INTRODUCTION TO

REAL ANALYSIS

INTRODUCTION TO REAL ANALYSIS

Liviu I Nicolaescu
University of Notre Dame, USA

World Scientific

NEW JERSEY • LONDON • SINGAPORE • BEIJING • SHANGHAI • HONG KONG • TAIPEI • CHENNAI • TOKYO

Published by

World Scientific Publishing Co. Pte. Ltd.
5 Toh Tuck Link, Singapore 596224
USA office: 27 Warren Street, Suite 401-402, Hackensack, NJ 07601
UK office: 57 Shelton Street, Covent Garden, London WC2H 9HE

Library of Congress Control Number: 2019952307

British Library Cataloguing-in-Publication Data
A catalogue record for this book is available from the British Library.

INTRODUCTION TO REAL ANALYSIS

ISBN 978-981-121-038-9
ISBN 978-981-121-075-4 (pbk)

For any available supplementary material, please visit
https://www.worldscientific.com/worldscibooks/10.1142/11553#t=suppl

To my students, for keeping me young.

Preface

This book started as notes for the Freshman Honors Calculus course at the University of Notre Dame. The word "calculus" is a misnomer since this course is intended to be an introduction to real analysis or, if you like, "calculus with proofs". For most students this class is the first encounter with mathematical rigor and it can be a bit disconcerting. In my view the best way to overcome this is to confront rigor head on and adopt it as standard operating procedure early on. This makes for a bumpy early going, but with a rewarding payoff.

A proof is an argument that uses the basic rules of Aristotelian logic and relies on facts everyone agrees to be true we mathematicians call axioms or postulates. The course is based on these basic rules of the mathematical discourse. It starts from a meagre collection of postulates and ends up constructing the main contours of the impressive edifice called real analysis.

In writing these notes I have benefitted immensely from the students who took the Honors Calculus course during the academic years 2013–2019. Their questions and reactions in class and to the notes, and their impressive ability of detecting typos have improved the original product. I asked a lot of them and I got a lot in return. Their hard work, curiosity and enthusiasm made my job so much more enjoyable.

The first nine chapters correspond essentially to the topics covered in the yearlong freshman course. Some years I could also squeeze in Chapter 10 on complex numbers. The first nine chapters cover the basics of real analysis of one real variable functions. In the process the reader is introduced to the fundamental facts about the topology of the real axis. No prior knowledge of calculus is assumed, but being comfortable performing algebraic manipulations is something that will make this journey more rewarding.

Chapters 11 through 16 deal with several variables calculus topics, corresponding to the yearlong sophomore Honors Calculus offered at the University of Notre Dame: Fréchet derivative, multidimensional Riemann integral and integrals over curves and surfaces. For this part, familiarity with the basic notions of linear algebra makes a great difference. By this I mean that the student is familiar with the concepts of vector spaces, linear operators, linear combinations, linear dependence, basis, rank and especially determinants and their applications.

When teaching this class, time constraints frequently forced me to decide the benefits of discussing in class all the details of one proof or another. The reactions of my students informed this decision. The fine print parts in this book cover details or technical proofs that, in my experience, are best skipped in class, or at the first reading of the text.

The trickiest to teach is integration on curves and surfaces: spending too much time on proofs leaves too little time for examples which is where the "rubber meets the road". Chapter 16 is my attempt of a compromise between these two competing goals.

Computations are an important part of Calculus, and the fact that many are essentially algorithmic is one reason for the versatility of Calculus. The exercises at the end of each chapter contain some computations but not in sufficient amount. This is by design. When I was teaching this course the typical weekly homework assignments consisted of two types of exercises: "theoretical", as my students would call them, and computational. The exercises at the end of each chapter in this book are mostly "theoretical". For the computational parts of the homework assignments I have used the venerable text by Demidovich [4]. This is less familiar to Western audience, but it was widely and very successfully used behind the Iron Curtain, when the Iron Curtain existed.

During an academic year I would ask my students to perform hundreds of computations from [4] until they reached the point where the process of computation became automatic and they could concentrate on concepts and ideas, without the distractions of the minutiae of an intervening computation. For multiple integrals the computations are essential in the understanding of this concept. I strongly encourage the would-be instructor of such a class to have a copy of [4] on her desk or in his hard drive.

As you might have noticed, some chapters have a section called "More challenging problems". In my class these were problems that students could try for extra credit. Some of these "more challenging" problems are indeed very challenging.

The present course notes are undoubtedly challenging, suitable for an Honors course. At the University of Notre Dame the topics in this book are covered during the first two years in college. The approach I propose is demanding to both the students and the instructor. However, as I learned from my own teachers, when you are challenged you will surprise yourself.

I found teaching this class to be extremely rewarding. To most of my freshmen this was the first encounter with mathematics as mathematicians know it, and it left an indelible impression in their young minds. A few discovered that math is a career that they want to pursue. The informal polls I conducted among my students showed that after one semester their perception of math had radically changed, in my view, for the better. More importantly, even the students that decided to pursue non-mathematical endeavors left with an appreciation of this subject and its practitioners.

My hope is that the lines to follow will help young minds get accustomed with the world of modern and abstract mathematics.

Notre Dame, September 30, 2019.

The Greek Alphabet

A	α	Alpha	N	ν	Nu
B	β	Beta	Ξ	ξ	Xi
Γ	γ	Gamma	O	o	Omicron
Δ	δ	Delta	Π	π	Pi
E	ε	Epsilon	P	ρ	Rho
Z	ζ	Zeta	Σ	σ	Sigma
H	η	Eta	T	τ	Tau
Θ	θ	Theta	Υ	υ	Upsilon
I	ι	Iota	Φ	φ	Phi
K	κ	Kappa	X	χ	Chi
Λ	λ	Lambda	Ψ	ψ	Psi
M	μ	Mu	Ω	ω	Omega

Contents

Preface vii

1. The Basics of Mathematical Reasoning 1

1.1 Statements and predicates . . . 1
1.2 Quantifiers . . . 4
1.3 Sets . . . 7
1.4 Functions . . . 10
1.5 Exercises . . . 14
1.6 More challenging problems . . . 16

2. The Real Number System 17

2.1 The algebraic axioms of the real numbers . . . 18
2.2 The order axiom of the real numbers . . . 20
2.3 The completeness axiom . . . 25
2.4 Visualizing the real numbers . . . 27
2.5 Exercises . . . 30

3. Special Classes of Real Numbers 33

3.1 The natural numbers and the induction principle . . . 33
3.2 Applications of the induction principle . . . 37
3.3 Archimedes' Principle . . . 41
3.4 Rational and irrational numbers . . . 44
3.5 Exercises . . . 50
3.6 More challenging problems . . . 53

4. Limits of Sequences 57

4.1 Sequences . . . 57
4.2 Convergent sequences . . . 59
4.3 The arithmetic of limits . . . 64
4.4 Convergence of monotone sequences . . . 69

4.5 Fundamental sequences and Cauchy's characterization of convergence . . 74
4.6 Series . . . 76
4.7 Power series . . . 87
4.8 Some fundamental sequences and series . . . 89
4.9 Exercises . . . 90
4.10 More challenging problems . . . 97

5. Limits of Functions 101

5.1 Definition and basic properties . . . 101
5.2 Exponentials and logarithms . . . 105
5.3 Limits involving infinities . . . 113
5.4 One-sided limits . . . 116
5.5 Some fundamental limits . . . 118
5.6 Trigonometric functions: a less than completely rigorous definition . . . 120
5.7 Useful trig identities . . . 126
5.8 Landau notation . . . 126
5.9 Exercises . . . 128
5.10 More challenging problems . . . 131

6. Continuity 133

6.1 Definition and examples . . . 133
6.2 Fundamental properties of continuous functions . . . 137
6.3 Uniform continuity . . . 145
6.4 Exercises . . . 149
6.5 More challenging problems . . . 152

7. Differential Calculus 155

7.1 Linear approximation and derivative . . . 155
7.2 Fundamental examples . . . 160
7.3 The basic rules of differential calculus . . . 164
7.4 Fundamental properties of differentiable functions . . . 172
7.5 Table of derivatives . . . 182
7.6 Exercises . . . 183
7.7 More challenging problems . . . 188

8. Applications of Differential Calculus 191

8.1 Taylor approximations . . . 191
8.2 L'Hôpital's Rule . . . 197
8.3 Convexity . . . 200
8.3.1 Basic facts about convex functions . . . 201
8.3.2 Some classical applications of convexity . . . 207
8.4 How to sketch the graph of a function . . . 217

8.5 Antiderivatives 221
8.6 Exercises 232
8.7 More challenging problems 237

9. Integral Calculus 239

9.1 The integral as area: a first look 239
9.2 The Riemann integral 241
9.3 Darboux sums and Riemann integrability 245
9.4 Examples of Riemann integrable functions 253
9.5 Basic properties of the Riemann integral 262
9.6 How to compute a Riemann integral 265
9.6.1 Integration by parts 267
9.6.2 Change of variables 273
9.7 Improper integrals 281
9.7.1 Euler's Gamma and Beta functions 291
9.8 Length, area and volume 292
9.8.1 Length 292
9.8.2 Area 295
9.8.3 Solids of revolution 298
9.9 Exercises 301
9.10 More challenging problems 308

10. Complex Numbers and Some of Their Applications 311

10.1 The field of complex numbers 311
10.1.1 The geometric interpretation of complex numbers 313
10.2 Analytic properties of complex numbers 316
10.3 Complex power series 323
10.4 Exercises 327

11. The Geometry and the Topology of Euclidean Spaces 329

11.1 Basic affine geometry 329
11.2 Basic Euclidean geometry 347
11.3 Basic Euclidean topology 354
11.4 Convergence 359
11.5 Normed vector spaces 366
11.6 Exercises 368
11.7 More challenging problems 375

12. Continuity 377

12.1 Limits and continuity 379
12.2 Connectedness and compactness 386
12.2.1 Connectedness 386

12.2.2 Compactness 387
12.3 Topological properties of continuous maps 394
12.4 Continuous partitions of unity 397
12.5 Exercises 401
12.6 More challenging problems 408

13. Multi-variable Differential Calculus 411

13.1 The differential of a map at a point 412
13.2 Partial derivatives and Fréchet differentials 416
13.3 The chain rule 426
13.4 Higher order partial derivatives 437
13.5 Exercises 442
13.6 More challenging problems 446

14. Applications of Multi-variable Differential Calculus 449

14.1 Taylor formula 449
14.2 Extrema of functions of several variables 451
14.3 Diffeomorphisms and the inverse function theorem 456
14.4 The implicit function theorem 464
14.5 Submanifolds of $\mathbb{R}^n$ 471
14.5.1 Definition and basic examples 472
14.5.2 Tangent spaces 481
14.5.3 Lagrange multipliers 488
14.6 Exercises 491
14.7 More challenging problems 497

15. Multidimensional Riemann Integration 499

15.1 Riemann integrable functions of several variables 499
15.1.1 The Riemann integral over a box 499
15.1.2 A conditional Fubini theorem 511
15.1.3 Functions Riemann integrable over $\mathbb{R}^n$ 515
15.1.4 Volume and Jordan measurability 518
15.1.5 The Riemann integral over arbitrary regions 521
15.2 Fubini theorem and iterated integrals 522
15.2.1 An unconditional Fubini theorem 523
15.2.2 Some applications 526
15.3 Change in variables formula 529
15.3.1 Formulation and some classical examples 529
15.3.2 Proof of the change of variables formula 545
15.4 Improper integrals 551
15.4.1 Locally integrable functions 551
15.4.2 Absolutely integrable functions 554

15.4.3 Examples . . . 557
15.5 Exercises . . . 562
15.6 More challenging problems . . . 571

16. Integration over Submanifolds 575

16.1 Integration along curves . . . 575
16.1.1 Integration of functions along curves . . . 575
16.1.2 Integration of differential 1-forms over paths . . . 583
16.1.3 Integration of 1-forms over oriented curves . . . 589
16.1.4 The 2-dimensional Stokes' formula: a baby case . . . 593
16.2 Integration over surfaces . . . 599
16.2.1 The area of a parallelogram . . . 599
16.2.2 Compact surfaces (with boundary) . . . 601
16.2.3 Integrals over surfaces . . . 604
16.2.4 Orientable surfaces in $\mathbb{R}^3$ . . . 614
16.2.5 The flux of a vector field through an oriented surface in $\mathbb{R}^3$ . . . 616
16.2.6 Stokes' Formulæ . . . 620
16.3 Differential forms and their calculus . . . 623
16.3.1 Differential forms on Euclidean spaces . . . 623
16.3.2 Orientable submanifolds . . . 633
16.3.3 Integration along oriented submanifolds . . . 635
16.3.4 The general Stokes' formula . . . 639
16.3.5 What are these differential forms anyway . . . 645
16.4 Exercises . . . 648
16.5 More challenging problems . . . 654

Bibliography 655

Index 657

Chapter 1

The Basics of Mathematical Reasoning

1.1. Statements and predicates

Mathematics deals in *statements*. These are sentences that have a definite truth value. What does this mean? The classical text [14] does a marvelous job explaining this point of view. I will not even attempt a rigorous or exhaustive explanation. Instead, I will try to suggest it to you through examples.

Example 1.1. (a) Consider the following sentence: "*if you walk in the rain without an umbrella, you will get wet*". This is a true sentence and we say that its truth value is $TRUE$ or T. This is an example of a *statement.*

(b) Consider the sentence: "*the number x is bigger than the number y*" or, in mathematical notation, $x > y$. This sentence could be $TRUE$ or $FALSE$ (F), depending on the choice of x and y. This is not a statement because it does not have a definitive truth value. It is a type of sentence called *predicate* that is encountered often in mathematics.

A *predicate* or *formula* is a sentence that depends on some parameters (or variables). In the above example the parameters were x and y. For some choices of parameters (or variables) it becomes a $TRUE$ statement, while for other values it could be $FALSE$.

When expressed in everyday language, statements and predicates must contain a verb.

Often a predicate comes in the guise of a *property*. For example the property "*the integer n is even*" stands for the predicate "*the integer n is twice an integer m*".

(c) Consider the following statement: "*This sentence is false.*" Is this sentence true? Clearly it cannot be true because if it were, then we would conclude that the sentence is false. Thus the sentence is false so the opposite must be true, i.e., the sentence is true. Something is obviously amiss. This type of sentence is *not* a statement because it does not have a truth value, and it is also *not* a predicate. It is a *paradox*. Paradoxes are to be avoided in mathematics. □

✍ **Notation.** It is time to explain the usage of the notation $:=$. For example an expression such as

$$x := \text{bla-bla-bla}$$

reads "*x is defined to be bla-bla-bla*", or "*x is short-hand for bla-bla-bla*".

The manipulations of statements and predicates are governed by the rules of *Aristotelian logic*. This and the following section will provide you with a very sparse introduction to logic. For more details and examples I refer to [20].

All the predicates used in mathematics are obtained from simpler ones called *atomic predicates* using the following *logical operators*.

- $NEGATION$ $\neg$ (read as *not*).
- $CONJUNCTION$ $\wedge$ (read as *and*).
- $DISJUNCTION$ $\vee$ (read as *or*).
- $IMPLICATION$ $\Rightarrow$ (read as *implies*).

To describe the effect of these operations we need to look at Table 1.1 describing the *truth tables* of these operations.

Table 1.1 The truth tables of $\neg, \wedge, \vee, \Rightarrow$.

p	T	F
$\neg p$	F	T

p	q	$p \wedge q$
T	T	T
T	F	F
F	T	F
F	F	F

p	q	$p \vee q$
T	T	T
T	F	T
F	T	T
F	F	F

p	q	$p \Rightarrow q$
T	T	T
T	F	F
F	T	T
F	F	T

Here is how one reads Table 1.1. When p is true (T), then $\neg p$ must be false (F), and when p is false, then $\neg p$ is true. To put it in simpler terms

$$\neg T = F, \quad \neg F = T.$$

The truth table for $\wedge$ can be presented in the simplified form

$$T \wedge T = T, \quad T \wedge F = F \wedge T = F \wedge F = F.$$

Observe that the predicate $p \vee q$ is true when at least one of the predicates p and q is true. *It is NOT an exclusive OR.* Another way of saying this is

$$T \vee T = T \vee F = F \vee T = T, \quad F \vee F = F.$$

The equivalence $\Leftrightarrow$ is the operation

$$p \Leftrightarrow q := (p \Rightarrow q) \wedge (q \Rightarrow p).$$

Its truth table is described in Table 1.2.

Table 1.2 The truth table of $\Leftrightarrow$.

p	q	$p \Leftrightarrow q$
T	T	T
T	F	F
F	T	F
F	F	T

Remark 1.2. (a) In everyday language, when we say that p *implies* q we mean that the statement $p \Rightarrow q$ is true. This signifies that either both p and q are true, or that p is false. Often we express this in the conditional form *if* p*, then* q.

If the implication $p \Rightarrow q$ is true, then we say that q *is a necessary condition for* p and that p *is a sufficient condition for* q. In everyday language, the implications are the "*if* bla-bla, *then* bla-bla" statements.

The truth table for $\Rightarrow$ hides certain subtleties best illustrated by the following example. Consider the statement

$$s := \text{if an elephant can fly, then it can also drive a car.}$$

This statement is composed of two simpler statements

$$p := \text{an elephant can fly}, \quad q := \text{an elephant can drive a car.}$$

We note that the statement s coincides with the implication $p \Rightarrow q$. Obviously, both statements p and q are false, but according to the truth table for $\Rightarrow$, the implication $p \Rightarrow q$ is true, and thus s is true as well. This conclusion is rather unsettling. It may be easier to accept it if we rephrase s as follows:

> *if you can show me a flying elephant, then I can show you that it can also drive a car.*

Fig. 1.1 *The elusive flying elephant.*

(b) In everyday language when we say that p *is equivalent to* q we mean that the statement $p \Leftrightarrow q$ is true. This signifies that either both p and q are true, or both are false. If p is equivalent to q, we say that q *is a necessary and sufficient condition for* p and that p *is a necessary and sufficient condition for* q.

We often express this in one of the following forms: p *if and only if* q. The mathematicians' abbreviation for the oft encountered construct "*if and only if*" is *iff.* □

Example 1.3. Consider the following predicate.

$s :=$ *if you do not clean your room, then you will not go to the movies.*

This is composed of two simpler predicates

- $p :=$ you do not clean the room.
- $q :=$ you do not go to the movies.

Observe that s is the compound predicate $p \Rightarrow q$. For s to be true, one of the following two mutually exclusive situations must happen

- either you do not clean your room AND you do not go to the movies A
- or you clean the room.

Note that there is no implied guarantee that if you clean your room, then you go to the movies. □

Example 1.4. Consider the following *true* statement: mathematicians like to be precise.

First, let us phrase this in a less ambiguous way. The above statement can be equivalently rephrased as: if you are a mathematician, then you are precise. To put it in symbolic terms

$$\underbrace{\text{you are a mathematician}}_{p} \Rightarrow \underbrace{\text{you are precise}}_{q}.$$

Thus, to be a mathematician it is necessary to be precise and to be precise it suffices to be a mathematician. However, to be precise it is not necessary to be a mathematician. □

A *tautology* is a compound predicate which is true no matter what the truth values of its atoms are.

Example 1.5. The predicate $p \vee \neg p$ is a tautology. In other words, in mathematics, a statement is either true, or false. There is no in-between. □

Two compound predicates P and Q are called *equivalent*, and we indicate this with the notation $P \longleftrightarrow Q$, if they have identical truth tables. In other words, P and Q are equivalent if the compound predicate $P \Longleftrightarrow Q$ is a tautology.

Example 1.6. Let us observe that the compound predicate $p \Rightarrow q$ is equivalent to the compound statement $(\neg p) \vee q$, i.e.

$$p \Rightarrow q \longleftrightarrow (\neg p) \vee q. \tag{1.1}$$

Indeed, if p is false then $p \Rightarrow q$ and $\neg p$ are true, no matter what q. In particular $(\neg p) \vee q$ is also true, no matter what q. If p is true, then $\neg p$ is false, and we deduce that $p \Rightarrow q$ and $(\neg p) \vee q$ are either simultaneously true, or simultaneously false. □

1.2. Quantifiers

Example 1.7. Consider the following property of a person x

$$x \textit{ is at least } 6ft \textit{ tall.}$$

This does not have a definite truth value because the truth value depends on the person x. However, the claims

$$C_1 := \textit{there exists a } \text{person } x \text{ that is at least 6ft tall,}$$

and

$$C_2 := \textit{any } \text{person } x \text{ is at least 6ft tall}$$

have definite truth values. The claim C_1 is true, while the claim C_2 is false. □

Example 1.8. Consider the following property involving the numbers x, y

$$x > y.$$

This does not have a definite truth value. However, we can modify it to obtain statements that have definite truth values. Here are several possible modifications. (Below and in the sequel the abbreviation s.t. stands for *such that*)

$$S_1 := \underline{\text{For any}}\ x,\ \underline{\text{for any}}\ y,\ x > y.$$

$$S_2 := \underline{\text{For any}}\ x,\ \underline{\text{there exists}}\ y \text{ s.t. } x > y.$$

$$S_3 := \underline{\text{There exists}}\ y \text{ s.t. } \underline{\text{for any}}\ x, \quad x > y.$$

Observe that the statements S_1 and S_3 are false, while S_2 is a true statement. Notice a *very important fact*. The statement S_3 is obtained from S_2 by a seemingly innocuous transformation: we changed the order of some words. However, in doing so, we have dramatically altered the meaning of the statement. *Let this be a warning!* □

The expressions *for any*, *for all*, *there exists*, *for some* appear very frequently in mathematical communications and for this reason they were given a name and special abbreviations. These expressions are called *quantifiers* and they are abbreviated as follows.

$$\forall := \text{for any, for all,}$$

$$\exists := \text{there exists, there exist, for some.}$$

The symbol $\forall$ is also called *the universal quantifier*, while the symbol $\exists$ is called *the existential quantifier*. There is another quantifier encountered quite frequently, namely

$$\exists! := \text{there exists a unique.}$$

The above examples illustrate the roles of the quantifiers: they are used to convert predicates, which have no definite truth value, to statements which have definite truth value. To achieve this, we need to attach a quantifier to each variable in the predicate. In Example 1.8 we used a quantifier for the variable x and a quantifier for the variable y. We cannot overemphasize the following fact.

☞ ***The meaning of a statement is sensitive to the order of the quantifiers in that statement!***

Example 1.9. Let us put to work the above simple principles in a concrete situation. Consider the statement:

$S :=$ *there is a person in this class such that, if he or she gets an* A *in the final, then everyone will get an* A *in the final.*

Is this a true statement or a false statement? There are two ways to decide this. The fastest way is to think of the persons who get the lowest grade in the final. If those persons get A's, then, obviously, everybody else will get A's.

We can use a more formal way of deciding the truth value of the above statement. Consider the predicate $P(x) :=$*the person* x *gets an* A *in the final.* The quantified form of S is then

$$\exists x: \; \big(P(x) \Rightarrow \forall y P(y) \big).$$

As we know, an implication $p \Rightarrow q$ is equivalent to the disjunction $\neg p \vee q$; see (1.1). We can rewrite the above statement as

$$\exists x: \; \big(\neg P(x) \vee \forall y P(y) \big).$$

In everyday language the above statement says that either there is a person who did not get an A or everybody gets an A. This is a *Duh!* statement or, as mathematicians like to call it, a *tautology*. □

Let us discuss how to concretely describe the negation of a statement involving quantifiers. Take for example the statements S_1, S_2, S_3 in Example 1.8. Their opposites are

$$\neg S_1 := \underline{\text{There exists}}\ x,\ \underline{\text{there exists}}\ y \text{ s.t. } x \leq y,$$

$$\neg S_2 := \underline{\text{There exists}}\ x \text{ s.t. } \underline{\text{for any}}\ y\text{: } x \leq y,$$

$$\neg S_3 := \underline{\text{For any}}\ \ y,\ \underline{\text{there exists}}\ x \text{ s.t. } x \leq y.$$

Observe that all the opposites were obtained by using the following simple operations.

- Globally replacing the existential quantifier $\exists$ with its opposite $\forall$.
- Globally replace the universal quantifier $\forall$ with its opposite, $\exists$.
- Replace the predicate $x > y$ with its opposite, $x \leq y$.

When dealing with more complex statements it is very useful to remember the above rules. We summarize them below.

☞ ***The opposite of a statement that contains quantifiers is obtained by replacing each quantifier with its opposite, and each predicate with its opposite.***

Example 1.10. Consider the following portion of a famous Abraham Lincoln quote: *you can fool all of the people some of the time.* There are two conceivable ways of phrasing this rigorously.

1. *For any person y there exists a moment of time t when y can be fooled by you at time t.*

2. *There exists a moment of time t such that any person y there can be fooled by you at time t.*

We can now easily transform these into *quantified statements.*

1. $S_1 := \forall$ *person y, $\exists$ moment t, s.t., y can be fooled by you at time t.*

2. $S_2 := \exists$ *moment t, s.t., $\forall$ person y, y can be fooled by you at time t.*

Note that the two statements carry different meanings. Which do you think was meant by Lincoln? Observe also

$\neg S_1 := \exists$ *person y s.t. $\forall$ moment t: y cannot be fooled by you at time t.*

In plain English this reads: *some people cannot be fooled at any time.* □

1.3. Sets

Now that we have learned a bit about the language of mathematics, let us mention a few fundamental concepts that appear in all the mathematical discourses. The most important concept is that of *set.*

Any attempt at a rigorous definition of the concept of set unavoidably leads to treacherous logical and philosophical marshes.[1] A more productive approach is not to define what a set is, but agree on a list of "uncontroversial" properties (or *axioms*) our intuition tells us the sets ought to satisfy.[2] Once these axioms are adopted, then the entire edifice of mathematics should be built on them. I refer to [15] for a detailed description of this point of view.

The axiomatic approach mentioned above is very labor intensive, and would send us far astray. Our goal for now is a bit more modest. We will adopt a more elementary (or naive) approach relying on the intuition of a set X as a collection of objects, usually referred to as the *elements of* X. In mathematics, a set is described by the "list" of its elements enclosed by brackets. In this list, no two objects are identical. For example, the set

$$\{winter, spring, summer, fall\}$$

is the set of seasons in a temperate region such as in Indiana. However, the list

$$\{winter, winter, spring\},$$

is not a set.

We will use the notation $x \in X$ (or $X \ni x$) to indicate that the *object x belongs to* the *set* X, i.e., the object x is an element of X. The notation $x \notin X$ indicates that x is not an element of X. Two sets A and B are considered identical if they consist of the same elements, i.e., the following (quantified) statement is true

$$\forall x \ \big(x \in A \Longleftrightarrow x \in B\big).$$

[1] For more details on the possible traps; see Wikipedia's article on set theory.

[2] See the above footnote.

In words, an object belongs to A iff it also belongs to B. For example, we have the equality of sets

$$\{winter, spring, summer, fall\} = \{spring, summer, fall, winter\}.$$

There exists a distinguished set, called the *empty set* and denoted by $\emptyset$. Intuitively, $\emptyset$ is the set with no elements.

Remark 1.11. The nature of the elements of a set is not important in set theory. In fact, the elements of a set can have varied natures. For example we have the set

$$\{\,1,\ \emptyset,\ apple\,\}$$

which consists of three elements: of the number 1, the empty set, and the word *apple*. Another more subtle example is the set $\{\emptyset\}$ which consists of the single element, the empty set $\emptyset$. Let us observe that $\emptyset \neq \{\emptyset\}$. □

We say that a set A is a *subset* of B, and we write this $A \subset B$, if any element of A is also an element of B. In other words, $A \subset B$ signifies that the following statement is true

$$\forall x\ \big(\, x \in A \Rightarrow x \in B\,\big).$$

A *proper subset* of B is a subset $A \subset B$ such that $A \neq B$. We will use the notation $A \subsetneq B$ to indicate that A is a proper subset of B.

The *union* of two sets A, B is a new set denoted by $A \cup B$. More precisely,

$$x \in A \cup B \iff (x \in A) \vee (x \in B).$$

The *intersection* of two sets A, B is a new set denoted by $A \cap B$. More precisely,

$$x \in A \cap B \iff (x \in A) \wedge (x \in B).$$

The sets A and B are said to be *disjoint* if $A \cap B = \emptyset$.

More generally, if $(A_i)_{i\in I}$ is a collection of sets, then we can define their union

$$\bigcup_{i\in I} A_i := \{\, x;\ \ \exists i \in I:\ \ x \in A_i\,\},$$

and their intersection

$$\bigcap_{i\in I} A_i := \{\, x;\ \ \forall i \in I;\ \ x \in A_i\,\}.$$

The *difference* between a set A and a set B is a new set $A \setminus B$ defined by

$$x \in A \setminus B \iff (x \in A) \wedge (x \notin B).$$

If A is a subset of X, then we will use the alternative notation $C_X A$ when referring to the difference $X \setminus A$. The set $C_X A$ is called the *complement of A in X*. Observe that

$$C_X\big(C_X A\big) = A.$$

It is sometimes convenient to visualize sets using *Venn diagrams*. A Venn diagram identifies a set with a region in the plane.

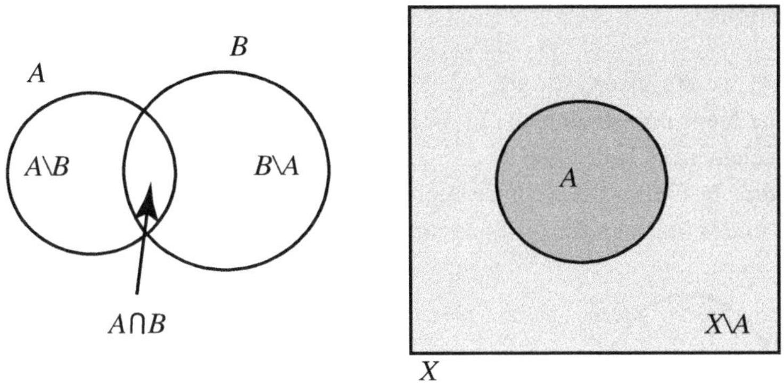

Fig. 1.2 Venn diagrams.

Proposition 1.12 (De Morgan Laws). *If A, B are subsets of a set X then*

$$C_X(A \cup B) = (C_X A) \cap (C_X B), \quad C_X(A \cap B) = (C_X A) \cup (C_X B). \qquad \square$$

Given two sets A and B we can form a new set $A \times B$ which consists of all ordered pairs of objects (a, b) where $a \in A$ and $b \in B$. The set $A \times B$ is called the *Cartesian product* of A and B.

Remark 1.13. As a curiosity, and to give you a sense of the intricacies of the axiomatic set theory, let us point out that above the concept of *ordered pair*, while intuitively clear, it is not rigorous. One rigorous definition of an ordered pair is due to *Norbert Wiener* who defined the ordered pair (a, b) to be the set consisting of two elements that are themselves sets: one element is the set $\{a, \emptyset\}$ and the other element is the set $\{\,\{b\}\,\}$, i.e.,

$$(a, b) := \{\,\{a, \emptyset\}, \{\,\{b\}\,\}\,\}. \qquad \square$$

Most of the time sets are defined by properties. For example, the interval $[0, 1]$ consists of the real numbers x satisfying the property

$$P(x) := (x \geq 0) \wedge (x \leq 1).$$

As we discussed in the previous section, a synonym for the term *property* is the term *predicate*. Proving that an object satisfying a property P also satisfies a property Q is tantamount to showing that the set of objects satisfying property P is contained in the set of objects satisfying property Q.

Remark 1.14. To prove that two sets A and B are equal one has to prove two inclusions: $A \subset B$ and $B \subset A$. In other words one has to prove two facts:

- If x is in A, then x is also in B.
- If x is in B, then x is also in A.

$\square$

1.4. Functions

Suppose that we are given two sets X, Y. Intuitively, a *function* f from X to Y is a "device" that feeds on elements of X. Once we feed this machine an element $x \in X$ it spits out an element of Y denoted by $f(x)$. The elements of X are called *inputs*, and those of Y, *outputs*. In Figure 1.3 we have depicted such a device. Each arrow starts at some input and its head indicates the resulting output when we feed that input to the function f.

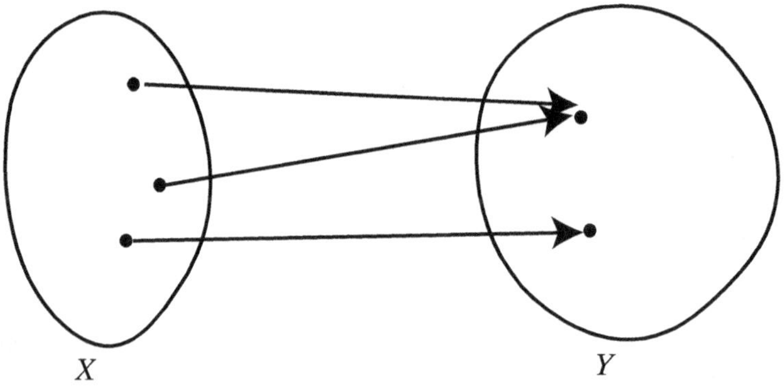

Fig. 1.3 *A Venn diagram depiction of a function from X to Y.*

The above definition may not sound too academic, but at least it gives an idea of what a function is supposed to do. Mathematically, a function is described by listing its effect on each and every one of the inputs $x \in X$. The result is a list G which consists of pairs $(x, y) \in X \times Y$, where the appearance of a pair (x, y) in the list indicates the fact that when the device is fed the input x, the output will be y. Note that the list G is a subset of $X \times Y$ and has two properties.

- For any $x \in X$ there exists $y \in Y$ such that $(x, y) \in G$. Symbolically

$$\forall x \in X \ \exists y \in Y : \ (x, y) \in G. \tag{F_1}$$

- For any $x \in X$ and any $y_1, y_2 \in Y$, if $(x, y_1), (x, y_2) \in G$, then $y_1 = y_2$. Symbolically

$$\forall x \in X, \ \forall y_1, y_2 \in Y, \Big((x, y_1) \in G \wedge (x, y_2) \in G \Big) \Rightarrow (y_1 = y_2). \tag{F_2}$$

Property F_1 states that to any input there corresponds at least one output, while property F_2 states that each input has at most one output.

We can use any symbol to name functions. The notation $f : X \to Y$ indicates that f is a function from X to Y. Often we will use the alternate notation $X \xrightarrow{f} Y$ to indicate that f is a function from X to Y. In mathematics there are many synonyms for the term function. They are also called *maps*, *mappings*, *operators*, or *transformations*.

Given a function $f : X \to Y$ we will refer to the set of inputs X as the *domain* of the function. The set Y is called the *codomain* of f. The result of feeding f the input $x \in X$ is denoted by $f(x)$. By definition $f(x) \in Y$. We say that x *is mapped to* $f(x)$ by f. The set

$$G_f := \{ (x, f(x));\ \ x \in X \} \subset X \times Y$$

lists the effect of f on each possible input $x \in X$, and it is usually referred to as the *graph* of f.

The *range* or *image* of a function $f : X \to Y$ is the set of all outputs of f. More precisely, it is the subset $f(X)$ of F defined by

$$f(X) := \{ y \in Y;\ \exists x \in X :\ \ y = f(x) \}.$$

The range of f is also denoted by $\boldsymbol{R}(f)$. More generally, for any subset $A \subset X$ we define

$$f(A) = \{ y \in Y;\ \exists a \in A;\ \ f(a) = y \} \subset Y. \tag{1.2}$$

The set $f(A)$ is called the *image* of A via f.

For a subset $S \subset Y$, we define the *preimage* of S via f to be the set of all inputs that are mapped by f to an element in S. More precisely the preimage of S is the set

$$f^{-1}(S) := \{ x \in X;\ \ f(x) \in S \} \subset X. \tag{1.3}$$

When S consists of a single point, $S = \{y_0\}$ we use the simpler notation $f^{-1}(y_0)$ to denote the preimage of $\{y_0\}$ via f. The set $f^{-1}(y_0)$ is a subset of X called the *fiber of f over y_0*.

A function $f : X \to Y$ is called *surjective*, or *onto*, if $f(X) = Y$. Using the visual description of a function given in Figure 1.3 we see that a function is onto if any element in Y is hit by an arrow originating at some element $x \in X$. Symbolically

$$f : X \to Y \text{ is surjective} \iff \forall y \in Y,\ \exists x \in X :\ \ y = f(x).$$

A function $f : X \to Y$ is called *injective*, or *one-to-one*, if different inputs have different outputs under f. More precisely

$$f : X \to Y \text{ is injective} \iff \forall x_1, x_2 \in X :\ \ x_1 \neq x_2 \Rightarrow f(x_1) \neq f(x_2)$$

$$\iff \forall x_1, x_2 \in X :\ \ f(x_1) = f(x_2) \Rightarrow x_1 = x_2.$$

A function $f : X \to Y$ is called *bijective* if it is both injective and surjective. We see that

$$f : X \to Y \text{ is bijective} \iff \forall y \in Y\ \ \exists!\, x \in X :\ \ y = f(x).$$

Example 1.15. (a) For any set X we denote by $\mathbb{1}_X$ or by e_X the function $X \to X$ which maps any $x \in X$ to itself. The function $\mathbb{1}_X$ is called the *identity map*. The identity map is clearly injective.

(b) Suppose that X, Y are two sets. We denote π_X the mapping $X \times Y \to X$ which sends a pair (x, y) to x. We say that π_X is the *natural projection* of $X \times Y$ onto X.

(c) Given a function $f : X \to Y$ and a subset $A \subset X$ we can construct a new function $f|_A : A \to Y$ called the *restriction* of f to A and defined in the obvious way

$$f|_A(a) = f(a),\ \ \forall a \in A.$$

(d) If X is a set and $A \subset X$, then we denote by i_A the function $A \to X$ defined as the restriction of $\mathbb{1}_X$ to A. More precisely

$$i_A(a) = a, \quad \forall a \in A.$$

The function i_A is called the *natural inclusion map* associated to the subset $A \subset X$. □

Given two functions

$$X \xrightarrow{f} Y, \quad Y \xrightarrow{g} Z$$

we can form their *composition* which is a function $g \circ f : X \to Z$ defined by

$$g \circ f(x) := g\big(\, f(x)\,\big).$$

Intuitively, the action of $g \circ f$ on an input x can be described by the diagram

$$x \overset{f}{\mapsto} f(x) \overset{g}{\mapsto} g\big(\, f(x)\,\big).$$

In words, this means the following: take an input $x \in X$ and drop it in the device $f : X \to Y$; out comes $f(x)$, which is an element of Y. Use the output $f(x)$ as an input for the device $g : Y \to Z$. This yields the output $g\big(\, f(x)\,\big)$.

Proposition 1.16. *Let $f : X \to Y$ be a function. The following statements are equivalent.*

(i) The function f is bijective.
(ii) There exists a function $g : Y \to X$ such that

$$f \circ g = \mathbb{1}_Y, \quad g \circ f = \mathbb{1}_X. \tag{1.4}$$

(iii) There exists a ***unique*** *function $g : Y \to X$ satisfying (1.4).*

Proof. (i) ⇒ (ii) Assume (i), so that f is bijective. Hence, for any $y \in Y$ there exists a unique $x \in X$ such that $f(x) = y$. This unique x depends on y and we will denote it by $g(y)$; see Figure 1.4.

The correspondence $y \mapsto g(y)$ defines a function $g : Y \to X$. By construction, if $x = g(y)$, then

$$y = f(x) = f\big(\, g(y)\,\big) = f \circ g(y) \;\; \forall y \in Y$$

so that $f \circ g = \mathbb{1}_Y$. Also, if $y = f(x)$, then

$$x = g(y) = g\big(\, f(x)\,\big) = g \circ f(x), \;\; \forall x \in X.$$

Hence $g \circ f = \mathbb{1}_X$. This proves the implication (i) ⇒ (ii).

(ii) ⇒ (iii) Assume (i). We need to show that if $g_1, g_2 : Y \to X$ are two functions satisfying (1.4), then $g_1 = g_2$, i.e., $g_1(y) = g_2(y)$, $\forall y \in Y$.

Let $y \in Y$. Set $x_1 = g_1(y)$. Then

$$f(x_1) = f\big(\, g_1(y)\,\big) = f \circ g_1(y) \overset{(1.4)}{=} y.$$

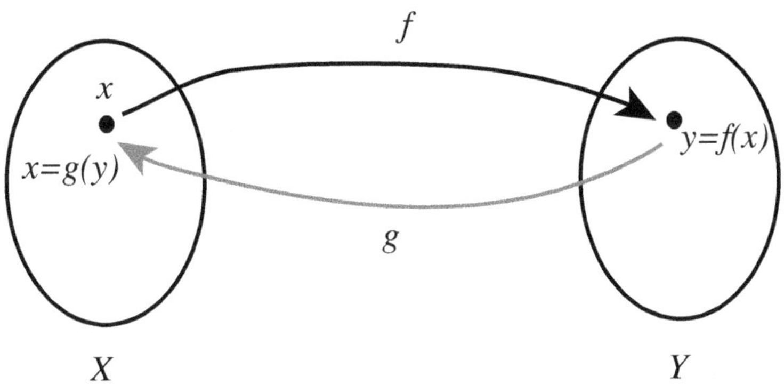

Fig. 1.4 *Constructing the inverse of a bijective function* $X \to Y$.

On the other hand,

$$g_2(y) = g_2\big(f(x_1)\big) = g_2 \circ f(x_1) \stackrel{(1.4)}{=} x_1 = g_1(y).$$

This proves the implication (ii) $\Rightarrow$ (iii).

(iii) $\Rightarrow$ (i). We assume that there exists a function $g : Y \to X$ satisfying (1.4) and we will show that f is bijective. We first prove that f is injective, i.e.,

$$\forall x_1, x_2 \in X : f(x_1) = f(x_2) \Rightarrow x_1 = x_2.$$

Indeed, if $f(x_1) = f(x_2)$, then

$$x_1 \stackrel{(1.4)}{=} g \circ f(x_1) = g\big(f(x_1)\big) = g\big(f(x_2)\big) = g \circ f(x_2) \stackrel{(1.4)}{=} x_2.$$

To prove surjectivity we need to show that for any $y \in Y$, there exists $x \in X$ such that $f(x) = y$. Let $y \in Y$. Set $x = g(y)$. Then

$$y \stackrel{(1.4)}{=} f \circ g(y) = f\big(g(y)\big) = f(x).$$

This proves the surjectivity of f and completes the proof of Proposition 1.16. □

Definition 1.17. Let $f : X \to Y$ be a bijective function. The *inverse* of f is the unique function $g : Y \to X$ satisfying (1.4). The inverse of a bijective function f is denoted by f^{-1}. □

1.5. Exercises

Exercise 1.1. Show that

$$\neg(p \vee q) \longleftrightarrow (\neg p \wedge \neg q), \quad \neg(p \wedge q) \longleftrightarrow \neg p \vee \neg q. \qquad \square$$

Exercise 1.2. (a) Show that

$$(p \Rightarrow q) \longleftrightarrow (\neg q \Rightarrow \neg p), \quad \neg(p \Rightarrow q) \longleftrightarrow (p \wedge \neg q).$$

(b) Consider the predicates

$$p := \text{the elephant } x \text{ can fly}, \quad q := \text{the elephant } x \text{ can drive}.$$

Let us stipulate that p is false. Show that the predicate $p \Rightarrow q$ is true by showing that its negation $\neg(p \Rightarrow q)$ is false. $\square$

Exercise 1.3. Consider the exclusive-OR operation $\vee^*$ with truth table.

Table 1.3 The truth table of "$\vee^*$".

p	q	$p \vee^* q$
T	T	F
T	F	T
F	T	T
F	F	F

Show that

$$(p \vee^* q) \longleftrightarrow (p \wedge \neg q) \vee (\neg p \wedge q) \longleftrightarrow (p \Longleftrightarrow \neg q) \longleftrightarrow (p \Rightarrow \neg q) \wedge (\neg p \Rightarrow q). \quad \square$$

Exercise 1.4 (Modus ponens). Show that the compound predicate

$$\big((p \Rightarrow q) \wedge p \big) \Rightarrow q$$

is a tautology. $\square$

Exercise 1.5 (Modus tollens). Show that the compound predicate

$$\big((p \Rightarrow q) \wedge \neg q \big) \Rightarrow \neg p$$

is a tautology. $\square$

Exercise 1.6. Translate each of the following propositions into a *quantified statement* in standard form, write its symbolic negation, and then state its negation in words. (Use Example 1.10 as guide.)

(i) You can fool some of the people all of the time.
(ii) Everybody loves somebody sometime.
(iii) You cannot teach an old dog new tricks.
(iv) When it rains, it pours.

$\square$

Exercise 1.7. Consider the following predicates.

$$P := \text{I will attend your party.}$$

$$Q := \text{I go to a movie.}$$

Rephrase the predicate

I will attend your party unless I go to a movie

using the predicates P, Q and the logical operators $\neg, \vee, \wedge, \Rightarrow$. □

Exercise 1.8. Give an example of three sets A, B, C satisfying the following properties

$$A \cap B \neq \emptyset, \quad B \cap C \neq \emptyset, \quad C \cap A \neq \emptyset, \quad A \cap B \cap C = \emptyset.$$ □

Exercise 1.9. Suppose that A, B, C are three arbitrary sets. Show that

$$A \cap (B \cup C) = (A \cap B) \cup (A \cap C),$$

$$A \cup (B \cap C) = (A \cup B) \cap (A \cup C),$$

and

$$A \setminus (B \cup C) = (A \setminus B) \cap (A \setminus C).$$

(In the above equalities it should be understood that the operations enclosed by parentheses are to be performed first.)

Hint. Use Remark 1.14. □

Exercise 1.10. Suppose that $f : X \to Y$ is a function and $A, B \subset Y$ are subsets of the codomain. Prove that

$$f^{-1}(A \cup B) = f^{-1}(A) \cup f^{-1}(B), \quad f^{-1}(A \cap B) = f^{-1}(A) \cap f^{-1}(B).$$

Hint. Take into account (1.3) and Remark 1.14. □

Exercise 1.11. Let $f : X \to Y$ be a map between the sets X, Y. Prove that f is one-to-one *if and only if* for any subsets $A, B \subset X$ we have

$$f(A \cap B) = f(A) \cap f(B).$$ □

Exercise 1.12. Suppose A, B are sets and $f : A \to B$ is a map.[3] Define the maps

$$\varphi : A \to A \times B, \quad \rho : A \times B \to B$$

by setting

$$\varphi(a) := \big(a, f(a)\big), \quad \forall a \in A, \quad \rho(a, b) := b, \quad \forall (a, b) \in A \times B.$$

Prove that the following hold.

[3]Recall that a map is a function.

(i) The map φ is injective.
(ii) The map ρ is surjective.
(iii) $f = \rho \circ \varphi$.

□

Exercise 1.13. Suppose that $f : X \to Y$ and $g : Y \to Z$ are two bijective maps. Prove that the composition $g \circ f$ is also bijective and

$$(g \circ f)^{-1} = f^{-1} \circ g^{-1}.$$ □

Exercise 1.14. Suppose that $f : X \to Y$ is a function. Prove that the following statements are equivalent.

(i) The function f is injective.
(ii) There exists a function $g : Y \to X$ such that $g \circ f = \mathbb{1}_X$.

Exercise 1.15. Suppose that $f : X \to Y$ is a function. Prove that the following statements are equivalent.

(i) The function f is surjective.
(ii) There exists a function $g : Y \to X$ such that $f \circ g = \mathbb{1}_Y$.

□

1.6. More challenging problems

Problem 1.1. Two old ladies left from A to B and from B to A at dawn heading towards one another along the same road. They met at noon, but did not stop, each carried on walking with the same speed as before they met. The first lady arrives at B at 4 pm, and the second lady arrives at A at 9 pm. What time was the dawn that day? □

Problem 1.2. A farmer must take a wolf, a goat and a cabbage across a river in a boat. However the boat is so small that he is able to take only one of the three on board with him. How should he transport all three across the river? (The wolf cannot be left alone with the goat, and the goat cannot be left alone with the cabbage.) □

Chapter 2

The Real Number System

Any attempt to define the concept of number is fraught with perils of a logical kind: we will eventually end up chasing our tails. Instead of trying to explain *what numbers are*, it is more productive to explain *what numbers do*, and *how they interact with each other.*

In this section we gather in a coherent way some of the basic properties our intuition tells us that real numbers[1] ought to satisfy. We will formulate them precisely and we will declare, by fiat, that *these are true statements*. We will refer to these as the *axioms of the real number system*. (Things are a bit more subtle, but that is the gist of our approach.) All the other properties of the real numbers follow from these axioms. Such deductible properties are known in mathematics as *Proposition*s or *Theorem*s. The term *Theorem* is used sparingly and it is reserved to the more remarkable properties.

The process of deducing new properties from the already established ones is called a mathematical *proof*. Intuitively, a proof is a complete, precise and coherent explanation of a fact. In this course we will prove all of the calculus facts you are familiar with, and much more.

The first thing that we observe is that the real numbers, whatever their nature, form a set. We will encounter this set so often in our mathematical discourse that it deserves a short name and symbol. We will denote the set of real numbers by $\mathbb{R}$. More importantly this set of mysterious objects called numbers satisfy certain properties that we use every day. We take them for granted, and do not bother to prove them. These are the axioms of the real numbers and they are of three types.

- Algebraic axioms.
- Order axioms.
- The completeness axiom.

In this chapter we discuss these axiom in some details and then we show some of their immediate consequences.

Remark 2.1. There is one rather delicate issue that we do not address in these notes. We introduce a set of objects whose nature we do not explain and then we take for granted that they satisfy certain properties.

Naturally, one should ask if such things exist, because, for all we know, we might be investigating the set of flying elephants. This is a rather subtle question, and answering it would force us to dig deep at the foundations of mathematics. Historically, this

[1] You may know them as *decimal numbers* or *decimals*.

question was settled relatively recently during the twentieth century but, mercifully, science progressed for two millennia before people thought of formulating and addressing this issue. To cut to the chase, no, we are not investigating flying elephants. □

2.1. The algebraic axioms of the real numbers

Another thing we know from experience is that we can operate with numbers. More precisely we can add, subtract, multiply and divide real numbers. Of these four operations, the addition and the multiplication are the fundamental ones. These are special instances of a more general mathematical concept, that of *binary operation*.

A binary operation on a set S is, by definition, a function $S \times S \to S$. Loosely, a binary operation is a gizmo that feeds on *ordered* pairs of elements of S, processes such a pair in some fashion, and produces a single element of S. We list the first axioms describing the set of real numbers.

Axiom 1. The set of real numbers $\mathbb{R}$ is equipped with two binary operations,

- *addition*

$$+ : \mathbb{R} \times \mathbb{R} \to \mathbb{R}, \quad (x, y) \mapsto x + y,$$

- and *multiplication*

$$\cdot : \mathbb{R} \times \mathbb{R} \to \mathbb{R}, \quad (x, y) \mapsto x \cdot y.$$

□

The operation of multiplication is sometimes denoted by the symbol $\times$.

Axiom 2. Addition is *associative*, i.e.,

$$\forall x, y, z \in \mathbb{R}; \quad (x + y) + z = x + (y + z).$$ □

The usage of parentheses ($-$) indicates that we first perform the operation enclosed by them.

Axiom 3. Addition is *commutative*, i.e.,

$$\forall x, y \in \mathbb{R} : \quad x + y = y + x.$$ □

Axiom 4. An *additive identity element* exists. This means that there exists at least one real number u such that

$$x + u = u + x = x, \quad \forall x \in \mathbb{R}. \tag{2.1}$$

□

Before we proceed to our next axiom, let us observe that there exists precisely one additive identity element.

Proposition 2.2. *If* $u_0, \hat{u}_0 \in \mathbb{R}$ *are additive identity elements, then* $u_0 = \hat{u}_0$.

Proof. Since u_0 is an identity element, if we choose $x = \hat{u}_0$ in (2.1) we deduce that

$$\hat{u}_0 + u_0 = u_0 + \hat{u}_0 = \hat{u}_0.$$

On the other hand, $\hat{u}_0$ is also an identity element and if we let $x = u_0$ in (2.1) we conclude that

$$u_0 + \hat{u}_0 = \hat{u}_0 + u_0 = u_0.$$

Thus $u_0 = \hat{u}_0$. □

Definition 2.3. The unique additive identity element of $\mathbb{R}$ is denoted by 0. □

Axiom 5. *Additive inverses exist.* More precisely, this means that for any $x \in \mathbb{R}$ there exists at least one real number $y \in \mathbb{R}$ such that

$$x + y = y + x = 0.$$ □

We have the following result whose proof is left to you as an exercise.

Proposition 2.4. *Additive inverses are unique. This means that if x, y, y' are real numbers such that*

$$x + y = y + x = 0 = x + y' = y' + x,$$

then $y = y'$. □

Definition 2.5. The unique additive inverse of a real number x is denoted by $-x$. Thus

$$x + (-x) = (-x) + x = 0, \quad \forall x \in \mathbb{R}.$$ □

Axiom 6. The multiplication is *associative*, i.e.,

$$\forall x, y, z \in \mathbb{R}; \quad (x \cdot y) \cdot z = x \cdot (y \cdot z).$$ □

Axiom 7. The multiplication is *commutative*, i.e.,

$$\forall x, y \in \mathbb{R}: \quad x \cdot y = y \cdot x.$$ □

Axiom 8. A *multiplicative identity element* exists. This means that there exists at least one *nonzero* real number u such that

$$x \cdot u = u \cdot x = x, \quad \forall x \in \mathbb{R}. \tag{2.2}$$

□

Arguing as in the proof of Proposition 2.2 we deduce that there exists precisely one multiplicative identity element. We denote it by 1. We define

$$2 := 1 + 1, \quad x^2 := x \cdot x, \quad \forall x \in \mathbb{R}.$$ (✎)

Axiom 9. *Multiplicative inverses exist.* More precisely, this means that for any $x \in \mathbb{R}$, $x \neq 0$, there exists at least one real number $y \in \mathbb{R}$ such that

$$x \cdot y = y \cdot x = 1.$$ □

Proposition 2.4 has a multiplicative counterpart that states that multiplicative inverses are unique. The multiplicative inverse of the *nonzero* real number x is denoted by x^{-1}, or $1/x$, or $\frac{1}{x}$. Also, we will frequently use the notation

$$\frac{x}{y} := x \cdot y^{-1}, \quad y \neq 0.$$

☞ ***The real number zero <u>does not have an inverse</u>. For this reason division by zero is an illegal and very dangerous operation. NEVER DIVIDE BY ZERO!***

Axiom 10. *Distributivity.*

$$\forall x, y, z \in \mathbb{R}: \quad x \cdot (y + z) = x \cdot y + x \cdot z.$$ □

✍ *To save energy and time we agree to replace the notation $x \cdot y$ with the simpler one, xy, whenever no confusion is possible.*

Definition 2.6. A set satisfying Axioms 1 through 10 is called a *field*. □

The above axioms have a number of "obvious" consequences.

Proposition 2.7.

(i) $\forall x \in \mathbb{R}$, $x \cdot 0 = 0$.
(ii) $\forall x, y \in \mathbb{R}$, $(xy = 0) \Rightarrow (x = 0) \vee (y = 0)$.
(iii) $\forall x \in \mathbb{R}$, $-x = (-1) \cdot x$.
(iv) $\forall x \in \mathbb{R}$, $(-1) \cdot (-x) = x$.
(v) $\forall x, y \in \mathbb{R}$, $(-x) \cdot (-y) = xy$.

Proof. We will prove only part (i). The rest are left as exercises. Since 0 is the additive identity element we have $0 + 0 = 0$ and

$$x \cdot 0 = x \cdot (0 + 0) = x \cdot 0 + x \cdot 0.$$

If we add $-(x \cdot 0)$ to both sides of the equality $x \cdot 0 = x \cdot 0 + x \cdot 0$ we deduce $0 = x \cdot 0$. □

2.2. The order axiom of the real numbers

Experience tells us that we can compare two real numbers, i.e., given two real numbers we can decide which is smaller than the other. In particular, we can decide whether a number is positive or not. In more technical terms we say that we can *order* the set of real numbers. The next axiom formalizes this intuition.

Axiom 11. There exists a subset $\boldsymbol{P} \subset \mathbb{R}$ called the *subset of positive real numbers* satisfying the following two conditions.

(i) If x and y are in $\boldsymbol{P}$, then so are their sum and product, $x + y \in \boldsymbol{P}$ and $xy \in \boldsymbol{P}$.
(ii) If $x \in \mathbb{R}$, then *exactly one* of the following statements is true:

$$x \in \boldsymbol{P}, \text{ or } x = 0, \text{ or } -x \in \boldsymbol{P}.$$ □

Definition 2.8. Let $x, y \in \mathbb{R}$.

(i) We say that x is *negative* if $-x \in \boldsymbol{P}$.
(ii) We say that x is *greater than* y, and we write this $x > y$ if $x - y$ is positive. We say that x is *less than* y, written $x < y$, if y is greater than x.
(iii) We say that x is *greater than or equal to* y, and we write this $x \geq y$, if $x > y$ or $x = y$. We say that x is *less than or equal to* y, and we write this $x \leq y$, if $y \geq x$.
(iv) A real number x is called *nonnegative* if $x \geq 0$.

□

Observe that $x > 0$ signifies that $x \in \boldsymbol{P}$.

Proposition 2.9.

(i) $1 > 0$, *i.e.,* $1 \in \boldsymbol{P}$.
(ii) If $x > y$ *and* $y > z$, *then* $x > z$, $x, y, z \in \mathbb{R}$.
(iii) If $x > y$, *then for any* $z \in \mathbb{R}$, $x + z > y + z$.
(iv) If $x > y$ *and* $z > 0$, *then* $xz > yz$.
(v) If $x > y$ *and* $z < 0$, *then* $xz < yz$.

Proof. We will prove only (i) and (ii). The proofs of the other statements are left to you as exercises. To prove (i) we argue by contradiction. Thus we assume that $1 \notin \boldsymbol{P}$. By Axiom 8, $1 \neq 0$, so Axiom 11 implies that $-1 \in \boldsymbol{P}$ and $(-1) \cdot (-1) \in \boldsymbol{P}$. Using Proposition 2.7(v) we deduce that

$$1 = (-1) \cdot (-1) \in \boldsymbol{P}.$$

We have reached a contradiction which proves (i).

To prove (ii) observe that

$$x > y \Rightarrow x - y \in \boldsymbol{P}, \quad y > z \Rightarrow y - z \in \boldsymbol{P}$$

so that

$$x - z = (x - y) + (y - z) \in \boldsymbol{P} \Rightarrow x > z.$$

□

Definition 2.10 (Intervals). Let $a, b \in \mathbb{R}$. We define the following sets.

(i) $(a, b) =]a, b[:= \{ x \in \mathbb{R};\ a < x < b \}$.
(ii) $(a, b] =]a, b] := \{ x \in \mathbb{R};\ a < x \leq b \}$.
(iii) $[a, b) = [a, b[:= \{ x \in \mathbb{R};\ a \leq x < b \}$.
(iv) $[a, b] := \{ x \in \mathbb{R};\ a \leq x \leq b \}$.
(v) $[a, \infty) = [a, \infty[:= \{ x \in \mathbb{R};\ a \leq x \}$.
(vi) $(a, \infty) =]a, \infty[:= \{ x \in \mathbb{R};\ a < x \}$.
(vii) $(-\infty, a) =]-\infty, a[:= \{ x \in \mathbb{R};\ x < a \}$.

(viii) $(-\infty, a] =]-\infty, a] := \{ x \in \mathbb{R};\;\; x \leq a \}$.

The above sets are generically called *intervals*. The intervals of the form $[a, b]$, $[a, \infty)$, or $(-\infty, a]$ are called *closed*, while the intervals of the form (a, b), (a, ∞), or $(-\infty, a)$ are called *open*. □

I would like to emphasize that in the above definition we made no claim that any or some of the intervals are nonempty. This is indeed the case, but this fact requires a proof.

Definition 2.11. For any $x \in \mathbb{R}$ we define the *absolute value* of x to be the quantity

$$|x| := \begin{cases} x & \text{if } x \geq 0, \\ -x & \text{if } x < 0. \end{cases}$$ □

Proposition 2.12.

(i) Let $\varepsilon > 0$. *Then* $|x| < \varepsilon$ *if and only if* $-\varepsilon < x < \varepsilon$, *i.e.,*

$$(-\varepsilon, \varepsilon) = \{ x \in \mathbb{R};\;\; |x| < \varepsilon \}.$$

(ii) $x \leq |x|$, $\forall x \in \mathbb{R}$.
(iii) $|xy| = |x| \cdot |y|$, $\forall x, y \in \mathbb{R}$. *In particular,* $|-x| = |x|$.
(iv) $|x + y| \leq |x| + |y|$, $\forall x, y \in \mathbb{R}$.

Proof. We prove only (i) leaving the other parts as an exercise. We have to prove two things,

$$|x| < \varepsilon \Rightarrow -\varepsilon < x < \varepsilon, \tag{2.3}$$

and

$$-\varepsilon < x < \varepsilon \Rightarrow |x| < \varepsilon. \tag{2.4}$$

To prove (2.3) let us assume that $|x| < \varepsilon$. We distinguish two cases. If $x \geq 0$, then $|x| = x$ and we conclude that $-\varepsilon < 0 \leq x < \varepsilon$. If $x < 0$, then $|x| = -x$ and thus $0 < -x = |x| < \varepsilon$. This implies $-\varepsilon < -(-x) = x < 0 < \varepsilon$.

Conversely, let us assume that $-\varepsilon < x < \varepsilon$. Multiplying this inequality by -1 we deduce that $-\varepsilon < -x < \varepsilon$. If $0 \leq x$, then $|x| = x < \varepsilon$. If $x < 0$ then $|x| = -x < \varepsilon$. □

Definition 2.13. The distance between two real numbers x, y is the nonnegative number $\operatorname{dist}(x, y)$ defined by

$$\operatorname{dist}(x, y) := |x - y|.$$ □

Very often in calculus we need to solve *inequalities*. The following examples describe some simple ways of doing this.

Example 2.14. (a) Suppose that we want to find all the real numbers x such that

$$(x - 1)(x - 2) > 0.$$

To solve this inequality we rely on the following simple principle: the product of two real numbers is positive if and only if both numbers are positive or both numbers are negative; see Exercise 2.8. In this case the answer is simple: the numbers $(x-1)$ and $(x-2)$ are both positive iff $x>2$ and they are both negative iff $x<1$. Hence

$$(x-1)(x-2)>0 \iff x \in (-\infty,1)\cup(2,\infty).$$

(b) Consider the more complicated problem: find all the real numbers x such that

$$(x-1)(x-2)(x-3)>0.$$

The answer to this question is also decided by the multiplicative rule of signs, but it is convenient to organize or work in a table. In each row we read the sign of the quantity listed at the beginning of the row. The signs in the bottom row are obtained by multiplying the signs in the column above them. We read

$$(x-1)(x-2)(x-3)>0 \iff x \in (1,2)\cup(3,\infty).$$

Table 2.1

x	$-\infty$		1		2		3		∞
$(x-1)$	$-\infty$	$----$	0	$+++$	$+$	$+++$	$+$	$+++$	∞
$(x-2)$	$-\infty$	$----$	$-$	$---$	0	$+++$	$+$	$+++$	∞
$(x-3)$	$-\infty$	$----$	$-$	$---$	$-$	$---$	0	$+++$	∞
$(x-1)(x-2)(x-3)$	$-\infty$	$----$	0	$+++$	0	$---$	0	$+++$	∞

(c) Consider the related problem: find all the real numbers x such that

$$\frac{(x-1)}{(x-2)(x-3)} \geq 0.$$

Before we proceed we need to eliminate the numbers $x=2$ and $x=3$ from our considerations because the denominator of the above fraction vanishes for these values of x and the ***division by*** 0 ***is an illegal operation***. We obtain a similar table.

Table 2.2

x	$-\infty$		1		2		3		∞
$(x-1)$	$-\infty$	$----$	0	$+++$	$+$	$+++$	$+$	$+++$	∞
$(x-2)$	$-\infty$	$----$	$-$	$---$	0	$+++$	$+$	$+++$	∞
$(x-3)$	$-\infty$	$----$	$-$	$---$	$-$	$---$	0	$+++$	∞
$\frac{(x-1)}{(x-2)(x-3)}$	$-\infty$	$----$	0	$+++$	!	$---$	!	$+++$	∞

The exclamation signs at the bottom row are warning us that for the corresponding values of x the fraction has no meaning. We read

$$\frac{(x-1)}{(x-2)(x-3)} \geq 0 \iff x \in [1,2)\cup(3,\infty). \qquad \square$$

Example 2.15. We want to discuss a question involving inequalities frequently encountered in real analysis. Consider the statement

$$P(M): \ \forall x \in \mathbb{R}, \ \ x > M \Rightarrow \left| \frac{x^2}{x^2 + x - 2} - 1 \right| < \frac{1}{10}.$$

We want to show that there exists at least one positive number M such that $P(M)$ is true, i.e., we want to prove that the statement

$$\exists M > 0 \ \text{ such that}, \ \ \forall x \in \mathbb{R}, \ \ x > M \Rightarrow \left| \frac{x^2}{x^2 + x - 2} - 1 \right| < \frac{1}{10}.$$

Let us observe that if $P(M)$ is true and $M' \geq M$, then $P(M')$ is also true. Thus, once we find one M such that $P(M)$ is true, then $P(M')$ is true for all $M' \in [M, \infty)$.

We are content with finding only one M such that $P(M)$ is true and the above observation shows that in our search we can assume that M is very large. This is a bit vague, so let us see how this works in our special case.

First, we need to make sure that our algebraic expression is well defined so we need to require that the denominator $x^2 + x - 2 = (x - 1)(x + 2)$ is not zero. Thus we need to assume that $x \neq 1, -2$. In particular, we will restrict our search for M to numbers larger than 1. We have

$$\left| \frac{x^2}{x^2 + x - 2} - 1 \right| = \left| \frac{x^2 - (x^2 + x - 2)}{x^2 + x - 2} \right| = \left| \frac{-x + 2}{x^2 + x - 2} \right| = \left| \frac{x - 2}{x^2 + x - 2} \right|.$$

Since we are investigating the properties of the last expression for $x > M > 1$ we deduce that for $x > 2$ both quantities $x - 2$ and $(x - 1)(x + 2)$ are positive and thus

$$\left| \frac{x - 2}{x^2 + x - 2} \right| = \frac{x - 2}{x^2 + x - 2}.$$

We want this fraction to be small, smaller than $\frac{1}{10}$. Note that for $x > 2$ we have

$$\frac{x - 2}{x^2 + x - 2} \leq \frac{x - 1}{x^2 + x - 2} = \frac{x - 1}{(x - 1)(x + 2)} = \frac{1}{x + 2},$$

and

$$x > 2 \ \wedge \ \frac{1}{x + 2} < \frac{1}{10} \Longleftrightarrow x + 2 > 10 \Longleftrightarrow x > 8\,.$$

We deduce that if $x > 8$, then

$$\frac{1}{10} > \frac{1}{x + 2} > \left| \frac{x^2}{x^2 + x - 2} - 1 \right|.$$

Hence $P(8)$ is true. □

2.3. The completeness axiom

Definition 2.16. Let $X \subset \mathbb{R}$ be a nonempty set of real numbers.

(i) A real number M is called an *upper bound* for X if

$$\forall x \in X: \quad x \leq M. \tag{2.5}$$

(ii) The set X is said to be *bounded above* if it admits an upper bound.

(iii) A real number m is called a *lower bound* for X if

$$\forall x \in X: \quad x \geq m. \tag{2.6}$$

(iv) The set X is said to be *bounded below* if it admits a lower bound.

(v) The set X is said to be *bounded* if it is bounded both above and below.

□

Example 2.17. (a) The interval $(-\infty, 0)$ is bounded above, but not below. The interval $(0, \infty)$ is bounded below, but not above, while the interval $(0, 1)$ is bounded. □

(b) Consider the set R consisting of positive real numbers x such that $x^2 < 2$. This set is not empty because $1^2 = 1 < 2$ so that $1 \in R$. Let us show that this set is bounded above. More precisely, we will prove that

$$x^2 < 2 \Rightarrow x \leq 2.$$

We argue by contradiction. Suppose that $x \in R$ yet $x > 2$. Then

$$x^2 - 2^2 = (x - 2)(x + 2) > 0.$$

Hence $x^2 > 2^2 > 2$ which shows that $x \notin R$. This contradiction proves that 2 is an upper bound for R. □

Definition 2.18. Let $X \subset \mathbb{R}$ be a nonempty set of real numbers.

(i) A *least upper bound* for X is an upper bound M with the following additional property: if M' is another upper bound of X, then $M \leq M'$.

(ii) A *greatest lower bound* for X is a lower bound m with the following additional property: if m' is another lower bound of X, then $m \geq m'$.

□

Thus, M is a least upper bound for X if

- $\forall x \in X, x \leq M$, and
- if $M' \in \mathbb{R}$ is such that $\forall x \in X, x \leq M'$, then $M \leq M'$.

Proposition 2.19. *Any nonempty set $X \subset \mathbb{R}$ admits at most one least upper bound, and at most one greatest lower bound.*

Proof. We prove only the statement concerning upper bounds. Suppose that M_1, M_2 are two least upper bounds. Since M_1 is a least upper bound, and M_2 is an upper bound we have $M_1 \leq M_2$. Similarly, since M_2 is a least upper bound we deduce $M_2 \leq M_1$. Hence $M_1 \leq M_2$ and $M_2 \leq M_1$ so that $M_1 = M_2$. □

Definition 2.20. Let $X \subset \mathbb{R}$ be a nonempty set of real numbers.

(i) The least upper bound of X, when it exists, is called the *supremum* of X and it is denoted by $\sup X$.
(ii) The greatest lower bound of X, when it exists, is called the *infimum* of X and it is denoted by $\inf X$.

□

Example 2.21. Suppose that $X = [0, 1)$. Then $\sup X = 1$ and $\inf X = 0$. Note that $\sup X$ is not an element of X. □

Proposition 2.22. *Let $X \subset \mathbb{R}$ be a nonempty set of real numbers and $M \in \mathbb{R}$. The following statements are equivalent.*

(i) $M = \sup X$.
(ii) The number M is an upper bound for X and for any $\varepsilon > 0$ there exists $x \in X$ such that $x > M - \varepsilon$.

Proof. (i) $\Rightarrow$ (ii) Assume that M is the least upper bound of X. Then clearly M is an upper bound and we have to show that for any $\varepsilon > 0$ we can find a number $x \in X$ such that $x > M - \varepsilon$.

Because M is the least upper bound and $M - \varepsilon < M$, we deduce that $M - \varepsilon$ is *not* an upper bound for X. In other words, the opposite of (2.5) must be true, i.e., there must exist $x \in X$ such that x is not less or equal to $M - \varepsilon$.

(ii) $\Rightarrow$ (i) We have to show that if M' is another upper bound then $M \leq M'$. We argue by contradiction. Suppose that $M' < M$. Then $M' = M - \varepsilon$ for some positive number ε. The assumption (ii) implies that $x > M - \varepsilon$ for some number $x \in X$ so that $M' = M - \varepsilon$ is not an upper bound. We reached a contradiction which completes the proof. □

The Completeness Axiom. *Any nonempty set of real numbers that is bounded above* ***admits*** *a least upper bound.* □

From the completeness axiom we deduce the following result whose proof is left to you as Exercise 2.22.

Proposition 2.23. *If the nonempty set $X \subset \mathbb{R}$ is bounded below, then it admits a greatest lower bound.* □

Definition 2.24. Let $X \subset \mathbb{R}$ be a nonempty subset.

(i) We say that X admits a *maximal element* if X is bounded above and $\sup X \in X$. In this case we say that $\sup X$ is the maximum of X and it is denoted by $\max X$.
(ii) We say that X admits a *minimal element* if X is bounded below and $\inf X \in X$. In this case we say that $\inf X$ is called the *minimum* of X and it is denoted by $\min X$.

□

Note that the interval $I = [0, 1)$ has no maximal element, but it has a minimal element

$$\min I = 0.$$

2.4. Visualizing the real numbers

The approach we have adopted in introducing the real numbers differs from the historical course of things. For centuries scientists did not bother to ask what are the real numbers, often relying on intuition to prove things. This lead to various contradictory conclusions which prompted mathematicians to think more carefully about the concept of number and to treat the intuition more carefully.

This does not mean that the intuition stopped playing an important part in the modern mathematical thinking. On the contrary, intuition is still the first guide, but it always needs to be checked and backed by rigorous arguments.

For example, you learned to visualize the numbers as points on a line called the *real line*. We will not even attempt to explain what a line is. Instead we will rely on our physical intuition of this geometric concept. The real line is more than just a line, it is a line enriched with several attributes.

- It has a distinguished point called the *origin* which should be thought of as the real number 0.
- It is equipped with an *orientation*, i.e., a direction of running along the line visually indicated by an arrowhead at one end of the line; see Figure 2.1. Equivalently, the origin splits the line into two sides, and choosing an orientation is equivalent to declaring one side to be the *positive side* and the other side to be the *negative side*. Traditionally the above arrowhead points towards the positive side; see Figure 2.1.
- There is a way of measuring the distance between two points on the line.

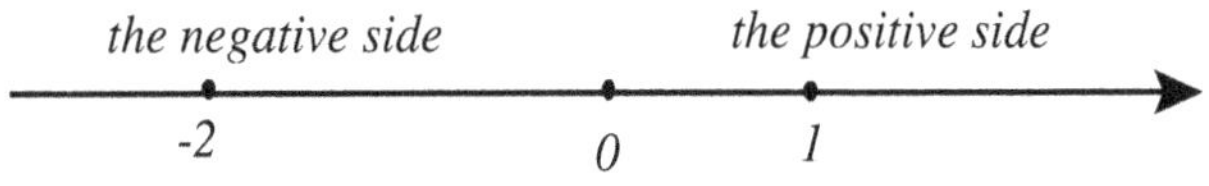

Fig. 2.1 *The real line.*

For example, the number -2 can be visualized as the point on the negative side situated at distance 2 from the origin; see Figure 2.1.

Now that we have identified the set $\mathbb{R}$ of real numbers with the set of points on a line, we can visualize the Cartesian product $\mathbb{R}^2 := \mathbb{R} \times \mathbb{R}$ with the set of points in a plane, called the *Cartesian plane*; see the top of Figure 2.2.

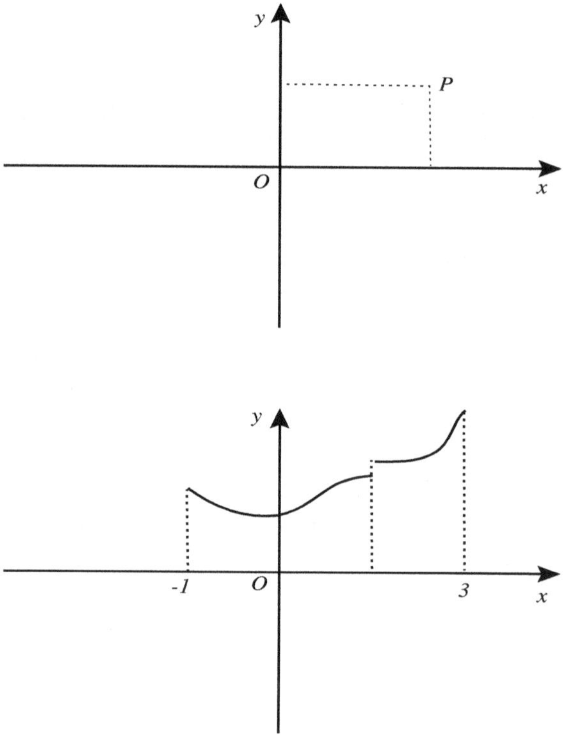

Fig. 2.2 *The real line.*

Just like the real line, the Cartesian plane is more than a plane: it is a plane enriched by several attributes.

- It contains a distinguished point, called the origin and denoted by O.
- It contains two distinguished perpendicular lines intersecting at O. These lines are called the *axes* of the Cartesian plane. One of the axes is declared to be horizontal and the other is declared to be vertical.
- Each of these two axes is a real line, i.e., it has the additional attributes of a real line: each has a distinguished point, O, each has an orientation, and each is equipped with a way measuring distances along that respective line. The horizontal axis is also known as the x-axis, while the vertical one is also known as the y-axis.

The position of a point P in that plane is determined by a pair of real numbers called the *Cartesian coordinates* of that point. These two numbers are obtained by intersecting the two axes with the lines through P which are perpendicular to the axes.

An interval of the real line can be visualized as a segment on the real line, possibly with one or both endpoints removed. If I is an interval of the real line and $f : I \to \mathbb{R}$ is a function, then its graph looks typically like a curve in the Cartesian plane. For example, the bottom of Figure 2.2 depicts the graph of a function $f : [-1, 3] \to \mathbb{R}$.

2.5. Exercises

Exercise 2.1. (a) Prove Proposition 2.4.
(b) State and prove the multiplicative counterpart of Proposition 2.2. □

Exercise 2.2. Prove parts (ii)–(v) of Proposition 2.7. □

Exercise 2.3. (a) Prove that

$$(x+y)+(z+t) = \big((x+y)+z\big)+t, \quad \forall x,y,z,t \in \mathbb{R}.$$

(b) Prove that for any $x, y, z, t, \in \mathbb{R}$ the sum $x+y+z+t$ is independent of the manner in which parentheses are inserted. □

Exercise 2.4. Prove parts (iii)–(v) of Proposition 2.9. □

Exercise 2.5. Show that for any real numbers x, y, z such that $y, z \neq 0$, we have

$$\frac{xz}{yz} = \frac{x}{y}.$$ □

Exercise 2.6. (a) Show that for any real numbers x, y, z, t such that $y, t \neq 0$ we have the equality

$$\frac{x}{y} + \frac{z}{t} = \frac{xt+yz}{yt}.$$

(b) Prove that for any real numbers x, y we have

$$x^2 - y^2 = (x-y)(x+y).$$

(c) Prove that the function $f : (0, \infty) \to \mathbb{R}$, $f(x) = x^2$, is injective but not surjective. □

Exercise 2.7. Prove that if $x \leq y$ and $y \leq x$, then $x = y$. □

Exercise 2.8. (a) Prove that if $xy > 0$, then either $x > 0$ and $y > 0$, or $x < 0$ and $y < 0$.□

(b) Prove that if $x > 0$, then $1/x > 0$.
(c) Let $x > 0$. Show that $x > 1$ if and only if $1/x < 1$.
(d) Prove that if $y > x \geq 1$, then

$$x + \frac{1}{x} < y + \frac{1}{y}.$$ □

Exercise 2.9. (a) Prove that $x^2 > 0$ for any $x \in \mathbb{R}$, $x \neq 0$.
(b) Consider the functions

$$f, g : \mathbb{R} \to \mathbb{R}, \quad f(x) = x^2 + 1, \quad g(x) = 2x + 1.$$

Decide if any of these two functions is injective or surjective.
(c) With f and g as above, describe the functions $f \circ g$ and $g \circ f$. □

Exercise 2.10. Using the technique described in Example 2.14 find all the real numbers x such that

$$\frac{x^2}{(x-1)(x+2)} \leq 1. \qquad \square$$

Exercise 2.11. (a) Find a positive number M with the following property:

$$\forall x : x > M \Rightarrow \frac{x^2}{x+1} > 10^5.$$

(b) Find a positive number M with the following property:

$$\forall x : \quad x > M \Rightarrow \frac{x^2}{x-1} > 10^6.$$

(c) Find a real number M with the following property:

$$\forall x : \quad x > M \Rightarrow \left| \frac{x^2}{(x-1)(x-2)} - 1 \right| < \frac{1}{100}. \qquad \square$$

Exercise 2.12. Let $a < b$. Show that

$$a < \frac{1}{2}(a+b) < b,$$

where 2 is the real number $2 := 1 + 1$. Conclude that the interval (a, b) is nonempty. $\square$

Exercise 2.13. Prove that $x^2 + y^2 \geq 2xy$, for any $x, y \in \mathbb{R}$. Use this inequality to prove that

$$x^2 + y^2 + z^2 \geq xy + yz + zx, \quad \forall x, y, z \in \mathbb{R}. \qquad \square$$

Exercise 2.14. Prove that if $0 \leq x \leq \varepsilon$, $\forall \varepsilon > 0$, then $x = 0$.[2] $\square$

Exercise 2.15. (a) Consider the function $f : [0, 2] \to \mathbb{R}$ given by

$$f(x) = \begin{cases} 0, & x \in [0, 1], \\ 1, & x \in (1, 2]. \end{cases}$$

Decide which of the following statements is true.

(i) $\exists L > 0$ such that $\forall x_1, x_2 \in [0, 2]$ we have $|f(x_1) - f(x_2)| \leq L|x_1 - x_2|$.
(ii) $\forall x_1, x_2 \in [0, 2]$, $\exists L > 0$ such that $|f(x_1) - f(x_2)| \leq L|x_1 - x_2|$.

(b) Same question, when we change the definition of f to $f(x) = x^2$, for all $x \in [0, 2]$. $\square$

Exercise 2.16. Show that for any $\delta > 0$ and any $a \in \mathbb{R}$ we have

$$(a - \delta, a + \delta) = \{ x \in \mathbb{R};\ |x - a| < \delta \}. \qquad \square$$

[2] The Greek letter ε (read *epsilon*) is ubiquitous in analysis and it is almost exclusively used to denote quantities that are extremely small.

Exercise 2.17. Prove the statements (ii)–(iv) of Proposition 2.12. □

Exercise 2.18. Prove that for any real numbers a, b, c we have

$$\operatorname{dist}(a, c) \leq \operatorname{dist}(a, b) + \operatorname{dist}(b, c).$$ □

Exercise 2.19. Prove that a set $X \subset \mathbb{R}$ is bounded if and only if there exists $C > 0$ such that $|x| \leq C, \forall x \in X$. □

Exercise 2.20. Fix two real numbers a, b such that $a < b$. Prove that for any $x, y \in [a, b]$ we have

$$|x - y| \leq b - a.$$ □

Exercise 2.21. State and prove the version of Proposition 2.22 involving the infimum of a bounded below set $X \subset \mathbb{R}$. □

Exercise 2.22. Let $X \subset \mathbb{R}$ be a nonempty set of real numbers. For $c \in \mathbb{R}$ define

$$cX := \{\, cx; \;\; x \in X \,\} \subset \mathbb{R}.$$

(i) Show that if $c > 0$ and X is bounded above, then cX is bounded above and

$$\sup cX = c \sup X.$$

(ii) Show that if $c < 0$ and X is bounded above, then cX is bounded below and

$$\inf cX = c \sup X.$$

□

Exercise 2.23. (a) Let

$$A := \left\{ \frac{a}{a+1}; \;\; a > 0 \right\}.$$

Compute $\inf A$ and $\sup A$.

(b) Let

$$B := \left\{ \frac{b}{b+1}; \;\; b \in \mathbb{R} \setminus \{-1\} \right\}.$$

Prove that the set B is not bounded below or above. □

Chapter 3

Special Classes of Real Numbers

3.1. The natural numbers and the induction principle

The numbers of the form

$$1, \quad 1+1, \quad (1+1)+1$$

and so forth are denoted respectively by $1, 2, 3, \ldots$ and are called *natural numbers*. The term *and so forth* is rather ambiguous and its rigorous justification is provided by the *principle of mathematical induction*.

Definition 3.1. A set $X \subset \mathbb{R}$ is called *inductive* if

$$\forall x : \quad (x \in X \Rightarrow x+1 \in X).$$

□

Example 3.2. The set $\mathbb{R}$ is inductive and so is any interval (a, ∞). If $(X_a)_{a \in A}$ is a collection of inductive sets, then so is their intersection

$$\bigcap_{a \in A} X_a.$$

□

Definition 3.3. The set of *natural numbers* is the smallest inductive set containing 1, i.e., the intersection of all inductive sets that contain 1. The set of natural numbers is denoted by $\mathbb{N}$. □

To unravel the above definition, the set $\mathbb{N}$ is the subset of $\mathbb{R}$ uniquely characterized by the following requirements.

- The set $\mathbb{N}$ is inductive and $1 \in \mathbb{N}$.
- If $S \subset \mathbb{R}$ is an inductive set that contains 1, then $\mathbb{N} \subset S$.

The set $\mathbb{N}$ consists of the numbers

$$1, \quad 2 := 1+1, \quad 3 := 2+1, \quad 4 := 3+1, \ldots .$$

Note that $0 \notin \mathbb{N}$. Indeed, the interval $[1, \infty)$ is an inductive set, containing 1 and thus must contain $\mathbb{N}$. On the other hand, this interval does not contain 0. The above argument proves that $\mathbb{N} \subset [1, \infty)$, i.e.,

$$n \geq 1, \ \ \forall n \in \mathbb{N}. \tag{3.1}$$

We set

$$\boxed{\mathbb{N}_0 := \{0\} \cup \mathbb{N} = \{\, 0, 1, 2, , 3, \dots, \}.}$$

☆ **The Principle of Mathematical Induction.** *If E is an inductive subset of the set of natural numbers such that $1 \in E$, then $E = \mathbb{N}$.*

In applications the set E consists of the natural numbers n satisfying a property $P(n)$. To prove that any natural number n satisfies the property $P(n)$ it suffices to prove two things.

- Prove $P(1)$. This is called the *initial step.*
- Prove that if $P(n)$ is true, then so is $P(n+1)$. This is called the *inductive step.*

Sometimes we need an alternate version of the induction principle.

☆ **The Principle of Mathematical Induction: alternate version.** Suppose that for any natural number n we are given a statement $P(n)$ and we know the following.

- The statement $P(1)$ is true.
- For any $n \in \mathbb{N}$, if $P(k)$ is true for any $k < n$, then $P(n)$ is true as well.

Then $P(n)$ is true for any $n \in \mathbb{N}$. □

We will spend the rest of this section presenting various instances of the induction principle at work.

Proposition 3.4. *The sum and the product of two natural numbers are also natural numbers.*

Proof.[1] Fix a natural number m. For each $n \in \mathbb{N}$ consider the statement

$$P(n) := \ \ m + n \text{ is a natural number.}$$

We have to prove that $P(n)$ is true for any $n \in \mathbb{N}$. We will achieve this using the principle of induction. We first need to check that $P(1)$ is true, i.e., that $m+1$ is a natural number. This follows from the fact that $m \in \mathbb{N}$ and $\mathbb{N}$ is an inductive set.

To complete the inductive step assume that $P(n)$ is true, i.e., $m + n \in \mathbb{N}$. Thus $(m+n)+1 \in \mathbb{N}$ and

$$m + (n+1) = (m+n) + 1 \in \mathbb{N}.$$

This shows that $P(n+1)$ is also true. □

[1]The proof can be omitted.

Lemma 3.5. $\forall n \in \mathbb{N}$, $(n \neq 1) \Rightarrow (n-1) \in \mathbb{N}$.

Proof. For $n \in \mathbb{N}$ consider the statement

$$P(n) := n \neq 1 \Rightarrow (n-1) \in \mathbb{N}.$$

We want to prove that this statement is true for any $n \in \mathbb{N}$. The initial step is obvious since for $n = 1$ the statement $n \neq 1$ is false and thus the implication is true.

For the inductive step assume that the statement $P(n)$ is true and we prove that $P(n+1)$ is also true. Observe that $n + 1 \neq 1$ because $n \in \mathbb{N}$ and thus $n \neq 0$. Clearly $(n+1) - 1 = n \in \mathbb{N}$. □

Lemma 3.6. *The set*

$$I_1 = \{x \in \mathbb{N};\ \ x > 1\}$$

admits a minimal element and $\min I_1 = 2$.

Proof. Consider the set

$$E := \{x \in \mathbb{N};\ \ x = 1 \vee x \geq 2\} \subset \mathbb{N}.$$

We will prove by induction that

$$E = \mathbb{N}. \tag{3.2}$$

Thus we need to show that $1 \in E$ and $x \in E \Rightarrow x + 1 \in E$. Clearly $1 \in E$.

If $x \in E$, then

- either $x = 1$ so that $x + 1 = 2 \geq 2$ so that $x + 1 \in E$,
- or $x \geq 2$ which implies $x + 1 \geq 2$ and thus $x + 1 \in E$.

The equality $E = \mathbb{N}$ implies that a natural number n is either equal to 1, or it is ≥ 2. Thus

$$x \in \mathbb{N} \wedge x > 1 \Rightarrow x \geq 2.$$

This shows that

$$x \geq 2,\ \ \forall x \in I_1.$$

Clearly $2 \in I_1$ so that $2 = \min I$. □

Corollary 3.7. *For any* $n \geq 1$ *the set*

$$H_n = \{x \in \mathbb{N};\ \ x > n\}$$

admits a minimal element and

$$\min H_n = n + 1.$$

Proof. We will prove that for any $n \in \mathbb{N}$ the statement

$$P(n):\ \ \min H_n = n + 1$$

is true. Lemma 3.6 shows that $P(1)$ is true.

Let us show that $P(n) \Rightarrow P(n+1)$. Since $n + 2 \in H_{n+1}$ it suffices to show that $x \geq n + 2$, $\forall x \in H_{n+1}$. Let $x \in H_{n+1}$. Lemma 3.5 implies that $x - 1 \in \mathbb{N}$ and $x - 1 > n$ so that $x - 1 \in H_n$. Since $P(n)$ is true, we deduce $x - 1 \geq n + 1$, i.e., $x \geq n + 2$. □

Corollary 3.8. *Suppose that* n *is a natural number. Any natural number* x *such that* $x > n$ *satisfies* $x \geq n + 1$. □

Corollary 3.9. *For any natural number* n, *the open interval* $(n, n+1)$ *contains no natural number.*

Proof. From Corollary 3.8 we deduce that if x is a natural number such that $x > n$, then $x \geq n+1$. Thus there cannot exist any natural number x such that $n < x < n+1$. □

The above results imply the following important theorem.

Theorem 3.10 (Well Ordering Principle). *Any set of natural numbers $S \subset \mathbb{N}$ has a minimal element.* □

For a proof of this theorem we refer to [23, §2.2.1].

Definition 3.11. For any $n \in \mathbb{N}$ we denote by $\mathbb{I}_n$ the set

$$\mathbb{I}_n := \{x \in \mathbb{N};\ 1 \leq x \leq n\} = [1, n] \cap \mathbb{N}.$$ □

Definition 3.12. We say two sets X and Y are said to have the *same cardinality*, and we write this $X \sim Y$, if and only if there exists a bijection $f : X \to Y$. A set X is called *finite* if there exists a natural number n such that $X \sim \mathbb{I}_n$. □

Let us observe that if X, Y, Z are three sets such that $X \sim Y$ and $Y \sim Z$, then $X \sim Z$; see Exercise 3.1. This implies that any set X equivalent to a finite set Y is also finite. Indeed, if $X \sim Y$ and $Y \sim \mathbb{I}_n$, then $X \sim \mathbb{I}_n$.

At this point we want to invoke (without proof) the following result.

Proposition 3.13. *For any $m, n \in \mathbb{N}$ we have*

$$\mathbb{I}_n \sim \mathbb{I}_m \Longleftrightarrow m = n.$$ □

The above result implies that if X is a finite set, then there exists a ***unique*** natural number n such that $X \sim \mathbb{I}_n$. This unique natural number is called the *cardinality* of X and it is denoted by $|X|$ or $\#X$. You should think of the cardinality of a finite set as the number of elements in that set.

An *infinite* set is a set that is not finite. We have the following *highly nontrivial* result. Its proof is too complex to present here.

Theorem 3.14. *A set X is infinite if and only if it is equivalent to one of its* proper[2] *subsets.* □

Theorem 3.15. *The set of natural numbers $\mathbb{N}$ is infinite.*

Proof. Consider the proper subset

$$H := \{n \in \mathbb{N};\ n > 1\} \subset \mathbb{N}.$$

Lemma 3.5 implies that if $n \in H$, then $(n-1) \in \mathbb{N}$. Consider the map

$$f : H \to \mathbb{N},\ \ f(n) = n - 1.$$

[2] We recall that a subset $S \subset X$ is called proper if $S \neq X$.

Observe that this map is injective. Indeed, if $f(n_1) = f(n_2)$, then $n_1 - 1 = n_2 - 1$ so that $n_1 = n_2$. This map is also surjective. Indeed, if $m \in \mathbb{N}$, then, according to (3.1) the natural number $n := m + 1$ is greater than 1 so it belongs to H. Clearly $f(n) = (m+1) - 1 = m$ which proves that f is also surjective. □

Definition 3.16. A set X is called *countable* if it is equivalent with the set of natural numbers. □

Example 3.17. The set $\mathbb{N} \times \mathbb{N}$ is countable. To see this arrange the elements of $\mathbb{N} \times \mathbb{N}$ in a sequence as follows:

$$(1,1),\ \underbrace{(2,1),(2,2)}_{S_2},\ \underbrace{(3,1),(3,2),(3,3)}_{S_3},\ \underbrace{(4,1),(4,2),(4,3),(4,4)}_{S_4}.$$

Now denote by $\phi(m,n)$ the location of the pair (m,n) in the above string. For example, $\phi(1,1) = 1$ since $(1,1)$ is the first term in the above sequence. Note that

$$\phi(4,2) = 1 + 2 + 3 + 2 = 8,$$

i.e., $(4,2)$ occupies the 8-th position in the above string. More precisely ϕ is the function

$$\phi : \mathbb{N} \times \mathbb{N} \to \mathbb{N},\ \ \phi(m,n) = \#S_1 + \cdots \#S_{m-1} + n.$$

It should be clear that ϕ is bijective proving that $\mathbb{N} \times \mathbb{N}$ has the same cardinality as $\mathbb{N}$. □

3.2. Applications of the induction principle

In this section we discuss some traditional applications of the induction principle. This serves two purposes: first, it familiarizes you with the usage of this principle, and second, some of the results we will discuss here will be needed later on in this class.

First let us introduce some notations. If n is a natural number, $n > 1$, and we are given n real numbers $a_1, \dots, a_n$, then define inductively

$$a_1 + \cdots + a_n := (a_1 + \cdots + a_{n-1}) + a_n,$$

$$a_1 \cdots a_n = (a_1 \cdots a_{n-1}) a_n.$$

We will use the following notations for the sum and products of a string of real numbers. Thus

$$\sum_{k=1}^{n} a_k := a_1 + \cdots + a_n,\ \ \prod_{k=1}^{n} a_k := a_1 \cdots a_n.$$

Similarly, given real numbers $a_0, a_1, \dots, a_n$ we define

$$\sum_{k=0}^{n} a_k = a_0 + a_1 + \cdots + a_n,\ \ \prod_{k=0}^{n} a_k := a_0 \cdots a_n.$$

For any natural number n and any real number x we define inductively

$$x^n := \begin{cases} x & \text{if } n = 1 \\ (x^{n-1}) \cdot x & \text{if } n > 1. \end{cases}$$

Intuitively

$$x^n = \underbrace{x \cdot x \cdots x}_{n}.$$

If x is a *nonzero* real number we set

$$x^0 := 1.$$

Let us observe that for any natural numbers m, n and any real number x we have the equality

$$x^{m+n} = (x^m) \cdot (x^n).$$

Exercise 3.2 asks you to prove this fact.

Example 3.18. Let us prove that

$$\sum_{k=1}^{n} k = \frac{n(n+1)}{2}, \quad \forall n \in \mathbb{N}. \tag{3.3}$$

The expanded form of the last equality is

$$1 + 2 + \cdots + n = \frac{n(n+1)}{2}, \quad \forall n \in \mathbb{N}.$$

Let us denote by S_n the sum $1+2+\cdots+n$. We argue by induction. The initial case $n = 1$ is trivial since

$$\frac{1 \cdot (1+1)}{2} = 1 = S_1.$$

For the inductive case we assume that

$$S_n = \frac{n(n+1)}{2},$$

and we have to prove that

$$S_{n+1} = \frac{(n+1)(\,(n+1)+1)}{2} = \frac{(n+1)(n+2)}{2}.$$

Indeed we have

$$S_{n+1} = S_n + (n+1) = \frac{n(n+1)}{2} + n + 1 = \frac{n(n+1)}{2} + \frac{2(n+1)}{2} = \frac{n(n+1) + 2(n+1)}{2}$$

(factor out $(n+1)$)

$$= \frac{(n+1)(n+2)}{2}.$$

□

Example 3.19 (Bernoulli's inequality). We want to prove a simple but very versatile inequality that goes by the name of *Bernoulli's inequality.* More precisely it states that inequality

$$\forall x \geq -1, \ \forall n \in \mathbb{N}: \ (1+x)^n \geq 1+nx. \tag{3.4}$$

We argue by induction. Clearly, the inequality is obviously true when $n = 1$ and the initial case is true. For the inductive case, we assume that

$$(1+x)^n \geq 1+nx, \ \forall x \geq -1 \tag{3.5}$$

and we have to prove that

$$(1+x)^{n+1} \geq 1+(n+1)x, \ \forall x \geq -1.$$

Since $x \geq -1$ we deduce $1+x \geq 0$. Multiplying both sides of (3.5) with the nonnegative number $1+x$ we deduce

$$(1+x)^{n+1} \geq (1+x)(1+nx) = 1+nx+x+nx^2 \geq 1+nx+x = 1+(n+1)x. \ \square$$

Example 3.20 (*Newton's Binomial Formula*). Before we state this very important formula we need to introduce several notations widely used in mathematics. For $n \in \mathbb{N} \cup \{0\}$ we define $n!$ (read n factorial) as follows

$$0! := 1, \ 1! := 1, \ 2! = 1 \cdot 2, \ 3! = 1 \cdot 2 \cdot 3, \cdots n! = 1 \cdot 2 \cdots n.$$

Given $k, n \in \mathbb{N} \cup \{0\}$, $k \leq n$ we define the *binomial coefficient* $\binom{n}{k}$ (read n *choose* k)

$$\binom{n}{k} := \frac{n!}{k! \cdot (n-k)!}.$$

We record below the values of these binomial coefficients for small values of n

$$\binom{0}{0} = 1, \ \binom{1}{0} = \binom{1}{1} = 1,$$

$$\binom{2}{0} = \frac{2!}{(0!)(2!)} = 1, \ \binom{2}{1} = \frac{2!}{(1!)(1!)} = 2, \ \binom{2}{2} = \frac{2!}{(2!)(0!)} = 1,$$

$$\binom{3}{0} = \frac{3!}{(0!)(3!)} = \binom{3}{3} = 1, \ \binom{3}{1} = \frac{(3!)}{(1!)(2!)} = \frac{3!}{(2!)(1!)} = \binom{3}{2} = 3.$$

Here is a more involved example

$$\binom{7}{3} = \frac{7!}{(3!)(4!)} = \frac{7 \cdot 6 \cdot 5 \cdot 4 \cdot 3 \cdot 2 \cdot 1}{(3!)1 \cdot 2 \cdot 3 \cdot 4} = \frac{7 \cdot 6 \cdot 5}{3!} = \frac{7 \cdot 6 \cdot 5}{6} = 35.$$

The binomial coefficients can be conveniently arranged in the so-called *Pascal triangle*

$$\begin{array}{ll}
\binom{0}{k}: & 1 \\
\binom{1}{k}: & 1 \quad 1 \\
\binom{2}{k}: & 1 \quad 2 \quad 1 \\
\binom{3}{k}: & 1 \quad 3 \quad 3 \quad 1 \\
\binom{4}{k}: & 1 \quad 4 \quad 6 \quad 4 \quad 1 \\
\vdots & \vdots \ \vdots \ \vdots \ \vdots \ \vdots \ \vdots \ \vdots \ \vdots \ \vdots
\end{array}$$

Observe that each entry in the Pascal triangle is the sum of the closest neighbors above it.

The binomial coefficients play an important role in mathematics. One reason behind their usefulness is *Newton's binomial formula* which states that, for any natural number n, and any real numbers x, y, we have the equality below

$$\begin{aligned}(x+y)^n &= \binom{n}{0}x^n + \binom{n}{1}x^{n-1}y + \binom{n}{2}x^{n-2}y^2 + \cdots + \binom{n}{n-1}xy^{n-1} + \binom{n}{n}y^n \\ &= \sum_{k=0}^{n}\binom{n}{k}x^{n-k}y^k.\end{aligned} \tag{3.6}$$

We will prove this equality by induction on n. For $n = 1$ we have

$$(x+y)^1 = x + y = \binom{1}{0}x + \binom{1}{1}y,$$

which shows that the case $n = 1$ of (3.6) is true.

As for the inductive steps, we assume that (3.6) is true for n and we prove that it is true for $n + 1$. We have

$$(x+y)^{n+1} = (x+y)(x+y)^n$$

(use the inductive assumption)

$$\begin{aligned}
&= (x+y)\left(\binom{n}{0}x^n + \binom{n}{1}x^{n-1}y + \binom{n}{2}x^{n-2}y^2 + \cdots + \binom{n}{n-1}xy^{n-1} + \binom{n}{n}y^n\right) \\
&= x\left(\binom{n}{0}x^n + \binom{n}{1}x^{n-1}y + \binom{n}{2}x^{n-2}y^2 + \cdots + \binom{n}{n-1}xy^{n-1} + \binom{n}{n}y^n\right) \\
&\quad + y\left(\binom{n}{0}x^n + \binom{n}{1}x^{n-1}y + \binom{n}{2}x^{n-2}y^2 + \cdots + \binom{n}{n-1}xy^{n-1} + \binom{n}{n}y^n\right) \\
&= \binom{n}{0}x^{n+1} + \binom{n}{1}\boxed{x^ny} + \binom{n}{2}\boxed{x^{n-1}y^2} + \cdots + \binom{n}{n-1}x^2y^{n-1} + \binom{n}{n}\boxed{xy^n} \\
&\quad + \binom{n}{0}\boxed{x^ny} + \binom{n}{1}\boxed{x^{n-1}y^2} + \binom{n}{2}x^{n-2}y^3 + \cdots + \binom{n}{n-1}\boxed{xy^n} + \binom{n}{n}y^{n+1} \\
&= \binom{n}{0}x^{n+1} + \left(\binom{n}{1} + \binom{n}{0}\right)\boxed{x^ny} + \left(\binom{n}{2} + \binom{n}{1}\right)\boxed{x^{n-1}y^2} + \cdots \\
&\quad + \left(\binom{n}{k} + \binom{n}{k-1}\right)x^{n+1-k}y^k + \cdots + \left(\binom{n}{n} + \binom{n}{n-1}\right)\boxed{xy^n} + \binom{n}{n}y^{n+1}.
\end{aligned}$$

Clearly

$$\binom{n}{0} = 1 = \binom{n+1}{0}, \quad \binom{n}{n} = 1 = \binom{n+1}{n+1}.$$

We want to show $1 \leq k \leq n$ we have the *Pascal's formula*

$$\binom{n+1}{k} = \binom{n}{k} + \binom{n}{k-1}. \tag{3.7}$$

Indeed, we have

$$\begin{aligned}\binom{n}{k}+\binom{n}{k-1} &= \frac{n!}{k!(n-k)!}+\frac{n!}{(k-1)!(n-k+1)!}\\ &= \frac{n!}{k(k-1)!(n-k)!}+\frac{n!}{(k-1)!(n-k)!(n-k+1)}\\ &= \frac{n!}{(k-1)!(n-k)!}\left(\frac{1}{k}+\frac{1}{n-k+1}\right)\\ &= \frac{n!}{(k-1)!(n-k)!}\cdot\left(\frac{(n-k+1)}{k(n-k+1)}+\frac{k}{k(n-k+1)}\right)\\ &= \frac{n!}{(k-1)!(n-k)!}\cdot\frac{n+1}{k(n-k+1)}\\ &= \frac{(n+1)n!}{(k(k-1)!)\cdot((n-k+1)(n-k)!)}=\frac{(n+1)!}{k!(n+1-k)!}=\binom{n+1}{k}.\end{aligned}$$

This completes the inductive step. □

3.3. Archimedes' Principle

We begin with a simple but fundamental observation.

Proposition 3.21. *Suppose that the nonempty subset $E \subset \mathbb{N}$ is bounded above. Then E has a maximal element n_0, i.e., $n_0 \in E$ and $n \leq n_0$, $\forall n \in E$.*

Proof. From the completeness axiom we deduce that E has a least upper bound $M = \sup E \in \mathbb{R}$. We want to prove that $M \in E$. We argue by contradiction. Suppose that $M \notin E$. In particular, this means that any number in E is strictly smaller than M.

From the definition of the *least* upper bound we deduce that there must exist $n_0 \in E$ such that

$$M - 1 < n_0 \leq M.$$

On the other hand, any natural number n greater than n_0 must be greater than or equal to n_0+1, $n \geq n_0+1$. Observing that $n_0+1 > M$, we deduce that any natural number $> n_0$ is also $> M$. Since $M \notin E$, then $n_0 < M$, and the above discussion show that the interval (n_0, M) contains no natural numbers, thus no elements of E. Hence, any real number in (n_0, M) will be an upper bound for E, contradicting that M is the *least* upper bound. □

Theorem 3.22 (Archimedes' Principle). *Let ε be a positive real number. Then for any $x > 0$ there exists $n \in \mathbb{N}$ such that $n\varepsilon > x$.*[3]

[3] A popular formulation of Archimedes' principle reads: one can fill an ocean with grains of sand.

Proof. Consider the set

$$E := \{ n \in \mathbb{N};\ \ n\varepsilon \leq x \}.$$

If $E = \emptyset$, then this means that $n\varepsilon > x$ for any $n \in \mathbb{N}$ and the conclusion of the theorem is guaranteed. Suppose that $E \neq \emptyset$. Observe that

$$n \leq \frac{x}{\varepsilon},\quad \forall n \in E.$$

Hence, the set E is bounded above, and the previous proposition shows that it has a maximal element n_0. Then $n_0 + 1 \notin E$, so that $(n_0 + 1)\varepsilon > x$. □

Definition 3.23. The set of *integers* is the subset $\mathbb{Z} \subset \mathbb{R}$ consisting of the natural numbers, the negatives of natural numbers and 0. □

The proof of the following results are left to you as an exercise.

Proposition 3.24. *If $m, n \in \mathbb{Z}$, then $m + n, mn \in \mathbb{Z}$.* □

Proposition 3.25. *For any real number x the interval $(x-1, x]$ contains exactly one integer.* □

Corollary 3.26. *For any real number x there exists a unique integer n such that*

$$n \leq x < n + 1.$$

This integer is called the ***integer part*** *of x and it is denoted by $\lfloor x \rfloor$.*

Proof. Observe that the inequalities $n \leq x < n + 1$ are equivalent to the inequalities

$$x - 1 < n \leq x.$$

By Proposition 3.25, the interval $(x - 1, x]$ contains exactly one integer. This proves the existence and uniqueness of the integer with the postulated properties. □

Observe for example that

$$\left\lfloor \frac{1}{2} \right\rfloor = 0,\quad \left\lfloor -\frac{1}{2} \right\rfloor = -1,$$

Theorem 3.27 (Division with remainder). *Let $m, n \in \mathbb{Z}$, $n > 0$. There exists a unique pair of integers $(q, r) \in \mathbb{Z} \times \mathbb{Z}$ satisfying the following properties.*

(i) $m = qn + r$.
(ii) $0 \leq r < n$.

Proof. *Uniqueness.* Suppose that there exist two pairs of integers (q_1, r_1) and (q_2, r_2) satisfying (i) and (ii). Then

$$nq_1 + r_1 = m = nq_2 + r_2,$$

so that,

$$nq_1 - nq_2 = r_2 - r_1 \Rightarrow n(q_1 - q_2) = r_2 - r_1 \Rightarrow n \cdot |q_1 - q_2| = |r_2 - r_1|.$$

The natural numbers r_1, r_2 satisfy $0 \leq r_1, r_2 < |n|$ so that $r_1, r_2 \in [0, n-1]$. Using Exercise 2.20 we deduce $|r_2 - r_1| \leq n - 1$. Hence $n \cdot |q_1 - q_2| \leq n - 1$ which implies

$$|q_1 - q_2| \leq \frac{n-1}{n} < 1.$$

The quantity $|q_1 - q_2|$ is a nonnegative integer < 1 so that it must equal 0. This implies $q_1 = 2$ and

$$r_2 - r_1 = n(q_1 - q_2) = 0.$$

This proves the uniqueness.

Existence. Let

$$q := \left\lfloor \frac{m}{n} \right\rfloor \in \mathbb{Z}.$$

Then

$$q \leq \frac{m}{n} < q + 1 \Rightarrow nq \leq m < n(q+1) = nq + n \Rightarrow 0 \leq m - nq < n.$$

We set $r := m - nq$ and we observe that the pair (q, r) satisfies all the required properties. □

Definition 3.28. (a) Let $m, n \in \mathbb{Z}$, $m \neq 0$. We say that m *divides* n, and we write this $m|n$ if there exists an integer k such that $n = km$. When m divides n we also say that m is a *divisor* of n, or that n *is a multiple of* m, or that n *is divisible by* m.
(b) A *prime number* is a natural number $p > 1$ whose only divisors are ± 1 and $\pm p$. □

Observe that if d is a divisor of m, then $-d$ is also a divisor of m. An *even integer* is an integer divisible by 2. An *odd integer* is an integer not divisible by 2.

Given two integers m, n consider the set of common positive divisors of m and n, i.e., the set

$$D_{m,n} := \{ d \in \mathbb{N};\ \ d|m \wedge d|n \}.$$

This set is not empty because 1 is a common positive divisor. This is bounded above because any divisor of m is $\leq |m|$. Thus the set $D_{m,n}$ has a maximal element called the *greatest common divisor* of m and n and denoted by $\gcd(m, n)$. Two integers are called *coprime* if $\gcd(m, n) = 1$, i.e., 1 is their only positive common divisor.

The next result describes on the most important property of the set $\mathbb{Z}$ of integers. We will not include its rather elaborate and tricky proof. The curious reader can find the proof in any of the books [2, 3, 17].

Theorem 3.29 (Fundamental Theorem of Arithmetic). *(a) If p is a prime number that divides a product of integers mn, then $p|m$ or $p|n$.*
(b) Any natural number n can be written in a unique fashion as a product

$$n = p_1^{\alpha_1} \cdots p_k^{\alpha_k},$$

where $p_1 < p_2 < \cdots < p_k$ are prime numbers and $\alpha_1, \ldots, \alpha_k$ are natural numbers. □

3.4. Rational and irrational numbers

We want to isolate another important subclasses of real numbers.

Definition 3.30. The set of *rationals* (or *rational numbers*) is the subset $\mathbb{Q} \subset \mathbb{R}$ consisting of real numbers of the form m/n where $m, n \in \mathbb{Z}$, $n \neq 0$. □

If q is a rational number, then it can be written as a fraction of the form $q = \frac{m}{n}$, $n \neq 0$. We denote by d the $\gcd(m, n)$. Thus there exist integers m_1 and n_1 such that

$$m = dm_1, \quad n = dn_1.$$

Clearly the numbers m_1, n_1 are coprime, and we have

$$q = \frac{dm_1}{dn_1} = \frac{m_1}{n_1}.$$

We have thus proved the following result.

Proposition 3.31. *Every rational number is the ratio of two coprime integers.* □

The proof of the following result is left to you as an exercise.

Proposition 3.32. *If $q, r \in \mathbb{Q}$, then $q + r, qr \in \mathbb{Q}$.* □

We have a sequence of inclusions

$$\mathbb{N} \subset \mathbb{Z} \subset \mathbb{Q} \subset \mathbb{R}.$$

Clearly $\mathbb{N} \neq \mathbb{Z}$ because $-1 \in \mathbb{Z}$, but $-1 \notin \mathbb{N}$. Note that, although $\mathbb{Z}$ contains $\mathbb{N}$, the set of integers $\mathbb{Z}$ is countable, i.e., it has the same cardinality as $\mathbb{N}$.

Next observe that $\mathbb{Z} \neq \mathbb{Q}$. Indeed, the rational number $1/2$ is not an integer, because it is positive and smaller than any natural number.

Similarly, although $\mathbb{Q}$ strictly contains $\mathbb{Z}$, these two sets have the same cardinality: they are both countable. However, the following **very important result** shows that, loosely speaking, there are "many more rational numbers".

Proposition 3.33 (Density of rationals). *Any open interval $(a, b) \subset \mathbb{R}$, no matter how small, contains at least one rational number.*

Proof. From Archimedes' Principle we deduce that there exists at least one natural number n such that $n > \frac{1}{b-a}$. Observe that $(b-a)$ is the length of the interval (a, b). This inequality is obviously equivalent to the inequality

$$\frac{1}{n} < b - a \Longleftrightarrow n(b-a) > 1.$$

(This last equality codifies a rather intuitive fact: one can divide a stick of length one into many equal parts so that the subparts are as small as we please.)

We will show that we can find an integer m such that $\frac{m}{n} \in (a,b)$. Observe that

$$a < \frac{m}{n} < b \Longleftrightarrow na < m < nb \Longleftrightarrow m \in (na, nb).$$

This shows that the interval (a,b) contains a rational number if the interval (na, nb) contains an integer.

Since $n(b-a) > 1$, we deduce $nb > na + 1$. In particular, this shows that the interval $(na, na+1]$ is contained in the interval (na, nb). From Proposition 3.25 we deduce that the interval $(na, na+1]$ contains an integer m. □

This abundance of rational numbers leads people to believe for quite a long while that all real numbers must be rational. Then the ancient Greeks showed that there must exist real numbers that cannot be rational. These numbers were called *irrational.* In the remainder of this section we will describe how one can produce a large supply of irrational numbers. We start with a baby case.

Proposition 3.34. *There exists a unique positive number r such that $r^2 = 2$. This number is called the square root of 2 and it is denoted by $\sqrt{2}$.*

Proof. We begin by observing the following *useful fact*:

$$\forall x, y > 0: \;\; x < y \Longleftrightarrow x^2 < y^2. \tag{3.8}$$

Indeed

$$y^2 - x^2 > 0 \Longleftrightarrow (y-x)(y+x) > 0 \Longleftrightarrow y > x.$$

This useful fact takes care of the uniqueness because, if r_1, r_2 are two *positive* real numbers such that $r_1^2 = r_2^2 = 2$, then $r_1 = r_2 \Longleftrightarrow r_1^2 = r_2^2$.

To establish the existence of a positive r such that $r^2 = 2$ consider as in Example 2.17(b) the set

$$R = \{ x > 0;\;\; x^2 < 2 \}.$$

We have seen that this set is bounded above and thus it admits a least upper bound

$$r := \sup R.$$

We want to prove that $r^2 = 2$. We argue by contradiction and we assume that $r^2 \neq 2$. Thus, either $r^2 < 2$ or $r^2 > 2$.

Case 1. $r^2 < 2$. We will show that there exists ε_0 such that $(r + \varepsilon_0)^2 < 2$. This would imply that $r + \varepsilon_0 \in R$ and would contradict the fact that r is an upper bound for R because r would be smaller than the element $r + \varepsilon_0$ of R.

Set $\delta := 2 - r^2$. For any $\varepsilon \in (0,1)$ we have

$$(r+\varepsilon)^2 - r^2 = \big((r+\varepsilon) - r \big)\big((r+\varepsilon) + r \big) = \varepsilon(2r+\varepsilon) < \varepsilon(2r+1).$$

Now choose a number $\varepsilon_0 \in (0,1)$ such that

$$\varepsilon_0 < \frac{\delta}{2r+1}.$$

Then

$$(r+\varepsilon_0)^2 - r^2 < \varepsilon_0(2r+1) < \delta$$

$$\Rightarrow (r+\varepsilon_0)^2 < r^2 + \delta = r^2 + 2 - r^2 = 2 \Rightarrow r+\varepsilon_0 \in R.$$

Case 2. $r^2 > 2$. We will prove that under this assumption

$$\exists \varepsilon_0 \in (0,1) \text{ such that } r - \varepsilon_0 > 0 \text{ and } (r-\varepsilon_0)^2 > 2. \tag{3.9}$$

Let us observe that (3.9) leads to a contradiction. Note that $(r - \varepsilon_0)$ is an upper bound for R. Indeed, if $x \in R$, then

$$x^2 < 2 < (r-\varepsilon_0)^2 \overset{(3.8)}{\Rightarrow} x < r_0 - \varepsilon.$$

Thus, $r - \varepsilon_0$ is an upper bound of R and this upper bound is obviously strictly smaller than r, the *least* upper bound of R. This is a contradiction which shows that the situation $r^2 > 2$ is also not possible. Let us now prove (3.9).

Denote by δ the difference $\delta = r^2 - 2 > 0$. For any $\varepsilon \in (0,r)$ we have

$$r^2 - (r-\varepsilon)^2 = \Big(r - (r-\varepsilon)\Big)\Big(r + (r-\varepsilon)\Big) = \varepsilon(2r-\varepsilon) < 2r\varepsilon.$$

We have thus shown that for any $\varepsilon \in (0,r)$ we have $(r-\varepsilon) > 0$ and

$$r^2 - (r-\varepsilon)^2 \leq 2r\varepsilon \iff (r-\varepsilon)^2 \geq r^2 - 2r\varepsilon.$$

Now choose $\varepsilon_0 \in (0,r)$ small enough so that $\varepsilon_0 < \frac{\delta}{2r}$. Hence $2r\varepsilon_0 < \delta$ so that $-2r\varepsilon_0 > -\delta$ and

$$(r-\varepsilon_0)^2 > r^2 - 2r\varepsilon_0 > r^2 - \delta = r^2 - (r^2-2) = 2.$$

We deduce again that the situation $r^2 > 2$ is not possible so that $r^2 = 2$. □

The result we have just proved can be considerably generalized.

Theorem 3.35. *Fix a natural number $n \geq 2$. Then for any positive real number a* there exists a unique, positive *real number r such that $r^n = a$.*

Proof. *Existence.* Consider the set

$$S := \big\{ s \in \mathbb{R};\ s \geq 0 \wedge s^n \leq a \big\}.$$

Observe that this is a nonempty set since $0 \in S$. We want to prove that S is also bounded. To achieve this we need a few auxiliary results.

Lemma 3.36 (A very handy identity). *For any real numbers x, y and any natural number n we have the equality*

$$x^n - y^n = (x-y)\cdot\big(x^{n-1} + x^{n-2}y + \cdots + xy^{n-2} + y^{n-1}\big). \tag{3.10}$$

Proof. We have

$$\begin{aligned}
&(x-y)\cdot\left(x^{n-1}+x^{n-2}y+\cdots+xy^{n-2}+y^{n-1}\right)\\
&=x\left(x^{n-1}+x^{n-2}y+\cdots+xy^{n-2}+y^{n-1}\right)-y\left(x^{n-1}+x^{n-2}y+\cdots+xy^{n-2}+y^{n-1}\right)\\
&=x^n+x^{n-1}y+x^{n-2}y^2+\cdots+x^2y^{n-2}+xy^{n-1}\\
&\quad -x^{n-1}y-x^{n-2}y^2-\cdots-x^2y^{n-2}-xy^{n-1}-y^n\\
&=x^n-y^n.
\end{aligned}$$

□

Here is an immediate useful consequence of this identity.

$$\forall n\in\mathbb{N},\ \ \forall x,y>0:\ \ x<y \Longleftrightarrow x^n<y^n. \tag{3.11}$$

Indeed

$$y^n-x^n>0 \Longleftrightarrow (y-x)\left(y^{n-1}+y^{n-2}x+\cdots+x^{n-1}\right)>0 \Longleftrightarrow y-x>0.$$

Lemma 3.37. *Any positive real number x such that $x^n\geq a$ is an upper bound for S. In particular, any natural number $k>a$ is an upper bound for S so that S is a bounded set.*

Proof. Let x be a positive real number such that $x^n\geq a$. We want to prove that $x\geq s$ for any $s\in S$. Indeed

$$s\in S\Rightarrow s^n\leq a\leq x^n \overset{(3.11)}{\Rightarrow} s\leq x.$$

This proves the first part of the lemma.

Suppose now that k is a natural number such that $k>a$. Observe first that

$$k^n>k^{n-1}>\cdots>k>a.$$

From the first part of the lemma we deduce that k is an upper bound for S. □

The nonempty set S is bounded above. The Completeness Axiom implies that it admits a least upper bound

$$r:=\sup S.$$

We will show that $r^n=a$. We argue by contradiction and we assume that $r^n\neq a$. Thus, either $r^n<a$, or $r^n>a$.

Case 1. $r^n<a$. We will show that we can find $\varepsilon_0\in(0,1)$ such that $(r+\varepsilon_0)^n<a$. This would imply that $r+\varepsilon_0\in S$ and it would contradict the fact that r is an upper bound for S because r is less than the number $r+\varepsilon_0\in S$.

Denote by δ the difference $\delta := a - r^n > 0$. For any $\varepsilon \in (0,1)$ we have

$$(r+\varepsilon)^n - r^n = \Big((r+\varepsilon) - r\Big)\Big((r+\varepsilon)^{n-1} + (r+\varepsilon)^{n-2}r^2 + \cdots + r^{n-1}\Big)$$

$(r+\varepsilon < r+1)$

$$\leq \varepsilon \underbrace{\Big((r+1)^{n-1} + (r+1)^{n-2}r + \cdots + r^{n-1}\Big)}_{=:q}.$$

We have thus proved that

$$(r+\varepsilon)^n \leq r^n + \varepsilon q, \quad \forall \varepsilon \in (0,1).$$

Choose $\varepsilon_0 \in (0,1)$ small enough so that

$$\varepsilon_0 < \frac{\delta}{q} \iff \varepsilon_0 q < \delta.$$

Then

$$(r+\varepsilon_0)^n \leq r^n + \varepsilon_0 q < r^n + \delta = a \Rightarrow r + \varepsilon_0 \in S.$$

This contradicts the fact that r is an upper bound for S and shows that the inequality $r^n < a$ is impossible.

Case 2. $r^n > a$. We will prove that under this assumption

$$\exists \varepsilon_0 \in (0,1) \text{ such that } r - \varepsilon_0 > 0 \text{ and } (r-\varepsilon_0)^n > a. \tag{3.12}$$

Let us observe that (3.12) leads to a contradiction. Indeed, Lemma 3.37 implies that $r - \varepsilon_0$ is an upper bound of S and this upper bound is obviously strictly smaller than r, the *least* upper bound of S. This is a contradiction which shows that the situation $b^n > a$ is also not possible. Let us now prove (3.12).

Denote by δ the difference $\delta = r^n - a > 0$. For any $\varepsilon \in (0,r)$ we have

$$r^n - (r-\varepsilon)^n = \Big(r - (r-\varepsilon)\Big)\Big(r^{n-1} + r^{n-2}(r-\varepsilon) + \cdots + \cdots + (r-\varepsilon)^{n-1}\Big)$$

$((r-\varepsilon) < b)$

$$\leq \varepsilon \underbrace{\left(r^{n-1} + r^{n-2}r + \cdots + r^{n-1}\right)}_{=:q}.$$

We have thus shown that for any $\varepsilon \in (0,r)$ we have $(r-\varepsilon) > 0$ and

$$r^n - (r-\varepsilon)^n \leq \varepsilon q \iff (r-\varepsilon)^n \geq r^n - \varepsilon q.$$

Now choose $\varepsilon_0 \in (0,c)$ small enough so that $\varepsilon_0 < \frac{\delta}{q}$. Hence $\varepsilon_0 q < \delta$ so that $-\varepsilon_0 q > -\delta$ and

$$(r-\varepsilon_0)^n > r^n - \varepsilon_0 q > r^n - \delta = r^n - (r^n - a) = a.$$

We deduce again that the situation $r^n > a$ is not possible so that $r^n = a$. This completes the existence part of the proof.

Uniqueness. Suppose that r_1, r_2 are two positive numbers such that $r_1^n = r_2^n = a$. Using (3.11) we deduce that $r_1 = r_2$. This completes the proof of Theorem 3.35. □

The above result leads to the following important concept.

Definition 3.38. Let a be a positive real number and $n \in \mathbb{N}$. The *n-th root* of a, denoted by $a^{\frac{1}{n}}$ or $\sqrt[n]{a}$ is the ***unique positive real number*** r such that $r^n = a$. □

Theorem 3.39. *The positive number* $\sqrt{2}$ *is not rational.*

Proof. We argue by contradiction and we assume that $\sqrt{2}$ is rational. It can therefore be represented as a fraction,

$$\sqrt{2} = \frac{m}{n}, \quad m, n \in \mathbb{N}, \quad \gcd(m, n) = 1.$$

Thus $2 = \frac{m^2}{n^2}$ and we deduce

$$2n^2 = m^2. \tag{3.13}$$

Since 2 is a prime number and $2|m^2$ we deduce that $2|m$, i.e., $m = 2m_1$ for some natural number m_1. Using this last equality in (3.13) we deduce

$$2n^2 = (2m_1)^2 = 4m_1^2 \Rightarrow n^2 = 2m_1^2.$$

Thus $2|n^2$, and arguing as above we deduce that $2|n$. Hence 2 is a common divisor of both m and n. This contradicts the starting assumption that $\gcd(m, n) = 1$ and proves that $\sqrt{2}$ cannot be rational. □

Now that we know there exist irrational numbers, we can ask, how many there are. It turns out that most real numbers are irrational, but we will not prove this fact now.

3.5. Exercises

Exercise 3.1. (a) Suppose that X, Y are two sets such that $X \sim Y$. Prove that $Y \sim X$.
(b) Prove that if X, Y, Z are sets such that $X \sim Y$ and $Y \sim Z$, then $X \sim Z$. □

Exercise 3.2. Prove by induction that for any natural numbers m, n and any real number x we have the equality

$$x^{m+n} = (x^m) \cdot (x^n).$$ □

Exercise 3.3. (a) Prove that for any natural number n and any real numbers

$$a_1, a_2, \dots, a_n, b_1, \dots, b_n, c$$

we have the equalities

$$\sum_{k=1}^{n}(a_k + b_k) = \sum_{i=1}^{n} a_i + \sum_{j=1}^{n} b_j, \; \sum_{k=1}^{n}(c a_k) = c\left(\sum_{k=1}^{n} a_k\right).$$ □

(b) Using (a) and (3.3) prove that for any natural number n and any real numbers a, r we have the equality

$$\sum_{k=0}^{n}(a + kr) = a + (a + r) + (a + 2r) \cdots + (a + nr) = (n + 1)a + \frac{rn(n + 1)}{2}.$$

(c) Use (b) to compute

$$3 + 7 + 11 + 15 + 19 + \cdots + 999,999.$$

Express the above using the symbol $\sum$.
(d) Prove that for any natural number n we have the equality

$$1 + 3 + 5 + \cdots + (2n - 1) = n^2.$$ □

Exercise 3.4. Prove that for any natural number n and any positive real numbers x, y such that $x < 1 < y$ we have

$$x^n \leq x, \; y \leq y^n.$$ □

Exercise 3.5. Prove that if $0 < a < b$, and $n \geq 2$, then

$$\sqrt[n]{a} < \sqrt[n]{b}, \; a < \sqrt{ab} < \frac{a + b}{2} < b.$$ □

Exercise 3.6. Find a natural number N_0 with the following property: for any $n > N_0$ we have

$$0 < \frac{n}{n^2 + 1} < \frac{1}{10^6} = \frac{1}{1,000,000}.$$ □

Exercise 3.7. Prove that for any natural number n and any real number $x \neq 1$ we have the equality.

$$\frac{1-x^n}{1-x} = 1 + x + x^2 + \cdots + x^{n-1}. \qquad \square$$

Exercise 3.8. (a) Compute

$$\binom{11}{2}, \binom{11}{3}, \binom{11}{8}, \binom{15}{4}, \binom{15}{11}.$$

(b) Show that for any $n, k \in \mathbb{N} \cup \{0\}$, $k \leq n$ we have

$$\binom{n}{k} = \binom{n}{n-k}.$$

(c) Use Newton's binomial formula to show that for any natural number n we have the equalities

$$\binom{n}{0} + \binom{n}{1} + \binom{n}{2} + \cdots + \binom{n}{n} = 2^n,$$

$$\binom{n}{0} - \binom{n}{1} + \binom{n}{2} + \cdots + (-1)^n \binom{n}{n} = 0.$$

Deduce that

$$\binom{n}{0} + \binom{n}{2} + \binom{n}{4} + \cdots = 2^{n-1}. \qquad \square$$

Exercise 3.9. Show that for any real number x, the interval $(x-1, x]$ contains exactly one integer.

Hint: For uniqueness use the Corollaries 3.8 and 3.9. To prove existence consider separately the cases

- $x \in \mathbb{Z}$.
- $(x \in \mathbb{R} \setminus \mathbb{Z}) \wedge (x > 0)$.
- $(x \in \mathbb{R} \setminus \mathbb{Z}) \wedge (x < 0)$.

$\square$

Exercise 3.10. Let a, b, c be real numbers, $a \neq 0$.

(a) Show that

$$ax^2 + bx + c = a\left(x + \frac{b}{2a}\right)^2 - \left(\frac{b^2 - 4ac}{4a}\right), \quad \forall x \in \mathbb{R}.$$

(b) Prove that the following statements are equivalent.

(i) There exist $r_1, r_2 \in \mathbb{R}$ such that

$$ax^2 + bx + c = a(x - r_1)(x - r_2).$$

(ii) There exists $r \in \mathbb{R}$ such that $ar^2 + br + c = 0$.
(iii) $b^2 - 4ac \geq 0$.

□

Exercise 3.11. Find the ranges of the functions

$$f : (-\infty, 5) \to \mathbb{R}, \;\; f(x) = \frac{x+1}{x-5},$$

and

$$g : \mathbb{R} \to \mathbb{R}, \;\; g(x) = \frac{x}{x^2+1}.$$

□

Exercise 3.12. (a) Show that the equation $x^2 - x - 1 = 0$ has two solutions $r_1, r_2 \in \mathbb{R}$ and then prove that r_1, r_2 satisfy the equalities

$$r_1 + r_2 = 1, \;\; r_1 r_2 = -1.$$

(b) For any nonnegative integer n we set

$$F_n = \frac{r_1^{n+1} - r_2^{n+1}}{r_1 - r_2},$$

where r_1, r_2 are as in (a). Compute F_0, F_1, F_2.
(c) Prove by induction that for any nonnegative integer n we have

$$r_1^{n+2} = r_1^{n+1} + r_1^n, \;\; r_2^{n+2} = r_2^{n+1} + r_2^n,$$

and

$$F_{n+2} = F_{n+1} + F_n.$$

(d) Use the above equality to compute $F_3, \dots, F_9$. □

Exercise 3.13. Prove Propositions 3.24 and 3.32 . □

Exercise 3.14. (a) Verify that for any $a, b > 0$ and any $m, n \in \mathbb{N}$ we have the equalities

$$(ab)^{\frac{1}{n}} = a^{\frac{1}{n}} \cdot b^{\frac{1}{n}}, \;\; \left(a^{\frac{1}{n}}\right)^{\frac{1}{m}} = a^{\frac{1}{mn}},$$

$$(a^m)^{\frac{1}{n}} = \left(a^{\frac{1}{n}}\right)^m =: a^{\frac{m}{n}},$$

$$a^{\frac{km}{kn}} = a^{\frac{m}{n}}, \;\; \forall k \in \mathbb{N}.$$

$$(a^{\frac{m}{n}})^{-1} = \left(a^{-1}\right)^{\frac{m}{n}} =: a^{-\frac{m}{n}}.$$

$$a^{-\frac{km}{kn}} = a^{-\frac{m}{n}}, \;\; \forall k \in \mathbb{N}.$$

☞ *Recall that an expression of the form "bla-bla-bla $=: x$" signifies that the quantity x is defined to be whatever bla-bla-bla means. In particular the notation*

$$\left(a^{\frac{1}{n}}\right)^m =: a^{\frac{m}{n}}$$

indicates that the quantity $a^{\frac{m}{n}}$ is defined to be the m-th power of the n-th root of a.
(b) Prove that if $a > 0$, then for any $m, m' \in \mathbb{Z}$ and $n, n' \in \mathbb{N}$ such that

$$\frac{m}{n} = \frac{m'}{n'},$$

then

$$a^{\frac{m}{n}} = a^{\frac{m'}{n'}}.$$

Any rational number r admits a nonunique representation as a fraction

$$r = \frac{m}{n}, \quad m \in \mathbb{Z}, \quad n \in \mathbb{N}.$$

Part (b) allows us to give a well defined meaning to a^r, > 0, $r \in \mathbb{Q}$.
(c) Show that for all $r_1, r_2 \in \mathbb{Q}$ and any $a > 0$ we have

$$a^{r_1} \cdot a^{r_2} = a^{r_1 + r_2}.$$

(d) Suppose that $a > b > 0$. Prove that for any rational number $r > 0$ we have

$$a^r > b^r.$$

(e) Suppose that $a > 1$. Prove that for any rational numbers r_1, r_2 such that $r_1 < r_2$ we have

$$a^{r_1} < a^{r_2}.$$

(f) Suppose that $a \in (0, 1)$. Prove that for any rational numbers r_1, r_2 such that $r_1 < r_2$ we have

$$a^{r_1} > a^{r_2}.$$

□

3.6. More challenging problems

Problem 3.1. There are 5 heads and 14 legs in a family. How many people and how many dogs are there in the family? □

Problem 3.2. You have two vessels of volumes 5 liters and 3 litters respectively. Measure one liter, producing it in one of the vessels. □

Problem 3.3. Each number from 1 to 10^{10} is written out in formal English (e.g., "two hundred and eleven", "one thousand and forty-two") and then listed in alphabetical order (as in a dictionary, where spaces and hyphens are ignored). What is the first odd number in the list? □

Problem 3.4. Consider the map $f : \mathbb{N} \to \mathbb{Z}$ defined by

$$f(n) = (-1)^{n+1} \left\lfloor \frac{n}{2} \right\rfloor.$$

(i) Compute $f(1), f(2), f(3), f(4), f(5), f(6), f(7)$.
(ii) Given a natural number k, compute $f(2k)$ and $f(2k-1)$.
(iii) Prove that f is a bijection.

□

Problem 3.5. (a) Let p be a prime number and n a natural number > 1. Prove that $\sqrt[n]{p}$ is irrational.
(b) Let m, n be natural numbers and p, q prime numbers. Prove that

$$p^{1/m} = q^{1/n} \iff (p = q) \wedge (m = n).$$ □

Problem 3.6. Start with the natural numbers $1, 2, \ldots, 999$ and change it as follows: select any two numbers, and then replace them by a single number, their difference. After 998 such changes you are left with a single number. Show that this number must be even. □

Problem 3.7. Let $S \subset [0, 1]$ be a set satisfying the following two properties.

(i) $0, 1 \in S$.
(ii) For any $n \in \mathbb{N}$ and any pairwise distinct numbers $s_1, \ldots, s_n \in S$ we have

$$\frac{s_1 + \cdots + s_n}{n} \in S.$$

Show that $S = \mathbb{Q} \cap [0, 1]$. □

Problem 3.8. Given 25 positive real numbers, prove that you can choose two of them x, y so none of the remaining numbers is equal to the sum $x+y$ or the differences $x-y$, $y-x$.□

Problem 3.9. At a stockholders' meeting, the board presents the month-by-month profit (or losses) since the last meeting. "Note," says the CEO, "that we made a profit over every consecutive eight-month period."

"Maybe so," a shareholder complains, "but I also see that we *lost* over every consecutive *five*-month period!"

What is the maximum number of months that could have passed since the last meeting?

□

Problem 3.10 (Erdös–Szekeres).
Suppose we are given an injection $f : \{1, \ldots, 10001\} \to \mathbb{R}$. Prove that there exists a subset $I \subset \{1, \ldots, 10001\}$ of cardinality 101 such that, either

$$f(i_1) < f(i_2), \;\; \forall i_1, i_2 \in I, \;\; i_1 < i_2,$$

or

$$f(i_1) > f(i_2), \;\; \forall i_1, i_2 \in I, \;\; i_1 < i_2.$$ □

Problem 3.11 (Chebyshev). Suppose that $p_1, \dots, p_n$ are positive numbers such that

$$p_1 + \cdots + p_n = 1.$$

Prove that if $x_1, \dots, x_n$ and $y_1, \dots, y_n$ are real numbers such that

$$x_1 \le x_2 \le \cdots \le x_n \quad \text{and} \quad y_1 \le y_2 \le \cdots \le y_n,$$

then

$$\sum_{k=1}^{n} x_k y_k p_k \ge \left(\sum_{i=1}^{n} x_i p_i \right) \left(\sum_{j=1}^{n} y_j p_j \right). \qquad \square$$

Problem 3.12. Let $k \in \mathbb{N}$. We are given k pairwise disjoint intervals $I_1, \dots, I_k \subset [0, 1]$. Denote by S their union. We know that for any $d \in [0, 1]$ there exist two points $p, q \in S$ such that $\mathrm{dist}(p, q) = d$. Prove that

$$\text{length}\,(I_1) + \cdots + \text{length}\,(I_k) \ge \frac{1}{k}. \qquad \square$$

Chapter 4

Limits of Sequences

The concept of limit is the central concept of this course. This chapter deals with the simplest incarnation of this concept, namely, the notion of limit of a sequence of real numbers.

4.1. Sequences

Formally, a *sequence* of real numbers is a function $x : \mathbb{N} \to \mathbb{R}$. We typically describe a sequence $x : \mathbb{N} \to \mathbb{R}$ as a list $(x_n)_{n\in\mathbb{N}}$ consisting of one real number for each natural number n,

$$x_1, x_2, x_3, \dots, x_n, \dots .$$

Often we will allow lists that start at time 0, $(x_n)_{n\geq 0}$,

$$x_0, x_1, x_2, \dots .$$

If we use our intuition of a real number as corresponding to a point on a line, we can think of a sequence $(x_n)_{n\geq 1}$ as describing the motion of an object along the line, where x_n describes the position of that object at time n.

Example 4.1. (a) The natural numbers form a sequence $(n)_{n\in\mathbb{N}}$,

$$1, 2, 3, \dots .$$

(b) The *arithmetic progression* with initial term $a \in \mathbb{R}$ and ratio $r \in \mathbb{R}$ is the sequence

$$a, a + r, a + 2r, a + 3r, \dots .$$

For example, the sequence

$$3, 7, 11, 15, 19, \dots$$

is an arithmetic progression with initial term 3 and ratio 4. The constant sequence

$$a, a, a, \dots ,$$

is an arithmetic progression with initial term a and ratio 0.
(c) The *geometric progression* with initial term $a \in \mathbb{R}$ and ratio $r \in \mathbb{R}$ is the sequence

$$a, ar, ar^2, ar^3, \dots .$$

For example, the sequence

$$1, -1, 1, -1,$$

is the geometric progression with initial term 1 and ratio -1.

(d) The *Fibonacci sequence*is the sequence $F_0, F_1, F_2, \dots$ given by the initial condition

$$F_0 = F_1 = 1,$$

and the recurrence relation

$$F_{n+2} = F_{n+1} + F_n, \quad \forall n \geq 0.$$

For example

$$F_2 = 1 + 1 = 2, \quad F_3 = 2 + 1 = 3, \quad F_4 = 3 + 2 = 5, \quad F_5 = 5 + 3 = 8, \dots .$$

In Exercise 3.12 we gave an alternate description to the Fibonacci sequence. □

Definition 4.2. Let $(x_n)_{n \in \mathbb{N}}$ be a sequence of real numbers.

(i) The sequence $(x_n)_{n \in \mathbb{N}}$ is called *increasing* if

$$x_n < x_{n+1}, \quad \forall n \in \mathbb{N}.$$

(ii) The sequence $(x_n)_{n \in \mathbb{N}}$ is called *decreasing* if

$$x_n > x_{n+1}, \quad \forall n \in \mathbb{N}.$$

(iii) The sequence $(x_n)_{n \in \mathbb{N}}$ is called *nonincreasing* if

$$x_n \geq x_{n+1}, \quad \forall n \in \mathbb{N}.$$

(iv) The sequence $(x_n)_{n \in \mathbb{N}}$ is called *nondecreasing* if

$$x_n \leq x_{n+1}, \quad \forall n \in \mathbb{N}.$$

(v) A sequence $(x_n)_{n \in \mathbb{N}}$ is called *monotone* if it is either non-decreasing, or non-increasing. It is called *strictly monotone* if it is either increasing, or decreasing.

(vi) The sequence $(x_n)_{n \in \mathbb{N}}$ is called *bounded* if there exist real numbers m, M such that

$$m \leq x_n \leq M, \quad \forall n \in \mathbb{N}.$$ □

Note that an arithmetic progression is increasing if and only if its ratio is positive, while a geometric progression with positive initial term and positive ratio is monotone: it is increasing if the ratio is > 1, decreasing if the ratio < 1 and constant if the ratio is $= 1$. A geometric progression is bounded if and only if its ratio r satisfies $|r| \leq 1$.

A *subsequence* of a sequence $x : \mathbb{N} \to \mathbb{R}$ is a restriction of x to an infinite subset $S \subset \mathbb{N}$. An infinite subset $S \subset \mathbb{N}$ can itself be viewed as an *increasing* sequence of natural numbers

$$n_1 < n_2 < n_3 < \dots,$$

where

$$n_1 := \min S, \quad n_2 := \min S \setminus \{n_1\}, \dots, n_{k+1} := \min S \setminus \{n_1, \dots, n_k\}, \dots .$$

Thus a subsequence of a sequence $(x_n)_{n \in \mathbb{N}}$ can be described as a sequence $(x_{n_k})_{k \in \mathbb{N}}$, where $(n_k)_{k \in \mathbb{N}}$ is an increasing sequence of natural numbers.

4.2. Convergent sequences

Definition 4.3. We say that the sequence of real numbers (x_n) *converges to the number* $x \in \mathbb{R}$ if

$$\forall \varepsilon > 0: \ \exists N = N(\varepsilon) \in \mathbb{N} \ \text{ such that } \forall n > N(\varepsilon), \ \text{ we have } |x_n - x| < \varepsilon. \tag{4.1}$$

A sequence (x_n) is called *convergent* if it converges to some number x. More precisely, this means

$$\exists x \in \mathbb{R}, \ \forall \varepsilon > 0: \ \exists N = N(\varepsilon) \in \mathbb{N} \ \text{ such that } \forall n > N(\varepsilon), \ \text{ we have } |x_n - x| < \varepsilon. \tag{4.2}$$

The number x is called *a limit* of the sequence (a_n). A sequence is called *divergent* if it is not convergent. □

Observe that condition (4.1) can be rephrased as follows

$$\forall \varepsilon > 0: \ \exists N = N(\varepsilon) \in \mathbb{N} \ \text{ such that } \forall n > N(\varepsilon), \ \text{ we have } \operatorname{dist}(x_n, x) < \varepsilon. \tag{4.3}$$

Before we proceed further, let us observe the following simple fact.

Proposition 4.4. *Given a sequence (x_n) there exists at most one real number x satisfying the convergence property (4.1).*

Proof. Suppose that x, x' are two real numbers satisfying (4.1). Thus,

$$\forall \varepsilon > 0: \ \exists N = N(\varepsilon) \in \mathbb{N} \ \text{ such that } \forall n > N, \ \text{ we have } |x_n - x| < \varepsilon,$$

and

$$\forall \varepsilon > 0: \ \exists N' = N'(\varepsilon) \in \mathbb{N} \ \text{ such that } \forall n > N', \ \text{ we have } |x_n - x'| < \varepsilon.$$

Thus, if $n > N_0(\varepsilon) := \max(\, N(\varepsilon), N'(\varepsilon)\,)$, then

$$|x_n - x|, \ |x_n - x'| < \varepsilon.$$

We observe that if $n > N_0(\varepsilon)$, then

$$|x - x'| = |(x - x_n) + (x_n - x')| \leq |x - x_n| + |x_n - x'| < 2\varepsilon.$$

In other words

$$\forall \varepsilon > 0: \ |x - x'| < 2\varepsilon, \ \forall n > N_0(\varepsilon).$$

In the above statement the variable n really plays no role: if $|x - x'| < 2\varepsilon$ for some n, then clearly $|x - x'| < 2\varepsilon$ for any n. We conclude that

$$\forall \varepsilon > 0: \ |x - x'| < 2\varepsilon.$$

In other words, the distance $\operatorname{dist}(x, x') = |x - x'|$ between x and x' is smaller than any positive real number, so that this distance must be zero (Exercise 2.14) and hence $x = x'$. □

Definition 4.5. Given a convergent sequence (x_n), the unique real number x satisfying the convergence condition (4.1) is called *the limit* of the sequence (x_n) and we will indicate this using the notations

$$x = \lim_{n\to\infty} x_n \text{ or } x = \lim_n x_n.$$

We will also say that (x_n) *tends (or converges) to* x *as* n *goes to* ∞. □

Observe that

$$\lim_{n\to\infty} x_n = x \Longleftrightarrow \lim_{n\to\infty} |x_n - x| = 0. \tag{4.4}$$

The next example shows that convergent sequences do exist.

Example 4.6. (a) If (x_n) is the constant sequence, $x_n = x$, for all n, then (x_n) is convergent and its limit is a.
(b) We want to show that

$$\boxed{\lim_{n\to\infty} \frac{C}{n} = 0, \ \forall C > 0}. \tag{4.5}$$

Let $\varepsilon > 0$ and set $N(\varepsilon) := \lfloor \frac{C}{\varepsilon} \rfloor + 1 \in \mathbb{N}$. We deduce

$$N(\varepsilon) > \frac{C}{\varepsilon}, \text{ i.e., } \frac{N(\varepsilon)}{C} > \frac{1}{\varepsilon}.$$

For any $n > N(\varepsilon)$ we have

$$\frac{n}{C} > \frac{N(\varepsilon)}{C} > \frac{1}{\varepsilon} \Rightarrow \frac{C}{n} < \varepsilon.$$

Hence for any $n > N(\varepsilon)$ we have

$$|x_n| = \frac{C}{n} < \varepsilon.$$ □

Definition 4.7. (a) A *neighborhood* of a real number x is defined to be an *open* interval (α, β) that contains x, i.e., $x \in (\alpha, \beta)$.
(b) A *neighborhood of* ∞ is an interval of the form (M, ∞), while a *neighborhood of* $-\infty$ is an interval of the form $(-\infty, M)$. □

We have the following equivalent description of convergence. Its proof is left to you as an exercise.

Proposition 4.8. *Let (x_n) be a sequence of real numbers. Prove that the following statements are equivalent.*

(i) The sequence (x_n) converges to $x \in \mathbb{R}$ as $n \to \infty$.
(ii) For any neighborhood U of x there exists a natural number N such that

$$\forall n (n > N \Rightarrow x_n \in U).$$ □

The proof of the following result is left to you as an exercise.

Proposition 4.9. *Suppose that* $(x_n)_{n\in\mathbb{N}}$ *is a convergent sequence and* $x = \lim_{n\to\infty} x_n$.

(i) If $(x_{n_k})_{k\geq 1}$ *is a subsequence of* (x_n)*, then*

$$\lim_{k\to\infty} x_{n_k} = x.$$

(ii) Suppose that $(x'_n)_{n\in\mathbb{N}}$ *is another sequence with the following property*

$$\exists N_0 \in \mathbb{N}: \quad \forall n > N_0 \;\; x'_n = x_n.$$

Then

$$\lim_{n\to\infty} x'_n = x. \qquad \square$$

Part (ii) of the above proposition shows that the convergence or divergence of a sequence is not affected if we modify only finitely many of its terms. The next result is very intuitive.

Proposition 4.10 (Squeezing Principle). *Let* (a_n) (x_n), (y_n) *be sequences such that*

$$\exists N_0 \in \mathbb{N}: \forall n > N_0, \quad x_n \leq a_n \leq y_n.$$

If

$$\lim_{n\to\infty} x_n = \lim_{n\to\infty} y_n = a,$$

then

$$\lim_{n\to\infty} a_n = a.$$

Proof. We have

$$\text{dist}(a_n, a) \leq \text{dist}(a_n, x_n) + \text{dist}(x_n, a).$$

Since a_n lies in the interval $[x_n, y_n]$ for $n > N_0$ we deduce that

$$\text{dist}(a_n, x_n) \leq \text{dist}(y_n, x_n), \quad \forall n > N_0,$$

so that

$$\text{dist}(a_n, a) \leq \text{dist}(y_n, x_n) + \text{dist}(x_n, a), \quad \forall n > N_0.$$

Now observe that

$$\text{dist}(y_n, x_n) \leq \text{dist}(y_n, a) + \text{dist}(a, x_n).$$

Hence,

$$\begin{aligned} \text{dist}(a_n, a) &\leq \text{dist}(y_n, a) + \text{dist}(a, x_n) + \text{dist}(x_n, a) \\ &= \text{dist}(y_n, a) + 2\,\text{dist}(x_n, a), \quad \forall n > N_0. \end{aligned} \tag{4.6}$$

Let $\varepsilon > 0$. Since $x_n \to a$ there exists $N_x(\varepsilon) \in \mathbb{N}$ such that

$$\forall n > N_x(\varepsilon): \quad \text{dist}(x_n, a) < \frac{\varepsilon}{3}.$$

Since $y_n \to a$ there exists $N_y(\varepsilon) \in \mathbb{N}$ such that

$$\forall n > N_y(\varepsilon): \ \ \text{dist}(y_n, a) < \frac{\varepsilon}{3}.$$

Set $N(\varepsilon) := \max\{N_0, N_x(\varepsilon), N_y(\varepsilon)\}$. For $n > N(\varepsilon)$ we have

$$\text{dist}(x_n, a) < \frac{\varepsilon}{3}, \ \ \text{dist}(y_n, a) < \frac{\varepsilon}{3}$$

and thus

$$\text{dist}(y_n, a) + 2\,\text{dist}(x_n, a) < \varepsilon.$$

Using this in (4.6) we conclude that

$$\forall n > N(\varepsilon) \ \ \text{dist}(a_n, a) < \varepsilon.$$

This proves that $a_n \to a$ as $n \to \infty$. □

Corollary 4.11. *Suppose that $a \in \mathbb{R}$ and (a_n), (x_n) are sequences of real numbers such that*

$$|a_n - a| \leq x_n \ \forall n, \ \ \lim_{n\to\infty} x_n = 0.$$

Then

$$\lim_{n\to\infty} a_n = a.$$

Proof. We have squeezed the sequence $|a_n - a|$ between the sequences (x_n) and the constant sequence 0, both converging to 0. Hence $|a_n - a| \to 0$ and, in view of (4.4), we deduce that also $a_n \to a$. □

Example 4.12. We want to show that

$$\boxed{\forall M > 0, \ \ \forall r \in (-1, 1) \ \lim_{n\to\infty} Mr^n = 0}. \tag{4.7}$$

Clearly, it suffices to show that $M|r|^n \to 0$. This is clearly the case if $r = 0$. Assume $r \neq 0$. Set

$$R := \frac{1}{|r|}.$$

Then $R > 1$ so that $R = 1 + \delta$, $\delta > 0$. Bernoulli's inequality (3.4) implies that $\forall n \in \mathbb{N}$ we have $R^n \geq 1 + n\delta$ so that

$$M|r|^n = \frac{M}{R^n} \leq \frac{M}{1 + n\delta} \leq \frac{M}{n\delta} = \frac{C}{n}, \ \ C := \frac{M}{\delta}.$$

From Example 4.6(b) we deduce that

$$\lim_n \frac{C}{n} = 0.$$

The desired conclusion now follows from the Squeezing Principle. □

Example 4.13. We want to prove that

$$\boxed{\lim_n \frac{r^n}{n!} = 0, \ \ \forall r \in \mathbb{R}}. \tag{4.8}$$

We will rely again on the Squeezing Principle. Fix $N_0 \in \mathbb{N}$ such that $N_0 > 2|r|$. Then for any $n > N_0$ we have

$$\begin{aligned}\left|\frac{r^n}{n!}\right| = \frac{|r|^n}{n!} &= \frac{|r|^{N_0} r^{n-N_0}}{1\cdot 2\cdots N_0\cdot(N_0+1)(N_0+2)\cdots n}\\ &= \underbrace{\frac{|r|^{N_0}}{N_0!}}_{=:C_0}\cdot \underbrace{\frac{|r|}{N_0+1}\cdot\frac{|r|}{N_0+2}\cdots\frac{|r|}{n}}_{(n-N_0)\text{ terms}}.\end{aligned}$$

Now observe that

$$\frac{|r|}{N_0+1}, \frac{|r|}{N_0+2}, \dots, \frac{|r|}{n} < \frac{|r|}{N_0} < \frac{1}{2},$$

and we deduce

$$\left|\frac{r^n}{n!}\right| < C_0\left(\frac{1}{2}\right)^{n-N_0} = C_0\left(\frac{1}{2}\right)^{-N_0}\left(\frac{1}{2}\right)^{n} = 2^{N_0}C_0 2^{-n}.$$

If we denote by M the constant $2^{N_0}C_0$ and we set $x_n := M2^{-n}$, $n \in \mathbb{N}$, we deduce that

$$\forall n > N_0 : \ \left|\frac{r^n}{n!}\right| < x_n.$$

Example 4.12 shows that $x_n \to 0$ and the conclusion (4.8) now follows from the Squeezing Principle. □

Proposition 4.14. *Any convergent sequence of real numbers is bounded.*

Proof. Suppose that $(a_n)_{n\geq 1}$ is a convergent sequence

$$a = \lim_{n\to\infty} a_n.$$

There exists $N \in \mathbb{N}$ such that, for any $n > N$ we have

$$|a_n - a| < 1.$$

Thus, for any $n > N$ we have $a_n \in (a-1, a+1)$. Now set

$$m := \min\{a_1, a_2, \dots, a_N, a-1\}, \ \ M := \max\{a_1, a_2, \dots, a_N, a+1\}.$$

Then for any $n \geq 1$ we have

$$m \leq a_n \leq M,$$

i.e., the sequence (a_n) is bounded. □

4.3. The arithmetic of limits

This section describes a few simple yet basic techniques that reduce the study of the convergence of a sequence to a similar study of potentially simpler sequences. Thus, we will prove that the sum of two convergent sequences is a convergent sequence, etc.

Proposition 4.15 (Passage to the limit). *Suppose that $(a_n)_{n\geq 1}$ and $(b_n)_{n\geq 1}$ are two convergent sequences,*

$$a := \lim_{n\to\infty} a_n, \quad b = \lim_{n\to\infty} b_n.$$

The following hold.

(i) The sequence $(a_n + b_n)_{n\geq 1}$ is convergent and

$$\lim_{n\to\infty}(a_n + b_n) = \lim_{n\to\infty} a_n + \lim_{n\to\infty} b_n = a + b.$$

(ii) If $\lambda \in \mathbb{R}$ then

$$\lim_{n\to\infty}(\lambda a_n) = \lambda \lim_{n\to\infty} a_n = \lambda a.$$

(iii)

$$\lim_{n\to\infty}(a_n \cdot b_n) = \big(\lim_{n\to\infty} a_n\big)\cdot\big(\lim_{n\to\infty} b_n\big) = ab.$$

(iv) Suppose that $b \neq 0$. Then there exists $N_0 > 0$ such that $b_n \neq 0$, $\forall N > N_0$ and

$$\lim_{n\to\infty}\frac{a_n}{b_n} = \frac{a}{b}.$$

(v) Suppose that m, M are real numbers such that $m \leq a_n \leq M$, $\forall n$. Then

$$m \leq \lim_{n\to\infty} a_n = a \leq M.$$

Proof. (i) Because (a_n) and (b_n) are convergent, for any $\varepsilon > 0$ there exist $N_a(\varepsilon), N_b(\varepsilon) \in \mathbb{N}$ such that

$$|a_n - a| < \frac{\varepsilon}{2}, \quad \forall n > N_a(\varepsilon), \tag{4.9a}$$

$$|b_n - b| < \frac{\varepsilon}{2}, \quad \forall n > N_b(\varepsilon). \tag{4.9b}$$

Let

$$N(\varepsilon) := \max\big\{N_a(\varepsilon),\ N_b(\varepsilon)\big\}.$$

Then for any $n > N(\varepsilon)$ we have $n > N_a(\varepsilon)$, $n > N_b(\varepsilon)$ and therefore

$$\big|(a_n + b_n) - (a + b)\big| = \big|(a_n - a) + (b_n - b)\big| \leq |a_n - a| + |b_n - b| \overset{(4.9a),(4.9b)}{<} \frac{\varepsilon}{2} + \frac{\varepsilon}{2} = \varepsilon.$$

This proves that $\lim_{n\to\infty}(a_n + b_n) = a + b$.

(ii) If $\lambda = 0$, then the sequence (λa_n) is the constant sequence $0, 0, 0, \ldots$ and the conclusion is obvious. Assume that $\lambda \neq 0$. The sequence (a_n) is convergent so for any $\varepsilon > 0$ there exists $N = N(\varepsilon) \in \mathbb{N}$ such that

$$|a_n - a| < \frac{\varepsilon}{|\lambda|}, \quad \forall n > N(\varepsilon).$$

Hence for any $n > N(\varepsilon)$ we have

$$|\lambda a_n - \lambda a| = |\lambda| \cdot |a_n - a| < |\lambda| \cdot \frac{\varepsilon}{|\lambda|} = \varepsilon.$$

(iii) The sequences (a_n), (b_n) are convergent and thus, according to Proposition 4.14 they are bounded so that

$$\exists M > 0 : \quad |a_n|, |b_n| \leq M, \quad \forall n.$$

We have

$$\begin{aligned} |a_n b_n - ab| &= |(a_n b_n - a b_n) + (a b_n - ab)| \leq |a_n b_n - a b_n| + |a b_n - ab| \\ &= |b_n| \cdot |a_n - a| + |a| \cdot |b_n - b| \leq M|a_n - a| + |a| \cdot |b_n - b|. \end{aligned}$$

Part (ii) coupled with the convergence of (a_n) and (b_n) show that

$$\lim_{n\to\infty} M|a_n - a| = \lim_{n\to\infty} |a| \cdot |b_n - b| = 0.$$

Using (i) we deduce

$$\lim_{n\to\infty} \big(M|a_n - a| + |a| \cdot |b_n - b| \big) = 0.$$

The squeezing principle shows that $|a_n b_n - ab| \to 0$.

(iv) Let us first show that if $b \neq 0$, then $b_n \neq 0$ for n sufficiently large. Since $b_n \to b$ there exists $N_0 \in \mathbb{N}$ such that

$$\forall n > N_0 \ \ |b_n - b| < \frac{|b|}{2}.$$

Thus, for any $n > N_0$, we have

$$\operatorname{dist}(b_n, b) = |b_n - b| < \frac{1}{2}|b| = \frac{1}{2}\operatorname{dist}(b, 0).$$

This shows that for $n > N_0$ we cannot have $b_n = 0$. In fact

$$|b_n| > \frac{|b|}{2}, \quad \forall n > N_0. \tag{4.10}$$

Thus, the ratio $\frac{b_n}{b_n}$ is well defined at least for $n > N_0$. We have

$$\left| \frac{1}{b_n} - \frac{1}{b} \right| = \frac{|b_n - b|}{|b_n| \cdot |b|}.$$

The inequality (4.10) implies

$$\frac{1}{|b_n|} < \frac{2}{|b|}, \quad \forall n > N_0.$$

Hence, for $n > N_0$ we have

$$\left|\frac{1}{b_n} - \frac{1}{b}\right| < \frac{2}{|b|^2}|b_n - b| \to 0.$$

This implies

$$\lim_{n\to\infty} \frac{1}{b_n} = \frac{1}{b}.$$

Thus

$$\lim_{n\to\infty} \frac{a_n}{b_n} = \lim_{n\to\infty} a_n \cdot \lim_{n\to\infty} \frac{1}{b_n} = \frac{a}{b}.$$

(v) We argue by contradiction. Suppose that $a > M$ or $a < m$. We discuss what happens if $a > M$, the other situation being entirely similar. Then $\delta = a - M = \text{dist}(a, M) > 0$. Since $a_n \to a$, there exists $N \in \mathbb{N}$ such that if $n > N$, then

$$\text{dist}(a_n, a) = |a_n - a| < \frac{\delta}{2}.$$

Thus, for $n > N_0$ we have

$$a - \frac{\delta}{2} < a_n < a + \frac{\delta}{2}.$$

Clearly $M = a - \delta < a - \frac{\delta}{2}$ and thus, a fortiori, $a_n > M$ for $n > N_0$. Contradiction! □

Corollary 4.16. *Suppose that (a_n) and (b_n) are convergent sequences such that $a_n \geq b_n$, $\forall n$. Then*

$$\lim_{n\to\infty} a_n \geq \lim_{n\to\infty} b_n.$$

Proof. Let $c_n = a_n - b_n$. Then $c_n \geq 0\ \forall n$ and thus

$$\lim_{n\to\infty} a_n - \lim_{n\to\infty} b_n = \lim_{n\to\infty} c_n \geq 0.$$

□

Let us see how the above simple principles work in practice.

Example 4.17. We already know that

$$\lim_{n\to\infty} \frac{1}{n} = 0.$$

We deduce that for any $k \in \mathbb{N}$ we have

$$\lim_{n\to\infty} \frac{1}{n^k} = 0.$$

Consider the sequence

$$a_n := \frac{5n^2 + 3n + 2}{3n^2 - 2n + 1}.$$

We have

$$a_n = \frac{n^2(5 + \frac{3}{n} + \frac{2}{n^2})}{n^2(3 - \frac{2}{n} + \frac{1}{n^2})} = \frac{(5 + \frac{3}{n} + \frac{2}{n^2})}{(3 - \frac{2}{n} + \frac{1}{n^2})}.$$

Now observe that as $n \to \infty$

$$5 + \frac{3}{n} + \frac{2}{n^2} \to 5, \; 3 - \frac{2}{n} + \frac{1}{n^2} \to 3,$$

so that

$$\lim_{n\to\infty} a_n = \frac{5}{3}.$$

More generally, given $k \in \mathbb{N}$ and real numbers $a_0, b_0, \dots, a_k, b_k$ such that $b_k \neq 0$ then

$$\boxed{\lim_{n\to\infty} \frac{a_k n^k + \cdots + a_1 n + a_0}{b_k n^k + \cdots + b_1 n + b_0} = \frac{a_k}{b_k}}. \tag{4.11}$$

The proof is left to you as an exercise. □

Example 4.18. We want to show that

$$\boxed{\forall r > 1 \;\; \lim_n \frac{n}{r^n} = 0}. \tag{4.12}$$

We plan to use the Squeezing Principle and construct a sequence $(x_n)_{n\geq 1}$ of positive numbers such that

$$\frac{n}{r^n} \leq x_n \;\; \forall n \geq 2,$$

and

$$\lim_n x_n = 0.$$

Observe that since $r > 1$, we have $r - 1 > 0$. Set $a := r - 1$ so that $r = 1 + a$. Then, using Newton's binomial formula we deduce that if $n \geq 2$ then

$$\begin{aligned} r^n = (1+a)^n &= 1 + \binom{n}{1}a + \binom{n}{2}a^2 + \cdots \geq 1 + \binom{n}{1}a + \binom{n}{2}a^2 \\ &= 1 + na + \frac{n(n-1)}{2}a^2 = 1 + na + \frac{a^2}{2}(n^2 - n). \end{aligned}$$

Hence for $n \geq 2$ we have

$$\frac{1}{r^n} \leq \frac{1}{\frac{1}{2}(n^2 - n)a^2 + na + 1}$$

so that

$$\frac{n}{r^n} \leq \frac{n}{\frac{a^2}{2}(n^2 - n) + na + 1} =: x_n.$$

Now observe that

$$x_n = \frac{n}{n^2\left(\frac{a^2}{2}(1 - \frac{1}{n}) + \frac{a}{n} + \frac{1}{n^2}\right)} = \frac{\frac{1}{n}}{\frac{a^2}{2}(1 - \frac{1}{n}) + \frac{a}{n} + \frac{1}{n^2}} \stackrel{n\to\infty}{\longrightarrow} 0.$$ □

Example 4.19. We want to show that

$$\boxed{\lim_n \sqrt[n]{n} = 1}. \tag{4.13}$$

Let $\varepsilon > 0$. The number $r_\varepsilon = 1 + \varepsilon$ is > 1. Since $\frac{n}{r_\varepsilon^n} \to 0$ we deduce that there exists $N = N(\varepsilon) \in \mathbb{N}$ such that

$$\frac{n}{r_\varepsilon^n} < 1, \;\; \forall n > N(\varepsilon).$$

In other words,

$$n < r_\varepsilon^n = (1+\varepsilon)^n, \;\; \forall n > N(\varepsilon).$$

In particular

$$1 \leq \sqrt[n]{n} < \sqrt[n]{(1+\varepsilon)^n} = 1 + \varepsilon.$$

We have thus proved that for any $\varepsilon > 0$ we can find $N = N(\varepsilon) \in \mathbb{N}$ so that, as soon as $n > N(\varepsilon)$, we have

$$1 \leq \sqrt[n]{n} < 1 + \varepsilon.$$

Clearly this proves the equality (4.13). □

Definition 4.20 (Infinite limits). Let $(a_n)_{n\in\mathbb{N}}$ be a sequence of real numbers.

(i) We say that a_n tends to ∞ as $n \to \infty$, and we write this

$$\lim_{n\to\infty} a_n = \infty$$

if

$$\forall C > 0 \;\; \exists N = N(C) \in \mathbb{N}: \;\; \forall n (n > N \Rightarrow a_n > C).$$

(ii) We say that a_n tends to $-\infty$ as $n \to \infty$, and we write this

$$\lim_{n\to\infty} a_n = -\infty$$

if

$$\forall C > 0 \;\; \exists N = N(C) \in \mathbb{N}: \;\; \forall n (n > N \Rightarrow a_n < -C). \qquad \square$$

Proposition 4.15 continues to hold if one or both of limits a, b are $\pm\infty$ provided we use the following conventions

$$\boxed{\infty + \infty = \infty \cdot \infty = \infty, \;\; \frac{C}{\infty} = 0, \;\; \forall C \in \mathbb{R}},$$

$$\boxed{C \cdot \infty = \begin{cases} \infty, & C > 0 \\ -\infty, & C < 0 \\ \text{undefined}, & C = 0, \end{cases}}$$

$$\infty - \infty = \text{undefined}, \;\; 0 \cdot \infty = \text{undefined}, \;\; \frac{\infty}{\infty} = \text{undefined}.$$

Example 4.21. (a) If we let $a_n = n$ and $b_n = \frac{1}{n}$, then Archimedes' Principle shows that $a_n \to \infty$ and $b_n \to 0$. We observe that $a_n b_n = 1 \to 1$. In this case $\infty \cdot 0 = 1$. On the other hand, if we let

$$a_n = n, \;\; b_n = \frac{1}{2^n}$$

then $a_n \to \infty$, $b_n \to 0$ and (4.12) shows that $a_n b_n \to 0$. In this case $\infty \cdot 0 = 0$.

(b) Consider the sequences $a_n = n$, $b_n = 2n$, $c_n = 3n$, $\forall n \in \mathbb{N}$. Observe that

$$\lim_{n\to\infty} a_n = \lim_{n\to\infty} b_n = \lim_{n\to\infty} c_n = \infty.$$

However

$$\lim_{n\to\infty} \frac{a_n}{b_n} = \frac{\infty}{\infty} = \frac{1}{2}, \;\; lim_{n\to\infty} \frac{a_n}{c_n} = \frac{\infty}{\infty} = \frac{1}{3}.$$

□

☛ ***Important Warning!*** When investigating limits of sequences you should keep in mind that the following arithmetic operations are treacherous and should be dealt with using *extreme care*.

$$\frac{\text{anything}}{0}, \;\; 0 \cdot \infty, \;\; \infty - \infty, \;\; \frac{\infty}{\infty}.$$

□

4.4. Convergence of monotone sequences

The definition of convergence has one drawback: to verify that a sequence is convergent using the definition we need to a priori know its limit. In most cases this is a nearly impossible job. In this section and the next we will discuss techniques for proving the convergence of a sequence without knowing the precise value of its limit.

Theorem 4.22 (Weierstrass[1]). *Any bounded and monotone sequence is convergent.*

Proof. Suppose that (a_n) is a bounded and monotone sequence, i.e., it is either nondecreasing, or nonincreasing. We investigate only the case when (a_n) is nondecreasing, i.e.,

$$a_1 \leq a_2 \leq a_3 \leq \cdots .$$

The situation when (a_n) is nonincreasing is completely similar.

The set of real numbers

$$A := \{ a_n; \;\; n \geq 1 \}$$

is bounded because the sequence (a_n) is bounded. The Completeness Axiom implies it has a least upper bound

$$a := \sup A.$$

[1]Karl Weierstrass (1815–1897) was a German mathematician often cited as the "father of modern analysis"; see Wikipedia.

We will prove that

$$\lim_{n\to\infty} a_n = a. \tag{4.14}$$

Since a is an upper bound for the sequence we have

$$a_n \leq a, \ \ \forall n. \tag{4.15}$$

Since a is the *least* upper bound of A we deduce that for any $\varepsilon > 0$ the number $a-\varepsilon$ cannot be an upper bound of A. Hence, for any $\varepsilon > 0$ there exists $N(\varepsilon) \in \mathbb{N}$ such that

$$a - \varepsilon < a_{N(\varepsilon)}.$$

Since (a_n) is nondecreasing we deduce that

$$a - \varepsilon < a_{N(\varepsilon)} \leq a_n, \ \ \forall n > N(\varepsilon). \tag{4.16}$$

Putting together (4.15) and (4.16) we deduce that

$$\forall \varepsilon > 0 \ \exists N = N(\varepsilon) \in \mathbb{N}: \ \ \forall n \ (n > N(\varepsilon) \Rightarrow a - \varepsilon < a_n \leq a).$$

This implies the claimed convergence (4.14) because $a - \varepsilon < a_n \leq a \Rightarrow |a_n - a| < \varepsilon$. □

We will spend the rest of this section presenting applications of the above *very important* theorem.

Example 4.23 (L. Euler). Consider the sequence of positive numbers

$$x_n = \left(1 + \frac{1}{n}\right)^n, \ \ n \in \mathbb{N}.$$

We will prove that this sequence is convergent. Its limit is called the *Euler*[2] *number e*.

We plan to use Weierstrass' theorem applied to a new sequence of positive numbers

$$y_n = \left(1 + \frac{1}{n}\right)^{n+1}, \ \ n \in \mathbb{N}.$$

Note that

$$y_n = \left(\frac{n+1}{n}\right)^{n+1}$$

and for $n \geq 2$ we have

$$\begin{aligned}\frac{y_{n-1}}{y_n} &= \frac{\left(\frac{n}{n-1}\right)^n}{\left(\frac{n+1}{n}\right)^{n+1}} = \left(\frac{n}{n-1}\right)^n \cdot \left(\frac{n}{n+1}\right)^{n+1} \\ &= \frac{n^{2n+1}}{(n-1)^n(n+1)^n \cdot (n+1)} = \frac{n^{2n}}{(n^2-1)^n} \cdot \frac{n}{n+1}\end{aligned}$$

[2]Leonhard Euler (1707–1783) was a Swiss mathematician, physicist, astronomer, logician and engineer who made important and influential discoveries in many branches of mathematics. He is considered to be the most prolific mathematician of all time; see Wikipedia.

$$= \left(\frac{n^2}{n^2-1}\right)^n \cdot \frac{n}{n+1} = \underbrace{\left(1+\frac{1}{n^2-1}\right)^n}_{=:q_n} \cdot \frac{n}{n+1}.$$

Bernoulli's inequality implies that

$$q_n := \left(1+\frac{1}{n^2-1}\right)^n \geq 1 + \frac{n}{n^2-1} > 1 + \frac{n}{n^2} = 1 + \frac{1}{n} = \frac{n+1}{n}.$$

Hence

$$\frac{y_{n-1}}{y_n} = q_n \cdot \frac{n}{n+1} > \frac{n+1}{n} \cdot \frac{n}{n+1} = 1.$$

Hence $y_{n-1} > y_n \ \forall n \geq 2$, i.e., the sequence (y_n) is decreasing. Since it is bounded below by 1 we deduce that the sequence (y_n) is convergent.

Now observe that $y_n = x_n \cdot \left(1+\frac{1}{n}\right) = x_n \cdot \frac{n+1}{n}$ so that

$$x_n = y_n \cdot \frac{n}{n+1}.$$

Since

$$\lim_n \frac{n}{n+1} = 1,$$

we deduce that (x_n) is convergent and has the same limit as the sequence (y_n). □

Definition 4.24. The *Euler number*, denoted e is defined to be

$$\boxed{e := \lim_{n\to\infty} \left(1+\frac{1}{n}\right)^n}.$$ □

The arguments in Example 4.23 show that

$$4 = y_1 \geq e \geq 2.$$

Using more sophisticated methods one can show that

$$e = 2.71828182845905\dots.$$

Example 4.25 (Babylonians and I. Newton). Consider the sequence $(x_n)_{n\in\mathbb{N}}$ defined recursively by the requirements

$$x_1 = 1, \ \ x_{n+1} = \frac{1}{2}\left(x_n + \frac{2}{x_n}\right), \ \ \forall n \in \mathbb{N}.$$

Thus

$$x_2 = \frac{1}{2}\left(1+\frac{2}{1}\right) = \frac{3}{2},$$

$$x_3 = \frac{1}{2}\left(\frac{3}{2}+\frac{2}{\frac{3}{2}}\right) = \frac{1}{2}\left(\frac{3}{2}+\frac{4}{3}\right) = \frac{17}{12}, \ \ \text{etc.}$$

We want to prove that this sequence converges to $\sqrt{2}$. We proceed gradually.

Lemma 4.26.

$$x_n \geq \sqrt{2}, \ \ \forall n \geq 2. \tag{4.17}$$

Proof. Multiplying with $2x_n$ both sides of the equality

$$x_{n+1} = \frac{1}{2}\left(x_n + \frac{2}{x_n}\right)$$

we deduce $2x_n x_{n+1} = x_n^2 + 2$, or equivalently

$$x_n^2 - 2x_{n+1}x_n + 2 = 0. \tag{4.18}$$

This shows that the quadratic equation

$$t^2 - 2x_{n+1}t + 2 = 0$$

has at least one real solution, $t = x_{n+1}$ so that (see Exercise 3.10)

$$\Delta = 4x_{n+1}^2 - 8 \geq 0,$$

i.e., $x_{n+1}^2 \geq 2$, or $x_{n+1} \geq \sqrt{2}$, $\forall n \in \mathbb{N}$. □

Lemma 4.27. *For any $n \geq 2$ we have $x_{n+1} \leq x_n$.*

Proof. Let $n \geq 2$. We have

$$x_n - x_{n+1} = x_n - \frac{1}{2}\left(x_n + \frac{2}{x_n}\right) = \frac{1}{2}\frac{x_n^2 - 2}{x_n} \overset{(4.17)}{\geq} 0.$$

□

Thus the sequence $(x_n)_{n\geq 2}$ is decreasing and bounded below and thus it is convergent. Denote by $\bar{x}$ the limit. The inequality (4.17) implies that $\bar{x} \geq \sqrt{2}$. Letting $n \to \infty$ in (4.18) we deduce

$$\bar{x}^2 - 2\bar{x}^2 + 2 = 0 \Rightarrow 2 = \bar{x}^2 \Rightarrow \bar{x} = \sqrt{2}.$$

For example

$$x_2 = 1.5, \quad x_3 = 1.4166..., \quad x_4 := 1.4142..., \quad x_5 := 1.4142....$$

Note that

$$(1.4142)^2 = 1.99996164.$$

□

Theorem 4.28 (Nested Intervals Theorem). *Consider a* nested sequence *of* **closed** *intervals $[a_n, b_n]$, $n \in \mathbb{N}$, i.e.,*

$$[a_1, b_1] \supset [a_2, b_2] \supset [a_3, b_3] \supset \cdots.$$

Then there exists $x \in \mathbb{R}$ that belongs to all the intervals, i.e.,

$$\bigcap_{n\in\mathbb{N}} [a_n, b_n] \neq \emptyset.$$

Proof. The nesting condition implies that for any $n \in \mathbb{N}$ we have

$$a_n \leq a_{n+1} \leq b_{n+1} \leq b_n.$$

This shows that the sequence (a_n) is non-decreasing and bounded while the sequence (b_n) is non-increasing. Therefore, these sequences are convergent and we set

$$a := \lim_n a_n, \quad b := \lim_n b_n$$

the condition $a_n \leq b_n$, $\forall n$ implies that

$$a_n \leq a \leq b \leq b_n, \quad \forall n.$$

Hence $[a, b] \subset [a_n, b_n]$, $\forall n$. □

Theorem 4.29 (Bolzano–Weierstrass). *Any bounded sequence has a convergent subsequence.*

Proof. Let (x_n) be a bounded sequence of real numbers. Thus, there exist real numbers a_1, b_1 such that $x_n \in [a_1, b_1]$, for all n. We set

$$n_1 := 1.$$

Divide the interval $[a_1, b_1]$ into two intervals of equal length. At least one of these intervals will contain infinitely many terms of the sequence (x_n). Pick such an interval and denote it by $[a_2, b_2]$. Thus

$$[a_1, b_1] \supset [a_2, b_2], \quad b_2 - a_2 = \frac{1}{2}(b_1 - a_1).$$

Choose $n_2 > 1$ such that $x_{n_2} \in [a_2, b_2]$. We now proceed inductively.

Suppose that we have produced the intervals

$$[a_1, b_1] \supset [a_2, b_2] \supset \cdots \supset [a_k, b_k]$$

and the natural numbers $n_1 < n_2 < \cdots < n_k$ such that

$$b_2 - a_2 = \frac{1}{2}(b_1 - a_1), \quad b_3 - a_3 = \frac{1}{2}(b_2 - a_2), \quad b_k - a_k = \frac{1}{2}(b_{k-1} - a_{k-1}),$$

$$x_{n_1} \in [a_1, b_1], \quad x_{n_2} \in [a_2, b_2], \cdots x_{n_k} \in [a_k, b_k],$$

and the interval $[a_k, b_k]$ contains infinitely many terms of the sequence (x_n). We then divide $[a_k, b_k]$ into two intervals of equal lengths. One of them will contain infinitely many terms of (x_n). Denote that interval by $[a_{k+1}, b_{k+1}]$. We can then find a natural number $n_{k+1} > n_k$ such that $x_{n_{k+1}} \in [a_{k+1}, b_{k+1}]$. By construction

$$b_{k+1} - a_{k+1} = \frac{1}{2}(b_k - a_k) = \cdots = \frac{1}{2^k}(a_1 - b_1).$$

We have thus produced sequences (a_k), (b_k) (x_{n_k}) with the properties

$$a_1 \leq a_2 \leq \cdots \leq a_k \leq x_{n_k} \leq b_k \leq \cdots \leq b_2 \leq b_1, \tag{4.19a}$$

$$b_k - a_k = \frac{1}{2^{k-1}}(b_1 - a_1). \tag{4.19b}$$

The inequalities (4.19a) show that the sequences (a_k) and (b_k) are monotone and bounded, and thus have limits which we denote by a and b respectively. By letting $k \to \infty$ in (4.19b) we deduce that $a = b$.

The subsequence (x_{n_k}) is squeezed between two sequences converging to the same limit so the squeezing theorem implies that it is convergent. □

Definition 4.30. A *limit point* of a sequence of real numbers (x_n) is a real number which is the limit of some subsequence of the original sequence (x_n). □

Example 4.31. Consider the sequence

$$x_n = (-1)^n + \frac{1}{n}, \quad n \in \mathbb{N}.$$

Thus

$$x_{2n} = 1 + \frac{1}{2n}, \quad x_{2n+1} = -1 + \frac{1}{2n+1}.$$

Then the numbers 1 and -1 are limit points of this sequence because

$$\lim_{n\to\infty} x_{2n} = \lim_{n\to\infty}\left(1 + \frac{1}{2n}\right) = 1,$$

$$\lim_{n\to\infty} x_{2n+1} = \lim_{n\to\infty}\left(-1 + \frac{1}{2n+1}\right) = -1.$$ □

4.5. Fundamental sequences and Cauchy's characterization of convergence

We know that any convergent sequence is bounded. In other words, so boundedness is a necessary condition for a sequence to be convergent. However, it is not also a sufficient condition. For example, the sequence

$$1, -1, 1, -1, \ldots$$

is bounded, but it is not convergent.

Weierstrass's theorem on bounded monotone sequences shows that monotonicity is a sufficient condition for a bounded sequence to be convergent. However, monotonicity is not a necessary condition for convergence. Indeed, the sequence

$$x_n = \frac{(-1)^n}{n}, \quad n \in \mathbb{N}$$

converges to zero, yet it is not monotone because the even order terms are positive while the odd order terms are negative. In this subsection we will present a fundamental necessary and sufficient condition for a sequence to be convergent that makes no reference to the precise value of the limit. We begin by defining a very important concept.

Definition 4.32. A sequence of real numbers $(a_n)_{n\in\mathbb{N}}$ is called *fundamental* (or *Cauchy*[3]) if the following holds:

$$\forall \varepsilon > 0 \ \exists N = N(\varepsilon) \in \mathbb{N} \ \text{such that} \ \forall m, n > N(\varepsilon): \ |a_m - a_n| < \varepsilon. \tag{4.20}$$

□

[3]Named after August-Louis Cauchy (1789–1857), French mathematician, reputed as a pioneer of analysis. He was one of the first to state and prove theorems of calculus rigorously, rejecting the heuristic principle of the generality of algebra of earlier authors; see Wikipedia.

Theorem 4.33 (Cauchy). *Let $(a_n)_{n\in\mathbb{N}}$ be a sequence of real numbers. Then the following statements are equivalent.*

(i) The sequence (a_n) is convergent.
(ii) The sequence (a_n) is fundamental.

Proof. (i) $\Rightarrow$ (ii). We know that there exists $a \in \mathbb{R}$ such that

$$\forall \varepsilon > 0 \;\; \exists N = N(\varepsilon) \in \mathbb{N} : \;\; \forall n > N(\varepsilon) \;\; |a_n - a| < \varepsilon.$$

Observe that for any $m, n > N(\varepsilon/2)$ we have

$$|a_m - a_n| \leq |a_m - a| + |a - a_n| < \frac{\varepsilon}{2} + \frac{\varepsilon}{2} = \varepsilon.$$

This proves that (a_n) is fundamental.

(ii) $\Rightarrow$ (i) This is the "meatier" part of the theorem. We will reach the conclusion in three conceptually distinct steps.

1. Using the fact that the sequence (a_n) is fundamental we will prove that it is bounded.
2. Since (a_n) is bounded, the Bolzano–Weierstrass Theorem implies that it has a subsequence that converges to a real number a.
3. Using the fact that the sequence (a_n) is fundamental we will prove that it converges to the real number a found above.

Here are the details. Since (a_n) is fundamental, there exists $n_1 > 0$ such that, for any $m, n \geq n_1$ we have $|a_m - a_n| < 1$. Hence if we let $m = n_1$ we deduce that for any $n \geq n_1$ we have

$$|a_{n_1} - a_n| < 1 \Rightarrow a_{n_1} - 1 < a_n < a_{n_1} + 1, \;\; \forall n \geq n_1.$$

Now let

$$m := \min\{a_1, a_2, \dots, a_{n_1-1}, a_{n_1} - 1\}, \;\; M := \max\{a_1, a_2, \dots, a_{n_1-1}, a_{n_1} + 1\}.$$

Clearly

$$m \leq a_n \leq M, \;\; \forall n \in \mathbb{N}$$

so that the sequence (a_n) is bounded.

Invoking the Bolzano–Weierstrass Theorem we deduce that there exists a subsequence $(a_{n_k})_{k\geq 1}$ and a real number a such that

$$\lim_{k\to\infty} a_{n_k} = a.$$

Let $\varepsilon > 0$. Since $a_{n_k} \to a$ as $k \to \infty$ we deduce that

$$\exists K = K(\varepsilon) \in \mathbb{N} \;\text{ such that }\; \forall k > K(\varepsilon) : \;\; |a_{n_k} - a| < \frac{\varepsilon}{2}.$$

On the other hand, the sequence $(a_n)_{n\in\mathbb{N}}$ is fundamental so that

$$\exists N' = N'(\varepsilon) \in \mathbb{N} \;\text{ such that }\; \forall m, n > N'(\varepsilon) : \;\; |a_m - a_n| < \frac{\varepsilon}{2}.$$

Now choose a natural number $k_0(\varepsilon) > K(\varepsilon)$ such that $n_{k_0(\varepsilon)} > N'(\varepsilon)$. Define

$$N(\varepsilon) = n_{k_0(\varepsilon)}.$$

If $n > N(\varepsilon)$ then $n, n_{k_0} > N'(\varepsilon)$ and thus

$$|a_n - a_{n_{k_0}}| < \frac{\varepsilon}{2}.$$

On the other hand, since $k_0(\varepsilon) > K(\varepsilon)$ we deduce that

$$|a_{n_{k_0}} - a| < \frac{\varepsilon}{2}.$$

Hence, for any $n > N(\varepsilon)$ we have

$$|a_n - a| \leq |a_n - a_{n_{k_0}}| + |a_{n_{k_0}} - a| < \frac{\varepsilon}{2} + \frac{\varepsilon}{2} < \varepsilon.$$

Since ε was arbitrary we conclude that (a_n) converges to a. □

4.6. Series

Often one has to deal with sums of infinitely many terms. Such a sum is called a *series*. Here is the precise definition.

Definition 4.34. The *series* associated to a sequence $(a_n)_{n\geq 0}$ of real numbers is the ***new*** sequence $(s_n)_{n\geq 0}$ defined by the *partial sums*

$$s_0 = a_0,\ \ s_1 = a_0 + a_1,\ \ s_2 = a_0 + a_1 + a_2, \dots, s_n = a_0 + a_1 + \cdots + a_n = \sum_{i=0}^{n} a_i, \ \dots .$$

The series associated to the sequence $(a_n)_{n\geq 0}$ is denoted by the symbol

$$\sum_{n=0}^{\infty} a_n \ \text{ or } \ \sum_{n\geq 0} a_n.$$

The series is called *convergent* if the sequence of partial sums $(s_n)_{n\geq 0}$ is convergent. The limit $\lim_{n\to\infty} s_n$ is called *the sum* series. We will use the notation

$$\sum_{n\geq 0} a_n = S$$

to indicate that the series is convergent and its sum is the real number S. □

Example 4.35 (Geometric series. Part 1). Let $r \in (-1, 1)$. The *geometric series*

$$\sum_{n=0}^{\infty} r^n = 1 + r + r^2 + \cdots$$

is convergent and we have the following *very useful equality*

$$\boxed{\sum_{n=0}^{\infty} r^n = \frac{1}{1-r}}. \tag{4.21}$$

Indeed, the n-th partial sum of this series is

$$s_n = 1 + r + \cdots + r^n = \frac{1 - r^{n+1}}{1 - r}.$$

Example 4.12 shows that when $|r| < 1$ we have $\lim_n r^{n+1} = 0$ so that

$$\sum_{n=0}^{\infty} r^n = \lim_n s_n = \frac{1}{1 - r}.$$

Observe that if we set $r = \frac{1}{2}$ in (4.21) we deduce

$$\sum_{n=0}^{\infty} \frac{1}{2^n} = 2. \qquad \square$$

The proof of the following result is left to you as an exercise.

Proposition 4.36. *Consider two series*

$$\sum_{n\geq 0} a_n \ \textit{and} \ \sum_{n\geq 0} a'_n$$

such that there exists $N_0 > 0$ with the property

$$a_n = a'_n \ \forall n > N_0.$$

Then

$$\sum_{n\geq 0} a_n \ \textit{is convergent} \Longleftrightarrow \sum_{n\geq 0} a'_n \ \textit{is convergent}. \qquad \square$$

Proposition 4.37. *If the series $\sum_{n=0}^{\infty} a_n$ is convergent, then*

$$\lim_{n\to\infty} a_n = 0.$$

Proof. Observe that for $n \geq 1$

$$s_n = a_0 + a_1 + \cdots + a_{n-1} + a_n = s_{n-1} + a_n.$$

Hence

$$a_n = s_n - s_{n-1}.$$

The sequences $(s_n)_{n\geq 1}$ and $(s_{n-1})_{n\geq 1}$ converge to the same finite limit so that

$$\lim_n a_n = \lim_n s_n - \lim_n s_{n-1} = 0.$$

$\square$

Example 4.38 (Geometric series. Part 2). Let $|r| \geq 1$. Then the geometric series

$$1 + r + r^2 + \cdots + r^n + \cdots = \sum_{n=0}^{\infty} r^n$$

is divergent. Indeed, if it were convergent, then the above proposition would imply that $r^n \to 0$ as $n \to \infty$. This is not the case when $|r| \geq 1$. $\square$

Proposition 4.39. *A series of positive numbers*

$$\sum_{n\geq 0} a_n, \quad a_n > 0 \ \forall n$$

is convergent if and only if the sequence of partial sums

$$s_n = a_0 + \cdots + a_n$$

is bounded.

Proof. Observe that the sequence of partial sums is increasing since

$$s_{n+1} - s_n = a_{n+1} > 0, \quad \forall n.$$

If the sequence (s_n) is also bounded, then Weierstrass' Theorem on monotone sequences implies that it must be convergent.

Conversely, if the sequence (s_n) is convergent, then Proposition 4.14 shows that it must also be bounded. □

Example 4.40. (a) The *harmonic series*

$$\sum_{n=1}^{\infty} \frac{1}{n} = 1 + \frac{1}{2} + \frac{1}{3} + \cdots$$

is *divergent*. Here is why.

This is a series with positive terms. Observe that

$$s_1 = 1 \geq 1, \quad s_2 = 1 + \frac{1}{2} \geq 1 + \frac{1}{2},$$

$$s_{2^2} = s_4 = s_2 + \frac{1}{3} + \frac{1}{4} > s_2 + \frac{1}{4} + \frac{1}{4} = s_2 + \frac{1}{2} = 1 + \frac{2}{2}$$

$$s_{2^3} = s_8 = s_4 + \underbrace{\frac{1}{5} + \frac{1}{6} + \frac{1}{7} + \frac{1}{8}}_{\text{4 terms}} > 1 + \frac{2}{2} + \underbrace{\frac{1}{5} + \frac{1}{6} + \frac{1}{7} + \frac{1}{8}}_{\text{4 terms}}$$

$$> 1 + \frac{2}{2} + \underbrace{\frac{1}{8} + \frac{1}{8} + \frac{1}{8} + \frac{1}{8}}_{\text{4 terms}} = 1 + \frac{3}{2}.$$

Thus

$$s_{2^3} > 1 + \frac{3}{2}.$$

We want to prove that

$$s_{2^n} > 1 + \frac{n}{2}, \quad \forall n \geq 2. \tag{4.22}$$

We have shown this for $n = 2$ and $n = 3$. The general case follows inductively. Observe that $2^{n+1} = 2 \cdot 2^n = 2^n + 2^n$ and thus

$$s_{2^{n+1}} = s_{2^n} + \underbrace{\frac{1}{2^n + 1} + \cdots + \frac{1}{2^{n+1}}}_{2^n\text{-terms}}$$

$$> s_{2^n} + \underbrace{\frac{1}{2^{n+1}} + \cdots + \frac{1}{2^{n+1}}}_{2^n\text{-terms}} = s_{2^n} + \frac{2^n}{2^{n+1}} = s_{2^n} + \frac{1}{2}$$

(use the inductive assumption)

$$> 1 + \frac{n}{2} + \frac{1}{2}.$$

This proves (4.22) which shows that the sequence s_{2^n} is not bounded. Invoking Proposition 4.39 we conclude that the harmonic series is not convergent.

(b) Let $r > 1$ be a rational number and consider the series

$$\sum_{n=1}^{\infty} \frac{1}{n^r}.$$

We want to show that this series is *convergent*.

We have

$$s_2 = 1 + \frac{1}{2^r},$$

$$s_4 = s_2 + \frac{1}{3^r} + \frac{1}{4^r} < s_2 + \frac{1}{2^r} + \frac{1}{2^r} < s_2 + \frac{2}{2^r} = \frac{1}{2^r} + 1 + \frac{1}{2^{(r-1)}},$$

$$s_{2^3} = s_8 = s_4 + \frac{1}{5^r} + \frac{1}{6^r} + \frac{1}{7^r} + \frac{1}{8^r} < s_4 + \frac{4}{4^r} = \frac{1}{2^r} + 1 + \frac{1}{2^{(r-1)}} + \frac{1}{2^{2(r-1)}}.$$

We claim that for any $n \geq 1$ we have

$$s_{2^{n+1}} < \frac{1}{2^r} + 1 + \frac{1}{2^{(r-1)}} + \frac{1}{2^{2(r-1)}} + \cdots + \frac{1}{2^{n(r-1)}}. \tag{4.23}$$

We argue inductively. The result is clearly true for $n = 1, 2$. We assume it is true for n and we prove it is true for $n + 1$. We have

$$s_{2^{n+1}} = s_{2^n} + \underbrace{\frac{1}{(2^n + 1)^r} + \frac{1}{(2^n + 2)^r} + \cdots + \frac{1}{(2^{n+1})^r}}_{2^n \text{ terms}}$$

$$< s_{2^n} + \underbrace{\frac{1}{(2^n)^r} + \frac{1}{(2^n)^r} + \cdots + \frac{1}{(2^n)^r}}_{2^n \text{ terms}} = s_{2^n} + \frac{1}{2^{n(r-1)}}$$

(use the induction assumption)

$$< \frac{1}{2^r} + 1 + \frac{1}{2^{(r-1)}} + \frac{1}{2^{2(r-1)}} + \cdots + \frac{1}{2^{(n-1)(r-1)}} + \frac{1}{2^{n(r-1)}}.$$

If we set

$$q := \frac{1}{2^{r-1}} = \left(\frac{1}{2}\right)^{r-1},$$

then we observe that the condition $r > 1$ implies $q \in (0, 1)$ and we can rewrite (4.23) as

$$s_{2^{n+1}} < \frac{1}{2^r} + 1 + q + \cdots + q^n < \frac{1}{2^r} + \frac{1}{1-q}, \quad \forall n \in \mathbb{N}.$$

This implies that the sequence (s_{2^n}) is bounded and thus the series

$$\sum_{n=1}^{\infty} \frac{1}{n^r}$$

is convergent. Its sum is denoted by $\zeta(r)$ and it is called *Riemann zeta function*. For most r's, the actual value $\zeta(r)$ is not known. However, L. Euler has computed the values $\zeta(r)$ when r is an even natural number. For example

$$\zeta(2) = \sum_{n=1}^{\infty} \frac{1}{n^2} = \frac{\pi^2}{6}.$$

All the known proofs of the above equality are very ingenious. □

Theorem 4.41 (Comparison Principle). *Suppose that*

$$\sum_{n\geq 0} a_n \text{ and } \sum_{n\geq 0} b_n$$

are two series of positive real numbers such that

$$\exists N_0 \in \mathbb{N} \text{ such that } \forall n > N_0 : \; a_n < b_n.$$

Then the following hold.
(a) $\sum_{n\geq 0} a_n$ *divergent* $\Rightarrow \sum_{n\geq 0} b_n$ *divergent.*
(b) $\sum_{n\geq 0} b_n$ *convergent* $\Rightarrow \sum_{n\geq 0} a_n$ *convergent.*

Proof. We set

$$s_n(a) = \sum_{k=0}^{n} a_n, \quad s_n(b) = \sum_{k=1}^{n} b_n.$$

In view of Proposition 4.36 the convergence or divergence of a series is not affected if we modify finitely many of its terms. Thus, we may assume that

$$a_n \leq b_n, \quad \forall n \geq 0.$$

In particular, we have

$$s_n(a) \leq s_n(b), \quad \forall n \geq 0. \tag{4.24}$$

Note that since the terms a_n are *positive*

$$\sum_{n\geq 0} a_n \text{ divergent} \Rightarrow s_n(a) \to \infty \Rightarrow s_n(b) \to \infty \Rightarrow \sum_{n\geq 0} b_n \text{ divergent}$$

and

$$\sum_{n\geq 0} b_n \text{ convergent} \Rightarrow s_n(b) \text{ bounded} \Rightarrow s_n(a) \text{ bounded} \Rightarrow \sum_{n\geq 0} a_n \text{ convergent.}$$

□

The above result has an immediate and very useful consequence whose proof is left to you as an exercise.

Corollary 4.42. *Suppose that*

$$\sum_{n\geq 0} a_n \text{ and } \sum_{n\geq 0} b_n$$

are two series with positive terms.

(a) If the sequence $(\frac{a_n}{b_n})_{n\geq 0}$ *is convergent and the series* $\sum_{n\geq 0} b_n$ *is convergent, then the series* $\sum_{n\geq 0} a_n$ *is also convergent.*

(b) If the sequence $(\frac{a_n}{b_n})_{n\geq 0}$ *has a limit* r *which is either positive,* $r > 0$*, or* $r = \infty$ *and the series* $\sum_{n\geq 0} b_n$ *is divergent, then the series* $\sum_{n\geq 0} a_n$ *is also divergent.* □

Example 4.43 (L. Euler). Consider the series

$$\sum_{n=0}^{\infty} \frac{1}{n!} = 1 + \frac{1}{1!} + \frac{1}{2!} + \frac{1}{3!} + \cdots . \tag{4.25}$$

Observe that if $n \geq 2$, then

$$\frac{1}{n!} = \frac{1}{2} \cdot \frac{1}{3} \cdots \frac{1}{n} \leq \underbrace{\frac{1}{2} \cdots \frac{1}{2}}_{(n-1)-\text{times}} = \frac{1}{2^{n-1}} = \frac{2}{2^n}.$$

Since the series

$$\sum_{n\geq 0} \frac{2}{2^n}$$

is convergent we deduce from the Comparison Principle that the series (4.25) is also convergent. Its sum is the Euler number

$$\sum_{n=0}^{\infty} \frac{1}{n!} = e = \lim_n \left(1 + \frac{1}{n}\right)^n . \tag{4.26}$$

This is a nontrivial result. We will describe a more conceptual proof in Corollary 8.8. However, that proof relies on the full strength of differential calculus.

Here is an elementary proof. We set

$$e_n := \left(1 + \frac{1}{n}\right)^n, \quad s_n = 1 + \frac{1}{1!} + \frac{1}{2!} + \cdots + \frac{1}{n!}, \quad \forall n \in \mathbb{N}.$$

We will prove two things.

$$e_n < s_n, \quad \forall n \geq 1 \tag{4.27a}$$

$$s_k \leq e, \quad \forall k \geq 1. \tag{4.27b}$$

Assuming the validity of the above inequalities, we observe that by letting $n \to \infty$ in (4.27a) we deduce that

$$e \leq \lim_n s_n .$$

On the other hand, if we let $k \to \infty$ in (4.27b), then we conclude that

$$\lim_k s_k \leq e.$$

Hence (4.27a, 4.27b) imply that

$$e = \lim_n s_n = \sum_{n=0}^{\infty} \frac{1}{n!}.$$

Proof of (4.27a). Using Newton's binomial formula we deduce

$$\begin{aligned}
e_n &= \left(1+\frac{1}{n}\right)^n = 1 + \binom{n}{1}\frac{1}{n} + \binom{n}{2}\frac{1}{n^2} + \cdots + \binom{n}{n}\frac{1}{n^n} \\
&= 1 + \frac{n}{1!}\frac{1}{n} + \frac{n(n-1)}{2!}\frac{1}{n^2} + \frac{n(n-1)(n-2)}{3!}\frac{1}{n^3} + \cdots + \frac{n(n-1)\cdots 1}{n!}\frac{1}{n^n} \\
&= 1 + \frac{n}{n}\frac{1}{1!} + \underbrace{\frac{n(n-1)}{n^2}}_{<1}\cdot\frac{1}{2!} + \underbrace{\frac{n(n-1)(n-2)}{n^3}}_{<1}\cdot\frac{1}{3!} + \cdots + \underbrace{\frac{n(n-1)\cdots 1}{n^n}}_{<1}\cdot\frac{1}{n!} \\
&< 1 + \frac{1}{1!} + \frac{1}{2!} + \frac{1}{3!} + \cdots + \frac{1}{n!} = s_n.
\end{aligned}$$

Proof of (4.27b). Fix $k \in \mathbb{N}$. Then from the same formula above we deduce that if $k \leq n$, then

$$e_n = 1 + \frac{n}{1!}\frac{1}{n} + \frac{n(n-1)}{2!}\frac{1}{n^2} + \frac{n(n-1)(n-2)}{3!}\frac{1}{n^3} + \cdots + \frac{n(n-1)\cdots 1}{n!}\frac{1}{n^n}$$

(neglect the terms containing the powers $\frac{1}{n^j}$, $j > k$)

$$\begin{aligned}
&> 1 + \frac{n}{1!}\frac{1}{n} + \frac{n(n-1)}{2!}\frac{1}{n^2} + \frac{n(n-1)(n-2)}{3!}\frac{1}{n^3} + \cdots + \frac{n(n-1)\cdots(n-k+1)}{k!}\frac{1}{n^k} \\
&= 1 + \frac{1}{1!} + \frac{n-1}{n}\frac{1}{2!} + \frac{n-1}{n}\frac{n-2}{n}\cdot\frac{1}{3!} + \cdots + \frac{n-1}{n}\cdots\frac{n-k+1}{n}\cdot\frac{1}{k!} \\
&= 1 + \frac{1}{1!} + \left(1-\frac{1}{n}\right)\frac{1}{2!} + \left(1-\frac{1}{n}\right)\left(1-\frac{2}{n}\right)\frac{1}{3!} + \cdots + \left(1-\frac{1}{n}\right)\left(1-\frac{2}{n}\right)\cdots\left(1-\frac{k-1}{n}\right)\frac{1}{k!}.
\end{aligned}$$

If we let $n \to \infty$, while keeping k fixed we deduce

$$\begin{aligned}
e &= \lim_{n\to\infty} e_n \\
&\geq 1 + \frac{1}{1!} + \lim_{n\to\infty}\left(1-\frac{1}{n}\right)\frac{1}{2!} + \lim_{n\to\infty}\left(1-\frac{1}{n}\right)\left(1-\frac{2}{n}\right)\frac{1}{3!} + \cdots \\
&\quad + \lim_{n\to\infty}\left(1-\frac{1}{n}\right)\left(1-\frac{2}{n}\right)\cdots\left(1-\frac{k-1}{n}\right)\frac{1}{k!} = s_k.
\end{aligned}$$

Let us now estimate the error

$$\begin{aligned}
\varepsilon_n &= e - s_n = \left(1 + \frac{1}{1!} + \frac{1}{2!} + \cdots\right) - \left(1 + \frac{1}{1!} + \frac{1}{2!} + \cdots + \frac{1}{n!}\right) \\
&= \frac{1}{(n+1)!} + \frac{1}{(n+2)!} + \cdots.
\end{aligned}$$

Clearly $\varepsilon_n > 0$ and

$$\begin{aligned}
\varepsilon_n &= \frac{1}{(n+1)!}\left(1 + \frac{1}{n+2} + \frac{1}{(n+2)(n+3)} + \cdots\right) \\
&< \frac{1}{(n+1)!}\left(1 + \frac{1}{n+2} + \frac{1}{(n+2)^2} + \frac{1}{(n+2)^3} + \cdots\right) \\
&= \frac{1}{(n+1)!}\cdot\frac{1}{1-\frac{1}{n+2}} = \frac{1}{(n+1)!}\cdot\frac{n+2}{n+1}.
\end{aligned}$$

For example, if we let $n = 6$, then we deduce that

$$0 < \varepsilon_6 < \frac{8}{7\cdot 6!} = \frac{8}{7\cdot 720} \approx 0.0002\ldots.$$

This shows that s_5 computes e with a 2-decimal precision. We have

$$s_5 = 1 + 1 + \frac{1}{2} + \frac{1}{6} + \frac{1}{24} + \frac{1}{120} = 2.71\ldots,$$

so that

$$e = 2.71\ldots.$$

□

Given a series $\sum_{n=0}^{\infty} a_n$ and natural numbers $m < n$ we have

$$s_n - s_m = (a_{m+1} + a_{m+1} + \cdots + a_n) = \sum_{k=m+1}^{n} a_k.$$

Cauchy's Theorem 4.33 implies the following useful result.

Theorem 4.44 (Cauchy). *Let $\sum_{n=0}^{\infty} a_n$ be a series of real numbers. Then the following statements are equivalent.*

(i) The series $\sum_{n=0}^{\infty} a_n$ is convergent.
(ii) $\forall \varepsilon > 0\ \exists N = N(\varepsilon) \in \mathbb{N}$ such that $\forall n > m > N(\varepsilon)\ |a_{m+1} + \cdots + a_n| < \varepsilon$.

□

Definition 4.45. The series of real numbers

$$\sum_{n\geq 0} a_n$$

is called *absolutely convergent* if the series of absolute values

$$\sum_{n\geq 0} |a_n|$$

is convergent. □

Theorem 4.46 (Absolute Convergence Theorem). *If the series*

$$\sum_{n\geq 0} a_n$$

is absolutely convergent, then it is also convergent.

Proof. Since

$$\sum_{n\geq 0} |a_n|$$

is convergent, then Theorem 4.44 implies that

$$\forall \varepsilon > 0\ \exists N = N(\varepsilon) \in \mathbb{N}:\ \ \forall n > m > N(\varepsilon):\ \ |a_{m+1}| + \cdots + |a_n| < \varepsilon.$$

On the other hand, we observe that

$$|a_{m+1} + \cdots + a_n| \leq |a_{m+1}| + \cdots + |a_n|$$

so that

$$\forall \varepsilon > 0\ \exists N = N(\varepsilon) \in \mathbb{N}:\ \ \forall n > m > N(\varepsilon):\ \ |a_{m+1} + \cdots + a_n| < \varepsilon.$$

Invoking Theorem 4.44 again we deduce that the series $\sum_{n\geq 0} a_n$ is convergent as well. □

The Comparison Principle has the following immediate consequence.

Corollary 4.47 (Weierstrass M-test). *Consider two series*

$$\sum_{n\geq 0} a_n, \ \sum_{n\geq 0} b_n$$

such that $b_n > 0$ *for any* n *and there exists* $N_0 \in \mathbb{N}$ *such that*

$$|a_n| < b_n, \ \ \forall n > N_0.$$

If the series $\sum_{n\geq 0} b_n$ *is convergent, then the series* $\sum_{n\geq 0} a_n$ *converges absolutely.* □

The Weierstrass M-test leads to a simple but very useful convergence test, called *the d'Alembert test* or the *ratio test.*

Corollary 4.48 (Ratio Test). *Let*

$$\sum_{n\geq 0} a_n$$

be a series such that $a_n \neq 0 \ \forall n$ *and the limit*

$$L = \lim_{n\to\infty} \frac{|a_{n+1}|}{|a_n|} \geq 0$$

exists, but it could also be infinite. Then the following hold.

(i) If $L < 1$, *then the series* $\sum_{n\geq 0} a_n$ *is absolutely convergent.*
(ii) If $L > 1$ *then the series* $\sum_{n\geq 0} a_n$ *is not convergent.*

Proof. (i) We know that $L < 1$. Choose r such that $L < r < 1$. Since

$$\frac{|a_{n+1}|}{|a_n|} \to L$$

there exists $N_0 \in \mathbb{N}$ such that

$$\frac{|a_{n+1}|}{|a_n|} \leq r, \ \ \forall n > N_0 \iff |a_{n+1}| \leq |a_n|r, \ \ \forall n > N_0.$$

We deduce that

$$|a_{N_0+1}| \leq |a_{N_0}|r, \ \ |a_{N_0+2}| \leq |a_{N_0+1}|r \leq |a_{N_0}|r^2,$$

and, inductively

$$|a_{N_0+k}| \leq r^k |a_{N_0}|, \ \ \forall k \in \mathbb{N}.$$

If we set $n = N_0 + k$ so that $k = n - N_0$, then we conclude from above that for any $n > N_0$ we have

$$|a_n| \leq |a_{N_0}|\, r^{n-N_0} = \underbrace{\frac{|a_{N_0}|}{r^{N_0}}}_{=:C} r^n.$$

In other words

$$|a_n| \leq Cr^n, \ \ \forall n \geq N_0.$$

The geometric series $\sum_{n\geq 0} b_n$, $b_n = Cr^n$, is convergent for $r \in (0,1)$ and we deduce from Weierstrass' Test that the series $\sum_{n\geq 0} |a_n|$ is also convergent.

(ii) We argue by contradiction and assume that the series $\sum_{n\geq 0} |a_n|$ is convergent. Since $L > 1$ we deduce that there exists a $N_0 \in \mathbb{N}$ such that

$$\frac{|a_{n+1}|}{|a_n|} > 1, \ \ \forall n > N_0 \iff |a_{n+1}| > |a_n|, \ \ \forall n > N_0.$$

Since the series $\sum_{n\geq 0} |a_n|$ is convergent we deduce that $\lim_n a_n = 0$. On the other hand, $|a_n| > |a_{N_0}|$ for $n > N_0$ so that

$$0 = \lim_n |a_n| \geq |a_{N_0}| > 0.$$

This contradiction shows that the series $\sum_{n\geq 0} |a_n|$ cannot be convergent. □

Example 4.49. (a) Consider the series

$$\sum_{n\geq 1} (-1)^n \frac{n^2}{2^n}.$$

Then

$$\frac{|a_{n+1}|}{|a_n|} = \frac{\frac{(n+1)^2}{2^{n+1}}}{\frac{n^2}{2^n}} = \frac{1}{2}\left(\frac{n+1}{n}\right)^2 \to \frac{1}{2} \to \frac{1}{2} \ \text{ as } n \to \infty.$$

The Ratio Test implies that the series is absolutely convergent.

(b) Consider the series

$$\sum_{n\geq 1} \frac{1}{\sqrt{n(n+1)}}.$$

We observe that

$$\frac{\frac{1}{\sqrt{n(n+1)}}}{\frac{1}{n}} = \frac{n}{\sqrt{n(n+1)}} = \frac{n}{\sqrt{n^2(1+\frac{1}{n})}} = \frac{1}{\sqrt{1+\frac{1}{n}}}.$$

Hence

$$\lim_{n\to\infty} \frac{\frac{1}{\sqrt{n(n+1)}}}{\frac{1}{n}} = 1$$

so that there exists $N_0 > 0$ such that

$$\frac{\frac{1}{\sqrt{n(n+1)}}}{\frac{1}{n}} > \frac{1}{2} \ \ \forall n > N_0,$$

i.e.,

$$\frac{1}{\sqrt{n(n+1)}} > \frac{1}{2n}, \ \ \forall n > N_0.$$

In Example 4.40(a) we have shown that the series $\sum_{n\geq 1} \frac{1}{2n}$ is divergent. Invoking the Comparison Principle we deduce that the series $\sum_{n\geq 1} \frac{1}{\sqrt{n(n+1)}}$ is also divergent. □

Definition 4.50. A series is called *conditionally convergent* if it is convergent, but not absolutely convergent. □

Example 4.51. Consider the series

$$\sum_{n\geq 0} \frac{(-1)^n}{n+1} = 1 - \frac{1}{2} + \frac{1}{3} - \frac{1}{4} + \cdots .$$

Example 4.40(a) shows that this series is not absolutely convergent. However, it is a convergent series. To see this observe first that

$$s_0 = 1, \;\; s_2 = s_0 - \frac{1}{2} + \frac{1}{3} = s_0 - \left(\frac{1}{2} - \frac{1}{3}.\right) < s_0,$$

$$s_{2n+2} = s_{2n} - \frac{1}{(2n+2)} + \frac{1}{2n+3} = s_{2n} - \left(\frac{1}{2n+2} - \frac{1}{2n+3}\right) < s_{2n}.$$

Thus the subsequence $s_0, s_2, s_4, \ldots,$ is decreasing.

Next observe that

$$s_1 = 1 - \frac{1}{2} > 0, \;\; s_3 = s_1 + \frac{1}{3} - \frac{1}{4} > s_1,$$

$$s_{2n+3} = s_{2n+1} + \frac{1}{2n+3} - \frac{1}{2n+4} > s_{2n+1}.$$

Thus, the subsequence $s_1, s_3, s_5, \ldots,$ is increasing. Now observe that

$$s_{2n+2} - s_{2n+1} = \frac{1}{2n+3} > 0.$$

Hence

$$s_0 > s_{2n+2} > s_{2n+1} \geq s_1.$$

This proves that the increasing subsequence (s_{2n+1}) is also bounded above and the decreasing sequence (s_{2n+2}) is bounded below. Hence these two subsequences are convergent and since

$$\lim_n (s_{2n+2} - s_{2n+1}) = \lim_n \frac{1}{2n+3} = 0$$

we deduce that they converge to the same real number. This implies that the full sequence $(s_n)_{n\geq 0}$ converges to the same number; see Exercise 4.23.

The sum of this alternating series is $\ln 2$, but the proof of this fact is more involved and requires the full strength of the calculus techniques; see Example 9.52. □

4.7. Power series

Definition 4.52. A *power series* in the variable x and real coefficients $a_0, a_1, a_2, \ldots$ is a series of the form

$$s(x) = a_0 + a_1 x + a_2 x^2 + \cdots .$$

The *domain of convergence*of the power series is the set of real numbers x such that the corresponding series $s(x)$ is convergent. □

Example 4.53. (a) The geometric series

$$1 + x + x^2 + \cdots$$

is a power series. It converges for $|x| < 1$ and diverges for $|x| \geq 1$.

(b) Consider the power series

$$\sum_{n\geq 1} \frac{x^n}{n} = x + \frac{x^2}{2} + \frac{x^3}{3} + \cdots .$$

Note that

$$\left| \frac{\frac{x^{n+1}}{n+1}}{\frac{x^n}{n}} \right| = |x| \frac{n}{n+1} \to |x| \text{ as } n \to \infty.$$

The Ratio Test shows that this series converges absolutely for $|x| < 1$ and diverges for $|x| > 1$.

When $x = 1$ the series becomes the harmonic series

$$1 + \frac{1}{2} + \frac{1}{3} + \cdots$$

which is divergent. When $x = -1$ the series becomes the alternating series

$$-1 + \frac{1}{2} - \frac{1}{3} + \cdots = -\sum_{n\geq 1} \frac{(-1)^n}{n}.$$

As explained in Example 4.51, this series is convergent.

(c) Consider the power series

$$\sum_{n\geq 0} \frac{x^n}{n!} = 1 + \frac{x}{1!} + \frac{x^2}{2!} + \cdots .$$

Note that

$$\left| \frac{\frac{x^{n+1}}{(n+1)!}}{\frac{x^n}{n!}} \right| = \frac{|x|}{n+1} \to 0 \text{ as } n \to \infty.$$

The Ratio Test implies that this series converges absolutely for any $x \in \mathbb{R}$. □

Proposition 4.54. *Consider a power series in the variable x with real coefficients*

$$s(x) = a_0 + a_1 x + a_2 x^2 + \cdots .$$

Suppose that the nonzero real number x_0 is in the domain of convergence of the series. Then for any real number x such that $|x| < |x_0|$ the series $s(x)$ is absolutely *convergent.*

Proof. Since the series

$$a_0 + a_1 x_0 + a_2 x_0^2 + \cdots$$

is convergent, the sequence $(a_n x_0^n)$ converges to zero. In particular, this sequence is bounded and thus there exists a positive constant C such that

$$|a_n x_0^n| < C, \ \ \forall n = 0, 1, 2, \dots .$$

We set

$$r := \frac{|x|}{|x_0|}$$

and we observe that $0 \leq r < 1$. Next we notice that

$$|a_n x^n| = |a_n x_0^n| \frac{|x|^n}{|x_0|^n} = |a_n x_0^n| r^n < C r^n, \ \ \forall n.$$

Since $0 \leq r < 1$ we deduce that the positive geometric series

$$C + Cr + Cr^2 + \cdots$$

is convergent. The Comparison Principle then implies that the series

$$|a_0| + |a_1 x| + |a_2 x^2| + \cdots$$

is also convergent. □

The above result has a very important consequence whose proof is left to you as an exercise.

Corollary 4.55. *Consider a power series in the variable x and real coefficients*

$$s(x) = a_0 + a_1 x + a_2 x^2 + \cdots .$$

We denote by D the domain of convergence of the series. We set

$$R := \begin{cases} \sup D, & \text{if } D \text{ is bounded above}, \\ \infty, & \text{if } D \text{ is not bounded above}. \end{cases} \tag{4.28}$$

Then the following hold.

(i) $R \geq 0$.
(ii) *If x is a real number such that $|x| < R$, then the series $s(x)$ is absolutely convergent.*
(iii) *If x is a real number such that $|x| > R$, then the series $s(x)$ is divergent.*

□

Definition 4.56. The quantity R defined in (4.28) is called the *radius of convergence* of the power series $s(x)$. □

Example 4.57. The power series in Example 4.53(a),(b) have radii of convergence 1, while the power series in Example 4.53(c) has radius of convergence ∞. □

4.8. Some fundamental sequences and series

$$\lim_{n\to\infty} \frac{C}{n} = 0, \quad \forall C > 0.$$

$$\lim_{n\to\infty} Cn = \infty, \quad \forall C > 0.$$

$$\lim_{n\to\infty} r^n = \lim_{n\to\infty} \frac{1}{a^n} = 0, \quad \forall r \in (0,1), \quad \forall a > 1.$$

$$\lim_{n\to\infty} \frac{n}{r^n} = 0, \quad \forall r > 1.$$

$$\lim_{n\to\infty} r^{\frac{1}{n}} = 1, \quad \forall r > 0.$$

$$\lim_{n\to\infty} n^{\frac{1}{n}} = 1.$$

$$\lim_{n\to\infty} \frac{r^n}{n!} = 0, \quad \forall r \in \mathbb{R}.$$

$$\lim_{n\to\infty} \left(1 + \frac{1}{n}\right)^n = e.$$

$$1 + r + r^2 + \cdots + r^n + \cdots = \frac{1}{1-r}, \quad \forall |r| < 1.$$

$$1 + \frac{1}{1!} + \frac{1}{2!} + \frac{1}{3!} + \cdots = e.$$

$$\sum_{n\geq 1} \frac{1}{n^s} = \begin{cases} \text{convergent}, & s > 1 \\ \text{divergent}, & s \leq 1. \end{cases}$$

4.9. Exercises

Exercise 4.1. Prove, *using the definition*, the following equalities.

$$\lim_{n\to\infty} \frac{n}{n^2+1} = 0, \tag{a}$$

$$\lim_{n\to\infty} \frac{3n+1}{2n+5} = \frac{3}{2}, \tag{b}$$

$$\lim_{n\to\infty} \frac{1}{\sqrt{n}} = 0. \tag{c}$$

Exercise 4.2. Prove Proposition 4.8. □

Exercise 4.3. Prove Proposition 4.9. □

Exercise 4.4. Let $(x_n)_{n\geq 0}$ be a sequence of real numbers and $x \in \mathbb{R}$. Consider the following statements.

(i) $\forall \varepsilon > 0$, $\exists N \in \mathbb{N}$ such that, $n > N \Rightarrow |x_n - x| < \varepsilon$.
(ii) $\exists N \in \mathbb{N}$ such that, $\forall \varepsilon > 0$, $n > N \Rightarrow |x_n - x| < \varepsilon$.

Prove that (ii) $\Rightarrow$ (i) and construct an example of sequence $(x_n)_{n\geq 1}$ and real number x satisfying (i) but not (ii). □

Exercise 4.5. (a) Prove that for any real numbers a, b we have

$$\big|\,|a| - |b|\,\big| \leq |a - b|.$$

(b) Let $(x_n)_{n\geq 0}$ be a sequence of real numbers that converges to $x \in \mathbb{R}$. Prove that

$$\lim_{n\to\infty} |x_n| = |x|.$$

□

Exercise 4.6. Compute

$$\lim_{n\to\infty} \left(\frac{1}{2} + \frac{1}{2^2} + \cdots + \frac{1}{2^n}\right).$$

Hint. Observe that

$$\frac{1}{2} + \frac{1}{2^2} + \cdots + \frac{1}{2^n} = \frac{1}{2}\left(1 + \frac{1}{2} + \cdots + \frac{1}{2^{n-1}}\right).$$

At this point you might want to use Exercise 3.7. □

Exercise 4.7. Compute

$$\lim_{n\to\infty} \frac{2^1 + 2^3 + 2^5 + \cdots + 2^{2n+1}}{2^{2n+3}}.$$

Hint. Use Exercise 3.7. □

Exercise 4.8. Let $X \subset \mathbb{R}$ be a bounded above set of real numbers. Denote by x^* the supremum of X. (The existence of the least upper bound of X is guaranteed by the Completeness Axiom.) Prove that there exists a sequence of real numbers $(x_n)_{n\in\mathbb{N}}$ satisfying the following properties.

(i) $x_n \in X, \forall n \in \mathbb{N}$.
(ii) $\lim_{n\to\infty} x_n = x^*$.

Hint. Use Proposition 2.22 and Corollary 4.11. □

Exercise 4.9. Prove the equality (4.11). □

Exercise 4.10. Let $0 < a < b$. Compute

$$\lim_{n\to\infty} \frac{a^{n+1} + b^{n+1}}{a^n + b^n}.$$ □

Exercise 4.11. (a) Let (a_n) be a sequence of positive real numbers such that $\lim_n a_n = 1$. Prove that

$$\lim_n \sqrt{a_n} = 1.$$

(b) Compute

$$\lim_{n\to\infty} \sqrt{n}\big(\sqrt{n+1} - \sqrt{n}\big).$$

Hint. Prove first that

$$\sqrt{x+1} - \sqrt{x} = \frac{1}{\sqrt{x+1} + \sqrt{x}}, \quad \forall x > 0.$$

□

Exercise 4.12. Prove that if $a > 0$, then

$$\lim_{n\to\infty} a^{\frac{1}{n}} = 1.$$

Hint. Consider first the case $a > 1$. Write $a^{\frac{1}{n}} = 1 + \varepsilon_n$ and then use Bernoulli's inequality. Show that the case $a < 1$ follows from the case $a > 1$. □

Exercise 4.13. Prove that for any real number x there exists an *increasing* sequence of *rational* numbers that converges to x and also a *decreasing* sequence of *rational* numbers that converges to x.

Hint. Use Proposition 3.33. □

Exercise 4.14. Let $(a_n)_{n\in\mathbb{N}}$ be a sequence of positive numbers that converges to a *positive* number a. Prove that

$$\exists c > 0 \text{ such that } \forall n \in \mathbb{N}\ a_n > c.$$

Hint. Argue by contradiction. □

Exercise 4.15. Let $k \in \mathbb{N}$ and suppose that $(a_n)_{n\in\mathbb{N}}$ is a sequence of positive numbers that converges to a *positive* number a.
(a) Using Exercise 4.14 prove that there exists $r > 0$ such that $a_n > r, \forall n$, so that $a_n^{\frac{1}{k}} > r^{\frac{1}{k}}$, $\forall n$.
(b) Prove that there exists a constant $M > 0$ such that

$$\left| a_n^{\frac{1}{k}} - a^{\frac{1}{k}} \right| \leq M|a_n - a|, \ \ \forall n \in \mathbb{N}.$$

Hint. Set $b_n := a_n^{\frac{1}{k}}$, $b := a^{\frac{1}{k}}$ and use the equality (3.10) to deduce.

$$a_n - a = b_n^k - b^k = (b_n - b)(b_n^{k-1} + b_n^{k-2}b + \cdots + b_n b^{k-2} + b^{k-1})$$

which implies

$$|b_n - b| = \frac{|a_n - a|}{b_n^{k-1} + b_n^{k-2}b + \cdots + b_n b^{k-2} + b^{k-1}}.$$

Now use part (a).

(c) Show that

$$\lim_n a_n^{\frac{1}{k}} = a^{\frac{1}{k}}.$$

(d) Show that if $r \in \mathbb{Q}$, then

$$\lim_n a_n^r = a^r.$$

□

Exercise 4.16. Let $r > 1$ and $k \in \mathbb{N}$. Prove that

$$\lim_{n\to\infty} r^n = \infty$$

and

$$\lim_{n\to\infty} \frac{n^k}{r^n} = 0.$$

Hint. Let $a = r^{\frac{1}{k}}$. Then

$$\frac{n^k}{r^n} = \left(\frac{n}{a^n}\right)^k.$$

□

Exercise 4.17. Compute

$$\lim_{n\to\infty} \left(1 + \frac{1}{2n}\right)^n.$$

□

Exercise 4.18. (a) Using Example 4.23 as inspiration prove that the sequence

$$x_n = \left(1 + \frac{1}{n}\right)^n$$

is increasing.
(b) Prove that the Euler number e satisfies the inequalities

$$\left(1 + \frac{1}{n}\right)^n < e < \left(1 + \frac{1}{n}\right)^{n+1}, \ \ \forall n \in \mathbb{N}.$$

Deduce from the above inequalities that $2 < e < 3$. □

Exercise 4.19. Consider the sequence (x_n) defined by the recurrence

$$x_1 = \sqrt{2}, \quad x_{n+1} = \sqrt{2 + x_n}, \quad \forall n \in \mathbb{N}.$$

Thus

$$x_2 = \sqrt{2 + \sqrt{2}}, \quad x_3 = \sqrt{2 + \sqrt{2 + \sqrt{2}}}, \quad x_4 = \sqrt{2 + \sqrt{2 + \sqrt{2 + \sqrt{2}}}}, \dots .$$

(a) Prove by induction that the sequence (x_n) is increasing.
(b) Prove by induction that $x_n < \sqrt{2} + 1$, $\forall n \in \mathbb{N}$.
(c) Find $\lim_{n\to\infty} x_n$.
Hint. Consider the function $f : (0, \infty) \to (0, \infty)$, $f(x) = \sqrt{2 + x}$ and prove that

$$0 < x < y \Rightarrow f(x) < f(y) \quad \text{and} \quad x > 0 \wedge x = f(x) \Longleftrightarrow x = 2.$$

□

Exercise 4.20. Fix $a > 0$, $a \neq 1$ and define $f : (0, \infty) \to (0, \infty)$ by

$$f(x) = \frac{1}{2}\left(x + \frac{a}{x}\right) = \frac{x^2 + a}{2x}.$$

Consider the sequence of positive real numbers $(x_n)_{n\geq 1}$ defined by the recurrence

$$x_1 = 1, \quad x_{n+1} = f(x_n), \quad \forall n \in \mathbb{N}.$$

Use the strategy employed in Example 4.25 to show that

$$\lim_{n\to\infty} x_n = \sqrt{a}.$$

□

Exercise 4.21 (Gauss). Let a_0, b_0 be two real numbers such that

$$0 < a_0 < b_0.$$

Define inductively

$$a_1 := \sqrt{a_0 b_0}, \quad b_1 = \frac{a_0 + b_0}{2},$$

$$a_{n+1} = \sqrt{a_n b_n}, \quad b_{n+1} = \frac{a_n + b_n}{2}.$$

(a) Prove by induction that

$$a_1 \leq a_2 \leq \cdots \leq a_n \leq b_n \leq \cdots \leq b_2 \leq b_1.$$

(b) Prove that the sequences (a_n) and (b_n) are convergent and

$$\lim_{n\to\infty} a_n = \lim_{n\to\infty} b_n.$$

Hint: For part (a) use Exercise 3.5. For part (b) use Weierstrass' Theorem on the convergence of bounded monotone sequences, Theorem 4.22.

□

Exercise 4.22. Establish the convergence or divergence of the sequence

$$a_n = \frac{1}{n+1} + \frac{1}{n+2} + \cdots + \frac{1}{2n}, \quad n \in \mathbb{N}.$$

□

Exercise 4.23. Let (a_n) be a sequence of real numbers. For each $n \in \mathbb{N}$ we set

$$b_n := a_{2n-1}, \quad c_n := a_{2n}.$$

Prove that the following statements are equivalent.

(i) The sequence (a_n) is convergent and its limit is $a \in \mathbb{R}$.
(ii) The subsequences $(b_n)_{n\in\mathbb{N}}$ and $(c_n)_{n\in\mathbb{N}}$ converge to the same limit a.

□

Exercise 4.24. Suppose $(a_n)_{n\in\mathbb{N}}$ is a *contractive* sequence of real numbers, i.e., there exists $r \in (0,1)$ such that

$$|a_n - a_{n+1}| < r|a_n - a_{n-1}|, \quad \forall n \in \mathbb{N}, n \geq 2.$$

Prove that the sequence $(a_n)_{n\in\mathbb{N}}$ is convergent.

Hint. Set $x_1 := a_1, \;\; x_2 := a_2 - a_1, \;\; x_3 := a_3 - a_2, \ldots$. Observe that $x_1 + x_2 + \cdots + x_n = a_n$, $\forall n \in \mathbb{N}$ so that the sequence $(a_n)_{n\in\mathbb{N}}$ is the sequence of partial sums of the series

$$x_1 + x_2 + x_3 + \cdots.$$

Use the Comparison Principle to show that this series is absolutely convergent. □

Exercise 4.25. Consider the sequence of positive real numbers $(x_n)_{n\geq 1}$ defined by the recurrence

$$x_1 = 1, \quad x_{n+1} = 1 + \frac{1}{x_n}, \quad \forall n \in \mathbb{N}.$$

Thus

$$x_2 = 1+1, \quad x_3 = 1 + \frac{1}{1+1} = \frac{3}{2}, \quad x_4 = 1 + \frac{1}{1+\frac{1}{1+1}} = 1 + \frac{2}{3} = \frac{5}{3},$$

$$x_5 = 1 + \cfrac{1}{1+\cfrac{1}{1+\frac{1}{1+1}}}, \quad x_6 = 1 + \cfrac{1}{1+\cfrac{1}{1+\cfrac{1}{1+\frac{1}{1+1}}}} \ldots .$$

(a) Prove that

$$x_1 < x_3 < \cdots < x_{2n+1} < x_{2n+2} < x_{2n} < \cdots < x_2, \quad \forall n \geq 1.$$

(b) Prove that for $n \geq 3$ we have

$$|x_{n+1} - x_n| = \frac{|x_n - x_{n-1}|}{x_n x_{n-1}} \leq \frac{4}{9}|x_n - x_{n-1}|.$$

(c) Conclude that the sequence (x_n) is convergent and find its limit. (**Hint:** Use Exercise 4.24.) □

Exercise 4.26. If $a_1 < a_2$

$$a_{n+2} = \frac{1}{2}\big(a_{n+1} + a_n \big), \ \ \forall n \in \mathbb{N}$$

show that the sequence $(a_n)_{n\in\mathbb{N}}$ is convergent.

Hint. Use Exercise 4.24. □

Exercise 4.27. Consider a sequence of positive numbers $(x_n)_{n\geq 1}$ satisfying the recurrence relation

$$x_{n+1} = \frac{1}{2+x_n}, \quad \forall n \in \mathbb{N}.$$

Show that $(x_n)_{n\in\mathbb{N}}$ is a contractive sequence (Exercise 4.24) and then compute its limit. □

Exercise 4.28. Find all the limit points (see Definition 4.30) of the sequence

$$a_n = (-1)^n \frac{n-1}{n}.$$ □

Exercise 4.29. Let $(a_n)_{n\in\mathbb{N}}$ be a bounded sequence of real numbers, i.e.,

$$\exists C \in \mathbb{R}: \ \ |a_n| \leq C, \ \ \forall n.$$

For any $k \in \mathbb{N}$ we set

$$b_k := \sup\{a_n; \ \ n \geq k\}.$$

(i) Show that the sequence $(b_k)_{k\in\mathbb{N}}$ is *nonincreasing* and conclude that it is convergent. Denote by b its limit.

(ii) Show that b is a limit point of the sequence $(a_n)_{n\in\mathbb{N}}$, i.e., there exists a subsequence $(a_{n_k})_{k\geq 1}$ of $(a_n)_{n\geq 1}$ such that

$$\lim_{k\to\infty} a_{n_k} = b.$$

(iii) Show that if α is a limit point of the sequence (a_n), then $\alpha \leq b$.

(iv) Show that the sequence

$$c_k = \inf_{n\geq k} a_n.$$

Show that the sequence (c_k) is nondecreasing and its limit c is the smallest limit point of the sequence (a_n).

The number b is called the *superior limit* of the sequence (a_n) and it is denoted by $\limsup_n a_n$. The number c is called the *inferior limit* of the sequence a_n and it is denoted by $\liminf a_n$. □

Exercise 4.30. Prove Proposition 4.36. □

Exercise 4.31. Prove that if $\sum_{n\geq 0} a_n$ and $\sum_{n\geq 0} b_n$ are convergent series of real numbers and $\alpha, \beta \in \mathbb{R}$, then the series $\sum_{n\geq 0}(\alpha a_n + \beta b_n)$ is convergent and

$$\sum_{n\geq 0}(\alpha a_n + \beta b_n) = \alpha \sum_{n\geq 0} a_n + \beta \sum_{n\geq 0} b_n.$$ □

Exercise 4.32. Can you give an example of convergent series $\sum_{n\geq 0} a_n$ and a divergent series $\sum_{n\geq 0} b_n$ such that $\sum_{n\geq 0}(a_n + b_n)$ is convergent? Explain. □

Exercise 4.33. Prove Corollary 4.42. □

Exercise 4.34. Consider the sequence

$$a_n = \frac{n^3 + 2n^2 + 2n + 4}{n^5 + n^4 + 7n^2 + 1}, \quad n \geq 0.$$

Prove that the series

$$\sum_{n\geq 0} a_n$$

is absolutely convergent.

Hint. Example 4.40(b) and Corollary 4.42. □

Exercise 4.35 (Leibniz). Suppose that (a_n) is a decreasing sequence of positive real numbers such that

$$\lim_{n\to\infty} a_n = 0.$$

Prove that the series

$$\sum_{n\geq 0} (-1)^n a_n$$

is convergent.

Hint. Imitate the strategy in Example 4.51. □

Exercise 4.36 (Cauchy). Suppose that $(a_n)_{n\geq 0}$ is a decreasing sequence of positive numbers that converges to 0. Prove that the series

$$\sum_{n\geq 0} a_n$$

converges if and only if the series

$$\sum_{k=0}^{\infty} 2^k a_{2^k} = a_1 + 2a_2 + 4a_4 + 8a_8 + 16a_{16} + \cdots$$

converges.

Hint. Imitate the strategy employed in Example 4.40. □

Exercise 4.37. We consider the power series

$$\sum_{n\geq 0} a_n x^n = a_0 + a_1 x + a_2 x^2 + \cdots .$$

Suppose that there exists $C > 0$ such that $|a_n| \leq C$, $\forall n$. Show that the radius of convergence of the series

$$\sum_{n\geq 0} a_n x^n$$

is ≥ 1. □

Exercise 4.38. Suppose that $(a_n)_{n\in\mathbb{N}}$ is a sequence of integers such that $0 \leq a_n \leq 9$ for any $n \in \mathbb{N}$, i.e.,

$$a_n \in \{0, 1, 2, \ldots, 9\}, \;\; \forall n \in \mathbb{N}.$$

Show that the series

$$\sum_{n\geq 1} a_n 10^{-n} = \frac{a_1}{10} + \frac{a_2}{10^2} + \cdots$$

is convergent.
(b) Compute the sum of the above series in the two special cases

$$a_n = 7, \;\; \forall n \in \mathbb{N},$$

and

$$a_n = \begin{cases} 1, & n \text{ is odd} \\ 2, & n, \text{ is even}. \end{cases}$$

In each case, express the sum in decimal form.
(c) Prove that for any $x \in [0, 1]$ there exists a sequence of real numbers $(a_n)_{n\in\mathbb{N}}$ such that

$$a_n \in \{0, 1, 2, \ldots, 9\}, \;\; \forall n \in \mathbb{N},$$

and

$$x = \sum_{n\geq 1} a_n 10^{-n}.$$

□

Exercise 4.39. Prove Corollary 4.55. □

4.10. More challenging problems

Problem 4.1. Fix rational numbers a, b such that $1 < a < b$.
(a) Prove that

$$\lim_{n\to\infty} \frac{(2n)^b}{(2n+1)^a} = \infty.$$

(b) Prove that the series

$$\frac{1}{1^a} + \frac{1}{2^b} + \frac{1}{3^a} + \frac{1}{4^b} + \cdots$$

is convergent. □

Problem 4.2. Consider two series of real numbers $\sum_{n\geq 0} a_n$ and $\sum_{n\geq 0} b_n$. For each non-negative integer n define

$$c_n := a_0 b_n + a_1 b_{n-1} + \cdots + a_n b_0 = \sum_{k=0}^{n} a_k b_{n-k}.$$

Prove that if the series $\sum_{n\geq 0} a_n$ and $\sum_{n\geq 0} b_n$ are *absolutely* convergent, then the series

$$\sum_{n\geq 0} c_n$$

is *absolutely* convergent and its sum is the product of the sums of the series $\sum_{n\geq 0} a_n$ and $\sum_{n\geq 0} b_n$, i.e.,

$$\lim_{n\to\infty}\sum_{k=0}^{n} c_n = \left(\lim_{n\to\infty}\sum_{j=0}^{n} a_j\right)\cdot\left(\lim_{n\to\infty}\sum_{k=0}^{n} b_k\right).$$

The series $\sum_{n\geq 0} c_n$ constructed above is called the *Cauchy product* of the series $\sum_{n\geq 0} a_n$ and $\sum_{n\geq 0} b_n$.

Hint: Consider first the special case $a_n, b_n \geq 0$, $\forall n$. Set

$$A_n := \sum_{j=0}^{n} a_j,\;\; B_n := \sum_{k=0}^{n} b_k,\;\; C_n = \sum_{\ell=0}^{n} c_\ell.$$

Prove that

$$\lim_{n\to\infty}(C_n - A_nB_n) = 0.$$

□

Problem 4.3. Let $(a_n)_{n\geq 0}$ and $(b_n)_{n\geq 0}$ be two sequences of real numbers. For any non-negative integer n we set

$$B_n := b_0 + b_1 + \cdots + b_n,\;\; C_n = a_0b_0 + a_1b_1 + \cdots + a_nb_n.$$

(a) (Abel's trick) Show that, for any $n\in\mathbb{N}$, we have

$$C_n = a_nB_n - \sum_{k=1}^{n-1}(a_{k+1}-a_k)B_k. \tag{4.29}$$

(b) Show that if the series

$$\sum_{n\geq 0} b_n$$

is convergent and the sequence $(a_n)_{n\geq 0}$ is monotone and bounded, then the series

$$\sum_{n\geq 0} a_nb_n$$

is convergent. □

Problem 4.4. Let (a_n) be a convergent sequence of real numbers. Form the new sequence (c_n) such that

$$c_n := \frac{a_1+\cdots+a_n}{n},\;\; n\geq 1.$$

Show that (c_n) is convergent and

$$\lim_{n\to\infty} c_n = \lim_{n\to\infty} a_n.$$ □

Problem 4.5. Let the two given sequences

$$a_0, a_1, a_2, \ldots,$$

$$b_0, b_1, b_2, \ldots$$

satisfy the conditions

$$b_n > 0, \quad \forall n \geq 0, \tag{4.30a}$$

$$b_0 + b_1 + b_2 + \cdots + b_n + \cdots = \infty, \tag{4.30b}$$

$$\lim_{n\to\infty} \frac{a_n}{b_n} = s. \tag{4.30c}$$

Prove that

$$\lim_{n\to\infty} \frac{a_0 + a_1 + \cdots + a_n}{b_0 + b_1 + \cdots + b_n} = s.$$

□

Problem 4.6. Suppose that $(p_n)_{n\geq 1}$ is a sequence of *positive* real numbers, and $(x_n)_{n\geq 1}$ is a sequence of real numbers. For $n \in \mathbb{N}$ we set

$$b_n := p_1 + \cdots + p_n, \quad s_n := x_1 + \cdots + x_n.$$

Suppose that

$$\lim_{n\to\infty} b_n = \infty.$$

Prove that if the series

$$\sum_{n\geq 1} \frac{x_n}{b_n}$$

is convergent, then

$$\lim_{n\to\infty} \frac{s_n}{b_n} = 0.$$

□

Problem 4.7 (Doob). Let $(x_n)_{n\in\mathbb{N}}$ be a sequence of real numbers. To any real numbers a, b such that $a < b$ we associate the sequences $(S_k(a,b))_{k\in\mathbb{N}}$ and $(T_k(a,b))_{k\in\mathbb{N}}$ in $\mathbb{N} \cup \{\infty\}$ defined inductively as follows

$$S_1(a,b) := \inf\{n \geq 1;\ x_n \leq a\},\ T_1(a,b) := \inf\{\, n \geq S_1(a,b);\ x_n \geq b\,\},$$

$$S_{k+1}(a,b) := \inf\{n \geq T_k(a,b);\ x_n \leq a\,\},\ T_{k+1}(a,b) := \inf\{\, n \geq S_k(a,b);\ x_n \geq b\,\},$$

where we set $\inf \emptyset = \infty$. We set

$$U_n(a,b) := \#\{k \leq n;\ T_k(a,b) \leq n\,\}.$$

(a) Prove that for any $a, b \in \mathbb{R}$, $a < b$, the sequence $(U_n(a,b))_{n\in bN}$ is nondecreasing. Set

$$U_\infty(a,b) := \lim_{n\to\infty} U_n(a,b).$$

(b) Prove that the following statements are equivalent.

(i) The sequence (x_n) has a limit as $n \to \infty$.

(ii) For any $a, b \in \mathbb{Q}$ such that $a < b$ we have $U_\infty(a, b) < \infty$.

□

Problem 4.8 (Fekete). Suppose that the sequence of real numbers $(a_n)_{n\in\mathbb{N}}$ satisfies the subadditivity condition

$$a_{m+n} \le a_m + a_n, \quad \forall m, n \in \mathbb{N}.$$

Prove that

$$\lim_{n\to\infty} \frac{a_n}{n} = \inf_{n\in\mathbb{N}} \frac{a_n}{n}.$$ □

Problem 4.9. Let $(x_n)_{n\ge 0}$ be a sequence of nonzero real numbers such that

$$x_n^2 - x_{n+1}x_{n-1} = 1, \quad \forall n \in \mathbb{N}.$$

Prove that there exists $a \in \mathbb{R}$ such that

$$x_{n+1} = ax_n - x_{n-1}, \quad \forall n \in \mathbb{N}.$$ □

Problem 4.10. Suppose that a sequence of real numbers $(a_n)_{n\in\mathbb{N}}$ satisfies

$$0 < a_n < a_{2n} + a_{2n+1}, \quad \forall n \in \mathbb{N}.$$

Prove that the series $\sum_{n\ge 1} a_n$ is divergent. □

Problem 4.11. Suppose that $(x_n)_{n\in\mathbb{N}}$ is a sequence of positive real numbers such that the series

$$\sum_{n\in\mathbb{N}} x_n$$

is convergent and its sum is S. Prove that for any bijection $\varphi : \mathbb{N} \to \mathbb{N}$ the series

$$\sum_{n\in\mathbb{N}} x_{\varphi(n)}$$

is also convergent and its sum is also S. □

Problem 4.12. Suppose that the series of real numbers

$$\sum_{n\in\mathbb{N}} x_n$$

is convergent, but *not absolutely convergent*. Prove that for any real number S there exists a bijection $\varphi : \mathbb{N} \to \mathbb{N}$ such that the series

$$\sum_{n\in\mathbb{N}} x_{\varphi(n)}$$

is convergent and its sum is S. □

Problem 4.13. Suppose that $(a_n)_{n\ge 1}$ is a decreasing sequence of positive real numbers that converges to 0 and satisfies the inequalities

$$a_n \le a_{n+1} + a_{n^2}, \quad \forall n \ge 1.$$

Prove that the series

$$\sum_{n\ge 1} a_n$$

is divergent. □

Chapter 5

Limits of Functions

5.1. Definition and basic properties

Let X be a nonempty subset of $\mathbb{R}$. A real number c is called a *cluster point* of X if there exists a sequence (x_n) of real numbers with the following properties.

(i) $x_n \in X, \forall n \in \mathbb{N}$.
(ii) $x_n \neq c, \forall n \in \mathbb{N}$.
(iii) $\lim_n x_n = c$.

Example 5.1. (a) If $A = (0, 1)$, then 0 and 1 are cluster points of A, although they are not in A. Indeed, the sequence $a_n = \frac{1}{n+1}$, $n \in \mathbb{N}$ consists of elements of $(0, 1)$ and $a_n \to 0$. Similarly, the sequence $b_n = 1 - \frac{1}{n+1}$ consists of points in $(0, 1)$ and $b_n \to 1$. Observe that every point in $(0, 1)$ is also a cluster point of $(0, 1)$.
(b) Any real number is a cluster point of the set $\mathbb{Q}$ of rational numbers. □

Definition 5.2. Let $X \subset \mathbb{R}$. Suppose that c is a cluster point of X and $f : X \to \mathbb{R}$ is a real valued function defined on X. We say that the limit of f at c is the real number A, and we write this

$$\lim_{x \to c} f(x) = A,$$

if the following holds:

$$\boxed{\forall \varepsilon > 0 \;\; \exists \delta = \delta(\varepsilon) > 0 : \;\; \forall x \in X : \;\; 0 < |x - c| < \delta \Rightarrow |f(x) - A| < \varepsilon.} \tag{5.1}$$

□

An alternate viewpoint. Recall that a *neighborhood* of a point a is an open interval that contains a inside. For example, the open interval $(0, 3)$ is a neighborhood of 1. We denote by $\mathcal{N}_a$ the collection of all neighborhoods of a. Thus, a statement of the form $U \in \mathcal{N}_a$ signifies that U is an open interval that contains a. A *symmetric neighborhood* of a is a neighborhood of the form $(a - \delta, a + \delta)$, where δ is some positive number. Observe that

$$x \in (a - \delta, a + \delta) \Longleftrightarrow \operatorname{dist}(a, x) < \delta \Longleftrightarrow |x - a| < \delta.$$

Thus, to describe a symmetric neighborhood of a, it suffices to indicate a positive real number δ, and then the symmetric neighborhood is described by the condition $\operatorname{dist}(x, a) < \delta$. We denote by $\mathcal{SN}_a$ the collection of symmetric neighborhoods of a. Clearly, any symmetric neighborhood of a is also a neighborhood of a so that

$$\mathcal{SN}_a \subset \mathcal{N}_a.$$

A *deleted neighborhood* of a is a set obtained from a neighborhood of a by removing the point a. For example

$$(0,2)\setminus\{1\} = (0,1)\cup(1,2)$$

is a deleted neighborhood of 1. We denote by $\mathcal{N}_a^*$ the collection of all deleted neighborhoods of a. A *symmetric deleted neighborhood* of a is a deleted neighborhood of the form

$$(a-r,a+r)\setminus\{a\} = (a-r,a)\cup(a,a+r).$$

We denote by $\mathcal{SN}_a^*$ the collection of deleted symmetric neighborhoods of a. Clearly

$$\mathcal{SN}_a^* \subset \mathcal{S}_a^*.$$

Observe that the definition (5.1) is equivalent with the following statement

$$\forall U \in \mathcal{SN}_a \;\; \exists V \in \mathcal{SN}_c^* \;\; \forall x \in X: \;\; x \in V \Rightarrow f(x) \in U. \tag{5.2}$$

Indeed, we can rephrase (5.1) in the following equivalent way: for any symmetric neighborhood U of A of the form $(A-\varepsilon, A+\varepsilon)$, there exists a deleted symmetric neighborhood V of c of the form $(c-\delta, c+\delta)\setminus\{c\}$ such that for any $x \in V$ we have $f(x) \in U$. That is precisely the content of (5.2).

The proof of the next result is left to you as an exercise.

Proposition 5.3. *Let $f : X \to \mathbb{R}$ be a function defined on a set $X \subset \mathbb{R}$ and c a cluster point of X. Then the following statements are equivalent.*

(i) $\lim_{x\to c} f(x) = A$, *i.e., f satisfies (5.1) or (5.2).*
(ii)

$$\forall U \in \mathcal{N}_A, \;\; \exists V \in \mathcal{N}_c^* \;\; \text{such that} \;\; \forall x \in X : x \in V \Rightarrow f(x) \in U. \tag{5.3}$$

□

The following very useful result reduces the study of limits of functions to the study of a concept we are already familiar with, namely the concept of limits of sequences.

Theorem 5.4. *Let c be a cluster point of the set $X \subset \mathbb{R}$ and $f : X \to \mathbb{R}$ a real valued function on X. The following statements are equivalent.*

(i) $\lim_{x\to c} f(x) = A \in \mathbb{R}$.
(ii) For any sequence $(x_n)_{n\in\mathbb{N}}$ in $X\setminus\{c\}$ such that $x_n \to c$, we have $\lim_n f(x_n) = A$.

Proof. (i) ⇒ (ii). We know that $\lim_{x\to c} f(x) = A$ and we have to show that if (x_n) is a sequence in $X\setminus\{c\}$ that converges to c then the sequence $(f(x_n))$ converges to A. In other words, given the above sequence (x_n) we have to show that

$$\forall \varepsilon > 0 \;\; \exists N = N(\varepsilon) \;\; \forall n \in \mathbb{N}: \;\; n > N(\varepsilon) \Rightarrow |f(x_n) - A| < \varepsilon.$$

Let $\varepsilon > 0$. We deduce from (5.1) that there exists $\delta(\varepsilon) > 0$ such that

$$\forall x \in X: \;\; 0 < |x - c| < \delta \Rightarrow |f(x) - A| < \varepsilon. \tag{5.4}$$

Since $x_n \to c$, there exists $N = N(\delta(\varepsilon))$ such that

$$0 < |x_n - c| < \delta, \;\; \forall n > N.$$

Using (5.4) we deduce that for any $n > N(\delta(\varepsilon))$ we have $|f(x_n) - A| < \varepsilon$. This proves the implication (i) ⇒ (ii).

(ii) $\Rightarrow$ (i). We know that for any sequence (x_n) in $X \setminus \{c\}$ that converges to c, the sequence $(f(x_n))$ converges to A and we have to prove (5.1), i.e.,

$$\forall \varepsilon > 0 \ \exists \delta = \delta(\varepsilon) > 0 : \ \forall x \in X : \ 0 < |x - c| < \delta \Rightarrow |f(x) - A| < \varepsilon. \tag{5.5}$$

We argue by contradiction and we assume that (5.5) is false, so that its opposite is true, i.e.,

$$\exists \varepsilon_0 > 0 : \ \forall \delta > 0, \ \exists x = x(\delta) \in X, \ 0 < |x(\delta) - c| < \delta \ \text{ and } |f(x(\delta)) - A| \geq \varepsilon_0. \tag{5.6}$$

From (5.6) we deduce that for any δ of the form $\delta = \frac{1}{n}$, $n \in \mathbb{N}$, there exists $x_n = x(1/n) \in X$ such that

$$0 < |x_n - c| < \frac{1}{n} \ \wedge \ |f(x_n) - A| \geq \varepsilon_0.$$

We have thus produced a sequence (x_n) in X such that

$$0 < \operatorname{dist}(x_n, c) < \frac{1}{n} \ \wedge \ \operatorname{dist}(f(x_n), A) \geq \varepsilon_0.$$

Thus, (x_n) is a sequence in $X \setminus \{c\}$ that converges to c, but the sequence $(f(x_n))$ does not converge to A. □

Using Proposition 4.15 we obtain the following immediate consequence.

Corollary 5.5. *Let $f, g : X \to \mathbb{R}$ be two functions defined on the same subset $X \subset \mathbb{R}$ and c a cluster point of X. Suppose additionally that*

$$\lim_{x \to c} f(x) = A \ \text{ and } \ \lim_{x \to c} g(x) = B.$$

Then the following hold.

(i)

$$\lim_{x \to c} \big(f(x) + g(x) \big) = A + B, \ \lim_{x \to c} \lambda f(x) = \lambda A, \ \forall \lambda \in \mathbb{R}.$$

(ii)

$$\lim_{x \to c} f(x) g(x) = AB.$$

(iii) If $B \neq 0$ and $g(x) \neq 0$, $\forall x \in X$, then

$$\lim_{x \to c} \frac{f(x)}{g(x)} = \frac{A}{B}.$$

□

Example 5.6. (a) Let $f : \mathbb{R} \to \mathbb{R}$, $f(x) = x$. Then for any $c \in \mathbb{R}$ we have

$$\lim_{x \to c} f(x) = \lim_{x \to c} x = c.$$

(b) Let $m \in \mathbb{N}$ and define $f : \mathbb{R} \to \mathbb{R}$, $f(x) = x^m$. Corollary 5.5 implies that

$$\lim_{x \to c} f(x) = \lim_{x \to c} x^m = c^m.$$

Thus,

$$\lim_{x\to 3} x^2 = 3^2 = 9.$$

(c) Let $m \in \mathbb{N}$ and define $f : (0,\infty) \to \mathbb{R}$, $f(x) = x^{-m} = \frac{1}{x^m}$. Corollary 5.5 implies that for any $c > 0$ we have

$$\lim_{x\to c} x^{-m} = \lim_{x\to c} \frac{1}{x^m} = c^{-m}.$$

(d) Let $m, k \in \mathbb{N}$ and define $f : (0,\infty) \to \mathbb{R}$, $f(x) = x^{\frac{m}{k}}$. We want to show that for any $c > 0$ we have

$$\lim_{x\to c} x^{\frac{m}{k}} = c^{\frac{m}{k}}. \tag{5.7}$$

We rely on Theorem 5.4. Suppose that (x_n) is a sequence of positive numbers such that $x_n \to c$ and $x_n \neq c$, $\forall n$. We have to show that

$$\lim_n x_n^{\frac{m}{k}} = c^{\frac{m}{k}}.$$

Using Exercise 4.15, we deduce that

$$\lim_n x_n^{\frac{1}{k}} = c^{\frac{1}{k}}.$$

Thus,

$$\lim_n x_n^{\frac{m}{k}} = \lim_n \big(x_n^{\frac{1}{k}} \big)^m = \big(c^{\frac{1}{k}} \big)^m = c^{\frac{m}{k}}.$$

Then,

$$\lim_{x\to c} x^r = c^r, \ \ \forall r \in \mathbb{Q}, \ \ r > 0.$$

The above equality obviously holds if $r = 0$. If $r < 0$, then $x^{-r} = \frac{1}{x^r}$ and we deduce

$$\lim_{x\to c} x^r = c^r, \ \ \forall c > 0, \ r \in \mathbb{Q}. \tag{5.8}$$

□

Proposition 5.7. *Let $f, g : X \to \mathbb{R}$ be two functions defined on the same subset $X \subset \mathbb{R}$. Suppose that c is a cluster point of X and*

$$\lim_{x\to c} f(x) = A, \ \ \lim_{x\to c} g(x) = B \ \ \textit{and} \ \ A < B.$$

Then there exists a $\delta_0 > 0$ such that $f(x) < g(x)$, $\forall x \in X$, $0 < |x - c| < \delta_0$.

Proof. Fix a positive number ε such that $3\varepsilon < B - A$. In other words, ε is smaller than one third of the distance from A to B. In particular, $A + \varepsilon < B - \varepsilon$ because

$$B - \varepsilon - (A + \varepsilon) = B - A - 2\varepsilon > 3\varepsilon - 2\varepsilon > 0.$$

Since $\lim_{x\to c} f(x) = A$, there exists $\delta = \delta_f(\varepsilon) > 0$ such that

$$\forall x \in X: \ \ 0 < |x - c| < \delta_f \Rightarrow A - \varepsilon < f(x) < A + \varepsilon.$$

Since $\lim_{x\to c} g(x) = B$, there exists $\delta = \delta_g(\varepsilon) > 0$ such that

$$\forall x \in X: \ \ 0 < |x - c| < \delta_g \Rightarrow B - \varepsilon < f(x) < B + \varepsilon.$$

Let $\delta_0 < \min\{\delta_f, \delta_g\}$ and define

$$U := (c - \delta_0, c + \delta_0).$$

If $x \in U \cap X$, $x \neq c$, then

$$0 < |x - c| < \delta_0 < \min\{\delta_f, \delta_g\} \Rightarrow f(x) < A + \varepsilon < B - \varepsilon < g(x).$$

□

5.2. Exponentials and logarithms

In this section we want to give a meaning to the exponential a^x where a is a positive real number and x is an arbitrary real number. The case $a = 1$ is trivial: we define $1^x = 1$, $\forall x \in \mathbb{R}$.

We consider next the case $a > 1$. In Exercise 3.14 we defined a^r for any $r \in \mathbb{Q}$ and we showed that

$$a^{r_1+r_2} = a^{r_1} \cdot a^{r_2}, \quad a^{r_1-r_2} = \frac{a^{r_1}}{a^{r_2}}, \quad \left(a^{r_1}\right)^{r_2} = a^{r_1 r_2}, \quad \forall r_1, r_2 \in \mathbb{Q}. \tag{5.9}$$

We will use these facts to define a^x for any $x \in \mathbb{R}$. This will require several auxiliary results.

Lemma 5.8. *If $a > 1$, then for any rational numbers r_1, r_2 we have*

$$r_1 < r_2 \Rightarrow a^{r_1} < a^{r_2}.$$

Proof. We will use the fact that if $x, y > 0$ and $n \in \mathbb{N}$, then

$$x < y \Longleftrightarrow x^n < y^n.$$

Since $a > 1$ we deduce that $a^{\frac{1}{n}} > 1$ because

$$\left(a^{\frac{1}{n}}\right)^n = a > 1 = 1^n.$$

Thus,

$$a^{\frac{m}{n}} > 1, \quad \forall m, n \in \mathbb{N}$$

that is,

$$a^r > 1, \quad \forall r \in \mathbb{Q}, \quad r > 0.$$

Suppose that $r_1 < r_2$. Then the above inequality implies that

$$\frac{a^{r_2}}{a^{r_1}} \overset{(5.9)}{=} a^{r_2 - r_1} > 1$$

because $r = r_2 - r_1$ is a positive rational number. □

Lemma 5.9. *Let $a > 1$ and $r_0 \in \mathbb{Q}$. Then*

$$\lim_{\mathbb{Q} \ni r \to r_0} a^r = a^{r_0}.$$

Proof. We first consider the case $r_0 = 0$, i.e., we first prove that

$$\lim_{\mathbb{Q} \ni r \to 0} a^r = 1. \tag{5.10}$$

We have to prove that, given $\varepsilon > 0$, we can find $\delta = \delta(\varepsilon) > 0$ such that

$$0 < |r| < \delta \text{ and } r \in \mathbb{Q} \Rightarrow |a^r - 1| < \varepsilon.$$

Observe first that Exercise 4.12 implies that

$$\lim_{n \to \infty} a^{\frac{1}{n}} = \lim_{n \to \infty} a^{-\frac{1}{n}} = 1.$$

In particular, this implies that there exists $n_0 = n_0(\varepsilon) > 0$ such that, for all $n \geq n_0$, we have

$$1 - \varepsilon < a^{-\frac{1}{n}} < a^{\frac{1}{n}} < 1 + \varepsilon.$$

We set $\delta(\varepsilon) = \frac{1}{n_0(\varepsilon)}$. If $0 < |r| < \delta(\varepsilon)$ and $r \in \mathbb{Q}$, then $-\frac{1}{n_0(\varepsilon)} < r < \frac{1}{n_0(\varepsilon)}$ and we deduce from Lemma 5.8 that

$$1 - \varepsilon < a^{-\frac{1}{n_0(\varepsilon)}} < a^r < a^{\frac{1}{n_0(\varepsilon)}} < 1 + \varepsilon \Rightarrow 1 - \varepsilon < a^r < 1 + \varepsilon \Rightarrow |a^r - 1| < \varepsilon.$$

This proves (5.10). To deal with the general case, let $r_0 \in \mathbb{Q}$. If r_n is a sequence of rational numbers $r_n \to r_0$, then

$$a^{r_n} = a^{r_0} a^{r_n - r_0}.$$

Since $r_n - r_0 \to 0$, we deduce from (5.10) that $a^{r_n - r_0} \to 1$ and thus, $a^{r_n} = a^{r_0} a^{r_n - r_0} \to a^{r_0}$. The conclusion now follows from Theorem 5.4. □

Proposition 5.10. *Let $a > 1$ and $x \in \mathbb{R}$. We set*

$$\mathbb{Q}_{<x} := \{ r \in \mathbb{Q},\ \ r < x \},\ \ \mathbb{Q}_{>x} := \{ r \in \mathbb{Q},\ \ r > x \}$$

$$s_x = \sup_{r \in \mathbb{Q}_{<x}} a^r,\ \ i_x = \inf_{r \in \mathbb{Q}_{>x}} a^r.$$

Then $s_x = i_x$. Moreover, if x is rational, then $s_x = i_x = a^x$.

Proof. Observe first that the set $\{a^r;\ \ r \in \mathbb{Q}_{<x}\}$ is bounded above. Indeed, if we choose a rational number $R > x$, then Lemma 5.8 implies that $a^r < a^R$ for any rational number $r < x$. A similar argument shows that the set $\{a^r;\ \ r \in \mathbb{Q}_{>x}\}$ is bounded below and we have

$$s_x \leq i_x.$$

Observe that for any rational numbers r_1, r_2 such that $r_1 < x < r_2$, we have

$$a^{r_1} \leq s_x \leq i_x \leq a^{r_2}.$$

Hence,

$$1 \leq \frac{i_x}{s_x} \leq \frac{a^{r_2}}{s_x} \leq \frac{a^{r_2}}{a^{r_1}} = a^{r_2 - r_1}.$$

Now choose two sequences $(r'_n) \subset \mathbb{Q}_{<x}$ and $(r''_n) \subset \mathbb{Q}_{>x}$ such that $r'_n \to x$ and $r''_n \to x$.[1] Then

$$1 \leq \frac{s_x}{i_x} \leq a^{r''_n - r'_n}.$$

If we let $n \to \infty$ and observe that $r''_n - r'_n \to 0$, we deduce from Lemma 5.9 that

$$1 \leq \frac{s_x}{i_x} \leq \lim_{n \to \infty} a^{r''_n - r'_n} = 1 \Rightarrow s_x = i_x.$$

If $x \in \mathbb{Q}$, then the sequences r'_n and r''_n above converge to x. Invoking Lemma 5.9 we deduce

$$s_x = \lim_n a^{r'_n} = a^x = \lim_n a^{r''_n} = i_x.$$

□

Definition 5.11. For any $a > 1$ and $x \in \mathbb{R}$ we set

$$\boxed{a^x := \sup\{a^r;\ \ r \in \mathbb{Q},\ \ r < x\} = \inf\{a^r;\ \ r \in \mathbb{Q},\ \ r > x\}}.$$

If $b \in (0, 1)$, then $\frac{1}{b} > 1$ and we set

$$b^x := \left(\frac{1}{b}\right)^{-x}.$$

□

Lemma 5.12. *Let $a > 1$. If $x < y$, then $a^x < a^y$.*

Proof. We can find rational numbers r_1, r_2 such that

$$x < r_1 < r_2 < y.$$

Then $r_1 \in \mathbb{Q}_{>x}$ and $r_2 \in \mathbb{Q}_{<y}$ so that

$$a^x \leq a^{r_1} < a^{r_2} \leq a^y.$$

□

[1] The existence of such sequences was left to you as Exercise 4.13.

Lemma 5.13. *Let $a > 1$ and $x \in \mathbb{R}$. If the sequence $(r_n) \subset \mathbb{Q}_{<x}$ converges to x, then*

$$\lim_{n\to\infty} a^{r_n} \to a^x.$$

Proof. We have

$$a^x = \sup_{r\in\mathbb{Q}_{<x}} a^r.$$

Thus, for any $\varepsilon > 0$, there exists $r_\varepsilon \in \mathbb{Q}_{<x}$ such that

$$a^x - \varepsilon < a^{r_\varepsilon} \le a^x.$$

Since $r_n \to x$ and $r_n \in \mathbb{Q}_{<x}$, we deduce that there exists $N = N(\varepsilon)$ such that, $\forall n > N(\varepsilon)$ we have $r_\varepsilon < r_n < x$. We deduce that for all $n > N(\varepsilon)$, we have

$$a^x - \varepsilon < a^{r_\varepsilon} < a^{r_n} < a^x.$$

□

Lemma 5.14. *Let $a > 0$ and $x, y > 0$. Then*

$$a^x \cdot a^y = a^{x+y}.$$

Proof. Choose sequences $(r'_n) \subset \mathbb{Q}_{<x}$ and $(r''_n) \subset \mathbb{Q}_{<y}$ such that $r'_n \to x$ and $r''_n \to y$. Lemma 5.13 implies that

$$a^{r'_n} \to a^x \quad \wedge \quad a^{r''_n} \to a^y.$$

Hence,

$$\lim_n a^{r'_n + r''_n} = \lim_n \big(a^{r'_n} \cdot a^{r''_n} \big) = \big(\lim_n a^{r'_n} \big) \cdot \big(\lim_n a^{r''_n} \big) = a^x \cdot a^y.$$

Now observe that $r'_n + r''_n \in \mathbb{Q}_{<x+y}$ and $r'_n + r''_n \to x + y$. Lemma 5.13 implies

$$\lim_n a^{r'_n + r''_n} = a^{x+y}.$$

□

The proofs of our next two results are left to you as an exercise.

Lemma 5.15. *Let $a > 0$ and $x \in \mathbb{R}$. Then for any sequence of real numbers (x_n) such that $x_n \to x$ we have*

$$\lim_{n\to\infty} a^{x_n} = a^x.$$ □

Lemma 5.16. *Suppose that $a, b > 0$. Then for any $x \in \mathbb{R}$ we have*

$$a^x \cdot b^x = (ab)^x. \tag{5.11}$$

□

Definition 5.17. Let $X \subset \mathbb{R}$ and $f : X \to \mathbb{R}$ be a real valued function defined on X.

(i) The function f is called *increasing* if

$$\forall x_1, x_2 \in X \;\; (x_1 < x_2) \Rightarrow \big(f(x_1) < f(x_2) \big).$$

(ii) The function f is called *decreasing* if

$$\forall x_1, x_2 \in X \;\; (x_1 < x_2) \Rightarrow \big(f(x_1) > f(x_2) \big).$$

(iii) The function f is called *nondecreasing* if

$$\forall x_1, x_2 \in X \;\; (x_1 < x_2) \Rightarrow \big(f(x_1) \le f(x_2) \big).$$

(iv) The function f is called *nonincreasing* if

$$\forall x_1, x_2 \in X \ \ (x_1 < x_2) \Rightarrow \big(f(x_1) \geq f(x_2) \big).$$

(v) The function is called *strictly monotone* if it is either increasing or decreasing. It is called *monotone* if it is either nondecreasing or nonincreasing.

□

Theorem 5.18. *Let $a > 0$, $a \neq 1$. Consider the function $f_a : \mathbb{R} \to (0, \infty)$ given by $f(x) = a^x$. Then the following hold.*

(i) $a^{x+y} = a^x \cdot a^y$, $\forall x, y \in \mathbb{R}$.
(ii) $(a^x)^y = a^{xy}$, $\forall x, y \in \mathbb{R}$.
(iii) The function f_a is increasing if $a > 1$, and decreasing if $a < 1$.
(iv) The function f is bijective.
(v) For any sequence of real numbers (x_n) such that $x_n \to x$ we have

$$\lim_{n\to\infty} a^{x_n} = a^x.$$

Proof. Part (v) above is Lemma 5.15. We thus have to prove (i)–(iv). We consider first the case $a > 1$. The equality (i) is Lemma 5.14. The statement (iii) follows from Lemma 5.12.

We first prove (ii) in the special case $y \in \mathbb{Q}$. Choose a sequence $r_n \in \mathbb{Q}$ such that $r_n \to x$, $r_n \neq x$. Then (5.9) implies

$$(a^{r_n})^y = a^{r_n y}.$$

Clearly $r_n y \to xy$ and Lemma 5.15 implies that

$$\lim_n a^{r_n y} = a^{xy}.$$

On the other hand, y is rational and $a^{r_n} \to a^x$, and using (5.8) we deduce that

$$\lim_n (a^{r_n})^y = (a^x)^y.$$

Thus,

$$(a^x)^y = \lim_n (a^{r_n})^y = \lim_n a^{r_n y} = a^{xy}, \ \ \forall x \in \mathbb{R}, \ \ y \in \mathbb{Q}. \tag{5.12}$$

Now fix $x, y \in \mathbb{R}$ and choose a sequence of *rational* numbers $y_n \to y$, $y_n \neq y$. Then

$$(a^x)^{y_n} \stackrel{(5.12)}{=} a^{xy_n}, \ \ \forall n.$$

Using Lemma 5.15, we deduce

$$(a^x)^y = \lim_n (a^x)^{y_n} = \lim_n a^{xy_n} = a^{xy}, \ \ \forall x, y \in \mathbb{R}.$$

This proves (ii).

To prove (iv) observe that f_a is injective because it is increasing. (We recall that we are working under the assumption $a > 1$.) To prove surjectivity, fix $y \in (0, \infty)$. We have to show that there exists $x \in \mathbb{R}$ such that $a^x = y$. Define

$$S := \{ s \in \mathbb{R}; \ a^s \leq y \}.$$

Observe first that $S \neq \emptyset$. Indeed

$$\lim_n a^{-n} = \lim_n \frac{1}{a^n} = 0$$

so that there exists $n_0 \in \mathbb{N}$ such that $a^{-n_0} < y$, i.e., $-n_0 \in S$. Observe that S is also bounded above. Indeed

$$\lim_n a^n = \infty.$$

Hence there exists $n_1 \in \mathbb{N}$ such that $a^{n_1} > y$. If $x \geq n_1$, then $a^x \geq a^{n_1} > y$ so that $S \cap [n_1, \infty) = \emptyset$. Thus $S \subset (-\infty, n_1)$ and therefore n_1 is an upper bound for S. Set

$$x := \sup S.$$

Note that if $x' > x$, then $a^{x'} \geq y$. Indeed, if $a^{x'} < y$ then for any $s < x'$ we have $a^s < a^{x'} < y$ and thus $(-\infty, x'] \subset S$. This contradicts the fact that x is an upper bound for S.

Consider now two sequences $s_n' \to x$ and $s_n'' \to x$, where $s_n' < x$ and $s_n'' > x$, then

$$a^{s_n'} \leq y \leq a^{s_n''}, \quad \forall n.$$

Letting $n \to \infty$ in the above inequalities we obtain, from Lemma 5.15, that

$$a^x \leq y \leq a^x \Longleftrightarrow a^x = y.$$

The case $a < 1$ follows from the case $a > 1$ by observing that

$$a^x = \left(\frac{1}{a}\right)^{-x}.$$

□

Definition 5.19. Let $a \in (0, \infty)$, $a \neq 1$. The bijective function

$$\mathbb{R} \ni x \mapsto a^x \in (0, \infty)$$

is called the *exponential function with base a*. Its inverse is called the *logarithm to base a* and it is a function

$$\log_a : (0, \infty) \to \mathbb{R}.$$

When $a = e =$ the Euler number, we will refer to $\log_e$ as the *natural logarithm* and we will use the simpler notation log or ln. Also, we will use the notation lg for $\log_{10}$. □

We have depicted below the graphs of the functions a^x and $\log_a x$ for $a = 2$ and $a = \frac{1}{2}$.

The meaning of the logarithm function answers the following question: given $a, y > 0$, $a \neq 1$, to what power do we need to raise a in order to obtain y? The answer: we need to raise a to the power $\log_a y$ in order to get y. Equivalently, $\log_a$ is uniquely determined by the following two fundamental identities

$$\log_a a^x = x \text{ and } a^{\log_a y} = y, \quad \forall x \in \mathbb{R}, \quad y > 0.$$

For example, $\log_2 8 = 3$ because $2^3 = 8$. Similarly $\lg 10,000 = 4$ since $10^4 = 10,000$.

Theorem 5.20. *Let $a > 0$, $a \neq 1$. Then the following hold.*

(i) $\log_a(y_1 y_2) = \log_a y_1 + \log_a y_2$, $\log_a \frac{y_1}{y_2} = \log_a y_1 - \log_a y_2$, $\forall y_1, y_2 > 0$.
(ii) $\log_a y^\alpha = \alpha \log_a y$, $\forall y > 0$, $\alpha \in \mathbb{R}$.

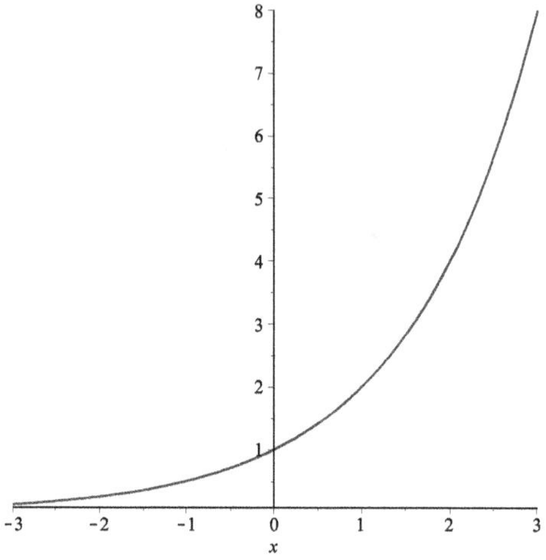

Fig. 5.1 *The graph of* 2^x.

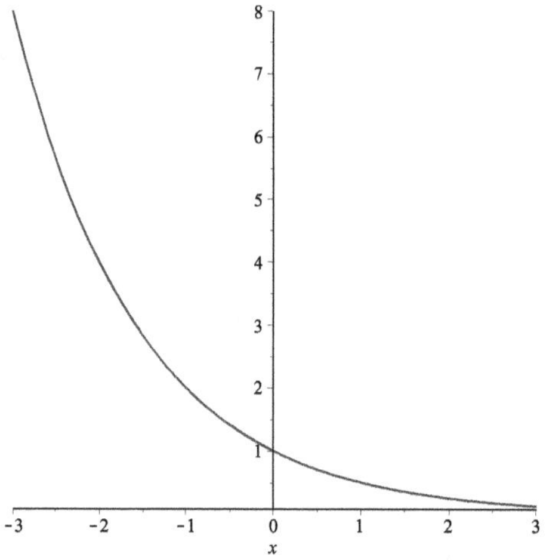

Fig. 5.2 *The graph of* $\left(\frac{1}{2}\right)^x$.

(iii) *If* $b > 0$ *and* $b \neq 1$, *then*

$$\log_b y = \frac{\log_a y}{\log_a b}, \quad \forall y > 0.$$

(iv) *If* $a > 1$, *then the function* $y \mapsto \log_a y$ *is increasing, while if* $a \in (0, 1)$, *then the function* $y \mapsto \log_a y$ *is decreasing.*

(v) *If* $y > 0$, *then for any sequence of positive numbers* (y_n) *that converges to* y *we have*

$$\lim_{n \to \infty} \log_a y_n = \log_a y.$$

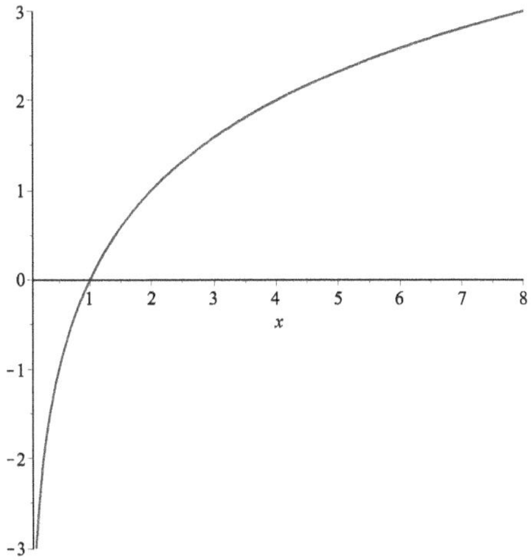

Fig. 5.3 *The graph of* $\log_2 x$.

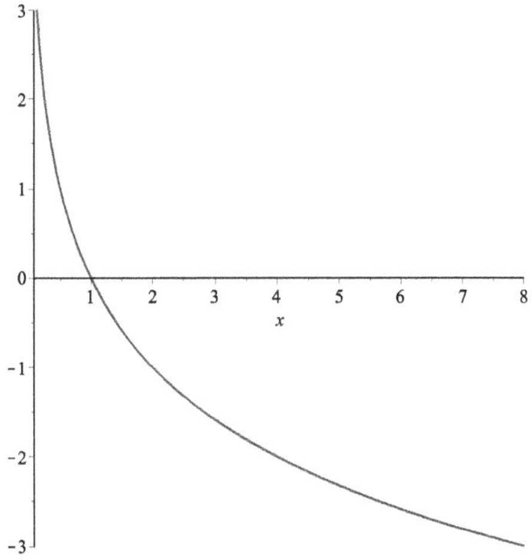

Fig. 5.4 *The graph of* $\log_{1/2} x$.

Proof. (i) Let $y_1, y_2 > 0$. Set $x_1 = \log_a y_1$, $x_2 = \log_a y_2$, i.e., $a^{x_1} = y_1$ and $y_2 = a^{x_2}$. We have to show that

$$\log_a(y_1 y_2) = x_1 + x_2, \quad \log_a \frac{y_1}{y_2} = x_1 - x_2.$$

We have

$$y_1 y_2 = a^{x_1} a^{x_2} = a^{x_1+x_2} \Rightarrow \log_a(y_1 y_2) = \log_a a^{x_1+x_2} = x_1 + x_2,$$

$$\frac{y_1}{y_2} = \frac{a^{x_1}}{a^{x_2}} = a^{x_1-x_2} \Rightarrow \log_a \frac{y_1}{y_2} = \log_a a^{x_1-x_2} = x_1 - x_2.$$

(ii) Let $x \in \mathbb{R}$ such that $a^x = y$, i.e., $\log_a y = x$. We have to prove that

$$\log_a y^\alpha = \alpha x.$$

We have

$$y^{\alpha} = (a^{x})^{\alpha} = a^{\alpha x} \Rightarrow \log_a y^{\alpha} = \log_a a^{\alpha x} = \alpha x.$$

(iii) Let $\beta, x, t \in \mathbb{R}$ such that $a^{\beta} = b$, $y = a^{x} = b^{t}$. Then

$$y = b^{t} = (a^{\beta})^{t} = a^{t\beta} = a^{x}.$$

Hence,

$$\log_a y = x = t\beta = (\log_b y)(\log_a b) \Rightarrow \log_b y = \frac{\log_a y}{\log_a b}.$$

(iv) Assume first that $a > 1$. Consider the numbers $y_2 > y_1 > 0$, and set

$$x_1 := \log_a y_1, \quad x_2 = \log_a y_2.$$

We have to show that $x_2 > x_1$. We argue by contradiction. If $x_1 \geq x_2$, then

$$y_1 = a^{x_1} \geq a^{x_2} = y_2 \Rightarrow y_1 \geq y_2.$$

This contradiction proves the statement (iv) in the case $a > 1$. The case $a \in (0, 1)$ is dealt with in a similar fashion.
(v) Assume first that $a > 1$ so that the function $y \mapsto \log_a y$ is increasing. Since $y_n \to y$, we deduce that $\frac{y_n}{y} \to 1$. Hence, for any $\varepsilon > 0$ there exists $N = N(\varepsilon) > 0$ such that $\forall n > N(\varepsilon) : \frac{y_n}{y_0} \in (a^{-\varepsilon}, a^{\varepsilon})$. We deduce that $\forall n > N(\varepsilon)$

$$-\varepsilon = \log_a a^{-\varepsilon} < \underbrace{\log_a \left(\frac{y_n}{y_0}\right)}_{= \log_a y_n - \log_a y_0} < \log_a a^{\varepsilon} = \varepsilon \iff |\log_a y_n - \log_a y_0| < \varepsilon.$$

□

Theorem 5.21. *Fix a real number s and consider $f : (0, \infty) \to (0, \infty)$ given by $f(x) = x^s$. Then for any $c > 0$ any sequence of positive numbers (x_n), and any sequence of real numbers (s_n) such that $x_n \to c$, and $s_n \to s$, we have*

$$\boxed{\lim_{n\to\infty} x_n^{s_n} = c^s.}$$

Proof. Set

$$y_n := \log x_n^{s_n} = s_n \log x_n.$$

Theorem 5.20(v) implies that

$$\lim_n y_n = (\lim_n s_n)(\lim_n \log x_n) = s \log c.$$

Using Theorem 5.18(v), we deduce that

$$\lim_n e^{y_n} = e^{s \log c} = (e^{\log c})^s = c^s.$$

Now observe that

$$e^{y_n} = e^{\log x_n^{s_n}} = x_n^{s_n}.$$

This proves Theorem 5.21. □

5.3. Limits involving infinities

Suppose that we are given a subset $X \subset \mathbb{R}$ and a function $f : X \to \mathbb{R}$.

Definition 5.22. Let c be a cluster point of X.
(a) We say that the limit of f as $x \to c$ is ∞, and we write this

$$\lim_{x \to c} f(x) = \infty$$

if for any $M > 0$, $\exists \delta = \delta(M) > 0$ such that

$$\forall x \in X \ \ \big(0 < |x - c| < \delta \Rightarrow f(x) > M\big).$$

(b) We say that the limit of f as $x \to c$ is $-\infty$, and we write this

$$\lim_{x \to c} f(x) = -\infty$$

if for any $M > 0$, $\exists \delta = \delta(M) > 0$ such that

$$\forall x \in X \ \ \big(0 < |x - c| < \delta \Rightarrow f(x) < -M\big).$$

□

We have the following version of Proposition 5.3. The proof is left to you.

Proposition 5.23. *Let $f : X \to \mathbb{R}$ be a function defined on a set $X \subset \mathbb{R}$ and c a cluster point of X. Then the following statements are equivalent.*

(i) $\lim_{x \to c} f(x) = \infty$, *i.e., f satisfies (5.1) or (5.2).*
(ii)

$$\forall M > 0, \ \ \exists V \in \mathcal{N}_c^* \ \text{such that} \ \forall x \in X : x \in V \Rightarrow f(x) \in (M, \infty). \tag{5.13}$$

□

Arguing as in the proof of Theorem 5.4 we obtain the following result. The details are left to you.

Theorem 5.24. *Let c be a cluster point of the set $X \subset \mathbb{R}$ and $f : X \to \mathbb{R}$ a real valued function on X. The following statements are equivalent.*

(i) $\lim_{x \to c} f(x) = \infty \in \mathbb{R}$.
(ii) For any sequence $(x_n)_{n \in \mathbb{N}}$ in $X \setminus \{c\}$ such that $x_n \to c$, we have $\lim_n f(x_n) = \infty$.

□

Observe that if $X \subset \mathbb{R}$ is not bounded above, then for any $M > 0$ the intersection $X \cap (M, \infty)$ is nonempty, i.e., for any number $M > 0$ there exists at least one number $x \in X$ such that $x > M$. Equivalently, this means that there exists a sequence $(x_n)_{n \in \mathbb{N}}$ of numbers in X such that

$$\lim_{n \to \infty} x_n = \infty.$$

Definition 5.25. Suppose $X \subset \mathbb{R}$ is a subset not bounded above and $f : X \to \mathbb{R}$ is a real function defined on X.

(a) We say that the limit of f as $x \to \infty$ is the real number A, and we write this $\lim_{x\to\infty} f(x) = A$, if

$$\forall \varepsilon > 0 \ \exists M = M(\varepsilon) > 0 \ \forall x \in X \ \ (x > M \Rightarrow |f(x) - A| < \varepsilon).$$

(b) We say that the limit of f as $x \to \infty$ is ∞,and we write this $\lim_{x\to\infty} f(x) = \infty$, if

$$\forall C > 0 \ \exists M = M(C) > 0 \ \forall x \in X \ \ (x > M \Rightarrow f(x) > C).$$

(c) We say that the limit of f as $x \to \infty$ is $-\infty$, and we write this $\lim_{x\to\infty} f(x) = -\infty$, if

$$\forall C > 0 \ \exists M = M(C) > 0 \ \forall x \in X \ \ (x > M \Rightarrow f(x) < -C). \qquad \square$$

Observe that if $X \subset \mathbb{R}$ is not bounded below, then for any $M > 0$ the intersection $X \cap (-\infty, -M)$ is nonempty, i.e., for any number $M > 0$ there exists at least one number $x \in X$ such that $x < -M$. Equivalently, this means that there exists a sequence $(x_n)_{n\in\mathbb{N}}$ of numbers in X such that

$$\lim_{n\to\infty} x_n = -\infty.$$

Definition 5.26. Suppose $X \subset \mathbb{R}$ is a subset not bounded below and $f : X \to \mathbb{R}$ is a real function defined on X.
(a) We say that the limit of f as $x \to -\infty$ is the real number A, and we write this $\lim_{x\to-\infty} f(x) = A$, if

$$\forall \varepsilon > 0 \ \exists M = M(\varepsilon) > 0 \ \forall x \in X \ \ (x < -M \Rightarrow |f(x) - A| < \varepsilon).$$

(b) We say that the limit of f as $x \to -\infty$ is ∞, and we write this $\lim_{x\to-\infty} f(x) = \infty$, if

$$\forall C > 0 \ \exists M = M(C) > 0 \ \forall x \in X \ \ (x < -M \Rightarrow f(x) > C).$$

(c) We say that the limit of f as $x \to -\infty$ is $-\infty$, and we write this $\lim_{x\to-\infty} f(x) = -\infty$, if

$$\forall C > 0 \ \exists M = M(C) > 0 \ \forall x \in X \ \ (x < -M \Rightarrow f(x) < -C). \qquad \square$$

The limits involving infinities have an alternate description involving sequences. Thus, if $X \subset \mathbb{R}$ is not bounded above and $f : X \to \mathbb{R}$ is a real function defined on X, then the equality

$$\lim_{x\to\infty} f(x) = A$$

can be given a characterization as in Theorem 5.4. More precisely, it means that for any sequence of real numbers $x_n \in X$ such that $x_n \to \infty$, the sequence $f(x_n)$ converges to A.

Example 5.27. (a) We want to prove that

$$\boxed{\lim_{x\to\infty} \left(1 + \frac{1}{x}\right)^x = e.} \tag{5.14}$$

We will use the fundamental result in Example 4.23 which states that the sequence

$$x_n := \left(1 + \frac{1}{n}\right)^n, \quad n \in \mathbb{N},$$

converges to the Euler number e. In particular, we deduce that

$$\lim_{n\to\infty}\left(1+\frac{1}{n+1}\right)^n=\lim_{n\to\infty}\left(1+\frac{1}{n}\right)^{n+1}=e. \tag{5.15}$$

Recall that for any real number x we denote by $\lfloor x\rfloor$ the integer part of the real number x, i.e., the largest integer which is $\leq x$. Thus $\lfloor x\rfloor$ is an integer and

$$\lfloor x\rfloor\leq x<\lfloor x\rfloor+1.$$

For $x\geq 1$ we have

$$1\leq\lfloor x\lfloor\leq x\leq\lfloor x\rfloor+1$$

and we deduce

$$1+\frac{1}{\lfloor x\rfloor+1}\leq 1+\frac{1}{x}\leq 1+\frac{1}{\lfloor x\rfloor}.$$

In particular, we deduce that

$$\left(1+\frac{1}{\lfloor x\rfloor+1}\right)^{\lfloor x\rfloor}\leq\left(1+\frac{1}{x}\right)^{\lfloor x\rfloor}\leq\left(1+\frac{1}{x}\right)^{x}\leq\left(1+\frac{1}{\lfloor x\rfloor}\right)^{x}\leq\left(1+\frac{1}{\lfloor x\rfloor}\right)^{\lfloor x\rfloor+1}. \tag{5.16}$$

From (5.15) we deduce that for any $\varepsilon>0$ there exists $N=N(\varepsilon)>0$ such that

$$\left(1+\frac{1}{n+1}\right)^n,\ \left(1+\frac{1}{n}\right)^{n+1}\in(e-\varepsilon,e+\varepsilon),\quad\forall n>N(\varepsilon).$$

If $x>N(\varepsilon)+1$, then $\lfloor x\rfloor>N(\varepsilon)$ and we deduce from the above that

$$e-\varepsilon<\left(1+\frac{1}{\lfloor x\rfloor+1}\right)^{\lfloor x\rfloor}<e+\varepsilon\ \text{ and }\ e-\varepsilon<\left(1+\frac{1}{\lfloor x\rfloor}\right)^{\lfloor x\rfloor+1}<e+\varepsilon.$$

The inequalities (5.16) now imply that for $x>N(\varepsilon)+1$ we have

$$e-\varepsilon<\left(1+\frac{1}{\lfloor x\rfloor+1}\right)^{\lfloor x\rfloor}\leq\left(1+\frac{1}{x}\right)^{x}\leq\left(1+\frac{1}{\lfloor x\rfloor}\right)^{\lfloor x\rfloor+1}<e+\varepsilon.$$

This proves (5.14).
(b) We want to prove that

$$\boxed{\lim_{x\to-\infty}\left(1+\frac{1}{x}\right)^{x}=e.} \tag{5.17}$$

We will prove that for any sequence of nonzero real numbers (x_n) such that $x_n\to-\infty$, we have

$$\lim_n\left(1+\frac{1}{x_n}\right)^{x_n}=e.$$

Consider the new sequence $y_n:=-x_n$. Clearly $y_n\to\infty$. We have

$$\left(1+\frac{1}{x_n}\right)^{x_n}=\left(1-\frac{1}{y_n}\right)^{-y_n}=\left(\frac{y_n-1}{y_n}\right)^{-y_n}=\left(\frac{y_n}{y_n-1}\right)^{y_n}.$$

Now set $z_n := y_n - 1$ so that $y_n = z_n + 1$ and

$$\left(\frac{y_n}{y_n - 1}\right)^{y_n} = \left(\frac{z_n + 1}{z_n}\right)^{z_n+1} = \left(1 + \frac{1}{z_n}\right)^{z_n+1} = \left(1 + \frac{1}{z_n}\right)^{z_n} \times \left(1 + \frac{1}{z_n}\right).$$

Clearly $z_n \to \infty$ so that

$$\lim_n \left(1 + \frac{1}{z_n}\right) = 1.$$

Invoking (5.14) we deduce

$$\lim_{n\to\infty} \left(1 + \frac{1}{z_n}\right)^{z_n} = e.$$

Hence,

$$\lim_n \left(1 + \frac{1}{x_n}\right)^{x_n} = \lim_n \left(1 + \frac{1}{z_n}\right)^{z_n} \times \lim_n \left(1 + \frac{1}{z_n}\right) = e.$$

This proves (5.17). □

5.4. One-sided limits

Suppose $X \subset \mathbb{R}$ is a set of real numbers. For any $c \in \mathbb{R}$ we define

$$X_{<c} := \{x \in X;\ x < c\} = X \cap (-\infty, c),\quad X_{>c} := \{x \in X;\ x > c\} = X \cap (c, \infty).$$

Definition 5.28. Let $f : X \subset \mathbb{R}$ and $c \in \mathbb{R}$. We say that L is the *left limit* of f at c, and we write this

$$L = \lim_{x \nearrow c} f(x) = \lim_{x\to c-} f(x),$$

if

- c is a cluster point of $X_{<c}$ and
- for any $\varepsilon > 0$ there exists $\delta = \delta(\varepsilon) > 0$ such that

$$\forall x \in X :\ \ x \in (c - \delta, c) \Rightarrow |f(x) - L| < \varepsilon.$$

We say that R is the *right limit* of f at c, and we write this

$$R = \lim_{x \searrow c} f(x) = \lim_{x\to c+} f(x),$$

if

- c is a cluster point of $X_{>c}$ and
- for any $\varepsilon > 0$ there exists $\delta = \delta(\varepsilon) > 0$ such that

$$\forall x \in X :\ \ x \in (c, c + \delta) \Rightarrow |f(x) - R| < \varepsilon.$$

□

The next result follows immediately from Theorem 5.4. The details are left to you.

Theorem 5.29. *Let $f : X \to \mathbb{R}$ be a real valued function defined on the set $X \subset \mathbb{R}$. Fix $c \in \mathbb{R}$.*
(a) Suppose that c is a cluster point of $X_{<c}$ and $L \in \mathbb{R}$. Then the following statements are equivalent.

(i)

$$\lim_{x \nearrow c} f(x) = L.$$

(ii) For any sequence of real numbers (x_n) in X such that $x_n \to c$ and $x_n < c \ \forall n$ we have

$$\lim_n f(x_n) = L.$$

(iii) For any nondecreasing sequence of real numbers (x_n) in X such that $x_n \to c$ and $x_n < c \ \forall n$ we have

$$\lim_n f(x_n) = L.$$

(b) Suppose that c is a cluster point of $X_{>c}$ and $L \in \mathbb{R}$. Then the following statements are equivalent.

(i)

$$\lim_{x \searrow c} f(x) = L.$$

(ii) For any sequence of real numbers (x_n) in X such that $x_n \to c$ and $x_n > c \ \forall n$ we have

$$\lim_n f(x_n) = L.$$

(iii) For any nonincreasing sequence of real numbers (x_n) in X such that $x_n \to c$ and $x_n > c \ \forall n$ we have

$$\lim_n f(x_n) = L.$$

□

The next result describes one of the reasons why the one-sided limits are useful. Its proof is left to you as an exercise.

Theorem 5.30. *Consider three real numbers $a < c < b$, a real valued function*

$$f : (a, b) \setminus \{c\} \to \mathbb{R}$$

and suppose that $A \in (-\infty, \infty)$. Then the following statements are equivalent.

(i)

$$\lim_{x \to c} f(x) = A.$$

(ii)

$$\lim_{x \nearrow c} f(x) = \lim_{x \searrow c} f(x).$$

□

5.5. Some fundamental limits

In this section we present a collection of examples that play a fundamental role in the development of real analysis.

Example 5.31. We want to prove that

$$\boxed{\lim_{x\to 0}\big(1+x\big)^{\frac{1}{x}}=e.} \tag{5.18}$$

We invoke Theorem 5.30, so we will prove that

$$\lim_{x\searrow 0}\big(1+x\big)^{\frac{1}{x}}=\lim_{x\nearrow 0}\big(1+x\big)^{\frac{1}{x}}=e.$$

We prove first the equality

$$\lim_{x\searrow 0}\big(1+x\big)^{\frac{1}{x}}=e.$$

We have to prove that if (x_n) is a sequence of *positive* numbers such that $x_n\to 0$, then

$$\lim_n\big(1+x_n\big)^{\frac{1}{x_n}}=e.$$

Set

$$y_n:=\frac{1}{x_n}.$$

Then $y_n\to\infty$ and

$$\big(1+x_n\big)^{\frac{1}{x_n}}=\left(1+\frac{1}{y_n}\right)^{y_n},$$

and, according to (5.15), we have

$$\left(1+\frac{1}{y_n}\right)^{y_n}=e.$$

The equality

$$\lim_{x\nearrow 0}\big(1+x\big)^{\frac{1}{x}}=e$$

is proved in a similar fashion invoking (5.17) instead of (5.15). □

Example 5.32. We have ($\log=\log_e$)

$$\boxed{\lim_{x\to 0}\frac{\log\big(1+x\big)}{x}=1.} \tag{5.19}$$

Indeed, consider a sequence of nonzero numbers (x_n) such that $x_n\to 0$. Set

$$y_n=(1+x_n)^{\frac{1}{x_n}}.$$

From (5.19) we deduce that $y_n\to e$. Using Theorem 5.20(v), we deduce that $\log y_n\to\log e=1$. □

Example 5.33. We have

$$\boxed{\lim_{x\to 0}\frac{e^x-1}{x}=1.} \tag{5.20}$$

Let $x_n \to 0$. Set $y_n := e^{x_n}$ so that $x_n = \log y_n$ and $y_n \to e^0 = 1$. Next, set $h_n := y_n - 1$ so that $h_n \to 0$. Then

$$\frac{e^{x_n}-1}{x_n} = \frac{y_n-1}{\log y_n} = \frac{h_n}{\log(1+h_n)} = \frac{1}{\frac{\log(1+h_n)}{h_n}} \overset{(5.19)}{\longrightarrow} 1. \qquad \square$$

Example 5.34. Suppose that $\alpha \in \mathbb{R}$, $\alpha \neq 0$. We have

$$\boxed{\lim_{x\to 0}\frac{(1+x)^\alpha-1}{x}=\alpha.} \tag{5.21}$$

Let $x_n \to 0$. Then

$$(1+x_n)^\alpha = e^{\alpha\log(1+x_n)}.$$

Set $y_n := \alpha\log(1+x_n)$ so that $y_n \to 0$. Then

$$\frac{(1+x_n)^\alpha-1}{x_n} = \frac{e^{y_n}-1}{y_n}\cdot\frac{y_n}{x_n} = \frac{e^{y_n}-1}{y_n}\cdot\frac{\alpha\log(1+x_n)}{x_n}.$$

Using (5.20) we deduce

$$\frac{e^{y_n}-1}{y_n} \to 1,$$

and using (5.19) we deduce

$$\frac{\alpha\log(1+x_n)}{x_n} \to \alpha.$$

This shows that

$$\frac{(1+x_n)^\alpha-1}{x_n} \to \alpha. \qquad \square$$

Here is a typical application of the equality (5.18).

Example 5.35. Let us compute

$$\lim_{x\to\infty}\left(1+\frac{x}{x^2+1}\right)^{2x}.$$

For any sequence $x_n \to \infty$ we have to compute

$$\lim_{n\to\infty}\left(1+\frac{x_n}{x_n^2+1}\right)^{2x_n}.$$

Set

$$y_n := \frac{x_n}{x_n^2+1}.$$

Note that $y_n \to 0$ as $n \to \infty$ so that

$$e = \lim_{y \to 0}(1+y)^{\frac{1}{y}} = \lim_{n \to \infty}(1+y_n)^{\frac{1}{y_n}}.$$

We first seek to express $2x_n$ in the form

$$2x_n = \frac{s_n}{y_n} \iff s_n = 2x_n y_n = \frac{2x_n^2}{x_n^2+1}.$$

Note that $s_n \to 2$ as $n \to \infty$. We deduce

$$\left(1+\frac{x_n}{x_n^2+1}\right)^{2x_n} = (1+y_n)^{\frac{s_n}{y_n}} = \left((1+y_n)^{\frac{1}{y_n}}\right)^{s_n},$$

so that

$$\lim_{n \to \infty}\left(1+\frac{x_n}{x_n^2+1}\right)^{2x_n} = \lim_{n \to \infty}\left((1+y_n)^{\frac{1}{y_n}}\right)^{s_n}$$

(use Theorem 5.21)

$$= \left(\lim_n (1+y_n)^{\frac{1}{y_n}}\right)^{\lim_n s_n} = e^2.$$

□

5.6. Trigonometric functions: a less than completely rigorous definition

Recall that the Cartesian product $\mathbb{R}^2 := \mathbb{R} \times \mathbb{R}$ is called the Cartesian plane and can be visualized as an Euclidean plane equipped with two perpendicular coordinate axes, the x-axis and the y-axis; see Figure 5.5. We can locate a point P in this plane if we can locate its projections P_x and P_y respectively, on the x- and the y-axis respectively; see Figure 5.5.

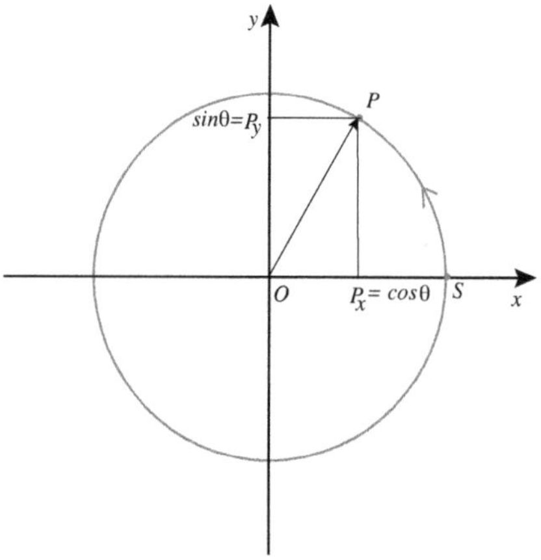

Fig. 5.5 *The trigonometric circle. The distance of the journey from S to P in the counterclockwise direction is θ.*

The locations of these projections are indicated by two numbers, the x-coordinate and the y-coordinate respectively, of P. The point with coordinates $(0,0)$ is called the origin and it is denoted by O.

The trigonometric circle is the circle of radius 1 centered at the origin; see Figure 5.5. More precisely, a point with coordinates (x, y) lies on this circle if and only if

$$x^2 + y^2 = 1. \tag{5.22}$$

Additionally, we agree that this circle is given an *orientation*, i.e., a prescribed way of traveling around it. In mathematics, the agreed upon orientation is *counterclockwise orientation* indicated by the arrow along the circle in Figure 5.5.

The starting point of the trigonometric circle is the point S with coordinates $(1, 0)$. It can alternatively be described as the intersection of the circle with the positive side of the x-axis. The length[2] of the upper semi-circle is a positive number known by its famous name, π. In particular, the total length of the circle is 2π.

Suppose that we start at the point S and we travel along the circle, in the counterclockwise direction a distance $\theta \geq 0$. We denote by P the final point of this journey. The coordinates of this point depend only on the distance θ traveled. The x-coordinate of P is denoted by $\cos\theta$, and the y-coordinate of P is denoted by $\sin\theta$. The equality (5.22) implies that

$$\cos^2\theta + \sin^2\theta = 1, \ \ \forall\theta \geq 0. \tag{5.23}$$

Observe that if we continue our journey from P in the counterclockwise direction for a distance 2π then we are back at P. This shows that

$$\cos(\theta + 2\pi) = \cos\theta, \ \ \sin(\theta + 2\pi) = \sin\theta, \ \ \forall\theta \geq 0. \tag{5.24}$$

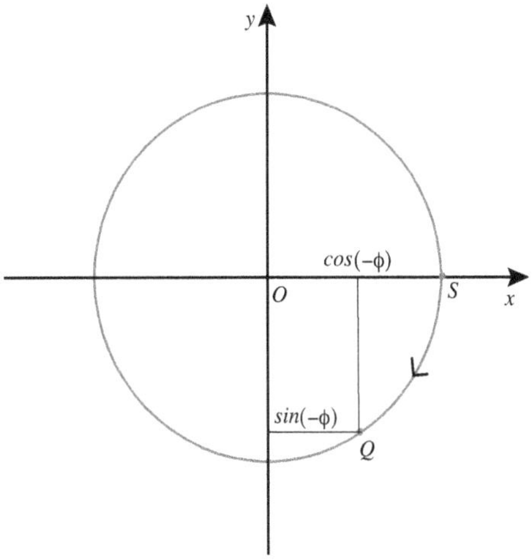

Fig. 5.6 *The trigonometric circle. The distance of the journey from S to Q in the* ***clockwise*** *direction is ϕ.*

We can define $\cos\theta$ and $\sin\theta$ for negative θ's as well. Suppose that $\theta = -\phi$, $\phi \geq 0$. If we start at S and travel along the circle in the *clockwise* direction a distance ϕ, then we reach a point Q. By definition, its coordinates are $\cos(-\phi)$ and $\sin(-\phi)$; see Figure 5.6.

[2]We have surreptitiously avoided explaining what length means.

From the description it is easily seen that

$$\cos(-\phi) = \cos\phi, \quad \sin(-\phi) = -\sin\phi, \quad \forall \phi \geq 0. \tag{5.25}$$

We have thus constructed two functions

$$\cos, \sin : \mathbb{R} \to \mathbb{R},$$

called *trigonometric functions*. Their graphs are depicted in Figures 5.7 and 5.8.

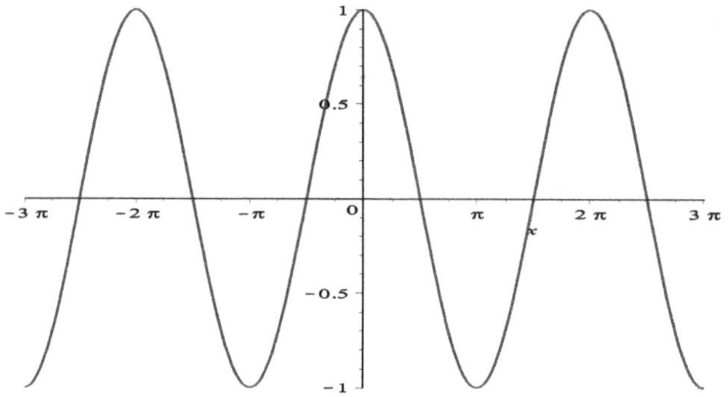

Fig. 5.7 *The graph of* $\cos x$.

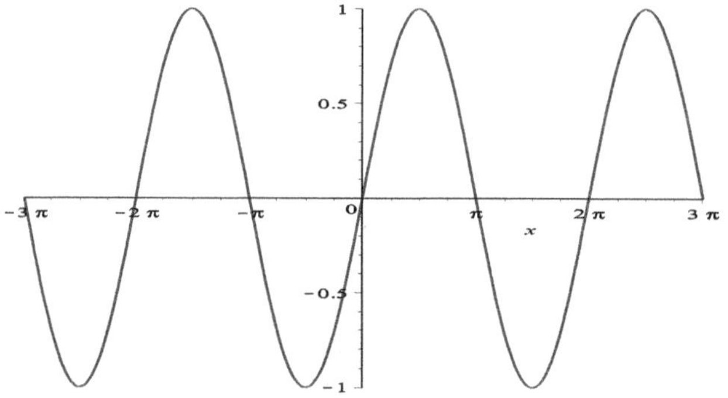

Fig. 5.8 *The graph of* $\sin x$.

Let us record a few important values of these functions.

Table 5.1 Some important values of trig functions.

θ	0	$\frac{\pi}{6}$	$\frac{\pi}{4}$	$\frac{\pi}{3}$	$\frac{\pi}{2}$	π	2π
$\cos\theta$	1	$\frac{\sqrt{3}}{2}$	$\frac{\sqrt{2}}{2}$	$\frac{1}{2}$	0	-1	1
$\sin\theta$	0	$\frac{1}{2}$	$\frac{\sqrt{2}}{2}$	$\frac{\sqrt{3}}{2}$	1	0	0

We list below some of the more elementary, but very important, properties of the trigonometric functions sin and cos.

$$\cos^2 x + \sin^2 x = 1, \ \ \forall x \in \mathbb{R}. \tag{5.26a}$$

$$\cos(x + 2\pi) = \cos x, \ \ \sin(x + 2\pi) = \sin x, \ \ \forall x \in \mathbb{R}. \tag{5.26b}$$

$$\cos(-x) = \cos x, \ \ \sin(-x) = -\sin(x), \ \ \forall x \in \mathbb{R}. \tag{5.26c}$$

$$\cos(x + \pi) = -\cos(x), \ \ \sin(x + \pi) = -\sin(x), \ \ \forall x \in \mathbb{R}, \tag{5.26d}$$

$$\sin\left(x + \frac{\pi}{2}\right) = \cos x, \ \ \forall x \in \mathbb{R}. \tag{5.26e}$$

$$|\cos x| \leq 1, \ \ |\sin x| \leq 1, \ \ \forall x \in \mathbb{R}. \tag{5.26f}$$

$$\cos x > 0, \ \ \forall x \in (-\frac{\pi}{2}, \frac{\pi}{2}) \ \text{ and } \ \sin x > 0, \ \ \forall x \in (0, \pi). \tag{5.26g}$$

$$\cos x = 0 \Longleftrightarrow x \text{ is an odd multiple of } \tfrac{\pi}{2}, \ \ \sin x = 0 \Longleftrightarrow x \text{ is a multiple of } \pi. \tag{5.26h}$$

Definition 5.36. Let $f : \mathbb{R} \to \mathbb{R}$ be a real valued function defined on the real axis $\mathbb{R}$.

(i) The function f is called *even* if

$$f(-x) = f(x), \ \ \forall x \in \mathbb{R}.$$

(ii) The function f is called *odd* if

$$f(-x) = -f(x), \ \ \forall x \in \mathbb{R}.$$

(iii) Suppose P is a positive real number. We say that f is *P-periodic* if

$$f(x + P) = f(x), \ \ \forall x \in \mathbb{R}.$$

(iv) The function f is called periodic if there exists $P > 0$ such that f is P-periodic. Such a number P is called a *period* of f.

□

We see that the functions $\cos x$ and $\sin x$ are 2π-periodic, $\cos x$ is even, and $\sin x$ is odd.

In applications, we often rely on other trigonometric functions derived from sin and cos. We define

$$\tan x = \frac{\sin x}{\cos x}, \ \ \text{whenever } \cos x \neq 0,$$

$$\cot x = \frac{\cos x}{\sin x}, \ \ \text{whenever } \sin x \neq 0.$$

The graphs of $\tan x$ and $\cot x$ are depicted in Figures 5.9 and 5.10.

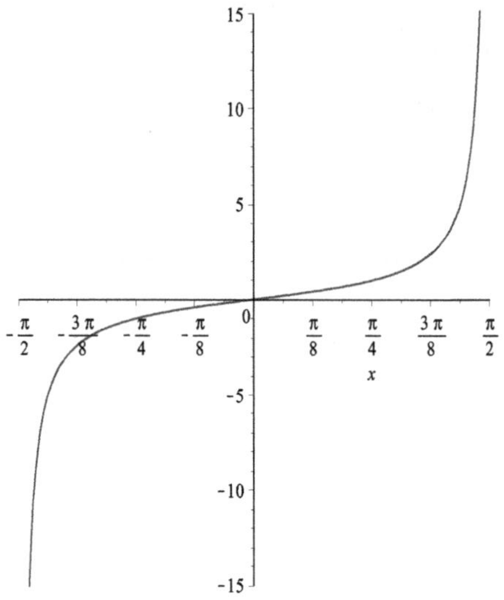

Fig. 5.9 *The graph of* $\tan x$ *for* $x \in (-\pi/2, \pi/2)$.

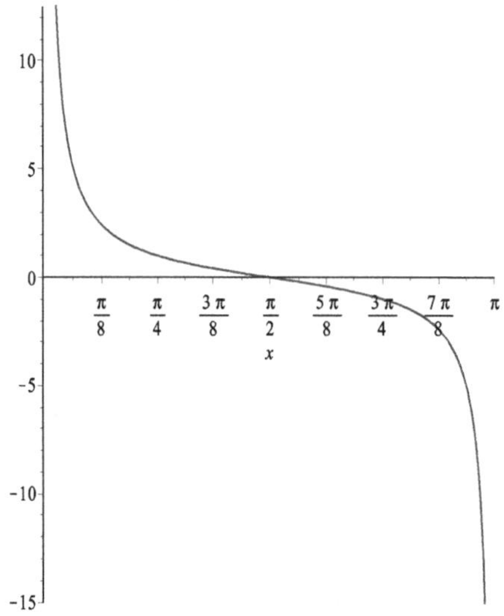

Fig. 5.10 *The graph of* $\cot x$ *for* $x \in (0, \pi)$.

Example 5.37. We want to outline a geometric explanation for an important limit.

$$\boxed{\lim_{x \to 0} \frac{\sin x}{x} = 1.} \tag{5.27}$$

We will prove that

$$\lim_{x \nearrow 0} \frac{\sin x}{x} = \lim_{x \searrow 0} \frac{\sin x}{x} = 1.$$

Since

$$\frac{\sin x}{x} = \frac{\sin(-x)}{-x}$$

it suffices to prove only that

$$\lim_{x \searrow 0} \frac{\sin x}{x} = 1. \tag{5.28}$$

This will follow immediately from the following fundamental inequalities

$$\theta \cos^2 \theta \leq \sin \theta \leq \theta, \ \ \forall 0 < \theta < \frac{\pi}{2}. \tag{5.29}$$

Let us temporarily take for granted these inequalities and show how they imply (5.28).

Observe that (5.29) implies that

$$0 \leq \sin \theta \leq \theta, \ \ \forall 0 < \theta < \frac{\pi}{2}.$$

The Squeezing Principle shows that

$$\lim_{\theta \searrow 0} \sin \theta = 0. \tag{5.30}$$

This shows that the limit

$$\lim_{\theta \searrow 0} \frac{\sin \theta}{\theta}$$

is a bad limit of the type $\frac{0}{0}$. We can rewrite (5.29) as

$$1 - \sin^2 \theta = \cos^2 \theta \leq \frac{\sin \theta}{\theta} \leq 1. \tag{5.31}$$

The equality (5.30) shows that

$$\lim_{\theta \searrow 0} (1 - \sin^2 \theta) = 1.$$

The equality (5.29) now follows by applying the Squeezing Principle to the inequalities (5.31).

"Proof" of (5.29). Fix θ, $0 < \theta < \frac{\pi}{2}$. We denote by P the point on the trigonometric circle reached from S by traveling a distance θ in the counterclockwise direction; see Figure 5.11. Denote by Q the projection of P onto the x-axis. We have

$$|OQ| = \cos \theta, \ \ |PQ| = \sin \theta.$$

Denote by M the intersection of the line OP with the circle centered at O and radius $|OP| = \cos \theta$. We distinguish three regions in Figure 5.12: the circular sector (OQM), the triangle $\triangle OSP$, and the circular sector (OSP). Clearly

$$(OQM) \subset \triangle OSP \subset (OSP)$$

so that we obtain inequalities between their areas[3]

$$\text{area}\,(OQM) \leq \text{area}\,\triangle OSP \leq \text{area}\,(OSP)$$

The area of a circular sector is[4]

$$\frac{1}{2} \times \text{square of the radius of the sector} \times \text{the size of the angle of the sector}. \tag{5.32}$$

We have

$$\text{area}\,(OQM) = \frac{1}{2}|OQ|^2\theta = \frac{\theta \cos^2 \theta}{2}, \ \ \text{area}\,(OSP) = \frac{1}{2}|OS|^2\theta = \frac{\theta}{2},$$

$$\text{area}\,\triangle OSP = \frac{1}{2}|PQ| \times |OS| = \frac{1}{2}\sin \theta.$$

Hence,

$$\frac{\theta \cos^2 \theta}{2} \leq \frac{1}{2}\sin \theta \leq \frac{\theta}{2}.$$

□

[3] At this point we do not have a rigorous definition of the area of a planar region.

[4] The equality (5.32) needs a justification.

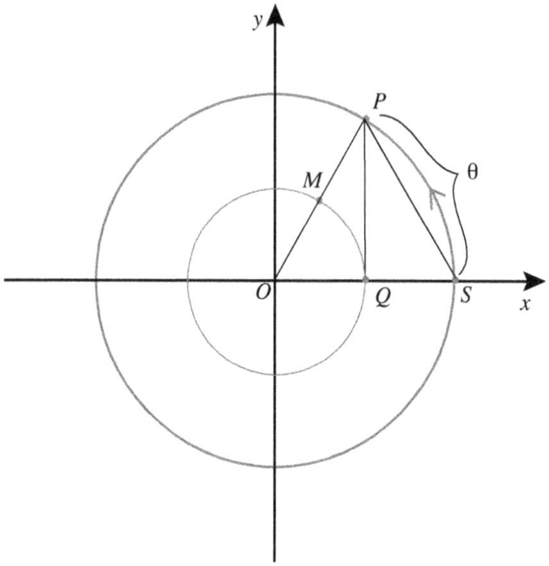

Fig. 5.11 *The trigonometric circle. The distance of the journey from S to P in the* **counterclockwise** *direction is θ.*

5.7. Useful trig identities

We list here a few important trigonometric identities that we will use in the future.

$$\sin(x \pm y) = \sin x \cos y \pm \sin y \cos x, \quad \cos(x \pm y) = \cos x \cos y \mp \sin x \sin y. \tag{5.33a}$$

$$\sin 2x = 2 \sin x \cos x, \quad \cos 2x = \cos^2 x - \sin^2 x. \tag{5.33b}$$

$$\frac{1 + \cos x}{2} = \cos^2(x/2), \quad \frac{1 - \cos x}{2} = \sin^2(x/2). \tag{5.33c}$$

$$\cos x \cos y = \frac{1}{2}\big(\cos(x-y) + \cos(x+y)\big), \quad \sin x \sin y = \frac{1}{2}\big(\cos(x-y) - \cos(x+y)\big). \tag{5.33d}$$

$$\tan(x \pm y) = \frac{\tan x + \tan y}{1 \mp \tan x \tan y}. \tag{5.33e}$$

5.8. Landau notation

Let $c \in [-\infty, \infty]$ and consider two real valued functions f, g defined on the same set $X \subset \mathbb{R}$ which admits c as a cluster point. We say that

$$f(x) = O\big(g(x)\big) \text{ as } x \to c \tag{5.34}$$

if there exists a positive constant C and a neighborhood U of c such that

$$\forall x \in X, \quad x \in (X \cap U) \setminus \{c\} \Rightarrow |f(x)| \le C|g(x)|.$$

For example,

$$\frac{x}{x^2+1} = O\left(\frac{1}{x}\right) \quad \text{as} \quad x \to \infty.$$

We say that

$$f(x) = o\big(g(x)\big) \quad \text{as } x \to c, \tag{5.35}$$

if, for any $\varepsilon > 0$, there exists a neighborhood U_ε of c such that

$$\forall x \in X, \quad x \in U_\varepsilon \setminus \{c\} \Rightarrow |f(x)| \leq \varepsilon |g(x)|.$$

If $g(x) \neq 0$ for any x in a neighborhood U of c, then

$$f(x) = o\big(g(x)\big) \quad \text{as } x \to c \Longleftrightarrow \lim_{x \to c} \frac{f(x)}{g(x)} = 0.$$

Loosely speaking, this means that $f(x)$ is much, much smaller than $g(x)$ as x approaches c. For example,

$$e^{-x} = o(x^{-25}) \quad \text{as} \quad x \to \infty,$$

and

$$x^3 = o(x^2) \quad \text{as} \quad x \to 0.$$

However,

$$x^2 = o(x^3) \quad \text{as } x \to \infty.$$

Finally, we say that f is *similar* to $g(x)$ as $x \to c$, and we write this

$$f(x) \sim g(x) \quad \text{as } x \to c$$

if

$$\lim_{x \to c} \frac{f(x)}{g(x)} = 1.$$

For example,

$$x^3 - 39x^2 + 17 \sim x^3 + 3x^2 + 2x + 1 \quad \text{as} \quad x \to \infty,$$

and

$$e^x - 1 \sim x \quad \text{as} \quad x \to 0.$$

5.9. Exercises

Exercise 5.1. Prove that any real number is a cluster point of the set of rational numbers. □

Exercise 5.2. Prove Proposition 5.3. □

Exercise 5.3 (Squeezing principle). Let $f, g, h : X \to \mathbb{R}$ be three functions defined on the same subset $X \subset \mathbb{R}$ and c a cluster point of X. Suppose that U is a deleted neighborhood of c such that

$$f(x) \leq h(x) \leq g(x), \ \ \forall x \in U \cap X.$$

Show that if

$$\lim_{x \to c} f(x) = A = \lim_{x \to c} g(x),$$

then

$$\lim_{x \to c} h(x) = A.$$

□

Exercise 5.4. Consider a subset $X \subset \mathbb{R}$, a function $f : X \to \mathbb{R}$, and a cluster point c of the set X. Prove that the following statements are equivalent.

(i) The limit $\lim_{x \to c} f(x)$ exists and it is finite.
(ii) For any sequence $(x_n)_{n \in \mathbb{N}} \subset X$ such that $x_n \to c$ and $x_n \neq c$, $\forall n \in \mathbb{N}$ the sequence $\big(f(x_n)\big)_{n \in \mathbb{N}}$ is convergent.

□

Exercise 5.5. Let $I \subset \mathbb{R}$ be an interval and $f : I \to \mathbb{R}$ a function. Suppose that f is a *Lipschitz function*, i.e., there exists a constant L such that

$$|f(x) - f(y)| \leq L|x - y|, \ \ \forall x, y \in I.$$

Show that for any $y \in Y$ we have

$$\lim_{x \to y} f(x) = f(y).$$

□

Exercise 5.6. We already know that the series

$$\sum_{n \geq 1} \frac{1}{n^s}$$

converges for any *rational number* $s > 1$. Prove that it converges for any *real number* $s > 1$. □

Exercise 5.7. (a) Prove that for any $n \in \mathbb{N}$ we have

$$\lim_{x \to \infty} \frac{1}{x^n} = 0.$$

(b) Let $k \in \mathbb{N}$ and consider the function $f : \mathbb{R} \setminus \{0\} \to \mathbb{R}$, $f(x) = \frac{1}{x^{2k}}$. Show that

$$\lim_{x \to 0} f(x) = \infty.$$

□

Exercise 5.8. Fix a natural number n. Consider the polynomial

$$P(x) = x^n + a_{n-1}x^{n-1} + \cdots + a_1x + a_0.$$

Show that

$$\lim_{x\to\infty} P(x) = \infty, \quad \lim_{x\to-\infty} P(x) = \begin{cases} \infty, & n \text{ is even} \\ -\infty, & n \text{ is odd.} \end{cases}$$

□

Exercise 5.9. Consider two convergent sequences of real numbers $(x_n)_{n\geq 0}$, $(y_n)_{n\geq 0}$. We set

$$x := \lim_{n\to\infty} x_n, \quad y := \lim_{n\to\infty} y_n.$$

Show that if $x_n > 0$, $\forall n \geq 0$ and $x > 0$ then

$$\lim_{n\to\infty} x_n^{y_n} = x^y.$$

Hint. Use the same strategy as in the proof of Theorem 5.21. □

Exercise 5.10. Prove that

$$\lim_{n\to\infty}\left(1 + \frac{x}{n}\right)^n = e^x, \quad \forall x \in \mathbb{R}.$$

Hint. Use the result in Example 5.27 and Theorem 5.21. □

Exercise 5.11. Fix an arbitrary number $a > 1$.
(a) Prove that for any $x > 1$ we have

$$a^x \geq a^{\lfloor x \rfloor} \geq 1 + (a-1)\lfloor x \rfloor + \binom{\lfloor x \rfloor}{2}(a-1)^2.$$

(b) Prove that

$$\lim_{x\to\infty} \frac{x}{a^x} = 0, \quad \lim_{x\to\infty} \frac{a^x}{x} = \infty.$$

Hint. Use (a) and Example 4.18.
(c) Let $r > 0$. Prove that

$$\lim_{x\to\infty} \frac{x^r}{a^x} = 0.$$

Hint. Reduce to (b).

(d) Prove that

$$\lim_{x\to\infty} \frac{\log_a x}{x} = 0.$$

Hint. Reduce to (c). □

Exercise 5.12. Fix a positive real number s and consider the function $f : (0, \infty) \to \mathbb{R}$, $f(x) = x^s$.
(a) Show that f is an increasing function.
(b) Show that

$$\lim_{x\searrow 0} x^s = 0.$$

Exercise 5.13. Let $a, b \in \mathbb{R}$, $a < b$. Prove that if $f : (a, b) \to \mathbb{R}$ is a nondecreasing function and $x_0 \in (a, b)$, then the one-sided limits

$$\lim_{x \nearrow x_0} f(x) \text{ and } \lim_{x \searrow x_0} f(x)$$

exist and

$$\lim_{x \nearrow x_0} f(x) = \sup_{x < x_0} f(x), \quad \lim_{x \searrow x_0} f(x) = \inf_{x > x_0} f(x). \qquad \square$$

Exercise 5.14. Consider the function

$$f : \mathbb{R} \setminus \{0\} \to \mathbb{R}, \quad f(x) = x \sin\left(\frac{1}{x}\right).$$

Prove that

$$\lim_{x \to 0} f(x) = 0. \qquad \square$$

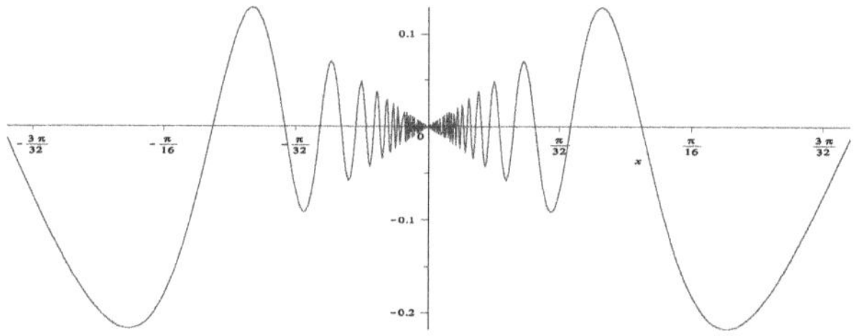

Fig. 5.12 *The graph of* $x\sin(1/x)$ *for* $|x| < \pi/10$.

Exercise 5.15. Consider $f : [0, \infty) \to \mathbb{R}$,

$$f(x) = \frac{2x^3 + x^2}{x^3 + x^2 + 1}.$$

Show that

$$f(x) = O(1) \text{ as } x \to \infty.$$

Above, we used Landau's notation introduced in section 5.8. $\square$

Exercise 5.16. (a) Prove Lemma 5.15.

Hint. The case $a = 1$ is trivial. In the case $a > 1$ show that there exists a sequence of positive rational numbers (r_n) such that

$$-r_n \le x_n - x \le r_n, \quad \forall n.$$

Now use Lemma 5.9, Lemma 5.12, and the Squeezing Principle to conclude. The case $a < 1$ follows from the case $a > 1$.

(b) Prove the equality (5.11).

Hint. First prove that (5.11) holds for any $x \in \mathbb{Q}$. Then conclude using Lemma 5.15. $\square$

Exercise 5.17. Prove Proposition 5.23. $\square$

Exercise 5.18. Prove Theorem 5.24. $\square$

Exercise 5.19. Prove Theorem 5.29. $\square$

Exercise 5.20. Prove Theorem 5.30. $\square$

5.10. More challenging problems

Problem 5.1 (Viète). Consider the sequence $(x_n)_{n\geq 0}$ defined by

$$x_0 = 0, \;\; x_{n+1} = \sqrt{\frac{1+x_n}{2}}, \;\; \forall n \geq 0.$$

(a) Prove that

$$x_n = \cos\frac{\pi}{2^{n+1}}, \;\; \forall n \geq 0.$$

(b) Prove that

$$\lim_{n\to\infty}(x_1 \cdot x_2 \cdots x_n) = \frac{2}{\pi}.$$

□

Problem 5.2. Suppose that

$$\sum_{n\geq 0} a_n$$

is a convergent series of real numbers. We denote by a its sum.

(i) Show that for any $x \in (-1, 1)$ the series

$$\sum_{n\geq 0} a_n x^n$$

is convergent. For $x \in (-1, 1)$ we denote by $A(x)$ the sum of the above series.

(ii) Prove that

$$\lim_{x\nearrow 1} A(x) = a.$$

Hint: At some point you will need to use Abel's trick (4.29).

□

Problem 5.3. Suppose that $U : (0, \infty) \to (0, \infty)$ is an *increasing function* such that the limit

$$\lim_{t\to\infty}\frac{U(tx)}{U(x)}$$

exists and it is positive for any $x > 0$. We denote by $\psi(x)$ the above limit.

(a) Prove that $\psi(x) \leq \psi(y)$, $\forall x, y \in (0, \infty)$, $x < y$.
(b) Prove that $\psi(xy) = \psi(x)\psi(y)$, $\forall x, y > 0$.
(c) Prove that there exists $p \geq 0$ such that $\psi(x) = x^p$, $\forall x > 0$. □

Chapter 6

Continuity

6.1. Definition and examples

The concept of continuity is a fundamental mathematical concept with a wide range of applications.

Definition 6.1. Suppose that $X \subset \mathbb{R}$ and $f : X \to \mathbb{R}$ is a real valued function defined on X. We say that the function f is *continuous at a point* $x_0 \in X$ if

$$\forall \varepsilon > 0 \ \exists \delta = \delta(\varepsilon) > 0 \ \text{ such that } \ \forall x \in X \ \ |x - x_0| < \delta \Rightarrow |f(x) - f(x_0)| < \varepsilon.$$

We say that the function f is *continuous (on X)* if it is continuous at every point $x_0 \in X$.□

Arguing as in the proof of Theorem 5.4 we obtain the following very useful alternate characterization of continuity. The details are left to you as an exercise.

Theorem 6.2. *Let $X \subset \mathbb{R}$, $x_0 \in X$, and $f : X \to \mathbb{R}$ a real valued function on X. The following statements are equivalent.*

(i) The function f is continuous at x_0.
(ii) For any sequence $(x_n)_{n \in \mathbb{N}}$ in X such that $x_n \to x_0$, we have $\lim_n f(x_n) = f(x_0)$.

□

We have the following useful consequence which relates the concept of continuity to the concept of limit. Its proof is left to you as an exercise.

Corollary 6.3. *Let $X \subset \mathbb{R}$ and $f : X \to \mathbb{R}$. Suppose that $x_0 \in X$ is a cluster point of X. Then the following statements are equivalent.*

(i) The function f is continuous at x_0.
(ii) $\lim_{x \to x_0} f(x) = f(x_0)$.

□

We have already encountered many examples of continuous functions.

Example 6.4. (a) Let $k \in \mathbb{N}$. Then the function

$$f : \mathbb{R} \to \mathbb{R}, \;\; f(x) = x^k, \;\; \forall x \in \mathbb{R},$$

is continuous on its domain $\mathbb{R}$. Indeed, if $x_0 \in \mathbb{R}$ and $(x_n)_{n\in\mathbb{R}}$ is a sequence of real numbers such that $x_n \to x_0$, then Proposition 4.15 implies that

$$x_n^k \to x_0^k,$$

thus proving the continuity of f at an arbitrary point $x_0 \in \mathbb{R}$.
(b) A similar argument shows that if $k \in \mathbb{N}$, then the function

$$f : \mathbb{R} \setminus \{0\} \to \mathbb{R}, \;\; f(x) = \frac{1}{x^k}, \;\; \forall x \in \mathbb{R} \setminus \{0\},$$

is continuous.
(c) Fix $s \in \mathbb{R}$. Then the function

$$f : (0, \infty) \to \mathbb{R}, \;\; f(x) = x^s, \;\; \forall x > 0,$$

is continuous. Indeed, this follows by invoking Theorems 5.21 and 6.2.
(d) Let $a > 0$. Then the functions

$$f : \mathbb{R} \to (0, \infty), \;\; f(x) = a^x,$$

and

$$g : (0, \infty) \to \mathbb{R}, \;\; g(x) = \log_a x,$$

are continuous on their domains. Indeed, the continuity of f follows from Lemma 5.15, while the continuity of g follows from Theorem 5.20.
(e) The trigonometric functions

$$\sin, \cos : \mathbb{R} \to \mathbb{R}$$

are continuous.

Let us first prove that these functions are continuous at $x_0 = 0$. The continuity of sin at $x_0 = 0$ follows immediately from (5.30) and Corollary 6.3. To prove the continuity of cos at $x_0 = 0$ we have to show that if (x_n) is a sequence of real numbers such that $x_n \to 0$, then $\cos x_n \to \cos 0 = 1$.

Let (x_n) be a sequence of real numbers converging to zero. Then

$$\cos^2 x_n = 1 - \sin^2 x_n$$

and we deduce that

$$\lim_n \cos^2 x_n = 1 - \sin^2 x_n = \lim_n (1 - \sin^2 x_n) = 1.$$

Since $x_n \to 0$, we deduce that there exists $N_0 > 0$ such that $|x_n| < \frac{\pi}{2}, \forall n > N_0$. The inequalities (5.26g) imply that

$$\cos x_n > 0, \;\; \forall n >_0,$$

so that,

$$\cos x_n = \sqrt{1 - \sin^2 x_n}, \;\; \forall n > N_0.$$

Exercise 4.15 now implies that

$$\lim_n \cos x_n = \sqrt{\lim_n (1 - \sin^2 x_n)} = \sqrt{1} = \cos 0.$$

We can now prove the continuity of sin and cos at an arbitrary point x_0. Suppose that x_n is a sequence of real numbers such that $x_n \to x_0$. We have to show that

$$\lim_n \sin x_n = \sin x_0 \quad \text{and} \quad \lim_n \cos x_n = \cos x_0.$$

We set $h_n = x_n - x_0$, so that, $x_n = x_0 + h$. Then

$$\sin x_n = \sin(x_0 + h_n) \stackrel{(5.33a)}{=} \sin x_0 \cos h_n + \sin h_n \cos x_0$$

and

$$\cos x_n = \cos(x_0 + h_n) \stackrel{(5.33a)}{=} \cos x_0 \cos h_n - \sin x_0 \sin h_n.$$

Observe that $h_n \to 0$ and, since sin and cos are continuous at 0, we have $\sin h_n \to 0$ and $\cos h_n \to 1$. We deduce

$$\lim_n \sin x_n = \lim_n \sin(x_0 + h_n) = \sin x_0 \lim_n \cos h_n + \cos x_0 \lim_n \sin h_n = \sin x_0$$

and

$$\lim_n \cos x_n = \lim_n \cos(x_0 + h_n) = \cos x_0 \lim_n \cos h_n - \sin x_0 \lim_n \sin h_n = \cos x_0.$$

(f) Recall that a function $f : X \to \mathbb{R}$, $X \subset \mathbb{R}$ is called *Lipschitz* if

$$\exists L > 0 : \ \forall x_1, x_2 \in X \ \ |f(x_1) - f(x_2)| \leq L|x_1 - x_2|.$$

Observe that a Lipschitz function is necessarily continuous. Indeed, if $x_0 \in X$ and (x_n) is a sequence in X such that $x_n \to x_0$ then

$$|f(x_n) - f(x_0)| \leq L|x_n - x_0| \to 0,$$

and the Squeezing Principle implies that $f(x_n) \to f(x_0)$.

Observe that the absolute value function $f : \mathbb{R} \to [0, \infty)$, $f(x) = |x|$ is Lipschitz because of the following elementary inequality (see Exercise 4.5)

$$|f(x) - f(y)| = \big|\,|x| - |y|\,\big| \leq |x - y|, \ \ \forall x, y \in \mathbb{R}. \tag{6.1}$$

Thus the absolute value function $f : \mathbb{R} \to \mathbb{R}$, $f(x) = |x|$ is a continuous function. □

Proposition 6.5. *Let $X \subset \mathbb{R}$, $c \in \mathbb{R}$, and suppose that $f, g : X \to \mathbb{R}$ are two continuous functions. Then the functions*

$$f + g, cf, f \cdot g : X \to \mathbb{R},$$

are continuous. Additionally, if $\forall x \in X$ $g(x) \neq 0$, then the function

$$\frac{f}{g} : X \to \mathbb{R}$$

is also continuous.

Proof. This is an immediate consequence of Proposition 4.15 and Theorem 6.2. □

Example 6.6. Polynomial function $p : \mathbb{R} \to \mathbb{R}$ defined by

$$p(x) = c_n x^n + \cdots + c_1 x + c_0,$$

$n \in \mathbb{Z}$, $n \geq 0$, $c_0, \ldots, c_n \in \mathbb{R}$ are continuous. For example, the function $p(x) = x^3 - 2x + 5$, $x \in \mathbb{R}$, is continuous on $\mathbb{R}$.

We can easily get more complicated examples. Thus, the function $(x^3 - 2x + 5)\sin x$, $x \in \mathbb{R}$, is continuous, the function $e^x + e^{-x}$, $x \in \mathbb{R}$, is continuous and nowhere zero, so the quotient

$$\frac{(x^3 - 2x + 5)\sin x}{e^x + e^{-x}}, \quad x \in \mathbb{R}$$

is also continuous on $\mathbb{R}$. □

Proposition 6.7. *Suppose that $X, Y \subset \mathbb{R}$ and that $f : X \to \mathbb{R}$ and $g : Y \to \mathbb{R}$ are continuous functions such that*

$$f(X) \subset Y.$$

Then the composition $g \circ f : X \to \mathbb{R}$, $g \circ f(x) = g(f(x))$, $\forall x \in X$ is also a continuous function.

Proof. Theorem 6.2 implies that we have to prove that for any $x_0 \in X$ and any sequence (x_n) in X such that $x_n \to x_0$ we have

$$g(f(x_n)) \to g(f(x_0)).$$

Set $y_0 := f(x_0)$, $y_n := f(x_n)$. Since f is continuous at x_0, Theorem 6.2 shows that $f(x_n) \to f(x_0)$, i.e., $y_n \to y_0$. Since g is continuous at y_0, Theorem 6.2 implies that $g(y_n) \to g(y_0)$, i.e.,

$$g(f(x_n)) \to g(f(x_0)).$$

□

Example 6.8. Consider the continuous functions

$$f, g, h : \mathbb{R} \to \mathbb{R}, \quad f(x) = \sin x, \quad g(x) = e^x, \quad h(x) = |x|.$$

Then $g \circ f(x) = e^{\sin x}$ is continuous on $\mathbb{R}$, and so is the function $f \circ g(x) = \sin e^x$. Similarly $f \circ h(x) = \sin |x|$ is a continuous function on $\mathbb{R}$. □

Definition 6.9. Let $X \subset \mathbb{R}$ be a set of real numbers and $f : X \to \mathbb{R}$ a real valued function on X.

(a) The sequence of functions $f_n : X \to \mathbb{R}, \quad n \in \mathbb{N}$ is said to converge *pointwisely* to the function $f : X \to \mathbb{R}$ if

$$\lim_{n\to\infty} f_n(x) = f(x), \quad \forall x \in X,$$

i.e.,

$$\forall \varepsilon > 0, \ \ \forall x \in X \ \ \exists N = N(\varepsilon, x) : \ \ \forall n > N(\varepsilon, x) \ \ |f_n(x) - f(x)| < \varepsilon. \tag{6.2}$$

(b) The sequence of functions $f_n : X \to \mathbb{R}, \quad n \in \mathbb{N}$ is said to *converge uniformly* to the function $f : X \to \mathbb{R}$ if

$$\forall \varepsilon > 0 \ \exists N = N(\varepsilon) > 0 \text{ such that } \forall n > N(\varepsilon), \ \ \forall x \in X : \ \ |f_n(x) - f(x)| < \varepsilon. \tag{6.3}$$

□

Theorem 6.10 (Continuity of uniform limits). *Let $X \subset \mathbb{R}$ be a set of real numbers. If the sequence of* continuous *functions $f_n : X \to \mathbb{R}$, $n \in \mathbb{N}$, converges uniformly to the function $f : X \to \mathbb{R}$, then the limit function f is also continuous on X.*

Proof. We have to prove that given $x_0 \in X$ the function f is continuous at x_0, i.e., we have to show that

$$\forall \varepsilon > 0 \ \exists \delta = \delta(\varepsilon) > 0 \ \forall x \in X \ \ |x - x_0| < \delta \Rightarrow |f(x) - f(x_0)| < \varepsilon. \tag{6.4}$$

Let $\varepsilon > 0$. The uniform convergence implies that

$$\exists N(\varepsilon) > 0 : \ \ \forall x \in X, \ \ \forall n > N(\varepsilon) \ \ |f_n(x) - f(x)| < \frac{\varepsilon}{3}. \tag{6.5}$$

Fix $n_0 > N(\varepsilon)$. The function f_{n_0} is continuous at x_0 and thus

$$\exists \delta(\varepsilon) > 0 \ \forall x \in X : \ \ |x - x_0| < \delta(\varepsilon) \Rightarrow |f_{n_0}(x) - f_{n_0}(x_0)| < \frac{\varepsilon}{3}. \tag{6.6}$$

We deduce that if $|x - x_0| < \delta(\varepsilon)$, then

$$|f(x) - f(x_0)| \le |f(x) - f_{n_0}(x)| + |f_{n_0}(x) - f_{n_0}(x_0)| + |f_{n_0}(x_0) - f(x_0)|. \tag{6.7}$$

From (6.5) we deduce that since $n_0 > N(\varepsilon)$ we have

$$|f(x) - f_{n_0}(x)|, \ \ |f_{n_0}(x_0) - f(x_0)| < \frac{\varepsilon}{3}, \ \ \forall x \in X.$$

From (6.6) we deduce that if $|x - x_0| < \delta(\varepsilon)$, then

$$|f_{n_0}(x) - f_{n_0}(x_0)| < \frac{\varepsilon}{3}.$$

Using these facts in (6.7) we deduce that if $|x - x_0| < \delta(\varepsilon)$, then

$$|f(x) - f(x_0)| < \varepsilon.$$

□

6.2. Fundamental properties of continuous functions

In this section we will discuss several fundamental properties of continuous functions, which hopefully will explain the usefulness of the concept of continuity.

Theorem 6.11. *Suppose that c is an arbitrary real number,* $X \subset \mathbb{R}$ *and* $f : X \to \mathbb{R}$ *is a function continuous at* $x_0 \in X$.
(a) If $x_0 \in X$ *satisfies* $f(x_0) < c$*, then there exists* $\delta > 0$ *such that*

$$\forall x \in X, \ \ |x - x_0| < \delta \Rightarrow f(x) < c.$$

In other words, if $f(x_0) < c$*, then for any* $x \in X$ *sufficiently close to* x_0 *we also have* $f(x) < c$.
(b) If $x_0 \in X$ *satisfies* $f(x_0) > c$*, then there exists* $\delta > 0$ *such that*

$$\forall x \in X, \ \ |x - x_0| < \delta \Rightarrow f(x) > c.$$

In other words, if $f(x_0) > c$*, then for any* $x \in X$ *sufficiently close to* x_0 *we also have* $f(x) > c$.

Proof. Fix $\varepsilon_0 > 0$, such that $f(x_0) + \varepsilon_0 < c$. (For example, we can choose $\varepsilon_0 = \frac{1}{2}(c - f(x_0))$.)

The continuity of f at x_0 (Definition 6.1) implies that there exists $\delta_0 > 0$ such that for any $x \in X$ satisfying $|x - x_0| < \delta_0$ we have

$$|f(x) - f(x_0)| < \varepsilon_0,$$

so that

$$f(x_0) - \varepsilon_0 < f(x) < f(x_0) + \varepsilon_0 < c.$$

□

Corollary 6.12. *Suppose that $X \subset \mathbb{R}$, $x_0 \in X$ and $f : X \to \mathbb{R}$ is a continuous function such that $f(x_0) \neq 0$. Then there exists $\delta > 0$ such that*

$$\forall x \in X \ \ (|x - x_0| < \delta \Rightarrow f(x) \neq 0).$$

In other words, if $f(x_0) \neq 0$, then for any $x \in X$ sufficiently close to x_0 we also have $f(x) \neq 0$.

Proof. Consider the function $g : X \to \mathbb{R}$, $g(x) = |f(x)|$. The function g is continuous because it is the composition of the absolute-value-function with the continuous function f. Additionally, $|g(x_0)| > 0$. The desired conclusion now follows from Theorem 6.11(b).

□

To state and prove our next result we need to make a small digression. Recall that the *Completeness Axiom* states that if the set $X \subset \mathbb{R}$ is *bounded above*, then it admits a least upper bound which is a *real number* denoted by $\sup X$. If the set X is not bounded above, then we define $\sup X := \infty$. Thus, we have given a meaning to $\sup X$ *for any* subset $X \subset \mathbb{R}$. Moreover,

$$\sup X < \infty \iff \text{the set } X \text{ is bounded above.}$$

Similarly, we define $\inf X = -\infty$ for any set X that is not bounded below. Thus we have given a meaning to $\inf X$ *for any* subset $X \subset \mathbb{R}$. Moreover,

$$\inf X > -\infty \iff \text{the set } X \text{ is bounded below.}$$

Lemma 6.13. *(a) If Y is a set of real numbers and $M = \sup Y \in (-\infty, \infty]$, then there exists an increasing sequence of real numbers $(M_n)_{n \geq 1}$ and a sequence (y_n) in Y such that*

$$M_n \leq y_n \leq M, \ \ \forall n, \ \ \lim_n M_n = M.$$

(b) If Y is a set of real numbers and $m = \inf Y \in [-\infty, \infty)$, then there exists a decreasing sequence of real numbers $(m_n)_{n \geq 1}$ and a sequence (y_n) in Y such that

$$m \leq y_n \leq m_n, \ \ \forall n, \ \ \lim_n m_n = m.$$

Proof. We prove only (a). The proof of (b) is very similar and it is left to you as an exercise. We distinguish two cases.

A. $M < \infty$. Since M is the least upper bound of Y, for any $n > 0$ there exists $y_n \in X$ such that

$$M - \frac{1}{n} \leq y_n \leq M.$$

The sequences (y_n) and $M_n = M - \frac{1}{n}$ have the desired properties.

B. $M = \infty$. Hence, the set Y is not bounded above. Thus, for any $n \in \mathbb{N}$ there exists $y_n \in Y$ such that $y_n \geq n$. The sequences (y_n) and $M_n = n$ have the desired properties. □

Theorem 6.14 (Weierstrass). *Consider a continuous real valued function f defined on a* ***closed and bounded*** *interval $[a, b]$, i.e., $f : [a, b] \to \mathbb{R}$. Then the following hold.*

(i)

$$M := \sup\{ f(x); \;\; x \in [a, b] \} < \infty.$$

(ii) $\exists x^* \in [a, b]$ such that $f(x^*) = M$.

(iii)

$$m := \inf\{ f(x); \;\; x \in [a, b] \} > -\infty.$$

(iv) $\exists x_* \in [a, b]$ such that $f(x_*) = m$.

Proof. We prove only (i) and (ii). The proofs of statements (iii) and (iv) are similar. Denote by Y the range of the function f,

$$Y = \{f(x); \;\; x \in [a, b] \}.$$

Hence $M = \sup Y$. From Lemma 6.13 we deduce that there exists a sequence (y_n) in Y and an increasing sequence (M_n) such that

$$M_n \leq y_n \leq M, \;\; \lim_n M_n = M.$$

The Squeezing Principle implies that

$$\lim_n y_n = M. \tag{6.8}$$

Since y_n is in the range of f there exists $x_n \in [a, b]$ such that $f(x_n) = y_n$. The sequence (x_n) is obviously bounded because it is contained in the bounded interval $[a, b]$. The Bolzano–Weierstrass Theorem (Theorem 4.29) implies that (x_n) admits a subsequence (x_{n_k}) which converges to some number x^*

$$\lim_k x_{n_k} = x^*.$$

Since $a \leq x_{n_k} \leq b$, $\forall k$, we deduce that $x^* \in [a, b]$. The continuity of f implies that

$$\lim_k y_{n_k} = \lim_k f(x_{n_k}) = f(x^*).$$

On the other hand,

$$\lim_k y_{n_k} = \lim_n y_n \overset{(6.8)}{=} M.$$

Hence

$$M = f(x^*) < \infty.$$

□

Definition 6.15. Let $f : X \to \mathbb{R}$ be a function defined on a nonempty set $X \subset \mathbb{R}$.
(a) A point $\boldsymbol{x}_* \in X$ is called a *global minimum* of f if

$$f(x_*) \leq f(x), \quad \forall x \in X.$$

(b) A point $\boldsymbol{x}^* \in X$ is called a *global maximum* of f if

$$f(x) \leq f(x^*), \quad \forall x \in X.$$ □

We can rephrase Theorem 6.14 as follows.

Corollary 6.16. *A continuous function $f : [a,b] \to \mathbb{R}$ admits a global minimum and a global maximum.* □

Remark 6.17. The conclusions of Theorem 6.14 *do not necessarily hold* for continuous functions defined on *non-closed* intervals. Consider for example the continuous function

$$f : (0,1] \to \mathbb{R}, \quad f(x) = \frac{1}{x}.$$

Note that $f(1/n) = n$, $\forall n \in \mathbb{N}$ so that

$$\sup\{\, f(x); \;\; x \in (0,1] \,\} = \infty.$$ □

Theorem 6.18 (The Intermediate Value Theorem). *Suppose that $f : [a,b] \to \mathbb{R}$ is a continuous function and $c, d \in [a,b]$ are real numbers such that*

$$c < d \;\; \text{and} \;\; f(c) \cdot f(d) < 0.$$

Then there exists a real number $r \in (c,d)$ such that $f(r) = 0$.

Proof. We distinguish two cases: $f(c) < 0$ or $f(c) > 0$. We discuss only the case $f(c) < 0$ depicted in Figure 6.1. The second case follows from the first case applied to the continuous function $-f$. Observe that the assumption $f(c)f(d) < 0$ implies that if $f(c) < 0$, then $f(d) > 0$.

Consider the set

$$X := \{x \in [c,d]; \;\; f(x) < 0\}.$$

Clearly X is nonempty because $c \in X$. By construction, the set X is bounded above by d. Define

$$r := \sup X.$$

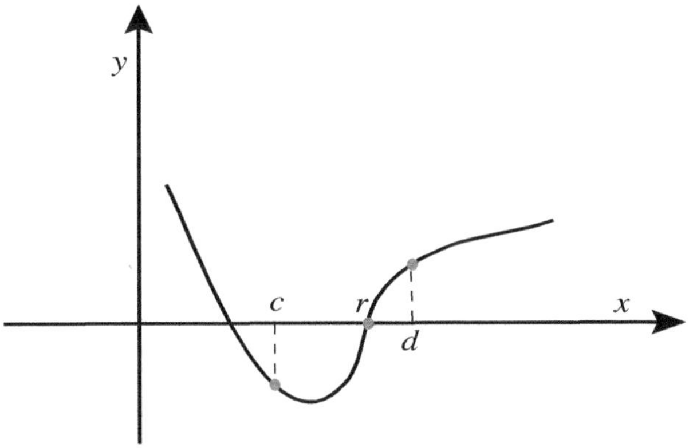

Fig. 6.1 *If the graph of a continuous functions has points both below and above the x-axis, then the graph must intersect the x-axis.*

We will prove that $f(r) = 0$. Since $r = \sup X$, we deduce from Lemma 6.13 that there exists a sequence (x_n) in X such that $x_n \to r$ as $n \to \infty$. The function f is continuous at r so that

$$f(r) = \lim_{x \to r} f(x) = \lim_{n \to \infty} f(x_n).$$

On the other hand, $f(x_n) < 0$ for any n because $x_n \in X$. Hence $f(r) \leq 0$. In particular $r \neq d$ because $f(d) > 0$.

Fig. 6.2 *The function f would be negative on $[r, r+\delta]$ if $f(r)$ were negative.*

To prove that $f(r) = 0$ it suffices to show that $f(r) \geq 0$. We argue by contradiction and we assume that $f(r) < 0$. Theorem 6.11 implies that there exists $\delta > 0$ such that if $x \in [a, b]$ and $|x - r| < \delta$, then $f(x) < 0$. Thus $f(x) < 0$ for any $x \in [a, b] \cap [r, r+\delta]$; Figure 6.2.

Choose $h > 0$ such that

$$h < \min\{\, \delta, \operatorname{dist}(r, d) \,\}.$$

Then $r + h \in [r, d]$ and $r + h \in [r, r+\delta]$. Hence $r + h \in [c, d]$ and $f(r+h) < 0$ so that $r + h \in X$. This contradicts the fact that $r = \sup X$. □

The Intermediate Value Theorem has many useful consequences. We present a few of them.

Corollary 6.19. *Suppose that $f : [a, b] \to \mathbb{R}$ is a continuous function, $y_0 \in \mathbb{R}$ and $c \leq d$ are real numbers in the interval $[a, b]$ such that*

- *either $f(c) \leq y_0 \leq f(d)$, or*

- $f(c) \geq y_0 \geq f(d)$.

Then there exists $x_0 \in [c,d]$ *such that* $f(x_0) = y_0$.

Proof. If $f(c) = y_0$ or $f(d) = y_0$, then there is nothing to prove so we assume that $f(c), f(d) \neq y_0$. Consider the function $g : [a,b] \to \mathbb{R}$, $g(x) = f(x) - y_0$. Then $g(c)g(d) < 0$, and the Intermediate Value Theorem implies that there exists $x_0 \in (c,d)$ such that $g(x_0) = 0$, i.e., $f(x_0) = y_0$. □

Corollary 6.20. *Suppose that* $f : [a,b] \to \mathbb{R}$ *is a continuous function and* $c < d$ *are real numbers in the interval* $[a,b]$ *such that*

$$f(x) \neq 0, \ \ \forall x \in (c,d).$$

Then the function f *does not change sign in the interval* (c,d), *i.e., either*

$$f(x) > 0, \ \ \forall x \in (c,d),$$

or

$$f(x) < 0, \ \ \forall x \in (c,d).$$

Proof. If f did change sign in the interval (c,d), then we could find two numbers $c', d' \in (c,d)$ such that $f(c') < 0$ and $f(d') > 0$. The Intermediate Value Theorem will then imply that f must equal zero at some point r situated between c' and d'. This would contradict the assumptions on f. □

Corollary 6.21. *Suppose that* $f : [a,b] \to \mathbb{R}$ *is a continuous function,*

$$M = \sup\{\, f(x); \ \ x \in [a,b] \,\}, \ \ m = \inf\{\, f(x); \ \ x \in [a,b] \,\}.$$

Then the range of the function f *is the interval* $[m,M]$.

Proof. Observe first that

$$m \leq f(x) \leq M, \ \ \forall x \in [a,b].$$

This shows that the range of f is contained in the interval $[m,M]$. Let us now prove the opposite inclusion, i.e., $[m,M]$ is contained in the range of f. More precisely, we need to show that for any $y_0 \in [m,M]$ there exists $x_0 \in [a,b]$ such that $f(x_0) = y_0$.

Observe first that Weierstrass' Theorem 6.14 implies that m, M belong to the range of f. In particular, there exist $c, d \in [a,b]$ such that $f(c) = m$ and $f(d) = M$. In particular,

$$f(c) \leq y_0 \leq f(d).$$

Corollary 6.19 implies that there exists a number x_0 situated between c and d such that $f(x_0) = y_0$. □

Corollary 6.22. *Suppose that $f : \mathbb{R} \to \mathbb{R}$ is a continuous function such that*

$$\lim_{x\to\infty} f(x) \in (0,\infty] \quad \lim_{x\to-\infty} f(x) \in [-\infty, 0).$$

Then there exists $r \in \mathbb{R}$ such that $f(r) = 0$. □

The proof of this corollary is left to you as an exercise.

Corollary 6.23. *Suppose that $a < b$ and $f : [a,b] \to \mathbb{R}$ is a continuous function. Then the following statements are equivalent,*

(i) The function f is injective.
(ii) The function f is strictly monotone; see Definition 5.17(v).

Proof. The implication (ii) $\Rightarrow$ (i) is immediate. Indeed, suppose $x_1, x_2 \in [a,b]$ and $x_1 \neq x_2$. One of the numbers x_1, x_2 is smaller than the other and we can assume $x_1 < x_2$. If f is strictly increasing, then $f(x_1) < f(x_2)$, thus $f(x_1) \neq f(x_2)$. If f is strictly decreasing, then $f(x_1) > f(x_2)$ and again we conclude that $f(x_1) \neq f(x_2)$.

Let us now prove (i) $\Rightarrow$ (ii). Since $a < b$ and f is injective we deduce that either $f(a) < f(b)$, or $f(a) > f(b)$. We discuss only the first situation, $f(a) < f(b)$. The second case follows from the first case applied to the continuous injective function $g = -f$. We will prove in several steps that f is strictly increasing.

Step 1. Suppose that $d \in [a,b)$ is such that $f(d) < f(b)$. Then

$$f(d) < f(c), \quad \forall c \in (d,b). \tag{6.9}$$

We argue by contradiction. Assume that there exists $c \in (d,b)$ such that $f(c) \leq f(d)$. Since f is injective and $d \neq c$ we deduce $f(d) \neq f(c)$ so that $f(c) < f(d)$; see Figure 6.3.

We observe that on the interval $[c,b]$ the function f has values both $< f(d)$ and $> f(d)$ because

$$f(c) < f(d) < f(b).$$

The Intermediate Value Theorem implies that there must exist a point r in the interval (c,b) such that $f(r) = f(d)$; see Figure 6.3. This contradicts the injectivity of f and completes the proof of Step 1.

Step 2. We will show that

$$f(c) < f(b), \quad \forall c \in (a,b). \tag{6.10}$$

Again we argue by contradiction. Assume that there exists $c \in (a,b)$ such that $f(c) \geq f(b)$. Since f is injective, $f(c) > f(b)$; see Figure 6.4.

We observe that on the interval $[a,c]$ the function f has values both $< f(b)$ and $> f(b)$ because

$$f(c) > f(b) > f(a).$$

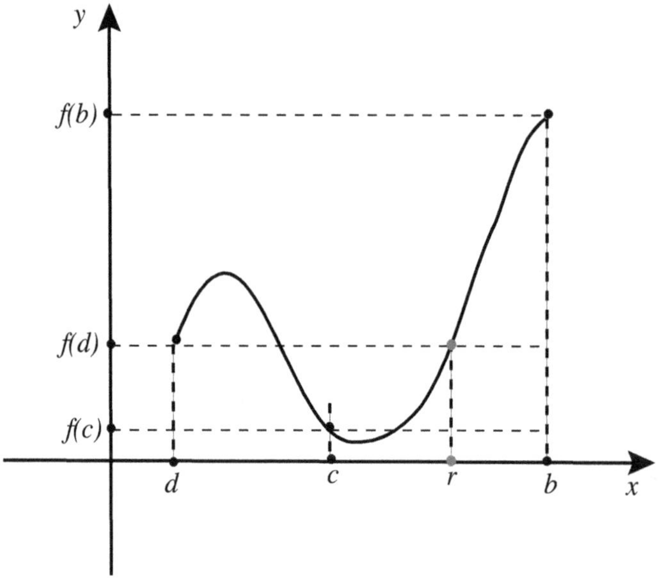

Fig. 6.3 A continuous injective function has to be monotone.

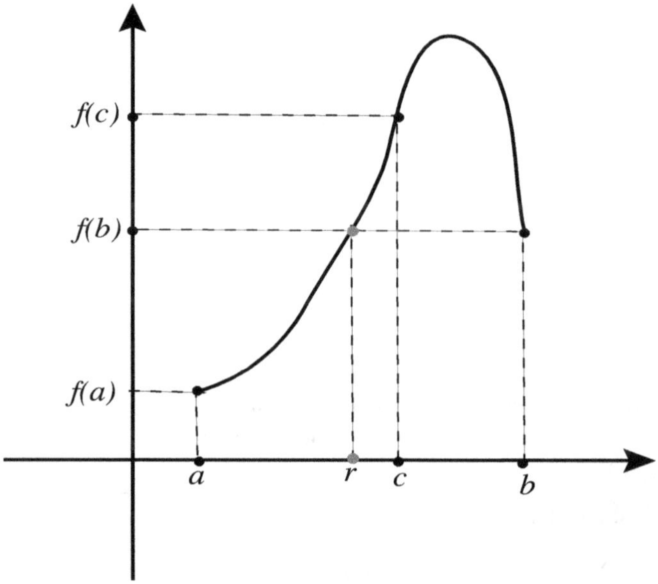

Fig. 6.4 A continuous injective function has to be monotone.

The Intermediate Value Theorem implies that there must exist a point r in the interval (a, c) such that $f(r) = f(b)$; see Figure 6.4. This contradicts the injectivity of f and completes the proof of Step 2.

Step 3. Suppose that $d < d'$ are points in the interval (a, b). We want to show that $f(d) < f(d')$. Note that since $d \in (a, b)$ we deduce from (6.9) and (6.10) that $f(a) < f(d) < f(b)$. Since $d' \in (d, b)$ and $f(d) < f(b)$ we deduce from Step 1 that $f(d) < f(d')$. □

Example 6.24. Consider the function

$$\sin : \bigl[-\pi/2, \pi/2\bigr] \to \mathbb{R}.$$

Using the trigonometric-circle definition of sin we deduce that the above function is strictly increasing. Note that

$$\sin(-\pi/2) = -1 = \min_{x\in\mathbb{R}} \sin x, \quad \sin(\pi/2) = 1 = \max_{x\in\mathbb{R}} \sin x.$$

Using Corollary 6.21 we deduce that the range of this function is $[-1, 1]$ so that the resulting function

$$\sin[-\pi/2, \pi/2] \to [-1, 1]$$

is bijective. Its inverse is the function

$$\arcsin : [-1, 1] \to [-\pi/2, \pi/2].$$

We want to emphasize that, by construction, the range of arcsin is $[-\pi/2, \pi/2]$.

Similarly, the function

$$\cos : [0, \pi] \to \mathbb{R}$$

is strictly decreasing and its range is $[-1, 1]$. Its inverse is the function

$$\arccos : [-1, 1] \to [0, \pi]. \qquad \square$$

Finally, consider the function

$$\tan : (-\pi/2, \pi/2) \to \mathbb{R}.$$

Exercise 6.12 asks you to prove that the above function is bijective. Its inverse is the function

$$\arctan : \mathbb{R} \to (-\pi/2, \pi/2). \qquad \square$$

6.3. Uniform continuity

We want to discuss a more subtle concept of continuity that will play an important role in our investigation of integrability.

Definition 6.25. Suppose that X is a nonempty subset of the real axis and $f : X \to \mathbb{R}$ is a real valued function defined on X. The *oscillation* of the function f on the set $S \subset X$ is the quantity

$$\operatorname{osc}(f, S) := \sup_{s\in S} f(s) - \inf_{s\in S} f(s) \in [0, \infty]. \qquad \square$$

Let us observe that

$$\operatorname{osc}(f,S) = \sup_{s',s''\in S} |f(s') - f(s'')|. \tag{6.11}$$

Exercise 6.11 asks you to prove this equality.

Definition 6.26. Let $J \subset \mathbb{R}$ be an interval and $f : J \to \mathbb{R}$ a function. We say that f is *uniformly continuous* on J if, for any $\varepsilon > 0$, there exists $\delta = \delta(\varepsilon) > 0$ such that, for any closed interval $I \subset J$ of length $\ell(I) \leq \delta$, we have

$$\operatorname{osc}(f,I) \leq \varepsilon.$$

□

Remark 6.27. The uniform continuity of $f : J \to \mathbb{R}$ can be alternatively characterized by the following quantized statement

$$\forall \varepsilon > 0 \;\; \exists \delta = \delta(\varepsilon) > 0 \;\text{ such that }\; \forall x, y \in J \;\; |x - y| < \delta \Rightarrow |f(x) - f(y)| < \varepsilon.$$

□

Proposition 6.28. *Let $J \subset \mathbb{R}$ be an interval and $f : J \to \mathbb{R}$ a function. If f is uniformly continuous, then f is continuous at any point $x_0 \in J$.*

Proof. Let $x_0 \in J$. We have to prove that $\forall \varepsilon > 0$ there exists $\delta > 0$ such that

$$\forall x \;\; |x - x_0| \leq \delta \Rightarrow |f(x) - f(x_0)| < \varepsilon.$$

Since f is uniformly continuous, there exists $\delta_0 = \delta_0(\varepsilon) > 0$ such that, for any interval $I \subset J$ of length $\leq \delta_0(\varepsilon)$ we have $\operatorname{osc}(f,I) < \varepsilon$. Consider now the interval

$$I_{x_0} := \left\{ x \in J;\;\; |x - x_0| < \frac{\delta_0}{2} \right\}.$$

Clearly I_{x_0} has length $< \delta_0$ so that $\operatorname{osc}(f, I_{x_0}) < \varepsilon$. In particular (6.11) implies that for any $x \in I_{x_0}$ we have

$$|f(x) - f(x_0)| < \varepsilon.$$

Hence

$$|x - x_0| < \delta(\varepsilon) := \frac{\delta_0(\varepsilon)}{2} \Rightarrow x \in I_{x_0} \Rightarrow |f(x) - f(x_0)| < \varepsilon.$$

□

Theorem 6.29 (Uniform Continuity). *Suppose that $a < b$ are two real numbers and $f : [a,b] \to \mathbb{R}$ is a continuous function. Then f is uniformly continuous, i.e., for any $\varepsilon > 0$ there exists $\delta = \delta(\varepsilon) > 0$ such that for any interval $I \subset [a,b]$ of length $\ell(I) \leq \delta$ we have*

$$\operatorname{osc}(f,I) \leq \varepsilon.$$

Proof. We have to prove that

$$\forall \varepsilon > 0 \ \exists \delta > 0 \ \forall I \subset [a, b] \ \text{interval}, \ \ \ell(I) \leq \delta \Rightarrow \operatorname{osc}(f, I) \leq \varepsilon.$$

We argue by contradiction and we assume that the opposite is true

$$\exists \varepsilon_0 > 0 \ \forall \delta > 0 \ \exists I = I_\delta \subset [a, b] \ \text{interval}, \ \ \ell(I_\delta) \leq \delta \wedge \operatorname{osc}(f, I_\delta) > \varepsilon_0.$$

We deduce that ***for any*** $n \in \mathbb{N}$ there exists a closed interval $I_n = [a_n, b_n] \subset [a, b]$ of length $\leq \frac{1}{n}$ such that

$$\operatorname{osc}(f, [a_n, b_n]) > \varepsilon_0. \tag{6.12}$$

Since the length of $[a_n, b_n]$ is $\leq \frac{1}{n}$ we deduce

$$a_n < b_n \leq a_n + \frac{1}{n}.$$

The Bolzano–Weierstrass Theorem 4.29 implies that the sequence (a_n) admits a convergent subsequence (a_{n_k}). We set

$$a_* := \lim_{k\to\infty} a_{n_k}.$$

Since $a \leq a_n \leq b$, we deduce $a_* \in [a, b]$. Since

$$a_{n_k} < b_{n_k} \leq a_{n_k} + \frac{1}{n_k}$$

we deduce from the Squeezing Principle that

$$\lim_{k\to\infty} b_{n_k} = \lim_{k\to\infty} a_{n_k} = a_*.$$

On the other hand, since $a_* \in [a, b]$, the function f is continuous at a_*. Thus there exists $\delta > 0$ such that

$$|x - a_*| < \delta \Rightarrow |f(x) - f(a_*)| < \frac{\varepsilon_0}{4}.$$

In other words,

$$\operatorname{dist}(x, a_*) < \delta \Rightarrow f(a_*) - \frac{\varepsilon_0}{4} < f(x) < f(a_*) + \frac{\varepsilon_0}{4}.$$

Since $a_{n_k}, b_{n_k} \to a_*$ there exists k_0 such that

$$[a_{n_{k_0}}, b_{n_{k_0}}] \subset (a_* - \delta, a_* + \delta) \Rightarrow f(a_*) - \frac{\varepsilon_0}{4} < f(x) < f(a_*) + \frac{\varepsilon_0}{4}, \ \ \forall x \in [a_{n_{k_0}}, b_{n_{k_0}}].$$

Thus

$$f(a_*) - \frac{\varepsilon_0}{4} \leq \inf_{x\in[a_{n_{k_0}}, b_{n_{k_0}}]} f(x) \leq \sup_{x\in[a_{n_{k_0}}, b_{n_{k_0}}]} f(x) \leq f(a_*) + \frac{\varepsilon_0}{4}.$$

This shows that

$$\operatorname{osc}\big(f, \ [a_{n_{k_0}}, b_{n_{k_0}}]\big) \leq \left(f(a_*) + \frac{\varepsilon_0}{4}\right) - \left(f(a_*) - \frac{\varepsilon_0}{4}\right) = \frac{\varepsilon_0}{2}.$$

This contradicts (6.12) and completes the proof of the theorem. □

Remark 6.30. The above result is no longer valid for continuous functions defined on *non-closed* or *unbounded* intervals. Consider for example the continuous function $f:(0,1)\to\mathbb{R}$, $f(x)=\frac{1}{x}$. For each $n\in\mathbb{N}$, $n>1$ we define

$$I_n=\Big[\frac{1}{n+1},\frac{1}{n}\Big].$$

Since f is decreasing we deduce that

$$\sup_{x\in I_n} f(x)=f\Big(\frac{1}{n+1}\Big)=n+1,\quad \inf_{x\in I_n} f(x)=f\Big(\frac{1}{n}\Big)=n$$

so that $\operatorname{osc}(f,I_n)=1$. On the other hand, $\ell(I_n)=\frac{1}{n(n+1)}\to 0$ as $n\to\infty$. We have thus produced arbitrarily short intervals over which the oscillation is 1.

Exercise 6.10 describes an example of continuous function over an *unbounded* interval that is *not* uniformly continuous on that interval. □

6.4. Exercises

Exercise 6.1. Prove Theorem 6.2. □

Exercise 6.2. Suppose that $f, g : \mathbb{R} \to \mathbb{R}$ are two continuous functions such that $f(q) = g(q)$, $\forall q \in \mathbb{Q}$. Prove that $f(x) = g(x)$, $\forall x \in \mathbb{R}$.

Hint. You may want to invoke Proposition 3.33. □

Exercise 6.3. Prove Corollary 6.3. □

Exercise 6.4. Prove the inequality (6.1). □

Exercise 6.5. Suppose that $f, g : [a, b] \to \mathbb{R}$ are continuous functions.
(a) Prove that the function $|f|$ is continuous.
(b) Prove that for any $x \in [a, b]$ we have

$$\max\{ f(x), g(x) \} = \frac{1}{2}\big(f(x) + g(x) + |f(x) - g(x)| \big).$$

(c) Prove that the function $h : [a, b] \to \mathbb{R}$, $h(x) = \max\{f(x), g(x)\}$ is continuous. □

Exercise 6.6 (Weierstrass). Suppose that X is a nonempty set of real numbers, $f_n : X \to \mathbb{R}$, $n \in \mathbb{N}$ is a sequence of functions, and $f : X \to \mathbb{R}$ a function on X. Suppose that for any $n \in \mathbb{N}$ we have

$$M_n := \sup_{x \in X} |f_n(x) - f(x)| < \infty.$$

Prove that the following statements are equivalent.

(i) The sequence (f_n) converges uniformly to f on X.
(ii) $\lim_{n\to\infty} M_n = 0$.

□

Exercise 6.7. Suppose that $f_n : [0, 1] \to \mathbb{R}$, $n \in \mathbb{N}$, is a sequence of continuous functions that converges uniformly to the function $f : [0, 1] \to \mathbb{R}$.
(a) Show that there exists $M > 0$ such that

$$|f_n(x)| \leq M, \;\; \forall x \in [0, 1], \;\; ; n \in \mathbb{N}.$$

(b) Show that the sequence f_n^2 converges uniformly to f^2. □

Exercise 6.8. (a) Prove Corollary 6.22.
(b) Let $f(x)$ be a polynomial of *odd* degree. Prove that there exists $r \in \mathbb{R}$ such that $f(r) = 0$. □

Exercise 6.9. Suppose that $f : [0, 1] \to [0, 1]$ is a continuous function. Prove that there exists $c \in [0, 1]$ such that $f(c) = c$. Can you give a geometric interpretation of this result? □

Exercise 6.10. (a) Find the oscillation of the function $f : [0, \infty) \to \mathbb{R}$, $f(x) = x^2$, over an interval $[a, b] \subset (0, \infty)$.
(b) Prove that for any $n \in \mathbb{N}$ one can find an interval $[a, b] \subset [0, \infty)$ of length $\leq \frac{1}{n}$ over which the oscillation of f is ≥ 1. □

Exercise 6.11. (a) Suppose that $f : X \to \mathbb{R}$ is a function defined on a set X, and $Y \subset X$. Prove that

$$\operatorname{osc}(f, X) = \sup_{x', x'' \in X} |f(x') - f(x'')| \text{ and } \operatorname{osc}(f, Y) \leq \operatorname{osc}(f, X).$$

(b) Consider a function $f : (a, b) \to \mathbb{R}$. Prove that f is continuous at a point $x_0 \in (a, b)$ if and only if

$$\lim_{\delta \searrow 0} \operatorname{osc}\big(\, f, [x_0 - \delta, x_0 + \delta]\,\big) = 0.$$

(c) Suppose that $f : [a, b] \to \mathbb{R}$ is a continuous function. Prove that

$$\operatorname{osc}(f,\ (a, b)\,) = \operatorname{osc}(f,\ [a, b]\,).$$

Exercise 6.12. Consider the function

$$f : (-\pi/2, \pi/2) \to \mathbb{R},\quad f(x) = \tan x = \frac{\sin x}{\cos x}.$$

Prove that f is strictly increasing and

$$\lim_{x \to \pm\pi/2} f(x) = \pm\infty.$$

Conclude that f is bijective. □

Exercise 6.13 (Weierstrass). Consider a sequence of functions $f_n : [a, b] \to \mathbb{R}$, $n \geq 0$, where a, b are real numbers $a < b$. Suppose that there exists a sequence of positive real numbers $(c_n)_{n \geq 0}$ with the following properties.

(i) $|f_n(x)| \leq c_n$, $\forall n \geq 0$, $\forall x \in [a, b]$.
(ii) The series $\sum_{n \geq 0} c_n$ is convergent.

(a) Prove that for any $x \in [a, b]$, the series of real numbers $\sum_{n \geq 0} f_n(x)$ is absolutely convergent. Denote by $s(x)$ its sum.
(b) Denote by $s_n(x)$ the n-th partial sum

$$s_n(x) = f_0(x) + f_1(x) + \cdots + f_n(x).$$

Prove that the sequence of functions $s_n : [a, b] \to \mathbb{R}$ converges uniformly on $[a, b]$ to the function $s : [a, b] \to \mathbb{R}$ defined in (a).

Hint. Use Exercise 6.6. □

Exercise 6.14. Consider the power series

$$\sum_{n\geq 0} a_n x^n, \quad a_n \in \mathbb{R}. \tag{6.13}$$

Suppose that for some $R > 0$ the series

$$\sum_{n\geq 0} a_n R^n$$

is absolutely convergent.

(a) Prove that the series (6.13) converges absolutely for any $x \in [-R, R]$. Denote by $s(x)$ its sum.

(b) Denote by $s_n(x)$ the n-th partial sum

$$s_n(x) = a_0 + a_1 x + \cdots + a_n x^n.$$

Prove that the resulting sequence of functions $s_n : [-R, R] \to \mathbb{R}$ converges uniformly to $s(x)$. Conclude that the function $s(x)$ is continuous on $[-R, R]$.

Hint. Use the results in Exercise 6.13. □

Exercise 6.15. Consider the sequence of functions

$$f_n : [0,1] \to \mathbb{R}, \quad f_n(x) = x^n, \quad n \in \mathbb{N}.$$

(a) Prove that for any $x \in [0, 1]$ the sequence $(f_n(x))_{n\in\mathbb{N}}$ is convergent. Compute its limit $f(x)$.

(b) Given $n \in \mathbb{N}$ compute

$$\sup_{x\in[0,1]} |f_n(x) - f(x)|.$$

(c) Prove that the sequence of functions $f_n(x)$ does *not* converge uniformly to the function $f(x)$ defined in (a). □

Exercise 6.16 (Cauchy). Suppose $X \subset \mathbb{R}$ is a nonempty set of real numbers and $f_n : X \to \mathbb{R}$ is a sequence of real valued functions defined on X. Prove that the following statements are equivalent.

(i) There exists a function $f : X \to \mathbb{R}$ such that the sequence $f_n : X \to \mathbb{R}$ converges uniformly on X to $f : X \to \mathbb{R}$.

(ii) $\forall \varepsilon > 0, \exists N = N(\varepsilon) \in \mathbb{N}$ such that

$$\forall n, m > N(\varepsilon), \quad \forall x \in X : \quad |f_n(x) - f_m(x)| < \varepsilon.$$

□

Exercise 6.17. Construct a sequence of functions $f_n : [0, 1] \to \mathbb{R}$, $n \in \mathbb{N}$, satisfying the following properties:

(i)

$$\sup_{x\in[0,1]} f_n(x) = 1, \quad \inf_{x\in[0,1]} f_n(x) = 0, \quad \forall x \in [0,1],$$

(ii)

$$\lim_{n\to\infty} f_n(x) = 0, \quad \forall x \in [0,1].$$

□

Exercise 6.18. Construct a sequence of continuous functions $f_n : (0,1] \to \mathbb{R}$ that converges uniformly to a function $f : (0,1] \to \mathbb{N}$, yet the sequence f_n^2 does *not* converge *uniformly* to f^2. □

6.5. More challenging problems

Problem 6.1. Suppose that $f : [a,b] \to \mathbb{R}$ is a continuous function. For any $x \in [a,b]$ we define

$$m(x) = \inf_{t\in[a,x]} f(x), \quad M(x) = \sup_{t\in[a,x]} f(x).$$

Prove that the functions $x \mapsto m(x)$ and $x \mapsto M(x)$ are continuous. □

Problem 6.2. Suppose that $f : \mathbb{R} \to \mathbb{R}$ is a continuous function satisfying

$$f(0) = 0, \quad f(1) = 1$$

and

$$f(x+y) = f(x) + f(y), \quad \forall x \in \mathbb{R}.$$

Prove that $f(x) = x$, $\forall x \in \mathbb{R}$. □

Problem 6.3. Suppose that $f : \mathbb{R} \to \mathbb{R}$ is a *continuous* function satisfying the following properties.

(i) $f(x) > 0$, $\forall x \in \mathbb{R}$.
(ii) $f(x+y) = f(x)f(y)$, $\forall x, y \in \mathbb{R}$.

Set $a := f(1)$. Prove that $f(x) = a^x$, $\forall x \in \mathbb{R}$. □

Problem 6.4. Suppose that $f : \mathbb{R} \to \mathbb{R}$ is a function satisfying the following conditions

$$f(x+y) = f(x) + f(y), \quad \forall x, y \in \mathbb{R}. \tag{6.14a}$$

$$f(xy) = f(x)f(y), \quad \forall x, y \in \mathbb{R}. \tag{6.14b}$$

$$f(1) \neq 0. \tag{6.14c}$$

Prove that the following hold.

(i) $f(0) = 0$, $f(1) = 1$.
(ii) $f(n) = n$, $\forall n \in \mathbb{N}$.
(iii) $f(m) = m$, $\forall m \in \mathbb{Z}$.
(iv) $f(q) = q$, $\forall q \in \mathbb{Q}$.
(v) If $x, y \in \mathbb{R}$ and $x < y$, then $f(x) < f(y)$.
(vi) $f(x) = x$, $\forall x \in \mathbb{R}$.

Problem 6.5 (Dini). Suppose that $f_n : [0,1] \to \mathbb{R}$, $n \in \mathbb{N}$ is a sequence of continuous functions with the following properties.

(i) For any $t \in [0,1]$ we have

$$f_1(t) \leq f_2(t) \leq f_3(t) \leq \cdots.$$

(ii) There exists a continuous function $f : [0,1] \to \mathbb{R}$ such that

$$\lim_{n\to\infty} f_n(t) = f(t).$$

Prove that the sequence of functions (f_n) converges *uniformly* to f on $[0,1]$. □

Problem 6.6. Suppose that $f : [0,1] \to \mathbb{R}$ is a continuous function. For $n \in \mathbb{R}$ define

$$f_n : [0,1] \to \mathbb{R}$$

by setting

$$f_n(x) := \begin{cases} f(0), & \text{if } x = 0 \\ \min\{ f(x); \frac{k-1}{n} \leq x \leq \frac{k}{n} \}, & \text{if } \frac{k-1}{n} < x \leq \frac{k}{n}, \quad k = 1, \dots, n. \end{cases}$$

Prove that the sequence of functions (f_n) converges *uniformly* to the function f on $[0,1]$. □

Chapter 7

Differential Calculus

7.1. Linear approximation and derivative

The differential calculus is one of the most consequential scientific discoveries in the history of mankind. Surprisingly, this revolutionary theory is based on a very simple principle: often one can learn nontrivial things about complicated objects by approximating them with simpler ones.

In the case at hand, the complicated object is a function $f : (a, b) \to \mathbb{R}$ and one would like to understand its behavior near a point $x_0 \in (a, b)$. To achieve this, we try to approximate f with a simpler function, and the linear functions are the simplest nontrivial candidates.

Definition 7.1. Suppose that I is an interval[1] on the real axis, $f : I \to \mathbb{R}$ is a function and $x_0 \in I$. A *linear approximation* or *linearization* of f at x_0 is a linear function

$$L : \mathbb{R} \to \mathbb{R}, \;\; L(x) = b + m(x - x_0)$$

such that

$$L(x_0) = f(x_0) \tag{7.1}$$

and

$$f(x) - L(x) = o(x - x_0) \;\; \text{as } x \to x_0. \tag{7.2}$$

Above, we used Landau's symbol o defined in (5.35) signifying that

$$\lim_{x \in I,\, x \to x_0} \frac{f(x) - L(x)}{x - x_0} = 0.$$

The function is said to be *linearizable* at x_0 if it admits a linearization at x_0. □

Suppose that L is a linearization of the function $f : I \to \mathbb{R}$ at x_0. By (7.1), the value of L at x_0 is equal to the value of f at x_0, $L(x_0) = f(x_0)$. On the other hand

$$L(x_0) = b + m(x_0 - x_0) = b$$

[1] The interval I could be closed, could be open, could be neither, could be bounded or not.

and we deduce that $L(x)$ has the form

$$L(x) = f(x_0) + m(x - x_0).$$

The linear function $L(x)$ is meant to approximate the function $f(x)$ for x not too far for x_0. The *error* of this linear approximation of $f(x)$ is the difference $r(x) = f(x) - L(x)$ which by definition is $o(x - x_0)$ as $x \to x_0$. In less rigorous terms, $r(x)$ is a tiny fraction of $(x - x_0)$ when x is close to x_0. Note that

$$f(x) - L(x) = f(x) - \big(m(x - x_0) + f(x_0)\big) = f(x) - f(x_0) - m(x - x_0),$$

$$\frac{f(x) - L(x)}{x - x_0} = \frac{f(x) - f(x_0)}{x - x_0} - m.$$

Since

$$0 = \lim_{x \to x_0} \frac{f(x) - L(x)}{x - x_0} = \lim_{x \to x_0} \frac{f(x) - f(x_0)}{x - x_0} - m$$

we deduce that

$$m = \lim_{x \to x_0} \frac{f(x) - f(x_0)}{x - x_0}. \tag{7.3}$$

Thus if f is linearizable at x_0, then there exists a *unique* linearization $L(x)$ described by

$$L(x) = f(x_0) + m(x - x_0),$$

where the slope m is given by (7.3).

Definition 7.2. Suppose that I is an interval of the real axis, $f : I \to \mathbb{R}$ is a function and $x_0 \in I$.

(i) We say that f is *differentiable at* x_0 if the limit (7.3)

$$\lim_{\substack{x \to x_0 \\ x \in I}} \frac{f(x) - f(x_0)}{x - x_0} \tag{7.4}$$

exists and it is finite. If this is the case, we denote the limit by $f'(x_0)$ or $\frac{df}{dx}|_{x=x_0}$ and we will refer to it as the *derivative* of f at x_0.

(ii) We say that f is differentiable on I if it is differentiable at *any* point $x \in I$. The function $f' : I \to \mathbb{R}$ that assigns to $x \in I$ the derivative $f'(x)$ of f at x is called the *derivative of the function f on the interval I.* □

Remark 7.3. In concrete computations it is often convenient to describe the derivative of f at x_0 as the limit

$$f'(x_0) = \lim_{h \to 0} \frac{f(x_0 + h) - f(x_0)}{h}.$$

This is obtained from (7.4) if we denote by h the "displacement" $x - x_0$. With this notation we have $x = x_0 + h$ and

$$\frac{f(x) - f(x_0)}{x - x_0} = \frac{f(x_0 + h) - f(x_0)}{h}.$$ □

The next result summarizes the observations we have made so far.

Proposition 7.4. *Suppose that I is an interval of the real axis, $f : I \to \mathbb{R}$ is a function and $x_0 \in I$. Then the following statements are equivalent.*

(i) The function f is differentiable at x_0.
(ii) The function f is linearizable at x_0.
(iii) The function f is differentiable at x_0 and the function $L(x) = f(x_0)+f'(x_0)(x-x_0)$ is the linearization of f at x_0, i.e.,

$$f(x) = f(x_0) + f'(x_0)(x - x_0) + r(x), \quad \lim_{x\to x_0} \frac{r(x)}{|x - x_0|} = 0. \tag{7.5}$$

□

We should perhaps give a geometric interpretation to the linear approximation of f at x_0. The graph of f is the curve

$$G_f := \Big\{ \big(x, f(x)\big) \in \mathbb{R}^2;\ \ x \in (a, b) \Big\}.$$

The point $x_0 \in I$ determines a point $P_0 = (x_0, f(x_0)\,)$ on the curve G_f; see Figure 7.1.

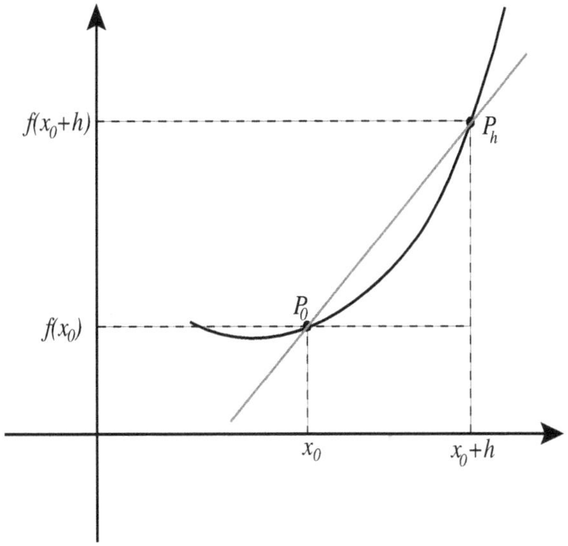

Fig. 7.1 *A tangent line to the graph of a function is a limit of secant lines.*

The graph of a linear function $L(x)$ is a line in the plane and since we are interested in approximating the behavior of f near x_0 it makes sense to look only at lines $\ell_{P_0,P}$ determined by two points P_0, P on the graph G_f. Since we are interested only in the behavior of f *near* x_0, we may assume that the point P is not too far from P_0. Thus we assume that the coordinates of P are $(\,x_0 + h, f(x_0 + h)\,)$, where h is very small.

In more concrete terms, we look at the lines ℓ_{P_0,P_h} determined by the two points

$$P_0 := (\,x_0, f(x_0)\,),\quad P_h := (\,x_0+h, f(x_0+h)\,),$$

where h very small. The slope of the line ℓ_{P_0,P_h} is

$$m(h) := \frac{f(x_0+h)-f(x_0)}{(x_0+h)-x_0} = \frac{f(x_0+h)-f(x_0)}{h},$$

so its equation is

$$y - f(x_0) = m(h)(x-x_0).$$

This is the graph of the linear function

$$L_{x_0,h}(x) = f(x_0) + m(h)(x-x_0).$$

Suppose that as $h \to 0$ the line ℓ_{P_0,P_h} stabilizes to some limiting position. This limit line goes through the point P_0 and therefore its position is determined by its slope

$$\lim_{h\to 0} m(h) = \lim_{h\to 0}\frac{f(x_0+h)-f(x_0)}{h} = \lim_{x\to x_0}\frac{f(x)-f(x_0)}{x-x_0}.$$

We see that this limit exists and it is finite if and only if f is differentiable at x_0. In this case, the limit line is the graph of the linear approximation of f at x_0.

Definition 7.5. Suppose $I \subset \mathbb{R}$ is an interval of the real axis and $f : I \to \mathbb{R}$ is a function differentiable at x_0. The *tangent line to the graph* of f at x_0 is the graph of the linearization of f at x_0. □

Remark 7.6. (a) The quantities

$$\frac{f(x)-f(x_0)}{x-x_0},\quad \frac{f(x_0+h)-f(x_0)}{h}$$

are called *difference quotients* of f at x_0. You should think of such a difference quotient as measuring the average rate of change of the quantity f over the interval $[x_0, x]$.

In physics, the numerator $f(x) - f(x_0)$ is denoted by Δf while the denominator is denoted Δx. The symbol Δ is shorthand for "*variation of*". Thus

$$\frac{df}{dx} = \lim_{\Delta x\to 0}\frac{\Delta f}{\Delta x}.$$

From the equality

$$f'(x) = \frac{df}{dx}$$

we deduce formally

$$df = f'(x)dx. \tag{7.6}$$

The expression $f'(x)dx$ is called the *differential* of f and as the above equality suggests, it is denoted by df.

(b) Often a function $f : [a, b] \to \mathbb{R}$ has a physical meaning. For example, the interval $[a, b]$ can signify a stretch of highway between mile a and mile b and $f(x)$ could be the temperature at mile x and thus it is measured in $^\circ F$. The difference quotient

$$\frac{f(x) - f(x_0)}{x - x_0}$$

has a different meaning. The numerator $f(x) - f(x_0)$ describes the change in temperature from mile x_0 to mile x and it is again measured in $^\circ F$, while the numerator $x - x_0$ is the "distance" (could be negative) from mile x_0 to mile x and thus it is measured in miles. We deduce that the quotient is measured in different units, degrees-per-mile, and should be viewed as the average rate of change in temperature per mile. When $x \to x_0$ we are measuring the rate of change in temperature over shorter and shorter stretches of highway. For this reason, the limit $f'(x_0)$ is sometimes referred to as the *infinitesimal rate of change*. □

The differentiability of a function at a point x_0 imposes restrictions on the behavior of the function near that point. Our next elementary result describes one such restriction. Its proof is left to you as an exercise.

Proposition 7.7. *Suppose I is an interval of the real axis $\mathbb{R}$ and $f : I \to \mathbb{R}$ is a function that is differentiable at a point $x_0 \in I$. Then f is continuous at x_0, i.e.,*

$$\lim_{I \ni x \to x_0} f(x) = f(x_0).$$ □

Remark 7.8. The converse of the above result is not true. There exist continuous functions $f : [0, 1] \to \mathbb{R}$ which are nowhere differentiable. For example, the function

$$f : [0,1] \to \mathbb{R}, \;\; f(t) = \sum_{n=0}^{\infty} \frac{\cos(5^n t)}{2^n},$$

is continuous and nowhere differentiable. Its graph, depicted in Figure 7.2, may convince you of the validity of this claim. The rigorous proof of this fact is rather ingenious and for details and generalizations we refer to [12]. □

Suppose that $I \subset \mathbb{R}$ is an interval and $f : I \to \mathbb{R}$ is a differentiable function. We say that f is *twice* differentiable if its derivative f', viewed as a function $f' : I \to \mathbb{R}$, is also differentiable. The *second derivative* of f denoted by f'' or $\frac{d^2 f}{dx^2}$ is the derivative of f'

$$f'' := \frac{d}{dx}(f').$$

Recursively, for any natural number $n > 1$, we say that f is *n-times differentiable* if its derivative is $(n-1)$-times differentiable. The n-th derivative of f is the function $f^{(n)} : I \to \mathbb{R}$ defined recursively as

$$f^{(n)} := \frac{d}{dx}\big(f^{(n-1)}\big).$$

Often we will use the alternate notation $\frac{d^n f}{dx^n}$ to denote the n-th derivative of f.

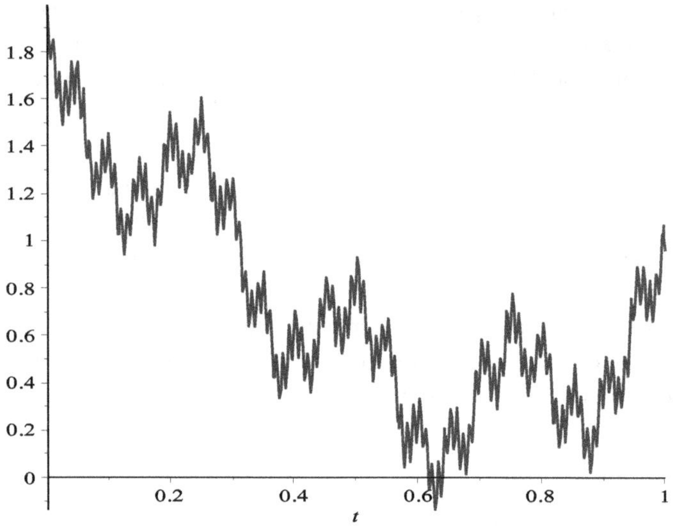

Fig. 7.2 *Weierstrass's example of continuous, nowhere differentiable function.*

Definition 7.9. Let $I \subset \mathbb{R}$ be an interval.

(i) We denote by $C^0(I)$ the set consisting of all the continuous functions $f : I \to \mathbb{R}$.

(ii) If n is a natural number, then we denote by $C^n(I)$ the space of functions $f : I \to \mathbb{R}$ which are

- n-times differentiable and
- the n-th derivative $f^{(n)}$ is a continuous function.

We will refer to the functions in $C^n(I)$ as C^n-*functions.*

(iii) We denote by $C^\infty(I)$ the space of functions $I \to \mathbb{R}$ which are infinitely many times differentiable. We will refer to such functions as *smooth.*

□

7.2. Fundamental examples

In this section we describe a *very important* collection of differentiable functions.

Example 7.10 (Constant functions). Suppose that $f : \mathbb{R} \to \mathbb{R}$ is the function which is identically equal to a fixed real number c,

$$f(x) = c, \quad \forall x \in \mathbb{R}.$$

Then f is differentiable and $f'(x) = 0$, $\forall x \in \mathbb{R}$. Indeed, for any $x_0 \in \mathbb{R}$

$$\frac{f(x_0 + h) - f(x_0)}{h} = 0, \quad \forall h \neq 0.$$

□

Example 7.11 (Monomials). Suppose that $n \in \mathbb{N}$ and consider the *monomial function* $\mu_n : \mathbb{R} \to \mathbb{R}$, $\mu_n(x) = x^n$. Then μ_n is differentiable on $\mathbb{R}$ and its derivative is

$$\mu_n'(x) = nx^{n-1}, \;\; \forall x \in \mathbb{R} \Longleftrightarrow \frac{d}{dx}(x^n) = nx^{n-1}. \tag{7.7}$$

To prove this claim we investigate the difference quotients of μ_n at $x_0 \in \mathbb{R}$. We have

$$\mu_n(x_0 + h) - \mu_n(x_0) = (x_0 + h)^n - x_0^n$$

(use Newton's binomial formula (3.6))

$$= x_0^n + \binom{n}{1} x_0^{n-1} h + \binom{n}{2} x_0^{n-2} h^2 + \cdots + \binom{n}{n} h^n - x_0^n$$
$$= h\left(\binom{n}{1} x_0^{n-1} + \binom{n}{2} x_0^{n-2} h + \cdots + \binom{n}{n} h^{n-1} \right),$$

so that

$$\frac{\mu_n(x_0 + h) - \mu_n(x_0)}{h} = \binom{n}{1} x_0^{n-1} + \binom{n}{2} x_0^{n-2} h + \cdots + \binom{n}{n} h^{n-1}.$$

Now observe that

$$\mu_n'(x_0) = \lim_{h \to 0} \frac{\mu_n(x_0 + h) - \mu_n(x_0)}{h}$$
$$= \lim_{h \to 0} \left(\binom{n}{1} x_0^{n-1} + \binom{n}{2} x_0^{n-2} h + \cdots + \binom{n}{n} h^{n-1} \right) = \binom{n}{1} x_0^{n-1} = n x_0^{n-1}.$$

For example

$$(x^2)' = 2x, \;\; d(x^2) = 2x dx.$$

□

Example 7.12 (Power functions). Fix a real number α and consider the power function

$$f : (0, \infty) \to \mathbb{R}, \;\; f(x) = x^\alpha.$$

Then f is differentiable and its derivative is

$$\boxed{f'(x) = \alpha x^{\alpha - 1}, \;\; \forall x > 0 \Longleftrightarrow \frac{d}{dx}(x^\alpha) = \alpha x^{\alpha - 1}}. \tag{7.8}$$

To prove this claim we investigate the difference quotients of $f(x)$ at $x_0 \in (0, \infty)$. We have

$$(x_0 + h)^\alpha - x_0^\alpha = \left(x_0 \left(1 + \frac{h}{x_0} \right) \right)^\alpha - x_0^\alpha = x_0^\alpha \left(\left(1 + \frac{h}{x_0} \right)^\alpha - 1 \right),$$

$$\frac{f(x_0 + h) - f(x_0)}{h} = x_0^\alpha \frac{\left(1 + \frac{h}{x_0} \right)^\alpha - 1}{h} = x_0^\alpha \frac{\left(1 + \frac{h}{x_0} \right)^\alpha - 1}{x_0 \frac{h}{x_0}}$$

$$= x_0^{\alpha - 1} \frac{\left(1 + \frac{h}{x_0} \right)^\alpha - 1}{\frac{h}{x_0}}.$$

We set $t := \frac{h}{x_0}$ and we observe that $t \to 0$ as $h \to 0$ and

$$\frac{f(x_0+h)-f(x_0)}{h} = x_0^{\alpha-1}\frac{(1+t)^\alpha - 1}{t}.$$

Invoking the fundamental limit (5.21) we deduce

$$\lim_{t\to 0}\frac{(1+t)^\alpha - 1}{t} = \alpha$$

so that

$$f'(x_0) = \lim_{h\to 0}\frac{f(x_0+h)-f(x_0)}{h} = \alpha x_0^{\alpha-1}.$$

Note that if $\alpha = \frac{1}{2}$, then $f(x) = \sqrt{x}$ and we deduce

$$\frac{d}{dx}(\sqrt{x}) = \frac{1}{2\sqrt{x}},\quad d(\sqrt{x}) = \frac{dx}{2\sqrt{x}}. \tag{7.9}$$

□

Example 7.13 (The exponential function). Consider the exponential function

$$f : \mathbb{R} \to \mathbb{R},\quad f(x) = e^x.$$

This function is differentiable and its derivative is

$$f'(x) = e^x,\quad \forall x \in \mathbb{R} \iff \frac{d}{dx}(e^x) = e^x. \tag{7.10}$$

To prove this claim we investigate the difference quotients of f at $x_0 \in \mathbb{R}$. We have

$$f(x_0+h) - f(x_0) = e^{x_0+h} - e^{x_0} = e^{x_0}(e^h - 1),$$

$$\frac{f(x_0+h)-f(x_0)}{h} = e^{x_0}\frac{e^h-1}{h}.$$

On the other hand, the fundamental limit (5.20) implies that

$$\lim_{h\to 0}\frac{e^h-1}{h} = 1.$$

Hence

$$f'(x_0) = \lim_{h\to 0}\frac{f(x_0+h)-f(x_0)}{h} = e^{x_0}.$$

These computations show that the exponential function is smooth, i.e., infinitely many times differentiable and

$$\boxed{\frac{d^n}{dx^n}e^x = e^x,\quad d(e^x) = e^x\,dx}. \tag{7.11}$$

□

Example 7.14 (The natural logarithm). Consider the natural logarithm

$$f:(0,\infty)\to\mathbb{R},\ \ f(x)=\ln x=\log x.$$

Then f is differentiable and its derivative is

$$\boxed{f'(x)=\frac{1}{x},\ \ \forall x>0\iff\frac{d}{dx}(\ln x)=\frac{1}{x},\ \ d(\ln x)=\frac{dx}{x}}. \tag{7.12}$$

To prove this claim we investigate the difference quotients of f at $x_0>0$. We have

$$\begin{aligned}f(x_0+h)-f(x_0)&=\ln(x_0+h)-\ln x_0=\ln\left(x_0\Big(1+\frac{h}{x_0}\Big)\right)-\ln x_0\\&=\ln x_0+\ln\Big(1+\frac{h}{x_0}\Big)-\ln x_0=\ln\Big(1+\frac{h}{x_0}\Big).\end{aligned}$$

$$\frac{f(x_0+h)-f(x_0)}{h}=\frac{\ln\Big(1+\frac{h}{x_0}\Big)}{h}=\frac{\ln\Big(1+\frac{h}{x_0}\Big)}{x_0\frac{h}{x_0}}=\frac{1}{x_0}\frac{\ln\Big(1+\frac{h}{x_0}\Big)}{\frac{h}{x_0}}.$$

We set $t=\frac{h}{x_0}$ and we conclude from above that

$$\frac{f(x_0+h)-f(x_0)}{h}=\frac{1}{x_0}\frac{\ln(1+t)}{t}.$$

Note that t goes to zero when $h\to 0$. We can now invoke (5.19) to conclude that

$$\lim_{t\to 0}\frac{\ln(1+t)}{t}=1.$$

This proves

$$\lim_{h\to 0}\frac{f(x_0+h)-f(x_0)}{h}=\frac{1}{x_0}.$$

□

Example 7.15 (Trigonometric functions). The trigonometric functions

$$\sin,\cos:\mathbb{R}\to\mathbb{R}$$

are differentiable and

$$\boxed{\frac{d}{dx}(\sin x)=\cos x,\ \ \frac{d}{dx}(\cos x)=-\sin x}. \tag{7.13}$$

Fix $x_0\in\mathbb{R}$. We have

$$\begin{aligned}&\sin(x_0+h)-\sin x_0\overset{(5.33a)}{=}\sin x_0\cos h+\cos x_0\sin h-\sin x_0\\&=\sin x_0\big(\cos h-1\big)+\cos x_0\sin h=-2\sin^2(h/2)\sin x_0+\cos x_0\sin h.\end{aligned}$$

Hence

$$\frac{\sin(x_0+h)-\sin x_0}{h}=-2\sin x_0\frac{\sin^2(h/2)}{h}+\cos x_0\frac{\sin h}{h}.$$

$$=-\sin x_0\frac{\sin^2(h/2)}{\frac{h}{2}}+\cos x_0\frac{\sin h}{h}=-\frac{h}{2}\sin x_0\left(\frac{\sin(\frac{h}{2})}{\frac{h}{2}}\right)^2+\cos x_0\frac{\sin h}{h}.$$

From the fundamental identity (5.27) we deduce that

$$\lim_{t\to 0}\frac{\sin t}{t}=1.$$

Hence

$$\lim_{h\to 0}\frac{h}{2}\sin x_0\left(\frac{\sin(\frac{h}{2})}{\frac{h}{2}}\right)^2=0,\;\;\lim_{h\to 0}\cos x_0\frac{\sin h}{h}=\cos x_0,$$

and thus

$$\lim_{h\to 0}\frac{\sin(x_0+h)-\sin x_0}{h}=\cos x_0.$$

The equality

$$\lim_{h\to 0}\frac{\cos(x_0+h)-\cos x_0}{h}=-\sin x_0$$

is proved in a similar fashion and the details are left to you as an exercise. □

7.3. The basic rules of differential calculus

In the previous section we have computed the derivatives of a few important functions. In this section we describe a few basic rules which will allow us to easily compute the derivatives of almost any function.

Theorem 7.16 (Arithmetic rules of differentiation). *Suppose that $I\subset\mathbb{R}$ is an interval, and $f,g:I\to\mathbb{R}$ are two functions differentiable at x_0. Then the following hold.*

Addition. *The sum $f+g$ is differentiable at x_0 and*

$$\boxed{(f+g)'(x_0)=f'(x_0)+g'(x_0).}$$

Scalar multiplication. *If c is a real number, then the function cf is differentiable at x_0 and*

$$\boxed{(cf)'(x_0)=cf'(x_0).}$$

Product. *The product $f\cdot g$ is differentiable at x_0 and its derivative is given by the* product rule *or* Leibniz rule

$$\boxed{(f\cdot g)'(x_0)=f'(x_0)g(x_0)+f(x_0)g'(x_0).}$$

Quotient. *If $g(x_0)\neq 0$, then there exists $\delta>0$ such that*

$$\forall x\in I\;\;|x-x_0|<\delta\Rightarrow g(x)\neq 0.$$

Set

$$I_{x_0,\delta}:=\big\{x\in I;\;\;|x-x_0|<\delta\big\}.$$

The quotient $\frac{f}{g}$ is a well defined function on $I_{x_0,\delta}$ which is differentiable at x_0 and its derivative at x_0 is determined by the quotient rule

$$\boxed{\left(\frac{f}{g}\right)'(x_0)=\frac{f'(x_0)g(x_0)-f(x_0)g'(x_0)}{g(x_0)^2}.}$$

Proof. Addition. We have

$$\begin{aligned}\frac{(f+g)(x_0+h)-(f+g)(x_0)}{h} &= \frac{f(x_0+h)-f(x_0)+g(x_0+h)-g(x_0)}{h}\\ &= \frac{f(x_0+h)-f(x_0)}{h}+\frac{g(x_0+h)-g(x_0)}{h}.\end{aligned}$$

Hence

$$\begin{aligned}&\lim_{h\to 0}\frac{(f+g)(x_0+h)-(f+g)(x_0)}{h}\\ &= \lim_{h\to 0}\frac{f(x_0+h)-f(x_0)}{h}+\lim_{h\to 0}\frac{g(x_0+h)-g(x_0)}{h}\\ &= f'(x_0)+g'(x_0).\end{aligned}$$

Scalar multiplication. We have

$$\frac{(cf)(x_0+h)-(cf)(x_0)}{h} = c\frac{f(x_0+h)-f(x_0)}{h}$$

so that

$$\lim_{h\to 0}\frac{(cf)(x_0+h)-(cf)(x_0)}{h} = c\lim_{h\to 0}\frac{f(x_0+h)-f(x_0)}{h} = cf'(x_0).$$

Product. We have

$$\begin{aligned}&(f\cdot g)(x_0+h)-(f\cdot g)(x_0) = f(x_0+h)g(x_0+h)-f(x_0)g(x_0)\\ &= f(x_0+h)g(x_0+h)-f(x_0)g(x_0+h)+f(x_0)g(x_0+h)-f(x_0)g(x_0)\\ &= \big(f(x_0+h)-f(x_0)\big)g(x_0+h)+f(x_0)\big(g(x_0+h)-g(x_0)\big),\end{aligned}$$

so that

$$\begin{aligned}&\frac{(f\cdot g)(x_0+h)-(f\cdot g)(x_0)}{h}\\ &= \frac{\big(f(x_0+h)-f(x_0)\big)}{h}g(x_0+h)+f(x_0)\frac{\big(g(x_0+h)-g(x_0)\big)}{h}.\end{aligned}$$

Since g is differentiable at x_0 it is also continuous at x_0 by Proposition 7.7. Hence

$$\lim_{h\to 0} g(x_0+h) = g(x_0),\quad \lim_{h\to 0} f(x_0)\frac{\big(g(x_0+h)-g(x_0)\big)}{h} = f(x_0)g'(x_0).$$

Since f is differentiable at x_0 we deduce

$$\begin{aligned}&\lim_{h\to 0}\frac{\big(f(x_0+h)-f(x_0)\big)}{h}g(x_0+h)\\ &= \lim_{h\to 0}\frac{\big(f(x_0+h)-f(x_0)\big)}{h}\cdot\lim_{h\to 0} g(x_0+h) = f'(x_0)g(x_0).\end{aligned}$$

Hence

$$\lim_{h\to 0}\frac{(f\cdot g)(x_0+h)-(f\cdot g)(x_0)}{h} = f'(x_0)g(x_0)+f(x_0)g'(x_0).$$

Quotient. The function g is differentiable at x_0, thus continuous at this point. From Theorem 6.11 we deduce that there exists $\delta > 0$ such that

$$\forall x\in I,\ \ |x-x_0|<\delta \Rightarrow g(x)\neq 0.$$

For $|h| < \delta$ such that $x_0 + h \in I$ we have

$$\left(\frac{1}{g}\right)(x_0 + h) - \left(\frac{1}{g}\right)(x_0) = \frac{1}{g(x_0 + h)} - \frac{1}{g(x_0)} = \frac{g(x_0) - g(x_0 + h)}{g(x_0)g(x_0 + h)}$$

so that

$$\frac{\left(\frac{1}{g}\right)(x_0 + h) - \left(\frac{1}{g}\right)(x_0)}{h} = \frac{g(x_0) - g(x_0 + h)}{h} \frac{1}{g(x_0)g(x_0 + h)}.$$

Hence

$$\begin{aligned}\left(\frac{1}{g}\right)'(x_0) &= \lim_{h\to 0} \frac{\left(\frac{1}{g}\right)(x_0 + h) - \left(\frac{1}{g}\right)(x_0)}{h} \\ &= \lim_{h\to 0} \frac{g(x_0) - g(x_0 + h)}{h} \cdot \lim_{h\to 0} \frac{1}{g(x_0)g(x_0 + h)} = -\frac{g'(x_0)}{g(x_0)^2}.\end{aligned}$$

To compute the derivative of $\frac{f}{g}$ at x_0 we use the product rule. We have

$$\begin{aligned}\frac{f}{g} = f \cdot \frac{1}{g} \Rightarrow \left(\frac{f}{g}\right)'(x_0) &= \left(f \cdot \frac{1}{g}\right)'(x_0) \\ &= f'(x_0)\frac{1}{g(x_0)} + f(x_0)\left(\frac{1}{g}\right)'(x_0) \\ &= f'(x_0)\frac{1}{g(x_0)} - f(x_0)\frac{g'(x_0)}{g(x_0)^2} = \frac{f'(x_0)g(x_0) - f(x_0)g'(x_0)}{g(x_0)^2}.\end{aligned}$$

□

Example 7.17. Let us see how the above rules work on some simple examples.

(a) Consider the polynomial function

$$p(x) = 5 - 3x^2 + 7x^5, \quad x \in \mathbb{R}.$$

From the scalar multiplication rule and the Examples 7.10, 7.11 we deduce that each of the functions 5, $-3x^2$ and $7x^5$ is differentiable and the addition rule implies that their sum is differentiable as well. We deduce

$$p'(x) = (5)' + (-3x^2)' + (7x^5)' = -6x + 35x^4.$$

(b) From the equalities

$$\frac{d}{dx}(\sin x) = \cos x, \quad \frac{d}{dx}(\cos x) = -\sin x$$

and the scalar multiplication rule we deduce that the trigonometric functions are smooth and we have

$$\frac{d^2}{dx^2}\sin x = -\sin x, \quad \frac{d^2}{dx^2}\cos x = -\cos x,$$

$$\frac{d^4}{dx^4}\sin x = \sin x, \quad \frac{d^4}{dx^4}\cos x = \cos x.$$

(c) If a is a positive real number, then

$$\log_a x = \frac{\ln x}{\ln a}$$

and we deduce

$$(\log_a x)' = \frac{1}{x \ln a}. \tag{7.14}$$

(d) If n is a natural number, then the function

$$f : \mathbb{R} \setminus \{0\} \to \mathbb{R}, \;\; f(x) = \frac{1}{x^n} = x^{-n}$$

is differentiable by the quotient rule and we have

$$(x^{-n})' = \left(\frac{1}{x^n}\right)' = -\frac{nx^{n-1}}{x^{2n}} = -\frac{n}{x^{n+1}} = -nx^{-n-1}.$$

(e) From the quotient rule we deduce

$$\frac{d}{dx}\tan x = \frac{d}{dx}\left(\frac{\sin x}{\cos x}\right) = \frac{\cos^2 x + \sin^2 x}{\cos x^2} = \frac{1}{\cos^2 x} = 1 + \tan^2 x.$$

Thus

$$(\tan x)' = 1 + \tan^2 x = \frac{1}{\cos^2 x}. \tag{7.15}$$

(f) Using the product rule we deduce

$$\frac{d}{dx}(e^x \sin x) = e^x \sin x + e^x \cos x.$$

The above simple rules are unfortunately not powerful enough to allow us to compute the derivative of simple functions such as $e^{\sqrt{x}}$, $x > 0$ or $\sqrt{2 + \sin x}$. For this we need a more powerful technology. □

Theorem 7.18 (Chain Rule). *Let I, J be two nontrivial intervals of the real axis. Suppose that we are given two functions $u : I \to \mathbb{R}$ and $f : J \to \mathbb{R}$ and a point $x_0 \in I$ with the following properties.*

(i) The range of the function u is contained in the interval J, i.e., $u(I) \subset J$.
(ii) The function u is differentiable at x_0.
(iii) The function f is differentiable at $u_0 := u(x_0)$.

Then the composition

$$f \circ u : I \to \mathbb{R}, \;\; f \circ u(x) = f\big(u(x)\big)$$

is differentiable at x_0 and

$$\boxed{(f \circ u)'(x_0) = f'(u_0)u'(x_0).}$$

Proof. Let us begin by giving a flawed proof. We have

$$\frac{f(u(x)) - f(u(x_0))}{x - x_0} = \frac{f(u(x)) - f(u(x_0))}{u(x) - u(x_0)} \cdot \frac{u(x) - u(x_0)}{x - x_0}.$$

Since u is differentiable at x_0 we have

$$\lim_{x \to x_0} u(x) = u(x_0).$$

Thus

$$\begin{aligned} &\lim_{x \to x_0} \frac{f(u(x)) - f(u(x_0))}{u(x) - u(x_0)} \cdot \frac{u(x) - u(x_0)}{x - x_0} \\ &= \left(\lim_{u(x) \to u(x_0)} \frac{f(u(x)) - f(u(x_0))}{u(x) - u(x_0)} \right) \cdot \left(\lim_{x \to x_0} \frac{u(x) - u(x_0)}{x - x_0} \right) \\ &= f'(u(x_0))u'(x_0) \end{aligned}$$

et voilà, we are done!

Unfortunately the above argument has one serious flaw. More precisely it is possible that $u(x) = u(x_0)$ for infinitely many values of x close to x_0. The quotient

$$\frac{f(u(x)) - f(u(x_0)}{u(x) - u(x_0)}$$

is ill-defined and thus the above argument is meaningless. Although problematic, the above argument displays the strategy of the proof. We need a bit of technical contortionism to avoid the problem of vanishing denominators. The details follow below.

Since f is differentiable at u_0 we deduce that it is linearly approximable at x_0. From (7.5) we deduce that

$$f(u) = f(u_0) + f'(u_0)(u - u_0) + r(u), \quad r(u) = o(u - u_0) \text{ as } u \to u_0.$$

Recall that the equality

$$r(u) = o(u - u_0) \text{ as } u \to u_0$$

signifies that

$$\lim_{u \to u_0} \frac{r(u)}{u - u_0} = 0. \tag{7.16}$$

In particular, we deduce that

$$f(u(x)) - f(u(x_0)) = f(u(x)) - f(u_0) = f'(u_0)(u(x) - u(x_0)) + r(u(x))$$

and

$$\frac{f(u(x)) - f(u(x_0))}{x - x_0} = f'(u_0)\frac{(u(x) - u(x_0))}{x - x_0} + \frac{r(u(x))}{x - x_0}.$$

Observe that if we prove that

$$\lim_{x \to x_0} \frac{r(u(x))}{x - x_0} = 0, \tag{7.17}$$

then we deduce

$$\lim_{x\to x_0} \frac{f\big(u(x)\big) - f\big(u(x_0)\big)}{x - x_0} = f'(u_0) \lim_{x\to x_0} \frac{\big(u(x) - u(x_0)\big)}{x - x_0} = f'(u_0)u'(x_0)$$

which is the claim of the theorem.

Why do we expect (7.17) to be true? We have $r(u(x)) = o(u(x) - u_0)$, i.e., $r(u(x))$ is a tiny fraction of $u(x) - u_0$ if $u(x)$ is close to x. When x is close to x_0, then $u(x)$ is close to u_0 so $r(u(x))$ is a tiny fraction of $u(x) - u_0$ when x is close to x_0.

On the other hand, when x is close to x_0 we have

$$\begin{aligned} u(x) - u_0 &= u'(x_0)(x - x_0) + o(x - x_0) = u'(x_0)(x - x_0) + \text{tiny fraction of } x - x_0 \\ &= (x - x_0)(u'(x_0) + \text{tiny number}). \end{aligned}$$

Thus when x is close to x_0 the remainder $r(u(x))$ is a tiny fraction of $(x - x_0)(u'(x_0) +$ tiny number$)$ which in turn is obviously a tiny fraction of $(x - x_0)$. The precise proof is presented below.

To prove (7.17) it suffices to show that

$$\forall \hbar > 0 \;\; \exists d = d(\hbar) > 0 : \;\; |x - x_0| < d(\hbar) \Rightarrow \frac{|r(u(x))|}{|x - x_0|} \leq \hbar. \tag{7.18}$$

The function u is differentiable at x_0 and it is linearizable at this point. Hence

$$u(x) - u_0 = u'(x_0)(x - x_0) + \rho(x), \;\; \rho(x) = o(x - x_0) \;\text{ as } x \to x_0.$$

Since $\rho(x) = o(x - x_0)$ as $x \to x_0$ we deduce that there exists a small $\gamma > 0$ such that

$$|x - x_0| < \gamma \Rightarrow |\rho(x)| \leq |x - x_0|.$$

Hence, for $|x - x_0| < \gamma$ we have

$$|u(x) - u_0| = |u'(x_0)(x - x_0) + \rho(x)|$$

$$\leq |u'(x_0)||x - x_0| + |\rho(x)| \leq (|u'(x_0)| + 1)|x - x_0|.$$

If we set $C := |u'(x_0)| + 1 > 0$, then we deduce

$$|x - x_0| < \gamma \Rightarrow |u(x) - u_0| \leq C|x - x_0|. \tag{7.19}$$

Note that (7.16) implies that

$$\forall \hbar > 0 \;\; \exists \varepsilon(\hbar) > 0 : \;\; |u - u_0| < \varepsilon(\hbar) \Rightarrow |r(u)| \leq \hbar|u - u_0|. \tag{7.20}$$

Observe that (7.19) implies

$$|x - x_0| < \delta(\hbar) := \min\Big\{\gamma, \frac{\varepsilon(\hbar)}{c}\Big\} \Rightarrow |u(x) - u_0| \leq C|x - x_0| < \varepsilon(\hbar).$$

Using this in (7.20) we deduce that

$$|x - x_0| < \delta(\hbar) \Rightarrow |u(x) - u_0| < \varepsilon(\hbar) \overset{(7.20)}{\Rightarrow} |r(\,u(x)\,)| \leq \hbar|u(x) - u_0| \overset{(7.19)}{\leq} C\hbar|x - x_0|.$$

We have thus proved that

$$\forall \hbar > 0 \;\; \exists \delta(\hbar) > 0 : \;\; |x - x_0| < \delta(\hbar) \Rightarrow \frac{|r(\,u(x)\,)|}{|x - x_0|} \leq C\hbar.$$

If we set

$$d(\hbar) := \delta(\hbar/C)$$

we obtain (7.18). □

Remark 7.19. Since the chain rule is without a doubt the key rule in differential calculus, it is perhaps appropriate to pause and provide a bit of intuition behind it. The classical point of view on this formula is in our view the most intuitive.

Before the modern concept of function (late 19th century) functions were regarded as quantities that depend on other quantities. In the chain rule we deal with three quantities denoted by x, u, f. The quantity u depends on the quantity x thus giving us the function $u = u(x)$. The quantity f depends on the quantity u thus giving us the function $f = f(u)$. Since u also depends on x, we deduce that through u as intermediary the function f also depends on x, this giving us the composition $f \circ u$.

The derivative of $f \circ u$ with respect to x measures the rate of change in the quantity f per unit of change in x. The classics would denote this rate of change by $\frac{df}{dx}$ instead of the more complete, but more cumbersome[2] $\frac{df\circ u}{dx}$. The quantity $\frac{df}{du}$ denotes the rate of change in f per unit of change in u. The quantity $\frac{du}{dx}$ is defined in a similar fashion and the chain rule takes the simpler form

$$\boxed{\frac{df}{dx} = \frac{df}{du} \cdot \frac{du}{dx}}. \tag{7.21}$$

A less rigorous but more intuitive way of phrasing the above equality is

$$\frac{\text{change in } f}{\text{change in } x} = \frac{\text{change in } f}{\text{change in } u} \cdot \frac{\text{change in } u}{\text{change in } x}.$$

□

Let us see the chain rule at work in some simple examples.

Example 7.20. (a) Consider the function

$$\sin\sqrt{x}, \quad x > 0.$$

It is the composition of the two functions

$$f(u) = \sin u, \quad u(x) = \sqrt{x}.$$

Then

$$\frac{d}{dx}\sin\sqrt{x} = \frac{df}{du} \cdot \frac{du}{dx} = (\cos u) \cdot \frac{1}{2\sqrt{x}} = \frac{\cos\sqrt{x}}{2\sqrt{x}}.$$

(b) Consider the function 2^x. We have

$$2^x = (e^{\ln 2})^x = e^{(\ln 2)x}.$$

It is the composition of two functions

$$f(u) = e^u, \quad u(x) = (\ln 2)x.$$

Then

$$\frac{d}{dx}2^x = \frac{df}{du} \cdot \frac{du}{dx} = e^u(\ln 2) = e^{(\ln 2)x}\ln 2 = 2^x \ln 2.$$

[2]The concept of composition of function was not clearly defined given that the concept of function was nebulous.

More generally, if a is a positive real number then

$$\frac{d}{dx}a^x = a^x \ln a. \tag{7.22}$$

Observe that for any $\lambda \in \mathbb{R}$ we have

$$\frac{d}{dx}e^{\lambda x} = \lambda e^{\lambda x},$$

and we conclude inductively that

$$\frac{d^n}{dx^n}e^{\lambda x} = \lambda^n e^{\lambda x}, \ \ \forall n \in \mathbb{N}. \tag{7.23}$$

(c) Consider now a trickier situation. Let $f : (0, \infty) \to \mathbb{R}$ be given by $f(x) = x^x$. We want to prove that f is differentiable and then compute its derivative. We set

$$g(x) = \ln f(x) = x \ln x.$$

Clearly g is differentiable since it is the product of differentiable functions. From the equality

$$f(x) = e^{g(x)}$$

we deduce that f is also differentiable because it is the composition of differentiable functions. Using the chain rule we deduce

$$f'(x) = e^{g(x)}g'(x) = (x^x)g'(x) = x^x(\ln x + 1).$$

□

Theorem 7.21 (Inverse function rule). *Suppose that I, J are two intervals of the real axis and $u : I \to J$ is a bijective function satisfying the following properties.*

(i) The function u is differentiable at the point $x_0 \in I$.
(ii) $u'(x_0) \neq 0$.
(iii) The inverse function u^{-1} is continuous at $y_0 = u(x_0)$.

Then the inverse function u^{-1} is differentiable at $y_0 = u(x_0)$ and

$$(u^{-1})'(y_0) = \frac{1}{u'(x_0)}.$$

Proof. Since u is bijective we deduce that for any $y \in J$, there exists a unique $x = x(y)$ in I such that $u(x) = y$. More precisely $x(y) = u^{-1}(y)$. Since u^{-1} is continuous at y_0 we have

$$\lim_{y \to y_0} x(y) = x(y_0) = x_0.$$

Then

$$\frac{u^{-1}(y) - u^{-1}(y_0)}{y - y_0} = \frac{x - x_0}{u(x) - u(x_0)} = \frac{1}{\frac{u(x)-u(x_0)}{x-x_0}},$$

so that

$$\lim_{y \to y_0} \frac{u^{-1}(y) - u^{-1}(y_0)}{y - y_0} = \lim_{x \to x_0} \frac{1}{\frac{u(x)-u(x_0)}{x-x_0}} = \frac{1}{u'(x_0)}.$$

□

Example 7.22. The inverse function rule is a bit tricky to use. We discuss a few classical examples.

(a) Consider the function

$$u : (-\pi/2, \pi/2) \to (-1, 1), \;\; u(x) = \sin x.$$

This function is bijective, differentiable, and the derivative $u'(x) = \cos x$ is nowhere zero. Its inverse is the continuous function

$$\arcsin : (-1, 1) \to (-\pi/2, \pi/2).$$

We have

$$\frac{d}{du} \arcsin u = \frac{1}{u'(x)} = \frac{1}{\cos x}, \quad u = \sin x.$$

Observe that on the interval $(-\pi/2, \pi/2)$ the function $\cos x$ is positive so that

$$\cos x = \sqrt{1 - \sin^2 x} = \sqrt{1 - u^2}.$$

Hence

$$\boxed{\frac{d}{du} \arcsin u = \frac{1}{\sqrt{1-u^2}}, \quad \forall u \in (-1, 1)}. \tag{7.24}$$

A similar argument shows that

$$\frac{d}{du} \arccos u = -\frac{1}{\sqrt{1-u^2}}, \quad \forall u \in (-1, 1). \tag{7.25}$$

(b) Consider the bijective differentiable function

$$u : (-\pi/2, \pi/2) \to \mathbb{R}, \;\; u(x) = \tan x.$$

Its inverse is the function $\arctan : \mathbb{R} \to (-\pi/2, \pi/2)$. It is continuous and

$$\frac{d}{du} \arctan u = \frac{1}{u'(x)} = \frac{1}{(\tan x)'}, \quad u = \tan x.$$

Using the equality $(\tan x)' = 1 + \tan^2 x$ we deduce

$$\boxed{\frac{d}{du} \arctan u = \frac{1}{1 + \tan^2 x} = \frac{1}{1 + u^2}}. \tag{7.26}$$

□

7.4. Fundamental properties of differentiable functions

The first fundamental result concerning differentiable functions is Fermat's Principle. Before we formulate it we need to introduce a new concept.

Definition 7.23. Suppose that $f : I \to \mathbb{R}$ is a function defined on an interval $I \subset \mathbb{R}$.

(i) A point $x_0 \in I$ is said to be a *local minimum* of f if there exists $\delta > 0$ with the following property

$$\forall x \in I, \ \ |x - x_0| < \delta \Rightarrow f(x) \geq f(x_0).$$

The point x_0 is called a *strict local minimum* if there exists $\delta > 0$ with the following property

$$\forall x \in I, \ \ 0 < |x - x_0| < \delta \Rightarrow f(x) > f(x_0).$$

(ii) A point $x_0 \in I$ is said to be a *local maximum* of f if there exists $\delta > 0$ with the following property

$$\forall x \in I, \ \ |x - x_0| < \delta \Rightarrow f(x) \leq f(x_0).$$

The point x_0 is called a *strict local maximum* if there exists $\delta > 0$ with the following property

$$\forall x \in I, \ \ 0 < |x - x_0| < \delta \Rightarrow f(x) < f(x_0).$$

(iii) A point $x_0 \in I$ is said to be a *(strict) local extremum* of f if it is either a (strict) local minimum, or a (strict) local maximum.

□

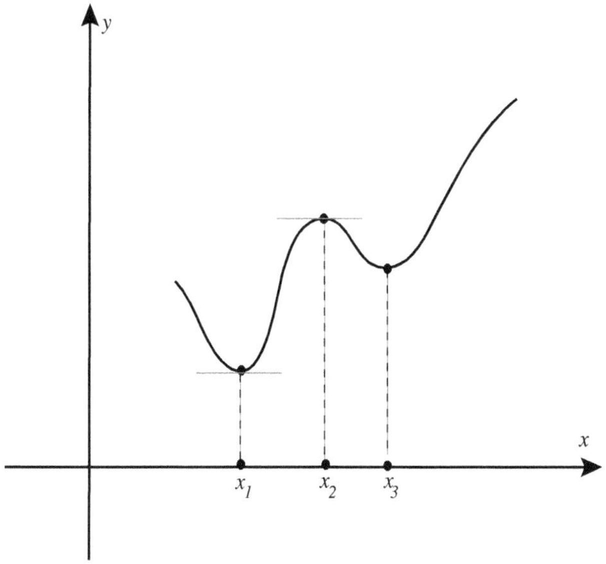

Fig. 7.3 *The points x_1 and x_3 are local minima, while the point x_2 is a local maximum.*

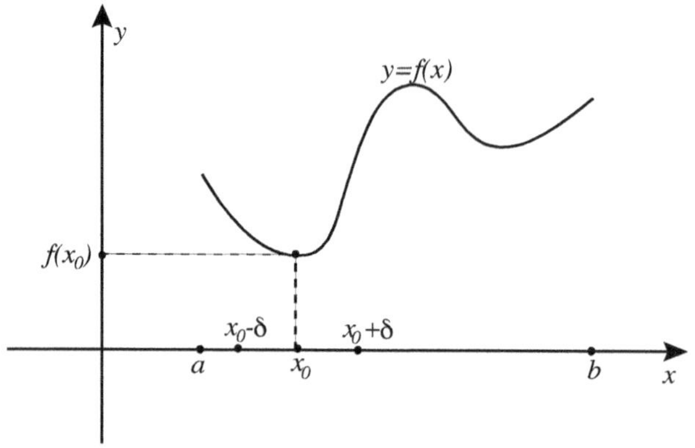

Fig. 7.4 *The point x_0 is an interior local minimum.*

Theorem 7.24 (Fermat's Principle). *Consider a function $f : [a,b] \to \mathbb{R}$ which is differentiable on the open interval (a,b). Suppose that x_0 is a local extremum of f* situated in the interior, *$x_0 \in (a,b)$. Then $f'(x_0) = 0$. In geometric terms, at an interior local extremum, the tangent line to the graph has zero slope, i.e., it is horizontal.*

Proof. Assume for simplicity that x_0 is a local minimum; see Figure 7.4. Since x_0 is in the *interior* of the interval $[a,b]$ we can find $\delta > 0$ such that

$$(x_0 - \delta, x_0 + \delta) \subset (a,b) \text{ and } f(x_0) \leq f(x), \ \forall x \in (x_0 - \delta, x_0 + \delta).$$

We have

$$\lim_{x \searrow x_0} \frac{f(x) - f(x_0)}{x - x_0} = f'(x_0) = \lim_{x \nearrow x_0} \frac{f(x) - f(x_0)}{x - x_0}.$$

Note that

$$x \in (x_0, x_0 + \delta) \Rightarrow f(x) - f(x_0) \geq 0 \wedge x - x_0 > 0 \Rightarrow \frac{f(x) - f(x_0)}{x - x_0} \geq 0$$
$$\Rightarrow \lim_{x \searrow x_0} \frac{f(x) - f(x_0)}{x - x_0} \geq 0 \Rightarrow f'(x_0) \geq 0.$$

Similarly

$$x \in (x_0 - \delta, x_0) \Rightarrow f(x) - f(x_0) \geq 0 \wedge x - x_0 < 0 \Rightarrow \frac{f(x) - f(x_0)}{x - x_0} \leq 0$$
$$\Rightarrow \lim_{x \nearrow x_0} \frac{f(x) - f(x_0)}{x - x_0} \leq 0 \Rightarrow f'(x_0) \leq 0.$$

This proves that $f'(x_0) = 0$. □

Remark 7.25. The importance of Fermat's Principle is difficult to overestimate. Locating the local extrema of a function is a problem with a huge number of applications beyond theoretical mathematics. Fermat's Principle states that the local extrema of a differentiable function $f : [a, b] \to \mathbb{R}$ are very special points: they are either endpoints of the interval, or points where the derivative of f vanishes.

This principle reduces the search of extrema to a set much much smaller than the interval $[a, b]$. Instead of looking for the needle in a haystack, we are looking for a needle hidden in a small matchbox. There is a caveat: the matchbox could be locked and it may take some ingenuity to unlock it. □

Definition 7.26. Suppose that $f : I \to \mathbb{R}$ is a differentiable function defined on an interval $I \subset \mathbb{R}$. A point $x_0 \in I$ is called a *critical* or *stationary* point of f if $f'(x_0) = 0$. □

We can thus rephrase Fermat's Principle as saying that ***interior*** *local extrema must be critical points.* We want to point out that not all critical points are necessarily local extrema. For example the point $x_0 = 0$ of $f(x) = x^3$, $x \in \mathbb{R}$, is a critical point of f. However it is not a local extremum because

$$f(x) > f(0) \;\; \forall x > 0 \;\wedge\;\; f(x) < f(0) \;\; \forall x < 0.$$

Fermat's Principle has several fundamental consequences. We describe a few of them.

Theorem 7.27 (Rolle). *Suppose that $f : [a, b] \to \mathbb{R}$ is a continuous function that is also differentiable on the open interval (a, b). If $f(a) = f(b)$, then there exists $\xi \in (a, b)$ such that $f'(\xi) = 0$.*

Proof. According to Weierstrass' Theorem 6.14 there exist $x_*, x^* \in [a, b]$ such that

$$f(x_*) = \inf_{x \in [a,b]} f(x), \;\; f(x^*) = \sup_{x \in [a,b]} f(x). \tag{7.27}$$

We distinguish two cases.

1. $f(x_*) = f(x^*)$. We deduce from (7.27) that f is the constant function $f(x) = f(x_*)$, $\forall x \in [a, b]$. In particular $f'(x) = 0$, $\forall x \in (a, b)$, proving the claim in the theorem.

2. $f(x_*) < f(x^*)$. Thus x_* and x^* cannot simultaneously be endpoints of the interval $[a, b]$ because $f(a) = f(b)$. Hence at least one of the points x_* or x^* is located in the interior of the interval. Suppose for that x_* is that point. Then x_* is a local minimum of f located in the interior of (a, b). Fermat's Principle implies that $f'(x_*) = 0$. □

Theorem 7.28 (Lagrange's Mean Value Theorem). *Suppose that $f : [a, b] \to \mathbb{R}$ is a continuous function that is also differentiable on the open interval (a, b). Then there exists a point $\xi \in (a, b)$ such that*

$$f'(\xi) = \frac{f(b) - f(a)}{b - a}.$$

Geometrically this signifies that somewhere on the graph of f there exists a point so that the tangent to the graph at that point is parallel to the line connecting the endpoints of the graph of f; see Figure 7.5.

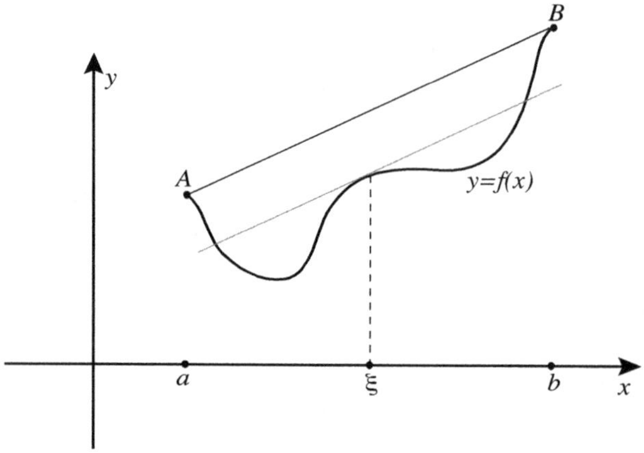

Fig. 7.5 The geometric interpretation of Theorem 7.28.

Proof. We set

$$m := \frac{f(b) - f(a)}{b - a}.$$

The line passing through the points $A = (a, f(a))$ and $B = (b, f(b))$ has slope m and is the graph of the linear function

$$L(x) = m(x - a) + f(a).$$

Observe that

$$L(a) = f(a), \;\; L(b) = f(b), \;\; L'(x) = m, \;\; \forall x.$$

Define

$$g : [a, b] \to \mathbb{R}, \;\; g(x) = f(x) - L(x).$$

Note that g is continuous on $[a, b]$ and differentiable on (a, b). Moreover

$$g(a) = f(a) - L(a) = 0 = f(b) - L(b) = g(b).$$

Rolle's Theorem implies that there exists $\xi \in (a, b)$ such that

$$0 = g'(\xi) = f'(\xi) - L'(\xi) = f'(\xi) - m \Rightarrow f'(\xi) = m.$$

□

Remark 7.29. In the Mean Value Theorem the requirement that f be continuous on the ***closed*** interval $[a, b]$ is essential and does not follow from the requirement that f be differentiable on the ***open*** interval (a, b).

In the theorem we have tacitly assumed that $a < b$. The result continues to be true even when $a > b$ because

$$\frac{f(b) - f(a)}{b - a} = \frac{f(a) - f(b)}{a - b}.$$

In this case ξ is a point in the open interval with endpoints a and b. □

Corollary 7.30. *Suppose that $f : [a, b] \to \mathbb{R}$ is a continuous function that is also differentiable on the open interval (a, b). Then the following statements are equivalent.*

(i) The function f is constant.
(ii) $f'(x) = 0$, $\forall x \in (a, b)$.

Proof. The implication (i) $\Rightarrow$ (ii) is immediate since the derivative of a constant function is 0.

To prove the implication (ii) $\Rightarrow$ (i) we argue by contradiction. Suppose that there exist $x_0, x_1 \in [a, b]$, such that $x_0 < x_1$ and $f(x_0) \neq f(x_1)$. The Mean Value Theorem implies that there exists $\xi \in (x_0, x_1)$ such that

$$f'(\xi) = \frac{f(x_1) - f(x_0)}{x_1 - x_0} \neq 0.$$

□

Corollary 7.31. *Suppose that $f : [a, b] \to \mathbb{R}$ is a continuous function that is differentiable on (a, b). If $f'(x) \neq 0$ for any (a, b), then f is injective.*

Proof. If $x_0, x_1 \in [a, b]$ and $x_0 \neq x_1$, say $x_0 < x_1$, then the Mean Value Theorem implies that there exists $\xi \in (x_0, x_1)$ such that

$$f(x_1) - f(x_0) = f'(\xi)(x_1 - x_0) \neq 0.$$

This proves the injectivity of f. □

Corollary 7.32. *Suppose that $f : [a, b] \to \mathbb{R}$ is a continuous function that is differentiable on (a, b). Then the following statements are equivalent.*

(i) The function f is nondecreasing.
(ii) $f'(x) \geq 0$, $\forall x \in (a, b)$.

Also, the following statements are equivalent.

(iii) The function f is nonincreasing.
(iv) $f'(x) \leq 0$, $\forall x \in (a, b)$.

Proof. (i) $\Rightarrow$ (ii). Let $x_0 \in (a, b)$. Then for $h > 0$ we have $f(x_0 + h) - f(x_0) \geq 0$ so that

$$\frac{f(x_0 + h) - f(x_0)}{h} \geq 0 \Rightarrow f'(x_0) = \lim_{h \searrow 0} \frac{f(x_0 + h) - f(x_0)}{h} \geq 0.$$

(ii) $\Rightarrow$ (i). Suppose that $x_0, x_1 \in [a, b]$ are such that $x_0 < x_1$. The Mean Value Theorem implies that there exists $\xi \in (x_0, x_1)$ such that

$$f'(\xi) = \frac{f(x_1) - f(x_0)}{x_1 - x_0} \Rightarrow f(x_1) - f(x_0) = f'(\xi)(x_1 - x_0) \geq 0.$$

□

Remark 7.33. If in the above result we replace (ii) with the stronger condition

$$f'(x) > 0, \quad \forall x \in (a, b),$$

then we obtain a stronger conclusion namely that f is (strictly) increasing. This follows by coupling Corollary 7.32 with Corollary 7.31. □

Example 7.34. (a) We want to prove that

$$e^x \geq x + 1, \quad \forall x \in \mathbb{R}. \tag{7.28}$$

To this aim consider the function $f : \mathbb{R} \to \mathbb{R}$, $f(x) = e^x - (x + 1)$. This function is differentiable and $f'(x) = e^x - 1$.

We see that the derivative is positive on $(0, \infty)$ and negative on $(-\infty, 0)$. Hence f is increasing on $(0, \infty)$ and thus $f(x) > f(0) = 0$, $\forall x > 0$ and $f(x) > 0$, $\forall x \in (-\infty, 0)$. In other words,

$$e^x - (x + 1) \geq 0, \quad \forall x \in \mathbb{R},$$

which is (7.28).
(b) We want to prove that

$$x \geq \sin x, \quad \forall x \geq 0. \tag{7.29}$$

Consider the function $f : [0, \infty) \to \mathbb{R}$, $f(x) = x - \sin x$. This function is differentiable and

$$f'(x) = 1 - \cos x \geq 0, \quad \forall x \geq 0.$$

Hence f is nondecreasing and thus

$$x - \sin x = f(x) \geq f(0) = 0, \quad \forall x \geq 0.$$

(c) We want to prove that

$$\cos x \geq 1 - \frac{x^2}{2}, \quad \forall x \in \mathbb{R}. \tag{7.30}$$

Consider the function

$$f : \mathbb{R} \to \mathbb{R}, \quad f(x) = \cos x - \left(1 - \frac{x^2}{2}\right), \quad \forall x \in \mathbb{R}.$$

We have to prove that $f(x) \geq 0$, $\forall x \in \mathbb{R}$. We observe that f is an even function, i.e., $f(-x) = f(x)$, $\forall x \in \mathbb{R}$ so it suffices to show that $f(x) \geq 0$, $\forall x \geq 0$. Note that f is differentiable and

$$f'(x) = -\sin x + x \overset{(7.29)}{\geq} 0 \ \forall x \geq 0.$$

Thus f is nondecreasing on the interval $[0, \infty)$ and we conclude that $f(x) \geq f(0) = 0$, $\forall x \geq 0$. □

Example 7.35 (Young's inequality). Suppose that $p \in (1, \infty)$. Define $q \in (1, \infty)$ by $\frac{1}{p} + \frac{1}{q} = 1$, i.e., $q = \frac{p}{p-1}$. Consider $f : (0, \infty) \to \mathbb{R}$

$$f(x) = x^{\alpha} - \alpha x + \alpha - 1, \quad \alpha := \frac{1}{p}.$$

We want to prove that $f(x) \leq 0, \forall x > 0$. We have

$$f'(x) = \alpha x^{\alpha-1} - \alpha = \alpha(x^{\alpha-1} - 1) = \alpha\left(\frac{1}{x^{1-\alpha}} - 1\right).$$

Observe that $f'(x) = 0$ if and only if $x = 1$. Moreover $f'(x) < 0$ for $x > 1$ and $f'(x) > 0$ for $x < 1$ because $1 - \alpha = 1 - \frac{1}{p} > 0$. Thus the function f increases on $(0, 1)$ and decreases on $(1, \infty)$ so that

$$0 = f(1) \geq f(x) \;\; \forall x > 0.$$

Thus

$$x^{\alpha} - \alpha x \leq 1 - \alpha = 1 - \frac{1}{p} > 0 = \frac{1}{q}.$$

If we choose $a, b > 0$ and set $x := \frac{a}{b}$, we deduce

$$\left(\frac{a}{b}\right)^{\frac{1}{p}} - \frac{1}{p}\left(\frac{a}{b}\right)^{\frac{1}{p}+\frac{1}{q}} \leq \frac{1}{q} \Rightarrow \left(\frac{a}{b}\right)^{\frac{1}{p}} \leq \frac{1}{p}\left(\frac{a}{b}\right)^{\frac{1}{p}+\frac{1}{q}} + \frac{1}{q}.$$

Multiplying both sides by $b = b^{\frac{1}{p}+\frac{1}{q}}$ we deduce

$$a^{\frac{1}{p}} b^{\frac{1}{q}} \leq \frac{a}{p} + \frac{b}{q}, \quad \forall a, b > 0. \tag{7.31}$$

If we set $u := a^{\frac{1}{p}}$, $v := b^{\frac{1}{q}}$, then we can rewrite the above inequality in the commonly encountered form

$$uv \leq \frac{u^p}{p} + \frac{v^q}{q}, \quad \forall u, v > 0, \;\; p, q > 1, \;\; \frac{1}{p} + \frac{1}{q} = 1. \tag{7.32}$$

The last inequality is known as *Young's inequality*. □

Corollary 7.36. *Suppose that $f : [a, b] \to \mathbb{R}$ is a continuous function that is twice differentiable on (a, b). Let $x_0 \in (a, b)$ be a critical point of f, i.e., $f'(x_0) = 0$. Then the following hold.*

(i) If $f''(x_0) > 0$, then x_0 is a strict local minimum of f.
(ii) If $f''(x_0) < 0$, then x_0 is a strict local maximum of f.

Proof. We prove only (i). Part (ii) follows by applying (i) to the new function $-f$. Suppose that

$$f'(x_0) = 0, \quad f''(x_0) > 0.$$

We have to prove that there exists $\delta > 0$ such that

$$0 < |x - x_0| < \delta \Rightarrow f(x) > f(x_0).$$

We have

$$\lim_{x \searrow x_0} \frac{f'(x)}{x - x_0} = \lim_{x \searrow x_0} \frac{f'(x) - f'(x_0)}{x - x_0} = f''(x_0) > 0.$$

Thus there exists $\delta_1 > 0$ such that,

$$x \in (x_0, x_0 + \delta_1) \Rightarrow \frac{f'(x)}{x - x_0} > 0 \Rightarrow f'(x) > 0.$$

The Mean Value Theorem implies that, for any $x \in (x_0, x_0 + \delta_1)$, there exists $\xi \in (x_0, x)$ such that

$$f(x) - f(x_0) = f'(\xi)(x - x_0).$$

Since $\xi \in (x_0, x_0 + \delta_1)$ we have $f'(\xi) > 0$ and thus $f'(\xi)(x - x_0) > 0$. Similarly

$$\lim_{x \nearrow x_0} \frac{f'(x)}{x - x_0} = \lim_{x \nearrow x_0} \frac{f'(x) - f'(x_0)}{x - x_0} = f''(x_0) > 0.$$

Thus there exists $\delta_2 > 0$ such that,

$$x \in (x_0 - \delta_2, x_0) \Rightarrow \frac{f'(x)}{x - x_0} > 0 \rightarrow f'(x) < 0.$$

Hence if $x \in (x_0 - \delta_2, x_0)$, then the Mean Value Theorem implies that there exists $\eta \in (x, x_0) \subset (x_0 - \delta_2, x_0)$ such that

$$f(x) - f(x_0) = f'(\eta)(x - x_0) > 0.$$

If we let $\delta := \min(\delta_1, \delta_2)$, then we deduce

$$0 < |x - x_0| < \delta \Rightarrow f(x) > f(x_0).$$

□

Example 7.37. Here is a simple application of the above corollary. Fix a positive number a. Consider the function

$$f : [0, a] \to \mathbb{R}, \quad f(x) = x(a - x)^2.$$

We want to find the maximum possible value of this function. It is achieved either at one of the end points $0, a$ or at some interior point x_0. Note that $f(0) = f(a) = 0$ and $f(x) \geq 0$, $\forall x \in [0, a]$, so there must exist an interior maximum which must be a critical point. To find the critical points of f we need to solve the equation $f'(x) = 0$. We have

$$f'(x) = (a - x)^2 - 2x(a - x) = x^2 - 2ax + a^2 - 2ax + 2x^2 = 3x^2 - 4ax + a^2.$$

The discriminant of the quadratic equation $3x^2 - 4ax + a^2 = 0$ is

$$\Delta = 16a^2 - 12a^2 = 4a^2 > 0.$$

Thus this quadratic equation has two roots

$$x_{\pm} = \frac{4a \pm 2a}{6} = a, \frac{a}{3}.$$

Only one of these roots is in the interval $(0, a)$, namely $\frac{a}{3}$. Note that

$$f''(x) = 6x - 4a, \quad f''(a/3) = 2a - 4a < 0.$$

Thus $a/3$ is the unique maximum point of f, and thus it is absolute maximum point. We have

$$f(x) \leq f(a/3) = \frac{4a^3}{27}, \quad \forall x \in [0, a].$$

□

Theorem 7.38 (Cauchy's Finite Increment Theorem). *Suppose that $f, g : [a,b] \to \mathbb{R}$ are two continuous functions that are differentiable on (a,b). Then there exists $\xi \in (a,b)$ such that*

$$f'(\xi)\big(\,g(b)-g(a)\,\big) = g'(\xi)\big(\,f(b)-f(a)\,\big). \tag{7.33}$$

In particular, if $g'(t) \neq 0$ for any $t \in (a,b)$, then $g(b) \neq g(a)$ and

$$\frac{f(b)-f(a)}{g(b)-g(a)} = \frac{f'(\xi)}{g'(\xi)}. \tag{7.34}$$

Proof. Consider the function $F : [a,b] \to \mathbb{R}$ defined by

$$F(x) = f(x)\underbrace{\big(\,g(b)-g(a)\,\big)}_{=:\Delta_g} - g(x)\underbrace{\big(\,f(b)-f(a)\,\big)}_{=:\Delta_f},\ \ \forall x \in [a,b].$$

This function is continuous on $[a,b]$ and differentiable on (a,b). Moreover

$$\begin{aligned} F(b)-F(a) &= \big(\,f(b)\Delta_g - g(b)\Delta_f\,\big) - \big(\,f(a)\Delta_g - g(a)\Delta_f\,\big) \\ &= \big(\,f(b)-f(a)\,\big)\Delta_g + \big(\,g(a)-g(b)\,\big)\Delta_f = 0. \end{aligned}$$

Rolle's Theorem implies that there exists $\xi \in (a,b)$ such that $F'(\xi) = 0$. This proves (7.33). To obtain (7.34) we observe that the assumption $g'(t) \neq 0$ for any $t \in (a,b)$ implies that g is injective and thus $g(b) \neq g(a)$. Dividing both sides of (7.33) by $g(b)-g(a)$ we deduce (7.34). □

Remark 7.39. In the above theorem we have tacitly assumed that $a < b$. The result continues to be true even when $a > b$ because

$$\frac{f(b)-f(a)}{g(b)-g(a)} = \frac{f(a)-f(b)}{g(a)-g(b)}.$$

In this case ξ is a point in the open interval with endpoints a and b. □

If $f : I \to \mathbb{R}$ is a function differentiable on the interval I, then its derivative $f' : I \to \mathbb{R}$ need not be continuous. However, the derivative is very close to being continuous in the sense that it satisfies the *intermediate value property*, just like continuous functions do.

Theorem 7.40 (Darboux). *Suppose that I is an interval of the real axis and $f : I \to \mathbb{R}$ is a differentiable function. Then the derivative f' satisfies the intermediate value property: given $a, b \in I$, $a < b$, and a number γ strictly between $f'(a)$ and $f'(b)$, there exists a number $\xi \in (a,b)$ such that $f'(\xi) = \gamma$.* □

Exercise 7.1 will guide you toward a proof of this theorem which is also a consequence of Fermat's Principle.

7.5. Table of derivatives

Table 7.1 summarizes the derivatives of the most frequently encountered functions.

Table 7.1 Table of derivatives.

$f(x)$	$f'(x)$
x^n, $(x \in \mathbb{R}, n \in \mathbb{N})$	nx^{n-1}
x^{-n} $(x \neq 0, n \in \mathbb{N})$	$-nx^{-n-1}$
x^α, $(\alpha \in \mathbb{R}, x > 0)$	$\alpha x^{\alpha-1}$
$\sqrt{x}$, $(x > 0)$	$\frac{1}{2\sqrt{x}}$
$\ln x$	$1/x$
e^x, $(x \in \mathbb{R})$	e^x
a^x, $(a > 0, x \in \mathbb{R})$	$a^x \ln a$
$\sin x$, $(x \in \mathbb{R})$	$\cos x$
$\cos x$, $(x \in \mathbb{R})$	$-\sin x$
$\tan x$, $(\cos x \neq 0)$	$1 + \tan^2 x = \frac{1}{\cos^2 x}$
$\arcsin x$, $x \in (-1, 1)$	$\frac{1}{\sqrt{1-x^2}}$
$\arccos x$, $x \in (-1, 1)$	$-\frac{1}{\sqrt{1-x^2}}$
$\arctan x$, $(x \in \mathbb{R})$	$\frac{1}{1+x^2}$
$\sinh x$, $(x \in \mathbb{R})$	$\cosh x$
$\cosh x$, $(x \in \mathbb{R})$	$\sinh x$

The *hyperbolic functions* $\sinh x$ and $\cosh x$ are defined by the equalities

$$\cosh x := \frac{e^x + e^{-x}}{2}, \quad \sinh x = \frac{e^x - e^{-x}}{2}.$$

The function sinh is called the *hyperbolic sine* while the function cosh is called the *hyperbolic cosine*.

7.6. Exercises

Exercise 7.1. Consider the function $f : \mathbb{R} \to \mathbb{R}$, $f(x) = |x|$.

(i) Sketch the graph of f.
(ii) Show that f is not differentiable at 0.
(iii) Show that f is differentiable at any point $x_0 \neq 0$ and then compute the derivative of f at x_0.

□

Exercise 7.2. Prove Proposition 7.7. □

Exercise 7.3. Imitate the strategy in Example 7.15 to prove

$$\lim_{h\to 0} \frac{\cos(x_0 + h) - \cos x_0}{h} = -\sin x_0.$$

Hint. You need to use the trigonometric identities (5.33a) and (5.33c). □

Exercise 7.4. Consider the function $f : (-\pi/2, \pi/2) \to \mathbb{R}$, $f(x) = \tan x$. Write the equation of the tangent line to the graph of f at the point $(\,\pi/4, f(\pi/4)\,)$. □

Exercise 7.5. Suppose that the functions $f, g : I \to \mathbb{R}$ are n-times differentiable. Prove that their product $f \cdot g$ is also n-times differentiable and satisfies the generalized product rule

$$\frac{d^n}{dx^n}(fg) = \sum_{k=0}^{n} \binom{n}{k} f^{(n-k)} g^{(k)} = \sum_{k=0}^{n} \binom{n}{k} f^{(k)} g^{(n-k)}, \tag{7.35}$$

where we defined $f^{(0)} := f$, $g^{(0)} = g$.

Hint. Argue by induction on n. At some point you need to use the Pascal formula (3.7),

$$\binom{n+1}{k} = \binom{n}{k} + \binom{n}{k-1},$$

also used in the proof of Newton's binomial formula (3.6). □

Exercise 7.6. Let n be a natural number. A real number r is said to be a *root* of order n of a polynomial $P(x)$ if there exists a polynomial $Q(x)$ with the following properties:

- $P(x) = (x - r)^n Q(x)$, $\forall x \in \mathbb{R}$.
- $Q(r) \neq 0$.

(a) Prove that if $n > 1$ and r is a root of $P(x)$ of order n, then r is also a root of order $(n-1)$ of $P'(x)$.
(b) Prove that for any natural numbers $k < n$ the real numbers ± 1 are roots of order $(n-k)$ of the polynomial

$$\frac{d^k}{dx^k}(x^2 - 1)^n.$$

(c) For any natural number n we define the n-th *Legendre polynomial* to be

$$P_n(x) := \frac{1}{2^n n!}\frac{d^n}{dx^n}\left(x^2-1\right)^n.$$

Use (7.35) to compute $P_n(\pm 1)$. □

Exercise 7.7. Consider the continuous function $f : [0,\infty) \to \mathbb{R}$, $f(x) = \sqrt{x}$. Show that f is not differentiable at 0. □

Exercise 7.8. Consider the function $f : \mathbb{R} \to \mathbb{R}$ given by

$$f(x) = \begin{cases} 0, & |x| \geq 1 \\ e^{-T(x)}, & |x| < 1, \end{cases} \quad \text{where } T(x) = \frac{1}{1-x^2}, \ \forall |x| < 1.$$

(a) Set

$$F_n(x) := \frac{d^n}{dx^n}\left(e^{-T(x)}\right), \ \forall |x| < 1.$$

Prove by induction that for any $n \in \mathbb{N}$ there exists a polynomial $P_n(x)$ and a natural number k_n such that

$$F_n(x) = P_n(x)T(x)^{k_n}e^{-T(x)}, \ \forall |x| < 1.$$

Hint. Observe that

$$T'(x) = 2xT(x)^2.$$

(b) Prove that f is a smooth function, i.e., infinitely many times differentiable.

Hint. Prove by induction that

$$f^{(n)}(x) = \begin{cases} 0, & |x| \geq 1, \\ F_n(x), & |x| < 1. \end{cases}$$

For the inductive step observe that for $|x| < 1$ we have

$$\frac{1}{x-1} = -(x+1)T(x),$$

$$\frac{f^{(n)}(x) - f^{(n)}(1)}{x-1} = \frac{F_n(x)}{x-1} = -(x+1)T(x)F_n(x) = (x+1)P_n(x)T(x)^{k_n+1}e^{-T(x)},$$

$$\frac{F_n(x)}{x+1} = (x-1)T(x)F_n(x) = -(x-1)P_n(x)T(x)^{k_n+1}e^{-T(x)}.$$

Then

$$\lim_{x\nearrow 1}\frac{f^{(n)}(x) - f^{(n)}(1)}{x-1} = -\lim_{x\nearrow 1}\frac{F_n(x)}{x-1} = \Big(\lim_{x\nearrow 1}(x+1)P_n(x)\Big)\cdot\Big(\lim_{x\nearrow 1}T(x)^{k_n+1}e^{-T(x)}\Big)$$

$$= -2P_n(1)\Big(\lim_{x\nearrow 1}T(x)^{k_n+1}e^{-T(x)}\Big).$$

Now observe that

$$\lim_{x\nearrow 1}T(x) = \infty.$$

Use the result in Exercise 5.11(b) to deduce

$$\lim_{x\nearrow 1}T(x)^{k_n+1}e^{-T(x)} = 0.$$

□

Exercise 7.9.[3] Fix a natural number n and real numbers p, q.

(a) Prove that for any $t \in \mathbb{R}$ we have

$$np(tp+q)^{n-1} = \sum_{k=1}^{n} k\binom{n}{k} t^{k-1}p^k q^{n-k},$$

$$n(n-1)p^2(tp+q)^{n-2} = \sum_{k=2}^{n} k(k-1)\binom{n}{k} t^{k-2}p^k q^{n-k}.$$

Hint. Consider the function

$$f_n : \mathbb{R} \to \mathbb{R}, \;\; f_n(t) = (\,tp+q\,)^n.$$

Compute the derivatives $f_n'(t)$, $f_n''(t)$. Then describe $f_n(t)$ using Newton's binomial formula and compute the same derivatives using the new description of $f_n(t)$.

(b) For any integer k, $0 \leq k \leq n$, and any $x \in \mathbb{R}$ set $w_k(x) := \binom{n}{k} x^k(1-x)^{n-k}$. Use part (a) to prove that for any $x \in \mathbb{R}$

$$1 = \sum_{k=0}^{n} w_k(x), \tag{7.36a}$$

$$nx = \sum_{k=0}^{n} kw_k(x), \tag{7.36b}$$

$$n(n-1)x^2 = \sum_{k=0}^{n} k(k-1)w_k(x) = \sum_{k=0}^{n} k^2 w_k(x) - nx, \tag{7.36c}$$

$$nx(1-x) = \sum_{k=0}^{n} (k-nx)^2 w_k(x). \tag{7.36d}$$

Hint. Use the results in (a) in the special case $p = x$, $q = 1-x$, $t = 1$. □

Exercise 7.10. Find the extrema and the intervals on which the following functions are increasing.

(i) $f(x) = \sqrt{x} - 2\sqrt{x+2}$, $x > 0$.
(ii) $g(x) = \frac{x}{x^2+1}$, $x \in \mathbb{R}$.

□

Exercise 7.11. Suppose that $f : [a,b] \to \mathbb{R}$ is continuous and differentiable on (a,b). Show that if

$$\lim_{x \to a} f'(x) = A,$$

then f is differentiable at a and $f'(a) = A$. □

[3] The results in this exercise are very useful in probability theory.

Exercise 7.12. Prove that if $f : I \to \mathbb{R}$ is a differentiable function defined on an interval I, and the derivative f' is bounded on I, then f is a Lipschitz function, i.e.,

$$\exists L > 0, \ \ \forall x, y \in I \ \ |f(x) - f(y)| \leq L|x - y|. \qquad \square$$

Exercise 7.13. Use the Mean Value Theorem to prove that

$$|\sin(x) - \sin(y)| \leq |x - y|, \ \ \forall x, y \in \mathbb{R}. \qquad \square$$

Exercise 7.14. Fix a real number λ and suppose that $u : \mathbb{R} \to \mathbb{R}$ is a differentiable function satisfying the differential equation

$$u'(t) = \lambda u(t), \ \ \forall t \in \mathbb{R}.$$

Prove that there exists a constant $c \in \mathbb{R}$ such that $u(t) = ce^{\lambda t}$, $\forall t \in \mathbb{R}$.

Hint. Show that the function $f(t) = e^{-\lambda t} u(t)$, $t \in \mathbb{R}$ is constant. □

Exercise 7.15. Suppose that b, c are real numbers and $u, v : \mathbb{R} \to \mathbb{R}$ are twice differentiable functions satisfying the differential equation

$$u''(t) + bu'(t) + cu(t) = 0 = v''(t) + bv'(t) + cv(t), \ \ \forall t \in \mathbb{R}.$$

Define the *Wronskian* to be the function

$$W(t) = u(t)v'(t) - u'(t)v(t), \ \ t \in \mathbb{R}.$$

Prove that

$$W'(t) + bW(t) = 0$$

and deduce that

$$W(t) = W(0)e^{-bt}.$$

Hint. You may want to use Exercise 7.14. □

Exercise 7.16. (a) Suppose that $u : \mathbb{R} \to \mathbb{R}$ is a twice differentiable function satisfying the differential equation

$$u''(t) + u(t) = 0, \ \ \forall t \in \mathbb{R}. \tag{7.37}$$

Prove that

$$u'(t)^2 + u(t)^2 = u'(0)^2 + u(0)^2, \ \ \forall t \in \mathbb{R}.$$

(b) Suppose that $u, v : \mathbb{R} \to \mathbb{R}$ are twice differentiable functions satisfying the differential equation (7.37), i.e.,

$$u''(t) + u(t) = 0 = v''(t) + v(t), \ \ \forall t \in \mathbb{R}.$$

Show that the difference $w(t) = u(t) - v(t)$ also satisfies the differential equation (7.37). Use part (a) to prove that if $u(0) = v(0)$ and $u'(0) = v'(0)$, then $u(t) = v(t)$, $\forall t \in \mathbb{R}$.

(c) Can you think of a function $u : \mathbb{R} \to \mathbb{R}$ satisfying (7.37) and the initial conditions

$$u(0) = 0, \ \ u'(0) = 1? \qquad \square$$

Exercise 7.17. (a) Prove that for any real number $\alpha \geq 1$ and any $x > -1$ we have

$$(1+x)^\alpha \geq 1 + \alpha x.$$

(b) Prove by induction that for any natural number n and any $x \geq 0$ we have

$$1 + x + \frac{x^2}{2!} + \cdots + \frac{x^n}{n!} \leq e^x.$$

Hint. Have a look at Example 7.34. □

Exercise 7.18. Prove that

$$\sin x \geq x - \frac{x^3}{6}, \quad \forall x \geq 0.$$

Hint. Have a look at Example 7.34. □

Exercise 7.19. Prove that the function

$$f : (0, \infty) \to \mathbb{R}, \quad f(x) = \left(1 + \frac{1}{x}\right)^x$$

is increasing. □

Exercise 7.20. Use Lagrange's Mean Value Theorem to show that for any $x > 0$ we have

$$\frac{1}{x+1} < \ln(x+1) - \ln x < \frac{1}{x}.$$

Conclude that

$$1 + \frac{1}{2} + \cdots + \frac{1}{n} > \ln(n+1), \quad \forall n \in \mathbb{N}.$$ □

Exercise 7.21. Fix a real number $s \in (0,1)$. Prove that for any $x > 0$ we have

$$(1+x)^{1-s} - x^{1-s} < \frac{1-s}{x^s}.$$

Conclude that

$$1 + \frac{1}{2^s} + \cdots + \frac{1}{n^s} > \frac{1}{1-s}\big((n+1)^{1-s} - 1\big).$$ □

Exercise 7.22. Find the maximum possible volume of an open rectangular box that can be obtained from a square sheet of cardboard with a 6 ft side by cutting squares at each of the corners and bending up the ends of the resulting cross-like figure; see Figure 7.6. □

Exercise 7.23. Prove that among all the rectangles with given perimeter P the square has the largest area. □

Exercise 7.24. Suppose that $f : [-1,1] \to \mathbb{R}$ is a differentiable function.
(a) Prove that if f is even, i.e., $f(x) = f(-x)$, $\forall x \in [-1,1]$, then $f'(x)$ is odd, $f'(-x) = -f'(x)$, $\forall x \in [-1,1]$. In particular, $f'(0) = 0$.
(b) Prove that if f is odd, then f' is even. □

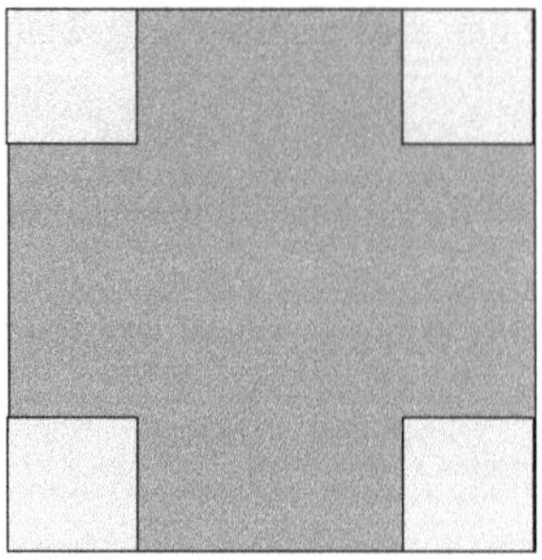

Fig. 7.6 *Cutting out a box.*

Exercise 7.25. Fix a natural number n and suppose that $f : (a,b) \to \mathbb{R}$ is a $2n$-times differentiable function. Prove the following statements.

(a) If $x_0 \in (a,b)$ satisfies

$$f'(x_0) = \cdots = f^{(2n-1)}(x_0) = 0, \quad f^{(2n)}(x_0) > 0,$$

then x_0 is a strict local minimum of f.

(b) If $x_0 \in (a,b)$ satisfies

$$f'(x_0) = \cdots = f^{(2n-1)}(x_0) = 0, \quad f^{(2n)}(x_0) < 0,$$

then x_0 is a strict local maximum of f.

Hint. Use proof of Corollary 7.36 as inspiration and prove (in case (a)) that there exists $\delta > 0$ such that for $x \in (x_0, x_0 + \delta)$ we have $f^{(k)}(x) > 0, \forall k = 1, \dots, 2n-1$ and for $x \in (x_0 - \delta, x_0)$ we have $f^{(k)}(x) < 0, \forall k = 1, \dots, 2n-1$. □

Exercise 7.26. (a) Prove that for any $n \in \mathbb{N}_0$ there exists a unique polynomial T_n of degree n such that

$$\cos n\theta = T_n\big(\cos\theta\big), \ \forall \theta \in \mathbb{R}.$$

(b) Prove that the above polynomials T_n satisfy the following properties.

(i) $T_n(-x) = (-1)^n T_n(x), \forall n \in \mathbb{N}_0, \forall x \in \mathbb{R}$.
(ii) $T_n(x) = xT_{n-1}(x) - T_{n-2}(x), \forall n \geq, \forall x \in \mathbb{R}$.
(iii) $T_n(x)T_m(x) = \frac{1}{2}\big(T_{n+m}(x) + T_{n-m}(x)\big)$.
(iv) $(1-x^2)T_n''(x) - xT_n'(x) + n^2 T_n(x) = 0, \forall n \in \mathbb{N}_0, \forall x \in \mathbb{R}$.

□

7.7. More challenging problems

Problem 7.1 (Intermediate value property of derivatives). Suppose that $f : [a,b] \to \mathbb{R}$ is a differentiable function.

(a) Prove that if $f'(a) < 0 < f'(b)$, then there exists $\xi \in (a,b)$ such that $f'(\xi) = 0$.

Hint. Think Fermat's Principle.

(b) More generally, prove that if $f'(a) < f'(b)$ and $m \in (f'(a), f'(b))$, then there exists $\xi \in (a, b)$ such that $f'(\xi) = m$. □

Problem 7.2. Suppose $f_n : [a, b] \to \mathbb{R}$, $n \in \mathbb{N}$ is a sequence of differentiable functions with the following properties.

(i) The sequence of derivatives $f_n' : [a, b] \to \mathbb{R}$ converges *uniformly* to a function $g : [a, b] \to \mathbb{R}$.
(ii) The sequence $f_n : [a, b] \to \mathbb{R}$ converges *pointwisely* to a function $f : [a, b] \to \mathbb{R}$.

Prove that the following hold.
(a) The sequence $f_n : [a, b] \to \mathbb{R}$ converges *uniformly* to $f : [a, b] \to \mathbb{R}$.
(b) The function f is differentiable and $f' = g$, i.e., the sequence $f_n' : [a, b] \to \mathbb{R}$ converges uniformly to f'.

Hint. Use Exercise 6.16 and the Mean Value Theorem. □

Problem 7.3. Suppose that $f : \mathbb{R} \to \mathbb{R}$ is a continuous function such that

$$\lim_{h \searrow 0} \frac{f(x + 2h) - f(x + h)}{h} = 0, \quad \forall x \in \mathbb{R}.$$

Prove that f is a constant function.

Hint. Argue by contradiction and assume there exist a, b such that $f(a) \neq f(b)$, say $f(a) < f(b)$. Consider the function $g(x) = f(x) - mx$, $m := \frac{f(b)-f(a)}{b-a}$. Note that $g(a) = g(b)$ and

$$\lim_{h \searrow 0} \frac{g(x + 2h) - g(x + h)}{h} = -m < 0,$$

and prove that g admits a local maximum in $[a, b)$. □

Problem 7.4. Suppose $f : \mathbb{R} \to \mathbb{R}$ is a C^2-function, i.e., twice differentiable and the second derivative is continuous. Show that if the functions f and $f^{(2)}$ are bounded on $\mathbb{R}$, then so is the function f'. □

Problem 7.5 (Bernstein). Let $f : [0, 1] \to \mathbb{R}$ be a continuous function. For any $n \in \mathbb{N}$ we denote by $B_n^f(x)$ the n-th Bernstein polynomial determined by f,

$$B_n(x) = \sum_{k=0}^{n} f(k/n) \binom{n}{k} x^k (1 - x)^{n-k}.$$

(a) Show that for any $x \in [0, 1]$ we have

$$f(x) - B_n^f(x) = \sum_{k=0}^{n} \big(f(x) - f(k/n) \big) \binom{n}{k} x^k (1 - x)^k.$$

(b) Show that for any $\delta \in (0, 1)$ and $x \in [0, 1]$ we have

$$\sum_{|k/n - x| \geq \delta} \binom{n}{k} x^k (1 - x)^k \leq \sum_{k=0} \frac{(k - nx)^2}{n^2 \delta^2} \leq \frac{x(1 - x)}{n\delta^2}.$$

(c) Use (a) and (b) to prove that as $n \to \infty$ the sequence $(B_n^f(x))$ converges to $f(x)$ uniformly in $x \in [0, 1]$.

Hint. Use the equalities in Exercise 7.9. □

Chapter 8

Applications of Differential Calculus

8.1. Taylor approximations

The concept of derivative is based on the idea of approximation. Thus, if $f : I \to \mathbb{R}$ is a differentiable function and $x_0 \in I$, then the linearization of f at x_0,

$$L(x) = f(x_0) + f'(x_0)(x - x_0),$$

is a good approximation for $f(x)$ when x is not too far from x_0. More precisely, the error

$$r(x) = f(x) - L(x)$$

is $o(x - x_0)$, much much smaller than $|x - x_0|$, which itself is small when x is close to x_0. In this section we want to refine and improve this observation.

Definition 8.1. Suppose that $f : I \to \mathbb{R}$ is an n-times differentiable function defined on an interval I. For $x_0 \in I$ we define the *degree n Taylor polynomial* of f at x_0 to be

$$T_n(x) = f(x_0) + \frac{f'(x_0)}{1!}(x - x_0) + \cdots + \frac{f^{(n)}(x_0)}{n!}(x - x_0)^n = \sum_{k=0}^{n} \frac{f^{(k)}(x_0)}{k!}(x - x_0)^k.$$

Often the Taylor polynomial of f at $x_0 = 0$ is referred to as the *Maclaurin polynomial.*

If $f : I \to \mathbb{R}$ is a smooth function, then the series

$$\sum_{k=0}^{\infty} \frac{f^{(k)}(x_0)}{k!}(x - x_0)^k$$

is called the *Taylor series* of the smooth function f at the point x_0. □

Example 8.2. (a) Consider a differentiable function $f : I \to \mathbb{R}$. Then the degree 1 Taylor polynomial of f at x_0 is

$$T_1(x) = f(x_0) + f'(x_0)(x - x_0).$$

Thus, $T_1(x)$ is the linearization of f at x_0.
(b) Consider the function $f : \mathbb{R} \to \mathbb{R}$, $f(x) = e^x$. We know that $f^{(n)}(x) = e^x$, $\forall n \in \mathbb{N}$, $x \in \mathbb{R}$ and we deduce that

$$f^{(k)}(0) = 1, \ \ \forall k \in \mathbb{N}.$$

In particular, the degree n Taylor polynomial of e^x at $x_0 = 0$ is

$$T_n(x) = 1 + \frac{x}{1!} + \cdots + \frac{x^n}{n!}.$$

The Taylor series of e^x at $x_0 = 0$ is

$$\sum_{k=0}^{\infty} \frac{x^k}{k!}.$$

(c) Consider the function $f : \mathbb{R} \to \mathbb{R}$, $f(x) = \sin x$. We have

$$f^{(4k)}(x) = \sin x, \quad f^{(4k+1)}(x) = \cos x, \quad f^{(4k+2)}(x) = -\sin x, \quad f^{(4k+3)}(x) = -\cos x,$$

$$\forall k \geq 0, \quad f^{(4k)}(0) = 0, \quad f^{(4k+1)}(0) = 1, \quad f^{(4k+2)}(0) = 0, \quad f^{(4k+3)}(0) = -1.$$

We deduce that the Taylor polynomials of $\sin x$ at $x_0 = 0$ are

$$T_1(x) = f(0) + \frac{f'(0)}{1!}x = x,$$

$$T_2(x) = f(0) + \frac{f'(0)}{1!}x + \frac{f''(0)}{2!}x^2 = x,$$

$$T_3(x) = f(0) + \frac{f'(0)}{1!}x + \frac{f''(0)}{2!}x^2 + \frac{f^{(3)}(0)}{3!}x^3 = x - \frac{x^3}{6},$$

$$T_n(x) = x - \frac{x^3}{3!} + \frac{x^5}{5!} - \frac{x^7}{7!} + \cdots.$$

The Taylor series of $\sin x$ at $x_0 = 0$ is

$$\sum_{k=0}^{\infty} (-1)^k \frac{x^{2k+1}}{(2k+1)!}$$

(d) Consider the function $f : \mathbb{R} \to \mathbb{R}$, $f(x) = \cos x$. We have

$$f^{(4k)}(x) = \cos x, \quad f^{(4k+1)}(x) = -\sin x, \quad f^{(4k+2)}(x) = -\cos x, \quad f^{(4k+3)}(x) = \sin x,$$

$$\forall k \geq 0, \quad f^{(4k)}(0) = 1, \quad f^{(4k+1)}(0) = 0, \quad f^{(4k+2)}(0) = -1, \quad f^{(4k+3)}(0) = 0.$$

We deduce that the Taylor polynomials of $\cos x$ at $x_0 = 0$ are

$$T_1(x) = f(0) + \frac{f'(0)}{1!}x = 1,$$

$$T_2(x) = f(0) + \frac{f'(0)}{1!}x + \frac{f''(0)}{2!}x^2 = 1 - \frac{x^2}{2!},$$

$$T_3(x) = f(0) + \frac{f'(0)}{1!}x + \frac{f''(0)}{2!}x^2 + \frac{f^{(3)}(0)}{3!}x^3 = 1 - \frac{x^2}{2!},$$

$$T_n(x) = 1 - \frac{x^2}{2!} + \frac{x^4}{4!} - \frac{x^6}{6!} + \cdots.$$

The Taylor series of $\cos x$ at $x_0 = 0$ is

$$\sum_{k=0}^{\infty}(-1)^k \frac{x^{2k}}{(2k)!}.$$

(e) Fix a real number α and define $f : (0,\infty) \to \mathbb{R}$, $f(x) = x^\alpha$. We have

$$f'(x) = \alpha x^{\alpha-1},\ \ f^{(2)}(x) = \alpha(\alpha-1)x^{\alpha-2},\ \ f^{(k)}(x) = \alpha(\alpha-1)\cdots(\alpha-(k-1)\,)x^{\alpha-k}.$$

We deduce that

$$f^{(k)}(1) = \alpha(\alpha-1)\cdots(\alpha-(k-1)\,)$$

and thus the degree n Taylor polynomial of x^α at $x_0 = 1$ is

$$T_n(x) = 1 + \frac{\alpha}{1!}(x-1) + \frac{\alpha(\alpha-1)}{2!}(x-1)^2 + \cdots + \frac{\alpha(\alpha-1)\cdots(\alpha-(n-1)\,)}{n!}(x-1)^n.$$

The coefficients of the above polynomial coincide with the binomial coefficients if α is a natural number. For this reason, for any $\alpha \in \mathbb{R}$ we introduce the notation

$$\binom{\alpha}{0} = 1,\ \ \binom{\alpha}{n} = \frac{\alpha(\alpha-1)\cdots(\alpha-(n-1)\,)}{n!},\ \ n \in \mathbb{N}.$$

The degree n Taylor polynomial of x^α at $x_0 = 1$ can then be described in the more compact form

$$T_n(x) = \sum_{k=0}^{n}\binom{\alpha}{k}(x-1)^{\alpha-k}.$$

□

Remark 8.3. The degree n Taylor polynomial of a function f at a point x_0 is the unique polynomial of degree $\leq n$ such that

$$T_n(x_0) = f(x_0),\ \ T_n'(x_0) = f'(x_0),\ \ T_n^{(k)}(x_0) = f^{(k)}(x_0),\ \ \forall k = 1,\dots,n.$$

Exercise 8.1 asks you to prove this fact. □

Example 8.2 shows that the degree 1 Taylor polynomial of a differentiable function at a point x_0 is the linear approximation of f at x_0, and we know that it provides a very good approximation for $f(x)$ if x is near x_0. The next result states that the same is true for the higher degree Taylor polynomials.

Theorem 8.4 (Taylor approximation). *Suppose that $f : [a,b] \to \mathbb{R}$ is $(n+1)$-times differentiable, $n \in \mathbb{N}$. Fix $x_0 \in [a,b]$. We form the degree n Taylor polynomial of f at x_0*

$$T_n(x) = f(x_0) + \frac{f'(x_0)}{1!}(x-x_0) + \cdots + \frac{f^{(n)}(x_0)}{n!}(x-x_0)^n$$

and we consider the remainder (or error)

$$R_n(x_0,x) = f(x) - T_n(x),\ \ x \in [a,b].$$

Fix $x \in [a,b]$, $x \neq x_0$, and a continuous function $\varphi : [x_0, x] \to \mathbb{R}$ which is differentiable on (x_0, x) and $\varphi'(t) \neq 0$, $\forall t \in (x_0, x)$. (Here we are deliberately a bit negligent and we think of $[x_0, x]$ as the closed interval with endpoints x_0, x, even in the case $x_0 > x$.)

Then there exists ξ in the open interval with endpoints x_0 and x such that

$$R_n(x_0, x) = \frac{\varphi(x) - \varphi(x_0)}{n!\varphi'(\xi)} f^{(n+1)}(\xi)(x-\xi)^n. \tag{8.1}$$

Proof. Consider the function $F : [x_0, x] \to \mathbb{R}$ given by

$$F(t) = f(x) - \left(f(t) + \frac{f'(t)}{1!}(x-t) + \cdots + \frac{f^{(n)}(t)}{n!}(x-t)^n \right), \quad \forall t \in [x_0, x].$$

Note that $F(x) = 0$, $F(x_0) = R_n(x_0, x)$. From Cauchy's finite increment theorem, Theorem 7.38, we deduce that there exists ξ in the interval (x_0, x) such that

$$\frac{F(x) - F(x_0)}{\varphi(x) - \varphi(x_0)} = \frac{F'(\xi)}{\varphi'(\xi)}.$$

Now observe that

$$\begin{aligned} -F'(t) &= f'(t) + \left(\frac{f''(t)}{1!}(x-t) - \frac{f'(t)}{1!} \right) + \left(\frac{f^{(3)}(t)}{2!}(x-t)^2 - \frac{f^{(2)}(t)}{1!}(x-t) \right) \\ &\quad + \cdots + \left(\frac{f^{(n+1)}(t)}{n!}(x-t)^n - \frac{f^{(n)}(t)}{(n-1)!}(x-t)^{n-1} \right) \\ &= \frac{f^{(n+1)}(t)}{n!}(x-t)^n. \end{aligned}$$

Thus

$$-\frac{R_n(x_0, x)}{\varphi(x) - \varphi(x_0)} = \frac{F(x) - F(x_0)}{\varphi(x) - \varphi(x_0)} = \frac{F'(\xi)}{\varphi'(\xi)} = -\frac{f^{(n+1)}(t)(x-\xi)^n}{n!\varphi'(\xi)}.$$

The last equality clearly implies (8.1). □

If we let $\varphi(t) = (x-t)^{n+1}$ in the above theorem, we obtain the following important consequence.

Corollary 8.5 (Lagrange Remainder Formula). *There exists $\xi \in (x_0, x)$ such that*

$$\boxed{f(x) - T_n(x) = R_n(x_0, x) = \frac{1}{(n+1)!} f^{(n+1)}(\xi)(x - x_0)^{n+1}.} \tag{8.2}$$

Proof. We have $\varphi(x) = 0$ and $\varphi(x) - \varphi(x_0) = -(x-x_0)^{n+1}$, $\varphi'(\xi) = -(n+1)(x-\xi)^n$. □

Remark 8.6. Let us explain how this works in applications. Suppose that $f : [a,b] \to \mathbb{R}$ is $(n+1)$-times differentiable and $x_0 \in [a,b]$. The degree n Taylor polynomial of f at x_0 is

$$T_n(x) = f(x_0) + \frac{f'(x_0)}{1!}(x-x_0) + \cdots + \frac{f^{(n)}(x_0)}{n!} .(x-x_0)^n.$$

It is convenient to introduce the notation $h = x - x_0$ so that $x = x_0 + h$ and we deduce

$$T_n(x_0 + h) = f(x_0) + \frac{f'(x_0)}{1!}h + \cdots + \frac{f^{(n)}(x_0)}{n!}h^n.$$

If h is sufficiently small, then $T_n(x_0 + h)$ is an approximation for $f(x_0 + h)$. The error of this approximation is given by the remainder $R_n(x_0, x) = f(x_0 + h) - T_n(x_0 + h)$. This remainder really depends only on the difference $h = x - x_0$ and, to emphasize this fact, we will write $R_n(x_0, h)$ instead of $R_n(x_0, x)$ in the argument below. Also, for simplicity, we will denote by $(x_0, x_0 + h)$ the open interval with endpoints x_0 and $x_0 + h$. (Note that $x_0 + h < x_0$ when $h < 0$.)

The Lagrange remainder formula tells us that there exists $\xi \in (x_0, x_0 + h)$ such that

$$R_n(x_0, h) = \frac{1}{(n+1)!}f^{(n+1)}(\xi)h^{n+1}.$$

If we define

$$M_{n+1}(x_0, h) := \sup_{\xi \in [x_0, x_0+h]} |f^{(n+1)}(\xi)|,$$

then we deduce

$$|\, R_n(x_0, h)\, | \leq \frac{M_{n+1}(x_0, h)|h|^{n+1}}{(n+1)!}. \tag{8.3}$$

If the right-hand side of the above inequality is small, then the error has to be small. The above result implies that

$$\big|\, f(x) - T_n(x)\, \big| = O\big(\, |x - x_0|^{n+1}\,\big) \text{ as } x \to x_0, \tag{8.4}$$

where O is Landau's symbol defined in (5.34). □

Example 8.7. Let us show how the above remark works in a rather concrete case. Suppose $f(x) = \sin x$. We use Taylor approximations of $\sin x$ at $x_0 = 0$. For example, the degree 4 Taylor polynomial of $\sin x$ at $x_0 = 0$ is

$$T_4(h) = \sin(0) + \frac{\cos(0)}{1!}h - \frac{\sin(0)}{2!}h^2 - \frac{\cos(0)}{3!}h^3 + \frac{\sin(0)}{4!}h^4 = h - \frac{h^3}{3!} = h - \frac{h^3}{6}.$$

We have

$$\sin h \approx h - \frac{h^3}{6}.$$

To estimate the error of this approximation we use (8.2). The 5th derivative of $\sin x$ is $\cos x$ so that $|\cos\xi| \leq 1$, $\forall x \in \mathbb{R}$. We deduce from (8.2) that for some ξ between 0 and x we have

$$\left|\sin h - \left(h - \frac{h^3}{6}\right)\right| = \frac{|\cos\xi|}{5!}h^5 \leq \frac{|h|^5}{5!} = \frac{|h|^5}{120}.$$

If for example $|h| \leq \frac{1}{2}$, then

$$\frac{|h|^5}{120} \leq \frac{1}{32 \cdot 120} = \frac{1}{3840} < \frac{1}{10^3}.$$

Thus for $|h| \leq \frac{1}{2}$ the expression $h - \frac{h^3}{6}$ approximates $\sin h$ up to two decimals. For example

$$0.5 - (0.5)^3/6 = 0.47916... \Rightarrow \sin 0.5 = 0.47...$$

If $h = \frac{1}{4}$, then

$$\frac{|h|^5}{120} = \frac{1}{4^5 \cdot 120} = \frac{1}{1024 \cdot 120} = \frac{1}{122880} \leq \frac{1}{10^5},$$

and $0.25 - (0.25)^3/6$ computes $\sin(0.25)$ up to four decimals. Thus

$$0.25 - (0.25)^3/6 = 0.248666... \Rightarrow \sin(0.25) = 0.2486....$$

In Figure 8.1 we have depicted side-by-side the graph of $\sin(x)$ for $|x| \leq 10$ and the graph of $T_7(x)$, its degree 7 Taylor approximation at $x_0 = 0$. While $T_7(x)$ takes large values for $|x|$ large, it matches very well the graph of $\sin x$ on the interval $[-3, 3]$. □

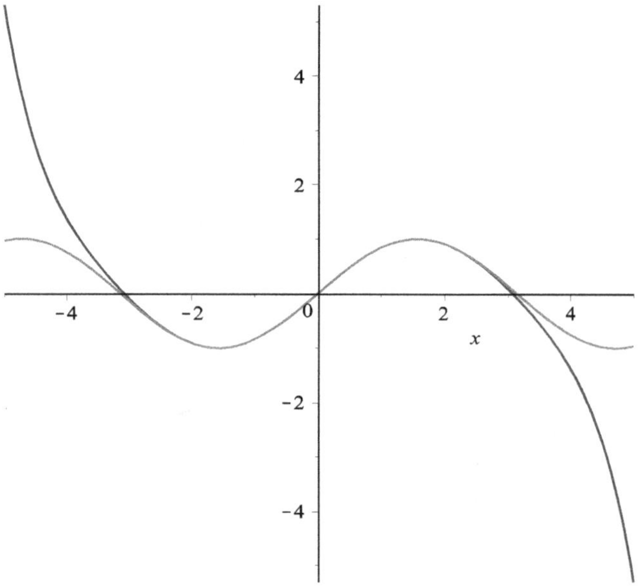

Fig. 8.1 *The graphs of* $\sin x$ *and its degree* 7 *Taylor approximation at the origin.*

Here is a nice consequence of Corollary 8.5.

Corollary 8.8. *For any* $x \in \mathbb{R}$ *we have*

$$e^x = \sum_{n=0}^{\infty} \frac{x^n}{n!}. \tag{8.5}$$

Note that for $x = 1$ *the above equality specializes to (4.26).*

Proof. Observe that for any natural number n the partial sum

$$s_n(x) = 1 + \frac{x}{1!} + \cdots + \frac{x^n}{n!}$$

is the n-th Taylor polynomial of e^x at $x_0 = 0$. Corollary 8.5 implies that there exists a real number ξ_n between 0 and x such that

$$e^x - s_n(x) = e^{\xi_n} \frac{x^{n+1}}{(n+1)!}.$$

Observe that since $-|x| \leq \xi_n \leq |x|$ we have $e^{\xi_n} \leq e^{|x|}$ so that

$$\left| e^x - s_n(x) \right| \leq e^{|x|} \frac{|x|^{n+1}}{(n+1)!}. \tag{8.6}$$

From (4.8) we deduce that

$$\lim_{n\to\infty} \frac{|x|^{n+1}}{(n+1)!} = 0.$$

The Squeezing Principle then implies that

$$\lim_{n\to\infty} \left| e^x - s_n(x) \right| = 0.$$

□

Remark 8.9. The above proof shows a bit more namely that for any $R > 0$, the partial sums $s_n(x)$ converge to e^x *uniformly* on $[-R, R]$. Indeed, if $x \in [-R, R]$ so that $|x| \leq R$, then (8.6) implies that

$$\left| e^x - s_n(x) \right| \leq e^R \frac{R^{n+1}}{(n+1)!}, \quad \forall |x| \leq R.$$

Note that the right-hand side of the above inequality is independent of x and converges to 0 as $n \to \infty$ according to (4.8). Weierstrass criterion in Exercise 6.6 implies the claimed uniform convergence. □

8.2. L'Hôpital's Rule

Differential calculus is also very useful in dealing with singular limits such as $\frac{0}{0}$, $\frac{\infty}{\infty}$.

Proposition 8.10 (L'Hôpital's Rule). *Let $a, b \in [-\infty, \infty]$, $a < b$. Suppose that the differentiable functions $f, g : (a, b) \to \mathbb{R}$ satisfy the following conditions.*

(i) $g'(x) \neq 0$, $\forall x \in (a, b)$.

(ii)

$$\lim_{x\nearrow b} \frac{f'(x)}{g'(x)} = A \in [-\infty, \infty].$$

(iii) Either

$$\lim_{x\nearrow b} f(x) = \lim_{x\nearrow b} g(x) = 0, \tag{iii$_0$}$$

or

$$\lim_{x\nearrow b} g(x) = \pm\infty. \tag{iii$_\infty$}$$

Then

$$\lim_{x \nearrow b} \frac{f(x)}{g(x)} = A.$$

Proof. Let us first observe that (i) and Rolle's Theorem imply that g is injective. Hence, there exists $a' \in [a, b)$ such that $g(x) \neq 0$, $\forall x \in (a', b)$. Without any loss of generality we can assume that $a = a'$ since we are interested in the behavior of f, g near b. We have to prove that for any sequence $x_n \in (a, b)$ such that $\lim x_n = b$ we have

$$\lim_{n\to\infty} \frac{f(x_n)}{g(x_n)} = A.$$

Fix one such sequence $(x_n)_{n\in\mathbb{N}}$. At this point we want to invoke the following auxiliary fact whose proof we postpone.

Lemma 8.11. *There exists a sequence (y_n) in (a, b) such that $x_n \neq y_n$, $\forall n$, $\lim_{n\to\infty} y_n = b$ and*

$$\lim \left(\frac{|f(y_n)| + |g(y_n)|}{|g(x_n)|} \right) = 0. \qquad \square$$

Choose a sequence (y_n) as in the above lemma so that

$$\lim_{n\to\infty} \frac{f(y_n)}{g(x_n)} = \lim_{n\to\infty} \frac{g(y_n)}{g(x_n)} = 0.$$

From Cauchy's Finite Increment Theorem 7.38 we deduce that there exists $\xi_n \in (x_n, y_n)$ such that

$$r_n = \frac{f(x_n) - f(y_n)}{g(x_n) - g(y_n)} = \frac{f'(\xi_n)}{g'(\xi_n)}.$$

Since $x_n \to b$ we deduce $\xi_n \to b$ so that

$$\lim_{n\to\infty} r_n = \lim_{n\to\infty} \frac{f'(\xi_n)}{g'(\xi_n)} = A. \tag{8.7}$$

On the other hand, for any n we have

$$r_n = \frac{f(x_n) - f(y_n)}{g(x_n) - g(y_n)} = \frac{f(x_n) - f(y_n)}{g(x_n)\left(1 - \frac{g(y_n)}{g(x_n)}\right)} = \frac{\frac{f(x_n)}{g(x_n)} - \frac{f(y_n)}{g(x_n)}}{1 - \frac{g(y_n)}{g(x_n)}}.$$

We deduce

$$\frac{f(x_n)}{g(x_n)} - \frac{f(y_n)}{g(x_n)} = r_n \left(1 - \frac{g(y_n)}{g(x_n)}\right) \Rightarrow \frac{f(x_n)}{g(x_n)} = \frac{f(y_n)}{g(x_n)} + r_n \left(1 - \frac{g(y_n)}{g(x_n)}\right).$$

Hence

$$\lim_{n\to\infty} \frac{f(x_n)}{g(x_n)} = \underbrace{\lim_{n\to\infty} \frac{f(y_n)}{g(x_n)}}_{=0} + \left(\lim_{n\to\infty} r_n\right) \cdot \underbrace{\lim_{n\to\infty} \left(1 - \frac{g(y_n)}{g(x_n)}\right)}_{=1}$$

$$= \lim_{n\to\infty} r_n \overset{(8.7)}{=} A.$$

All there is left to do is to prove Lemma 8.11.
Proof of Lemma 8.11. We consider two cases.

1. Suppose that (iii$_0$) holds, i.e.,

$$\lim_{x\nearrow b} f(x) = \lim_{\nearrow b} g(x) = 0.$$

Then for any n we can find $y_n \in (x_n, b)$ such that

$$|f(y_n)| + |g(y_n)| < \frac{1}{n}|g(x_n)|,$$

so that

$$\frac{|f(y_n)| + |g(y_n)|}{|g(x_n)|} < \frac{1}{n}, \quad \forall n,$$

and thus

$$\lim_{n\to\infty} \frac{|f(y_n)| + |g(y_n)|}{|g(x_n)|} = 0.$$

2. Suppose that (iii$_\infty$) holds, i.e.,

$$\lim_{n\to\infty} g(x_n) = \pm\infty.$$

For $t \in (a,b)$ we set $h(t) := |f(t)| + |g(t)|$. We construct inductively an increasing sequence of natural numbers (n_k) as follows.

A. Since $|g(x_n)| \to \infty$ there exists $n_0 \in \mathbb{N}$ such that

$$|g(x_n)| > h(x_1), \quad \forall n \geq n_0.$$

B. Since $|g(x_n)| \to \infty$, for any $k \in \mathbb{N}$, $k > 1$, we can find $n_k \in \mathbb{N}$ such that $n_k > n_{k-1}$ and

$$|g(x_n)| > 2^k h(x_{n_{k-1}}) \ \forall n \geq n_k. \tag{8.8}$$

Now define y_n by setting

$$y_n := \begin{cases} x_1, & 1 \leq n < n_1 \\ x_{n_{k-1}}, & n_k \leq n < n_{k+1}, \ k \in \mathbb{N}. \end{cases}$$

Observe that for $n \in [n_k, n_{k+1})$ we have

$$\frac{h(y_n)}{|g(x_n)|} = \frac{|h(x_{n_{k-1}})|}{g(x_n)} \overset{(8.8)}{<} \frac{1}{2^k}.$$

This proves that

$$\lim_{n\to\infty} \frac{h(y_n)}{|g(x_n)|} = 0. \qquad \square$$

$\square$

Remark 8.12. Proposition 8.10 has a counterpart involving the left limit $\lim_{x\searrow a}$. Its statement is obtained from the statement of Proposition 8.10 by globally replacing the limit at b with the limit at a. The proof is entirely similar. $\square$

Example 8.13. (a) We want to compute

$$\lim_{x\to 0}\frac{1-\cos x}{x^2}.$$

According to L'Hôpital's Theorem we have

$$\lim_{x\to 0}\frac{1-\cos x}{x^2}=\lim_{x\to 0}\frac{(1-\cos x)'}{(x^2)'}=\lim_{x\to 0}\frac{\sin x}{2x}=\frac{1}{2}.$$

(b) Consider the function $f:(0,\infty)\to\mathbb{R}$, $f(x)=x^x$. We want to investigate the limit

$$\lim_{x\to 0}x^x.$$

Formally the limit ought to be 0^0, but we do not know what 0^0 means. Consider a new function

$$g(x)=\ln x^x=x\ln x,\quad x>0.$$

In this case we have

$$\lim_{x\to 0+}g(x)=0\cdot(-\infty)$$

which is a degenerate limit. We rewrite

$$g(x)=\frac{\ln x}{\frac{1}{x}}$$

and we observe that in this case

$$\lim_{x\to 0+}g(x)=-\frac{\infty}{\infty}$$

which suggests trying L'Hôpital's Rule. We have

$$(\ln x)'=\frac{1}{x},\quad (1/x)'=-\frac{1}{x^2}$$

and

$$\frac{1/x}{-1/x^2}=-x\to 0\ \text{ as } x\to 0+.$$

Hence

$$\lim_{x\to 0+}g(x)=0\Rightarrow\lim_{x\to 0+}f(x)=e^0=1.$$ □

8.3. Convexity

In this section we discuss in some detail a concept that has find many applications.

8.3.1. *Basic facts about convex functions*

We begin with a simple geometric observation.

Proposition 8.14. *Let $x, x_1, x_2 \in \mathbb{R}$, $x_1 < x_2$. The following statements are equivalent.*

(i) $x \in [x_1, x_2]$.
(ii) There exist $t_1, t_2 \geq 0$ such that $t_1 + t_2 = 1$ and $x = t_1x_1 + t_2x_2$.

Proof. (i) $\Rightarrow$ (ii) Suppose $x \in [x_1, x_2]$. We set

$$t_1 := \frac{x_2 - x}{x_2 - x_1}, \quad t_2 := \frac{x - x_1}{x_2 - x_1}. \tag{8.9}$$

Since $x_1 \leq x \leq x_2$ we deduce that $t_1, t_2 \geq 0$. We observe that

$$t_1 + t_2 = \frac{x_2 - x}{x_2 - x_1} + \frac{x - x_1}{x_2 - x_1} = \frac{x_2 - x_1}{x_2 - x_1} = 1,$$

and

$$t_1x_1 + t_2x_2 = \frac{x_1(x_2 - x) + x_2(x - x_1)}{x_2 - x_1} = \frac{x_2x - x_1x}{x_2 - x_1} = x. \tag{8.10}$$

(ii) $\Rightarrow$ (i) Suppose that there exist $t_1, t_2 \geq 0$ such that $t_1 + t_2 = 1$ and $x = t_1x_1 + t_2x_2$. We have

$$x - x_1 = (t_1 - 1)x_1 + t_2x_2 = -t_2x_1 + t_2x_2 = t_2(x_2 - x_1) \geq 0,$$

$$x_2 - x = (1 - t_2)x_2 - t_1x_1 = t_1x_2 - t_1x_1 = t_1(x_2 - x_1) \geq 0.$$

Hence $x_1 \leq x \leq x_2$. □

Remark 8.15. The point $t_1x_1 + t_2x_2$ is interpreted as the center of mass of a system of two particles, one located at x_1 and of mass t_1 and the other located at x_2 and of mass t_2.

In general, given n particles of masses $m_1, \dots, m_n$ respectively located at $x_1, \dots, x_n$, then the *center of mass* of this system is the point

$$\bar{x} = \frac{m_1x_1 + \cdots + m_nx_n}{m_1 + \cdots + m_n}.$$

Note that if we define

$$t_k := \frac{m_k}{m_1 + m_2 + \cdots + m_n}, \quad k = 1, 2, \dots, n,$$

then

$$t_1 + t_2 + \cdots + t_n = 1 \text{ and } \bar{x} = t_1x_1 + \cdots + t_nx_n.$$

Thus, a point x lies between x_1 and x_2 if and only if it is the center of mass of a system of particles located at x_1 and x_2. □

Given a function $f : (a, b) \to \mathbb{R}$ and $x_1, x_2 \in (a, b)$, $x_1 < x_2$, we denote by $L^f_{x_1,x_2}$ the linear function whose graph contains the points $(x_1, f(x_1))$ and $(x_2, f(x_2))$ on the graph of f. The slope of this line is

$$m = \frac{f(x_2) - f(x_1)}{x_2 - x_1}$$

and thus the equation of this line is

$$\begin{aligned} L^f_{x_1,x_2}(x) &= f(x_1) + m(x - x_1) = f(x_1) + \frac{f(x_2) - f(x_1)}{x_2 - x_1}(x - x_1) \\ &= f(x_1)\left(1 - \frac{x - x_1}{x_2 - x_1}\right) + f(x_2)\frac{x - x_1}{x_2 - x_1} = \frac{x_2 - x}{x_2 - x_1}f(x_1) + f(x_2)\frac{x - x_1}{x_2 - x_1}. \end{aligned}$$

Hence

$$L^f_{x_1,x_2}(x) = \frac{x_2 - x}{x_2 - x_1}f(x_1) + \frac{x - x_1}{x_2 - x_1}f(x_2). \tag{8.11}$$

Above we recognize the numbers t_1, t_2 defined in (8.9).

Proposition 8.16. *Consider a function $f : (a, b) \to \mathbb{R}$ and $x_1, x_2 \in (a, b)$, $x_1 < x_2$. Denote by $L^f_{x_1,x_2}$ the linear function whose graph contains the points $(x_1, f(x_1))$ and $(x_2, f(x_2))$ on the graph of f. The following statements are equivalent.*

$$f(x) \leq L^f_{x_1,x_2}(x), \;\; \forall x \in [x_1, x_2]. \tag{8.12a}$$

$$f(x) \leq \frac{x_2 - x}{x_2 - x_1}f(x_1) + \frac{x - x_1}{x_2 - x_1}f(x_2), \;\; \forall x \in [x_1, x_2]. \tag{8.12b}$$

$$\forall t_1, t_2 \geq 0 \;\; \textit{such that} \;\; t_1 + t_2 = 1 \;\; f(t_1x_1 + t_2x_2) \leq t_1f(x_1) + t_2f(x_2), \tag{8.12c}$$

Proof. The equivalence (8.12a) $\Longleftrightarrow$ (8.12b) follows from (8.11). The equivalence (8.12b) $\Longleftrightarrow$ (8.12c) follows from (8.9) and (8.10). □

Remark 8.17. The part of the graph of $L^f_{x_1,x_2}$ over the interval $[x_1, x_2]$ is called the *chord* of the graph of f determined by the interval $[x_1, x_2]$. The condition (8.12a) is equivalent to saying that the part of the graph of f corresponding to the interval $[x_1, x_2]$ lies below the chord of the graph determined by this interval; see Figure 8.2. □

Definition 8.18. Let $f : I \to \mathbb{R}$ be a real valued function defined on an interval I.

(i) The function f is called *convex* if, for any $x_1, x_2 \in I$, and any $t_1, t_2 \geq 0$ such that $t_1 + t_2 = 1$, we have

$$f(t_1x_1 + t_2x_2) \leq t_1f(x_1) + t_2f(x_2).$$

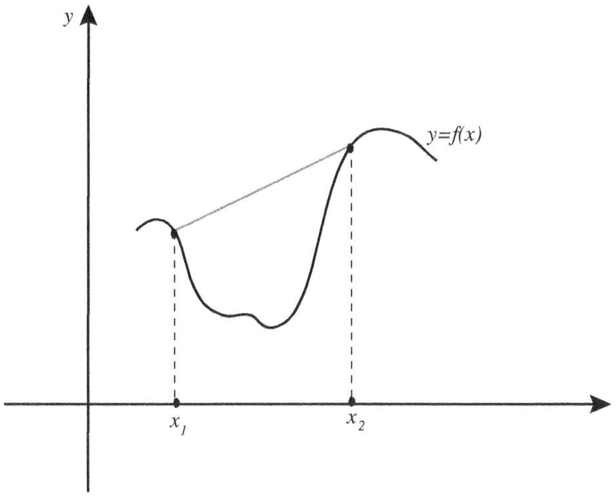

Fig. 8.2 *The graph lies below the chord.*

(ii) The function f is called *concave* if, for any $x_1, x_2 \in I$, and any $t_1, t_2 \geq 0$ such that $t_1 + t_2 = 1$, we have

$$f(t_1x_1 + t_2x_2) \geq t_1f(x_1) + t_2f(x_2).$$

□

Remark 8.19. (a) From Propositions 8.14 and 8.16 we deduce that a function $f : I \to \mathbb{R}$ is convex if and only if, for any interval $[x_1, x_2] \subset I$, the part of the graph of f determined by the interval $[x_1, x_2]$ is below the chord of the graph determined by this interval. It is concave if the graph is above the chords.
(b) Observe that if $t_1 = 1$ and $t_2 = 0$ we have $t_1x_1 + t_2x_2 = x_1$ and $t_1f(x_1) + t_2f(x_2) = f(x_1)$ and thus the inequality

$$f(t_1x_1 + t_2x_2) \leq t_1f(x_1) + t_2f(x_2)$$

is trivially satisfied. A similar thing happens when $t_1 = 0$ and $t_2 = 1$. Thus the definition of convexity is equivalent to the weaker requirement that for any $x_1, x_2 \in I$, and any *positive* t_1, t_2 such that $t_1 + t_2 = 1$, we have

$$f(t_1x_1 + t_2x_2) \leq t_1f(x_1) + t_2f(x_2).$$

(c) Observe that a function f is concave if and only if $-f$ is convex.
(d) In many calculus texts, convex functions are called *concave-up* and concave functions are called *concave-down*. □

Before we can give examples of convex functions we need to produce simple criteria for recognizing when a function is convex.

Proposition 8.14 implies that a function $f : I \to \mathbb{R}$ is convex if and only if for any $x_1, x_2 \in I$ and any $x \in (x_1, x_2)$ we have

$$f(x) \leq \frac{x_2 - x}{x_2 - x_1} f(x_1) + \frac{x - x_1}{x_2 - x_1} f(x_2).$$

Since

$$1 = \frac{x_2 - x}{x_2 - x_1} + \frac{x - x_1}{x_2 - x_1},$$

we deduce that f is convex if and only if

$$\begin{aligned} f(x)\left(\frac{x_2 - x}{x_2 - x_1} + \frac{x - x_1}{x_2 - x_1}\right) &\leq \frac{x_2 - x}{x_2 - x_1} f(x_1) + \frac{x - x_1}{x_2 - x_1} f(x_2) \\ \iff \frac{x_2 - x}{x_2 - x_1}\big(f(x) - f(x_1)\big) &\leq \frac{x - x_1}{x_2 - x_1}\big(f(x_2) - f(x)\big) \\ \iff (x_2 - x)\big(f(x) - f(x_1)\big) &\leq (x - x_1)\big(f(x_2) - f(x)\big) \\ \iff \frac{f(x) - f(x_1)}{x - x_1} &\leq \frac{f(x_2) - f(x)}{x_2 - x}. \end{aligned}$$

We have thus proved the following result.

Corollary 8.20. *Let $f : I \to \mathbb{R}$ be a function defined on the interval $I \subset \mathbb{R}$. The following statements are equivalent.*

(i) The function f is convex.
(ii) For any $x_1, x, x_2 \in I$ such that $x_1 < x < x_2$ we have

$$\frac{f(x) - f(x_1)}{x - x_1} \leq \frac{f(x_2) - f(x)}{x_2 - x}.$$

□

Let us observe that $\frac{f(x) - f(x_1)}{x - x_1}$ is the slope of the chord determined by $[x_1, x]$ while $\frac{f(x_2) - f(x)}{x_2 - x}$ is the slope of the chord determined by $[x, x_2]$. The above result states that f is convex if and only if for any $x_1 < x < x_2$ the chord determined by $[x_1, x]$ has a smaller inclination than the chord determined by $[x, x_2]$; see Figure 8.3.

Corollary 8.21. *Suppose that $f : I \to \mathbb{R}$ is a convex function. Then*

$$\frac{f(x_2) - f(x_1)}{x_2 - x_1} \leq \frac{f(x_4) - f(x_3)}{x_4 - x_3}, \quad \forall x_1, x_2, x_3, x_4 \in I, \;\; x_1 < x_2 < x_3 < x_4.$$

Proof. From Corollary 8.20 we deduce that the slope of the chord determined by $[x_1, x_2]$ is smaller than the slope of the chord determined by $[x_2, x_3]$ which in turn is smaller than the slope of the chord determined by $[x_3, x_4]$; see Figure 8.4. In other words,

$$\frac{f(x_2) - f(x_1)}{x_2 - x_1} \leq \frac{f(x_3) - f(x_2)}{x_3 - x_2} \leq \frac{f(x_4) - f(x_3)}{x_4 - x_3}.$$

□

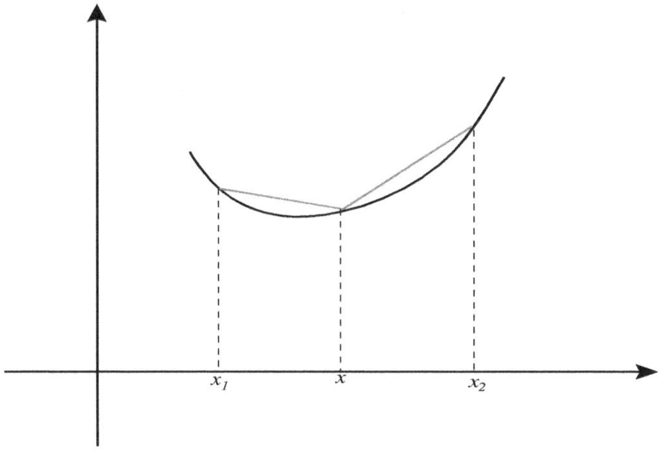

Fig. 8.3 *Chords of the graph of a convex function become less inclined as they move to the right.*

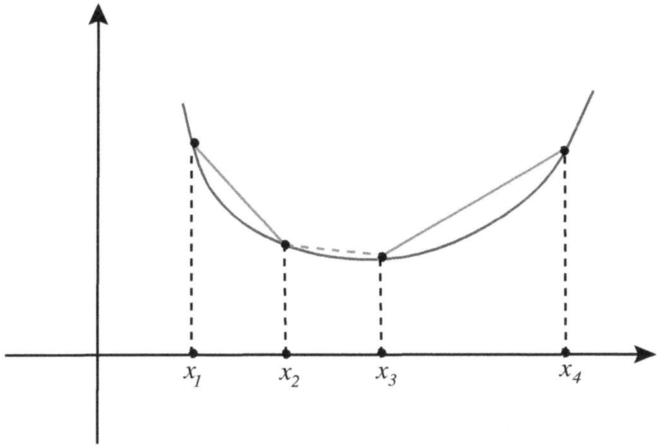

Fig. 8.4 *Chords of the graph of a convex function become more inclined as they move to the right.*

Corollary 8.22. *Suppose that $f : I \to \mathbb{R}$ is a differentiable function. Then the following statements are equivalent.*

(i) The function f is convex.
(ii) The derivative f' is a nondecreasing function.

Proof. (ii) $\Rightarrow$ (i) In view of Corollary 8.20 we have to prove that for any $x_1 < x_2 < x_3 \in I$ we have

$$\frac{f(x_2) - f(x_1)}{x_2 - x_1} \leq \frac{f(x_3) - f(x_2)}{x_3 - x_2}.$$

From Lagrange's Mean Value Theorem we deduce that there exist $\xi_1 \in (x_1, x_2)$ and $\xi_2 \in (x_2, x_3)$ such that

$$f'(\xi_1) = \frac{f(x_2) - f(x_1)}{x_2 - x_1}, \quad f'(\xi_2) = \frac{f(x_3) - f(x_2)}{x_3 - x_2}.$$

Since f' is nondecreasing and $\xi_1 < x_2 < \xi_2$, we deduce $f'(\xi_1) \le f'(\xi_2)$.

(i) $\Rightarrow$ (ii) We know that f is convex and we have to prove that f' is nondecreasing, i.e.,

$$x_1 < x_2 \Rightarrow f'(x_1) \le f'(x_2).$$

For $h > 0$ sufficiently small, $h < \frac{1}{2}(x_2 - x_1)$, we have

$$x_1 < x_1 + h < x_2 - h < x_2.$$

From Corollary 8.21 we deduce that slope of the chord determined by $[x_1, x_1+h]$ is smaller than the slope of the chord determined by $[x_2 - h, x_2]$, that is,

$$\frac{f(x_1 + h) - f(x_1)}{h} \le \frac{f(x_2) - f(x_2 - h)}{h} = \frac{f(x_2 - h) - f(x_2)}{-h}.$$

Hence

$$f'(x_1) = \lim_{h \to 0+} \frac{f(x_1 + h) - f(x_1)}{h} \le \lim_{h \to 0+} \frac{f(x_2 - h) - f(x_2)}{-h} = f'(x_2).$$

□

Since a differentiable function is nondecreasing iff its derivative is nonnegative, we deduce the following useful result.

Corollary 8.23. *Suppose that $f : I \to \mathbb{R}$ is a twice differentiable function. Then the following statements are equivalent.*

(i) The function f is convex.
(ii) The second derivative f'' is nonnegative, $f''(x) \ge 0$, $\forall x \in I$.

□

Since a function is concave if and only if $-f$ is convex we deduce the following result.

Corollary 8.24. *Suppose that $f : I \to \mathbb{R}$ is a twice differentiable function. Then the following statements are equivalent.*

(i) The function f is concave.
(ii) The second derivative f'' is nonpositive, $f''(x) \le 0$, $\forall x \in I$.

□

Example 8.25. The function $f : \mathbb{R} \to \mathbb{R}$, $f(x) = e^x$ is convex since $f''(x) = e^x > 0$ for any $x \in \mathbb{R}$. The function $f : (0, \infty) \to \mathbb{R}$, $f(x) = \ln x$ is concave since

$$f'(x) = \frac{1}{x}, \quad f''(x) = -\frac{1}{x^2} < 0, \quad \forall x > 0.$$

Fix $\alpha \in \mathbb{R}$ and consider the power function

$$p : (0, \infty) \to \mathbb{R}, \quad p(x) = x^\alpha.$$

Then

$$p''(x) = \alpha(\alpha - 1)x^{\alpha-2}.$$

Note that if $\alpha(\alpha - 1) > 0$ this function is convex, if $\alpha(\alpha - 1) < 0$ this function is concave, and if $\alpha = 0$ or $\alpha = 1$ this function is both convex and concave. Thus, the function $\sqrt{x}$ is concave, while the function $\frac{1}{\sqrt{x}} = x^{-\frac{1}{2}}$ is convex. □

8.3.2. *Some classical applications of convexity*

We start with a simple geometric consequence of convexity.

Proposition 8.26. *Suppose that $f : I \to \mathbb{R}$ is a differentiable convex function. Then the graph of f lies above any tangent to the graph; see Figure 8.5. If additionally f' is strictly increasing, then any tangent to the graph intersects the graph at a unique point.*

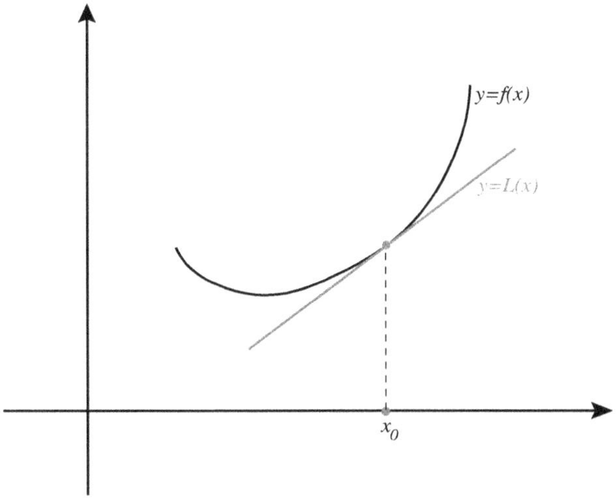

Fig. 8.5 *The graph of a convex function lies above any of its tangents.*

Proof. Let $x_0 \in I$. The tangent to the graph of f at the point $(x_0, f(x_0))$ is the graph of the linearization of f at x_0 which is the function

$$L(x) = f(x_0) + f'(x_0)(x - x_0).$$

We have to prove that

$$f(x) - L(x) \geq 0, \ \ \forall x \in I.$$

We have

$$f(x) - L(x) = f(x) - f(x_0) - f'(x_0)(x - x_0).$$

Suppose $x \neq x_0$. Lagrange's Mean Value Theorem implies that there exists ξ between x_0 and x such that $f(x) - f(x_0) = f'(\xi)(x - x_0)$. Hence

$$f(x) - L(x) = f'(\xi)(x - x_0) - f'(x_0)(x - x_0) = (\, f'(\xi) - f'(x_0)\,)(x - x_0).$$

We distinguish two cases.

1. $x > x_0$. Then $\xi > x_0$ and $(x - x_0) > 0$. Since f is convex, f' is increasing and thus $f'(\xi) \geq f'(x_0)$ so that

$$(\, f'(\xi) - f'(x_0)\,)(x - x_0) \geq 0.$$

Clearly if f' is strictly increasing, then $f'(\xi) > f'(x_0)$ and $(\, f'(\xi) - f'(x_0)\,)(x - x_0) > 0$.

2. $x < x_0$. Then $\xi < x_0$ and $(x - x_0) < 0$. Since f is convex, f' is increasing and thus $f'(\xi) \leq f'(x_0)$ so that

$$(\, f'(\xi) - f'(x_0)\,)(x - x_0) \geq 0.$$

Clearly if f' is strictly increasing, then $f'(\xi) < f'(x_0)$ and $(\, f'(\xi) - f'(x_0)\,)(x - x_0) > 0$.

□

Example 8.27 (Newton's Method). We want to describe an ingenious method devised by Isaac Newton[1] for approximating the solutions of an equation $f(x) = 0$.

Suppose that $f : (a, b) \to \mathbb{R}$ is a C^2-function such that

$$f'(x), \ \ f''(x) > 0, \ \ \forall x \in (a, b). \tag{8.13}$$

Suppose $z_0 \in (a, b)$ satisfies

$$f(z_0) = 0.$$

The condition (8.13) implies that f is strictly increasing and thus z_0 is the unique solution of the equation $f(x) = 0$. Newton's method described one way of constructing very accurate approximations for z_0.

Here is roughly the principle behind the method. Pick an arbitrary point $x_0 \in (z_0, b)$. The linearization $L(x)$ of f at x_0 is an approximation for $f(x)$. It is reasonable to hope that the solution of the equation $L(x) = 0$ approximates the solution of the equation $f(x) = 0$. Denote by $Z(x_0)$ the solution of the equation $L(x) = 0$, i.e., the point where the tangent to the graph of f at $(x_0, f(x_0))$ intersects the horizontal axis; see Figure 8.6.

More precisely, we have $L(x) = f(x_0) + f'(x_0)(x - x_0)$ and thus,

[1] Isaac Newton (1642–1726) was an English mathematician and physicist who is widely recognized as one of the most influential scientists of all time and a key figure in the scientific revolution; see Wikipedia.

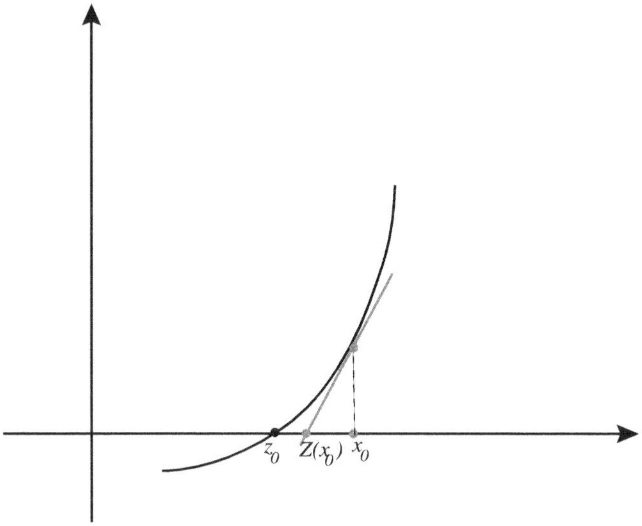

Fig. 8.6 *The geometry behind Newton's method.*

$$L(x) = 0 \iff f'(x_0)(x - x_0) = -f(x_0) \iff x - x_0 = -\frac{f(x_0)}{f'(x_0)}$$
$$\iff x = Z(x_0) = x_0 - \frac{f(x_0)}{f'(x_0)}.$$

Key Remark. *The point $Z(x_0)$ lies between z_0 and x_0, $z_0 < Z(x_0) < x_0$. In particular, $Z(x_0)$ is closer to z_0 than x_0.*

Clearly $Z(x_0) < x_0$ because $L(x_0) = f(x_0) > 0 = L(Z(x_0))$ and the linear function $L(x)$ is increasing. The assumption (8.13) implies that f is convex and f' is strictly increasing. Proposition 8.26 implies that the tangent lies below the graph, i.e.,

$$f\big(Z(x_0)\big) > L\big(Z(x_0)\big) = 0 = f(z_0).$$

Since f is strictly increasing we deduce $Z(x_0) > z_0$.

The correspondence $x_0 \mapsto Z(x_0)$ is thus a map $(z_0, b) \to (z_0, b)$ with the property that $z_0 < Z(x_0) < x_0$, $\forall x_0 \in (z_0, b)$.

We iterate this procedure. We set $x_1 = Z(x_0)$ so that $z_0 < x_1 < x_0$. Define next $x_2 = Z(x_1)$ so that $z_0 < x_2 < x_1$ and inductively

$$x_{n+1} := Z(x_n) = x_n - \frac{f(x_n)}{f'(x_n)}, \quad n \geq 0. \tag{8.14}$$

The above discussion shows that the sequence (x_n) is strictly decreasing and bounded below by z_0. It is therefore convergent and we set $\bar{x} = \lim x_n$. Observe that $\bar{x} \geq z_0$. Letting $n \to \infty$ in (8.14) and taking into account the continuity of f and f' we deduce

$$\bar{x} = \bar{x} - \frac{f(\bar{x})}{f'(\bar{x})} \Rightarrow \frac{f(\bar{x})}{f'(\bar{x})} = 0 \Rightarrow f(\bar{x}) = 0.$$

Since z_0 is the unique solution of the equation $f(x) = 0$ we deduce $\bar{x} = z_0$. Thus the sequence generated by Newton's iteration (8.14) converges to the unique zero of f.

Remarkably, the above sequence (x_n) converges to z_0 extremely quickly. Taylor's formula with Lagrange remainder implies that for any n there exists $\xi_n \in (z_0, x_n)$ such that

$$0 = f(z_0) = f(x_n) + f'(x_n)(z_0 - x_n) + \frac{1}{2}f''(\xi_n)(z_0 - x_n)^2.$$

Hence

$$0 = \frac{f(x_n)}{f'(x_n)} + z_0 - x_n + \frac{f''(\xi_n)}{2f'(x_n)}(z_0 - x_n)^2 \Rightarrow \underbrace{\frac{f(x_n)}{f'(x_n)} + z_0 - x_n}_{=z_0 - x_{n+1}} = -\frac{f''(\xi_n)}{2f'(x_n)}(z_0 - x_n)^2.$$

So

$$(z_0 - x_{n+1}) = -\frac{f''(\xi_n)}{2f'(x_n)}(z_0 - x_n)^2.$$

If we denote by ε_n the error, $\varepsilon_n := x_n - z_0$ we deduce

$$\varepsilon_{n+1} = \frac{f''(\xi_n)}{2f'(x_n)}\varepsilon_n^2. \tag{8.15}$$

Thus, the error at the $(n+1)$-th step is roughly the square of the error at the n-th step. If, e.g., the error ε_n is < 0.01, then we expect $\varepsilon_{n+1} < (0.1)^2 = 0.0001$.

Let us see how this works in a simple case. Let k be a natural number ≥ 2. Consider the function

$$f : (0, \infty) \to \mathbb{R}, \;\; f(x) = x^k - 2.$$

Then

$$f'(x) = kx^{k-1}, \;\; f''(x) = k(k-1)x^{k-2},$$

so the assumption (8.13) is satisfied. The unique solution of the equation $f(x) = 0$ is the number $\sqrt[k]{2}$ and Newton's method will produce approximations for this number.

We first need to choose a number $x_0 > \sqrt[k]{2}$. How do we do this when we do not know what the number $\sqrt[k]{2}$ is?

Observe we have to choose a number x_0 such that $f(x_0) > f(\sqrt[k]{2}) = 0$, or equivalently,

$$x_0^k > 2.$$

Let us pick $x_0 = \frac{3}{2}$. Then

$$\left(\frac{3}{2}\right)^k \geq \left(\frac{3}{2}\right)^2 = \frac{9}{4} > 2.$$

Note also that $f(1) = 1^k - 2 = -1 < 0$ so that

$$1 < \sqrt[k]{2} < \frac{3}{2},$$

and thus the error

$$\varepsilon_0 = x_0 = \sqrt[k]{2} < \frac{1}{2}.$$

In this case we have

$$Z(x) = x - \frac{f(x)}{f'(x)} = x - \frac{x^k - 2}{kx^{k-1}} = \frac{k-1}{k}x + \frac{2}{kx^{k-1}}.$$

Observe that for $k = 2$ we have

$$Z(x) = \frac{x}{2} + \frac{1}{x},$$

and the recurrence $x_{n+1} = Z(x_n)$ takes the form

$$x_{n+1} = \frac{x_n}{2} + \frac{1}{x_n}.$$

Above, we recognize the recurrence that we have investigated earlier in Example 4.25.

For $k = 3$ the recurrence $x_{n+1} = Z(x_n)$ takes the form

$$x_{n+1} = \frac{2x_n}{3} + \frac{2}{3x_n^2}, \quad x_0 = 1.5.$$

We have

$$x_1 = 1.296296..., \quad x_2 = 1.260932..., \quad x_3 = 1.25992186...,$$

$$x_4 = 1.25992104..., \quad x_5 = 1.25992104...$$

Note that, as predicted theoretically, this sequence displays a very rapid stabilization. Thus

$$\sqrt[3]{2} \approx 1.25992....$$

We can independently confirm the above claim by observing that

$$(1.25992)^3 = 1.999995. \qquad \square$$

Theorem 8.28 (Jensen's inequality). *Suppose that $f : I \to \mathbb{R}$ is a convex function defined on an interval I. Then for any $n \in \mathbb{N}$, any $x_1, \dots, x_n \in I$ and any $t_1, \dots, t_n \geq 0$ such that*

$$t_1 + \cdots + t_n = 1,$$

we have $t_1x_1 + \cdots + t_nx_n \in I$ and

$$f\big(t_1x_1 + \cdots + t_nx_n\big) \leq t_1f(x_1) + \cdots + t_nf(x_n). \tag{8.16}$$

Proof. We argue by induction on n. For $n = 1$ the inequality is trivially true, while for $n = 2$ it is the definition of convexity. We assume that the inequality is true for n and we prove it for $n + 1$.

Let $x_0, \dots, x_n \in I$ and $t_0, \dots, t_n \geq 0$ such that

$$t_0 + \cdots + t_n = 1.$$

We have to prove that

$$f(t_0x_0 + t_1x_1 + t_2x_2 + \cdots + t_nx_n) \leq t_0f(x_0) + t_1f(x_1) + t_2f(y_2) + \cdots + t_nf(y_n). \tag{8.17}$$

If one of the numbers $t_0, t_1, \dots, t_n$ is zero, then the above inequality reduces to the case n. We can therefore assume that $t_0, t_1, \dots, t_n > 0$. Consider now the real numbers

$$s_1 := t_0 + t_1, \quad s_2 := t_2, \dots, s_n := t_n,$$

$$y_1 := \frac{t_0}{t_0 + t_1}x_0 + \frac{t_1}{t_0 + t_1}x_1, \quad y_2 := x_2, \dots, y_n := x_n.$$

Note that

$$s_1, s_2, \dots, s_n \geq 0 \text{ and } s_1 + \cdots + s_n = 1$$

and since

$$\frac{t_0}{t_0+t_1}+\frac{t_1}{t_0+t_1}=1$$

the point y_1 lies between x_0 and x_1 and thus in the interval I. From the induction assumption we deduce

$$s_1y_1+\cdots+s_ny_n\in I,$$

and

$$\begin{aligned}f(s_1y_1+\cdots+s_ny_n)&\leq s_1f(y_1)+s_2f(y_2)+\cdots+s_nf(y_n)\\&=(t_0+t_1)f\left(\frac{t_0}{t_0+t_1}x_0+\frac{t_1}{t_0+t_1}x_1\right)+s_2f(y_2)+\cdots+s_nf(y_n).\end{aligned}$$

Now observe that

$$\begin{aligned}s_1y_1+\cdots+s_ny_n&=(t_0+t_1)\left(\frac{t_0}{t_0+t_1}x_0+\frac{t_1}{t_0+t_1}x_1\right)+t_2y_2+\cdots+t_ny_n\\&=t_0x_0+t_1x_1+t_2x_2+\cdots+t_nx_n,\end{aligned}$$

and, since f is convex,

$$f\left(\frac{t_0}{t_0+t_1}x_0+\frac{t_1}{t_0+t_1}x_1\right)\leq\frac{t_0}{t_0+t_1}f(x_0)+\frac{t_1}{t_0+t_1}f(x_1),$$

so that

$$(t_0+t_1)\left(\frac{t_0}{t_0+t_1}x_0+\frac{t_1}{t_0+t_1}x_1\right)\leq t_0f(x_0)+t_1f(x_1).$$

Putting together all of the above we deduce (8.17). □

Corollary 8.29. *If $f: I\to\mathbb{R}$ is a convex function defined on an interval I, then for any $n\in\mathbb{N}$ and any $x_1,\dots,x_n\in I$ we have*

$$\boxed{f\left(\frac{x_1+\cdots+x_n}{n}\right)\leq\frac{f(x_1)+\cdots+f(x_n)}{n}}. \tag{8.18}$$

Proof. Use (8.16) in which $t_1=t_2=\cdots=t_n=\frac{1}{n}$. □

Corollary 8.30. *Suppose that $g: I\to\mathbb{R}$ is a concave function defined on an interval I. Then for any $n\in\mathbb{N}$, any $x_1,\dots,x_n\in I$ and any $t_1,\dots,t_n\geq 0$ such that*

$$t_1+\cdots+t_n=1,$$

we have $t_1x_1+\cdots+t_nx_n\in I$ and

$$g\big(t_1x_1+\cdots+t_nx_n\big)\geq t_1g(x_1)+\cdots+t_ng(x_n). \tag{8.19}$$

In particular,

$$\boxed{g\left(\frac{x_1+\cdots+x_n}{n}\right)\geq\frac{g(x_1)+\cdots+g(x_n)}{n}}. \tag{8.20}$$

Proof. Apply Theorem 8.28 to the convex function $f=-g$. □

Corollary 8.31 (AM-GM inequality). *For any natural number n and any positive real numbers $x_1,\dots,x_n$ we have*

$$\boxed{\big(x_1\cdots x_n\big)^{\frac{1}{n}}\leq\frac{x_1+\cdots+x_n}{n}}. \tag{8.21}$$

The left-hand side of the above inequality is called the geometric mean *(GM) of the numbers $x_1,\dots,x_n$, while the right-hand side is called the* arithmetic mean *(AM) of the same numbers.*

Proof. Consider the function

$$f:(0,\infty)\to\mathbb{R},\ \ f(x)=\ln x.$$

This function is concave and (8.20) implies that

$$\ln\left(\frac{x_1+\cdots+x_n}{n}\right)\geq\frac{\ln x_1+\cdots+\ln x_n}{n}.$$

Exponentiating this inequality we deduce

$$\frac{x_1+\cdots+x_n}{n}=e^{\ln\left(\frac{x_1+\cdots+x_n}{n}\right)}$$
$$\geq e^{\frac{\ln x_1+\cdots+\ln x_n}{n}}=e^{\frac{\ln(x_1\cdots x_n)}{n}}=(x_1\cdots x_n)^{\frac{1}{n}}.$$

□

Corollary 8.32 (Hölder's inequality). *Fix a real number $p>1$ and define $q>1$ by the equality*

$$\frac{1}{q}=1-\frac{1}{p}=\frac{p-1}{p}.$$

Then for any natural number n and any nonnegative real numbers $a_1,\dots,a_n$, $b_1,\dots,b_n$ we have

$$a_1b_1+\cdots+a_nb_n\leq\left(a_1^p+\cdots+a_n^p\right)^{\frac{1}{p}}\left(b_1^q+\cdots+b_n^q\right)^{\frac{1}{q}},\tag{8.22}$$

or, using the summation notation,

$$\boxed{\sum_{k=1}^n a_kb_k\leq\left(\sum_{i=1}^n a_i^p\right)^{\frac{1}{p}}\left(\sum_{j=1}^n b_j^q\right)^{\frac{1}{q}}.}\tag{8.23}$$

Proof. Since $p>1$, the function $f:[0,\infty)\to\mathbb{R}$, $f(x)=x^p$, is convex. We define

$$B:=b_1^q+\cdots+b_n^q,$$

$$t_k:=\frac{b_k^q}{B},\ \ k=1,\dots,n,$$

$$x_k:=a_kb_k^{-\frac{1}{p-1}}B,\ \ k=1,\dots,n.$$

Observe that $t_k\geq 0$, $\forall k$ and

$$t_1+\cdots+t_k=1.$$

Using Jensen's inequality (8.18) we deduce that

$$\left(t_1x_1+\cdots+t_nx_n\right)^p\leq t_1x_1^p+\cdots+t_nx_n^p.$$

Observe that

$$\left(t_1x_1+\cdots+t_nx_n\right)^p=\left(a_1b_1^{q-\frac{1}{p-1}}+\cdots+a_nb_n^{q-\frac{1}{p-1}}\right)^p$$

$(q - \frac{1}{p-1} = 1)$

$$= (a_1 b_1 + \cdots + a_n b_n)^p .$$

Similarly

$$t_1 x_1^p + \cdots + t_n x_n^p = \frac{b_1^q}{B} a_1^p b_1^{-\frac{p}{p-1}} B^p + \cdots + \frac{b_n^q}{B} a_n^p b_n^{-\frac{p}{p-1}} B^p$$

$(q - \frac{p}{p-1} = 0)$

$$= B^{p-1} (a_1^p + \cdots + a_n^p).$$

Hence

$$(a_1 b_1 + \cdots + a_n b_n)^p \leq B^{p-1} (a_1^p + \cdots + a_n^p)$$

so that

$$\begin{aligned} a_1 b_1 + \cdots + a_n b_n &\leq B^{\frac{p-1}{p}} (a_1^p + \cdots + a_n^p)^{\frac{1}{p}} \\ &= (a_1^p + \cdots + a_n^p)^{\frac{1}{p}} (b_1^q + \cdots + b_n^q)^{\frac{1}{q}}. \end{aligned}$$

□

If in Hölder's inequality we let $p = 2$, then $q = 2$, and we obtain the following important result.

Corollary 8.33 (Cauchy–Schwarz inequality). *For any natural number n and any real numbers $x_1, \ldots, x_n, y_1, \ldots, y_n$ we have*

$$\boxed{\left| \sum_{k=1}^{n} x_k y_k \right| \leq \left(\sum_{i=1}^{n} x_i^2 \right)^{\frac{1}{2}} \left(\sum_{j=1}^{n} y_j^2 \right)^{\frac{1}{2}} .} \tag{8.24}$$

Proof. We define

$$a_k = |x_k|, \quad b_k = |y_k|, \quad k = 1, \ldots, n.$$

Note that $a_k^2 = x_k^2$, $b_k^2 = y_k^2$. Using Hölder's inequality with $p = q = 2$ we deduce

$$\sum_{k=1}^{n} |x_k y_k| \leq \left(\sum_{i=1}^{n} x_i^2 \right)^{\frac{1}{2}} \left(\sum_{j=1}^{n} y_j^2 \right)^{\frac{1}{2}} .$$

Now observe that

$$\left| \sum_{k=1}^{n} x_k y_k \right| \leq \sum_{k=1}^{n} |x_k y_k|.$$

□

Corollary 8.34 (Minkowski's inequality). *For any real number $p \in [1, \infty)$, any natural number n, and any real numbers $x_1, \dots, x_n, y_1, \dots, y_n$ we have*

$$\boxed{\left(\sum_{k=1}^{n} |x_k + y_k|^p\right)^{\frac{1}{p}} \leq \left(\sum_{k=1}^{n} |x_k|^p\right)^{\frac{1}{p}} + \left(\sum_{k=1}^{n} |y_k|^p\right)^{\frac{1}{p}}.} \tag{8.25}$$

Proof. We set

$$X := \left(\sum_{k=1}^{n} |x_k|^p\right)^{\frac{1}{p}}, \quad Y := \left(\sum_{k=1}^{n} |y_k|^p\right)^{\frac{1}{p}}, \quad Z := \left(\sum_{k=1}^{n} |x_k + y_k|^p\right)^{\frac{1}{p}}.$$

Clearly $X, Y, Z \geq 0$. We have to prove that $Z \leq X + Y$. This inequality is obviously true if $Z = 0$ so we assume that $Z > 0$. Note that we have

$$\begin{aligned} Z^p &= \sum_{k=1}^{n} |x_k + y_k|^p = \sum_{k=1}^{n} \underbrace{|x_k + y_k|}_{\leq |x_k| + |y_k|} \, |x_k + y_k|^{p-1} \\ &\leq \sum_{k=1}^{n} |x_k| \, |x_k + y_k|^{p-1} + \sum_{k=1}^{n} |y_k| \, |x_k + y_k|^{p-1}. \end{aligned} \tag{8.26}$$

This proves (8.25) in the special case $p = 1$ so in the sequel we assume that $p > 1$. Let $q = \frac{p}{p-1}$ so that

$$\frac{1}{p} + \frac{1}{q} = 1.$$

Using Hölder's inequality we deduce that, for any $k = 1, \dots, n$, we have

$$\sum_{k=1}^{n} |x_k| \, |x_k + y_k|^{p-1} \leq \underbrace{\left(\sum_{k=1}^{p} |x_k|^p\right)^{\frac{1}{p}}}_{X} \underbrace{\left(\sum_{k=1}^{n} |x_k + y_k|^p\right)^{\frac{p-1}{p}}}_{Z^{p-1}},$$

$$\sum_{k=1}^{n} |y_k| \, |x_k + y_k|^{p-1} \leq \underbrace{\left(\sum_{k=1}^{p} |y_k|^p\right)^{\frac{1}{p}}}_{Y} \underbrace{\left(\sum_{k=1}^{n} |x_k + y_k|^p\right)^{\frac{p-1}{p}}}_{Z^{p-1}}.$$

Using the last two inequalities in (8.26) we deduce

$$Z^p \leq (X + Y) Z^{p-1} \stackrel{Z>0}{\Rightarrow} Z \leq X + Y.$$

□

Remark 8.35. Minkowski's inequality has a very useful interpretation. For a natural number n we denote by $\mathbb{R}^n$ the n-dimensional Euclidean space whose points are called (n-dimensional) *vectors* and are defined to be n-tuples

$$\boldsymbol{x} = (x_1, \dots, x_n), \;\; x_i \in \mathbb{R}, \;\; 1 \leq i \leq n.$$

The space $\mathbb{R}^n$ has a rich algebraic structure. We mention here two operations. One such operation is the addition of vectors. Given

$$\boldsymbol{x} = (x_1, \dots, x_n), \ \ \boldsymbol{y} = (y_1, \dots, y_n) \in \mathbb{R}^n,$$

we define their sum $\boldsymbol{x} + \boldsymbol{y}$ to be the vector

$$\boldsymbol{x} + \boldsymbol{y} := (x_1 + y_1, \dots, x_n + y_n).$$

Another operation is the multiplication by a scalar. Given

$$\boldsymbol{x} = (x_1, \dots, x_n) \in \mathbb{R}^n, \ \ t \in \mathbb{R},$$

we define

$$t\boldsymbol{x} := (tx_1, \dots, tx_n).$$

For $p \in [1, \infty)$ and $\boldsymbol{x} \in \mathbb{R}^n$ we set

$$\|\boldsymbol{x}\|_p := \left(\sum_{k=1}^{n} |x_k|^p \right)^{\frac{1}{p}}.$$

Note that

$$\begin{aligned} &\|t\boldsymbol{x}\|_p = |t| \, \|\boldsymbol{x}\|_p, \ \ \forall t \in \mathbb{R}, \ \ \boldsymbol{x} \in \mathbb{R}^n, \\ &\|\boldsymbol{x}\|_p \geq 0, \ \ \forall \boldsymbol{x} \in \mathbb{R}^n, \\ &\|\boldsymbol{x}\|_p = 0 \Longleftrightarrow \boldsymbol{x} = (0, 0, \dots, 0). \end{aligned} \tag{8.27}$$

Minkowski's inequality is then equivalent to the *triangle inequality*

$$\|\boldsymbol{x} + \boldsymbol{y}\|_p \leq \|\boldsymbol{x}\|_p + \|\boldsymbol{y}\|_p, \ \ \forall \boldsymbol{x}, \boldsymbol{y} \in \mathbb{R}^n. \tag{8.28}$$

A function $\mathbb{R}^n \to \mathbb{R}$ that associates to a vector $\mathbb{R}$ a real number $\|\boldsymbol{x}\|$ satisfying (8.27) and (8.28) is called a *norm* on $\mathbb{R}^n$. Minkowski's inequality can be interpreted as saying that for any $p \in [1, \infty)$, the correspondence

$$\mathbb{R}^n \ni \boldsymbol{x} \mapsto \|\boldsymbol{x}\|_{\boldsymbol{p}} \in [0, \infty)$$

defines a norm on $\mathbb{R}^n$.

Note that (8.28) implies that for any $\boldsymbol{u}, \boldsymbol{v}, \boldsymbol{w} \in \mathbb{R}^n$ we have

$$\|\boldsymbol{u} - \boldsymbol{w}\|_p \leq \|\boldsymbol{u} - \boldsymbol{v}\|_p + \|\boldsymbol{v} - \boldsymbol{w}\|_p, \tag{8.29}$$

since

$$\underbrace{(\boldsymbol{u} - \boldsymbol{v})}_{\boldsymbol{x}} + \underbrace{(\boldsymbol{v} - \boldsymbol{w})}_{\boldsymbol{y}} = \underbrace{(\boldsymbol{u} - \boldsymbol{w})}_{\boldsymbol{x}+\boldsymbol{y}}.$$

□

8.4. How to sketch the graph of a function

Differential calculus can be quite useful in producing sketches of the graphs of functions. Instead of giving a detailed description of the steps that need to be taken to produce a sketch of a graph, we will outline a few general principles and illustrate them on a few examples.

In sketching the graph of a function $f(x)$, one needs to look at certain distinguishing features.

- Locate the intersections of f with the coordinate axes, if possible.
- Locate, if possible, the critical points of f, i.e., the points x such that $f'(x) = 0$.
- Locate the intervals where f is increasing and the intervals where f is decreasing, if possible.
- Locate the intervals where f is convex, and the intervals where f is concave, if possible. The endpoints of such intervals are found among the solutions of the equation.

  $$f''(x) = 0.$$

  Sometimes solving this equation explicitly may not be possible.
- Locate the asymptotes, if any.

Example 8.36 (Cubic polynomials). Consider an arbitrary cubic polynomial

$$p : \mathbb{R} \to \mathbb{R},\ \ p(x) = x^3 + a_2x^2 + a_1x + a_0,$$

where a_0, a_1, a_2 are given real numbers. We would like to describe the general appearance of the graph of p and analyze how it depends on the coefficients a_0, a_1, a_2. Observe first that

$$\lim_{x\to\pm\infty} p(x) = \pm\infty.$$

The graph intersects the y-axis at $y = a_0$. The intersection with the x-axis is difficult to find because the equation $p(x) = 0$ is difficult to solve. Instead, we will try to find the critical points of $p(x)$, i.e., the solutions of the equation $p'(x) = 0$.

$$3x^2 + 2a_2x + a_1 = 0. \tag{8.30}$$

The function $p'(x)$ has a global minimum achieved at the point μ defined by the equation

$$p''(\mu) = 0 \Longleftrightarrow 6\mu + 2a_2 = 0 \Longleftrightarrow \mu = -\frac{a_2}{3}.$$

The function $p'(x)$ is decreasing on the interval $(-\infty, \mu]$ and increasing on $[\mu, \infty)$. Thus $p(x)$ is concave on $(-\infty, \mu]$ and convex on $[\mu, \infty)$. The point μ is an inflection point of p.

The general theory of quadratic equations tells us that (8.30) can have zero, one or two solutions depending on whether $\Delta = 4a_2^2 - 12a_1$ is negative, zero or positive. These situations are depicted in Figure 8.7.

If p has no critical points, as in the left-hand side of Figure 8.7, then $p'(x) > 0$ for any $x \in \mathbb{R}$. This shows that p is increasing. Similarly, if p has a single critical point, then again $p(x)$ is increasing. In both cases, the graph of p looks like the left-hand side of Figure 8.9.

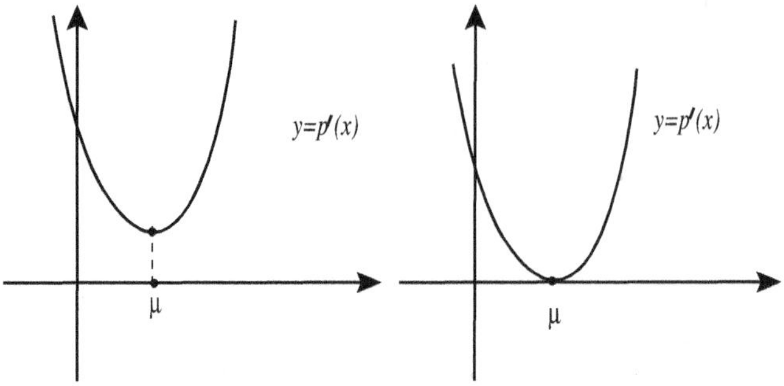

Fig. 8.7 $\Delta = 4a_2^2 - 12a_1 \leq 0.$

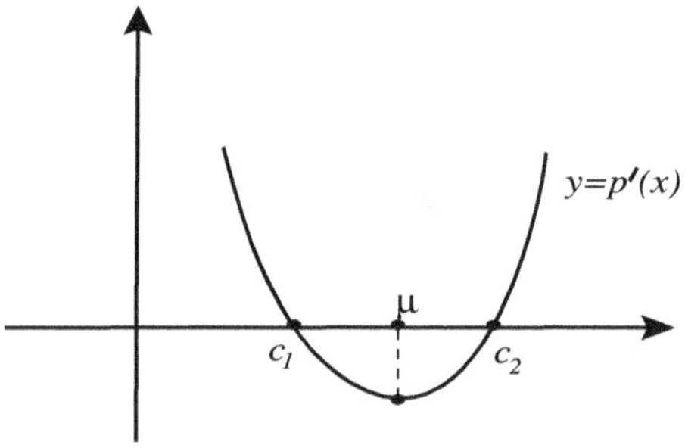

Fig. 8.8 $\Delta = 4a_2^2 - 12a_1 > 0.$

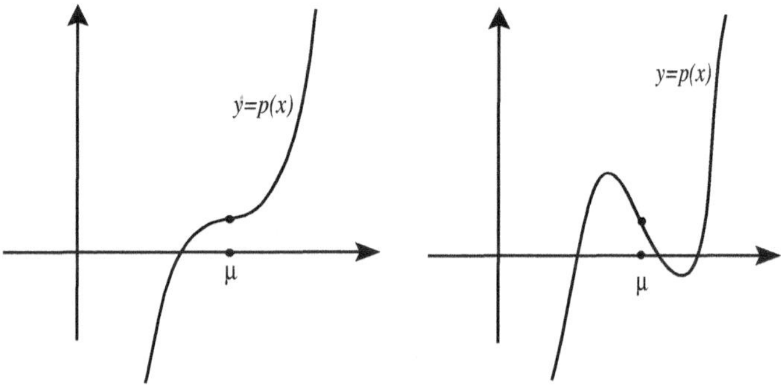

Fig. 8.9 *The graph of* $y = x^3 + a_2x^2 + a_1x + a_0.$

If $p(x)$ has two critical points $c_1 < c_2$, then $p'(x) < 0$ on (c_1, c_2) and positive on $(-\infty, c_1) \cup (c_2, \infty)$; see Figure 8.8. The point c_1 is a local max of p and c_2 is a local min of p. The inflection point μ is the midpoint of the interval $[c_1, c_2]$. The graph of p is depicted on the right-hand side of Figure 8.9. □

Example 8.37. Consider the function

$$f(x) = \frac{x^2+1}{x^2-3x+2}.$$

We have not specified its domain so it is understood to consist of all the x for which the fraction

$$\frac{x^2+1}{x^2-3x+2}$$

is well defined. The only problems are the points where the denominator vanishes,

$$x^2 - 3x + 2 = 0 \iff x = 1 \ \vee \ x = 2.$$

Thus the domain is

$$(-\infty, 1) \cup (1, 2) \cup (2, \infty).$$

The points 1 and 2 are also points where the vertical asymptotes could be located. We will investigate this issue later. We have

$$\begin{aligned} f'(x) &= \frac{(2x)(x^2-3x+2) - (x^2+1)(2x-3)}{(x^2-3x+2)^2} \\ &= \frac{2x^3 - 6x^2 + 4x - (2x^3 - 3x^2 + 2x - 3)}{(x^2-3x+2)^2} \\ &= \frac{-3x^2+2x+3}{(x^2-3x+2)^2}. \end{aligned}$$

The derivative vanishes when $3x^2 - 2x - 3 = 0$. The roots of this quadratic polynomial are

$$\frac{2 \pm \sqrt{4+36}}{6} = \frac{2 \pm \sqrt{40}}{6} = \frac{2 \pm 2\sqrt{10}}{6} = \frac{1 \pm \sqrt{10}}{3}.$$

One root is obviously negative. Since $3 < \sqrt{10} < 4$ we deduce

$$1 < \frac{1+\sqrt{10}}{3} < \frac{5}{3} < 2.$$

The intersection with the y-axis is obtained by computing $f(0) = \frac{1}{2}$. There is no intersection with the x axis since the numerator does not vanish. We have already detected several remarkable points

$$-\infty,\ c_1 = \frac{1-\sqrt{10}}{3},\ 1,\ c_2 = \frac{1+\sqrt{10}}{3},\ 2,\ \infty.$$

Observe that

$$\lim_{x\to\pm\infty} \frac{x^2+1}{x^2-3x+2},$$

so the horizontal line $y = 1$ is a horizontal asymptote for $f(x)$ at $\pm\infty$. We do not investigate the second derivative because it requires a substantial amount of work, with little payoff.

Table 8.1 *Organizing all the relevant data.*

x	$-\infty$		c_1		1		c_2		2		∞
(x^2-3x+2)	∞	$++$	$+$	$++$	0	$---$	$-$	$--$	0	$++$	∞
$-3x^2+2x+3$	$-\infty$	$--$	0	$++$	$+$	$++$	0	$--$	$-$	$--$	$-\infty$
$f'(x)$	$-$	$--$	0	$++$	!	$++$	0	$--$	!	$--$	$-$
$f(x)$	1	$\searrow$	min	$\nearrow$	!	$\nearrow$	max	$\searrow$	!	$\searrow$	1

Table 8.1 organizes the information we have collected. The exclamation signs indicate that the corresponding functions are not defined at those points. As x approaches 1 from the left, the function $f(x)$ is increasing and

$$\lim_{x\to 1-} f(x) = \infty.$$

Similarly, the table shows

$$\lim_{x\to 1+} f(x) = -\infty, \quad \lim_{x\to 2-} f(x) = -\infty, \quad \lim_{x\to 2+} f(x) = \infty.$$

This shows that the vertical lines $x = 1$ and $x = 2$ are asymptotes of $f(x)$. Figure 8.10 contains a sketch of the graph of the function $f(x)$.

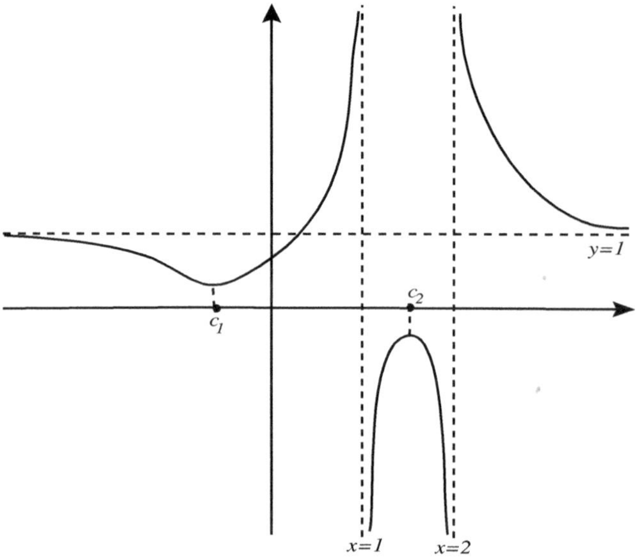

Fig. 8.10 *The graph of* $\frac{x^2+1}{x^2-3x+2}$.

□

Some functions admit *inclined asymptotes.*

Definition 8.38. (a) The line $y = mx + b$ is the asymptote of $f(x)$ at ∞ if

$$\lim_{x\to\infty} \frac{f(x)}{x} = m \text{ and } \lim_{x\to\infty} (f(x) - mx) = b.$$

(b) The line $y = mx + b$ is the asymptote of $f(x)$ at $-\infty$ if

$$\lim_{x\to-\infty}\frac{f(x)}{x} = m \text{ and } \lim_{x\to-\infty}(f(x) - mx) = b. \qquad \square$$

Example 8.39. The function

$$f(x) = \frac{x^5 + 2x^4 + 3x^3 + 4x + 5}{x^4 + 1}$$

admits an inclined asymptote $y = mx + b$ as $x \to \infty$. The slope m can be found from the equality

$$m = \lim_{x\to\infty}\frac{f(x)}{x} = 1,$$

and b can be found from the equality

$$\begin{aligned} b = \lim_{x\to\infty}\big(f(x) - x\big) &= \lim_{x\to\infty}\frac{x^5 + 2x^4 + 3x^3 + 4x + 5 - x(x^4 + 1)}{x^4 + 1} \\ &= \lim_{x\to\infty}\frac{2x^4 + 3x^3 + 3x + 5}{x^4 + 1} = 2. \end{aligned}$$

$\square$

8.5. Antiderivatives

Definition 8.40. Suppose that $f : I \to \mathbb{R}$ is a function defined on an interval $I \subset \mathbb{R}$. A function $F : I \to \mathbb{R}$ is called an *antiderivative* or *primitive* of f on I if F is differentiable, and

$$F'(x) = f(x), \quad \forall x \in I. \qquad \square$$

Example 8.41. (a) The function x^2 is an antiderivative of $2x$ on $\mathbb{R}$. Similarly, the function $\sin x$ is an antiderivative of $\cos x$ on $\mathbb{R}$. $\square$

Observe that if $F(x)$ is an antiderivative of a function $f(x)$ on an interval I, then for any constant $C \in \mathbb{R}$ the function $F(x) + C$ is also an antiderivative of $f(x)$ on I. The converse is also true.

Proposition 8.42. *If F_1, F_2 are antiderivatives of the function $f : I \to \mathbb{R}$, then $F_1 - F_2$ is constant.*

Proof. Observe that $(F_1 - F_2)' = F_1' - F_2' = f - f = 0$ and Corollary 7.30 implies that $F_1 - F_2$ is constant on I. $\square$

Definition 8.43. Given a function $f : I \to \mathbb{R}$ we denote by $\int f(x)dx$ the *collection* of all the antiderivatives of f on I. Usually $\int f(x)dx$ is referred to as the *indefinite integral* of f. $\square$

Table 8.2 Table of integrals.

$f(x)$	$\int f(x)dx$
$x^n, (x \in \mathbb{R}, n \in \mathbb{Z},\ n \geq 0)$	$\frac{x^{n+1}}{(n+1)} + C$
$\frac{1}{x^n}\ (x \neq 0, n \in \mathbb{N},\ \ n > 1)$	$-\frac{1}{(n-1)x^{n-1}} + C$
$x^\alpha, (\alpha \in \mathbb{R}, \alpha \neq -1, x > 0)$	$\frac{x^{\alpha+1}}{\alpha+1} + C$
$1/x, x \neq 0$	$\ln\lvert x\rvert + C$
$e^x, (x \in \mathbb{R})$	$e^x + C$
$\sin x, (x \in \mathbb{R})$	$-\cos x + C$
$\cos x, (x \in \mathbb{R})$	$\sin x + C$
$1/\cos^2 x$	$\tan x + C$
$\frac{1}{\sqrt{1-x^2}}, x \in (-1, 1)$	$\arcsin x + C$
$\frac{1}{1+x^2}, x \in \mathbb{R}$	$\arctan x + C$
$\int \frac{1}{\sqrt{x^2 \pm 1}} dx, x^2 \pm 1 > 0$	$\ln\lvert\, x + \sqrt{x^2 \pm 1}\,\rvert + C$

For example,

$$\int \cos x\, dx = \sin x + C, \quad \int 2x\, dx = x^2 + C.$$

Table 8.2 describes the antiderivatives of some basic functions.

Note that if $f :\to \mathbb{R}$ is a differentiable function, then f is an antiderivative of f' so that

$$\int f'(x)dx = f(x) + c. \tag{8.31}$$

Observing that $f'(x)dx = df$ we rewrite the above equality as

$$\int df = f + C. \tag{8.32}$$

In general, the computation of an antiderivative is a more challenging task that cannot always be completed. There are a few tricks and a few classes of functions for which this task is feasible. We will spend the remainder of this section discussing a few frequently encountered techniques for computing antiderivatives.

Proposition 8.44 (Linearity). *Suppose $f, g : I \to \mathbb{R}$ and $a, b \in \mathbb{R}$. If $F, G : I \to \mathbb{R}$ are antiderivatives of f and g respectively on I, then $aF + bG$ is an antiderivative of $af + bg$ on I. We write this in condensed form*

$$\int (af + bg)dx = a \int f dx + b \int g dx.$$

Proof.

$$(aF + bG)' = aF' + bG' = af + bg.$$

□

Example 8.45.

$$\int (3 + 5x + 7x^2)dx = 3 \int dx + 5 \int x dx + 7 \int x^2 dx = 3x + \frac{5}{2}x^2 + \frac{7}{3}x^3 + C. \quad \square$$

Proposition 8.46 (Integration by parts). *Suppose that $f, g : I \to \mathbb{R}$ are two differentiable functions. If the function $f(x)g'(x)$ admits antiderivatives on I, then so does the function $f'(x)g(x)$. Moreover*

$$\boxed{\int f(x)g'(x)dx = f(x)g(x) - \int g(x)f'(x)dx}. \tag{8.33}$$

Proof. The function $(fg)' = f'g + fg'$ admits antiderivatives and thus the difference

$$(fg)' - fg' = f'g$$

admits antiderivatives. Moreover,

$$fg = \int (fg)'dx = \int (f'g + fg')dx = \int f'gdx + \int fg'dx \Rightarrow \int fg'dx = fg - \int gf'dx.$$

□

Let us observe that we can rewrite (8.33) in a simpler form

$$\boxed{\int fdg = fg - \int gdf}. \tag{8.34}$$

Example 8.47. (a) We can use integration by parts to find the antiderivatives of $\ln x$, $x > 0$. We have

$$\begin{aligned}\int \ln x dx &= (\ln x)x - \int x d(\ln x) = x \ln x - \int x \frac{dx}{x} \\ &= x \ln x - \int dx = x \ln x - x + C.\end{aligned}$$

(b) For $a \in \mathbb{R}$ consider the indefinite integrals

$$\boxed{I_a = \int e^{ax} \cos x \, dx, \quad J_a = \int e^{ax} \sin x \, dx}.$$

We have

$$I_a = \int e^{ax} d(\sin x) = e^{ax}\sin x - \int \sin x d(e^{ax})$$
$$= e^{ax}\sin x - \int a e^{ax}\sin x dx = e^{ax}\sin x - aJ_a.$$

Similarly we have

$$J_a = \int e^{ax} d(-\cos x) = -e^{ax}\cos x + \int \cos x d(e^{ax})$$
$$= -e^{ax}\cos x + a\int e^{ax}\cos x dx = -e^{ax}\cos x + aI_a.$$

We deduce

$$I_a = e^{ax}\sin x - a(-e^{ax}\cos x + aI_a) = e^{ax}\sin x + ae^{ax}\cos x - a^2 I_a,$$

so that

$$(a^2+1)I_a = e^{ax}\sin x + ae^{ax}\cos x,$$

which shows that

$$\boxed{I_a = \frac{1}{a^2+1}\big(e^{ax}\sin x + ae^{ax}\cos x\big) + C}. \tag{8.35}$$

From this we deduce

$$J_a = aI_a - e^{ax}\cos x = \frac{a}{a^2+1}\big(e^{ax}\sin x + ae^{ax}\cos x\big) - e^{ax}\cos x + C,$$

so that

$$\boxed{J_a = \frac{1}{a^2+1}\big(ae^{ax}\sin x - e^{ax}\cos x\big) + C}. \tag{8.36}$$

(c) For any nonnegative integer n we consider the indefinite integral

$$\boxed{I_n = \int x^n e^x dx}.$$

Note that

$$I_0 = \int e^x dx = e^x + c.$$

In general, we have

$$I_{n+1} = \int x^{n+1} d(e^x) = x^{n+1}e^x - \int e^x d(x^{n+1}) = x^{n+1}e^x - (n+1)\int x^n e^x dx$$

so that

$$\boxed{I_{n+1} = x^{n+1}e^x - (n+1)I_n, \ \ \forall n = 0,1,2,\dots}. \tag{8.37}$$

If we let $n = 0$ in the above equality we deduce

$$I_1 = xe^x - I_0 = xe^x - e^x + C. \tag{8.38}$$

Using $n = 1$ in (8.37) we obtain

$$I_2 = x^2 e^x - 2I_1 = x^2 e^x - 2xe^x + 2e^x + C.$$

This suggests that in general $I_n = P_n(x)e^x + C$, where $P_n(x)$ is a polynomial of degree n. For example,

$$P_0(x) = 1, \ \ P_1(x) = (x-1), \ \ P_2(x) = x^2 - 2x + 2.$$

The equality (8.37) shows that

$$P_{n+1}(x) = x^{n+1} - (n+1)P_n(x), \ \ \forall n = 0, 1, 2, \dots. \tag{8.39}$$

(d) Let us now explain how to compute the integrals

$$\boxed{A_n = \int \frac{dx}{(x^2+1)^n}}.$$

Note that

$$A_1 = \int \frac{dx}{x^2+1} = \arctan x + C.$$

In general

$$\begin{aligned}
A_n &= \int (x^2+1)^{-n} dx = x(x^2+1)^{-n} - \int x d\Big((x^2+1)^{-n}\Big) \\
&= \frac{x}{(x^2+1)^n} - \int x \frac{-2nx}{(x^2+1)^{n+1}} dx = \frac{x}{(x^2+1)^n} + 2n \int \frac{x^2}{(x^2+1)^{n+1}} dx \\
&= \frac{x}{(x^2+1)^n} + 2n \int \frac{x^2+1-1}{(x^2+1)^{n+1}} dx \\
&= \frac{x}{(x^2+1)^n} + 2n \int \frac{1}{(x^2+1)^n} dx - 2n \int \frac{1}{(x^2+1)^{n+1}} dx \\
&= \frac{x}{(x^2+1)^n} + 2nA_n - 2nA_{n+1}.
\end{aligned}$$

Hence

$$A_n = \frac{x}{(x^2+1)^n} + 2nA_n - 2nA_{n+1},$$

so that

$$2nA_{n+1} = \frac{x}{(x^2+1)^n} + (2n-1)A_n,$$

and thus

$$\boxed{A_{n+1} = \frac{1}{2n}\frac{x}{(x^2+1)^n} + \frac{(2n-1)}{2n}A_n}. \tag{8.40}$$

For example,

$$\int \frac{1}{(x^2+1)^2} dx = \frac{1}{2}\frac{x}{x^2+1} + \frac{1}{2}\arctan x + C.$$

□

Proposition 8.48 (Integration by substitution). *Suppose that $u : I \to J$ and $f : J \to \mathbb{R}$ are differentiable functions. Then the function $f'(u(x))u'(x)$ admits antiderivatives on I and*

$$\int f'(u(x))u'(x)dx = \int f'(u)du = \int df = f(u) + C, \quad u = u(x). \tag{8.41}$$

Proof. The chain formula shows that $f'(u(x))u'(x)$ is the derivative of $f(u(x))$ so that $f(u(x))$ is an antiderivative of $f'(u(x))u'(x)$. □

Example 8.49. (a) To find an antiderivative of xe^{x^2} we use the change in variables $u = x^2$. Then

$$du = 2xdx \Rightarrow xdx = \frac{du}{2}$$

so that

$$\int e^{x^2} xdx = \int e^u \frac{du}{2} = \frac{1}{2}e^u + C = \frac{1}{2}e^{x^2} + C.$$

(b) Let us compute an antiderivative of $\tan x = \frac{\sin x}{\cos x}$ on an interval I where $\cos x \neq 0$. We distinguish two cases.

1. $\cos x > 0$ on I. We make the change in variables $u = \cos x$ so that $u > 0$, and $du = -\sin x dx$. We have

$$\int \frac{\sin x}{\cos x} dx = -\int \frac{du}{u} = -\ln u + C = -\ln \cos x + C = -\ln|\cos x| + C.$$

2. $\cos x < 0$ on I. We make the change in variables $v = -\cos x$ so that $v > 0$ and $dv = \sin x dx$. We have

$$\int \frac{\sin x}{\cos x} dx = \int \frac{dv}{-v} = -\ln v + C = -\ln(-\cos x) + C = -\ln|\cos x| + C.$$

Thus, in either case we have

$$\int \tan x dx = -\ln|\cos x| + C. \tag{8.42}$$

(c) To compute the integral

$$\int (ax+b)^n dx, \quad n \in \mathbb{N}, \quad a > 0,$$

we make the change in variables $u = ax + b$. Then $du = adx$ so that $dx = \frac{1}{a}du$ and we have

$$\int (ax+b)^n dx = \frac{1}{a}\int u^n du = \frac{1}{a(n+1)}u^{n+1} + C = \frac{1}{a(n+1)}(ax+b)^{n+1} + C.$$

(d) To compute the integral

$$\int \frac{1}{(ax+b)^n} dx, \quad a \neq 0, n \in \mathbb{N}$$

we again make the change in variables $u = (ax + b)$ and we deduce

$$\int \frac{1}{(ax+b)^n}dx = \frac{1}{a}\int \frac{1}{u^n}du = C + \begin{cases} \frac{1}{a}\ln|u|, & n = 1, \\ \\ \frac{1}{a(1-n)u^{n-1}}, & n > 1, \end{cases} \quad u = ax + b.$$

(e) To compute the integral

$$\boxed{B_n := \int \frac{x}{(x^2+1)^n}dx}$$

make the change in variables $u = x^2 + 1$. Then $du = 2xdx$ so that $xdx = \frac{1}{2}du$ and thus

$$\int \frac{x}{(x^2+1)^n}dx = \frac{1}{2}\int \frac{1}{u^n}du = C + \frac{1}{2} \times \begin{cases} \ln u, & n = 1, \\ \\ \frac{1}{(1-n)u^{n-1}}, & n > 1, \end{cases} \quad u = x^2 + 1.$$

(f) The integrals of the form

$$\boxed{\int (\sin x)^m (\cos x)^{2k+1} dx, \quad k, m \in \mathbb{Z}_{\geq 0}}$$

are found using the change in variables $u = \sin x$. Then

$$du = \cos x dx, \ (\cos x)^{2k+1}dx = (\cos^2 x)^k \cos x dx = (1-\sin^2 x)^k d(\sin x) = (1-u^2)^k du$$

and

$$\int (\sin x)^m (\cos x)^{2k+1} dx = \int u^m (1-u^2)^k du.$$

Similarly, the integrals of the form

$$\int (\cos x)^m (\sin x)^{2k+1} dx, \quad m, k \in \mathbb{Z}_{\geq 0}$$

are found using the change in variables $v = \cos x$. Then

$$\int (\cos x)^m (\sin x)^{2k+1} dx = -\int v^m (1-v^2)^k dv.$$

(g) The integrals of the form

$$\boxed{\int (\sin x)^{2m} (\cos x)^{2k} dx, \quad k, m \in \mathbb{Z}_{\geq 0}}$$

are a bit trickier to compute. There are two possible strategies.
One strategy is based on the trigonometric identities

$$\boxed{\sin^2 x = \frac{1-\cos 2x}{2}, \quad \cos^2 x = \frac{1+\cos 2x}{2}}.$$

Using the change in variables $u = 2x$, so that

$$du = 2dx \Rightarrow dx = \frac{1}{2}du,$$

we deduce

$$\int (\sin x)^{2m} (\cos x)^{2k} dx = \frac{1}{2^{m+k+1}} \int (1-\cos u)^m (1+\cos u)^k du.$$

The last integral involves <u>*smaller*</u> powers in $\cos u$. For example,

$$\int \cos^4 x dx \stackrel{u=2x}{=} \int \left(\frac{1+\cos u}{2}\right)^2 \frac{du}{2}$$

$$= \frac{1}{8}\int (1+2\cos u+\cos^2 u)du = \frac{1}{8}u + \frac{1}{4}\sin u + \frac{1}{8}\int \cos^2 u du$$

$$\stackrel{v=2u=4x}{=} \frac{1}{8}u + \frac{1}{4}\sin u + \frac{1}{8}\int \left(\frac{1+\cos v}{2}\right)\frac{dv}{2} = \frac{1}{8}u + \frac{1}{4}\sin u + \frac{1}{32}\int (1+\cos v)dv$$

$$= \frac{1}{4}x + \frac{1}{4}\sin(2x) + \frac{v}{32} + \frac{1}{32}\sin v + C$$

$$= \frac{x}{4} + \frac{1}{4}\sin(2x) + \frac{x}{8} + \frac{1}{32}\sin(4x) + C = \frac{3}{8}x + \frac{1}{4}\sin(2x) + \frac{1}{32}\sin(4x) + C.$$

One other possible strategy is to use the change in variables $u = \tan x$. Then

$$\cos^2 x = \frac{1}{1+\tan^2 x} = \frac{1}{1+u^2}, \quad \sin^2 x = \cos^2 x \tan^2 x = \frac{\tan^2 x}{1+\tan^2 x} = \frac{u^2}{1+u^2},$$

$$du = d(\tan x) = (1+\tan^2 x)dx = (1+u^2)dx \Rightarrow dx = \frac{du}{1+u^2}.$$

We deduce

$$\int (\sin x)^{2m} (\cos x)^{2k} dx = \int \left(\frac{u^2}{1+u^2}\right)^m \left(\frac{1}{1+u^2}\right)^k \frac{du}{1+u^2}$$

$$= \int \frac{u^{2m}}{(1+u^2)^{m+k+1}} du.$$

Thus we need to know how to compute integrals of the form

$$\boxed{J(m,n) = \int \frac{u^{2m}}{(1+u^2)^n} du, \;\; 0 \leq m < n, \;\; m,n \in \mathbb{Z}}.$$

Observe first that, when $m = 0$, the integrals $J(0,n)$ coincide with the integrals A_n of (8.40). The general case can be gradually reduced to the case $J(0,n)$ by observing that

$$J(m,n) = \int \frac{u^{2m} + u^{2m-2} - u^{2m-2}}{(1+u^2)^n} du = \int \frac{u^{2m-2}(1+u^2)}{(1+u^2)^n} - \int \frac{u^{2m-2}}{(1+u^2)^n}$$

$$= \int \frac{u^{2m-2}}{(1+u^2)^{n-1}} - J(m-1,n),$$

so that

$$\boxed{J(m,n) = J(m-1,n-1) - J(m-1,n)}.$$

For example

$$J(2,4) = J(1,3) - J(1,4) = J(0,2) - J(0,3) - \big(J(0,3) - J(0,4)\big)$$

$$= J(0,2) - 2J(0,3) + J(0,4) = A_2 - 2A_3 + A_4.$$

□

The examples discussed above will allow us to describe a procedure for computing the antiderivatives of any *rational function*, i.e., a function $f(x)$ of the form

$$f(x) = \frac{P(x)}{Q(x)}$$

where $P(x)$ and $Q(x)$ are polynomials. Theoretically, the procedure works for any rational function, but the practical implementation can lead to complex computations. Such computation is possible because any rational function can be written as a sum of rational functions of the following simpler types.

Type I.

$$ax^n, \quad a \in \mathbb{R}, \quad n = 0, 1, 2, \ldots .$$

Type II.

$$\frac{a}{(x-r)^n}, \quad c, r \in \mathbb{R}, \quad n \in \mathbb{N}.$$

Type III.

$$\frac{bx+c}{\big((x-r)^2+a^2\big)^n}, \quad a, b, c, r \in \mathbb{R}, \quad n \in \mathbb{N}.$$

If the degree of the numerator $P(x)$ is smaller than the degree of the denominator $Q(x)$, then only the Type II and Type III functions appear in the decomposition of $\frac{P(x)}{Q(x)}$. The functions of Type II and III are also known as *partial fractions* or *simple fractions*.

Actually finding the decomposition of a rational function as a sum of simple fractions requires a substantial amount of work and it is not very practical for more complicated rational functions. For this reason we will not discuss this technique in great detail.

The primitives of a function of Type I are known. More precisely

$$\int ax^n dx = \frac{a}{n+1}x^{n+1} + C.$$

The primitives of the functions of Type II were computed in Example 8.49(e). To deal with the Type III functions we make a change in variables

$$x - r = at \iff x = at + r.$$

Then

$$dx = adt, \quad bx + c = b(at+r) + \beta = abt + rb + c,$$

$$(x-r)^2 + a^2 = a^2t^2 + a^2 = a^2(t^2+1),$$

so that

$$\int \frac{bx+c}{\big((x-r)^2+a^2\big)^n}dx = \int \frac{abt+rb+c}{a^{2n}(t^2+1)^n}adt$$
$$= \frac{b}{a^{2n-2}}\int \frac{t}{(t^2+1)^n}dt + \frac{rb+c}{a^{2n-1}}\int \frac{1}{(t^2+1)^n}dt.$$

The computation of integral

$$\int \frac{1}{(t^2+1)^n}dt$$

is described in (8.40), while the computation of the integral

$$\int \frac{t}{(t^2+1)^n}dt$$

as described in Example 8.49(e).

Let us illustrate this strategy on a simple example.

Example 8.50. Consider the rational function

$$f(x) = \frac{1}{(x-1)^2(x^2+2x+2)}.$$

Let us observe that

$$x^2+2x+2 = (x+1)^2+1^2.$$

The function admits a decomposition of the form

$$\frac{1}{(x-1)^2(x^2+2x+2)} = f(x) = \frac{A_1}{x-1} + \frac{A_2}{(x-1)^2} + \frac{B_1x+C_1}{x^2+2x+2}.$$

Multiplying both sides by $(x-1)^2(x^2+2x+2)$ we deduce that for any $x \in \mathbb{R}$ we have

$$\begin{aligned}
1 &= A_1(x-1)(x^2+2x+2) + A_2(x^2+2x+2) + (B_1x+C_1)(x-1)^2 \\
&= A_1(x^3+2x^2+2x-x^2-2x-2) + A_2(x^2+2x+2) + (B_1x+C_1)(x^2-2x+1) \\
&= A_1(x^3+x^2-2) + A_2(x^2+2x+2) + (B_1x^3-2B_1x^2+B_1x+C_1x^2-2C_1x+C_1) \\
&= (A_1+B_1)x^3 + (A_1+A_2-2B_1+C_1)x^2 + (2A_2+B_1-2C_1)x - 2A_1+2A_2+C_1.
\end{aligned}$$

This implies

$$\begin{cases} A_1+B_1 = 0 \\ A_1+A_2-2B_1+C_1 = 0 \\ 2A_2+B_1-2C_1 = 0 \\ -2A_1+2A_2+C_1 = 1. \end{cases}$$

From the first equality we deduce $A_1 = -B_1$ and using this in the last three equalities above we deduce

$$\begin{cases} A_2-3B_1+C_1 = 0 \\ 2A_2+B_1-2C_1 = 0 \\ 2A_2+2B_1+C_1 = 1. \end{cases}$$

From the first equality we deduce $A_2 = 3B_1 - C_1$. Using this in the last two equalities we deduce

$$\begin{cases} 7B_1-4C_1 = 0 \\ 8B_1-C_1 = 1. \end{cases}$$

So,

$$\frac{7}{4}B_1 = C_1 = 8B_1 - 1 \Rightarrow \frac{25}{4}B_1 = 1 \Rightarrow B_1 = \frac{4}{25},\ C_1 = \frac{7}{25} \Rightarrow A_1 = -\frac{4}{25},$$

$$A_2 = 3B_1 - C_1 = \frac{12}{25} - \frac{7}{25} = \frac{1}{5}.$$

Hence

$$\frac{1}{(x-1)^2(x^2+2x+2)} = -\frac{4}{25(x-1)} + \frac{1}{5(x-1)^2} + \frac{4x+7}{25\big((x+1)^2+1^2\big)}.$$ □

Example 8.51 (First order linear differential equations). A quantity u that depends on time can be viewed as a function

$$u : I \to \mathbb{R}, \;\; t \mapsto u(t),$$

where $I \subset \mathbb{R}$ is a time interval. We say that u satisfies a *linear first order differential equation* if u is differentiable and it satisfies an equality of the form

$$u'(t) + r(t)u(t) = f(t), \;\; \forall t \in I, \tag{8.43}$$

where $r, f : I \to \mathbb{R}$ are some given functions. Solving a differential equation such as (8.43) means finding all the differentiable functions $u : I \to \mathbb{R}$ satisfying the above equality. Let us look at some special examples.

(a) If $r(t) = 0$ for any $t \in I$, then (8.43) has the simpler form $u'(t) = f(t)$, so that $u(t)$ must be an antiderivative of $f(t)$.

(b) The general case. Suppose that $r(t)$ admits antiderivatives on I. The differential equation (8.43) is solved as follows.

Step 1. Choose one antiderivative $R(t)$ of $r(t)$, i.e., a function $R(t)$ such that $R'(t) = r(t)$.

Step 2. Multiply both sides of (8.43) by $e^{R(t)}$. We obtain the equality

$$e^{R(t)}u'(t) + e^{R(t)}r(t)u(t) = f(t)e^{R(t)}.$$

Now observe that the left-hand side of the above equality is the derivative of $e^{R(t)}u(t)$,

$$\big(e^{R(t)}u(t)\big)' = e^{R(t)}u'(t) + e^{R(t)}R'(t)u(t) = e^{R(t)}u'(t) + e^{R(t)}r(t)u(t) = f(t)e^{R(t)}.$$

This shows that $e^{R(t)}u(t)$ is an antiderivative of $f(t)e^{R(t)}$.

Step 3. Find one antiderivative $G(t)$ of $f(t)e^{R(t)}$. We deduce that there exists a constant $C \in \mathbb{R}$ such that

$$e^{R(t)}u(t) = G(t) + C \Rightarrow u(t) = e^{-R(t)}G(t) + Ce^{-R(t)}.$$

Take for example the differential equation

$$u'(t) + 2tu(t) = t.$$

In this case

$$r(t) = 2t, \;\; f(t) = t.$$

We can choose $R(t) = t^2$ and we have

$$\frac{d}{dt}\big(e^{t^2}u(t)\big) = e^{t^2}u'(t) + 2te^{t^2}u(t) = e^{t^2}t,$$

so that

$$e^{t^2}u(t) = \int e^{t^2}tdt = \frac{1}{2}\int e^{t^2}d(t^2) = \frac{1}{2}e^{t^2} + C$$
$$\Rightarrow u(t) = e^{-t^2}\Big(C + \frac{1}{2}e^{t^2}\Big) = Ce^{-t^2} + \frac{1}{2}.$$

□

8.6. Exercises

Exercise 8.1. Let $n \in \mathbb{N}$, $x_0, c_0, c_1, \dots, c_n \in \mathbb{R}$ and

$$P(x) = c_0 + \frac{c_1}{1!}(x-x_0) + \frac{c_2}{2!}(x-x_0)^2 + \cdots + \frac{c_n}{n!}(x-x_0)^n = \sum_{k=0}^{n} \frac{c_k}{k!}(x-x_0)^k.$$

(a) Prove that for any $k = 0, 1, 2, \dots, n$ we have

$$P^{(k)}(x_0) = c_k.$$

(b) Prove that if $Q(x) = q_0 + q_1 x + \cdots q_n x^n$ is a polynomial of degree $\leq n$ such that

$$Q^{(k)}(x_0) = c_k, \;\; \forall k = 0, 1, 2, \dots, n,$$

then $Q(x) = P(x)$, $\forall x \in \mathbb{R}$.

Hint. Consider the difference $D(x) = P(x) - Q(x)$, observe that

$$D^{(k)}(x_0) = 0, \;\; \forall k = 0, 1, 2, \dots, n,$$

and conclude from the above that $D(x) = 0$, $\forall x \in \mathbb{R}$. To reach this conclusion write

$$D(x) = d_0 + d_1 x + \cdots + d_n x^n,$$

and observe first that $D^{(n)}(x) = n! d_n$, $\forall x \in \mathbb{R}$. □

Exercise 8.2. Suppose that $a, b \in \mathbb{R}$, $b \geq 0$ and consider $f : \mathbb{R} \to \mathbb{R}$

$$f(x) = \frac{1 + ax^2}{1 + bx^2}.$$

Find the degree 4 Taylor polynomial of f at $x_0 = 0$. For which values of a, b does this polynomial coincide with the degree 4 Taylor polynomial of $\cos x$ at $x_0 = 0$?

Hint. To simplify the computations of the derivatives of f at 0 use the following trick. Let $N(x) = 1 + ax^2$ be the numerator of the fraction, $D(x) = 1 + bx^2$ be the denominator. Then

$$N(0) = D(0) = 1, \;\; N'(0) = D'(0) = 0, \;\; N''(0) = 2a, \;\; D''(0) = 2b, \tag{8.44}$$

$$N^{(k)}(x) = D^{(k)}(x) = 0, \;\; \forall k \geq 3, \;\; x \in \mathbb{R}. \tag{8.45}$$

We have

$$\begin{aligned} N(x) &= D(x)f(x), \;\; N'(x) = D'(x)f(x) + D(x)f'(x), \\ N''(x) &= D''(x)f(x) + 2D'(x)f'(x) + D(x)f''(x), \\ N^{(n)}(x) &\overset{(7.35)}{=} \sum_{k=0}^{n} \binom{n}{k} D^{(k)}(x) f^{(n-k)}(x) \overset{(8.45)}{=} \sum_{k=0}^{2} \binom{n}{k} D^{(k)}(x) f^{(n-k)}(x) \\ &= D(x)f^{(n)}(x) + nD'(x)f^{(n-1)}(x) + \frac{n(n-1)}{2} D''(x) f^{(n-2)}(x), \;\; \forall n > 2. \end{aligned}$$

We deduce

$$\begin{aligned} f(0) &= D(0)f(0) = N(0) = 1, \;\; f'(0) = D(0)f'(0) = N'(0) - D'(0)f(0) \overset{(8.44)}{=} 0, \\ f''(0) &= D(0)f''(0) = N''(0) - 2D'(0)f'(0) - D''(0)f(0) \overset{(8.44)}{=} N''(0) - D''(0)f(0) = 2a - 2b, \\ f^{(n)}(0) &= D(0)f^{(n)}(0) = N^{(n)}(0) - nD'(0)f^{(n-1)}(0) - \frac{n(n-1)}{2} D''(0) f^{(n-2)}(0) \\ &\overset{(8.44)}{=} -bn(n-1)f^{(n-2)}(0), \;\; n > 2. \end{aligned}$$

□

Exercise 8.3. Use the inequality $2 < e < 3$ and the strategy outlined in Remark 8.6 to show that

$$\left| e^h - \left(1 + \frac{h}{1!} + \cdots + \frac{h^n}{n!}\right) \right| \leq \frac{3|h|^{n+1}}{(n+1)!}, \quad \forall |h| \leq 1.$$

□

Exercise 8.4. Using Example 8.7 as a guide, compute $\cos 1$ up to two decimals. □

Exercise 8.5. Approximate $\sqrt[3]{8.1}$ using the degree 3 Taylor polynomial of $f(x) = \sqrt[3]{x}$ at $x_0 = 8$. Estimate the error of this approximation using the Lagrange estimate (8.3). □

Exercise 8.6. Find the Taylor series of the function

$$f(x) = \frac{1}{1-x}, \quad x \neq 1$$

at $x_0 = 0$. For which values of x is this series convergent? □

Exercise 8.7. Prove that the Taylor series of $\ln(1-x)$ at $x_0 = 0$ is

$$-\sum_{n=1}^{\infty} \frac{x^n}{n}.$$

and then show that this series converges to $\ln(1-x)$ for any $x \in (-1, \frac{1}{2})$.

Hint. Use Corollary 8.5.[2] □

Exercise 8.8. (a) Prove that the Taylor series of $\sin x$ at $x_0 = 0$,

$$\sum_{k \geq 0} (-1)^k \frac{x^{2k+1}}{(2k+1)!},$$

is absolutely convergent for any $x \in \mathbb{R}$ and its sum is $\sin x$. Show that the convergence is uniform on any interval $[-R, R]$.

(b) Prove that the Taylor series of $\cos x$ at $x_0 = 0$,

$$\sum_{k \geq 0} (-1)^k \frac{x^{2k}}{(2k)!},$$

is absolutely convergent for any $x \in \mathbb{R}$ and its sum is $\cos x$. Show that the convergence is uniform on any interval $[-R, R]$.

Hint. Use Corollary 8.5. □

Exercise 8.9. Find

$$\lim_{x \to \infty} x \left[\frac{1}{e} - \left(\frac{x}{x+1} \right)^x \right].$$

□

[2]The Taylor series of $\ln(1-x)$ at $x_0 = 0$ converges to $\ln(1-x)$ *for all* $|x| < 1$. However, the Lagrange remainder formula is not strong enough to prove this. We need a different remainder formula (9.50) to prove this stronger statement. For details see Example 9.52.

Exercise 8.10. Using the fact that the function $\ln : (0, \infty) \to \mathbb{R}$ is concave prove *Young's inequality*: if $p, q \in (1, \infty)$ are such that

$$\frac{1}{p} + \frac{1}{q} = 1,$$

then

$$xy \leq \frac{x^p}{p} + \frac{y^q}{q}, \quad \forall x, y > 0. \tag{8.46}$$

□

Exercise 8.11. Use the AM-GM inequality to prove that if $x \in \mathbb{R}$, $n, m \in \mathbb{N}$ and $-x < n < m$, then

$$\left(1 + \frac{x}{n}\right)^n \leq \left(1 + \frac{x}{m}\right)^m.$$

□

Exercise 8.12. Let $x_1, \dots, x_n > 0$.

(i) Prove that

$$x_1^2 + \cdots + x_n^2 + \frac{1}{x_1 \cdots x_n} \geq n + 1.$$

(ii) Prove that

$$\sum_{1 \leq i \leq j \leq 1} x_i x_j + \sum_{k=1}^{n} \frac{1}{x_k^n} \geq \frac{n(n+3)}{2}.$$

□

Exercise 8.13. Suppose that $a < b$ are two real numbers and $f : (a, b) \to \mathbb{R}$ is a convex function.

(a) Prove that for any $x_1 < x_2 < x_3 \in (a, b)$ we have

$$\frac{f(x_2) - f(x_1)}{x_2 - x_1} \leq \frac{f(x_3) - f(x_1)}{x_3 - x_1} \leq \frac{f(x_3) - f(x_2)}{x_3 - x_2}.$$

Hint. Give a geometric interpretation to this statement and then think geometrically.

(b) Suppose that $x_0 \in (a, b)$. Prove that the one-sided limits

$$m_{\pm}(x_0) = \lim_{h \to 0\pm} \frac{f(x_0 + h) - f(x_0)}{h}$$

exist, are finite and $m_-(x_0) \leq m_+(x_0)$.

(c) Suppose $x_0 \in (a, b)$ and $m_{\pm}(x_0)$ are as above. Fix $m \in [m_-(x_0), m_+(x_0)]$. Show that

$$f(x) \geq f(x_0) + m(x - x_0), \quad \forall x \in (a, b).$$

Can you give a geometric interpretation of this fact?

(d) Prove that $f : \mathbb{R} \to \mathbb{R}$, $f(x) = |x|$ is convex. For $x_0 := 0$, compute the numbers $m_{\pm}(x_0)$ defined as in (b). □

Exercise 8.14.[3] Suppose that $f : [0,1] \to [0,\infty)$ is a C^2-function satisfying the following additional properties.

(i) $f'(x) \geq 0, \forall x \in [0,1]$.
(ii) $f''(x) > 0, \forall x \in (0,1)$.
(iii) $f(1) = 1$, $f'(1) > 1$ and $f(0) > 0$.

Prove that the following hold.

(a) $f(x) \in [0,1], \forall x \in [0,1]$.
(b) If $x_0 \in (0,1)$ is a *fixed point* of f, i.e., $f(x_0) = x_0$, then $f'(x_0) < 1$.
Hint. Argue by contradiction. Use the Mean Value Theorem with the quotient

$$\frac{f(1) - f(x_0)}{1 - x_0}.$$

(c) The function f has a *unique* fixed point x_* located in the *open* interval $(0,1)$.
Hint. Argue by contradiction. Suppose that f has two fixed points $x_* < y_*$ in $(0,1)$. Use the Mean Value Theorem for the quotient

$$\frac{f(y_*) - f(x_*)}{y_* - x_*}$$

and reach a contradiction using (b).
(d) Fix $s \in (0,1)$ and consider the sequence (x_n) defined by the recurrence

$$x_0 = s, \;\; x_{n+1} = f(x_n), \;\; \forall n \geq 0.$$

Prove that

$$\lim_n x_n = x_*,$$

where x_* is the unique fixed point of f located in the interval $(0,1)$.

Hint. The sequence is bounded since it lies in $[0,1]$. Show that the sequence is monotone and the limit lies in $(0,1)$. □

Exercise 8.15. Prove that for any $n \in \mathbb{N}$ and any numbers $x_1, x_2, \ldots, x_n \geq 0$ we have

$$\left(\frac{x_1 + \cdots + x_n}{n}\right)^2 \leq \frac{x_1^2 + \cdots + x_n^2}{n}.$$

Hint. Use the Cauchy–Schwarz inequality. □

Exercise 8.16. Consider the *Gauss bell*, i.e., the function

$$\gamma : \mathbb{R} \to \mathbb{R}, \;\; \gamma(x) = e^{-\frac{x^2}{2}}.$$

(a) Prove that for any $n \in \mathbb{N}$ there exists a polynomial $H_n(x)$ of degree n such that

$$\gamma^{(n)}(x) = (-1)^n H_n(x)\gamma(x).$$

(The polynomial $H_n(x)$ is called the *degree n Hermite polynomial*.)
(b) Prove that

$$H_{n+1}(x) = xH_n(x) - H_n'(x), \;\; \forall n \in \mathbb{N}.$$

(c) Compute $H_1(x)$, $H_2(x)$, $H_3(x)$.
(d) Find the intervals of convexity and concavity of $\gamma(x)$.
(e) Sketch the graph of the function $\gamma(x)$. □

[3]The results in this exercise are particularly useful in probability theory in the investigation of the so-called *branching processes*.

Exercise 8.17. Consider the *hyperbolic functions*

$$\cosh, \sinh : \mathbb{R} \to \mathbb{R}, \;\; \cosh x = \frac{e^x + e^{-x}}{2}, \;\; \sinh(x) = \frac{e^x - e^{-x}}{2}, \;\; \forall x \in \mathbb{R}.$$

(cosh=hyperbolic cosine, sinh= hyperbolic sine)
(a) Prove that

$$\cosh' x = \sinh x, \;\; \sinh' x = \cosh x,$$

$$\cosh^2 x - \sinh^2 x = 1, \;\; \cosh^2 x + \sinh^2 x = \cosh(2x), \;\; \forall x \in \mathbb{R}.$$

(b) Find the Taylor series of $\cosh x$ and $\sinh x$ at $x_0 = 0$.
(c) Prove that the function sinh is bijective and then find its inverse.
(d) Sketch the graphs of cosh and sinh. □

Exercise 8.18. Compute

$$\int xe^{2x}dx, \;\; \int xe^{2x}\cos x dx, \int xe^{2x}\sin x dx, \;\; \int \sin^3 x \cos^2 x dx.$$ □

Exercise 8.19. Compute

$$\int \frac{1}{(4+x^2)^5}dx$$

by reducing it to the computation in Example 8.47(d). □

Exercise 8.20. Compute

$$\int (\cos x)^{11} dx.$$ □

Exercise 8.21. Using the strategy outlined in Example 8.51 find the function $u(t)$, $v(t)$, $f(t)$ satisfying the differential equations

$$u'(t) + 2u(t) = t, \;\; v'(t) - v(t) = \cos t,$$

$$f'(t) - (\tan t)f(t) = t, \;\; -\frac{\pi}{2} < t < \frac{\pi}{2}.$$ □

Exercise 8.22. Suppose that we are given a huge container containing 200 liters of pure water. In this container, starting at $t = 0$, we continuously add 10 liters of salted water per minute containing 1.5 grams of salt per liter and, at the same time, the container is leaking salt-water mixture at a constant rate of 10 liters per minute. Denote by $m(t)$ the amount of salt (in grams) contained in the mixture after t minutes from the start.
(a) Prove that $m(t)$ satisfies the differential equation

$$\frac{dm}{dt} = 15 - \frac{m(t)}{20}.$$

(b) Recalling that initially there was no salt in the water, i.e., $m(0) = 0$, find $m(t)$ for any $t > 0$. □

8.7. More challenging problems

Problem 8.1. Suppose that $f:(0,\infty)\to\mathbb{R}$ is a differentiable function such that

$$\lim_{x\to\infty}\big(f(x)+f'(x)\big)=0.$$

Show that

$$\lim_{x\to\infty}f(x)=\lim_{x\to\infty}f'(x)=0. \qquad \square$$

Problem 8.2. (a) Prove that for any $n\in\mathbb{N}$ and any real numbers $a,r>0$ we have

$$a^{\frac{n}{n+1}}\leq\frac{1}{r}\left(\frac{r^{n+1}}{n+1}+\frac{na}{n+1}\right).$$

Hint: Use Young's inequality (8.46).

(b) Prove that if $\sum_{n\geq 0}a_n$ is a convergent series of positive numbers, then so is $\sum_{n\geq 0}a_n^{\frac{n}{n+1}}$. $\square$

Problem 8.3. Suppose that $f:\mathbb{R}\to\mathbb{R}$ is a C^3-function. Prove that there exists $a\in\mathbb{R}$ such that

$$f(a)\cdot f'(a)\cdot f''(a)\cdot f'''(a)\geq 0. \qquad \square$$

Problem 8.4. Suppose that $f:\mathbb{R}\to\mathbb{R}$ is a convex C^1 function. For $c\in\mathbb{R}$ we denote by $E_c(x)$ the function

$$E_c:\mathbb{R}\to\mathbb{R},\ \ E_c(x)=\frac{x^2}{2}-cx+f(x).$$

(a) Prove that E_c has a unique critical point.
(b) Prove that the function $g:\mathbb{R}\to\mathbb{R}$, $g(x)=x+f'(x)$ is bijective. $\square$

Problem 8.5. Suppose that $f:[a,b]\to\mathbb{R}$ is a continuous function satisfying

$$f\left(\frac{x+y}{2}\right)\leq\frac{f(x)+f(y)}{2},\ \ \forall x,y\in[a,b].$$

Prove that f is convex. $\square$

Problem 8.6. Suppose that $f:(a,b)\to\mathbb{R}$ is a convex function. Prove that f is continuous.

Hint. You need to use the facts proven in Exercise 8.13. $\square$

Problem 8.7. Show that for any positive real numbers a,b,c we have

$$a+b+c\leq\frac{a^3}{bc}+\frac{b^3}{ac}+\frac{c^3}{ab}. \qquad \square$$

Problem 8.8. Fix a natural number n and positive real numbers $x_1,\dots,x_n$. For any $\alpha>0$ we set

$$M_\alpha(x_1,\dots,x_n):=\left(\frac{x_1^\alpha+\cdots+x_n^\alpha}{n}\right)^{\frac{1}{\alpha}}.$$

(a) Show that

$$M_\alpha(x_1,\ldots,x_n) \le M_\beta(x_1,\ldots,x_n), \quad \forall 0<\alpha<\beta.$$

Hint. Use Hölder's inequality (8.22).

(b) Compute

$$\lim_{\alpha\to 0+} M_\alpha(x_1,\ldots,x_n). \qquad \square$$

Problem 8.9. (a) Prove that for any $n \in \mathbb{N}$ the equation $x^n + x = 1$ has a unique positive solution x_n.
(b) Prove that

$$\lim_{n\to\infty} x_n = 1. \qquad \square$$

Chapter 9

Integral Calculus

9.1. The integral as area: a first look

The Riemann integral is a very complicated infinite summation process that is often required when we want to compute areas or volumes of more irregular regions.

By way of motivation, let us consider a famous problem first solved by Archimedes by other means. Consider the arc of parabola in Figure 9.1 given by the equation

$$y = x^2, \quad 0 \leq x \leq 1.$$

We would like to compute the area of the region R between the x-axis, the parabola and the vertical line $x = 1$.

Let us observe that we do not have a precise definition of the concept of area. We only have an intuitive belief that

(i) the area of a rectangle is width $\times$ length, and
(ii) the area of a union of rectangles that intersect only along edges should be the sum of the area of the rectangles. We will refer to such regions as *simple type* regions.

We proceed by approximating R by a region of simple type. We subdivide the interval $[0, 1]$ into N equal parts, where N is a very large natural number. We obtain the points

$$x_0 = 0, \quad x_1 = \frac{1}{N}, \quad x_2 = \frac{2}{N}, \dots, x_N = \frac{N}{N}.$$

For each $k = 1, 2, \dots, N$ we denote by R_k the very thin slice of R of width $\frac{1}{N}$ delimited by the vertical lines $x = x_{k-1}$ and $x = x_k$. We have thus decomposed R into N thin slices $R_1, \dots, R_N$ and

$$\text{area}(R) = \sum_{k=1}^{n} \text{area}(R_k) = \text{area}(R_1) + \cdots + \text{area}(R_N).$$

Now observe that the slice R_k contains a thin rectangle $\underline{R}_k$ of height $f(x_{k-1})$ and is contained in a thin rectangle $\overline{R}_k$ of height $f(x_k)$; see Figure 9.1.

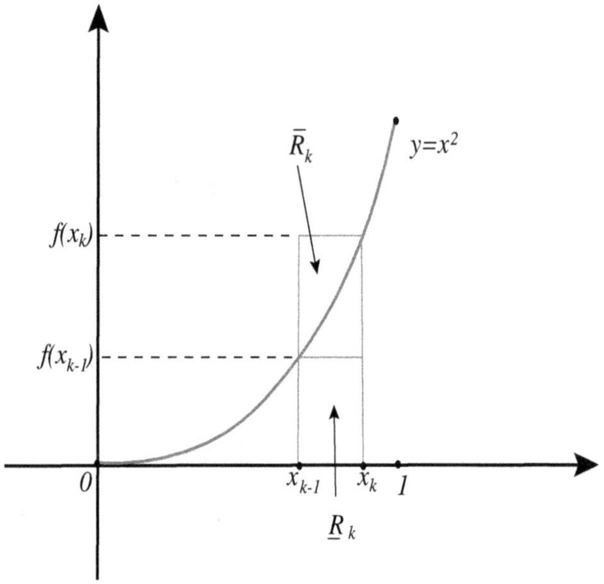

Fig. 9.1 *Computing the area underneath an arc of parabola.*

Thus

$$f(x_{k-1}) \times (x_k - x_{k-1}) = \text{area}(\underline{R}_k) \leq \text{area}(R_k) \leq \text{area}(\overline{R}_k) = f(x_k) \times (x_k - x_{k-1}).$$

Since $f(x_k) = \frac{k^2}{N^2}$ and $x_k - x_{k-1} = \frac{1}{N}$ we deduce

$$\frac{(k-1)^2}{N^3} \leq \text{area}(R_k) \leq \frac{k^2}{N^3},$$

and thus

$$\underbrace{\sum_{k=1}^{N} \frac{(k-1)^2}{N^3}}_{=:L_N} \leq \underbrace{\sum_{k=1}^{N} \text{area}(R_k)}_{=\text{area}(R)} \leq \underbrace{\sum_{k=1}^{N} \frac{k^2}{N^3}}_{=:U_N}. \tag{9.1}$$

Thus

$$L_N \leq \text{area}(R) \leq U_N. \tag{9.2}$$

Observe that

$$L_N = \frac{0^2}{N^3} + \frac{1^2}{N^3} + \cdots + \frac{(N-1)^2}{N^3} = \frac{1^2 + 2^2 + \cdots + (N-1)^2}{N^3},$$

$$U_N = \frac{1^2}{N^3} + \cdots + \frac{(N-1)^2}{N^3} + \frac{N^2}{N^3} = \frac{1^2 + 2^2 + \cdots + N^2}{N^3},$$

so that

$$U_N - L_N = \frac{N^2}{N^3} = \frac{1}{N}.$$

For N very large, the difference $U_N - L_N$ is very small and thus the sequence (L_N) converges if and only if the sequence (U_N) converges. Moreover, the inequality (9.2) shows that the common limit of these sequences, if it exists, must be equal to the area of R. To compute the limit of U_N we use the following famous identity whose proof is left to you as an exercise.

$$1^2 + 2^2 + \cdots + N^2 = \frac{N(N+1)(2N+1)}{6}. \tag{9.3}$$

We deduce that

$$U_N = \frac{N(N+1)(2N+1)}{6N^3} = \frac{1}{6}\frac{N}{N}\frac{N+1}{N}\frac{2N+1}{N} \to \frac{2}{6} \text{ as } N \to \infty.$$

Thus

$$\text{area}(R) = \frac{1}{3}.$$

This example describes the bare bones of the process called *integration*. As this simple example suggests, the integration involves a sophisticated infinite summation and a bit of good fortune, in the guise of (9.3), that allowed us to actually compute the result of this infinite summation.

We will spend the rest of this chapter describing rigorously and in great generality this process and we will show that in a large number of cases we can cleverly create our good fortune and succeed in carrying out explicit computations of the limits of infinite summations involved.

9.2. The Riemann integral

The process sketched in the previous section can be carried out in greater generality. We present the quite involved details in this section.

Definition 9.1 (Partitions). Fix an interval $[a, b]$, $a < b$.

(a) A *partition* $\boldsymbol{P}$ of $[a, b]$ is a finite collection of points $x_0, x_1, \dots, x_n$ of the interval such that

$$a = x_0 < x_1 < \cdots < x_n = b.$$

The natural number n is called the *order* of the partition, while the points $x_0, \dots, x_n$ are called the *nodes* of the partition. The intervals

$$[x_0, x_1], [x_1, x_2], \dots, [x_{n-1}, x_n]$$

are called the *intervals of the partition*. The interval $[x_{k-1}, x_k]$ is called the k-th interval of the partition and it is denoted by $I_k(\boldsymbol{P})$. Its length is denoted by $\Delta_k(\boldsymbol{P})$ or Δx_k. The largest of these lengths is called the *mesh size* of the partition and it is denoted by $\|\boldsymbol{P}\|$,

$$\boxed{\|\boldsymbol{P}\| := \max_{1 \le k \le n}(x_k - x_{k-1}) = \max_{1 \le k \le n} \Delta_k(\boldsymbol{P})}.$$

We denote by $\mathcal{P}_{[a,b]}$ the collection of all partitions of the interval $[a, b]$.

(b) A *sample* of a partition $\boldsymbol{P}$ of order n is a collection $\underline{\xi}$ consisting of n points $\xi_1, \dots, \xi_n$ such that

$$\xi_k \in I_k(\boldsymbol{P}), \ \ \forall k = 1, \dots, n.$$

The point ξ_k is called the *sample point* of the interval $I_k(\boldsymbol{P})$. We denote by $\mathcal{S}(\boldsymbol{P})$ the collection of all possible samples of the partition $\boldsymbol{P}$.

(c) A *sampled partition* of the interval $[a, b]$ is a pair $(\boldsymbol{P}, \underline{\xi})$, where $\boldsymbol{P}$ is a partition of $[a, b]$ and $\underline{\xi} \in \mathcal{S}(\boldsymbol{P})$ is a sample of $\boldsymbol{P}$. □

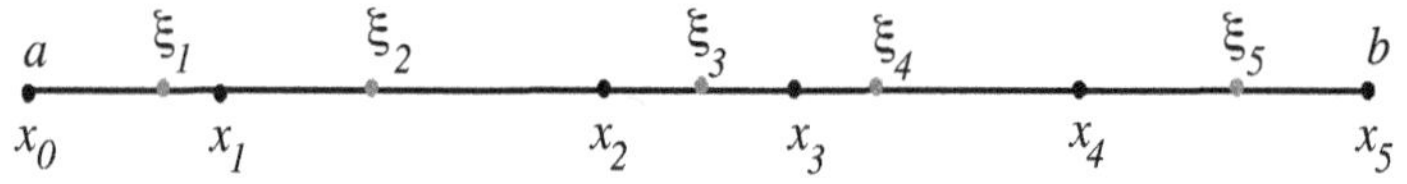

Fig. 9.2 *A sampled partition of order* 5 *of an interval* $[a, b]$. *Its longest interval is* $[x_1.x_2]$ *so its mesh size is* $(x_2 - x_1)$.

Example 9.2. Any compact interval $[a, b]$ has a natural partition $\boldsymbol{U}_n$ of order n corresponding to a subdivision of $[a, b]$ into n subintervals of order n. More precisely, $\boldsymbol{U}_n$ is defined by the points

$$x_0 = a, \ \ x_1 = a + \frac{1}{n}(b - a), \ \ x_k = a + \frac{k}{n}(b - a), \ \ k = 0, 1, \dots, n.$$

The partition $\boldsymbol{U}_n$ is called the *uniform partition of order* n of $[a, b]$. Note that

$$\|\boldsymbol{U}_n\| = \frac{b - a}{n}.$$ □

Definition 9.3. Let $f : [a, b] \to \mathbb{R}$ be a function defined on the closed and bounded interval $[a, b]$. Given a partition $\boldsymbol{P} = (x_0 < \cdots < x_n)$ of $[a, b]$, and a sample $\underline{\xi}$ of $\boldsymbol{P}$, we define the *Riemann*[1] *sum* of f associated to the sampled partition $(P, \underline{\xi})$ to be the number

$$\boldsymbol{S}(f, \boldsymbol{P}, \underline{\xi}) = \sum_{k=1}^{n} f(\xi_k)\Delta_k(\boldsymbol{P}) = \sum_{k=1}^{n} f(\xi_k)\Delta x_k = \sum_{k=1}^{n} f(\xi_k)(x_k - x_{k-1}).$$ □

As depicted in Figure 9.3, each term $f(\xi_k)(x_k - x_{k-1})$ in a Riemann sum is equal to the area of a "thin" rectangle of width $\Delta x_k = (x_k - x_{k-1})$, and height given by the altitude of the point on the graph of f determined by the sample point $\xi_k \in [x_{k-1}, x_k]$. The Riemann sum is therefore the area of the region formed by putting side by side each of these thin rectangles. The hope is that the area of this rather jagged looking region is an approximation for the area of the region under the graph of f. The next definition makes this intuition precise.

Definition 9.4. Suppose that $f : [a, b] \to \mathbb{R}$ is a function defined on the *closed and bounded* interval $[a, b]$. We say that f is *Riemann integrable on* $[a, b]$ if there exists a real number

[1]Named after Bernhardt Riemann (1826–1866), German mathematician who made lasting and revolutionary contributions to analysis, number theory, and differential geometry; see Wikipedia.

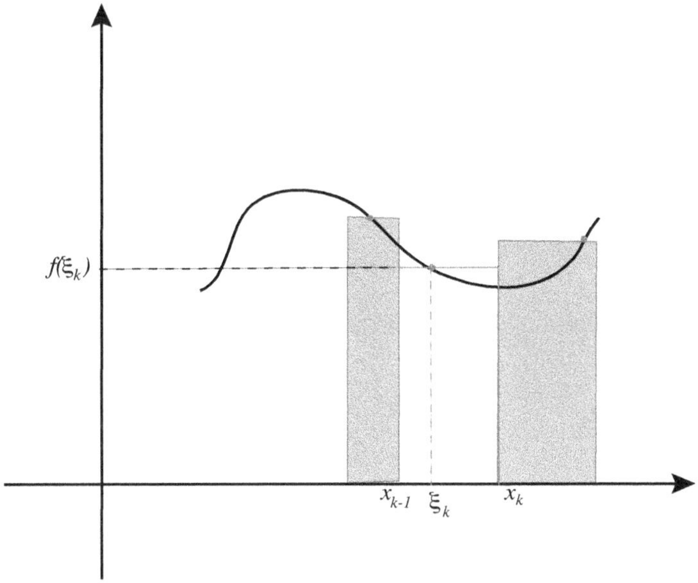

Fig. 9.3 *The term $f(\xi_k)\Delta x_k$ is the area of a rectangle.*

I with the following property: for any $\varepsilon > 0$ there exists $\delta = \delta(\varepsilon) > 0$ such that, *for any partition* $\boldsymbol{P}$ of $[a, b]$ with mesh size $\|\boldsymbol{P}\| < \delta$, *and any sample* $\underline{\xi}$ of $\boldsymbol{P}$ we have

$$\left| I - \boldsymbol{S}(f, \boldsymbol{P}, \underline{\xi}) \right| < \varepsilon.$$

Equivalently, as a quantified statement, the above reads

$$\begin{aligned} \exists I \in \mathbb{R}, \;\; \forall \varepsilon > 0, \;\; \exists \delta = \delta(\varepsilon) > 0, \;\; \forall \boldsymbol{P} \in \mathcal{P}_{[a,b]}, \;\; \forall \underline{\xi} \in \mathcal{S}(\boldsymbol{P}) : \\ \|\boldsymbol{P}\| < \delta \Rightarrow \left| I - \boldsymbol{S}(f, \boldsymbol{P}, \underline{\xi}) \right| < \varepsilon. \end{aligned} \tag{9.4}$$

We will denote by $\mathcal{R}[a, b]$ the collection of all Riemann integrable functions $f : [a, b] \to \mathbb{R}$.

□

Suppose that $f : [a, b] \to \mathbb{R}$ is Riemann integrable. For any $n \in \mathbb{N}$ we fix a sample $\underline{\xi}^{(n)}$ of $\boldsymbol{U}_n$, the uniform partition of order n of $[a, b]$. If I is *any* real number satisfying (9.4), then from the equality

$$\lim_{n\to\infty} \|\boldsymbol{U}_n\| = 0$$

we deduce that

$$I = \lim_{n\to\infty} \boldsymbol{S}\big(f, \boldsymbol{U}_n, \underline{\xi}^{(n)} \big).$$

Since a convergent sequence has a *unique* limit, we deduce that there exists precisely one real number I satisfying (9.4). This real number is called the *Riemann integral* of f on $[a, b]$ and it is denoted by

$$\int_a^b f(x)dx.$$

It bears repeating the definition of $\int_a^b f(x)dx$.

The Riemann integral of f over $[a,b]$, when it exists, is the **unique real number** $\int_a^b f(x)dx$ *with the following property: for any $\varepsilon > 0$ there exists $= \delta = \delta(\varepsilon) > 0$ such that for any partition $\boldsymbol{P}$ of $[a,b]$ with mesh $\|\boldsymbol{P}\| < \delta$, and for any sample $\underline{\xi}$ of $\boldsymbol{P}$, the Riemann sum $\boldsymbol{S}(f, \boldsymbol{P}, \underline{\xi})$ is within ε of $\int_a^b f(x)dx$, i.e.,*

$$\left| \int_a^b f(x)dx - \boldsymbol{S}(f, \boldsymbol{P}, \underline{\xi}) \right| < \varepsilon.$$

We can *loosely* rephrase this as follows

$$\int_a^b f(x)dx = \lim_{\substack{\|\boldsymbol{P}\| \to 0, \\ \underline{\xi} \in \mathcal{S}(\boldsymbol{P})}} \boldsymbol{S}(f, \boldsymbol{P}, \underline{\xi}). \tag{9.5}$$

Example 9.5. Consider the constant function $f : [a,b] \to \mathbb{R}$, $f(x) = C$, for all $x \in [a,b]$ where C is a fixed real number. Note that for any sampled partition of order n $(\boldsymbol{P}, \underline{\xi})$ of $[a,b]$ we have

$$\begin{aligned} \boldsymbol{S}(f, \boldsymbol{P}, \underline{\xi}) &= f(\xi_1)(x_1 - x_0) + f(\xi_2)(x_2 - x_1) + \cdots + f(\xi_n)(x_n - x_{n-1}) \\ &= C(x_1 - x_0) + C(x_2 - x_1) + \cdots + C(x_n - x_{n-1}). \\ &= C\big((x_1 - x_0) + (x_2 - x_1) + \cdots + (x_n - x_{n-1})\big) = C(x_n - x_0) = C(b-a). \end{aligned}$$

This shows that the constant function is integrable and

$$\int_a^b C dx = C(b-a). \qquad \square$$

It is natural to ask if there exist Riemann integrable functions more complicated than the constant functions. The next section will address precisely this issue. We will see that indeed, the world of integrable functions is very large. Until then, let us observe that not any function is Riemann integrable.

Proposition 9.6. *Suppose that $f : [a,b] \to \mathbb{R}$ is a Riemann integrable function. Then f is bounded, i.e.,*

$$-\infty < \inf_{x \in [a,b]} f(x) < \sup_{x \in [a,b]} f(x) < \infty.$$

Proof. We argue by contradiction. Suppose that $f : [a,b] \to \mathbb{R}$ is Riemann integrable and unbounded above, i.e.,

$$\sup_{x \in [a,b]} f(x) = \infty.$$

For any $n \in \mathbb{N}$ consider the uniform partition $\boldsymbol{U}_n$ of $[a, b]$. Then there exists $k = k(n)$ such that f is unbounded the interval $I_k = I_{k(n)}$ of this partition. For $j \neq k$ fix an arbitrary sample point $\xi_j \in I_j$. Since f is not bounded above on I_k, there exists $\xi_k \in I_k$ such that

$$f(\xi_k) > \frac{n}{\Delta x_k} - \sum_{j \neq k} f(\xi_j) \frac{\Delta x_j}{\Delta x_k} \iff f(\xi_k)\Delta x_k + \sum_{j \neq k} f(\xi_j)\Delta x_j > n.$$

We obtain a sample $\underline{\xi}^{(n)}$ of $\boldsymbol{U}_n$ and for this sample we have

$$\boldsymbol{S}\big(f, \boldsymbol{U}_n, \underline{\xi}^{(n)}\big) = f(x_k)\Delta x_k + \sum_{j \neq k} f(\xi_j)\Delta x_j > n, \;\; \forall n \in \mathbb{N}.$$

The Riemann integrability of f implies that the sequence of Riemann sums $\boldsymbol{S}\big(f, \boldsymbol{U}_n, \underline{\xi}^{(n)}\big)$ is convergent. This contradicts the last inequality which states that this sequence is unbounded. □

The above result shows that, e.g., the function

$$f : [0, 1] \to \mathbb{R}, \;\; f(x) = \begin{cases} 0, & x = 0, \\ \frac{1}{\sqrt{x}}, & x \in (0, 1], \end{cases}$$

is not Riemann integrable because it is not bounded.

9.3. Darboux sums and Riemann integrability

To be able to construct examples of integrable functions we need a criterion for recognizing such functions, more flexible than the definition. Fortunately there is one such criterion due to Gaston Darboux. To formulate it we need to introduce several new concepts.

Definition 9.7. Suppose that $f : [a, b] \to \mathbb{R}$ is a *bounded* function defined on the closed and bounded interval $[a, b]$. For any partition $\boldsymbol{P}$ of $[a, b]$ of order n we set

$$\boldsymbol{S}^*(f, \boldsymbol{P}) := \sum_{k=1}^{n} \sup_{x \in I_k(\boldsymbol{P})} f(x)\Delta x_k,$$

$$\boldsymbol{S}_*(f, \boldsymbol{P}) := \sum_{k=1}^{n} \inf_{x \in I_k(\boldsymbol{P})} f(x)\Delta x_k,$$

$$\omega(f, \boldsymbol{P}) := \sum_{k=1} \operatorname{osc}(f, I_k)\Delta x_k,$$

where

- $I_k = I_k(\boldsymbol{P})$ is the k-th interval of the partition $\boldsymbol{P}$,
- Δx_k is the length of I_k,
- $\operatorname{osc}(f, I_k)$ denotes the oscillation of f on I_k.

The quantity $\boldsymbol{S}^*(f, \boldsymbol{P})$ is called *the upper Darboux*[2] *sum* of the function f determined by the partition $\boldsymbol{P}$, while $\boldsymbol{S}_*(f, \boldsymbol{P})$ is called *the lower Darboux sum* of the function f determined by the partition $\boldsymbol{P}$. We will refer to $\omega(f, \boldsymbol{P})$ as the *mean oscillation of f* along $\boldsymbol{P}$. □

Proposition 9.8. *If $f : [a, b] \to \mathbb{R}$ is a bounded function, then for any partition $\boldsymbol{P}$ of $[a, b]$ and any sample $\underline{\xi}$ of $\boldsymbol{P}$ we have*

$$\boldsymbol{S}_*(f, \boldsymbol{P}) \leq \boldsymbol{S}(f, \boldsymbol{P}, \underline{\xi}) \leq \boldsymbol{S}^*(f, \boldsymbol{P}), \tag{9.6a}$$

$$\omega(f, \boldsymbol{P}) = \boldsymbol{S}^*(f, \boldsymbol{P}) - \boldsymbol{S}_*(f, \boldsymbol{P}). \tag{9.6b}$$

Proof. Suppose that $\boldsymbol{P}$ is a partition of order n of $[a, b]$ and $\underline{\xi}$ is a sample of $\boldsymbol{P}$. For $k = 1, \dots, n$ we denote by I_k the k-th interval of $\boldsymbol{P}$ and we set

$$M_k := \sup_{x \in I_k} f(x), \quad m_k := \inf_{x \in I_k} f(x).$$

Then $M_k - m_k = \operatorname{osc}(f, I_k)$ and

$$\begin{aligned} \boldsymbol{S}^*(f, \boldsymbol{P}) - \boldsymbol{S}_*(f, \boldsymbol{P}) &= \big(M_1 \Delta x_1 + \cdots + M_n \Delta x_n \big) - \big(m_1 \Delta x_1 + \cdots + m_n \Delta x_n \big) \\ &= (M_1 - m_1)\Delta x_1 + \cdots + (M_n - m_n)\Delta x_n \\ &= \operatorname{osc}(f, I_1)\Delta x_1 + \cdots + \operatorname{osc}(f, I_n)\Delta x_n = \omega(f, \boldsymbol{P}). \end{aligned}$$

This proves (9.6b). If $\underline{\xi}$ is a sample of $\boldsymbol{P}$, then

$$m_k \Delta x_k \leq f(\xi_k)\Delta x_k \leq M_k \Delta x_k, \quad \forall k = 1, \dots, n,$$

so that

$$\sum_{k=1}^n m_k \Delta x_k \leq \sum_{k=1}^n f(\xi_k)\Delta x_k \leq \sum_{k=1}^n M_k \Delta x_k.$$

This proves (9.6a). □

Corollary 9.9. *If $f : [a, b] \to \mathbb{R}$ is a bounded function, then for any partition P of $[a, b]$ and for any samples $\underline{\xi}, \underline{\xi}'$ of $\boldsymbol{P}$ we have*

$$\big| \boldsymbol{S}(f, \boldsymbol{P}, \underline{\xi}) - \boldsymbol{S}(f, \boldsymbol{P}, \underline{\xi}') \big| \leq \omega(f, \boldsymbol{P}).$$

Proof. According to (9.6a) the Riemann sums $\boldsymbol{S}(f, \boldsymbol{P}, \underline{\xi})$, $\boldsymbol{S}(f, \boldsymbol{P}, \underline{\xi}')$ are both contained in the interval $[\boldsymbol{S}_*(f, \boldsymbol{P}), \boldsymbol{S}^*(f, \boldsymbol{P})]$ so the distance between them must be smaller than the length of this interval which is equal to $\omega(f, \boldsymbol{P})$ according to (9.6b). □

[2]Named after Gaston Darboux (1842–1917), French mathematician who made several important contributions to geometry and mathematical analysis; see Wikipedia.

Proposition 9.10. *Suppose that $f : [a,b] \to \mathbb{R}$ is a bounded function and $\boldsymbol{P}$ is a partition of $[a,b]$. If $\boldsymbol{P}'$ is a partition of $[a,b]$ obtained from $\boldsymbol{P}$ by adding one extra node x' in the interior of some interval of $\boldsymbol{P}$, then*

$$\boldsymbol{S}_*(f,\boldsymbol{P}) \leq \boldsymbol{S}_*(f,\boldsymbol{P}') \leq \boldsymbol{S}^*(f,\boldsymbol{P}') \leq \boldsymbol{S}^*(f,\boldsymbol{P}).$$

Thus, by adding a node the upper Darboux sums decrease, while the lower Darboux sums increase.

Proof. The inequality (9.6a) shows that $\boldsymbol{S}_*(f,\boldsymbol{P}') \leq \boldsymbol{S}^*(f,\boldsymbol{P}')$. Suppose that the extra node x' is contained in (x_{k-1},x_k). We set

$$M_k := \sup_{x\in I_k} f(x), \quad m_k := \inf_{x\in I_k} f(x).$$

Then

$$\begin{aligned}
&\boldsymbol{S}_*(f,\boldsymbol{P}') \\
&= \sum_{j<k} m_j\Delta x_j + \underbrace{\inf_{x\in[x_{k-1},x']} f(x)}_{\geq m_k}\,(x'-x_{k-1}) + \underbrace{\inf_{[x',x_k]} f(x)}_{\geq m_k}\,(x_k-x') + \sum_{\ell>k} m_\ell\Delta x_\ell \\
&\geq \sum_{j<k} m_j\Delta x_j + \underbrace{m_k(x'-x_{k-1}) + m_k(x_k-x')}_{=m_k(x_k-x_{k-1})} + \sum_{\ell>k} m_\ell\Delta x_\ell \\
&= \sum_{j<k} m_j\Delta x_j + m_k\Delta x_k + \sum_{\ell>k} m_\ell\Delta x_\ell = \sum_{i=1}^{n} m_i\Delta x_i = \boldsymbol{S}_*(f,\boldsymbol{P}).
\end{aligned}$$

The inequality

$$\boldsymbol{S}^*(f,\boldsymbol{P}') \leq \boldsymbol{S}^*(f,\boldsymbol{P})$$

is proved in a similar fashion. □

Definition 9.11. Given two partitions $\boldsymbol{P},\boldsymbol{P}'$ of $[a,b]$, we say that $\boldsymbol{P}'$ is a *refinement* of $\boldsymbol{P}$, and we write this $\boldsymbol{P}' \succ \boldsymbol{P}$, if $\boldsymbol{P}'$ is obtained from $\boldsymbol{P}$ by adding a few more nodes. □

Since the addition of nodes increases lower Darboux sums and decreases upper Darboux sums we deduce the following result.

Proposition 9.12. *Suppose that $f : [a,b] \to \mathbb{R}$ is a bounded function and $\boldsymbol{P},\boldsymbol{P}'$ are partitions of $[a,b]$. If $\boldsymbol{P}' \succ \boldsymbol{P}$, then*

$$\boldsymbol{S}_*(f,\boldsymbol{P}) \leq \boldsymbol{S}_*(f,\boldsymbol{P}') \leq \boldsymbol{S}^*(f,\boldsymbol{P}') \leq \boldsymbol{S}^*(f,\boldsymbol{P}). \qquad □$$

Corollary 9.13. *Suppose that $f : [a,b] \to \mathbb{R}$ is a bounded function and $\boldsymbol{P},\boldsymbol{P}'$ are partitions of $[a,b]$. If $\boldsymbol{P}' \succ \boldsymbol{P}$,*

$$\omega(f,\boldsymbol{P}') \leq \omega(f,\boldsymbol{P}). \tag{9.7}$$

Proof. From (9.8) we deduce

$$S_*(f, P) \leq S_*(f, P') \leq S^*(f, P') \leq S^*(f, P),$$

so that,

$$\omega(f, P') = S^*(f, P') - S_*(f, P') \leq S^*(f, P) - S_*(f, P) = \omega(f, P).$$

□

Given two partitions P, P' of $[a, b]$ we denote by $P \vee P'$ the partition whose set of nodes is the union of the sets of nodes of the partitions P and P'. Clearly $P \vee P'$ is a refinement of both P and P'. From Proposition 9.12 we deduce the following important consequence.

Corollary 9.14. *Suppose that $f : [a, b] \to \mathbb{R}$ is a bounded function and P_0, P_1 are partitions of $[a, b]$. Then*

$$S_*(f, P_1) \leq S_*(f, P_0 \vee P_1) \leq S^*(f, P_0 \vee P_1) \leq S^*(f, P_0). \tag{9.8}$$

□

The above corollary shows that if $f : [a, b] \to \mathbb{R}$ is a bounded function, then the set

$$\{ S^*(f, P);\ \ P \in \mathcal{P}_{[a,b]} \}$$

is bounded below. Indeed, if we denote by U_1 the uniform partition of order 1 of $[a, b]$, then (9.8) shows that

$$S_*(f, U_1) \leq S^*(f, P),\ \ \forall P \in \mathcal{P}_{[a,b]}.$$

We set

$$S^*(f) := \inf\{ S^*(f, P);\ \ P \in \mathcal{P}_{[a,b]} \}.$$

Similarly, the set

$$\{ S_*(f, P);\ \ P \in \mathcal{P}_{[a,b]} \}$$

is bounded above and we define

$$S_*(f) := \sup\{ S_*(f, P);\ \ P \in \mathcal{P}_{[a,b]} \}.$$

Proposition 9.15. *If $f : [a, b] \to \mathbb{R}$ is a bounded function, then*

$$S_*(f) \leq S^*(f). \tag{9.9}$$

Proof. From (9.8) we deduce that $\forall P_0, P_1 \in \mathcal{P}_{[a,b]}$ we have

$$\begin{aligned} S_*(f, P_1) \leq S^*(f, P_0) &\Rightarrow S_*(f, P_1) \leq \inf_{P_0} S^*(f, P_0) = S^*(f) \\ &\Rightarrow S_*(f) = \sup_{P_1} S_*(f, P_1) \leq S^*(f). \end{aligned}$$

□

Definition 9.16. Let $f : [a, b] \to \mathbb{R}$ be a *bounded* function.

(a) The numbers $\boldsymbol{S}_*(f)$ and respectively $\boldsymbol{S}^*(f)$ are called the *lower* and respectively *upper Darboux integrals* of f.
(b) The function f is called *Darboux integrable* if $\boldsymbol{S}_*(f) = \boldsymbol{S}^*(f)$. □

Theorem 9.17 (Riemann–Darboux). *Suppose that $f : [a, b] \to \mathbb{R}$ is a bounded function. Then the following statements are equivalent.*

(i) The function f is Riemann integrable.
(ii) The function f is Darboux integrable, i.e., $\boldsymbol{S}_*(f) = \boldsymbol{S}^*(f)$.
(iii) $\inf_{\boldsymbol{P}} \omega(f, \boldsymbol{P}) = 0$, *i.e.,*

$$\forall \varepsilon > 0, \quad \exists \boldsymbol{P}_\varepsilon \in \mathcal{P}_{[a,b]} : \quad \omega(f, \boldsymbol{P}_\varepsilon) < \varepsilon. \tag{$\boldsymbol{\omega}_0$}$$

(iv) $\lim_{\|\boldsymbol{P}\| \to 0} \omega(f, \boldsymbol{P}) = 0$, *i.e.,*

$$\forall \varepsilon > 0 \ \exists \delta = \delta(\varepsilon) > 0 \ \forall \boldsymbol{P} \in \mathcal{P}_{[a,b]} : \quad \|\boldsymbol{P}\| < \delta \Rightarrow \omega(f, \boldsymbol{P}) < \varepsilon. \tag{$\boldsymbol{\omega}$}$$

Proof. We will prove these equivalences using the following logical successions

$$(iii) \Longleftrightarrow (ii), \quad (iv) \Rightarrow (iii), \quad (iv) \Longleftrightarrow (i), \quad (iii) \Rightarrow (iv).$$

(iii) ⇒ (ii). For any $\varepsilon > 0$ we can find a partition $\boldsymbol{P}_\varepsilon$ such that $\omega(f, \boldsymbol{P}_\varepsilon) < \varepsilon$. Now observe that

$$\boldsymbol{S}_*(f, \boldsymbol{P}_\varepsilon) \leq \boldsymbol{S}_*(f) \leq \boldsymbol{S}^*(f) \leq \boldsymbol{S}^*(f, \boldsymbol{P}_\varepsilon),$$

and

$$\boldsymbol{S}^*(f, \boldsymbol{P}_\varepsilon) - \boldsymbol{S}_*(f, \boldsymbol{P}_\varepsilon) = \omega(f, \boldsymbol{P}_\varepsilon) < \varepsilon.$$

Hence

$$0 \leq \boldsymbol{S}^*(f) - \boldsymbol{S}_*(f) \leq \boldsymbol{S}^*(f, \boldsymbol{P}_\varepsilon) - \boldsymbol{S}_*(f, \boldsymbol{P}_\varepsilon) < \varepsilon, \ \ \forall \varepsilon > 0,$$

so that

$$\boldsymbol{S}_*(f) = \boldsymbol{S}^*(f).$$

(ii) ⇒ (iii). We know that $\boldsymbol{S}_*(f) = \boldsymbol{S}^*(f)$. Denote by $\boldsymbol{S}(f)$ this common value. Since

$$\boldsymbol{S}(f) = \boldsymbol{S}_*(f) = \sup_{\boldsymbol{P}} \boldsymbol{S}_*(f, \boldsymbol{P}),$$

we deduce that for any $\varepsilon > 0$ there exists a partition P_ε^- such that

$$\boldsymbol{S}(f) - \frac{\varepsilon}{2} < \boldsymbol{S}_*(f, \boldsymbol{P}_\varepsilon^-) \leq \boldsymbol{S}(f).$$

Since

$$\boldsymbol{S}(f) = \boldsymbol{S}^*(f) = \inf_{\boldsymbol{P}} \boldsymbol{S}^*(f, \boldsymbol{P}),$$

we deduce that for any $\varepsilon > 0$ there exists a partition P_ε^+ such that

$$S(f) \leq S^*(f, P_\varepsilon^+) < S(f) + \frac{\varepsilon}{2}.$$

Hence

$$S(f) - \frac{\varepsilon}{2} < S_*(f, P_\varepsilon^-) \leq S^*(f, P_\varepsilon^+) < S(f) + \frac{\varepsilon}{2}.$$

Now set $P_\varepsilon := P_\varepsilon^- \vee P_\varepsilon^+$. We deduce from (9.8) that

$$S(f) - \frac{\varepsilon}{2} < S_*(f, P_\varepsilon^-) \leq S_*(f, P_\varepsilon) \leq S^*(f, P_\varepsilon) \leq S^*(f, P_\varepsilon^+) < S(f) + \frac{\varepsilon}{2}.$$

This proves that

$$\omega(f, P_\varepsilon) = S^*(f, P_\varepsilon) - S_*(f, P_\varepsilon) < \varepsilon.$$

(iv) $\Rightarrow$ (iii). This is obvious.

(iv) $\Rightarrow$ (i). From the above we deduce that (iv) $\Rightarrow$ (ii) $\wedge$ (iii), so $S_*(f) = S^*(f)$. We set

$$S(f) := S_*(f) = S^*(f).$$

We will show that f is integrable and its Riemann integral is $S(f)$.

Fix $\varepsilon > 0$. According to (ω), there exists $\delta = \delta(\varepsilon) > 0$ such that for any partition P of $[a, b]$ satisfying $\|P\| < \delta$ we have

$$\omega(f, P) < \varepsilon.$$

Given a partition P such that $\|P\| < \delta$ and $\underline{\xi}$ a sample of P we have

$$S_*(f, P) \leq S(f) \leq S^*(f, P),$$
$$S_*(f, P) \leq S(f, P, \underline{\xi}) \leq S^*(f, P).$$

Thus both numbers $S(f)$ and $S(f, P, \underline{\xi})$ lie in the interval $[S_*(f, P), S^*(f, P)]$ of length $\omega(f, P) < \varepsilon$. Hence

$$\left| S(f, P, \underline{\xi}) - S(f) \right| < \varepsilon, \quad \forall \|P\| < \delta(\varepsilon), \quad \forall \underline{\xi} \in \mathcal{S}(P).$$

This proves that f is Riemann integrable.

(i) $\Rightarrow$ (iv). We have to prove that if f is Riemann integrable, then f satisfies (ω). We first need an auxiliary result.

Lemma 9.18. *Suppose that $f : [a, b] \to \mathbb{R}$ is a bounded function. Then, for any partition P of $[a, b]$ we have*

$$S_*(f, P) = \inf_{\underline{\xi} \in \mathcal{S}(P)} S(f, P, \underline{\xi}),$$

$$S^*(f, P) = \sup_{\underline{\xi} \in \mathcal{S}(P)} S(f, P, \underline{\xi}).$$

In other words, for any $\varepsilon > 0$, and any partition P of $[a, b]$, there exist samples $\underline{\xi}'$ and $\underline{\xi}''$ of P such that

$$S_*(f, P) \leq S(f, P, \underline{\xi}') < S_*(f, P) + \varepsilon,$$

$$S^*(f, P) - \varepsilon < S(f, P, \underline{\xi}'') \leq S^*(f, P).$$

In particular

$$\omega(f, P) = S^*(f, P) - S_*(f, P) = \sup_{\underline{\xi} \in \mathcal{S}(P)} S(f, P, \underline{\xi}) - \inf_{\underline{\xi} \in \mathcal{S}(P)} S(f, P, \underline{\xi}). \tag{9.10}$$

Proof. We prove only the statement involving lower sums. The proof of the statement involving upper sums is similar. Denote by n the order of $\boldsymbol{P}$ and by I_k the k-th interval of $\boldsymbol{P}$ and, as usual, we set

$$m_k = \inf_{x \in I_k} f(x).$$

In particular, there exists $\xi_k' \in I_k$ such that

$$m_k \leq f(\xi_k') < m_k + \frac{\varepsilon}{b-a}.$$

The collection $\underline{\xi}' = (\xi_k')_{1 \leq k \leq n}$ is a sample of $\boldsymbol{P}$ satisfying

$$m_k(x_k - x_{k-1}) \leq f(\xi_k')(x_k - x_{k-1}) < m_k(x_k - x_{k-1}) + \frac{\varepsilon}{b-a}(x_k - x_{k-1}).$$

Hence

$$\begin{aligned}\boldsymbol{S}_*(f, \boldsymbol{P}) &= \sum_{k=1}^{n} m_k(x_k - x_{k-1}) \leq \underbrace{\sum_{k=1}^{n} f(\xi_k')(x_k - x_{k-1})}_{= \boldsymbol{S}(f, \boldsymbol{P}, \underline{\xi}')} \\ &< \underbrace{\sum_{k=1}^{n} m_k(x_k - x_{k-1})}_{= \boldsymbol{S}_*(f, \boldsymbol{P})} + \frac{\varepsilon}{b-a} \underbrace{\sum_{k=1}^{n} (x_k - x_{k-1})}_{= (b-a)} = \boldsymbol{S}_*(f, P) + \varepsilon.\end{aligned}$$

□

We can now complete the proof of ($\boldsymbol{\omega}$). Since f is Riemann integrable, there exists $S_f \in \mathbb{R}$ such that, for any $\varepsilon > 0$ we can find $\delta = \delta(\varepsilon) > 0$ with the property that for any partition $\boldsymbol{P}$ with mesh size $\|\boldsymbol{P}\| < \delta$ and any sample $\underline{\xi}$ of $\boldsymbol{P}$ we have

$$\big| S_f - \boldsymbol{S}(f, \boldsymbol{P}, \underline{\xi}) \big| < \frac{\varepsilon}{4}. \tag{9.11}$$

According to Lemma 9.18 we can find samples $\underline{\xi}'$ and $\underline{\xi}''$ such that

$$\big| \boldsymbol{S}_*(f, \boldsymbol{P}) - \boldsymbol{S}(f, \boldsymbol{P}, \underline{\xi}') \big|, \;\; \big| \boldsymbol{S}^*(f, \boldsymbol{P}) - \boldsymbol{S}(f, \boldsymbol{P}, \underline{\xi}'') \big| < \frac{\varepsilon}{4}. \tag{9.12}$$

If $\|\boldsymbol{P}\| < \delta$, then

$$\begin{aligned}\omega(f, \boldsymbol{P}) &= \big| \boldsymbol{S}^*(f, \boldsymbol{P}) - \boldsymbol{S}_*(f, \boldsymbol{P}) \big| \\ &\leq \big| \boldsymbol{S}_*(f, \boldsymbol{P}) - \boldsymbol{S}(f, \boldsymbol{P}, \underline{\xi}') \big| + \big| \boldsymbol{S}(f, \boldsymbol{P}, \underline{\xi}') - \boldsymbol{S}(f, \boldsymbol{P}, \underline{\xi}'') \big| \\ &\quad + \big| \boldsymbol{S}(f, \boldsymbol{P}, \underline{\xi}'') - \boldsymbol{S}^*(f, \boldsymbol{P}) \big| \\ &\stackrel{(9.12)}{<} \frac{\varepsilon}{4} + \big| \boldsymbol{S}(f, \boldsymbol{P}, \underline{\xi}') - \boldsymbol{S}(f, \boldsymbol{P}, \underline{\xi}'') \big| + \frac{\varepsilon}{4} \\ &\leq \frac{\varepsilon}{2} + \big| \boldsymbol{S}(f, \boldsymbol{P}, \underline{\xi}') - S_f \big| + \big| S_f - \boldsymbol{S}(f, \boldsymbol{P}, \underline{\xi}'') \big| \stackrel{(9.11)}{<} \frac{\varepsilon}{2} + \frac{\varepsilon}{4} + \frac{\varepsilon}{4} = \varepsilon.\end{aligned}$$

(iii) $\Rightarrow$ (iv). We have to show that if f satisfies ($\boldsymbol{\omega}_0$), then it also satisfies ($\boldsymbol{\omega}$). We need the following auxiliary result.

Lemma 9.19. *Suppose that $\boldsymbol{P}_0 = \{a = z_0 < z_1 < \cdots < z_{n_0} = b\}$ is a partition of $[a, b]$ of order n_0. Denote by λ_0 the length of the shortest intervals of the partition $\boldsymbol{P}_0$, i.e.,*

$$\lambda_0 := \min_{1 \leq j \leq n_0} (z_j - z_{j-1}).$$

For any partition $\boldsymbol{P}$ such that $\|\boldsymbol{P}\| < \lambda_0$ we have

$$\omega(f, \boldsymbol{P}) \leq (n_0 - 1)\|\boldsymbol{P}\| \operatorname{osc}(f, [a, b]) + \omega(f, \boldsymbol{P}_0). \tag{9.13}$$

Proof. Denote by $I_1, \ldots, I_{n_0}$ the intervals of $\boldsymbol{P}_0$. Denote by n the order of $\boldsymbol{P}$, and by $J_1, \ldots, J_n$ the intervals of $\boldsymbol{P}$. We will denote by $\ell(J_k)$ the length of J_k and by $\ell(I_j)$ the length of I_j.

Since $\ell(J_k) \le \ell(I_j)$, $\forall j = 1, \ldots, n_0$, $k = 1, \ldots, n$ we deduce that the intervals J_k of $\boldsymbol{P}$ are of only the following two types.

Type 1. The interval J_k is contained in an interval I_j of $\boldsymbol{P}_0$.
Type 2. The interval J_k contains in the *interior* a node $z_{j(k)}$ of $\boldsymbol{P}_0$.

We denote by $\mathcal{J}^1$ the collection of Type 1 intervals of $\boldsymbol{P}$, and by $\mathcal{J}^2$ the collection of Type 2 intervals of $\boldsymbol{P}$. We remark that $\mathcal{J}^2$ could be empty. Moreover, for any node z_j of $\boldsymbol{P}_0$ there exists at most one Type 2 interval of $\boldsymbol{P}$ that contains z_j in the interior. Thus $\mathcal{J}^2$ consist of at most $n_0 - 1$ intervals, i.e., its cardinality $|\mathcal{J}^2|$ satisfies

$$|\mathcal{J}^2| \le n_0 - 1.$$

We have

$$\omega(f, \boldsymbol{P}) = \sum_{k=1}^{n} \operatorname{osc}(f, J_k)\ell(J_k) = \underbrace{\sum_{j_k \in \mathcal{J}^1} \operatorname{osc}(f, J_k)\ell(J_k)}_{=:S_1} + \underbrace{\sum_{J_k \in \mathcal{J}^2} \operatorname{osc}(f, J_k)\ell(J_k)}_{=:S_2}.$$

We now estimate S_1 from above

$$S_1 = \sum_{j=1}^{n_0} \left(\sum_{J_k \subset I_j} \operatorname{osc}(f, J_k)\ell(J_k) \right)$$

$(\operatorname{osc}(f, J_k) \le \operatorname{osc}(f, I_j)$ whenever $J_k \subset I_j)$

$$\le \sum_{j=1}^{n_0} \left(\sum_{J_k \subset I_j} \operatorname{osc}(f, I_j)\ell(J_k) \right) = \sum_{j=1}^{n_0} \operatorname{osc}(f, I_j) \underbrace{\left(\sum_{J_k \subset I_j} \ell(J_k) \right)}_{\le \ell(I_j)} \le \sum_{j=1}^{n_0} \operatorname{osc}(f, I_j)\ell(I_j) = \omega(f, \boldsymbol{P}_0).$$

Now observe that if J_k is a Type 2 interval of $\boldsymbol{P}$, then $\ell(J_k) \le \|\boldsymbol{P}\|$ and $\operatorname{osc}(f, J_k) \le \operatorname{osc}(f, [a, b])$. Hence

$$S_2 \le \sum_{J_k \in \mathcal{J}^2} \operatorname{osc}(f, [a, b])\|\boldsymbol{P}\| \le |\mathcal{J}^2| \operatorname{osc}(f, [a, b])\|\boldsymbol{P}\| \le (n_0 - 1)\operatorname{osc}(f, [a, b])\|\boldsymbol{P}\|.$$

Thus

$$\omega(f, \boldsymbol{P}) = S_1 + S_2 \le (n_0 - 1)\operatorname{osc}(f, [a, b])\|\boldsymbol{P}\| + \omega(f, \boldsymbol{P}_0).$$

$\square$

Returning to our implication $(\omega_0) \Rightarrow (\omega)$, we observe that (ω_0) implies that for any $\varepsilon > 0$ there exists a partition $\boldsymbol{P}_\varepsilon$ such that

$$\omega(f, \boldsymbol{P}_\varepsilon) < \frac{\varepsilon}{2}.$$

Denote by n_ε the order of $\boldsymbol{P}_\varepsilon$ and by $x_0 < x_1 < \cdots < x_{n_\varepsilon}$ the nodes of $\boldsymbol{P}_\varepsilon$. We set

$$\lambda_\varepsilon := \min_{1 \le j \le n_\varepsilon} (x_j - x_{j-1}).$$

Now choose $\delta = \delta(\varepsilon) > 0$ such that

$$\delta < \lambda_\varepsilon \text{ and } (n_\varepsilon - 1)\operatorname{osc}(f, [a, b])\delta < \frac{\varepsilon}{2} \iff \delta < \min\left(\lambda_\varepsilon, \frac{\varepsilon}{2(n_\varepsilon - 1)\operatorname{osc}(f, [a, b])} \right).$$

If $\boldsymbol{P}$ is an arbitrary partition of $[a, b]$ such that $\|\boldsymbol{P}\| < \delta(\varepsilon)$, then Lemma 9.19 implies that

$$\omega(f, \boldsymbol{P}) \le (n_\varepsilon - 1)\operatorname{osc}(f, [a, b])\delta + \omega(f, \boldsymbol{P}_\varepsilon) < \varepsilon.$$

This proves that f satisfies (ω) and completes the proof of the Riemann–Darboux Theorem.

$\square$

We record here for later use a direct consequence of the above proof.

Corollary 9.20. *Suppose that* $f : [a, b] \to \mathbb{R}$ *is a Riemann integrable function. Then*

$$\int_a^b f(x)dx = \boldsymbol{S}_*(f) = \boldsymbol{S}^*(f). \tag{9.14}$$

In particular,

$$\boldsymbol{S}_*(f, \boldsymbol{P}) \le \int_a^b f(x)dx \le \boldsymbol{S}^*(f, \boldsymbol{P}'), \ \ \forall \boldsymbol{P}, \boldsymbol{P}' \in \mathcal{P}_{[a,b]}. \tag{9.15}$$

□

9.4. Examples of Riemann integrable functions

We are now going to collect the reward for the effort we spent proving the Riemann–Darboux Theorem.

Proposition 9.21. *Any continuous function* $f : [a, b] \to \mathbb{R}$ *is Riemann integrable.*

Proof. We will use the Riemann–Darboux Theorem to prove the claim. Note first that the Weierstrass Theorem 6.14 shows that f is bounded.

To prove that f satisfies $(\boldsymbol{\omega})$ we rely on the Uniform Continuity Theorem 6.29. According to this theorem, for any $\varepsilon > 0$ there exists $\delta = \delta(\varepsilon) > 0$ such that for any interval $I \subset [a, b]$ of length $< \delta$ we have

$$\operatorname{osc}(f, I) < \frac{\varepsilon}{b-a}.$$

If $\boldsymbol{P}$ is any partition of $[a, b]$ of order n and mesh size $\|\boldsymbol{P}\| < \delta(\varepsilon)$, then for any interval I_k of $\boldsymbol{P}$ we have

$$\operatorname{osc}(f, I_k) < \frac{\varepsilon}{b-a}.$$

Hence

$$\omega(f, \boldsymbol{P}) = \sum_{k=1}^{n} \operatorname{osc}(f, I_k)\Delta x_k < \frac{\varepsilon}{b-a} \underbrace{\sum_{k=1}^{n} \Delta x_k}_{=(b-a)} = \varepsilon.$$

This shows that f satisfies $(\boldsymbol{\omega})$ and thus it is Riemann integrable. □

Example 9.22. The function $f : [0, 1] \to \mathbb{R}$, $f(x) = x^2$ is continuous and thus integrable. Therefore

$$\int_0^1 x^2 dx = \lim_{N\to\infty} \boldsymbol{S}_*(f, \boldsymbol{U}_N),$$

where $\boldsymbol{U}_N$ denote the uniform partition of order N of $[0, 1]$. Since f is nondecreasing we deduce that $\boldsymbol{S}_*(f, \boldsymbol{U}_N)$ coincides with the sum L_N defined in (9.1). As explained in Section 9.1 the sum L_N converges to $\frac{1}{3}$ as $N \to \infty$. □

Proposition 9.23. *Any nondecreasing function $f : [a,b] \to \mathbb{R}$ is Riemann integrable.*

Proof. Clearly f is bounded since $f(a) \le f(x) \le f(b)$, $\forall x \in [a,b]$. If $\boldsymbol{P}$ is any partition of $[a,b]$ of order n, then for an interval $I_k = [x_{k-1}, x_k]$ of this partition we have

$$\operatorname{osc}(f, I_k) = f(x_k) - f(x_{k-1}),$$

$$\operatorname{osc}(f, I_k)\Delta x_k \le \operatorname{osc}(f, I_k)\|\boldsymbol{P}\| = \|\boldsymbol{P}\|\big(f(x_k) - f(x_{k-1})\big)$$

so that

$$\omega(f, \boldsymbol{P}) = \sum_{k=1}^{n} \operatorname{osc}(f, I_k)\Delta x_k \le \|\boldsymbol{P}\| \sum_{k=1}^{n}\big(f(x_k) - f(x_{k-1})\big) = \|\boldsymbol{P}\|\big(f(b) - f(a)\big).$$

This shows that f satisfies ($\boldsymbol{\omega}$) since

$$\lim_{\|\boldsymbol{P}\| \to 0} \|\boldsymbol{P}\|\big(f(b) - f(a)\big) = 0.$$

□

Proposition 9.24. *Suppose that $f : [a,b] \to \mathbb{R}$ is a bounded function which is continuous on (a,b). Then f is Riemann integrable.*

Proof. We will prove that f satisfies ($\boldsymbol{\omega}_0$). Fix $\varepsilon > 0$ and choose a positive real number $d(\varepsilon)$ such that

$$\operatorname{osc}(f, [a,b])d(\varepsilon) < \frac{\varepsilon}{4}. \tag{9.16}$$

Denote by J_ε the compact interval $J_\varepsilon := [a + d(\varepsilon), b - d(\varepsilon)]$; see Figure 9.4.

The restriction of f to J_ε is continuous. The Uniform Continuity Theorem 6.29 implies that there exists $\delta = \delta(\varepsilon) < d(\varepsilon)$ with the property that for any interval $I \subset J_\varepsilon$ of length $\ell(I) < \delta(\varepsilon)$ we have

$$\operatorname{osc}(f, I) < \frac{\varepsilon}{2(b-a)}. \tag{9.17}$$

Consider a partition $\boldsymbol{P}_\varepsilon$ of order n of J_ε satisfying $\|\boldsymbol{P}\| < \delta(\varepsilon)$. We denote by I_k, $k = 1\dots,n$, the intervals of $\boldsymbol{P}_\varepsilon$; see Figure 9.4. We set

$$I_* := [a, a + d(\varepsilon)], \;\; I^* = [b - d(\varepsilon), b].$$

Fig. 9.4 *Isolating the possible points of discontinuity of f.*

The collection of intervals

$$I_*, I_1, \dots, I_n, I^*$$

defines a partition $\widehat{\boldsymbol{P}}_\varepsilon$ of $[a,b]$; see Figure 9.4. We have

$$\omega(f,\widehat{\boldsymbol{P}}_\varepsilon)=\underbrace{\operatorname{osc}(f,I_*)\ell(I_*)}_{=:T_*}+\underbrace{\sum_{k=1}^n\operatorname{osc}(f,I_k)\ell(I_k)}_{=:T}+\underbrace{\operatorname{osc}(f,I^*)\ell(I^*)}_{=:T^*}.$$

Note that

$$\ell(I_*)=\ell(I^*)=d(\varepsilon),$$

so that

$$T_*=\operatorname{osc}(f,I_*)d(\varepsilon)\leq\operatorname{osc}(f,[a,b])d(\varepsilon)\overset{(9.16)}{<}\frac{\varepsilon}{4},$$

$$T^*=\operatorname{osc}(f,I^*)d(\varepsilon)\leq\operatorname{osc}(f,[a,b])d(\varepsilon)\overset{(9.16)}{<}\frac{\varepsilon}{4}.$$

Moreover,

$$T=\sum_{k=1}^n\operatorname{osc}(f,I_k)\ell(I_k)\overset{(9.17)}{<}\frac{\varepsilon}{2(b-a)}\sum_{k=1}^n\ell(I_k)=\frac{\varepsilon}{2(b-a)}(b-a)=\frac{\varepsilon}{2}.$$

Hence,

$$\omega(f,\widehat{\boldsymbol{P}}_\varepsilon)=T_*+T+T^*<\frac{\varepsilon}{4}+\frac{\varepsilon}{2}+\frac{\varepsilon}{4}=\varepsilon.$$

This proves that f satisfies $(\boldsymbol{\omega}_0)$ and thus it is Riemann integrable. □

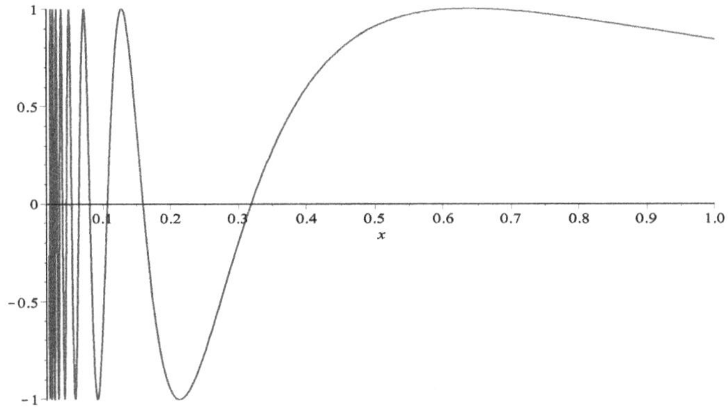

Fig. 9.5 *A wildly oscillating, yet Riemann integrable function.*

Remark 9.25. Proposition 9.24 has some surprising nontrivial consequences. For example, it shows that the wildly oscillating function (see Figure 9.5)

$$f:[0,1]\to\mathbb{R},\quad f(x)=\begin{cases}\sin\left(\frac{1}{x}\right), & x\in(0,1],\\ 0, & x=0,\end{cases}$$

is Riemann integrable. □

Proposition 9.26. *Suppose that $f : [a,b] \to \mathbb{R}$ is a bounded function and $c \in (a,b)$. The following statements are equivalent.*

(i) The function f is Riemann integrable on $[a,b]$.
(ii) The restrictions of $f|_{[a,c]}$ and $f|_{[c,b]}$ of f to $[a,c]$ and $[c,b]$ are Riemann integrable functions.

Moreover, if f satisfies either one of the two equivalent conditions above, then

$$\int_a^b f(x)dx = \int_a^c f(x)dx + \int_c^b f(x)dx. \tag{9.18}$$

Proof. (i) $\Rightarrow$ (ii). Suppose that f is Riemann integrable on $[a,b]$. Given a partition $\boldsymbol{P}'$ of $[a,c]$ and a partition $\boldsymbol{P}''$ of $[c,b]$ we obtain a partition $\boldsymbol{P}' * \boldsymbol{P}''$ of $[a,b]$ whose set of nodes is the union of the sets of nodes of $\boldsymbol{P}'$ and $\boldsymbol{P}''$. Note that

$$\|\boldsymbol{P}' * \boldsymbol{P}''\| \leq \max\{\, \|\boldsymbol{P}'\|,\ \|\boldsymbol{P}''\|\,\},$$

and

$$\omega(\, f, \boldsymbol{P}' * \boldsymbol{P}''\,) = \omega(\, f|_{[a,c]}, \boldsymbol{P}'\,) + \omega(\, f|_{[c,b]}, \boldsymbol{P}'\,).$$

Since f is Riemann integrable on $[a,b]$, it satisfies the property ($\boldsymbol{\omega}$) so, for any $\varepsilon > 0$, there exists $\delta = \delta(\varepsilon) > 0$ such that, for any partition $\boldsymbol{P}$ of $[a,b]$ with mesh size $\|\boldsymbol{P}\| < \delta(\varepsilon)$, we have

$$\omega(f, \boldsymbol{P}) < \varepsilon.$$

If the partitions $\boldsymbol{P}'$ and $\boldsymbol{P}''$ satisfy

$$\max\{\, \|\boldsymbol{P}'\|,\ \|\boldsymbol{P}''\|\,\} < \delta(\varepsilon),$$

then $\|\boldsymbol{P}' * \boldsymbol{P}''\| < \delta(\varepsilon)$ so that

$$\omega(f|_{[a,c]}, \boldsymbol{P}') + \omega(f|_{[c,b]}, \boldsymbol{P}'') = \omega(f, \boldsymbol{P}' * \boldsymbol{P}'') < \varepsilon.$$

This shows that both restrictions $f|_{[a,c]}$ and $f|_{[c,b]}$ satisfy ($\boldsymbol{\omega}$) and thus are Riemann integrable.

(ii) $\Rightarrow$ (i). We will prove that if $f|_{[a,c]}$ and $f|_{[c,b]}$ are Riemann integrable, then f is integrable on $[a,b]$. We invoke Theorem 9.17. It suffices to show that f satisfies ($\boldsymbol{\omega}_0$). Fix $\varepsilon > 0$. We have to prove that there exists a partition $\boldsymbol{P}_\varepsilon$ of $[a,b]$ such that $\omega(f, \boldsymbol{P}_\varepsilon) < \varepsilon$.

Since $f|_{[a,c]}$ and $f|_{[c,b]}$ are Riemann integrable, they satisfy ($\boldsymbol{\omega}_0$), and we deduce that there exist partitions $\boldsymbol{P}'_\varepsilon$ of $[a,c]$, and $\boldsymbol{P}''_\varepsilon$ of $[c,b]$ such that

$$\omega(f, \boldsymbol{P}'_\varepsilon),\ \ \omega(f, \boldsymbol{P}''_\varepsilon) < \frac{\varepsilon}{2}.$$

Then $\boldsymbol{P}_\varepsilon = \boldsymbol{P}'_\varepsilon * \boldsymbol{P}''_\varepsilon$ is a partition of $[a,b]$, and

$$\omega(f, \boldsymbol{P}_\varepsilon) = \omega(f, \boldsymbol{P}'_\varepsilon) + \omega(f, \boldsymbol{P}''_\varepsilon) < \varepsilon.$$

To prove (9.18) assume that f satisfies both (i) and (ii). Denote by $\boldsymbol{U}'_n$ the uniform partition of order n of $[a, c]$ and by $\boldsymbol{U}''_n$ the uniform partition of order n of $[c, b]$. Set

$$\boldsymbol{P}_n := \boldsymbol{U}'_n * \boldsymbol{U}''_n.$$

Note that

$$\|\boldsymbol{P}_n\| = \max\big(\|\boldsymbol{U}'_n\|,\ \|\boldsymbol{U}''_n\| \big) \to 0 \ \text{ as } n \to \infty. \tag{9.19}$$

Denote by $\underline{\xi}'_n$ the midpoint sample of $\boldsymbol{U}'_n$, and by $\underline{\xi}''_n$ the midpoint sample of $\boldsymbol{U}''_n$. Then $\underline{\xi}_n := \underline{\xi}_n \cup \underline{\xi}''_n$ is the midpoint sample of $\boldsymbol{P}_n$. We have

$$\boldsymbol{S}(f, \boldsymbol{P}_n, \underline{\xi}_n) = \boldsymbol{S}(f, \boldsymbol{U}'_n, \underline{\xi}'_n) + \boldsymbol{S}(f, \boldsymbol{U}''_n, \underline{\xi}''_n). \tag{9.20}$$

From (i), (9.19), and (9.5) we deduce that

$$\lim_{n\to\infty} \boldsymbol{S}(f, \boldsymbol{P}_n, \underline{\xi}_n) = \int_a^b f(x)dx.$$

From (ii), (9.19), and (9.5) we deduce that

$$\lim_{n\to\infty} \boldsymbol{S}(f, \boldsymbol{U}'_n, \underline{\xi}'_n) = \int_a^c f(x)dx,$$

$$\lim_{n\to\infty} \boldsymbol{S}(f, \boldsymbol{U}''_n, \underline{\xi}''_n) = \int_c^b f(x)dx.$$

The equality (9.18) now follows from the above three equalities after letting $n \to \infty$ in (9.20). □

Applying Proposition 9.26 iteratively we deduce the following consequence.

Corollary 9.27. *Suppose that $f : [a, b] \to \mathbb{R}$ is a bounded function and*

$$\boldsymbol{P} = (a = x_0 < x_1 < \cdots < x_n = b)$$

is a partition of $[a, b]$. Then the following statements are equivalent.

(i) The function f is Riemann integrable on $[a, b]$.
(ii) For any $k = 1, \dots, n$ the restriction of f to $[x_{k-1}, x_k]$ is Riemann integrable.

Moreover, if any of the above two equivalent conditions is satisfied, then

$$\int_a^b f(x)dx = \int_a^{x_1} f(x)dx + \int_{x_1}^{x_2} f(x)dx + \cdots + \int_{x_{n-1}}^b f(x)dx. \tag{9.21}$$

□

Corollary 9.28. *If $f : [a, b] \to \mathbb{R}$ is a bounded function and $D \subset [a, b]$ is a finite set such that f is continuous at any point in $[a, b] \setminus D$, then f is Riemann integrable.*

Proof. We add to D the endpoints a, b if they are not contained in D and we obtain a partition $\boldsymbol{P}$ of $[a,b]$ such that f is continuous in the *interior* of any interval $[x_{k-1}, x_k]$ of $\boldsymbol{P}$. Proposition 9.24 implies that f is Riemann integrable on each of the intervals $[x_{k-1}, x_k]$ and Corollary 9.27 implies that f is integrable on $[a,b]$. □

Proposition 9.29. *If $f, g : [a,b] \to \mathbb{R}$ are Riemann integrable, then for any constants $\alpha, \beta \in \mathbb{R}$ the sum $\alpha f + \beta g : [a,b] \to \mathbb{R}$ is also Riemann integrable and*

$$\int_a^b \big(\alpha f(x) + \beta g(x)\big)dx = \alpha \int_a^b f(x)dx + \beta \int_a^b g(x)dx. \tag{9.22}$$

Proof. We will show that $\alpha f + \beta g$ satisfies the definition of Riemann integrability, Definition 9.4. Observe first that if $(\boldsymbol{P}, \underline{\xi})$ is a sampled partition of $[a,b]$, then

$$\boldsymbol{S}\big(\alpha f + \beta g, \boldsymbol{P}, \underline{\xi}\big) = \alpha \boldsymbol{S}(f, \boldsymbol{P}, \underline{\xi}) + \beta \boldsymbol{S}(g, \boldsymbol{P}, \underline{\xi}). \tag{9.23}$$

Indeed, if the partition $\boldsymbol{P}$ is

$$\boldsymbol{P} = \{a = x_0 < x_1 < \cdots < x_{n-1} < x_n = b\},$$

and the sample $\underline{\xi}$ is $\underline{\xi} = (\xi_k)_{1 \leq k \leq n}$, then

$$\begin{aligned}\boldsymbol{S}\big(\alpha f + \beta g, \boldsymbol{P}, \underline{\xi}\big) &= \sum_k \big(\alpha f(\xi_k) + \beta g(\xi_k)\big)\Delta x_k = \sum_k \alpha f(\xi_k)\Delta x_k + \sum_k \beta g(\xi_k)\Delta x_k \\ &= \alpha \sum_k f(\xi_k)\Delta x_k + \beta \sum_k g(\xi_k)\Delta x_k = \alpha \boldsymbol{S}(f, \boldsymbol{P}, \underline{\xi}) + \beta \boldsymbol{S}(g, \boldsymbol{P}, \underline{\xi}).\end{aligned}$$

Set

$$K := (|\alpha| + |\beta| + 1).$$

Fix $\varepsilon > 0$. Since f is Riemann integrable, there exists $\delta_1 = \delta_1(\varepsilon) > 0$ such that $\forall \boldsymbol{P} \in \mathcal{P}_{[a,b]}$, $\forall \underline{\xi} \in \mathcal{S}(\boldsymbol{P})$, we have

$$\|\boldsymbol{P}\| < \delta_1 \Rightarrow \left| \boldsymbol{S}(f, \boldsymbol{P}, \underline{\xi}) - \int_a^b f(x)dx \right| < \frac{\varepsilon}{K}. \tag{9.24}$$

Since g is Riemann integrable, there exists $\delta_2 = \delta_2(\varepsilon) > 0$ such that $\forall \boldsymbol{P} \in \mathcal{P}_{[a,b]}$, $\forall \underline{\xi} \in \mathcal{S}(\boldsymbol{P})$, we have

$$\|\boldsymbol{P}\| < \delta_2 \Rightarrow \left| \boldsymbol{S}(g, \boldsymbol{P}, \underline{\xi}) - \int_a^b g(x)dx \right| < \frac{\varepsilon}{K}. \tag{9.25}$$

Set

$$\delta = \delta(\varepsilon) := \min\big(\delta_1(\varepsilon), \delta_2(\varepsilon)\big), \quad S := \alpha \int_a^b f(x)dx + \beta \int_a^b g(x)dx.$$

Let $\boldsymbol{P} \in \mathcal{P}_{[a,b]}$ be an arbitrary partition such that $\|\boldsymbol{P}\| < \delta$. Then for any sample $\underline{\xi} \in \mathcal{S}(\boldsymbol{P})$ we have

$$|\boldsymbol{S}(\alpha f + \beta g, \boldsymbol{P}, \underline{\xi}) - S|$$

$$\overset{(9.23)}{=} \left| \alpha\Big(\boldsymbol{S}(f,\boldsymbol{P},\underline{\xi}) - \int_a^b f(x)dx \Big) + \beta\Big(\boldsymbol{S}(g,\boldsymbol{P},\underline{\xi}) - \int_a^b g(x)dx \Big) \right|$$

$$\leq |\alpha| \cdot \left| \boldsymbol{S}(f,\boldsymbol{P},\underline{\xi}) - \int_a^b f(x)dx \right| + |\beta| \cdot \left| \boldsymbol{S}(g,\boldsymbol{P},\underline{\xi}) - \int_a^b g(x)dx \right|$$

(use (9.24) and (9.25))

$$\leq |\alpha| \frac{\varepsilon}{K} + |\beta| \frac{\varepsilon}{K} = \frac{|\alpha| + |\beta|}{|\alpha| + |\beta| + 1} \varepsilon < \varepsilon.$$

This proves that $\alpha f + \beta g$ is Riemann integrable and

$$\int_a^b f(x)dx = S = \alpha \int_a^b f(x)dx + \beta \int_a^b g(x)dx.$$

□

Corollary 9.30. *Suppose that $f, g : [a,b] \to \mathbb{R}$ are two functions such that*

$$f(x) = g(x), \quad \forall x \in (a,b).$$

If f is Riemann integrable, then so is g. Moreover,

$$\int_a^b f(x)dx = \int_a^b g(x)dx. \tag{9.26}$$

Proof. Consider the difference $h : [a,b] \to \mathbb{R}$, $h(x) = g(x) - f(x)$, $\forall x \in [a,b]$. Note that h is bounded on $[a,b]$ and continuous on (a,b) because $h(x) = 0$, $\forall x \in (a,b)$. Using Proposition 9.24 we deduce that h is Riemann integrable on $[a,b]$. Since $g = f + h$, we deduce from Proposition 9.29 that g is Riemann integrable on $[a,b]$ and

$$\int_a^b g(x)dx = \int_a^b f(x)dx + \int_a^b h(x)dx.$$

Thus, to prove (9.26) we have to show that

$$\int_a^b h(x)dx = 0.$$

To do this, denote by $\boldsymbol{U}_n$ the uniform partition of order n of $[a,b]$, and denote by $\underline{\xi}^{(n)}$ the sample of $\boldsymbol{U}_n$ consisting of the midpoints of the intervals of $\boldsymbol{U}_n$. Then

$$\boldsymbol{S}(h, \boldsymbol{U}_n, \underline{\xi}^{(n)}) = 0.$$

Since h is Riemann integrable, we have

$$\int_a^b h(x)dx = \lim_{n\to\infty} \boldsymbol{S}(h, \boldsymbol{U}_n, \underline{\xi}^{(n)}) = 0.$$

□

Example 9.31. We say that a function $f : [a, b] \to \mathbb{R}$ is *piecewise constant* if there exists a partition

$$\boldsymbol{P} = (a = x_0 < x_1 < \cdots < x_n = b)$$

and constants $c_1, \ldots, c_n$ such that for any $k = 1, \ldots, n$ the restriction of f to the open interval (x_{k-1}, x_k) is the constant function c_k. From the above corollary we deduce that f is Riemann integrable on each of the intervals $[x_{k-1}, x_k]$. Moreover, the computation in Example 9.5 implies that

$$\int_{x_{k-1}}^{x_k} f(t)dt = c_k(x_k - x_{k-1}).$$

Corollary 9.27 implies that f is Riemann integrable on $[a, b]$ and

$$\int_a^b f(x)dx = c_1(x_1 - x_0) + \cdots + c_n(x_n - x_{n-1}).$$

□

Proposition 9.32. *Suppose that $f : [a, b] \to \mathbb{R}$ is a Riemann integrable function, J is an interval containing the range of f and $G : J \to \mathbb{R}$ is a Lipschitz function. Then $G \circ f : [a, b] \to \mathbb{R}$ is Riemann integrable.*

Proof. Fix a positive constant L such that

$$|G(y_1) - G(y_2)| \leq L|y_1 - y_2|, \ \ \forall y_1, y_2 \in J.$$

Observe that for any $X \subset [a, b]$ and any $x', x'' \in X$ we have

$$\big|\, G \circ f(x') - G \circ f(x'') \,\big| \leq L|f(x') - f(x'')|.$$

Hence

$$\operatorname{osc}(G\circ f, X) = \sup_{x', x'' \in X} \big|\, G\circ f(x') - G\circ f(x'') \,\big| \leq L \sup_{x', x'' \in X} |f(x') - f(x'')| = L \operatorname{osc}(f, X).$$

We deduce as in the proof of Proposition 9.29 that for any partition $\boldsymbol{P}$ of $[a, b]$ we have

$$\omega(G \circ f, \boldsymbol{P}) \leq L\omega(f, \boldsymbol{P}).$$

Since f is Riemann integrable we deduce that

$$\lim_{\|\boldsymbol{P}\| \to 0} \omega(f, \boldsymbol{P}) = 0$$

so that

$$\lim_{\|\boldsymbol{P}\| \to 0} \omega(G \circ f, \boldsymbol{P}) = 0.$$

□

Corollary 9.33. *Suppose that $f : [a, b] \to \mathbb{R}$ is Riemann integrable. Then f^2 is also Riemann integrable on $[a, b]$.*

Proof. Since f is Riemann integrable it is bounded so its range is contained in some interval $[-M, M]$, $M > 0$. The function $G : [-M, M] \to \mathbb{R}$, $G(x) = x^2$ is Lipschitz on this interval because for any $x, y \in [-M, M]$ we have

$$|G(x) - G(y)| = |x^2 - y^2| = |x + y| \cdot |x - y| \leq (|x| + |y|)|x - y| \leq 2M|x - y|.$$

Proposition 9.32 implies that $G \circ f = f^2$ is Riemann integrable. □

Corollary 9.34. *If $f, g : [a, b] \to \mathbb{R}$ are Riemann integrable, then so is their product fg.*

Proof. The function $f + g$ is integrable according to Proposition 9.29. Invoking Corollary 9.33 we deduce that the functions $(f + g)^2, f^2, g^2$ are Riemann integrable. Proposition 9.29 now implies that the function

$$\frac{1}{2}\Big((f + g)^2 - f^2 - g^2 \Big) = \frac{1}{2}\big(f^2 + g^2 + 2fg - f^2 - g^2 \big) = fg$$

is Riemann integrable. □

Corollary 9.35. *Suppose that $f : [a, b] \to \mathbb{R}$ is Riemann integrable. Then the function $|f|$ is also Riemann integrable.*

Proof. The function $G : \mathbb{R} \to \mathbb{R}$, $G(y) = |y|$ is Lipschitz so the function $G \circ f = |f|$ is Riemann integrable. □

☞ ***A very useful convention.*** We denoted the Riemann integral of a function $f : [a, b] \to \mathbb{R}$ with the symbol

$$\int_a^b f(x)dx,$$

where the lower endpoint a is at the bottom of the integral sign $\int$ and the upper endpoint b is at the top of the integral sign. We define

$$\int_b^a f(x)dx := -\int_a^b f(x)dx, \quad \int_a^a f(x)dx = 0.$$

There are several arguments in favor of this convention. For example, we can rewrite (9.31) as

$$f(\xi) = \frac{1}{b - a}\int_a^b f(x)dx = \frac{1}{a - b}\int_b^a f(x)dx. \tag{9.27}$$

This formulation will be especially useful when we do not know whether $a < b$ or $b < a$. The above equality says that it does not matter.

Another advantage comes from the following additivity identity.

$$\int_a^c f(x)dx = \int_a^b f(x)dx + \int_b^c f(x)dx, \quad \forall a, b, c \in \mathbb{R}. \tag{9.28}$$

If $a < b < c$, then (9.28) is an immediate consequence of Corollary 9.27. When the numbers a, b, c are situated in a different order, the identity (9.28) is still a consequence of Corollary 9.27, but in a more roundabout way. For example, if $a = 0$, $b = 2$ and $c = 1$, then

$$\int_0^1 f(x)dx = \int_0^2 f(x)dx - \int_1^2 f(x)dx = \int_0^2 f(x)dx + \int_2^1 f(x)dx. \qquad \square$$

9.5. Basic properties of the Riemann integral

Now that we have seen how the concept of integrability interacts with the basic arithmetic operations on functions, we want to discuss a few simple techniques for estimating Riemann integrals. All these techniques are based on the following simple result.

Proposition 9.36 (Positivity). *Suppose that $f : [a,b] \to \mathbb{R}$ is Riemann integrable and $f(x) \geq 0$ for any $x \in [a,b]$. Then*

$$\int_a^b f(x)dx \geq 0.$$

Proof. Denote by $\boldsymbol{U}_1$ the partition of $[a,b]$ consisting of a single interval. Then

$$0 \leq \Big(\inf_{x\in[a,b]} f(x)\Big)(b-a) = \boldsymbol{S}_*(f,\boldsymbol{U}_1) \overset{(9.15)}{\leq} \int_a^b f(x)dx.$$

□

Corollary 9.37 (Monotonicity). *If $f,g : [a,b] \to \mathbb{R}$ are Riemann integrable functions and $f(x) \leq g(x)$, $\forall x \in [a,b]$, then*

$$\int_a^b f(x)dx \leq \int_a^b g(x)dx.$$

Proof. The function $(g-f)$ is integrable and nonnegative so

$$\int_a^b g(x)dx - \int_a^b f(x)dx = \int_a^b (g(x)-f(x))dx \geq 0.$$

□

Corollary 9.38. *If $f : [a,b] \to \mathbb{R}$ is Riemann integrable, then*

$$\boxed{\left|\int_a^b f(x)dx\right| \leq \int_a^b |f(x)|\,dx.} \tag{9.29}$$

Proof. We know that

$$f(x) \leq |f(x)| \text{ and } -f(x) \leq |f(x)|, \ \forall x \in [a,b].$$

Hence

$$\int_a^b f(x)dx \leq \int_a^b |f(x)|dx \text{ and } -\int_a^b f(x)dx \leq \int_a^b |f(x)|dx.$$

The last two inequalities imply (9.29). □

Corollary 9.39. *Suppose that $f : [a,b] \to \mathbb{R}$ is a Riemann integrable function. We set*

$$m := \inf_{x\in[a,b]} f(x), \quad M = \sup_{x\in[a,b]} f(x).$$

Then

$$m(b-a) \leq \int_a^b f(x)dx \leq M(b-a).$$

Proof. We have

$$m \leq f(x) \leq M, \quad \forall x \in [a,b],$$

so that

$$m(b-a) = \int_a^b mdx \leq \int_a^b f(x)dx \leq \int_a^b Mdx = M(b-a).$$

□

Definition 9.40. If $f : [a,b] \to \mathbb{R}$ is a Riemann integrable function, then the quantity

$$\frac{1}{b-a}\int_a^b f(x)dx$$

is called the *average value* of f, or the *mean* of f, or the *expectation* of f and we denote it by $\operatorname{Mean}(f)$. □

We see that we can rephrase the inequality in Corollary 9.39 as

$$\inf_{x\in[a,b]} f(x) \leq \operatorname{Mean}(f) \leq \sup_{x\in[a,b]} f(x). \tag{9.30}$$

Theorem 9.41 (Integral Mean Value Theorem). *Suppose that $f : [a,b] \to \mathbb{R}$ is a* ***continuous*** *function. Then there exists $\xi \in [a,b]$ such that*

$$f(\xi) = \operatorname{Mean}(f),$$

i.e.,

$$f(\xi) = \frac{1}{b-a}\int_a^b f(x)dx. \tag{9.31}$$

Proof. Let

$$m := \inf_{x\in[a,b]} f(x), \quad M = \sup_{x\in[a,b]} f(x).$$

Then (9.30) implies that $\operatorname{Mean}(f) \in [m,M]$.

On the other hand, since f is continuous we deduce from Weierstrass' Theorem 6.14 that there exist $x_*, x^* \in [a,b]$ such that

$$f(x_*) = m, \quad f(x^*) = M.$$

Since $\operatorname{Mean}(f) \in [f(x_*), f(x^*)]$ we deduce from the Intermediate Value Theorem that there exists ξ in the interval $[x_*, x^*]$ such that $f(\xi) = \operatorname{Mean}(f)$. □

Theorem 9.42. *Suppose that $f : [a,b] \to \mathbb{R}$ is a Riemann integrable function. We define*

$$F : [a,b] \to \mathbb{R}, \;\; F(x) := \int_a^x f(t)dt.$$

Then the following hold.

(i) The function F is Lipschitz. In particular, F is continuous.
(ii) If the function f is continuous, then the function $F(x)$ is differentiable on $[a,b]$ and

$$F'(x) = f(x), \;\; \forall x \in [a,b].$$

In other words, $F(x)$ is an antiderivative of f, more precisely the unique antiderivative on $[a,b]$ such that $F(a) = 0$.

Proof. (i) We set

$$M := \sup_{x\in[a,b]} |f(x)|.$$

If $x, y \in [a,b]$, $x < y$, then

$$|F(x) - F(y)| = |F(y) - F(x)| = \left| \int_a^y f(t)dt - \int_a^x f(t)dt \right|$$

$$= \left| \int_x^y f(t)dt \right| \leq \int_x^y |f(t)|dt \leq \int_x^y Mdt = M(y-x) = M|x-y|.$$

This proves that F is Lipschitz.
(ii) We have to prove that if $x_0 \in [a,b]$, then

$$\lim_{x\to x_0} \frac{F(x) - F(x_0)}{x - x_0} = f(x_0).$$

Using (9.28) we deduce

$$F(x) - F(x_0) = \int_a^x f(t)dt - \int_a^{x_0} f(t)dt = \int_{x_0}^x f(t)dt$$

so that we have to show that

$$\lim_{x\to x_0} \frac{1}{x - x_0} \int_{x_0}^x f(t)dt = f(x_0).$$

In other words, we have to prove that for any $\varepsilon > 0$ there exists $\delta = \delta(\varepsilon) > 0$ such that

$$\forall x \in [a,b], \;\; 0 < |x - x_0| < \delta \Rightarrow \left| \frac{1}{x - x_0} \int_{x_0}^x f(t)dt - f(x_0) \right| < \varepsilon. \tag{9.32}$$

Since f is continuous at x_0, given $\varepsilon > 0$ we can find $\delta = \delta(\varepsilon) > 0$ such that

$$\forall x \in [a,b], \;\; |x - x_0| < \delta \Rightarrow |f(x) - f(x_0)| < \varepsilon.$$

On the other hand, invoking the continuity of f again, we deduce from the Integral Mean Value Theorem that, for any $x \neq x_0$, there exists ξ_x between x_0 and x such that

$$f(\xi_x) = \frac{1}{x - x_0} \int_{x_0}^x f(t)dt.$$

In particular, if $|x - x_0| < \delta$, then $|\xi_x - x_0| < \delta$, and thus

$$\left| \frac{1}{x - x_0} \int_{x_0}^x f(t)dt - f(x_0) \right| = |f(\xi_x) - f(x_0)| < \varepsilon.$$

□

9.6. How to compute a Riemann integral

To this day, the best method of computing Riemann integrals, by hand, is the Fundamental Theorem of Calculus.

Theorem 9.43 (The Fundamental Theorem of Calculus: Part 1). *Suppose that* $f : [a,b] \to \mathbb{R}$ *is a function satisfying the following two conditions.*

(i) The function f is Riemann integrable.
(ii) The function f admits antiderivatives on $[a,b]$.

If $F : [a,b] \to \mathbb{R}$ is an antiderivative of f, then

$$\boxed{\int_a^x f(t)dt = F(x) - F(a), \;\; \forall x \in (a,b].} \tag{9.33}$$

In particular,

$$\boxed{\int_a^b f(t)dt = F(t)\Big|_{t=a}^{t=b} := F(b) - F(a).} \tag{9.34}$$

Proof. Fix $x \in (a,b]$. Denote by $\boldsymbol{U}_n$ the uniform partition of $[a,x]$ of order n. Since f is Riemann integrable we deduce that *for any choices of samples* $\underline{\xi}^{(n)}$ of $\boldsymbol{U}_n$ we have

$$\int_a^x f(t)dt = \lim_{n\to\infty} \boldsymbol{S}\big(f, \boldsymbol{U}_n, \underline{\xi}^{(n)}\big).$$

The miracle is that for any n we can cleverly choose a sample

$$\underline{\xi}^{(n)} = (\xi_1^n, \dots, \xi_n^n)$$

of $\boldsymbol{U}_n$ such that the Riemann sum $\boldsymbol{S}\big(f, \boldsymbol{U}_n, \underline{\xi}^{(n)}\big)$ has an extremely simple form. Here are the details.

The k-th node of $\boldsymbol{U}_n$ is $x_k^n = a + \frac{k}{n}(x-a)$ and the k-th interval is $I_k = [x_{k-1}^n, x_k^n]$. The function F is differentiable on the closed interval $[a,b]$ and, in particular, it is continuous on $[a,b]$. We can invoke Lagrange's Mean Value Theorem to conclude that, for any $k = 1, \dots, n$, there exists $\xi_k^n \in (x_{k-1}^n, x_k^n)$ such that

$$f(\xi_k^n) = F'(\xi_k^n) = \frac{F(x_k^n) - F(x_{k-1}^n)}{x_k^n - x_{k-1}^n},$$

i.e.,

$$f(\xi_k^n)(x_k^n - x_{k-1}^n) = F(x_k^n) - F(x_{k-1}^n).$$

The collection $(\xi_1^n, \dots, \xi_n^n)$ is a sample $\underline{\xi}^{(n)}$ of the partition $\boldsymbol{U}_n$. The associated Riemann sum satisfies

$$\boldsymbol{S}\big(f, \boldsymbol{U}_n, \underline{\xi}^{(n)}\big) = f(\xi_1^n)(x_1^n - x_0^n) + f(\xi_2^n)(x_2^n - x_1^n) + \cdots + f(\xi_n^n)(x_n^n - x_{n-1}^n)$$

$$= F(x_1^n) - F(x_0^n) + F(x_2^n) - F(x_1^n) + \cdots + F(x_n^n) - F(x_{n-1}^n)$$

(the above is a telescopic sum!!!)

$$= F(x_n^n) - F(x_0^n) = F(x) - F(a).$$

Thus the sequence of Riemann sums $\boldsymbol{S}(f, \boldsymbol{U}_n, \underline{\xi}^{(n)})$ is constant, equal to $F(x) - F(a)$. Hence

$$\int_a^x f(t)dt = \lim_{n\to\infty} \boldsymbol{S}\big(f, \boldsymbol{U}_n, \underline{\xi}^{(n)}\big) = F(x) - F(a).$$

The equality (9.34) follows from (9.33) by letting $x = b$. □

Corollary 9.44 (The Fundamental Theorem of Calculus: Part 2). *Suppose that $f : [a,b] \to \mathbb{R}$ is a continuous function. Then f admits antiderivatives on $[a,b]$ and, if $F(x)$ is any antiderivative of f on $[a,b]$, then*

$$\int_a^b f(x)dx = F\Big|_a^b := F(b) - F(a), \;\; F(x) = F(a) + \int_a^x f(t)dt, \;\; \forall x \in [a,b]. \quad (9.35)$$

Proof. The fact that f admits antiderivatives follows from Theorem 9.42(b). The rest follows from Theorem 9.43. □

Remark 9.45. (a) Theorem 9.43 shows that the computation of Riemann integral of a function can be reduced to the computation of the antiderivatives of that function, if they exist. As we have seen in the previous chapter, for many classes of continuous function this computation can be carried out successfully in a *finite number* of purely algebraic steps.

If we ponder for a little bit, the equality (9.34) is a truly remarkable result. The left-hand side of (9.34) is a Riemann integral defined by a very laborious limiting process which involves *infinitely many and computationally very punishing steps*. The right-hand side of (9.34) involves computing the values of an antiderivative at two points. Often this can be achieved in *finitely many arithmetic steps*!

The attribute *fundamental* attached to Theorem 9.43 is fully justified: it describes a *finite-time* shortcut to an *infinite-time* process.

(b) Both assumptions (i) and (ii) are needed in Theorem 9.43! Indeed, there exist functions that satisfy (i) but not (ii), and there exist function satisfying (ii), but not (i). Their constructions are rather ingenious and we refer to [10] for more details. Note that the continuous functions automatically satisfy both (i) and (ii). □

Example 9.46. For $k \in \mathbb{N}$ consider the continuous function $f : [0,1] \to \mathbb{R}$, $f(x) = x^k$. The function $F(x) = \frac{1}{k+1}x^{k+1}$ is an antiderivative of f and (9.35) implies

$$\int_0^1 x^k dx = \Big(\frac{1}{k+1}x^{k+1}\Big)\Big|_0^1 = \frac{1}{k+1}.$$

In particular, for $k = 2$ we deduce

$$\int_0^1 x^2 dx = \frac{1}{3}.$$

This agrees with the elementary computations in Section 9.1. □

The techniques for computing antiderivatives can now be used for computing Riemann integrals. As we have seen, there are basically two methods for computing antiderivatives: integration by parts, and change of variables. These lead to two basic techniques for computing Riemann integrals. In applications, most often one needs to use a blend of these techniques to compute a Riemann integral.

9.6.1. *Integration by parts*

We state a special case that covers most of the concrete situations.

Proposition 9.47. *Suppose that $u, v : [a,b] \to \mathbb{R}$ are two C^1 functions, i.e., they are differentiable and have continuous derivatives. Then uv' and $u'v$ are Riemann integrable and*

$$\boxed{\int_a^b u(x)v'(x)dx = u(x)v(x)\Big|_a^b - \int_a^b v(x)u'(x)dx}. \tag{9.36}$$

Proof. The functions $u'v$ and uv' are continuous since they are products of continuous functions. In particular these functions are integrable, and we have

$$\begin{aligned}\int_a^b u'(x)v(x)dx + \int_a^b u(x)v'(x)dx &= \int_a^b \big(\, u'(x)v(x) + u(x)v'(x)\,\big)dx\\ &= \int_a^b (uv)'(x)dx \overset{(9.34)}{=} u(x)v(x)\Big|_a^b.\end{aligned}$$

The equality (9.36) is now obvious. □

Remark 9.48. The integration-by-parts formula (9.36) is often written in the shorter form

$$\boxed{\int_a^b udv = uv\Big|_a^b - \int_a^b vdu.} \tag{9.37}$$

Observing that

$$uv\Big|_b^a = u(a)v(a) - u(b)v(b) = -\big(\,u(b)v(b) - u(a)v(a)\,\big) = -uv\Big|_a^b,$$

we deduce that

$$\int_b^a udv = uv\Big|_b^a - \int_b^a vdu,$$

even though the upper limit of integration a is smaller than the lower limit of integration b.
□

Example 9.49. For any nonnegative integers m, n we set

$$I_{m,n} = \int_{-1}^1 (x-1)^m(x+1)^n dx. \tag{9.38}$$

This integral is theoretically computable because $(x-1)^m(x+1)^n$ is a polynomial. Its precise form is obtained via Newton's binomial formula and the final result is rather complicated. For example

$(x-1)^2(x+1)^3 = (x^2-2x+1)(x^3+3x^2+3x+1) = x^5+x^4-2x^3-2x^2+x+1.$

In general, we need to multiply the two polynomials in the right-hand side of (9.38) to obtain the explicit form of $(x-1)^m(x+1)^n$. This is an elaborate process which becomes increasingly more complex as the powers m and n increase. However, an ingenious usage of the integration-by-parts trick leads to a much simpler way of computing $I_{m,n}$.

Let us first observe that

$$(x+1)^n = \frac{1}{n+1}\frac{d}{dx}(x+1)^{n+1},$$

from which we deduce

$$I_{0,n} = \int_{-1}^{1}(x+1)^n dx = \frac{1}{n+1}(x+1)^{n+1}\Big|_{-1}^{1} = \frac{2^{n+1}}{n+1}. \tag{9.39}$$

Observe now that if $m>0$, then

$$\begin{aligned} I_{m,n} &= \int_{-1}^{1}(x-1)^m(x+1)^n dx = \frac{1}{n+1}\int_{-1}^{1}(x-1)^m\frac{d}{dx}(x+1)^{n+1}dx \\ &= \underbrace{\frac{1}{n+1}(x-1)^m(x+1)^{n+1}\Big|_{-1}^{1}}_{=0} - \frac{m}{n+1}\int_{-1}^{1}(x-1)^{m-1}(x+1)^{n+1}dx. \end{aligned}$$

We obtain in this fashion the recurrence relation

$$I_{m,n} = -\frac{m}{n+1}I_{m-1,n+1},\ \ \forall m>0,\ \ n\geq 0. \tag{9.40}$$

If $m-1>0$, then we can continue this process and we deduce

$$I_{m-1,n+1} = -\frac{m-1}{n+2}I_{m-2,n+2} \Rightarrow I_{m,n} = \frac{m(m-1)}{(n+1)(n+2)}I_{m-2,n+2}.$$

Iterating this procedure we conclude that

$$\begin{aligned} I_{m,n} &= (-1)^m\frac{m(m-1)\cdots 2\cdot 1}{(n+1)(n+2)\cdots(n+m-1)(n+m)}I_{0,n+m} \\ &= (-1)^m\frac{m!}{(n+1)\cdots(n+m)}I_{0,n+m} = (-1)^m\frac{1}{\binom{n+m}{m}}I_{0,n+m}. \end{aligned}$$

Invoking (9.39) we deduce

$$\boxed{I_{m,n} = (-1)^m\frac{1}{\binom{n+m}{m}}\cdot\frac{2^{n+m+1}}{(n+m+1)}}. \tag{9.41}$$

When $m=n$ we have

$$I_{n,n} = \int_{-1}^{1}(x-1)^n(x+1)^n dx = \int_{-1}^{1}(x^2-1)^n dx$$

and we conclude that

$$\boxed{\int_{-1}^{1}(x^2-1)^n dx = I_{n,n} = \frac{(-1)^n}{\binom{2n}{n}}\cdot\frac{2^{2n+1}}{(2n+1)}}. \tag{9.42}$$

□

Example 9.50 (Wallis' formula). For nonnegative integer n we set

$$I_n := \int_0^{\frac{\pi}{2}} (\sin x)^n \, dx.$$

Note that

$$I_0 = \frac{\pi}{2}, \ I_1 = \int_0^{\frac{\pi}{2}} \sin x \, dx = (-\cos x)\Big|_{x=0}^{x=\frac{\pi}{2}} = 1.$$

In general, for $n > 0$, we have

$$I_{n+1} = \int_0^{\frac{\pi}{2}} (\sin x)^n d(-\cos x) = \underbrace{(\sin x)^n(-\cos x)\Big|_{x=0}^{x=\frac{\pi}{2}}}_{=0} + \int_0^{\frac{\pi}{2}} \cos x \, d(\sin x)^n$$

$$= n\int_0^{\frac{\pi}{2}} (\sin x)^{n-1} \cos^2 x \, dx = n\int_0^{\frac{\pi}{2}} (\sin x)^{n-1}(1-\sin^2 x) \, dx = nI_{n-1} - nI_{n+1}.$$

Hence

$$I_{n+1} = nI_{n-1} - nI_{n+1}$$

so that

$$(n+1)I_{n+1} = nI_{n-1}, \ I_{n+1} = \frac{n}{n+1} I_{n-1}. \tag{9.43}$$

We deduce

$$I_2 = \frac{1}{2} I_0 = \frac{1}{2}\frac{\pi}{2}, \ I_4 = \frac{3}{4} I_2 = \frac{3}{4}\frac{1}{2}\frac{\pi}{2},$$

and, in general,

$$\boxed{I_{2n} = \int_0^{\frac{\pi}{2}} (\sin x)^{2n} dx = \frac{2n-1}{2n} \cdots \frac{3}{4}\frac{1}{2}\frac{\pi}{2}}. \tag{9.44}$$

Similarly,

$$I_3 = \frac{2}{3} I_1 = \frac{2}{3}, \ I_5 = \frac{4}{5} I_3 = \frac{4}{5}\frac{2}{3},$$

and, in general,

$$\boxed{I_{2n+1} = \int_0^{\frac{\pi}{2}} (\sin x)^{2n+1} dx = \frac{2n}{2n+1} \cdots \frac{4}{5}\frac{2}{3}}. \tag{9.45}$$

If we introduce the notation

$$(2k)!! := 2 \cdot 4 \cdot 6 \cdots (2n), \ (2k-1)!! := 1 \cdot 3 \cdot 5 \cdots (2k-1), \tag{9.46}$$

then we can rewrite the equalities (9.44) and (9.45) in a more compact form

$$\boxed{I_{2j} = \frac{\pi}{2}\frac{(2j-1)!!}{(2j)!!}, \ I_{2j-1} = \frac{(2j-2)!!}{(2j-1)!!}}. \tag{9.47}$$

Since $\sin x \in [0,1]$, $\forall x \in [0,\pi/2]$ we deduce

$$(\sin x)^{n+1} \leq (\sin x)^n, \quad \forall x \in [0,\pi/2],$$

and thus,

$$I_{n+1} \leq I_n, \quad \forall n \in \mathbb{N}.$$

We deduce

$$\frac{2n}{2n+1} \stackrel{(9.43)}{=} \frac{I_{2n+1}}{I_{2n-1}} \leq \frac{I_{2n+1}}{I_{2n}} \leq 1.$$

From the above equalities we deduce

$$\lim_{n\to\infty} \frac{I_{2n+1}}{I_{2n}} = 1.$$

Using (9.44) and (9.45) we deduce

$$\frac{I_{2n+1}}{I_{2n}} = \frac{2}{\pi} \cdot \frac{1}{2n+1} \cdot \frac{2^2 4^2 \cdots (2n)^2}{1^2 3^2 \cdots (2n-1)^2}.$$

This implies the celebrated *Wallis' formula*

$$\boxed{\frac{\pi}{2} = \frac{\pi}{2} \lim_{n\to\infty} \frac{I_{2n+1}}{I_{2n}} = \lim_{n\to\infty} \frac{2^2 4^2 \cdots (2n)^2}{1^2 3^2 \cdots (2n-1)^2} \cdot \frac{1}{2n+1}.} \tag{9.48}$$

Later on, we will need an equivalent version of the above equality, namely

$$\boxed{\frac{\pi}{2} = \frac{\pi}{2} \lim_{n\to\infty} \frac{2n+1}{2n} \frac{I_{2n+1}}{I_{2n}} = \lim_{n\to\infty} \frac{2^2 4^2 \cdots (2n)^2}{1^2 3^2 \cdots (2n-1)^2} \cdot \frac{1}{2n}.} \tag{9.49}$$

□

Let us discuss another simple but useful application of the integration-by-parts trick.

Proposition 9.51 (Integral remainder formula). *Let $n \in \mathbb{N}$ and suppose that $f : [a,b] \to \mathbb{R}$ is a C^{n+1}-function, i.e., $(n+1)$-times differentiable and the $(n+1)$-th derivative is continuous. If $x_0 \in [a,b]$ and $T_n(x)$ is the degree n-Taylor polynomial of f at x_0,*

$$T_n(x) = f(x_0) + \frac{f'(x_0)}{1!}(x-x_0) + \cdots + \frac{f^{(n)}(x_0)}{n!}(x-x_0)^n,$$

then the remainder $R_n(x) := f(x) - T_n(x)$ admits the integral representation

$$\boxed{R_n(x) = \frac{1}{n!} \int_{x_0}^{x} f^{(n+1)}(t)(x-t)^n dt, \quad \forall x \in [a,b]}. \tag{9.50}$$

Proof. Fix $x \neq x_0$. We have

$$\begin{aligned}
f(x) - f(x_0) &= \int_{x_0}^{x} f'(t)dt = -\int_{x_0}^{x} f'(t)\frac{d}{dt}(x-t)\, dt \\
&= -\Big(f'(t)(x-t)\Big)\Big|_{t=x_0}^{t=x} + \int_{x_0}^{x} f''(t)(x-t)dt \\
&= f'(x_0)(x-x_0) - \int_{x_0}^{x} f''(t)\frac{d}{dt}\left(\frac{1}{2}(x-t)^2\right) dt \\
&= f'(x_0)(x-x_0) - \left(\frac{1}{2}f''(t)(x-t)^2\right)\Big|_{t=x_0}^{t=x} + \frac{1}{2}\int_{x_0}^{x} f^{(3)}(t)(x-t)^2 dt \\
&= f'(x_0)(x-x_0) + \frac{f''(x_0)}{2}(x-x_0)^2 - \frac{1}{3!}\int_{x_0}^{x} f^{(3)}(t)\frac{d}{dt}(x-t)^3 dt \\
&= f'(x_0)(x-x_0) + \frac{f''(x_0)}{2}(x-x_0)^2 - \frac{1}{3!}\big(f^{(3)}(t)(x-t)^3\big)\Big|_{t=x_0}^{t=x} + \frac{1}{3!}\int_{x_0}^{x} f^{(4)}(t)(x-t)^3 dt \\
&= f'(x_0)(x-x_0) + \frac{f''(x_0)}{2}(x-x_0)^2 + \frac{f^{(3)}(x_0)}{3!}(x-x_0)^3 + \frac{1}{3!}\int_{x_0}^{x} f^{(4)}(t)(x-t)^3 dt \\
&= f'(x_0)(x-x_0) + \frac{f''(x_0)}{2}(x-x_0)^2 + \frac{f^{(3)}(x_0)}{3!}(x-x_0)^3 - \frac{1}{4!}\int_{x_0}^{x} f^{(4)}(t)\frac{d}{dt}(x-t)^4 dt \\
&= \cdots\cdots\cdots\cdots\cdots = \\
&= f'(x_0)(x-x_0) + \frac{f''(x_0)}{2}(x-x_0)^2 + \cdots + \frac{f^{(n)}(x_0)}{n!}(x-x_0)^n + \frac{1}{n!}\int_{x_0}^{x} f^{(n+1)}(t)(x-t)^n dt.
\end{aligned}$$

Thus

$$\begin{aligned}
f(x) &= f(x_0) + f'(x_0)(x-x_0) + \frac{f''(x_0)}{2}(x-x_0)^2 + \cdots + \frac{f^{(n)}(x_0)}{n!}(x-x_0)^n \\
&\quad + \frac{1}{n!}\int_{x_0}^{x} f^{(n+1)}(t)(x-t)^n dt \\
&= T_n(x) + \frac{1}{n!}\int_{x_0}^{x} f^{(n+1)}(t)(x-t)^n dt.
\end{aligned}$$

This proves (9.50). □

Example 9.52. Let us show how we can use the integral remainder formula to strengthen the result in Exercise 8.7. Consider the function $f : (-1,1) \to \mathbb{R}$, $f(x) = \ln(1-x)$. Since

$$f'(x) = -\frac{1}{1-x} = (x-1)^{-1}, \quad f''(x) = \frac{d}{dx}(x-1)^{-1} = -(x-1)^{-2},$$

$$f^{(3)}(x) = -\frac{d}{dx}(x-1)^{-2} = 2(x-1)^{-3}, \ldots,$$

$$f^{(n)}(x) = (-1)^{n-1}(n-1)!(x-1)^{-n}, \quad \forall n \in \mathbb{N},$$

we deduce that

$$f(0) = 0, \quad f^{(n)}(0) = (-1)^n(n-1)!(-1)^{-n} = -(n-1)!, \quad \forall n \in \mathbb{N},$$

and thus, the Taylor series of f at $x_0 = 0$ is

$$-\sum_{k=1}^{\infty} \frac{x^k}{k}.$$

We denote by $T_n(x)$ the degree n Taylor polynomial of $f(x)$ at $x_0 = 0$,

$$T_n(x) = -\sum_{k=1}^{n} \frac{x^k}{k} = -x - \frac{x^2}{2} - \cdots - \frac{x^n}{n!}.$$

We want to prove that this series converges to $\ln(1-x)$ for any $x \in [-1, 1)$. To do this we have to show that

$$\lim_{n\to\infty} |f(x) - T_n(x)| = 0, \;\; \forall x \in [-1, 1).$$

We need to estimate the remainder $R_n(x) = f(x) - T_n(x)$. We distinguish two cases.

1. $x \in [0, 1)$. Using the integral remainder formula (9.50) we deduce

$$R_n(x) = \frac{1}{n!}\int_0^x f^{(n+1)}(t)(x-t)^n dt = (-1)^n \int_0^x (t-1)^{-n-1}(x-t)^n dt.$$

Hence

$$|R_n(x)| = \int_0^x \frac{(x-t)^n}{(1-t)^{n+1}} dt.$$

Observe that for $t \in [0, x]$ we have $1 - t \geq 1 - x > 0$ so that, for any $t \in [0, x]$ we have

$$(1-t)^{n+1} \geq (1-t)^n(1-x) > 0 \Longleftrightarrow 0 < \frac{1}{(1-t)^{n+1}} \leq \frac{1}{1-x} \cdot \frac{1}{(1-t)^n}.$$

Hence

$$|R_n(x)| \leq \frac{1}{1-x}\int_0^x \left(\frac{x-t}{1-t}\right)^n dt.$$

Now consider the function

$$g : [0, x] \to \mathbb{R}, \;\; g(t) = \frac{x-t}{1-t}.$$

We have

$$g'(t) = \frac{-(1-t) + (x-t)}{(1-t)^2} = \frac{x-1}{(1-t)^2} < 0.$$

Hence

$$0 = g(x) \leq g(t) \leq g(0) = x, \;\; \forall t \in [0, x],$$

and thus

$$|R_n(x)| \leq \frac{1}{1-x}\int_0^x g(t)^n dt \leq \frac{1}{1-x}\int_0^x x^n dt = \frac{x^{n+1}}{1-x}.$$

We deduce

$$|R_n(x)| \leq \frac{x^{n+1}}{1-x}, \;\; \forall x \in [0, 1),$$

so that

$$\lim_{n\to\infty} R_n(x) = \lim_{n\to\infty} \frac{x^{n+1}}{1-x} = 0, \ \ \forall x \in [0,1).$$

2. $x \in [-1,0)$. We estimate $R_n(x)$ using the Lagrange remainder formula. Hence, there exists $\xi \in (x,0)$ such that

$$R_n(x) = \frac{f^{(n+1)}(\xi)}{(n+1)!}x^{n+1} = (-1)^n \frac{n!(\xi-1)^{-(n+1)}}{(n+1)!}x^{n+1}$$
$$= (-1)^n \frac{1}{(n+1)(\xi-1)^{n+1}}x^{n+1}.$$

Hence, since $\xi \in (x,0)$, we have $|\xi - 1| = |\xi| + 1$ and

$$|R_n(x)| = \frac{|x|^{n+1}}{(n+1)(1+|\xi|)^{n+1}} \leq \frac{|x|^{n+1}}{n+1}.$$

Since $|x| \leq 1$ we deduce

$$\lim_{n\to\infty} |R_n(x)| = 0, \ \ \forall x \in [-1,0).$$

We have thus proved that

$$\ln(1-x) = -\sum_{n=1}^{\infty} \frac{x^n}{n}, \ \ \forall x \in [-1,1).$$

Note in particular that

$$f(-1) = \ln 2 = -\sum_{n=1}^{\infty} \frac{(-1)^n}{n} = 1 - \frac{1}{2} + \frac{1}{3} - \frac{1}{4} + \cdots. \tag{9.51}$$

□

9.6.2. *Change of variables*

The change of variables in the Riemann integral is very similar to the integration-by-substitution trick used in the computation of antiderivatives, but it has a few peculiarities. There are two versions of the change of variables formula.

Proposition 9.53 (Change of variables formula: version 1, $t = \phi(x)$). *Suppose that $f : [a,b] \to \mathbb{R}$ is a continuous function and $\phi : [\alpha,\beta] \to [a,b]$ is a C^1-function. Then the function $f\big(\phi(x)\big)\phi'(x)$ is integrable on $[\alpha,\beta]$ and*

$$\boxed{\int_\alpha^\beta f\big(\phi(x)\big)\phi'(x)dx = \int_{\phi(\alpha)}^{\phi(\beta)} f(t)dt.} \tag{9.52}$$

Proof. Since f is continuous it admits antiderivatives. Fix an antiderivative F of f. The chain rule shows that $F\big(\phi(x)\big)$ is an antiderivative of the continuous function $f\big(\phi(x)\big)\phi'(x)$. The Fundamental Theorem of Calculus then shows

$$\int_\alpha^\beta f\big(\phi(x)\big)\phi'(x)dt = F\big(\phi(x)\big)\Big|_{x=\alpha}^{x=\beta} = F\big(\phi(\beta)\big) - F\big(\phi(\alpha)\big) = \int_{\phi(\alpha)}^{\phi(\beta)} f(t)dt.$$

□

We can relax the continuity assumption of f, but to do so we need to make an additional assumption of the nature of the change in variables, $t = \phi(x)$.

Proposition 9.54 (Change of variables formula: version 2, $x = \varphi(t)$). *Suppose that $f : [a, b] \to \mathbb{R}$ is a Riemann integrable function and $\varphi : [\alpha, \beta] \to [a, b]$ is a C^1-function such that*

$$\varphi'(t) \neq 0, \quad \forall t \in (\alpha, \beta).$$

Then $f(\varphi(t))\varphi'(t)$ is Riemann integrable on $[\alpha, \beta]$ and

$$\boxed{\int_{\varphi(\alpha)}^{\varphi(\beta)} f(x)dx = \int_{\alpha}^{\beta} f(\varphi(t))\varphi'(t)dt.} \tag{9.53}$$

Proof. Set

$$M := \sup_{t \in [\alpha, \beta]} |\varphi'(t)|.$$

Note that $M > 0$. Since $|\varphi'(t)|$ is continuous, Weierstrass' Theorem implies that $M < \infty$. Since $\varphi'(t) \neq 0$ for any $t \in (\alpha, \beta)$ we deduce from the Intermediate Value Theorem that

$$\text{either } \varphi'(t) > 0, \ \forall t \in (\alpha, \beta) \ \text{ or } \ \varphi'(t) < 0, \ \forall t \in (\alpha, \beta).$$

Thus, either φ is strictly increasing and its range is $[\varphi(\alpha), \varphi(\beta)]$, or φ is strictly decreasing and its range is $[\varphi(\beta), \varphi(\alpha)]$. We need to discuss each case separately, but we will present the details only for the first case and leave the details for the second case for you as an exercise. In the sequel we will assume that φ is increasing and thus

$$0 < \varphi'(t) \leq M, \ \forall t \in (\alpha, \beta).$$

For simplicity we set

$$g(t) := f(\varphi(t))\varphi'(t), \ \ t \in [\alpha, \beta].$$

We will show that g is Riemann integrable on $[\alpha, \beta]$ and its Riemann integral is given by the left-hand side of (9.53). We will need the following technical result.

Lemma 9.55. *For any partition $\boldsymbol{P}$ of $[\alpha, \beta]$, there exists a partition $\boldsymbol{P}_\varphi$ of $[\varphi(\alpha), \varphi(\beta)]$ and samples $\underline{\xi}$ of $\boldsymbol{P}$ and $\underline{\eta}$ of $\boldsymbol{P}_\varphi$ such that*

$$\|\boldsymbol{P}_\varphi\| \leq M\|\boldsymbol{P}\|, \tag{9.54a}$$

$$\boldsymbol{S}(\boldsymbol{P}, g, \underline{\xi}) = \boldsymbol{S}(\boldsymbol{P}_\varphi, f, \underline{\xi}_\varphi). \tag{9.54b}$$

Let us first show that Lemma 9.55 implies that g is Riemann integrable and satisfies (9.53). Fix $\varepsilon > 0$. The function f is Riemann integrable on $[\varphi(\alpha), \varphi(\beta)]$ and thus there exists $\delta_0 = \delta_0(\varepsilon) > 0$ such that for any partition $\boldsymbol{Q}$ of $[\varphi(\alpha), \varphi(\beta)]$ and any sample $\underline{\eta}$ of $\boldsymbol{Q}$ we have

$$\left| \int_{\varphi(\alpha)}^{\varphi(\beta)} f(x)dx - \boldsymbol{S}(f, \boldsymbol{Q}, \underline{\eta}) \right| < \varepsilon. \tag{9.55}$$

Set

$$\delta = \delta(\varepsilon) := \frac{1}{M}\delta_0(\varepsilon).$$

For any partition $\boldsymbol{P}$ of $[\alpha, \beta]$ of mesh $\|\boldsymbol{P}\| < \delta(\varepsilon)$, and any sample $\underline{\xi}$ of $\boldsymbol{P}$, the sampled partition $(\boldsymbol{P}_\varphi, \underline{\xi}_\varphi)$ of $[\varphi(\alpha), \varphi(\beta)]$ associated to $(\boldsymbol{P}, \underline{\xi})$ by Lemma 9.55 satisfies

$$\boldsymbol{P}_\varphi\| < M\delta(\varepsilon) = \delta_0(\varepsilon) \ \text{ and } \ \boldsymbol{S}(g, \boldsymbol{P}, \underline{\xi}) = \boldsymbol{S}(f, \boldsymbol{P}_\varphi, \underline{\xi}_\varphi).$$

We deduce that

$$\left| \int_{\varphi(\alpha)}^{\varphi(\beta)} f(x)dx - \boldsymbol{S}(g, \boldsymbol{P}, \underline{\xi}) \right| = \left| \int_{\varphi(\alpha)}^{\varphi(\beta)} f(x)dx - \boldsymbol{S}(f, \boldsymbol{P}_\varphi, \underline{\xi}_\varphi) \right| \overset{(9.55)}{<} \varepsilon.$$

This proves that $g(t)$ is integrable on $[\alpha, \beta]$ and its integral is equal to $\int_{\varphi(\alpha)}^{\varphi(\beta)} f(x)dx$.

Proof of Lemma 9.55. Consider a partition $\boldsymbol{P} = (\alpha = t_0 < t_1 < \cdots < t_n = \beta)$ of $[\alpha, \beta]$. For $k = 0, 1, \dots, x_n$ we set

$$x_k := \varphi(t_k).$$

Since φ is increasing we have

$$x_{k-1} < x_k, \quad \forall k = 1, \dots, n.$$

Thus

$$\varphi(\alpha) = x_0 < x_1 < \cdots < x_n = \varphi(\beta)$$

is a partition of $[\varphi(\alpha), \varphi(\beta)]$ that we denote by $\boldsymbol{P}_\varphi$. Note that

$$x_k - x_{k-1} = \varphi(t_k) - \varphi(t_{k-1}).$$

Lagrange's Mean Value Theorem implies that there exists $\xi_k \in (t_{k-1}, t_k)$ such that

$$x_k - x_{k-1} = \varphi(t_k) - \varphi(t_{k-1}) = \varphi'(\xi_k)(t_k - t_{k-1}).$$

In particular, this shows that

$$|x_{k-1} - x_k| = |\varphi'(\xi_k)| \cdot |t_k - t_{k-1}| \leq M|t_k - t_{k-1}|, \quad \forall k = 1, \dots, k.$$

Hence

$$\|\boldsymbol{P}_\varphi\| \leq M\|\boldsymbol{P}\|.$$

This proves (9.54a).

Set $\eta_k := \varphi(\xi_k)$. Note that since φ is increasing we have $\eta_k \in (x_{k-1}, x_k)$. The collection $\underline{\xi} = (\xi_1, \dots, \xi_k)$ is a sample of $\boldsymbol{P}$, and the collection $\underline{\eta} = (\eta_1, \dots, \eta_n)$ is a sample of $\boldsymbol{P}_\varphi$. Observe that

$$f(\eta_k)(x_k - x_{k-1}) = f\big(\varphi(\xi_k)\big)\varphi'(\xi_k)(t_k - t_{k-1}) = g(\xi_k)(t_k - t_{k-1}).$$

Thus

$$\boldsymbol{S}(f, \boldsymbol{P}_\varphi, \underline{\eta}) = \sum_{k=1}^n f(\eta_k)(x_k - x_{k-1}) = \sum_{k=1}^n g(\xi_k)(t_k - t_{k-1}) = \boldsymbol{S}(g, \boldsymbol{P}, \underline{\xi}).$$

This proves (9.54b) and completes the proof of Proposition 9.54. □

Remark 9.56. In concrete examples, the right-hand sides of the equalities (9.52) and (9.53) are quantities that we know how to compute. The left-hand sides are the unknown quantities whose computations are sought. For this reason these two equalities play different roles in applications. □

Example 9.57. (a) Suppose that we want to compute

$$\int_{-1}^{2} \cos(x^2)x dx = \frac{1}{2}\int_{-1}^{2} \cos(x^2) d(x^2).$$

We make the change of variables $t = x^2$. Note that $x = -1 \Rightarrow t = 1$, $x = 2 \Rightarrow t = 4$ and we deduce

$$\int_{-1}^{2} \cos(x^2)x dx \stackrel{(9.52)}{=} \frac{1}{2}\int_{1}^{4} \cos t\, dt = \frac{\sin 4 - \sin 1}{2}.$$

Note that in this case (9.53) is not applicable.

(b) Suppose that we want to compute

$$\int_0^{\frac{\pi}{2}} e^{\sin x} \cos x\, dx = \int_0^{\frac{\pi}{2}} e^{\sin x} d(\sin x).$$

We make the change in variables $t = \sin x$. Note that $x = 0 \Rightarrow t = 0$, $x = \frac{\pi}{2} \Rightarrow t = 1$ and we deduce

$$\int_0^{\frac{\pi}{2}} e^{\sin x} \cos x dx \overset{(9.52)}{=} \int_0^1 e^t dt = e^t\Big|_{t=0}^{t=1} = e - 1.$$

(c) Suppose we want to compute

$$\int_{-1}^{1} \sqrt{1-x^2}dx.$$

We make a change of variables $x = \sin t$ so that $dx = d(\sin t) = \cos t dt$. Note that

$$x = -1 \Rightarrow t = -\frac{\pi}{2},\ \ x = 1 \Rightarrow t = \frac{\pi}{2},$$

and $\cos t > 0$ when $t \in (-\frac{\pi}{2}, \frac{\pi}{2})$. Hence

$$\sqrt{1-x^2} = \sqrt{1-\sin^2 t} = \sqrt{\cos^2 t} = \cos t,\ \ -\frac{\pi}{2} \le t \le \frac{\pi}{2}.$$

We deduce

$$\int_{-1}^{1} \sqrt{1-x^2}dx \overset{(9.53)}{=} \int_{-\frac{\pi}{2}}^{\frac{\pi}{2}} \cos^2 t dt = \int_{-\frac{\pi}{2}}^{\frac{\pi}{2}} \frac{1+\cos 2t}{2} dt$$

$$= \frac{1}{2}\Big(\frac{\pi}{2} + \frac{\pi}{2}\Big) + \frac{1}{2}\int_{-\frac{\pi}{2}}^{\frac{\pi}{2}} \cos 2t dt = \frac{\pi}{2} + \frac{1}{2}\int_{-\frac{\pi}{2}}^{\frac{\pi}{2}} \cos 2t dt.$$

To compute the last integral we use the change in variables $u = 2t$ so that $dt = \frac{1}{2}du$,

$$t = -\frac{\pi}{2} \Rightarrow u = -\pi,\ \ t = \frac{\pi}{2} \Rightarrow u = \pi.$$

Hence

$$\int_{-\frac{\pi}{2}}^{\frac{\pi}{2}} \cos 2t\, dt = \frac{1}{2}\int_{-\pi}^{\pi} \cos u\, du = \frac{1}{2}\big(\sin\pi - \sin(-\pi)\big) = 0.$$

We conclude that

$$\int_{-1}^{1} \sqrt{1-x^2}dx = \frac{\pi}{2}. \tag{9.56}$$

Let us observe that this equality provides a way of approximating $\frac{\pi}{2}$ by using Riemann sums to approximate the integral in the left-hand side. If we use the uniform partition $\boldsymbol{U}_{200}$ of order 200 of $[-1, 1]$ and as sample $\underline{\xi}$ the right endpoints of the intervals of the partition, then we deduce

$$\pi \approx 2\boldsymbol{S}\big(\sqrt{1-x^2}, \boldsymbol{U}_{200}, \underline{\xi}\big) \approx 3.14041.....$$

If we use the uniform partition of order $2,000$ and a similar sample, then we deduce

$$\pi \approx 2\boldsymbol{S}\big(\sqrt{1-x^2}, \boldsymbol{U}_{2,000}, \underline{\xi}\big) \approx 3.14157.....$$

(d) Suppose that we want to compute the integral.

$$\int_1^e \frac{\ln x}{x} dx.$$

We make the change in variables $x = e^t$ and we observe that $dx = e^t dt$,

$$x = 1 \Rightarrow t = 0,\ \ x = e \Rightarrow t = 1.$$

The derivative $\frac{dx}{dt} = e^t$ is everywhere positive and we deduce

$$\int_1^e \frac{\ln x}{x} dx \overset{(9.53)}{=} \int_0^1 \frac{\ln e^t}{e^t} e^t dt = \int_0^1 t dt = \frac{t^2}{2}\Big|_{t=0}^{t=1} = \frac{1}{2}. \qquad \square$$

Example 9.58 (Stirling's formula). In many applications we need to have a simpler way of understanding the size of $n!$ for n very large. This is what Stirling's formula accomplishes. More precisely, we want to prove the refined inequalities

$$\boxed{1 < \frac{n!}{n^n e^{-n}\sqrt{2\pi n}} < 1 + \frac{1}{4n}, \quad \forall n \in \mathbb{N}}. \tag{9.57}$$

The inequalities (9.57) imply the classical *Stirling formula*

$$\boxed{n! \sim \sqrt{2\pi n}\left(\frac{n}{e}\right)^n \quad \text{as } n \to \infty}, \tag{9.58}$$

where we recall that the notation $x_n \sim y_n$ as $n \to \infty$ (read x_n *is asymptotic to* y_n *as* $n \to \infty$) signifies

$$\lim_{n\to\infty} \frac{x_n}{y_n} = 1.$$

To prove the inequalities (9.57) we follow the very nice approach in [7, §2.6]. We set

$$F_n := \ln n! = \ln(1) + \ln(2) + \cdots + \ln(n) = \ln(2) + \cdots + \ln(n)$$

and we aim to find accurate approximations for F_n. We will find these by providing rather sharp approximations for the integral

$$I_n := \int_1^n \ln x dx = \big(x\ln x - x\big)\Big|_{x=1}^{x=n} = n\ln n - n + 1.$$

To see why such an integral might be relevant observe that

$$\ln\left(\frac{n}{e}\right)^n = n\ln n - n.$$

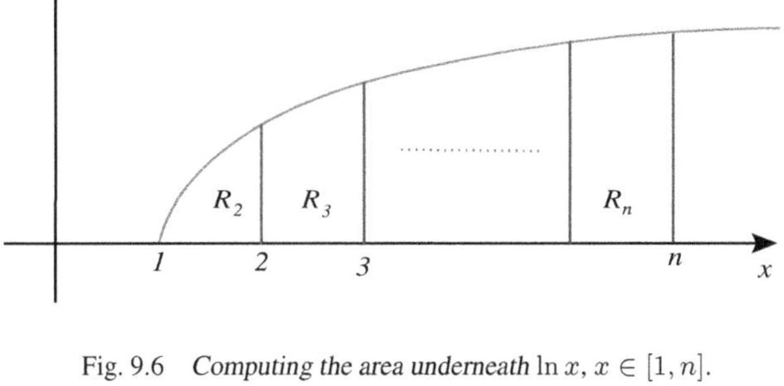

Fig. 9.6 *Computing the area underneath* $\ln x$, $x \in [1, n]$.

Observe that I_n is the area below the graph of $\ln x$ and above the interval $[1, n]$ on the x axis. For $k =, 2, 3, \ldots, n$ we denote by R_k the region below the graph of $\ln x$ and above the interval $[k-1, k]$ on the x-axis; see Figure 9.6. Then

$$I_n = \text{Area}\,(R_2) + \cdots + \text{Area}\,(R_n).$$

We will provide lower and upper estimates for I_n by producing lower and upper estimates for the areas of the regions R_k. To produce these bounds for the area of R_k we will take advantage of the fact that $\ln x$ is concave so its graph lies above any chord and below any tangent.

Denote by p_k the point on the graph of $\ln x$ corresponding to $x = k$, i.e., $p_k = (k, \ln k)$. Due to the concavity of $\ln x$ the region R_k contains the trapezoid A_k determined by the chord connecting the points p_{k-1} and p_k; see Figure 9.7.

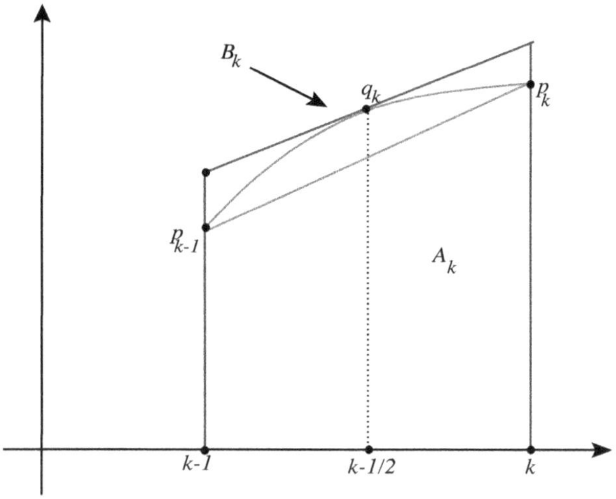

Fig. 9.7 *Approximating the region R_k by trapezoids.*

Denote by q_k the point on the graph of $\ln x$ above the midpoint of the interval $[k-1, k]$, i.e., $q_k = (k - 1/2, \ln(k - 1/2))$. The tangent to the graph of $\ln x$ at q_k determines a trapezoid B_k that contains the region R_k. Thus

$$\underbrace{\text{Area}\,(R_k) - \text{Area}\,(A_k)}_{=:s_k} < \text{Area}\,(B_k) - \text{Area}\,(A_k).$$

Hence

$$I_n = \sum_{k=2}^{n} \text{Area}\,(R_k) = \sum_{k=2}^{n} \text{Area}\,(A_k) + \underbrace{\sum_{k=2}^{n} s_k}_{=:S_n}. \tag{9.59}$$

Observe that

$$\text{Area}\,(A_k) = \frac{1}{2}\big(\ln(k-1) + \ln k\big),$$

so

$$\begin{aligned}\text{Area}\,(A_2) + \cdots + \text{Area}\,(A_n) &= \frac{1}{2}\log 2 + \frac{1}{2}\big(\ln 2 + \ln 3\big) + \cdots + \frac{1}{2}\big(\ln(n-1) + \ln n\big)\\ &= \ln 2 + \ln 3 + \cdots + \ln n - \frac{1}{2}\ln n = \ln n! - \frac{1}{2}\ln n.\end{aligned}$$

Using this in (9.59) we deduce

$$I_n + \frac{1}{2}\ln n = \ln n! + S_n.$$

Recalling that $I_n = n\ln n - n + 1$ we deduce

$$n\ln n - n + \frac{1}{2}\ln n + 1 - S_n = \ln n!,$$

or, equivalently

$$n! = C_n\sqrt{n}\left(\frac{n}{e}\right)^n, \quad C_n = e^{1-S_n}. \tag{9.60}$$

To progress further we need to gain some information about S_n. Observing that

$$\text{Area}\,(B_k) = \ln\left(k - \frac{1}{2}\right)$$

we deduce

$$\begin{aligned}s_k < \text{Area}\,(B_k) - \text{Area}\,(A_k) &= \ln\left(k - \frac{1}{2}\right) - \frac{1}{2}\big(\ln(k-1) + \ln k\big)\\ &= \frac{1}{2}\ln\left(\frac{k - \frac{1}{2}}{k-1}\right) - \frac{1}{2}\ln\left(\frac{k}{k - \frac{1}{2}}\right)\\ &= \frac{1}{2}\ln\left(1 + \frac{1}{2k-2}\right) - \frac{1}{2}\ln\left(1 + \frac{1}{2k-1}\right)\\ &< \frac{1}{2}\ln\left(1 + \frac{1}{2k-2}\right) - \frac{1}{2}\ln\left(1 + \frac{1}{2k}\right).\end{aligned}$$

We conclude

$$S_n = \sum_{k=2}^{n} s_k < \frac{1}{2}\sum_{k=2}^{n}\left(\ln\left(1 + \frac{1}{2k-2}\right) - \frac{1}{2}\ln\left(1 + \frac{1}{2k}\right)\right)$$

(the last sum is a telescoping sum)

$$= \frac{1}{2}\ln\left(1 + \frac{1}{2}\right) - \frac{1}{2}\ln\left(1 + \frac{1}{2n}\right) < \frac{1}{2}\ln\frac{3}{2}.$$

This shows that the sequence S_n is bounded above. Since this sequence is obviously increasing, we deduce that (S_n) is convergent. We denote by S its limit. Since the sequence S_n is increasing, the sequence $C_n = e^{1-S_n}$ is decreasing and converges to $C = e^{1-S}$. Using this in (9.60) we deduce

$$n! = C_n\sqrt{n}\left(\frac{n}{e}\right)^n > C\sqrt{n}\left(\frac{n}{e}\right)^n. \tag{9.61}$$

Observe next that

$$\frac{C_n}{C} = e^{S-S_n}$$

and

$$S - S_n = \sum_{k>n} s_k < \frac{1}{2}\sum_{k>n}\left(\ln\left(1+\frac{1}{2k-2}\right) - \frac{1}{2}\ln\left(1+\frac{1}{2k}\right)\right)$$
$$= \frac{1}{2}\ln\left(1+\frac{1}{2n}\right) = \ln\left(1+\frac{1}{2n}\right)^{\frac{1}{2}}.$$

Hence

$$\frac{C_n}{C} < \left(1+\frac{1}{2k}\right)^{\frac{1}{2}} < 1+\frac{1}{4n} \Rightarrow C_n < C\left(1+\frac{1}{4n}\right).$$

We deduce

$$C\sqrt{n}\left(\frac{n}{e}\right)^n < n! = C_n\sqrt{n}\left(\frac{n}{e}\right)^n < C\left(1+\frac{1}{4n}\right)\sqrt{n}\left(\frac{n}{e}\right)^n. \tag{9.62}$$

It remains to determine the constant C. We set

$$P_n := \sqrt{n}\left(\frac{n}{e}\right)^n.$$

From (9.62) we deduce

$$n! \sim CP_n \text{ as } n \to \infty. \tag{9.63}$$

To obtain C from the above equality we rely on Wallis' formula (9.49) which states that

$$\frac{\pi}{2} = \lim_{n\to\infty} \frac{2^2 4^2 \cdots (2n)^2}{1^2 3^2 \cdots (2n-1)^2} \cdot \frac{1}{2n}.$$

Now observe that

$$\frac{2^2 4^2 \cdots (2n)^2}{1^2 3^2 \cdots (2n-1)^2} \cdot \frac{1}{2n} = \frac{(n!)^2 2^{2n}}{1^2 3^2 \cdots (2n-1)^2} \cdot \frac{1}{2n} = \frac{(n!)^4 2^{4n}}{(\,(2n)!\,)^2 (2n)}.$$

Hence

$$\sqrt{\frac{\pi}{2}} = \lim_{n\to\infty} \frac{(n!)^2 2^{2n}}{(2n)!\sqrt{2n}},$$

i.e.,

$$\sqrt{\pi} = \lim_{n\to\infty} \frac{(n!)^2 2^{2n}}{(2n)!\sqrt{n}} = \lim_{n\to\infty} \frac{C^2 P_n^2 2^{2n}}{CP_{2n}\sqrt{n}} \cdot \frac{\left(\frac{n!}{CP_n}\right)^2 2^{2n}}{\frac{(2n)!}{CP_{2n}}}$$
$$\stackrel{(9.63)}{=} \lim_{n\to\infty} \frac{C^2 P_n^2 2^{2n}}{CP_{2n}\sqrt{n}} = C \lim_{n\to\infty} \frac{P_n^2 2^{2n}}{P_{2n}\sqrt{n}}.$$

Now observe that

$$P_n = \sqrt{n}\left(\frac{n}{e}\right)^n \Rightarrow P_n^2 = \frac{n^{2n+1}}{e^{2n}},\;\; P_{2n} = \sqrt{2n}\frac{(2n)^{2n}}{e^{2n}} = 2^{2n}\sqrt{2n}\frac{n^{2n}}{e^{2n}},$$

and thus

$$\frac{P_n^2 2^{2n}}{P_{2n}\sqrt{n}} = \frac{2^{2n}\frac{n^{2n+1}}{e^{2n}}}{2^{2n}\sqrt{2n}\cdot\frac{n^{2n}}{e^{2n}}\cdot\sqrt{n}} = \frac{1}{\sqrt{2}}.$$

Hence

$$\sqrt{\pi} = C\lim_{n\to\infty}\frac{P_{2n}\sqrt{n}}{2^{2n}P_n^2} = \frac{C}{\sqrt{2}} \Rightarrow C = \sqrt{2\pi}.$$

The inequalities (9.62) with $C = \sqrt{2\pi}$ are precisely the inequalities (9.57) that we wanted to prove. □

9.7. Improper integrals

The Riemann integral is an operation defined for certain *bounded* functions defined on *bounded* intervals. Sometimes, even when one or both of these boundedness requirements are violated we can still give a meaning to an integral. Before we proceed with rigorous definitions it is helpful to look at some guiding examples.

Example 9.59. (a) Let $\alpha \in (0,1)$ and consider the function

$$f:(0,1]\to\mathbb{R},\ \ f(x)=\frac{1}{x^\alpha}.$$

This function is continuous on $(0,1]$, but it is not bounded on this interval because

$$\lim_{x\to 0+}\frac{1}{x^\alpha}=\infty.$$

It is however continuous on any compact interval $[\varepsilon,1]$ and so it is Riemann integrable on such an interval. Note that

$$\int_\varepsilon^1 x^{-\alpha}dx=\frac{x^{1-\alpha}}{1-\alpha}\Big|_\varepsilon^1=\frac{1}{1-\alpha}(1-\varepsilon^{1-\alpha}).$$

Since $1-\alpha>0$ we deduce that $\varepsilon^{1-\alpha}\to 0$ as $\varepsilon\searrow 0$ and thus

$$\lim_{\varepsilon\searrow 0}\int_\varepsilon^1 x^{-\alpha}dx=\frac{1}{1-\alpha}.$$

We can define the *improper Riemann* integral of $x^{-\alpha}$ over $[0,1]$ to be

$$\int_0^1 x^{-\alpha}dx:=\lim_{\varepsilon\searrow 0}\int_\varepsilon^1 x^{-\alpha}dx=\frac{1}{1-\alpha}.$$

(b) Let $p>1$ and consider the function $g:[1,\infty)\to\mathbb{R}$, $g(x)=\frac{1}{x^p}$. The function g is bounded

$$0<g(x)\leq 1,\ \ \forall x\geq 1$$

but it is defined on the unbounded interval $[1,\infty)$. It is integrable on any interval $[1,L]$ and we have

$$\int_1^L x^{-p}dx=\frac{x^{1-p}}{1-p}\Big|_1^L=\frac{1}{1-p}(L^{1-p}-1).$$

Since $1-p<0$ we deduce that $L^{1-p}\to 0$ as $L\to\infty$ and thus

$$\lim_{L\to\infty}\int_1^L x^{-p}dx=-\frac{1}{1-p}=\frac{1}{p-1}.$$

We define the *improper Riemann* integral of x^{-p} over $[1,\infty)$ to be

$$\int_1^\infty x^{-p}dx:=\lim_{L\to\infty}\int_1^L x^{-p}dx=\frac{1}{p-1}.$$

□

The above examples gave meaning to integrals of *functions that are not defined on compact intervals*. Such integrals are called *improper*.

Definition 9.60 (Improper integrals). (a) Let $-\infty < a < \omega \leq \infty$. Given a function $f : [a, \omega) \to \mathbb{R}$ we say that the *improper integral*

$$\int_a^\omega f(x)\,dx$$

is *convergent* if

- the restriction of f to any interval $[a, x] \subset [a, \omega)$ is Riemann integrable and
- the limit

$$\lim_{x \nearrow \omega} \int_a^x f(t)\,dt$$

exists and it is finite.

When these happen we set

$$\int_a^\omega f(x)dx := \lim_{x \nearrow \omega} \int_a^x f(t)\,dt.$$

(b) Let $-\infty \leq \omega < b < \infty$. Given a function $f : (\omega, b] \to \mathbb{R}$ we say that the *improper integral*

$$\int_\omega^b f(x)dx$$

is *convergent* if

- the restriction of f to any interval $[x, b] \subset (\omega, b]$ is Riemann integrable and
- the limit

$$\lim_{x \searrow \omega} \int_x^b f(t)\,dt$$

exists and it is finite.

When these happen we set

$$\int_\omega^b f(x)dx := \lim_{x \searrow \omega} \int_x^b f(t)\,dt.$$ □

Remark 9.61. (a) We can rephrase the conclusion of Example 9.59(a) by saying that the integral

$$\int_0^1 \frac{1}{x^\alpha} dx$$

is convergent if $\alpha \in (0, 1)$. Example 9.59(b) shows that the integral

$$\int_1^\infty \frac{1}{x^p} dx$$

is convergent if $p > 1$.

(b) In the sequel, in order to keep the presentation within bearable limits, we will state and prove results only for the improper integrals of type (a) in Definition 9.60. These involve functions that have a "problem" at the *upper* endpoint ω of their domain: either that endpoint is infinite, or the function "explodes" as x approaches ω.

These results have obvious counterparts for the integrals of type (b) in Definition 9.60 that involve functions that have a "problem" at the *lower* endpoint of their domain. Their statements and proofs closely mimic the corresponding ones for type (a) integrals. □

Example 9.62. For any $a, b \in \mathbb{R}$, $a < b$, the improper integrals

$$\int_a^b \frac{1}{(x-a)^\alpha} dx, \quad \int_a^b \frac{1}{(b-x)^\alpha} dx$$

are convergent for $\alpha < 1$ and divergent if $\alpha \geq 1$. Indeed, if $\alpha \neq 1$ we have

$$\int_{a+\varepsilon}^b \frac{1}{(x-a)^\alpha} dx = \frac{1}{1-\alpha}(x-a)^{1-\alpha}\Big|_{x=a+\varepsilon}^{a=b} = \frac{1}{1-\alpha}\big((b-a)^{1-\alpha} - \varepsilon^{1-\alpha}\big),$$

If $\alpha = 1$ we have

$$\int_{a+\varepsilon}^b \frac{1}{(x-a)} dx = \ln(x-a)\Big|_{x=a+\varepsilon}^{x=b} = \ln(b-a) - \ln\varepsilon.$$

These computations show that

$$\lim_{\varepsilon\searrow 0} \int_{a+\varepsilon}^b \frac{1}{(x-a)^\alpha} dx = \begin{cases} \frac{1}{1-\alpha}(b-a)^{1-\alpha}, & \alpha < 1, \\ \infty, & \alpha \geq 1. \end{cases}$$

The convergence of the integral

$$\int_a^b \frac{1}{(b-x)^\alpha} dx$$

is analyzed in a similar fashion.

(b) The integral

$$\int_1^\infty \frac{1}{x^p} dx, \quad p \in \mathbb{R}.$$

is convergent for $p > 1$ and divergent if $p \leq 1$.

Indeed, if $p \neq 1$, then

$$\int_1^L x^{-p} dx = \frac{1}{1-p} x^{1-p}\Big|_{x=1}^{x=L} = \frac{1}{1-p}\big(L^{1-p} - 1\big).$$

Now observe that

$$\lim_{L\to\infty} L^{1-p} = \begin{cases} 0, & p > 1, \\ \infty, & p < 1. \end{cases}$$

When $p = 1$, we have

$$\int_1^L \frac{1}{x} dx = \ln L \to \infty \text{ as } L \to \infty.$$

Similarly, the integral

$$\int_{-\infty}^{-1} \frac{1}{|x|^p} dx$$

converges for $p > 1$ and diverges for $p \leq 1$. □

We have the following immediate result whose proof is left to you as an exercise.

Proposition 9.63. *Let $-\infty < a < \omega \le \infty$ and $f_1, f_2 : [a, \omega) \to \mathbb{R}$ be functions that are Riemann integrable on each of the intervals $[a, x]$, $x \in (a, \omega)$.*

(a) If $t_1, t_2 \in \mathbb{R}$, and the improper integrals

$$\int_a^\omega f_i(x)dx, \quad i = 1, 2$$

are convergent, then the integral

$$\int_a^\omega \big(t_1 f_1(x) + t_2 f_2(x) \big) dx$$

is convergent, and

$$\int_a^\omega \big(t_1 f_1(x) + t_2 f_2(x) \big) dx = t_1 \int_a^\omega f_1(x)dx + t_2 \int_a^\omega f_2(x)dx.$$

(b) Let $b \in (a, \omega)$. The improper integral

$$\int_a^\omega f_1(x)dx$$

is convergent if and only if the improper integral

$$\int_b^\omega f_1(x)dx$$

is convergent. Moreover, when these integrals are convergent we have

$$\int_a^\omega f_1(x)dx = \int_a^b f_1(x)dx + \int_b^\omega f_1(x)dx. \tag{9.64}$$

□

Theorem 9.64 (Cauchy). *Let $-\infty < a < \omega \le \infty$ and suppose that $f : [a, \omega) \to \mathbb{R}$ is a function which is Riemann integrable on each of the intervals $[a, x] \subset [a, \omega)$. Then the following statements are equivalent.*

(i) The integral $\int_a^\omega f(t)dt$ is convergent.
(ii) For any $\varepsilon > 0$ there exists $c = c(\varepsilon) \in (a, \omega)$ such that

$$\forall x, y : \ x, y \in (c(\varepsilon), \omega) \Rightarrow \left| \int_x^y f(t)dt \right| < \varepsilon.$$

Proof. We set

$$I(x) := \int_a^x f(t)dt, \ \forall x \in [a, \omega).$$

(i) ⇒ (ii). We know that the limit

$$I_\omega := \lim_{x \to \omega} I(x)$$

exists and it is finite. Let $\varepsilon > 0$. There exists $c = c(\varepsilon) \in [a, \omega)$ such that

$$\forall x, y: \; x, y \in (c, \omega) \Rightarrow |I(x) - I_\omega| < \frac{\varepsilon}{2} \text{ and } |I(y) - I_\omega| < \frac{\varepsilon}{2}.$$

Observe that for any $x, y \in (c, \omega)$ we have

$$\left| \int_x^y f(t)dt \right| = \left| I(y) - I(x) \right| \leq \left| I(y) - I_\omega \right| + \left| I_\omega - I(x) \right| < \varepsilon.$$

This proves (ii).
(ii) $\Rightarrow$ (i). We know that for any $\varepsilon > 0$ there exists $c = c(\varepsilon) \in [a, \omega)$ such that

$$\forall x < y: \; x, y \in (\, c(\varepsilon), \omega\,) \Rightarrow \left| \int_x^y f(t)dt \right| < \frac{\varepsilon}{2}. \tag{9.65}$$

Choose a sequence (x_n) in $[a, \omega)$ such that

$$\lim_n x_n = \omega.$$

We deduce that for any $\varepsilon > 0$ there exists $N = N(\varepsilon)$ such that

$$\forall n: \; n > N(\varepsilon) \Rightarrow x_n \in (\, c(\varepsilon), \omega).$$

Hence, for any $m, n > N(\varepsilon)$ we have

$$|I(x_m) - I(x_n)| < \frac{\varepsilon}{2} < \varepsilon, \;\; \forall m, n > N(\varepsilon) \tag{9.66}$$

proving that the sequence $(I(x_n))$ is Cauchy, thus convergent. Set

$$J := \lim_n I(x_n).$$

We will show that

$$\lim_{x \to \omega} I(x) = J.$$

Letting $m \to \infty$ in (9.66) we deduce that for any $\varepsilon > 0$ and any $n > N(\varepsilon)$ we have

$$x_n \in (\, c(\varepsilon), \omega) \text{ and } |J - I(x_n)| \leq \frac{\varepsilon}{2}. \tag{9.67}$$

Let $x \in (\, c(\varepsilon), \omega)$ and $n > N(\varepsilon/2)$. Then $x, x_n \in (\, c(\varepsilon), \omega)$ and (9.65) implies that

$$|I(x_n) - I(x)| < \frac{\varepsilon}{2} \tag{9.68}$$

We deduce

$$|I(x) - J| \leq \left| I(x) - I(x_n) \right| + \left| I(x_n) - J \right| \overset{(9.67),(9.68)}{<} \varepsilon, \;\; \forall x \in (c(\varepsilon), \omega).$$

This proves (i). □

Corollary 9.65 (Comparison Principle). *Let $-\infty < a < \omega \leq \infty$ and suppose that $f, g : [a, \omega) \to \mathbb{R}$ are two real functions satisfying the following properties.*

(i) For any $x \in [a, \omega)$ the restrictions of f, g to $[a, x]$ are Riemann integrable.
(ii) $\exists b \in [a, \omega)$, such that $0 \leq f(x) \leq g(x)$, $\forall x \in [b, \omega)$.

Then

$$\int_a^\omega g(x)dx \text{ is convergent} \Rightarrow \int_a^\omega f(x)dx \text{ is convergent.}$$

Proof. Since the improper integral

$$\int_a^\omega g(x)dx$$

is convergent we deduce from Proposition 9.63(b) that the integral

$$\int_b^\omega g(x)dx$$

is also convergent. Theorem 9.64 shows that for any $\varepsilon > 0$ there exists $c(\varepsilon) \in [b, \omega)$ such that

$$\forall x < y : \; x, y \in (c(\varepsilon), \omega) \Rightarrow \int_x^y g(t)dt = \left| \int_x^y g(t)dt \right| < \varepsilon.$$

Using the assumption (i) we deduce that

$$\forall x < y : \; x, y \in (c(\varepsilon), \omega) \Rightarrow \left| \int_x^y f(t)dt \right| = \int_x^y f(t)dt \leq \int_x^y g(t)dt.$$

We can invoke Theorem 9.64 to conclude that the integral

$$\int_b^\omega f(x)dx$$

is convergent. Proposition 9.63(b) now implies that

$$\int_a^\omega f(x)dx$$

is convergent. □

Remark 9.66. Using the logical tautology

$$p \Rightarrow q \longleftrightarrow \neg q \Rightarrow \neg p,$$

we see that if f and g are as in Corollary 9.65, then

$$\int_a^\omega f(x)dx \text{ is divergent} \Rightarrow \int_a^\omega g(x)dx \text{ is divergent.}$$ □

Corollary 9.67. *Let $-\infty < a < \omega \leq \infty$ and suppose that $f, g : [a, \omega) \to \mathbb{R}$ are two real functions satisfying the following properties.*

(i) $\exists b \in [a, \omega)$, such that $f(x) \geq 0$ and $g(x) > 0$, $\forall x \in [b, \omega)$.
(ii) There exists $C \geq 0$ such that

$$\lim_{x \to \omega} \frac{f(x)}{g(x)} = C.$$

(iii) For any $x \in [a, \omega)$ the restrictions of f and g to $[a, x]$ are Riemann integrable.

Then

$$\int_a^\omega g(x)dx \text{ is convergent } \Rightarrow \int_a^\omega f(x)dx \text{ is convergent.}$$

Proof. The integral

$$\int_a^\omega (C+1)g(x)dx$$

is convergent.
The assumption (ii) implies that there exists $b_0 \in (b, \omega)$ such that

$$f(x) < (C+1)g(x), \quad \forall x \in (b_0, \omega).$$

We can now invoke Corollary 9.65 to reach the desired conclusion. □

Example 9.68. (a) Consider the continuous function

$$f : [1, \infty) \to \mathbb{R}, \quad f(x) = \frac{x+2}{4x^3 + 3x^2 + 2x + 1}.$$

Note that $f(x) \geq 0$ for any $x \in [1, \infty)$. To decide the convergence of the integral

$$\int_1^\infty f(x)dx$$

we compare $f(x)$ with the function $g : [1, \infty) \to \mathbb{R}$, $g(x) = \frac{1}{x^2}$. Observe that

$$\frac{f(x)}{g(x)} = \frac{x^3 + 2x^2}{4x^3 + 3x^2 + 2x + 1} \to \frac{1}{4} \quad \text{as } x \to \infty.$$

Since the integral

$$\int_1^\infty \frac{1}{x^2}dx$$

is convergent we deduce from Corollary 9.67 that the integral

$$\int_1^\infty f(x)dx$$

is also convergent.
(b) Consider the function

$$f : (0, 1] \to \mathbb{R}, \quad f(x) = \frac{\sin\sqrt{x}}{x}.$$

Note that

$$\lim_{x \searrow 0} f(x) = \lim_{x \searrow 0} \frac{\sin\sqrt{x}}{\sqrt{x}} \frac{1}{\sqrt{x}} = \infty.$$

In particular, $f(x) > 0$ for $x > 0$ small. Since

$$\frac{f(x)}{\frac{1}{\sqrt{x}}} = \frac{\sin\sqrt{x}}{\sqrt{x}} \to 1 \text{ as } x \searrow 0$$

and the improper integral

$$\int_0^1 \frac{1}{\sqrt{x}}dx$$

is convergent, we deduce from Corollary 9.67 that the improper integral $\int_0^1 f(x)dx$ is also convergent.

(c) Consider the function $f : [0, \infty) \to \mathbb{R}$, $f(x) = xe^{-x^2}$. Note that $f(x) \geq 0$, $\forall x$ and

$$\frac{f(x)}{\frac{1}{x^2}} = x^3 e^{-x^2} = \frac{x^2}{e^{x^2}} \to 0 \ \text{ as } x \to \infty.$$

Thus the integral

$$\int_0^\infty xe^{-x^2} dx$$

is convergent. To evaluate this integral we begin by evaluating the integrals

$$\int_0^L xe^{-x^2} dx$$

where $L \to \infty$. We use the change in variables $u = x^2$ so that $du = 2xdx$

$$x = 0 \Rightarrow u = 0, \ \ x = L \Rightarrow u = L^2$$

and we deduce

$$\int_0^L xe^{-x^2} dx = \frac{1}{2}\int_0^L e^{-x^2}(2xdx) = \frac{1}{2}\int_0^{L^2} e^{-u}du = \frac{1}{2}\Big(-e^{-u}\Big)\Big|_{u=0}^{u=L^2} = \frac{1}{2}(1-e^{-L^2}).$$

Now observe that

$$\lim_{L\to\infty} \frac{1}{2}(1 - e^{-L^2}) = \frac{1}{2},$$

so that

$$\int_0^\infty xe^{-x^2} dx = \frac{1}{2}.$$

So far we have investigated improper integrals of functions that had a problem at ω, one of the endpoints of its domain: either $\omega = \infty$, or the function "explodes" as it approaches ω. Sometime we need to deal with functions that have problems at both endpoints of its domain. The next example explains how to proceed in this case.

(d) Consider the function

$$f : (-1, 1) \to \mathbb{R}, \ \ f(x) = \frac{1}{\sqrt{(1 - x^2)}}.$$

To decide the convergence of the integral

$$\int_{-1}^1 f(x)dx,$$

we must first locate the sources of the possible problems. We note that $f(x)$ "explodes" as $x \to \pm 1$, i.e.,

$$\lim_{x\to\pm 1} f(x) = \infty.$$

We split the integral into two parts,

$$I_{-1} = \int_{-1}^{0} f(x)dx, \;\; I_1 = \int_0^1 f(x)dx.$$

Each of the above integrals has only one problem point and, if both integrals are convergent, then the original integral will be convergent if and only if both integrals above are convergent and, when this happens, we have

$$\int_{-1}^{1} f(x)dx = \int_{-1}^{0} f(x)dx + \int_0^1 f(x)dx.$$

Now observe that

$$f(x) = \frac{1}{\sqrt{(1-x)(1+x)}}.$$

The term $(1-x)$ is responsible for the bad behavior near $x = 1$, while the term $(1+x)$ is responsible for the bad behavior near $x = -1$.

From Example 9.62 we deduce that both integrals

$$\int_{-1}^{0} \frac{1}{\sqrt{1+x}}dx, \;\; \int_0^1 \frac{1}{\sqrt{1-x}}$$

are convergent. Observe next that

$$\lim_{x\to -1} \frac{f(x)}{\frac{1}{\sqrt{1+x}}} = \lim_{x\to -1} \frac{\frac{1}{\sqrt{(1-x)(1+x)}}}{\frac{1}{\sqrt{1+x}}} = \lim_{x\to -1} \frac{1}{\sqrt{1-x}} = \frac{1}{\sqrt{2}},$$

$$\lim_{x\to 1} \frac{f(x)}{\frac{1}{\sqrt{1-x}}} = \lim_{x\to 1} \frac{\frac{1}{\sqrt{(1-x)(1+x)}}}{\frac{1}{\sqrt{1-x}}} = \lim_{x\to 1} \frac{1}{\sqrt{1+x}} = \frac{1}{\sqrt{2}}.$$

Using Corollary 9.67 we now deduce that both integrals $I_{\pm 1}$ are convergent. In particular, we deduce that the improper integral

$$\int_{-1}^{1} f(x)dx$$

is convergent. We can actually compute it. Let $-1 < a < 0 < b < 1$. We have

$$\int_a^b \frac{1}{\sqrt{1-x^2}}dx = \arcsin x \Big|_{x=a}^{x=b} = \arcsin b - \arcsin a.$$

Note that

$$\lim_{b\nearrow 1} \arcsin b = \arcsin 1 = \frac{\pi}{2}, \quad \lim_{a\searrow -1} \arcsin a = \arcsin(-1) = -\frac{\pi}{2}$$

so that

$$\int_{-1}^{1} \frac{1}{\sqrt{1-x^2}}dx = \frac{\pi}{2} - \left(-\frac{\pi}{2}\right) = \pi. \tag{9.69}$$

Definition 9.69. Let $-\infty < a < \omega \leq \infty$ and $f : [a, \omega) \to \mathbb{R}$ be a function that is Riemann integrable on any interval $[a, x]$, $x \in (a, \omega)$. We say that the improper integral

$$\int_a^\omega f(x)dx$$

is *absolutely convergent* if the improper integral

$$\int_a^\omega |f(x)|dx$$

is convergent. □

The next result is very similar to Theorem 4.46.

Proposition 9.70. *Let $-\infty < a < \omega \leq \infty$ and $f : [a, \omega) \to \mathbb{R}$ be a function that is Riemann integrable on any interval $[a, x]$, $x \in (a, \omega)$. Then*

$$\int_a^\omega f(x)dx \ \textit{absolutely convergent} \Rightarrow \int_a^\omega f(x)dx \ \textit{convergent}.$$

Proof. We rely on Cauchy's Theorem 9.64. Since the integral

$$\int_a^\omega |f(x)|dx$$

is convergent we deduce from Cauchy's Theorem that for any $\varepsilon > 0$ there exists $c(\varepsilon) \in (a, \omega)$ such that

$$\forall x, y; \quad x, y \in (c(\varepsilon), \omega) \Rightarrow \left| \int_x^y |f(t)|dt \right| < \varepsilon.$$

On the other hand, (9.29) shows that

$$\left| \int_x^y f(t)dt \right| \leq \left| \int_x^y |f(t)|dt \right|$$

and we deduce that

$$\forall x, y, \quad x, y \in (c(\varepsilon), \omega) \Rightarrow \left| \int_x^y f(t)dt \right| < \varepsilon.$$

Cauchy's Theorem now implies that

$$\int_a^\omega f(x)dx$$

is convergent. □

The Comparison Principle (Corollary 9.65) yields a Comparison Principle involving absolute convergence.

Corollary 9.71 (Comparison Principle). *Let $-\infty < a < \omega \leq \infty$ and suppose that $f, g : [a, \omega) \to \mathbb{R}$ are two real functions satisfying the following properties.*

(i) $\exists b \in [a, \omega)$, *such that* $|f(x)| \leq |g(x)|$, $\forall x \in [b, \omega)$.
(ii) For any $x \in [a, \omega)$ *the restrictions of* f, g *to* $[a, x]$ *are Riemann integrable.*

Then

$$\int_a^\omega g(x)dx \text{ is absolutely convergent } \Rightarrow \int_a^\omega f(x)dx \text{ is absolutely convergent.} \quad \square$$

Example 9.72. Consider the function

$$f : [1, \infty) \to \mathbb{R}, \;\; f(x) = \frac{\sin x}{x^2}.$$

Note that

$$|f(x)| \leq \frac{1}{x^2}, \;\; \forall x \geq 1$$

and since $\int_1^\infty \frac{1}{x^2}dx$ is convergent we deduce that $\int_a^\infty f(x)dx$ is absolutely convergent. $\square$

9.7.1. *Euler's Gamma and Beta functions*

For every $x > 0$ we set

$$\boxed{\Gamma(x) := \int_0^\infty t^{x-1}e^{-t}dt}. \tag{9.70}$$

For each fixed $x > 0$ this improper integral is convergent. To see this we split the above integral into two parts

$$I_0 = \int_0^1 t^{x-1}e^{-t}dt, \;\; I_\infty = \int_1^\infty t^{x-1}e^{-t}dt.$$

To prove the convergence of I_0 we observe that

$$0 < t^{x-1}e^{-t} \leq t^{x-1} \;\; \forall t \in (0, 1].$$

Since $x - 1 > -1$ the improper integral

$$\int_0^1 t^{x-1}dt$$

is convergent. The Comparison Principle then implies that I_0 is also convergent.

To prove the convergence of I_∞ we observe that and as $t \to \infty$ the function $t^{x-1}e^{-t}$ decays to zero faster, than any power t^{-n}, $n \in \mathbb{N}$. In particular

$$\lim_{t\to\infty} \frac{t^{x-1}e^{-t}}{t^{-2}}dt = 0.$$

Since the integral

$$\int_1^\infty t^{-2}dt$$

is convergent we deduce from the Comparison Principle that I_∞ is convergent as well.

The resulting function

$$(0,\infty) \ni x \mapsto \Gamma(x) \in (0,\infty)$$

is called *Euler's Gamma function*. Observe that

$$\Gamma(1) = \int_0^\infty e^{-t}dt = \left(-e^{-t}\right)\Big|_{t=0}^{t=\infty} = 1, \tag{9.71}$$

and, for $x > 0$,

$$\begin{aligned}\Gamma(x+1) &= \int_0^\infty t^x e^{-t}dt = -\int_0^\infty t^x d(e^{-t}) \\ &= \underbrace{-\left(t^x e^{-t}\right)\Big|_{t=0}^{\infty}}_{=0} + x\underbrace{\int_0^\infty t^{x-1}e^{-t}dt}_{=\Gamma(x)} = x\Gamma(x),\end{aligned}$$

so that

$$\boxed{\Gamma(x+1) = x\Gamma(x), \ \ \forall x > 0.} \tag{9.72}$$

From (9.71) and (9.72) we deduce inductively

$$\Gamma(2) = 1\Gamma(1) = 1, \ \Gamma(3) = 2\Gamma(2) = 2, \ \Gamma(4) = 3\Gamma(3) = 3\cdot 2 = 3!, \dots,$$

$$\boxed{\Gamma(n) = (n-1)!, \ \ \forall n \in \mathbb{N}.} \tag{9.73}$$

9.8. Length, area and volume

The concept of integral is involved in the definition of important geometric quantities such as length, area and volume. Their definition in the most general context is quite involved and we restrict ourselves to special cases that still have a wide range of applications.

9.8.1. *Length*

We will define the length of special curves in the plane, namely the curves defined by the graphs of differentiable functions.

Definition 9.73. Suppose that $-\infty \leq a < b \leq \infty$ and $f : (a,b) \to \mathbb{R}$ is a C^1-function. We say that its graph has *finite length* if the integral

$$\int_a^b \sqrt{1+f'(x)^2}dx$$

is convergent. The value of this integral is then declared to be the *length of the graph* Γ_f *of* f. We write this

$$\operatorname{length}(\Gamma_f) \int_a^b \sqrt{1+f'(x)^2}dx. \tag{9.74}$$

□

Here is the intuition behind the definition. If we are located at the point $(x_0, y_0) = (x_0, f(x_0))$ on the graph of f and we move a tiny bit, from x_0 to $x_0 + dx$, then the rise, that is the change in altitude is

$$dy = \frac{dy}{dx} \cdot dx = f'(x_0)dx.$$

The Pythagorean Theorem then shows that the distance covered along the graph is approximately

$$\sqrt{dx^2 + dy^2} = \sqrt{dx^2 + f'(x_0)^2 dx^2} = \sqrt{1 + f'(x_0)^2}\, dx.$$

The total distance traveled along the graph, i.e., the length of the trap is obtained by summing all these infinitesimal distances

$$\int_a^b \sqrt{1 + f'(x)^2}\, dx.$$

The next examples support the validity of the proposed formula for the length.

Example 9.74. Consider two points in the plane, P_1 with coordinates (x_1, y_1) and P_2 with coordinates (x_2, y_2). Assume further that $x_1 < x_2$; see Figure 9.8. We want to compute the length $|P_1P_2|$ of the line segment connecting P_1 to P_2.

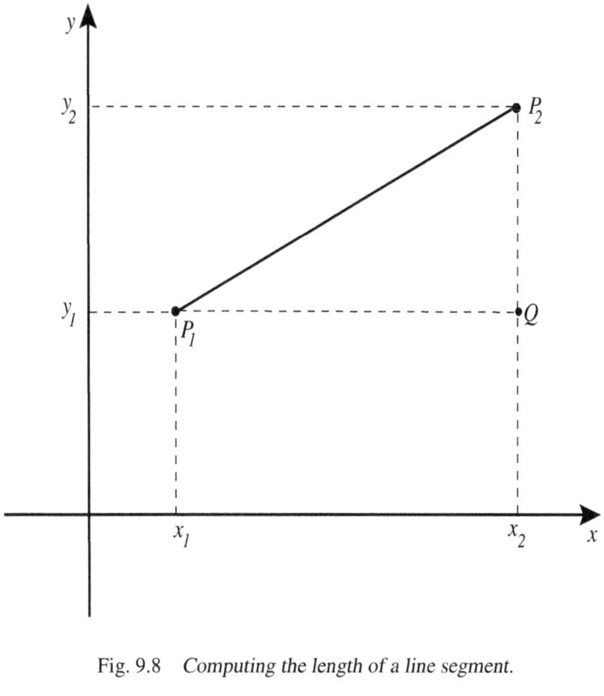

Fig. 9.8 *Computing the length of a line segment.*

Pythagoras' Theorem shows that

$$|P_1P_2|^2 = |P_1Q|^2 + |QP_2|^2 = (x_2 - x_1)^2 + (y_2 - y_1)^2. \tag{9.75}$$

Let us show that the formula proposed in Definition 9.73 yields the same result.

The line determined by the points P_1, P_2 has slope

$$m := \frac{y_2 - y_1}{x_2 - x_1},$$

and thus it is described by the equation

$$y = m(x - x_1) + y_1.$$

In other words, the line segment is the graph of the linear function

$$f : [x_1, x_2] \to \mathbb{R}, \quad f(x) = m(x - x_1) + y_1.$$

Note that $f'(x) = m$, $\forall x \in [x_1, x_2]$ and, according to Definition 9.73, we have

$$|P_1P_2| = \int_{x_1}^{x_2} \sqrt{1 + f'(x)^2} dx = \int_{x_1}^{x_2} \sqrt{1 + m^2} dx = \sqrt{1 + m^2}(x_2 - x_1).$$

Hence

$$|P_1P_2|^2 = (1+m^2)(x_2-x_1)^2 = \left(1 + \frac{(y_2 - y_1)^2}{(x_2 - x_1)^2}\right)(x_2-x_1)^2 = (x_2-x_1)^2+(y_2-y_1)^2.$$

This agrees with the Pythagorean prediction (9.75). □

Example 9.75. Consider the function

$$f : (-1, 1) \to \mathbb{R}, \quad f(x) = \sqrt{1 - x^2}.$$

The graph of this function is the upper semi-circle of radius 1 centered at the origin; see Figure 9.9. Indeed, a point (x, y) on this circle satisfies

$$x^2 + y^2 = 1, \quad y \geq 0 \Longleftrightarrow y = \sqrt{1 - x^2}.$$

The function $f(x)$ is differentiable on $(-1, 1)$ and we have

$$f'(x) = -\frac{x}{\sqrt{1 - x^2}}, \quad 1 + f'(x)^2 = 1 + \frac{x^2}{1 - x^2} = \frac{1}{1 - x^2}, \quad \forall x \in (-1, 1). \tag{9.76}$$

Hence the length of this semi-circle is

$$\int_{-1}^{1} \frac{1}{\sqrt{1 - x^2}} dx \overset{(9.69)}{=} \pi. \qquad \square$$

We can define the length of more complicated curves.

Definition 9.76. Let $-\infty < a < b \leq \infty$. A *continuous* function $(a, b) \to \mathbb{R}$ is called *piecewise* C^1 if there exist points $x_1, \dots, x_n \in (a, b)$ such that

$$a < x_1 < x_2 < \cdots < x_n < b$$

and the function f is C^1 on each of the subintervals

$$(a, x_1), \ (x_1, x_2), \dots, (x_n, b).$$

The length of its graph is then given by

$$\int_a^{x_1} \sqrt{1 + f'(x)^2}\, dx + \int_{x_1}^{x_2} \sqrt{1 + f'(x)^2}\, dx + \cdots + \int_{x_n}^{b} \sqrt{1 + f'(x)^2}\, dx.$$

Above, some of the integrals could be improper and for the length to be finite these integrals have to be convergent. □

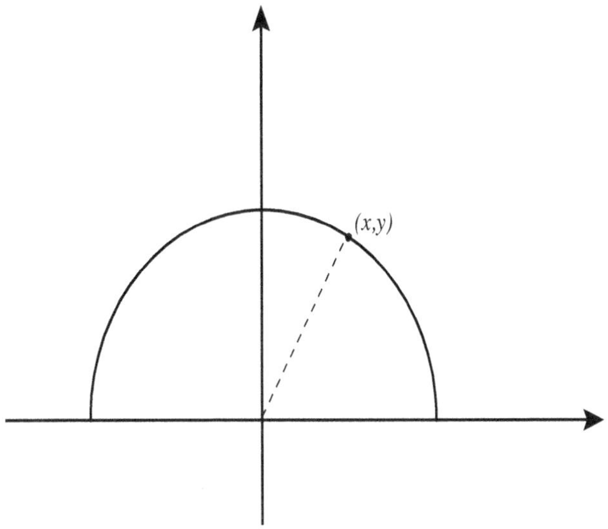

Fig. 9.9 *Computing the length of a semi-circle.*

9.8.2. *Area*

A region D of the cartesian plane $\mathbb{R}^2$ is said to be of *simple type* with respect to the x-axis if there exists an interval I and functions

$$F, C : I \to \mathbb{R}$$

such that

$$F(x) \leq C(x), \ \ \forall x \in I,$$

and

$$(x, y) \in D \Longleftrightarrow x \in I \ \wedge \ F(x) \leq y \leq C(x).$$

The function F is called the *floor* of the region D, while the function C is called the *ceiling* of the region; see Figure 9.10.

The *area* of the region D is given by the improper integral

$$\text{Area}\,(D) := \int_I \big(C(x) - F(x) \big) \, dx,$$

whenever this integral is well defined[3] and convergent.

A region D of the cartesian plane $\mathbb{R}^2$ is said to be of *simple type* with respect to the y-axis if there exists an interval J and functions

$$L, R : J \to \mathbb{R}$$

[3]The integral is well defined if the function $C(x) - F(x)$ is Riemann integrable on any compact interval $[\alpha, \beta] \subset (a, b)$.

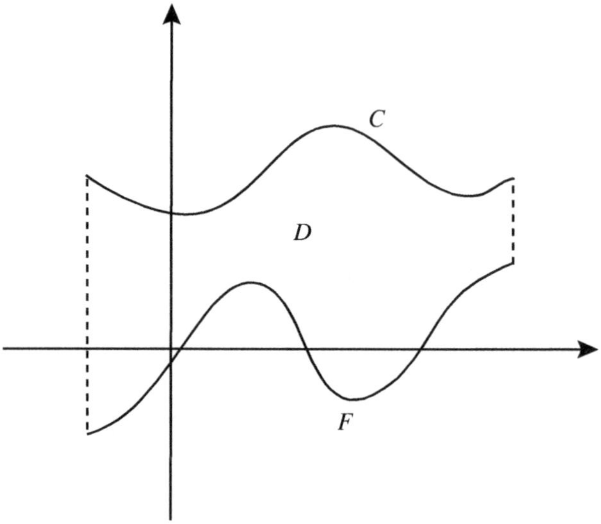

Fig. 9.10 *A planar region of simple type with respect to the x-axis.*

such that

$$L(y) \leq R(y), \ \ \forall y \in J$$

and

$$(x, y) \in R \Longleftrightarrow y \in J \ \wedge \ L(y) \leq x \leq R(y).$$

The function L is called the *left wall* of the region D, while the function R is called the *right wall* of the region; see Figure 9.11.

The *area* of the region D is given by the improper integral

$$\text{Area}\,(D) := \int_J \big(\, R(y) - L(y) \,\big)\, dy,$$

whenever this integral is well defined.

Remark 9.77. (a) We swept under the rug a rather subtle fact. A region in the plane can be simultaneously simple type with respect to the x-axis, and simple type with respect to the y-axis. In such situations there are two possible ways of computing the area and they had better produce the same result. This is indeed the case, but the proof in general is quite complicated, and the best approach relies on the concept of multiple integrals.

To see that this is not merely a theoretical possibility, consider the region (see Figure 9.12)

$$R = \big\{ (x, y) \in \mathbb{R}^2; \ \ x \in [0, 1], \ \ x^2 \leq y \leq x \,\big\}.$$

The above description shows that R is a region of simple type with respect to the x-axis. However, R can be given the alternate description as a region of simple type with respect to the y-axis,

$$R = \big\{ (x, t) \in \mathbb{R}^2; \ \ y \in [0, 1], \ \ y \leq x \leq \sqrt{y} \,\big\}.$$

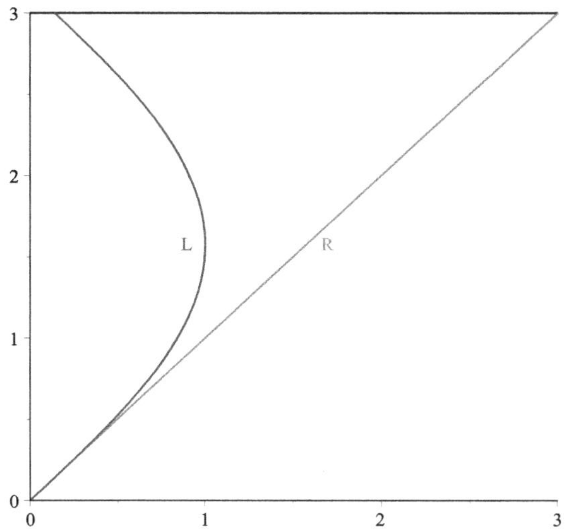

Fig. 9.11 *A planar region of simple type with respect to the y-axis,* $\sin y \leq x \leq y, 0 \leq y \leq 3$.

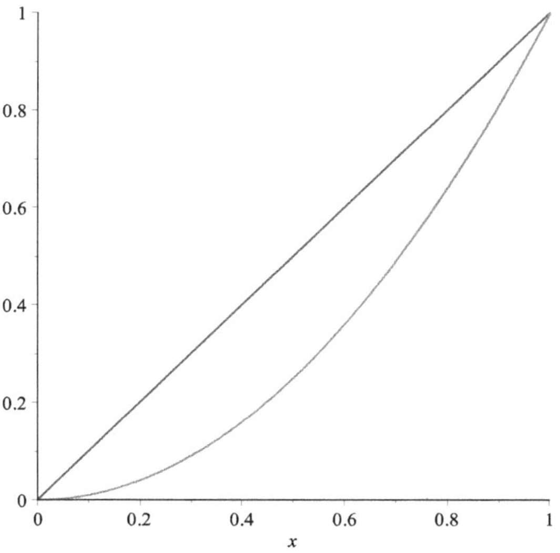

Fig. 9.12 *A planar region of simple type with respect to both axes:* $x^2 \leq y \leq x, 0 \leq x \leq 1$.

If we use the first description we deduce

$$\text{Area}\,(R) = \int_0^1 (x - x^2)dx = \left(\frac{x^2}{2} - \frac{x^3}{3}\right)\Big|_{x=0}^{x=1} = \frac{1}{2} - \frac{1}{3} = \frac{1}{6}.$$

If we use the second description we deduce

$$\text{Area}\,(R) = \int_0^1 (\sqrt{y} - y)dy = \left(\frac{2x^{3/2}}{3} - \frac{x^2}{2}\right)\Big|_{x=0}^{x=1} = \frac{2}{3} - \frac{1}{2} = \frac{1}{6}.$$

Many regions in the plane decompose into finitely many simple type regions that have overlaps only along boundary curves. For such a region, the area is defined as the sum of the areas of the simple-type sub-regions it decomposes into. This raises an even trickier question: why is the answer independent of the procedure we use to decompose the region into simple-type sub-regions? To answer this question one needs the full apparatus of multiple integrals.

(b) Let us observe that a simple-type region can have finite area, even if it is unbounded. Consider for example the region between the x-axis and the graph of the function

$$g : [0, \infty) \to \mathbb{R}, \;\; g(x) = e^{-x}.$$

The area of this region is

$$\int_0^\infty e^{-x} dx = \left(-e^{-x}\right)\Big|_0^\infty = -e^{-\infty} - (-1) = 0 + 1 = 1. \qquad \square$$

9.8.3. *Solids of revolution*

Suppose that we are given an open interval (a, b) and a function

$$g : (a, b) \to \mathbb{R}$$

called *generatrix* such that $g(x) \geq 0$, $\forall x \in (a, b)$. If we rotate the graph of g about the x-axis we get a surface of revolution Σ_g that surrounds a solid of revolution S_g; see Figure 9.13.

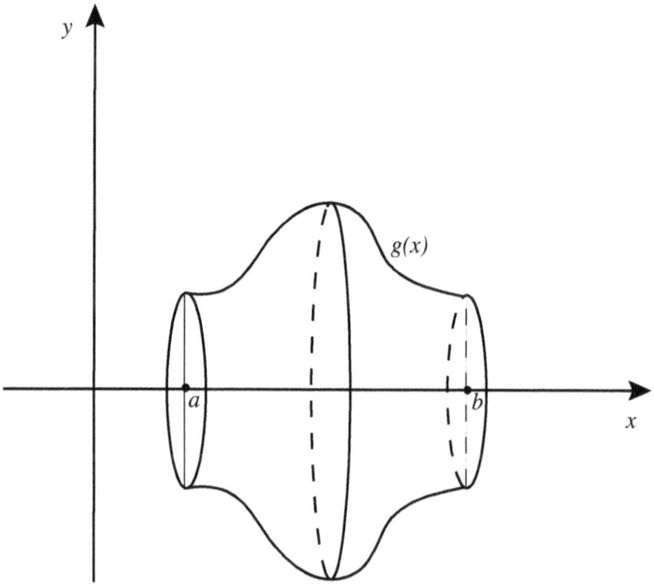

Fig. 9.13 *A surface of revolution.*

The *area* of the surface of revolution Σ_g is given by the improper integral

$$\boxed{\operatorname{area}(\Sigma_g) := 2\pi \int_a^b g(x)\sqrt{1+g'(x)^2}dx}, \tag{9.77}$$

whenever the integral is well defined. The *volume* of the solid of revolution S_g is given by the improper integral

$$\boxed{\operatorname{vol}(S_g) := \pi \int_a^b g(x)^2 dx}, \tag{9.78}$$

whenever the integral is well defined.

Example 9.78. (a) Suppose that the generatrix is the function $g : (-1,1) \to \mathbb{R}$, $g(x) = \sqrt{1-x^2}$. Its graph is the upper semi-circle of radius 1 depicted in Figure 9.9. When we rotate this semi-circle about the x-axis, the surface of revolution obtained is a sphere Σ_g of radius 1 that surrounds a solid ball S_g of radius 1.

The computations in (9.76) show that

$$\sqrt{1+g'(x)^2} = \frac{1}{\sqrt{1-x^2}}$$

so that

$$g(x)\sqrt{1+g'(x)^2} = 1.$$

We deduce that the area of the unit sphere is

$$2\pi \int_{-1}^{1} g(x)\sqrt{1+g'(x)^2}dx = 2\pi \in_{-1}^{1} dx = 4\pi.$$

The volume of the unit ball is

$$\pi \int_{-1}^{1} g(x)^2 dx = \pi \int_{-1}^{1} (1-x^2)dx = \pi \left(x\Big|_{-1}^{1} - \frac{x^3}{3}\Big|_{-1}^{1} \right) = \pi\Big(2 - \frac{2}{3}\Big) = \frac{4\pi}{3}.$$

These equalities confirm the classical formulæ taught in elementary solid geometry.

(b) Consider the cone depicted in Figure 9.14. It is obtained by rotating a line segment about the x-axis, more precisely, the line segment connecting the point $(0,r)$ on the y-axis with the point $(h,0)$ on the x-axis. Here $h, r > 0$.

This line segment lies on the line with slope $m = -r/h$ and y-intercept r. In other words, this line is given by the equation

$$g(x) = -\frac{r}{h}x + r.$$

Observe that

$$g'(x) = -\frac{r}{h}, \quad \sqrt{1+g'(x)^2} = \frac{\sqrt{h^2+r^2}}{h},$$

$$g(x)\sqrt{1+g'(x)^2} = \frac{h^2+r^2}{h}\Big(-\frac{r}{h}x + r\Big).$$

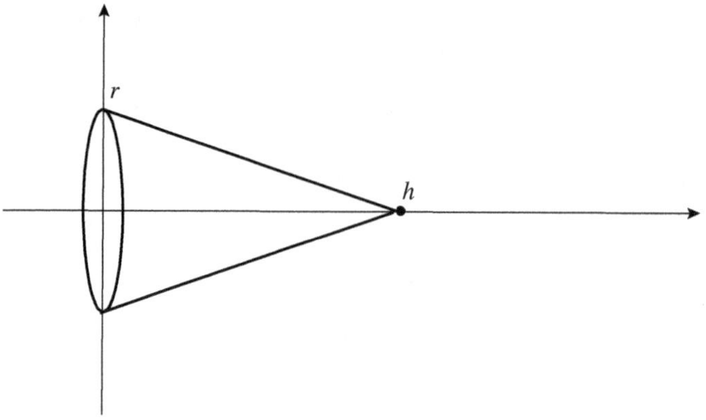

Fig. 9.14 A cone.

We deduce that the area of this cone (excluding its base) is

$$2\pi\int_0^h \frac{\sqrt{h^2+r^2}}{h}\left(-\frac{r}{h}x+r\right)dx = 2\pi\frac{\sqrt{h^2+r^2}}{h}\int_0^h\left(-\frac{r}{h}x+r\right)dx$$
$$= 2\pi r\frac{\sqrt{h^2+r^2}}{h}\int_0^h dx - 2\pi r\frac{\sqrt{h^2+r^2}}{h^2}\int_0^h xdx$$
$$= 2\pi r\sqrt{h^2+r^2} - \pi r\sqrt{h^2+r^2} = \pi r\sqrt{h^2+r^2}.$$

This agrees with the known formulæ in solid geometry.

The volume of the cone is

$$\pi\int_0^h\left(-\frac{r}{h}x+r\right)^2 dx = \frac{\pi r^2}{h^2}\int_0^h (h-x)^2dx = \frac{\pi r^2}{h^2}\times\frac{h^3}{3} = \frac{\pi r^2 h}{3}.$$

(c) Let $\alpha \in (\frac{1}{2}, 1)$ and consider the function

$$g : [1,\infty) \to \mathbb{R},\ \ g(x) = \frac{1}{x^\alpha}.$$

The surface of revolution obtained by rotating the graph of g about the x-axis has the bugle shape in Figure 9.15. The volume of this bugle is

$$\pi\int_1^\infty g(x)^2dx = \pi\int_1^\infty \frac{1}{x^{2\alpha}}dx.$$

Since $2\alpha > 1$, the above integral is convergent and in fact

$$\pi\int_1^\infty \frac{1}{x^{2\alpha}}dx = \frac{\pi}{2\alpha-1}.$$

On the other hand, the area of the bugle is

$$2\pi\lim_{L\to\infty}\int_1^L g(x)\sqrt{1+g'(x)^2}dx \geq 2\pi\lim_{L\to\infty}\int_1^L g(x)dx$$
$$= 2\pi\lim_{L\to\infty}\int_1^L\frac{1}{x^\alpha}dx = 2\pi\lim_{L\to\infty}\left(\frac{x^{1-\alpha}}{1-\alpha}\right)\Big|_1^L = \infty,$$

because $\alpha < 1$. This is surprising: you need a finite amount of water to fill the bugle, but an infinite amount of paint if you want to paint it!!! □

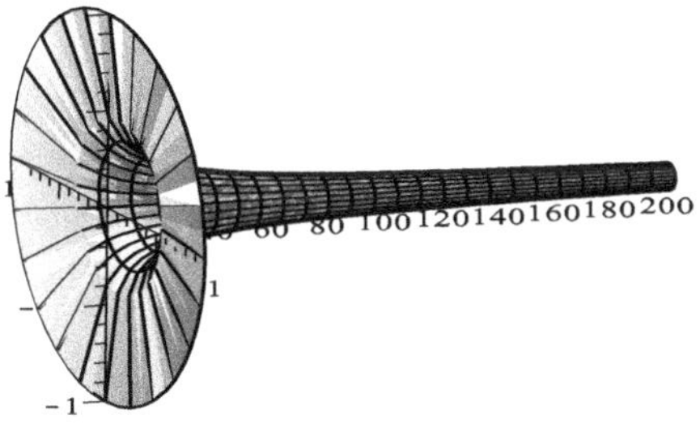

Fig. 9.15 *An infinite bugle.*

9.9. Exercises

Exercise 9.1. Prove by induction the equality (9.3). □

Exercise 9.2. Consider the function $f:[0,4]\to\mathbb{R}$, $f(x)=x^2$, and the partition

$$\boldsymbol{P}=(0,\ 0.5,\ 1,\ 1.5,\ 2,\ 2.5,\ 3,\ 3.5,\ 4)$$

of the interval $[0,4]$.
(a) Find the mesh size $\|\boldsymbol{P}\|$ of $\boldsymbol{P}$.
(b) Compute the Riemann sum $\boldsymbol{S}(f,\boldsymbol{P},\underline{\xi})$ when the sample $\underline{\xi}$ consists of the right endpoints of the subintervals of $\boldsymbol{P}$. □

Exercise 9.3. (a) Suppose that $f,g:[a,b]\to\mathbb{R}$ are two functions. Prove that for any sampled partition $(\boldsymbol{P},\underline{\xi})$ of $[a,b]$ and for any real numbers α,β we have

$$\boldsymbol{S}\big(\alpha f+\beta g,\boldsymbol{P},\underline{\xi}\big)=\alpha\boldsymbol{S}\big(f,\boldsymbol{P},\underline{\xi}\big)+\beta\boldsymbol{S}\big(g,\boldsymbol{P},\underline{\xi}\big).$$ □

(b) Let $f:[a,b]\to\mathbb{R}$. Prove that the following statements are equivalent.

(i) The function f is *not* Riemann integrable.
(ii) There exists ε_0 such that, for any $n\in\mathbb{N}$ there exists sampled partitions $(\boldsymbol{P}_n,\underline{\xi}^n)$ and $(\boldsymbol{Q}_n,\underline{\zeta}^n)$ satisfying

$$\|\boldsymbol{P}_n\|,\ \|\boldsymbol{Q}_n\|<\frac{1}{n}\ \text{ and }\ \big|\,\boldsymbol{S}(f,\boldsymbol{P}_n,\underline{\xi}^n)-\boldsymbol{S}(f,\boldsymbol{Q}_n,\underline{\zeta}^n)\,\big|>\varepsilon_0.$$

Hint. For the implication (ii) $\Rightarrow$ (i) use the Riemann–Darboux Theorem and the equality (9.10). □

Exercise 9.4. Consider the function $f:[-2,2]\to\mathbb{R}$, $f(x)=x^2$ and the partition

$$\boldsymbol{P}=(-2,\ -1.5,\ -1\ ,-0.5,\ 0,\ 0.5,\ 1,\ 1.5,\ 2)$$

of $[-2,2]$.

(a) Compute the upper and lower Darboux sums $\boldsymbol{S}^*(f, \boldsymbol{P})$, $\boldsymbol{S}_*(f, \boldsymbol{P})$.
(b) Compute $\omega(f, \boldsymbol{P})$. □

Exercise 9.5. Suppose that $f : [0, 1] \to \mathbb{R}$ is a C^1-function, i.e., it is differentiable on $[0, 1]$ and the derivative is continuous. We set

$$M := \sup_{x \in [0,1]} |f'(x)|.$$

(a) Suppose that $I \subset [0, 1]$ is an interval of length δ. Show that

$$\operatorname{osc}(f, I) \leq M\delta.$$

(b) For $n \in \mathbb{N}$ we denote by $\boldsymbol{U}_n$ the uniform partition of order n of $[0, 1]$. Show that

$$\omega(f, \boldsymbol{U}_n) \leq \frac{M}{n}, \quad \forall n \in \mathbb{N}.$$

(c) Fix $n \in \mathbb{N}$ and a sample $\underline{\xi}$ of $\boldsymbol{U}_n$. Show that

$$\left| \int_0^1 f(x)dx - \boldsymbol{S}(f, \boldsymbol{U}_n, \underline{\xi}) \right| \leq \frac{M}{n}.$$ □

Exercise 9.6. Let $a > 0$ and assume that $f : [-a, a] \to \mathbb{R}$ is a Riemann integrable function.
(a) Prove that if f is an *odd* function, i.e., $f(-x) = -f(x)$, $\forall x \in [-a, a]$, then

$$\int_{-a}^{a} f(x)dx = 0.$$

(b) Prove that if f is an *even* function, i.e., $f(-x) = f(x)$, $\forall x \in [-a, a]$, then

$$\int_{-a}^{a} f(x)dx = 2\int_0^a f(x)dx.$$ □

Exercise 9.7. (a) Suppose that $f, g : \mathbb{R} \to \mathbb{R}$ are two Lipschitz functions. Show that the composition $f \circ g$ is also Lipschitz.
(b) Suppose that the function $g : [a, b] \to \mathbb{R}$ is Riemann integrable and the function $f : \mathbb{R} \to \mathbb{R}$ is Lipschitz. Prove that $f \circ g$ is Riemann integrable.
(c) Suppose that the function $g : [a, b] \to \mathbb{R}$ is Riemann integrable and the function $f : \mathbb{R} \to \mathbb{R}$ is C^1, i.e., differentiable with continuous derivative. Prove that $f \circ g$ is Riemann integrable. □

Exercise 9.8. Suppose that the functions $f, g : [a, b] \to \mathbb{R}$ are Riemann integrable. Let $p, q > 1$ such that

$$\frac{1}{p} + \frac{1}{q} = 1.$$

(a) Prove that the functions $|f|^p$ and $|g|^q$ are Riemann integrable.

(b) Prove that

$$\int_a^b |f(x)g(x)|dx \leq \left(\int_a^b |f(x)|^p dx\right)^{\frac{1}{p}} \left(\int_a^b |g(x)|^q dx\right)^{\frac{1}{q}}.$$

(c) Prove that

$$\left(\int_a^b |f(x)+g(x)|^p dx\right)^{\frac{1}{p}} \leq \left(\int_a^b |f(x)|^p dx\right)^{\frac{1}{p}} + \left(\int_a^b |g(x)|^p dx\right)^{\frac{1}{p}}.$$

Hint. Approximate the integrals by Riemann sums and then use the inequalities (8.22) and (8.25). □

Exercise 9.9. (a) Suppose that $f : [a,b] \to \mathbb{R}$ is a continuous function such that $f(x) \geq 0$, $\forall x \in [a,b]$. Prove that

$$\int_a^b f(x)dx = 0 \iff f(x) = 0, \ \ \forall x \in [a,b].$$ □

(b) Show that for any $a < b$ there exists a continuous function $u : \mathbb{R} \to \mathbb{R}$ such that $u(x) > 0$, $\forall x \in (a,b)$, and $u(x) = 0$ $\forall x \in \mathbb{R} \setminus (a,b)$.

Hint. Think of a function u whose graph looks like a roof.

(c) Suppose that $f : [0,1] \to \mathbb{R}$ is a continuous function such that

$$\int_0^1 f(x)u(x)dx = 0,$$

for any continuous function $u : [0,1] \to \mathbb{R}$. Prove that $f(x) = 0$, $\forall x \in [0,1]$.

Hint. Argue by contradiction. Suppose that there exists $x_0 \in [0,1]$ such that $f(x_0) \neq 0$, say $f(x_0) > 0$. Reach a contradiction using Theorem 6.11, and the facts (a), (b) above. □

Exercise 9.10. Suppose that $f_n : [a,b] \to \mathbb{R}$, $n \in \mathbb{N}$ is a sequence of Riemann integrable functions that converges *uniformly* on $[a,b]$ to the function $f : [a,b] \to \mathbb{R}$. We set

$$d_n := \sup_{x \in [a,b]} |f(x) - f_n(x)|.$$

(a) (Compare with Exercise 6.6.) Prove that

$$\lim_{n \to \infty} d_n = 0.$$

(b) Let $X \subset [a,b]$ be a nonempty subset of $[a,b]$. Prove that, for any $n \in \mathbb{N}$, we have

$$\operatorname{osc}(f,X) \leq \operatorname{osc}(f_n,X) + 2d_n.$$

(c) Prove that, for any partition $\boldsymbol{P}$ of $[a,b]$, and any $n \in \mathbb{N}$, we have

$$\omega(f,\boldsymbol{P}) \leq \omega(f_n,\boldsymbol{P}) + 2d_n(b-a).$$

(d) Prove that f is Riemann integrable and

$$\lim_{n \to \infty} \int_a^b f_n(x)dx = \int_a^b f(x)dx.$$ □

Exercise 9.11. (a) Suppose that $f : [a,b] \to \mathbb{R}$ is a continuous and convex function. Prove that

$$\frac{1}{b-a}\int_a^b f(x)dx \leq \frac{f(a)+f(b)}{2}.$$

(b) Use (a) to show that for any $x > y > 0$ we have

$$\frac{1}{2y}\ln\frac{x+y}{x-y} \leq \frac{x}{x^2-y^2}.$$

□

Exercise 9.12. Consider the function $f : [0,1] \to \mathbb{R}$, $f(x) = \frac{1}{x+1}$.
(a) Compute $\int_0^1 f(x)dx$.
(b) For $n \in \mathbb{N}$ we denote by $\boldsymbol{U}_n$ the uniform partition of order n of $[0,1]$ and by $\underline{\xi}^{(n)}$ the sample of $\boldsymbol{U}_n$ given by

$$\underline{\xi}_k^{(n)} = \frac{k}{n},\ \ k = 1,\dots,n.$$

Describe explicitly the Riemann sum $\boldsymbol{S}(f, \boldsymbol{U}_n, \underline{\xi}^{(n)})$.
(c) Use parts (a) and (b) to compute the limit in Exercise 4.22. □

Exercise 9.13. Use Riemann sums for an appropriate Riemann integrable function to compute the limit

$$\lim_{n\to\infty}\frac{1}{\sqrt{n}}\left(\frac{1}{\sqrt{n+1}}+\frac{1}{\sqrt{n+2}}+\cdots+\frac{1}{\sqrt{2n}}\right).$$

□

Exercise 9.14. Fix a natural number k.
(a) Prove that for any $n \in \mathbb{N}$ we have

$$1^k + 2^k + \cdots + (n-1)^k \leq \int_0^n x^k dx \leq 1^k + 2^k + \cdots + n^k.$$

(b) Use (a) to prove that

$$\lim_{n\to\infty}\frac{1^k + 2^k + \cdots + n^k}{n^{k+1}} = \frac{1}{k+1}.$$

□

Exercise 9.15. Consider the function

$$F : [0,\infty) \to \mathbb{R},\ \ F(x) = \int_0^{\sqrt{x}} e^{\frac{t^2}{2}}dt.$$

Show that $F(x)$ is differentiable on $(0,\infty)$ and then compute $F'(x)$, $x > 0$. □

Exercise 9.16. Suppose $f_n : [a,b] \to \mathbb{R}$, $n \in \mathbb{N}$, is a sequence of C^1-functions with the following properties.

(i) The sequence of derivatives $f_n' : [a,b] \to \mathbb{R}$ converges *uniformly* to a function $g : [a,b] \to \mathbb{R}$.
(ii) The sequence $f_n : [a,b] \to \mathbb{R}$ converges *pointwisely* to a function $f : [a,b] \to \mathbb{R}$.

Prove that the following hold.

(a) The sequence $f_n : [a,b] \to \mathbb{R}$ converges *uniformly* to $f : [a,b] \to \mathbb{R}$.

Hint. Define $G : [a,b] \to \mathbb{R}$, $G(x) = f(a) + \int_a^x g(t)dt$. (The function g is continuous since it is a uniform limit of continuous functions.) Since f_n' is continuous, the Fundamental Theorem of Calculus shows that

$$f_n(x) = f_n(a) + \int_a^x f_n'(t)dt.$$

Then

$$f_n(x) - G(x) = f_n(a) - f(a) + \int_a^x \left(f_n'(t) - g(t) \right) dt.$$

Use the above equality to show that the sequence f_n converges uniformly on $[a,b]$ to G. Argue next that $G = f$.

(b) The function f is C^1 and $f' = g$, i.e., the sequence $f_n' : [a,b] \to \mathbb{R}$ converges uniformly to f'. □

Exercise 9.17. Let $L > 0$. Suppose that the power series with real coefficients

$$a_0 + a_1 x + a_2 x^2 + \cdots$$

converges absolutely for any $|x| < L$. For every $x \in (-L, L)$ we denote by $s(x)$ the sum of the above series.

(a) Show that the function $x \mapsto s(x)$ is continuous on $(-L, L)$ and, for any $R \in (0, L)$, we have

$$\int_0^R s(x)dx = a_0 R + \frac{a_1}{2} R^2 + \frac{a_2}{3} R^3 + \cdots.$$

Hint. Use the Exercises 6.14 and 9.10.

(b) Prove that the power series

$$a_1 + 2a_2 x + 3a_3 x^2 + \cdots$$

also converges absolutely for any $|x| < L$.

(c) Prove that $s(x)$ is differentiable on $(-L, L)$ and that

$$s'(x) = a_1 + 2a_2 x + 3a_3 x^2 + \cdots, \quad \forall |x| < L.$$

Hint. Use the Exercises 6.14 and 9.16. □

Exercise 9.18. Consider the power series

$$x - \frac{x^3}{3!} + \frac{x^5}{5!} - \frac{x^7}{7!} + \cdots,$$

and respectively,

$$1 - \frac{x^2}{2!} + \frac{x^4}{4!} - \frac{x^6}{6!} + \cdots.$$

(a) Prove that the above series converge absolutely for any $x \in \mathbb{R}$. Denote their sums by $a(x)$ and respectively $b(x)$.

(b) Show that the functions $a, b : \mathbb{R} \to \mathbb{R}$ are differentiable and satisfy the equalities

$$a'(x) = b(x), \quad b'(x) = -a(x).$$

Hint. Use Exercise 9.17.

(c) Show that $a(x)$ is the unique solution of the differential equation

$$a''(x) + a(x) = 0, \quad \forall x \in \mathbb{R}$$

satisfying the condition $a(0) = 0$, $a'(0) = 1$. (Compare with Exercise 7.16.) □

Exercise 9.19. Consider the function

$$f : \mathbb{R} \to \mathbb{R}, \ \ f(x) = \frac{1}{1+x^2}.$$

(a) Prove that

$$f(x) = \sum_{n=0}^{\infty} (-1)^n x^{2n}, \ \ \forall |x| < 1.$$

(b) Conclude from (a) that the Taylor series of $f(x)$ at $x_0 = 0$ is

$$1 - x^2 + x^4 - x^6 + \cdots.$$

Hint. Use Exercise 9.17.

(c) Deduce from (a) that

$$\arctan x = \sum_{k=0}^{\infty} (-1)^k \frac{x^{2k+1}}{2k+1} = x - \frac{x^3}{3} + \frac{x^5}{5} - \frac{x^7}{7} + \cdots, \ \ \forall |x| < 1.$$ □

Exercise 9.20. (a) Suppose that $f, w : [a, b] \to \mathbb{R}$ are two continuous functions satisfying the following properties.

(i) The function f is continuous.
(ii) The function w is Riemann integrable and nonnegative, i.e., $w(x) \geq 0$, $\forall x \in [a, b]$.
(iii) The integral

$$W := \int_a^b w(x)dx$$

is strictly positive.

We set

$$m := \inf_{x \in [a,b]} f(x), \ \ M := \sup_{x \in [a,b]} f(x).$$

Show that

$$m \leq \frac{1}{W} \int_a^b f(x)w(x)dx \leq M,$$

and then conclude that there exists a point ξ in the *open* interval (a, b) such that

$$f(\xi) = \frac{1}{W} \int_a^b f(x)w(x)dx.$$ □

(b) Use the result in (a) to show that the Integral Remainder Formula (9.50) implies the Lagrange Remainder Formula (8.2). □

Exercise 9.21. Consider the function $f : [0, 2] \to \mathbb{R}$, $f(x) = 1 - |x - 1|$, $\forall x \in [0, 2]$.
(a) Sketch the graph of f.
(b) Compute $\int_0^2 f(x)dx$. □

Exercise 9.22. For any natural number n we define the n-th *Legendre polynomial* to be

$$P_n(x) := \frac{1}{2^n n!}\frac{d^n}{dx^n}\left(x^2-1\right)^n.$$

We set $P_0(x) = 1, \forall x$.
(a) Compute $P_1(x), P_2(x), P_3(x)$.
(b) Compute

$$\int_{-1}^{1} P_1(x)^2 dx,\ \int_{-1}^{1} P_2(x)^2 dx,\ \int_{-1}^{1} P_3(x)^2 dx,\ \int_{-1}^{1} P_1(x)P_2(x) dx,$$

(c) Use integration-by-parts to compute

$$\int_{-1}^{1} P_m(x)P_n(x) dx,\ \int_{-1}^{1} P_n(x)^2 dx,\ \ m,n\in\mathbb{N},\ \ m\neq n.$$

Hint. You may want to use the results in Exercise 7.6 and Example 9.49. □

Exercise 9.23. Fix an integer k. Use Stirling's formula (9.58) to compute

$$\lim_{n\to\infty}\frac{\sqrt{2n}}{2^{2n}}\binom{2n}{n+k},\ \ \lim_{n\to\infty}\frac{\sqrt{2n+1}}{2^{2n+1}}\binom{2n+1}{n+k}.$$ □

Exercise 9.24. (a) For any integer $n\geq 0$ compute the numbers

$$\int_0^1 \sin^2(2\pi nt)dt\ \ \int_0^1 \cos^2(2\pi nt)dt.$$

(b) Consider the function

$$f:[0,1]\to\mathbb{R},\ \ f(x)=\frac{1}{2}-\left|x-\frac{1}{2}\right|.$$

Sketch its graph and then compute

$$\int_0^1 f^2(x)dx.$$

(c) Let f be as above. For any integer $n\geq 0$ compute the numbers

$$a_n=\int_0^1 f(x)\cos(2\pi nx)dx,\ \ b_n=\int_0^1 f(x)\sin(2\pi nx)dx.$$

(d) With a_n, b_n as in (c) prove that the series

$$\sum_{n\geq 1}(a_n^2+b_n^2)$$

is convergent.[4]

Hint. When computing the above integrals it is convenient to use the change in variables $u=x-\frac{1}{2}$, some of the trig identities in Section 5.6 and Exercise 9.6. □

[4]A nontrivial result in the theory of Fourier series shows that

$$\int_0^1 f^2(x)dx = a_0^2 + 2\sum_{n\geq 1}(a_n^2+b_n^2).$$

Exercise 9.25. Compute the area of the region depicted in Figure 9.11. □

Exercise 9.26. Prove Proposition 9.63. □

Exercise 9.27. Consider the function

$$f:[0,2]\to\mathbb{R},\;\; f(x)=\max\{2-x,x^2\}.$$

(a) Sketch the graph of the function.
(b) Compute the area of the region between the x-axis and the graph of f.
(c) Show that the function f is piecewise C^1 and then compute the length of its graph. □

Exercise 9.28. Prove that for any $a\in(-1,0)$ and any $b>0$ the integrals

$$\int_0^1 t^a|\ln t|^b dt,\;\; \int_1^\infty t^{a-1}|\ln t|^b dt$$

are convergent. □

Exercise 9.29. Prove that the Gamma function $\Gamma:(0,\infty)\to(0,\infty)$

$$\Gamma(x)=\int_0^\infty t^{x-1}e^{-t}dt$$

is continuous.

Hint. Fix $t>0$ and then use Lagrange's Mean Value Theorem for the function $f:(0,\infty)\to\mathbb{R}$, $f(x)=t^x$. Then use Exercise 9.28 to conclude. □

Exercise 9.30. Suppose that $f:[0,\infty)\to(0,\infty)$ is a decreasing function. Prove that the following statements are equivalent.

(i) The improper integral

$$\int_0^\infty f(x)dx$$

is convergent.

(ii) The series

$$f(0)+f(1)+f(2)+\cdots$$

is convergent.

□

9.10. More challenging problems

Problem 9.1. Suppose that $f:[a,b]\to\mathbb{R}$ is a continuous function and $\Phi:\mathbb{R}\to\mathbb{R}$ is a convex continuous[5] function. Prove *Jensen's inequality*

$$\Phi\left(\frac{1}{b-a}\int_a^b f(x)dx\right)\leq\frac{1}{b-a}\int_a^b\Phi\big(f(x)\big)dx. \tag{9.79}$$

□

[5] The continuity assumption is redundant since any convex function $\mathbb{R}\to\mathbb{R}$ is automatically continuous.

Problem 9.2. Show that the improper integrals

$$\int_0^\infty \frac{\sin x}{x} dx, \quad \int_0^\infty \sin(x^2) dx$$

are convergent. $\square$

Problem 9.3. Construct a continuous function $f : [0, \infty) \to \mathbb{R}$ satisfying the following properties.

(i) $f(x) \geq 0, \forall x \geq 0$.
(ii) $\sup_{x \geq 0} f(x) = \infty$.
(iii) The integral $\int_0^\infty f(x)dx$ is convergent.

$\square$

Problem 9.4. Suppose that $f : [0, \infty) \to \mathbb{R}$ is a C^2-function satisfying the following conditions

(i) $f'(0) = 0$.
(ii)

$$\lim_{x \to \infty} \frac{1}{\ln x} \big(f(x) + f'(x) \big) = 0.$$

Prove that for any $\alpha \in (0, 1)$ the integral $\int_0^\infty \frac{f'(x)}{x^\alpha} dx$ is convergent. $\square$

Problem 9.5. Suppose that $f : [1, \infty) \to \mathbb{R}$ is differentiable, the derivative $f' : [0, \infty) \to \mathbb{R}$ is increasing and

$$\lim_{x \to \infty} f'(x) = 0.$$

(For example $f(x) = \frac{1}{x}$ or $f(x) = \ln x$.) Prove that the sequence

$$S_n := \frac{1}{2} f(1) + f(2) + \cdots + f(n-1) + \frac{1}{2} f(n) - \int_1^n f(x)dx$$

is convergent and, if S is its limit, then for any $n \in \mathbb{N}$ we have

$$\frac{f'(n)}{n} < \frac{1}{2} f(1) + f(2) + \cdots + f(n-1) + \frac{1}{2} f(n) - \int_1^n f(x)dx - S < 0. \quad \square$$

Problem 9.6. Suppose that $f : [1, \infty) \to \mathbb{R}$ is differentiable, the derivative $f' : [0, \infty) \to \mathbb{R}$ is increasing and

$$\lim_{x \to \infty} f'(x) = \infty.$$

(For example, $f(x) = x^\alpha$, $\alpha > 1$.) Prove that there exists a constant $C > 0$ such that, for any $n \in \mathbb{N}$, we have

$$\left| \frac{1}{2} f(1) + f(2) + \cdots + f(n-1) + \frac{1}{2} f(n) - \int_1^n f(x)dx \right| \leq C |f'(n)|. \quad \square$$

Problem 9.7. (a) Suppose that $f : [a, b] \to [0, \infty)$ is a Riemann integrable function. For any natural numbers $k \leq n$ we set

$$\delta_n := \frac{b-a}{n}, \quad f_{n,k} = f\big(a + k\delta_n\big).$$

Prove that

$$\lim_{n\to\infty} \frac{1}{n}\sum_{k=1}^{n} f_{n,k} = \frac{1}{b-a}\int_a^b f(x)dx,$$

$$\lim_{n\to\infty}\left(\prod_{k=1}^{n} f_{n,k}\right)^{\frac{1}{n}} = \exp\left(\frac{1}{b-a}\int_a^b f(x)dx\right), \quad \exp(x) := e^x.$$

(b) Fix real numbers $c, r > 0$. Denote by A_n and G_n, the arithmetic and geometric, respectively, mean of the numbers

$$c + r, c + 2r, \ldots, c + nr.$$

Prove that

$$\lim_{n\to\infty} \frac{G_n}{A_n} = \frac{2}{e}.$$

□

Problem 9.8. Prove that the sequence

$$x_n = \frac{1^n + 2^n + \cdots + n^n}{n^{n+1}}$$

is convergent and then compute its limit. □

Chapter 10

Complex Numbers and Some of Their Applications

10.1. The field of complex numbers

It is well known that there exists no real number x such that $x^2 = -1$ because $x^2 \geq 0 > -1$, $\forall x \in \mathbb{R}$. Following L. Euler, we introduce an imaginary number $\boldsymbol{i}$ with the property that

$$\boldsymbol{i}^2 = -1. \tag{10.1}$$

Sometimes we write $\boldsymbol{i} = \sqrt{-1}$. The number $\boldsymbol{i}$ is called the *imaginary unit.* This bold and somewhat arbitrary move raises some troubling questions.

Can we really do this? Yes, we just did, by fiat. Where does the "number" $\boldsymbol{i}$ come from? As its name suggests, it comes from our imagination. Can we get into some sort of trouble? This vaguely formulated question is the more serious one, but let us just admit that we will not get in any trouble. This can be argued rigorously, but requires more advanced mathematics that did not even exist during Euler's time. It took more than a century to settle this issue. During that time mathematicians found convincing semi-rigorous arguments that this construction leads to no contradictions. As Euler and its followers, we will take it on faith that this construction will not lead us to shaky grounds.

What can we do with $\boldsymbol{i}$? Following Gauss, we define the *complex numbers.* These are quantities of the form

$$z := x + y\boldsymbol{i}, \quad x, y \in \mathbb{R}.$$

The *real part* of the complex number z is

$$\mathbf{Re}\, z := x,$$

while its *imaginary part* is

$$\mathbf{Im}\, z := y.$$

The set of all the complex numbers is denoted by $\mathbb{C}$.

The reason we are referring to the quantities $a+b\boldsymbol{i}$ as *numbers* is because we can operate with them, much like we do with real numbers. First, we can add complex numbers. If

$$z_1 := x_1 + y_1\boldsymbol{i}, \quad z_2 = x_2 + y_2\boldsymbol{i},$$

then we define

$$z_1 + z_2 = (x_1 + x_2) + (y_1 + y_2)\boldsymbol{i}.$$

This operation satisfies the same properties as the addition of real numbers, namely the Axioms 1–4 in Section 2.1. Note that the real numbers are special examples of complex numbers: they are the complex numbers whose imaginary part is zero.

We can also multiply complex numbers in a natural way, taking (10.1) into account. Thus

$$\begin{aligned}(x_1 + y_1\boldsymbol{i})(x_2 + y_2\boldsymbol{i}) &= x_1x_2 + x_1y_2\boldsymbol{i} + y_1x_2\boldsymbol{i} + y_1y_2\boldsymbol{i}^2\\ &= (x_1x_2 - y_1y_2) + (x_1y_2 + y_1x_2)\boldsymbol{i}.\end{aligned}$$

One can check that this multiplication is commutative, associative, and distributive with respect to the addition. Moreover, the real number 1 acts as a multiplicative unit for this operation as well, and every nonzero real number z has an inverse. The construction of the inverse requires a bit of ingenuity.

To a complex number $z = x + y\boldsymbol{i}$ we associate its conjugate

$$\bar{z} = x - y\boldsymbol{i}.$$

Observe that

$$z\bar{z} = (x + y\boldsymbol{i})(x - y\boldsymbol{i}) = x^2 - (y\boldsymbol{i})^2 = x^2 + y^2.$$

The quantity $\sqrt{x^2 + y^2}$ is called the *norm* of the complex number z and it is denoted by $|z|$,

$$|z| := \sqrt{x^2 + y^2}.$$

Thus

$$\bar{z}z = z\bar{z} = |z|^2.$$

In particular, if $z \neq 0$, then $|z| \neq 0$ and we have

$$\frac{1}{|z|^2}\bar{z} \cdot z = z \cdot \frac{1}{|z|^2}\bar{z} = 1.$$

Thus

$$z^{-1} = \frac{1}{z} = \frac{\bar{z}}{|z|^2}. \tag{10.2}$$

The operation of conjugation interacts well with the operations of addition and multiplication introduced above. More precisely,

$$\overline{z_1 + z_2} = \bar{z}_1 + \bar{z}_2, \quad \overline{z_1 z_2} = \bar{z}_1\bar{z}_2, \quad \forall z_1, z_2 \in \mathbb{C}. \tag{10.3}$$

Moreover

$$|z_1 z_2| = |z_1| \cdot |z_2|, \quad \forall z_1, z_2 \in \mathbb{C}. \tag{10.4}$$

The simple proofs of these equalities are left to you as an exercise.

10.1.1. *The geometric interpretation of complex numbers*

The complex numbers have a very useful geometric interpretation. More precisely, we identify the complex number $z = x + y\boldsymbol{i}$ with the point $Z = (x, y)$ in the Cartesian plane $\mathbb{R}^2$. In turn we can identify the point Z with its position vector $\overrightarrow{OZ}$. For this reason we will often refer to $\mathbb{C}$ as the *complex plane*.

Given two complex numbers $z_1 = x_1 + y_1\boldsymbol{i}$, $z_2 = x_2 + y_2\boldsymbol{i}$ represented in the plane by the position vectors $\overrightarrow{OZ_1}$ and $\overrightarrow{OZ_2}$, then their sum $z_3 = (x_1 + x_2) + (y_1 + y_2)\boldsymbol{i}$ is represented in the plane by the point Z_3 with position vector

$$\overrightarrow{OZ_3} = \overrightarrow{OZ_1} + \overrightarrow{OZ_2},$$

where the addition of vectors is performed via the parallelogram rule; see Figure 10.1.

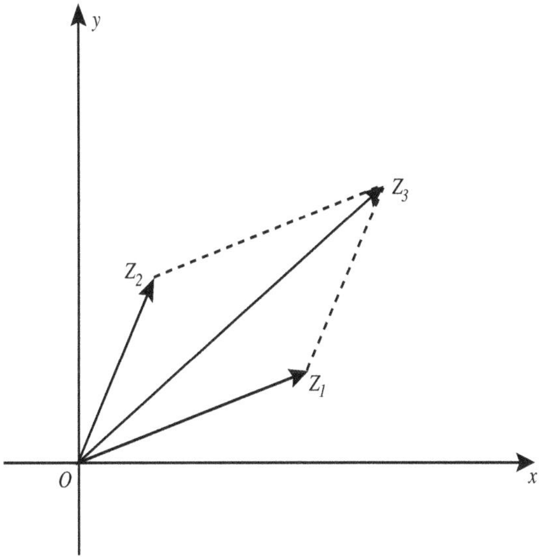

Fig. 10.1 *The geometric interpretation of the sum of complex numbers.*

If the complex number $z = x + y\boldsymbol{i}$ is described by the point $Z = (x, y)$ in $\mathbb{R}^2$, then its conjugate $\bar{z} = x - y\boldsymbol{i}$ is represented by the point $Z^- = (x, -y)$, the reflection of Z in the x-axis; see Figure 10.2. Note that the norm $|z| = \sqrt{x^2 + y^2}$ is equal to the length of the vector $\overrightarrow{OZ}$,

$$|z| = \left|\overrightarrow{OZ}\right|.$$

The vector $\overrightarrow{OZ}$ makes an angle θ with the x-axis measured in a counterclockwise fashion, starting on the x-axis. Measured in radians, it can be any number in $[0, 2\pi)$. This angle is called the *argument* of the complex number z and it is denoted by $\arg z$.

Denote by r the norm of z

$$r = |z| = \sqrt{x^2 + y^2}.$$

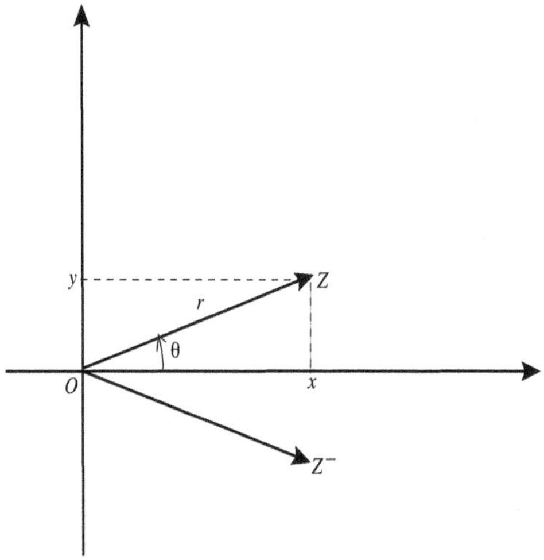

Fig. 10.2 *The geometric interpretation of the conjugation of complex numbers.*

From Figure 10.2 we deduce that the coordinates (x, y) of Z can be expressed in terms of r and θ via the equalities

$$x = r\cos\theta, \quad y = r\sin\theta,$$

so that

$$z = r\cos\theta + r\sin\theta\boldsymbol{i} = r(\cos\theta + \boldsymbol{i}\sin\theta), \quad r = |z|, \quad \theta = \arg z. \tag{10.5}$$

The equality (10.5) is usually referred to as the *trigonometric representation* of the complex number $z = x + y\boldsymbol{i}$.

Suppose that we have two complex numbers z_1, z_2 with trigonometric representations

$$z_k = r_k(\cos\theta_k + \boldsymbol{i}\sin\theta_k), \quad r_k \geq 0, \quad k = 1, 2.$$

Then

$$\mathbf{Re}\, z_k = r_k\cos\theta_k, \quad \mathbf{Im}\, z_k = r_k\sin\theta_k.$$

Moreover

$$\begin{aligned} z_1z_2 &= (r_1r_2)(\cos\theta_1 + \boldsymbol{i}\sin\theta_1)(\cos\theta_2 + \boldsymbol{i}\sin\theta_2) \\ &= r_1r_2\Big\{\underbrace{\big(\cos\theta_1\cos\theta_2 - \sin\theta_1\sin\theta_2\big)}_{=\cos(\theta_1+\theta_2)} + \boldsymbol{i}\underbrace{\big(\sin\theta_1\cos\theta_2 + \cos\theta_1\sin\theta_2\big)}_{=\sin(\theta_1+\theta_2)}\Big\}. \end{aligned}$$

We have thus proved that

$$r_1(\cos\theta_1 + \boldsymbol{i}\sin\theta_1) \times r_2(\cos\theta_2 + \boldsymbol{i}\sin\theta_2) = r_1r_2\big(\cos(\theta_1+\theta_2) + \boldsymbol{i}\sin(\theta_1+\theta_2)\big). \tag{10.6}$$

Applying the above equality iteratively we obtain the celebrated *Moivre's formula*

$$\big(\cos\theta + \boldsymbol{i}\sin\theta\big)^n = \cos(n\theta) + \boldsymbol{i}\sin(n\theta), \quad \forall n \in \mathbb{N}, \quad \theta \in \mathbb{R}. \tag{10.7}$$

If we combine Moivre's formula with Newton's binomial formula we can obtain many interesting consequences. We have

$$\cos n\theta + \boldsymbol{i}\sin n\theta = \sum_{k=0}^{n}\binom{n}{k}\boldsymbol{i}^k(\cos\theta)^k(\sin\theta)^{n-k}.$$

Separating the real and imaginary parts in the right-hand side of the above equality taking into account that

$$\boldsymbol{i}^2 = -1,\ \boldsymbol{i}^3 = -\boldsymbol{i},\ \boldsymbol{i}^4 = 1,$$

we deduce

$$\cos n\theta = (\cos\theta)^n - \binom{n}{2}(\cos\theta)^{n-2}(\sin\theta)^2 + \binom{n}{4}(\cos\theta)^n(\sin\theta)^4 - \cdots \tag{10.8a}$$

$$\sin n\theta = \binom{n}{1}(\cos\theta)^{n-1}\sin\theta - \binom{n}{3}(\cos\theta)^{n-3}(\sin\theta)^3 + \cdots. \tag{10.8b}$$

For example,

$$\cos 2\theta = \cos^2\theta - \sin^2\theta,\quad \sin 2\theta = 2\sin\theta\cos\theta,$$

$$\cos 3\theta = \cos^3\theta - 3\cos\theta\sin^2\theta,\quad \sin 3\theta = 3\cos^2\theta\sin\theta - \sin^3\theta,$$

$$\cos 4\theta = \cos^4\theta - \binom{4}{2}\cos^2\theta\sin^2\theta + \sin^4\theta = \cos^4\theta - 6\cos^2\theta\sin^2\theta + \cos^4\theta,$$

$$\sin 4\theta = 4\cos^3\theta\sin\theta - 4\cos\theta\sin^3\theta.$$

Example 10.1. Consider the complex number

$$z = \cos\frac{\pi}{4} + \boldsymbol{i}\sin\frac{\pi}{4} = \frac{1}{\sqrt{2}}(1+\boldsymbol{i}).$$

For any $n \in \mathbb{N}$ we have

$$z^{8n} = \cos 2n\pi + \boldsymbol{i}\sin 2n\pi = 1.$$

On the other hand we have

$$z^{8n} = \frac{1}{2^{4n}}(1+\boldsymbol{i})^{8n}$$

so that

$$2^{4n} = (1+\boldsymbol{i})^{4n} = \sum_{k=0}^{8n}\binom{8n}{k}\boldsymbol{i}^k.$$

Isolating the real and imaginary parts in the right-hand side and equating them with the real and imaginary parts in the left-hand side we deduce

$$2^{4n} = \binom{8n}{0} - \binom{8n}{2} + \binom{8n}{4} - \cdots,$$

$$0 = \binom{8n}{1} - \binom{8n}{3} + \binom{8n}{5} - \cdots.$$

□

Example 10.2. Fix a natural number $n \geq 2$. Observe that the numbers

$$\zeta_k = \cos\Big(\frac{2\pi}{k}n\Big) + \boldsymbol{i}\sin\Big(\frac{2\pi}{n}\Big), \quad k = 0, 1, \dots, n-1$$

satisfy the equation

$$\zeta_k^n = 1, \ \ \forall k.$$

Conversely, if z is a complex number such that $z^n = 1$, then we deduce

$$|z|^n = 1 \Rightarrow |z| = 1,$$

and thus there exists $\theta \in [0, 2\pi)$ such that

$$z = \cos\theta + \boldsymbol{i}\sin\theta.$$

Using Moivre's formula we deduce $\cos n\theta = 1$ and $\sin n\theta = 0$ which is possible if and only if $n\theta$ is a multiple of 2π. Thus θ can only be one of the numbers

$$\frac{2\pi k}{n}, \quad k = 0, 1, \dots, n-1.$$

In other words $z^n = 1$ if and only if z is equal to one of the numbers ζ_k. For this reason the numbers ζ_k are called *the n-th roots of unity.* □

10.2. Analytic properties of complex numbers

Most of the analysis we developed for real numbers carries over to complex numbers. The next result is crucial in this endeavor.

Proposition 10.3. *(a) For any complex numbers z_1, z_2 we have*

$$|z_1 + z_2| \leq |z_1| + |z_2|. \tag{10.9}$$

(b) If $z = x + y\boldsymbol{i} \in \mathbb{C}$ then

$$\frac{1}{2}(|x| + |y|) \leq |z| = \sqrt{x^2 + y^2} \leq |x| + |y|. \tag{10.10}$$

Proof. (a) Let

$$z_1 = x_1 + y_1\boldsymbol{i}, \ \ z_2 = x_2 + y_2\boldsymbol{i}.$$

Then

$$|z_1| = \sqrt{x_1^2 + y_1^2}, \ \ |z_2| = \sqrt{x_2^2 + y_2^2}.$$

The Cauchy–Schwarz inequality, Corollary 8.33, implies that

$$x_1x_2 + y_1y_2 \leq \left(\sqrt{x_1^2 + y_1^2}\right) \cdot \left(\sqrt{x_2^2 + y_2^2}\right) = |z_1| \cdot |z_2|.$$

We have

$$\begin{aligned} z_1 + z_2 &= (x_1 + x_2) + (y_1 + y_2)\boldsymbol{i}, \\ |z_1 + z_2|^2 &= (x_1 + y_1)^2 + (x_2 + y_2)^2 \\ &= x_1^2 + y_1^2 + 2x_1y_1 + x_2^2 + y_2^2 + 2x_2y_2 \\ &= |z_1|^2 + |z_2|^2 + 2(x_1y_1 + 2x_2y_2) \\ &\leq |z_1|^2 + |z_2|^2 + 2|z_1| \cdot |z_2| = (|z_1| + |z_2|)^2. \end{aligned}$$

This proves (10.9).

(b) Observe that

$$(|x| + |y|)^2 = |x|^2 + |y|^2 + 2|x| \cdot |y| \geq |x|^2 + |y|^2 = x^2 + y^2.$$

This shows that

$$|x| + |y| \geq \sqrt{x^2 + y^2}.$$

On the other hand,

$$\begin{aligned} &0 \leq (|x| - |y|)^2 = |x|^2 + |y|^2 - 2|xy| \Rightarrow 2|xy| \leq x^2 + y^2 \\ &\Rightarrow (|x| + |y|)^2 = |x|^2 + |y|^2 + 2|x| \cdot |y| \leq 2(x^2 + y^2) \\ &\Rightarrow \frac{1}{\sqrt{2}}(|x| + |y|) \leq \sqrt{x^2 + y^2}. \end{aligned}$$

This proves (10.10). □

Definition 10.4. We define the distance between two complex numbers z_1, z_2 to be the nonnegative real number

$$\operatorname{dist}(z_1, z_2) := |z_1 - z_2|.$$ □

Corollary 10.5 (The triangle inequality). *For any $z_1, z_2, z_3 \in \mathbb{C}$ we have*

$$\operatorname{dist}(z_1, z_3) \leq \operatorname{dist}(z_1, z_2) + \operatorname{dist}(z_2, z_3).$$

Proof. We have

$$\begin{aligned} \operatorname{dist}(z_1, z_3) &= |z_1 - z_3| = |(z_1 - z_2) + (z_2 - z_3)| \\ &\overset{(10.9)}{\leq} |z_1 - z_2| + |z_2 - z_3| = \operatorname{dist}(z_1, z_2) + \operatorname{dist}(z_2, z_3). \end{aligned}$$

□

Definition 10.6. (a) Let $z_0 \in \mathbb{C}$ and $r > 0$. The *open disk* of center z_0 and radius r is the set

$$D_r(z_0) := \{z \in \mathbb{C};\ \operatorname{dist}(z, z_0) < r\}.$$

(b) A subset $\mathcal{O} \subset \mathbb{C}$ is called *open* if for any $z_0 \in \mathcal{O}$ there exists $\varepsilon > 0$ such that

$$D_\varepsilon(z_0) \subset \mathcal{O}.$$ □

(c) A set $X \subset \mathbb{C}$ is called *closed* if the complement $\mathbb{C} \setminus X$ is open.
(d) A set $X \subset \mathbb{C}$ is called *bounded* if there exists $R > 0$ such that

$$X \subset D_R(0) \iff |z| < R,\ \forall z \in X.$$ □

Definition 10.7. (a) We say that a sequence of complex numbers $(z_n)_{n\geq 1}$ is *bounded* if the sequence of norms $(|z_n|)_{n\geq 1}$ is bounded as a sequence of real numbers.
(b) We say that a sequence of complex numbers $(z_n)_{n\geq 1}$ *converges to the complex number* z_*, and we denote this

$$\lim_n z_n = z_*,$$

if the sequence of nonnegative real numbers $\operatorname{dist}(z_n, z_*)$ converges to 0, i.e.,

$$\forall \varepsilon > 0\ \exists N = N(\varepsilon) > 0 \text{ such that } \forall n\ \ (n > N(\varepsilon) \Rightarrow |z_n - z_*| < \varepsilon).$$ □

Proposition 10.8. *Suppose that $(z_n)_{n\geq 1}$ is a sequence of complex numbers. We set $x_n =$ $\mathbf{Re}\, z_n$, $y_n = \mathbf{Im}\, z_n$. The following statements are equivalent.*

(i) The sequence (z_n) converges to the complex number $z_* = x_* + y_*\boldsymbol{i}$.
(ii) The sequences of real numbers $(x_n)_{n\geq 1}$ and $(y_n)_{n\geq 1}$ converge to x_* and respectively y_*.

Proof. (i) ⇒ (ii). From the first part of (10.10) we deduce that

$$\frac{1}{2}\left(|x_n - x_*| + |y_n - y_*|\right) \leq |z_n - z_*|.$$

Since $\lim_n z_n = z_*$ we deduce $\lim_n |z_n - z_*| = 0$ and the Squeezing Principle implies

$$\lim_n \left(|x_n - x_*| + |y_n - y_*|\right) = 0.$$

The last equality implies (ii).
(ii) ⇒ (i). From the second part of (10.10) we deduce that

$$|z_n - z_*| \leq |x_n - x_*| + |y_n - y_*|.$$

The assumption (ii) implies that

$$\lim_n \big(\, |x_n - x_*| + |y_n - y_*|\,\big) = 0.$$

From this we conclude that $\lim_n |z_n - z_*| = 0$, which is the statement (i). □

Corollary 10.9. *If the sequence of complex numbers* $(z_n)_{n\geq 1}$ *converges to* z*, then*

$$\lim_n |z_n| = |z|.$$

Proof. Let $x_n := \mathbf{Re}\, z_n$ and $y_n := \mathbf{Im}\, z_n$, $x = \mathbf{Re}\, z$, $y := \mathbf{Im}\, z$. Then

$$\lim_n z_n = z \Rightarrow \lim_n x_n = x \;\wedge\; \lim_n y_n = y$$
$$\Rightarrow \lim_n (x_n^2 + y_n^2) = x^2 + y^2 \Rightarrow \lim_n \sqrt{x_n^2 + y_n^2} = \sqrt{x^2+y^2} \iff \lim_n |z_n| = |z|.$$

□

Corollary 10.10. *Any convergent sequence of complex numbers is bounded.*

Proof. Given a convergent sequence of complex numbers, the associated sequence of norms is convergent according to Corollary 10.9. The sequence of norms is thus a convergent sequence of *real* numbers, hence bounded according to Proposition 4.14. □

Example 10.11. Suppose z is a complex number such that $|z| < 1$. Then

$$\lim_n z^n = 0.$$

We have to show that the sequence of nonnegative numbers $|z^n|$ goes to zero as $n \to \infty$. We set $r : |z|$ and we observe that

$$|z^n| \overset{(10.4)}{=} |z|^n = r^n.$$

As shown in Example 4.12

$$|r| < 1 \Rightarrow \lim_n r^n = 0 \Rightarrow \lim_n z^n = 0.$$

□

The convergent sequences of complex numbers satisfy many of the same properties of convergent sequences of real numbers. We summarize these facts in our next result whose proof is left to you as an exercise.

Proposition 10.12 (Passage to the limit). *Suppose that* $(a_n)_{n\geq 1}$ *and* $(b_n)_{n\geq 1}$ *are two convergent sequences of complex numbers,*

$$a := \lim_{n\to\infty} a_n, \quad b = \lim_{n\to\infty} b_n.$$

The following hold.

(i) The sequence $(a_n + b_n)_{n\geq 1}$ *is convergent and*

$$\lim_{n\to\infty} (a_n + b_n) = \lim_{n\to\infty} a_n + \lim_{n\to\infty} b_n = a + b.$$

(ii) If $\lambda \in \mathbb{C}$ *then*

$$\lim_{n\to\infty} (\lambda a_n) = \lambda \lim_{n\to\infty} a_n = \lambda a.$$

(iii)

$$\lim_{n\to\infty}(a_n \cdot b_n) = \big(\lim_{n\to\infty} a_n\big)\cdot\big(\lim_{n\to\infty} b_n\big) = ab.$$

(iv) Suppose that $b \neq 0$*. Then there exists* $N_0 > 0$ *such that* $b_n \neq 0$, $\forall N > N_0$ *and*

$$\lim_{n\to\infty}\frac{a_n}{b_n} = \frac{a}{b}.$$ □

Definition 10.13. A sequence of complex numbers $(z_n)_{n\geq 1}$ is called *Cauchy* if

$$\forall\varepsilon > 0\ \exists N = N(\varepsilon) > 0 \text{ such that } \forall m,n\ \ (m,n > N(\varepsilon) \Rightarrow |z_m - z_n| < \varepsilon).$$ □

The concept of Cauchy sequence of complex numbers is closely related to the notion of Cauchy sequence of real numbers. We state this in a precise form in our next result. Its proof is very similar to the proof of Proposition 10.8 and we leave the details to you as an exercise.

Proposition 10.14. *Suppose that* $(z_n)_{n\geq 1}$ *is a sequence of complex numbers. We set* $x_n := \mathbf{Re}\, z_n$, $y_n := \mathbf{Im}\, z_n$*. The following statements are equivalent.*

(i) The sequence $(z_n)_{n\geq 1}$ *is Cauchy.*
(ii) The sequences of real *numbers* $(x_n)_{n\geq 1}$ *and* $(y_n)_{n\geq 1}$ *are Cauchy.*

□

Definition 10.15. The *series* associated to a sequence $(z_n)_{n\geq 0}$ of complex numbers is the new sequence $(s_n)_{n\geq 1}$ defined by the *partial sums*

$$s_0 = z_0,\ \ s_1 = z_0 + z_1,\ \ s_2 = z_0 + z_1 + z_2, \dots, s_n = \sum_{j=0}^{n} a_j \ \dots .$$

The series associated to the sequence $(a_n)_{n\geq 0}$ is denoted by the symbol

$$\sum_{n\geq 0}^{\infty} z_n \ \ or\ \ \sum_{n\geq 0} z_n.$$

The series is called *convergent* if the sequence of partial sums $(s_n)_{n\geq 0}$ is convergent. The limit $\lim_{n\to\infty} s_n$ is called *the sum* series. We will use the notation

$$\sum_{n\geq 0} a_n = S$$

to indicate that the series is convergent and its sum is the real number S. □

Example 10.16. The geometric series

$$\sum_{n=0}^{\infty} z^n = 1 + z + z^2 + \cdots$$

is convergent for any complex number z of norm $|z| < 1$. Indeed, its n-th partial sum is

$$s_n = 1 + z + \cdots + z^n = \frac{1 - z^{n+1}}{1 - z}.$$

If $|z| < 1$, then we deduce from Example 10.11 and Proposition 10.12 that

$$\lim_n s_n = \lim_n \frac{1 - z^{n+1}}{1 - z} = \frac{1}{1 - z}.$$

This shows that the series is convergent and its sum is

$$1 + z + z^2 + \cdots + z^n + \cdots = \frac{1}{1 - z}, \quad \forall |z| < 1. \tag{10.11}$$

□

Proposition 10.17. *If the series of complex numbers*

$$\sum_{n \geq 0} z_n$$

is convergent, then its terms converge to zero, $\lim_n z_n = 0$.

Proof. Denote by s the sum of the series and by s_n its n-th partial sum,

$$s_n = z_0 + z_1 + \cdots + z_n.$$

Then $z_n = s_n - s_{n-1}$ and

$$\lim_n z_n = \lim_n (s_n - s_{n-1}) = \lim_n s_n - \lim_n s_{n-1} = s - s = 0.$$

□

Example 10.18. The geometric series

$$1 + z + z^2 + \cdots$$

is divergent if $|z| \geq 1$. Indeed, we have

$$|z^n| = |z|^n$$

and

$$\lim_n |z|^n = \begin{cases} 1, & |z| = 1, \\ \infty, & |z| > 1. \end{cases}$$

This shows that when $|z| \geq 1$ the sequence (z^n) does not converge to zero and thus, according to Proposition 10.17, the geometric series cannot be convergent. □

Definition 10.19. A series of complex numbers

$$\sum_{n\geq 0} z_n$$

is called *absolutely convergent* if the series of *nonnegative real numbers*

$$\sum_{n\geq 0} |z_n|$$

is convergent. □

Proposition 10.20. *If the series of complex numbers $\sum_{n\geq 0} z_n$ is absolutely convergent, then it is also convergent.*

Proof. We mimic the proof of Theorem 4.46. Denote by s_n the n-th partial sum of the series $\sum_{n\geq 0} z_n$ and by $\hat{s}_n$ the n-th partial sum of the series $\sum_{n\geq 0} |z_n|$,

$$s_n = z_0 + \cdots + z_n,\ \ \hat{s}_n = |z_0| + \cdots + |z_n|.$$

For $n > m$ we have

$$s_n - s_m = z_{m+1} + \cdots + z_n,\ \ \hat{s}_n - \hat{s}_m = |z_{m+1}| + \cdots + |z_n|.$$

Using (10.9) we deduce

$$|s_n - s_m| = |z_{m+1} + \cdots + z_n| \leq |z_{m+1}| + \cdots + |z_n| = \hat{s}_n - \hat{s}_m = |\hat{s}_n - \hat{s}_m|. \quad (10.12)$$

Since the series $\sum_{n\geq 0} |z_n|$ is convergent we deduce that the sequence of partial sums $(\hat{s}_n)_{n\geq 0}$ is Cauchy. Hence, for any $\varepsilon > 0$ there exists $N = N(\varepsilon) > 0$ such that for any $n > m > N(\varepsilon)$ we have

$$|\hat{s}_n - \hat{s}_m| < \varepsilon.$$

Using (10.12) we deduce that for any $n > m > N(\varepsilon)$ we have

$$|s_n - s_m| < \varepsilon.$$

This shows that the sequence (s_n) is Cauchy and thus convergent according to Proposition 10.14. □

The above result reduces the problem of deciding the absolute convergence of a series of complex numbers to deciding whether a series of nonnegative *real* numbers is convergent. We have investigated this issue in Section 4.6. We mention here one useful convergence test.

Corollary 10.21 (Ratio test). *Suppose that*

$$z_0 + z_1 + z_2 + \cdots$$

is a series of complex numbers such that

$$L = \lim_n \frac{|z_{n+1}|}{|z_n|}$$

exists, $L \in [0, \infty]$. Then the following hold.

(i) If $L < 1$, then the series $\sum_{n\geq 0} z_n$ is absolutely convergent.
(ii) If $L > 1$, then the series is divergent.

Proof. (i) The ratio test Corollary 4.48 implies that the series of positive real numbers

$$\sum_{n\geq 0} |z_n|$$

is convergent.

(ii) If

$$\lim_n \frac{|z_n|}{|z_n|} > 1,$$

then $|z_{n+1}| > |z_n|$ for n sufficiently large. In particular, the sequence (z_n) does not converge to 0 and thus the series $\sum_{n\geq 0} z_n$ is divergent. □

10.3. Complex power series

A complex *power series* is a series of the form

$$s(z) = a_0 + a_1 z + a_2 z^2 + a_3 z^3 + \cdots = \sum_{n\geq 0} a_n z^n,$$

where z and the numbers $a_0, a_1, \ldots$ are complex. The number z should be viewed as a quantity that is allowed to vary, while the numbers $a_0, a_1, \ldots$ should be viewed as fixed quantities. As such they are called the *coefficients* of the power series. Note that for different choices of z we obtain different series.

Example 10.22. Consider for example the power series

$$s(z) = 1 - 2z + 2^2 z^2 - 2^3 z^3 + \cdots .$$

The coefficients of this power series are

$$a_0 = 1, \quad a_1 = -2, \quad a_2 = 2^2, \ldots, a_n = (-2)^n, \ldots .$$

Note that we can rewrite the above series as

$$s(z) = 1 + (-2z) + (-2z)^2 + (-2z)^3 + \cdots = \sum_{n\geq 0} (-2z)^n.$$

If we make the substitution $\zeta := -2z$ we can further rewrite

$$s(z) = 1 + \zeta + \zeta^2 + \cdots .$$

We know that this series is absolutely convergent for $|\zeta| > 1$ and divergent for $|\zeta| > 1$. In other words the power series $s(z)$ converges absolutely if $|z| < \frac{1}{2}$ and diverges if $|z| > \frac{1}{2}$. Note that the set of complex numbers z such that $|z| < \frac{1}{2}$ is the open disk of center 0 and radius $\frac{1}{2}$. □

Proposition 10.23. *Consider a complex power series*

$$s(z) = \sum_{n \geq 0} a_n z^n.$$

(a) If for some $z_0 \neq 0$ the series $s(z_0)$ converges absolutely, *then for any $z \in \mathbb{C}$ such that $|z| \leq |z_0|$ the series $s(z)$ converges absolutely.*
(b) If for some $z_0 \neq 0$ the series $s(z_0)$ is convergent, *not necessarily absolutely, then for any $z \in \mathbb{C}$ such that $|z| < |z_0|$, the series $s(z)$ converges* absolutely.

Proof. (a) Since $|z| \leq |z_0|$ we deduce that

$$|a_n z^n| \leq |a_n z_0^n|, \;\; \forall n \geq 0.$$

The desired conclusion now follows from the Comparison Principle.
(b) Since $s(z_0)$ converges we deduce that

$$\lim_n a_n z_0^n = 0.$$

In particular, we deduce that the sequence $(a_n z_0^n)$ is bounded, i.e., there exists $C > 0$ such that

$$|a_n z_0^n| \leq C, \;\; \forall n \geq 0.$$

We set

$$r := \left| \frac{z}{z_0} \right| = \frac{|z|}{|z_0|} < 1.$$

We observe that

$$|a_n z^n| = |a_n z_0^n| \frac{|z|^n}{|z_0|^n} \leq C r^n.$$

Since $r < 1$ we deduce that the geometric series

$$\sum_{n \geq 0} C r^n$$

is convergent and the Comparison Principle implies that the series

$$\sum_{n \geq 0} |a_n z^n|$$

is also convergent. □

To a complex power series

$$s(z) = \sum_{n \geq 0} a_n z^n$$

associate the set of nonnegative numbers

$$\mathcal{R} = \{ r \geq 0; \;\; \exists z \in \mathbb{C} \text{ such that } |z| = r, \;\; s(z) \text{ is convergent} \} \subset \mathbb{R}.$$

Note that the set $\mathcal{R}$ is not empty because $0 \in \mathbb{R}$. Next observe that Proposition 10.23(b) implies that if $r_0 \in \mathcal{R}$, then $[0, r_0) \subset \mathcal{R}$. We set

$$R := \sup \mathcal{R} \in [0, \infty].$$

Proposition 10.23 shows that $s(z)$ converges absolutely for any $|z| < R$, and diverges for $|z| > R$. The number $R \in [0, \infty]$ is called the *radius of convergence* of the power series $s(z)$.

Example 10.24 (Complex exponential). Consider the power series

$$E(z) = 1 + \frac{z}{1!} + \frac{z^2}{2!} + \cdots = \sum_{n \geq 0} \frac{1}{n!} z^n.$$

This series is absolutely convergent for any $z \in \mathbb{C}$ because the series of positive numbers

$$\sum_{n \geq 0} \frac{|z|^n}{n!}$$

is convergent for any z. Thus the radius of convergence of this power series is ∞. For simplicity we will denote by $E(z)$ the sum of the series $E(z)$.

Observe that for a real number x the sum of the series $E(x)$ is e^x; see Exercise 8.7. We write this

$$E(x) = e^x, \quad \forall x \in \mathbb{R}. \tag{10.13}$$

The properties of the exponential show that

$$E(x+y) = e^{x+y} = e^x e^y = E(x)E(y), \quad \forall x, y \in \mathbb{R}. \tag{10.14}$$

A more general result is true, namely,

$$E(z+\zeta) = E(z)E(\zeta), \quad \forall z, \zeta \in \mathbb{C}. \tag{10.15}$$

To prove the above equality we denote by $E_n(z)$ the n-th partial sum of the series $E(z)$,

$$E_n(z) = 1 + \frac{z}{1!} + \cdots + \frac{z^n}{n!}.$$

The equality (10.15) is equivalent to the equality

$$\lim_n \big(E_{2n}(z+\zeta) - E_{2n}(z)E_{2n}(\zeta) \big) = 0. \tag{10.16}$$

Fix a real number $M > 1$ such that

$$|z|, \ |\zeta| < M.$$

We have

$$\begin{aligned} E_{2n}(z+\zeta) &= \sum_{m=0}^{2n} \frac{1}{m!}(z+\zeta)^m = \sum_{m=0}^{2n} \frac{1}{m!} \sum_{j=0}^{m} \binom{m}{j} z^{m-j}\zeta^j \\ &= \sum_{m=0}^{2n} \frac{1}{m!} \sum_{j=0}^{m} \frac{m! z^{m-j}\zeta^j}{(m-j)!j!} = \sum_{m=0}^{2n} \sum_{j=0}^{m} \frac{z^{m-j}\zeta^j}{(m-j)!j!} \end{aligned}$$

$(k := m - j)$

$$= \sum_{m=0}^{2n} \sum_{\substack{j+k=m \\ j,k \geq 0}} \frac{z^k \zeta^j}{k!j!} = \sum_{\substack{j+k \leq 2n \\ j,k \geq 0}} \frac{z^k \zeta^j}{k!j!}.$$

Similarly we have

$$E_{2n}(z)E_{2n}(\zeta) = \left(\sum_{k=0}^{2n} \frac{z^k}{k!} \right) \left(\sum_{j=0}^{2n} \frac{\zeta^j}{j!} \right) = \sum_{0 \leq j,k \leq 2n} \frac{z^k \zeta^j}{k!j!}.$$

We deduce

$$\begin{aligned} |E_{2n}(z+\zeta) - E_{2n}(z)E_{2n}(\zeta)| &= \left| \sum_{\substack{j+k>2n \\ 0 \leq j,k \leq 2n}} \frac{z^k \zeta^j}{k!j!} \right| \\ &\leq \sum_{\substack{j+k>2n \\ 0 \leq j,k \leq 2n}} \frac{|z|^k |\zeta|^j}{k!j!} \leq M^{4n} \sum_{\substack{j+k>2n \\ 0 \leq j,k \leq 2n}} \frac{1}{k!j!} \leq \frac{M^{4n}}{n!} \sum_{\substack{j+k>2n \\ 0 \leq j,k \leq 2n}} 1 \leq \frac{4n^2 M^{4n}}{n!}. \end{aligned}$$

From (4.8) we deduce that

$$\lim_n \frac{4n^2 M^{4n}}{n!} \to 0.$$

Because of the equalities (10.13) and (10.15), for any $z \in \mathbb{C}$ we set

$$e^z := E(z) = 1 + \frac{z}{1!} + \frac{z^2}{2!} + \frac{z^3}{3!} \cdots . \tag{10.17}$$

Suppose that in (10.17) the number z is purely imaginary, $z = \boldsymbol{i}t$, $t \in \mathbb{R}$. We deduce the celebrated *Euler's formula*

$$\begin{aligned} e^{\boldsymbol{i}t} &= 1 + \frac{\boldsymbol{i}t}{1!} + \frac{\boldsymbol{i}^2 t^2}{2!} + \frac{\boldsymbol{i}^3 t^3}{3!} + \cdots \\ &= \left(1 - \frac{t^2}{2!} + \frac{t^4}{4!} + \cdots\right) + \boldsymbol{i}\left(t - \frac{t^3}{3!} + \frac{t^5}{5!} + \cdots\right) \\ &= \cos t + \boldsymbol{i} \sin t. \end{aligned} \tag{10.18}$$

If we let $t = \pi$ in the above equality we deduce

$$e^{\boldsymbol{i}\pi} = \cos\pi + \boldsymbol{i}\sin\pi = -1,$$

i.e.,

$$e^{\boldsymbol{i}\pi} + 1 = 0. \tag{10.19}$$

The last very compact equality describes a deep connection between the five most important numbers in science: $0, 1, e, \pi, \boldsymbol{i}$. □

10.4. Exercises

Exercise 10.1. Prove the equalities (10.3) and (10.4). □

Exercise 10.2. (a) Consider the complex numbers

$$z_1 = 4 + 5\boldsymbol{i}, \quad z_2 = 5 + 12\boldsymbol{i}.$$

Compute

$$z_1 z_2, \; |z_2|, \; \frac{z_1}{z_2}.$$

(b) Show that if

$$z = \frac{1}{2}(1 + \sqrt{3}\boldsymbol{i}),$$

then

$$z^2 + z + 1 = \bar{z}^2 + \bar{z} + 1 = 0, \quad z^3 = \bar{z}^3 = 1.$$ □

Exercise 10.3. (a) Prove that if $z \in \mathbb{C}$, then

$$z^5 = 1 \wedge z \neq 1 \iff z^4 + z^3 + z^2 + z + 1 = 0 \iff z^2 + z + 1 + \frac{1}{z} + \frac{1}{z^2} = 0.$$

(b) Suppose that z satisfies the above equation, $z^4 + z^3 + z^2 + z + 1 = 0$. We set

$$\zeta := z + \frac{1}{z}.$$

Prove that

$$z^2 + \frac{1}{z^2} = \zeta^2 - 2,$$

and

$$\zeta^2 + \zeta - 1 = 0. \tag{10.20}$$

(c) Find the two roots ζ_1, ζ_2 of the quadratic equation (10.20).
(d) If ζ_1, ζ_2 are as above, find all the complex numbers z such that

$$z + \frac{1}{z} = \zeta_1 \;\vee\; z + \frac{1}{z} = \zeta_2.$$

(e) Use (d) to compute $\cos(2\pi/5)$, $\sin(2\pi/5)$. □

Exercise 10.4. (a) Let $z_0 \in \mathbb{C}$ and $r > 0$. Prove that the open disc $D_r(z_0)$ is an open set in the sense of Definition 10.6(b).
(b) Prove that if $\mathcal{O}_1, \mathcal{O}_2 \subset \mathbb{C}$ are open sets, then so are the sets $\mathcal{O}_1 \cap \mathcal{O}_2$, $\mathcal{O}_1 \cup \mathcal{O}_2$.
(c) Consider the set

$$S := \{ z \in \mathbb{C}; \; \mathbf{Im}\, z = 0, \; \mathbf{Re}\, z \in [0, 1] \}.$$

Draw a picture of S and then prove that it is a closed set in the sense of Definition 10.6(c). □

Exercise 10.5. Let S be a subset of the complex plane, $S \subset \mathbb{C}$. Prove that the following statements are equivalent.

(i) The set S is closed.
(ii) For any sequence $(z_n)_{n\geq 1}$ of points in S, $z_n \in S, \forall n$, if the sequence converges to z_*, then $z_* \in S$.

□

Exercise 10.6. Use the ideas in the proof of Proposition 10.8 to prove Proposition 10.14. □

Exercise 10.7. Prove Proposition 10.12 by imitating the proof of Proposition 4.15. □

Chapter 11

The Geometry and the Topology of Euclidean Spaces

The calculus of one-real-variable functions has a several-variable counterpart. To state and prove these results we need an appropriate language. The goal of this chapter is to introduce the terminology and the concepts required to make the jump into higher dimensions.

11.1. Basic affine geometry

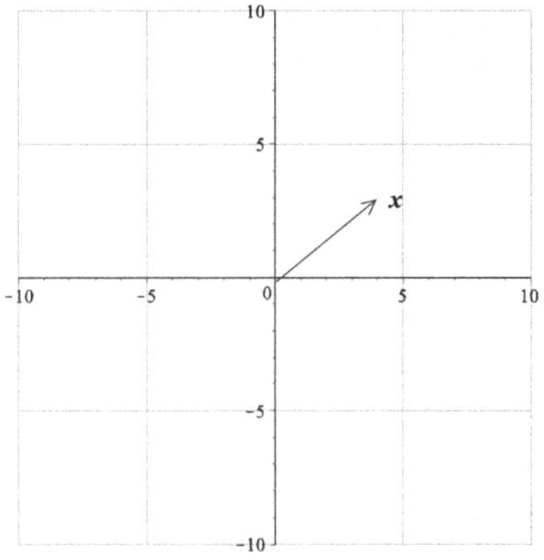

Fig. 11.1 *The point $\boldsymbol{x} \in \mathbb{R}^2$ with (Cartesian) coordinates $(4,3)$ is identified with the vector that starts at the origin and ends at $\boldsymbol{x}$.*

Let $n \in \mathbb{N}$. The *canonical n-dimensional real Euclidean space* is the Cartesian product

$$\mathbb{R}^n := \underbrace{\mathbb{R} \times \cdots \times \mathbb{R}}_{n \ times}.$$

The elements of $\mathbb{R}^n$ are called (n-dimensional) *vectors* or *points* and they are n-tuple of real numbers

$$\boldsymbol{x} := \begin{bmatrix} x^1 \\ \vdots \\ x^n \end{bmatrix}. \tag{11.1}$$

Above, the real numbers $x^1, \dots, x^n$ are called the *Cartesian coordinates* of the vector $\boldsymbol{x}$; see Figure 11.1.

☞ Several comments are in order. First, note that we represent the vector as a (vertical) *column*. To remind us of this, we use the *superscript* notation x^i rather than the *subscript* notation x_i. There are several other good reasons for this choice of notation, but explaining them is difficult at this time. This choice is part of a larger collection of conventions sometimes referred to as the *Einstein's conventions*. For now, accept and use this convention as a very good idea with a nebulous payoff that will reveal itself once your mathematical background is a bit more sophisticated.

For typographical reasons it is inconvenient to work with tall columns of numbers of the type appearing in (11.1) so we will use the notation $[x^1, \dots, x^n]^\top$ or $(x^1, \dots, x^n)$ to denote the *column* in the right-hand side of (11.1).

Also, when we refer to a point $\boldsymbol{x} \in \mathbb{R}^n$ as a vector we secretly think of $\boldsymbol{x}$ as the tip of an arrow that starts at the origin and ends at $\boldsymbol{x}$; see Figure 11.1.

The attribute *Euclidean space* attached to the set $\mathbb{R}^n$ refers to the additional structure this set is equipped with. First of all, $\mathbb{R}^n$ has a structure of *vector space*.[1] More precisely, it is equipped with two operations, *addition* and *multiplication by scalars* satisfying certain properties.

The addition is a function $\mathbb{R}^n \times \mathbb{R}^n \to \mathbb{R}^n$ that associates to a pair of vectors $(\boldsymbol{x}, \boldsymbol{y}) \in \mathbb{R}^n \times \mathbb{R}^n$ a third vector, its *sum* $\boldsymbol{x} + \boldsymbol{y} \in \mathbb{R}^n$, defined as follows: if

$$\boldsymbol{x} = \big(x^1, \dots, x^n\big), \;\; \boldsymbol{y} = \big(y^1, \dots, y^n\big),$$

then

$$\boldsymbol{x} + \boldsymbol{y} := \big(x^1 + y^1, \dots, x^n + y^n\big) \in \mathbb{R}^n.$$

The multiplication-by-scalars operation associates to a pair $(\lambda, \boldsymbol{x})$ consisting of a real number (or scalar) λ and a vector $\boldsymbol{x} \in \mathbb{R}^n$, a new vector denoted by $\lambda\boldsymbol{x}$ (or $\lambda \cdot \boldsymbol{x}$) and defined as follows: if $\boldsymbol{x} = \big(x^1, x^2, \dots, x^n\big)$, then

$$\lambda\boldsymbol{x} := \big(\lambda x^1, \dots, \lambda x^n\big) \in \mathbb{R}^n.$$

These operations satisfy the following properties.[2]

(i) (Associativity) For any $\boldsymbol{x}, \boldsymbol{y}, \boldsymbol{z} \in \mathbb{R}^n$

$$(\boldsymbol{x} + \boldsymbol{y}) + \boldsymbol{z} = \boldsymbol{x} + (\boldsymbol{y} + \boldsymbol{z}).$$

(ii) (Commutativity) For any $\boldsymbol{x}, \boldsymbol{y} \in \mathbb{R}^n$,

$$\boldsymbol{x} + \boldsymbol{y} = \boldsymbol{y} + \boldsymbol{x}.$$

(iii) (Neutral or identity element) The vector $\mathbf{0} := (0, \dots, 0) \in \mathbb{R}^n$ has the property: $\forall \boldsymbol{x} \in \mathbb{R}^n$ we have

$$\mathbf{0} + \boldsymbol{x} = \boldsymbol{x} + \mathbf{0} = \boldsymbol{x}.$$

[1]As we progress in this course I will assume increased knowledge of linear algebra. I recommend [21] as a linear algebra source very appropriate for the goals of this course.

[2]Compare them with the algebraic axioms of $\mathbb{R}$.

(iv) (Inverse or opposite element) For any $\boldsymbol{x} = \big(x_1, \ldots, x_n\big) \in \mathbb{R}^n$, the vector

$$-\boldsymbol{x} := \big(-x_1, \ldots, -x_n\big)$$

has the property:

$$\boldsymbol{x} + (-\boldsymbol{x}) = (-\boldsymbol{x}) + \boldsymbol{x} = \mathbf{0}.$$

(v) (Distributivity with respect to vector addition) For any $\lambda \in \mathbb{R}$, $\boldsymbol{x}, \boldsymbol{y} \in \mathbb{R}^n$,

$$\lambda(\boldsymbol{x} + \boldsymbol{y}) = \lambda\boldsymbol{x} + \lambda\boldsymbol{y}.$$

(vi) (Distributivity with respect to the scalar addition) For any $\lambda, \mu \in \mathbb{R}$, $\boldsymbol{x} \in \mathbb{R}^n$,

$$(\lambda + \mu)\boldsymbol{x} = \lambda\boldsymbol{x} + \mu\boldsymbol{x}, \quad (\lambda\mu)\boldsymbol{x} = \lambda(\mu\boldsymbol{x}).$$

(vii) For any $\boldsymbol{x} \in \mathbb{R}^n$,

$$1 \cdot \boldsymbol{x} = \boldsymbol{x}.$$

Note that $0 \cdot \boldsymbol{x} = \mathbf{0}$, $\forall \boldsymbol{x} \in \mathbb{R}^n$.

Definition 11.1. The canonical or natural basis of $\mathbb{R}^n$ is the set of vectors $\{\boldsymbol{e}_1, \ldots, \boldsymbol{e}_n\}$, where

$$\boldsymbol{e}_1 := \begin{bmatrix} 1 \\ 0 \\ 0 \\ \vdots \\ 0 \\ 0 \end{bmatrix}, \quad \boldsymbol{e}_2 := \begin{bmatrix} 0 \\ 1 \\ 0 \\ \vdots \\ 0 \\ 0 \end{bmatrix}, \ldots, \boldsymbol{e}_n := \begin{bmatrix} 0 \\ 0 \\ 0 \\ \vdots \\ 0 \\ 1 \end{bmatrix}. \tag{11.2}$$

□

Note that if $\boldsymbol{x} = \big(x^1, \ldots, x^n\big)$, then

$$\boxed{\boldsymbol{x} = \begin{bmatrix} x^1 \\ \vdots \\ x^n \end{bmatrix} = x^1\boldsymbol{e}_1 + \cdots + x^n\boldsymbol{e}_n = \sum_{i=1}^{n} x^i\boldsymbol{e}_i}\,.$$

For example, we have the following equality in $\mathbb{R}^3$,

$$\begin{bmatrix} 3 \\ -4 \\ 5 \end{bmatrix} = 3\begin{bmatrix} 1 \\ 0 \\ 0 \end{bmatrix} - 4\begin{bmatrix} 0 \\ 1 \\ 0 \end{bmatrix} + 5\begin{bmatrix} 0 \\ 0 \\ 1 \end{bmatrix} = 3\boldsymbol{e}_1 - 4\boldsymbol{e}_2 + 5\boldsymbol{e}_3. \tag{11.3}$$

At this point it is convenient to introduce the *Kronecker symbol* δ^i_j,

$$\delta^i_j := \begin{cases} 1, & i = j, \\ 0, & i \neq j. \end{cases} \tag{11.4}$$

Using the Kronecker symbol we observe that

$$e_k = \begin{bmatrix} \delta_k^1 \\ \delta_k^2 \\ \vdots \\ \delta_k^n \end{bmatrix}, \quad \forall k = 1, \ldots, n.$$

Remark 11.2. When $n = 2$, the coordinates x^1, x^2 are usually denoted by x and respectively y, and the vectors $\boldsymbol{e}_1, \boldsymbol{e}_2$ are usually denoted by $\boldsymbol{i}$ and respectively $\boldsymbol{j}$; see Figure 11.2.

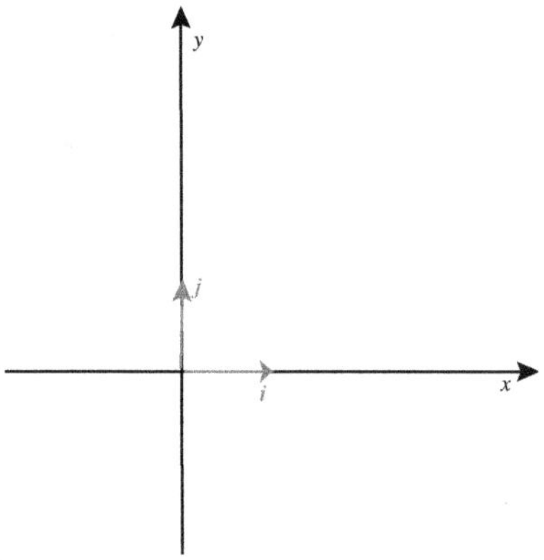

Fig. 11.2 *A Cartesian coordinate system in* $\mathbb{R}^2$.

When $n = 3$, the coordinates x^1, x^2, x^3 are usually denoted by x, y and respectively z, and the vectors $\boldsymbol{e}_1, \boldsymbol{e}_2, \boldsymbol{e}_3$ are usually denoted by $\boldsymbol{i}, \boldsymbol{j}$ and respectively $\boldsymbol{k}$; see Figure 11.3. Thus, in $\mathbb{R}^3$, the equality (11.3) could be rewritten as

$$\begin{bmatrix} 3 \\ -4 \\ 5 \end{bmatrix} = 3\boldsymbol{i} - 4\boldsymbol{j} + 5\boldsymbol{k}. \qquad \square$$

Definition 11.3. Two nonzero vectors $\boldsymbol{u}, \boldsymbol{v} \in \mathbb{R}^n$ are called *collinear* if one is a multiple of the other, i.e., there exists $t \in \mathbb{R}$, $t \neq 0$, such that $\boldsymbol{v} = t\boldsymbol{u}$ (and thus $\boldsymbol{u} = t^{-1}\boldsymbol{v}$). □

Definition 11.4. Let $\boldsymbol{p}, \boldsymbol{v} \in \mathbb{R}^n$, $\boldsymbol{v} \neq \boldsymbol{0}$. The *line* in $\mathbb{R}^n$ *through* $\boldsymbol{p}$ *and in the direction* $\boldsymbol{v}$ is the set

$$\boxed{\ell_{\boldsymbol{p},\boldsymbol{v}} := \left\{ \boldsymbol{p} + t\boldsymbol{v};\ \ t \in \mathbb{R} \right\} \subset \mathbb{R}^n}. \tag{11.5}$$

The vector $\boldsymbol{v}$ is called a *direction vector* of the line. □

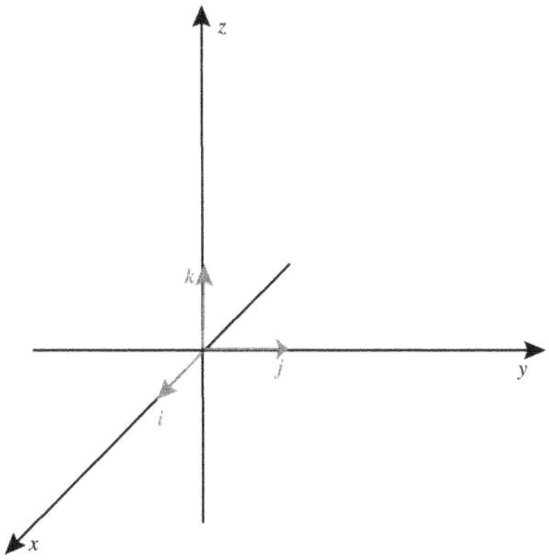

Fig. 11.3 *A Cartesian coordinate system in* $\mathbb{R}^3$.

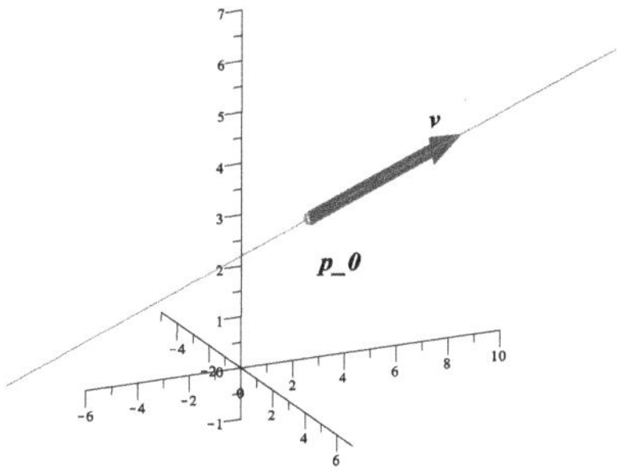

Fig. 11.4 *The line through the point* $\boldsymbol{p} = [1, 2, 3]^\top$ *and in the direction* $\boldsymbol{v} = [3, 4, 2]^\top$.

Let us point out that, if the two nonzero vectors $\boldsymbol{u}, \boldsymbol{v} \in \mathbb{R}^n$ are collinear, then, for any point $\boldsymbol{p} \in \mathbb{R}^n$, the line through $\boldsymbol{p}$ in the direction $\boldsymbol{u}$ coincides with the line through $\boldsymbol{p}$ in the direction $\boldsymbol{v}$, i.e.,

$$\ell_{\boldsymbol{p},\boldsymbol{u}} = \ell_{\boldsymbol{p},\boldsymbol{v}}.$$

Exercise 11.1 asks you to prove this fact.

Observe that the line through $\boldsymbol{p}$ and in the direction $\boldsymbol{v}$ is the image of the function

$$\boldsymbol{f} : \mathbb{R} \to \mathbb{R}^n, \ \ \boldsymbol{f}(t) = \boldsymbol{p} + t\boldsymbol{v}.$$

You can think of the map $\boldsymbol{f}$ as describing the motion of a point in $\mathbb{R}^n$ so that its location at time $t \in \mathbb{R}$ is $\boldsymbol{p} + t\boldsymbol{v}$. The line $\ell_{\boldsymbol{p},\boldsymbol{v}}$ is then the *trajectory* described by this point during its motion. If

$$\boldsymbol{p} = \begin{bmatrix} p^1 \\ \vdots \\ p^n \end{bmatrix}, \quad \boldsymbol{v} = \begin{bmatrix} v^1 \\ \vdots \\ v^n \end{bmatrix},$$

then

$$\boldsymbol{p} + t\boldsymbol{v} = \begin{bmatrix} p^1 + tv^1 \\ \vdots \\ p^n + tv^n \end{bmatrix}$$

and we can describe the line through $\boldsymbol{p}$ in the direction $\boldsymbol{v}$ using the *parametric equation* or *parametrization*

$$\begin{cases} x^1 = p^1 + tv^1 \\ \vdots \; \vdots \; \vdots \\ x^n = p^n + tv^n, \end{cases} \quad t \in \mathbb{R}. \tag{11.6}$$

Above, the variable t is called the *parameter* (of the parametric equations). As t varies, the right-hand side of (11.6) describes the coordinates of a moving point along the line. The parametric equations (11.6) should be interpreted as saying that

$$\boldsymbol{x} \in \ell_{\boldsymbol{p},\boldsymbol{v}} \Longleftrightarrow \exists t \in \mathbb{R}: \; x^i = p^i + tv^i, \; \forall i = 1, \dots, n.$$

Definition 11.5. The lines $\ell_{\boldsymbol{0},\boldsymbol{e}_1}, \dots, \ell_{\boldsymbol{0},\boldsymbol{e}_n}$ are called the *coordinate axes* of $\mathbb{R}^n$. □

Example 11.6. Figures 11.2 and 11.3 depict the coordinate axes in $\mathbb{R}^2$ and respectively $\mathbb{R}^3$. □

Suppose that we are given two *distinct* points $\boldsymbol{p}, \boldsymbol{q} \in \mathbb{R}^n$. These two points determine two collinear vectors, $\boldsymbol{v} = \boldsymbol{q} - \boldsymbol{p}$ and $-\boldsymbol{v} = \boldsymbol{p} - \boldsymbol{q}$; see Figure 11.5.[3]

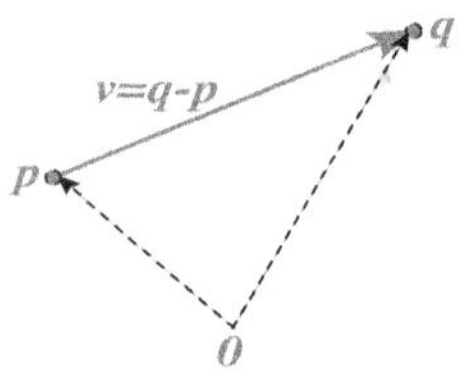

Fig. 11.5 *You should think of $\boldsymbol{v} = \boldsymbol{q} - \boldsymbol{p}$ as the vector described by the arrow that starts at $\boldsymbol{p}$ and ends at $\boldsymbol{q}$.*

The distinct points $\boldsymbol{p}, \boldsymbol{q}$ belong to both lines $\ell_{\boldsymbol{p},\boldsymbol{v}}$ and $\ell_{\boldsymbol{q},-\boldsymbol{v}}$. Since these two lines intersect in two distinct points they must coincide; see Exercise 11.2. Thus

$$\ell_{\boldsymbol{p},\boldsymbol{v}} = \ell_{\boldsymbol{q},-\boldsymbol{v}}.$$

[3]The old-fashioned notation for the vector $\boldsymbol{q} - \boldsymbol{p}$ is $\overrightarrow{\boldsymbol{pq}}$.

This line is called the *line determined by the (distinct) points* $\boldsymbol{p}$ *and* $\boldsymbol{q}$, and we will denote it by $\boldsymbol{pq}$. In other words, $\boldsymbol{pq}$ is the line through $\boldsymbol{p}$ in the direction $\boldsymbol{q}-\boldsymbol{p}$,

$$\boldsymbol{pq} = \ell_{\boldsymbol{p},\boldsymbol{q}-\boldsymbol{p}}.$$

By construction, either of the vectors $\boldsymbol{q}-\boldsymbol{p}$ or $\boldsymbol{p}-\boldsymbol{q}$ is a direction vector of the line $\boldsymbol{pq}$. Observe that this line consists of all the points in $\mathbb{R}^n$ of the form

$$\boldsymbol{p}+t\boldsymbol{v} = \boldsymbol{p}+t(\boldsymbol{q}-\boldsymbol{p}) = (1-t)\boldsymbol{p}+t\boldsymbol{q}, \quad t \in \mathbb{R}.$$

We thus have the important equality

$$\boxed{\boldsymbol{pq} = \Big\{ (1-t)\boldsymbol{p}+t\boldsymbol{q} \in \mathbb{R}^n, \quad t \in \mathbb{R} \Big\} = \boldsymbol{qp}}. \tag{11.7}$$

Example 11.7. Consider the points $\boldsymbol{p} = (1,2,3)$ and $\boldsymbol{q} = (4,5,6)$ in $\mathbb{R}^3$. Then the line through $\boldsymbol{p}$ and $\boldsymbol{q}$ is the subset of $\mathbb{R}^3$ described by

$$\begin{aligned}\boldsymbol{pq} &= \{ (1-t)\cdot(1,2,3)+t\cdot(4,5,6); \;\; t \in \mathbb{R} \} \\ &= \{ (1+3t, 2+3t, 3+3t); \;\; t \in \mathbb{R} \}.\end{aligned}$$

Equivalently, we say that the line $\boldsymbol{pq}$ is described by the equations

$$\begin{aligned} x &= 1+3t, \\ y &= 2+3t, \\ z &= 3+3t, \end{aligned}$$

$t \in \mathbb{R}$. □

Given $\boldsymbol{p},\boldsymbol{q} \in \mathbb{R}^n$, $\boldsymbol{p} \neq \boldsymbol{q}$, the line $\boldsymbol{pq}$ is the image of the function

$$\boldsymbol{f}_{\boldsymbol{p},\boldsymbol{q}} : \mathbb{R} \to \mathbb{R}^n, \quad \boldsymbol{f}_{\boldsymbol{p},\boldsymbol{q}}(t) = (1-t)\boldsymbol{p}+t\boldsymbol{q}.$$

Moreover,

$$\boldsymbol{f}_{\boldsymbol{p},\boldsymbol{q}}(0) = \boldsymbol{p}, \quad \boldsymbol{f}_{\boldsymbol{p},\boldsymbol{q}}(1) = \boldsymbol{q}.$$

Intuitively, the function $\boldsymbol{f}_{\boldsymbol{p},\boldsymbol{q}}$ describes the motion of a particle in the space $\mathbb{R}^n$ that is located at $\boldsymbol{f}_{\boldsymbol{p},\boldsymbol{q}}(t)$ at the moment of time t. The line $\boldsymbol{pq}$ is then the trajectory described by this moving particle. Note that at $t=0$ the particle is located at $\boldsymbol{p}$ while a second later, at $t=1$, the particle is located at $\boldsymbol{q}$. The *line segment* connecting $\boldsymbol{p}$ to $\boldsymbol{q}$ is defined to be the portion of the trajectory described by this particle during the time interval $[0,1]$. We denote this line segment by $[\boldsymbol{p},\boldsymbol{q}]$ and we observe that it has the algebraic description

$$\boxed{[\boldsymbol{p},\boldsymbol{q}] := \Big\{ (1-t)\boldsymbol{p}+t\boldsymbol{q}; \;\; t \in [0,1] \Big\}}. \tag{11.8}$$

Definition 11.8 (Convex sets). Let $n \in \mathbb{N}$. A subset $C \subset \mathbb{R}^n$ is called *convex* if for any two points in C, the segment connecting them is entirely contained in C. More formally, C is convex iff

$$\forall \boldsymbol{p},\boldsymbol{q} \in C, \;\; [\boldsymbol{p},\boldsymbol{q}] \subset C,$$

or, equivalently,

$$\boxed{\forall \boldsymbol{p},\boldsymbol{q} \in C, \;\; \forall t \in [0,1], \;\; (1-t)\boldsymbol{p}+t\boldsymbol{q} \in C}.$$ □

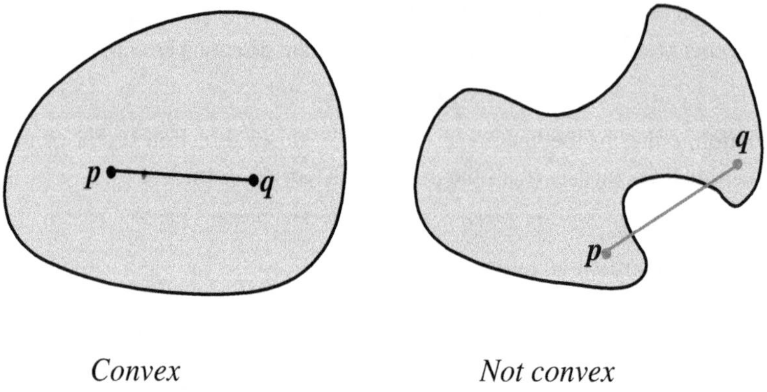

Fig. 11.6 *Examples of convex and non-convex planar sets.*

Definition 11.9 (Linear forms). A *linear form* or *linear functional*[4] on $\mathbb{R}^n$ is a map $\boldsymbol{\xi}$: $\mathbb{R}^n \to \mathbb{R}$ satisfying the following two properties.

(i) (Additivity.) For any $\boldsymbol{x}, \boldsymbol{y} \in \mathbb{R}^n$ we have $\boldsymbol{\xi}(\boldsymbol{x}+\boldsymbol{y}) = \boldsymbol{\xi}(\boldsymbol{x}) + \boldsymbol{\xi}(\boldsymbol{y})$.
(ii) (Homogeneity.) For any $t \in \mathbb{R}$ and any $\boldsymbol{x} \in \mathbb{R}^n$ we have $\boldsymbol{\xi}(t\boldsymbol{x}) = t\boldsymbol{\xi}(\boldsymbol{x})$.

We denote by $(\mathbb{R}^n)^*$ the set of linear forms on $\mathbb{R}^n$ and we will refer to it as the *dual* of $\mathbb{R}^n$. □

☞ We want to emphasize that the linear forms are "*beasts that eat vectors and spit out numbers*".

Example 11.10. (a) The set $(\mathbb{R}^n)^*$ is not empty. The trivial map $\mathbb{R}^n \to \mathbb{R}$ that sends every $\boldsymbol{x}$ to 0 is a linear functional. We will denote it by **0**.

(b) Consider *addition function* $\alpha : \mathbb{R}^2 \to \mathbb{R}$, $\alpha(\boldsymbol{x}) = x^1 + x^2$. Concretely, the function α "eats" a 2-dimensional vector $\boldsymbol{x} = (x^1, x^2)$ and returns the sum of its coordinates. Let us verify that α is a linear form. Indeed, we have

$$\begin{aligned}\alpha(\boldsymbol{x}+\boldsymbol{y}) &= \alpha\big((x^1+y^1, x^2+y^2)\big) = (x^1+y^1) + (x^2+y^2)\\ &= (x^1+x^2) + (y^1+y^2) = \alpha(\boldsymbol{x}) + \alpha(\boldsymbol{y}),\ \ \forall \boldsymbol{x}, \boldsymbol{y} \in \mathbb{R}^2,\\ \alpha(t\boldsymbol{x}) &= \alpha\big((tx^1, tx^2)\big) = tx^1 + tx^2 = t(x^1+x^2) = t\alpha(\boldsymbol{x}),\ \ \forall t \in \mathbb{R},\ \ \boldsymbol{x} \in \mathbb{R}^2.\end{aligned}$$

(c) For any $k = 1, \dots, n$, we define $\boldsymbol{e}^k : \mathbb{R}^n \to \mathbb{R}$ by

$$\boldsymbol{e}^k(\boldsymbol{x}) = x^k,\ \ \forall \boldsymbol{x} = (x^1, \dots, x^n) \in \mathbb{R}^n. \tag{11.9}$$

From the definition of the addition and multiplication by scalars we deduce immediately that the maps $\boldsymbol{e}^k$ are linear functionals. The linear forms $\boldsymbol{e}^1, \dots, \boldsymbol{e}^n$ are called the *basic linear forms on* $\mathbb{R}^n$. □

[4] We stick with the classical terminilogy of *functional* rather than *function*.

The proof of the next result is left to you as an exercise.

Proposition 11.11. *If $\boldsymbol{\xi}, \boldsymbol{\omega}$ are linear forms on $\mathbb{R}^n$ and t is a real number, then the sum $\boldsymbol{\xi} + \boldsymbol{\omega}$ and the multiple $t\boldsymbol{\xi}$ are linear functionals on $\mathbb{R}^n$.*[5] □

The linear forms on $\mathbb{R}^n$ have a very simple structure described in our next result.

Proposition 11.12. *Let $\boldsymbol{\xi} : \mathbb{R}^n \to \mathbb{R}$ be a linear form. For $i = 1, \dots, n$ we set*[6]

$$\xi_i := \xi(\boldsymbol{e}_i),$$

where $\boldsymbol{e}_1, \dots, \boldsymbol{e}_n$ is the canonical basis (11.2) of $\mathbb{R}^n$. Then,

$$\boldsymbol{\xi}(\boldsymbol{x}) = \xi_1 x^1 + \xi_2 x^2 + \cdots + \xi_n x^n = \sum_{i=1}^{n} \xi_i x^i, \;\; \forall \boldsymbol{x} = (x^1, \dots, x^n) \in \mathbb{R}^n. \tag{11.10}$$

Conversely, given any real numbers $\xi_1, \dots, \xi_n$, the linear form

$$\boldsymbol{\xi} = \xi_1 \boldsymbol{e}^1 + \cdots + \xi_n \boldsymbol{e}^n,$$

where $\boldsymbol{e}^k$ are defined by (11.9), satisfies (11.10).

Proof. To prove (11.10) let $\boldsymbol{x} = (x^1, \dots, x^n) \in \mathbb{R}^n$. Then

$$\boldsymbol{x} = x^1 \boldsymbol{e}_1 + \cdots + x^n \boldsymbol{e}_n.$$

From the additivity of $\boldsymbol{\xi}$ we deduce

$$\boldsymbol{\xi}(\boldsymbol{x}) = \boldsymbol{\xi}(x^1 \boldsymbol{e}_1 + \cdots + x^n \boldsymbol{e}_n) = \boldsymbol{\xi}(x^1 \boldsymbol{e}_1) + \cdots + \xi(x^n \boldsymbol{e}_n)$$

(use the homogeneity of ξ)

$$= x^1 \boldsymbol{\xi}(\boldsymbol{e}_1) + \cdots + x^n \boldsymbol{\xi}(\boldsymbol{e}_n) = \xi_1 x^1 + \xi_2 x^2 + \cdots + \xi_n x^n.$$

This proves (11.10).

Conversely, if $\boldsymbol{\xi} = \xi_1 \boldsymbol{e}^1 + \cdots + \xi_n \boldsymbol{e}^n$, then

$$\boldsymbol{\xi}(\boldsymbol{x}) = \xi_1 \boldsymbol{e}^1(\boldsymbol{x}) + \cdots + \xi_n \boldsymbol{e}^n(\boldsymbol{x}) \stackrel{(11.9)}{=} \xi_1 x^1 + \xi_2 x^2 + \cdots + \xi_n x^n.$$

□

The above proposition shows that a linear form $\boldsymbol{\xi}$ on $\mathbb{R}^n$ is completely and uniquely determined by its values on the basic vectors $\boldsymbol{e}_1, \dots, \boldsymbol{e}_n$. We will identify $\boldsymbol{\xi}$ with the *row*

$$[\xi_1, \dots, \xi_n], \;\; \xi_i = \xi(\boldsymbol{e}_i),$$

and *we will think of any length-n row of real numbers as defining a linear form on* $\mathbb{R}^n$. In the physics literature the linear forms are often referred to as *covectors*.

[5] In modern language this signifies that the space $(\mathbb{R}^n)^*$ of linear forms on $\mathbb{R}^n$ is a vector subspace of the vector space of functions on $\mathbb{R}^n \to \mathbb{R}$.

[6] Note that here we use the *subscript* notation, ξ_i, instead of the superscript notation ξ^i. This is part of Einstein's conventions I referred to at the beginning of this chapter.

The basic linear forms $\boldsymbol{e}^1, \ldots, \boldsymbol{e}^n$ defined in (11.9) are uniquely determined by the equalities

$$\boxed{\boldsymbol{e}^i(\boldsymbol{e}_j) = \delta^i_j, \ \ \forall i, j = 1, \ldots, n}, \tag{11.11}$$

where we recall that δ^i_j is the Kronecker symbol (11.4).

Example 11.13. Suppose that $n = 4$. Then the linear form $\boldsymbol{\xi} : \mathbb{R}^4 \to \mathbb{R}$ defined by the *row* vector $[3, 5, 7, 9]$ is given by

$$\boldsymbol{\xi}(x^1, x^2, x^3, x^4) = 3x^1 + 5x^2 + 7x^3 + 9x^4, \ \ \forall (x^1, x^2, x^3, x^4) \in \mathbb{R}^4. \qquad \square$$

Definition 11.14. A subset H of $\mathbb{R}^n$ is called a *hyperplane* if there exists a *nonzero* linear form $\boldsymbol{\xi} : \mathbb{R}^n \to \mathbb{R}$ and a real constant c such that H consists of all the points $\boldsymbol{x} \in \mathbb{R}^n$ satisfying $\boldsymbol{\xi}(\boldsymbol{x}) = c$. $\square$

Example 11.15. (a) A hyperplane in $\mathbb{R}^2$ is a line in $\mathbb{R}^2$. Indeed, any linear form on $\mathbb{R}^2$ has the form

$$\xi(x, y) = ax + by$$

where a, b are fixed real numbers and x, y denote the Cartesian coordinates on $\mathbb{R}^2$. An equation of the form

$$ax + by = c$$

describes a line in $\mathbb{R}^2$. For example, the equation $-2x + y = 3$ describes the line $y = 2x + 3$, with slope 2 and y-intercept 3; see Figure 11.7.

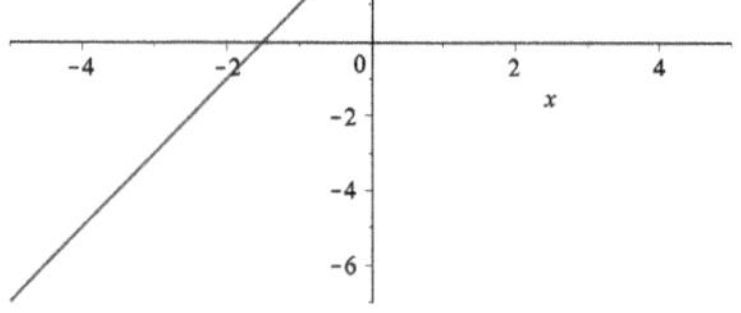

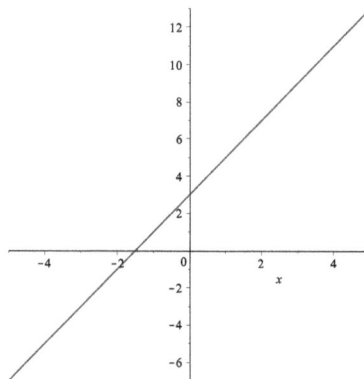

Fig. 11.7 *The planar line with slope 2 and y-intercept 3.*

(b) A hyperplane in $\mathbb{R}^3$ is a plane. For example, Figure 11.8 depicts the plane $x + 2y + 3z = 4$.

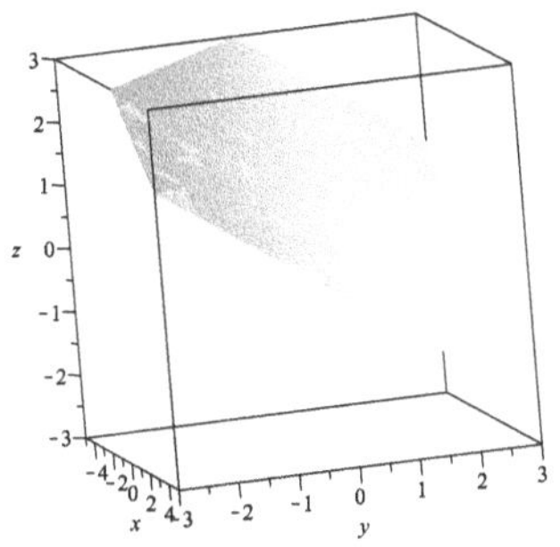

Fig. 11.8 *The plane* $x + 2y + 3z = 4$.

(c) A row vector $[\xi_1, \dots, \xi_n]$ and a constant c define the hyperplane in $\mathbb{R}^n$ consisting of all the points $\boldsymbol{x} = (x^1, \dots, x^n) \in \mathbb{R}^n$ satisfying the linear equation

$$\xi_1 x^1 + \cdots + \xi_n x^n = c.$$

All the hyperplanes in $\mathbb{R}^n$ are of this form. □

Definition 11.16 (Affine subspaces). (a) A nonempty subset $S \subset \mathbb{R}^n$ is called an *affine subspace* if it has the following property: for any points $\boldsymbol{p}, \boldsymbol{q} \in S$, $\boldsymbol{p} \neq \boldsymbol{q}$, the line $\boldsymbol{pq}$ is contained in S. In algebraic terms this means that S is an affine subspace if and only if, for any $\boldsymbol{p}, \boldsymbol{q} \in S$, $\boldsymbol{p} \neq \boldsymbol{q}$, and any $t \in \mathbb{R}$ we have $(1-t)\boldsymbol{p} + t\boldsymbol{q} \in S$.

(b) The subset S is called a *linear subspace* or *vector subspace* if it is an affine subspace and contains the origin. □

Example 11.17. (a) Any point in $\mathbb{R}^n$ is an affine subspace. The space $\mathbb{R}^n$ is obviously an affine subspace of itself.
(b) The lines and the hyperplanes in $\mathbb{R}^n$ are special examples of affine subspaces; see Exercise 11.8. When $n > 3$, there are examples of affine subspaces of $\mathbb{R}^n$ that are neither lines, nor hyperplanes.
(c) If nonempty, the intersection of two affine subspaces is an affine subspace. In particular, if two hyperplanes are not disjoint, then their intersection is an affine subspace. One can prove that if S is an affine subspace of $\mathbb{R}^n$ and $S \neq \mathbb{R}^n$, then S is the intersection of finitely many hyperplanes. □

Proposition 11.18. *Let S be a nonempty subset of $\mathbb{R}^n$. Then the following statements are equivalent.*

(i) The set S is a linear subspace, i.e., it is an affine subspace of $\mathbb{R}^n$ containing the origin.
(ii) For any $\boldsymbol{u}, \boldsymbol{v} \in S$ and any $t \in \mathbb{R}$ we have

$$t\boldsymbol{u} \in S \;\; \text{and} \;\; \boldsymbol{u} + \boldsymbol{v} \in S.$$

In other words, either of the conditions (i) or (ii) above can be used as definition of a linear subspace.

Proof. (i) $\Rightarrow$ (ii) We know that S is an affine subspace and $\mathbf{0} \in S$. Clearly $t\mathbf{0} = \mathbf{0}$, $\forall t \in \mathbb{R}$. For any $\boldsymbol{v} \in S$, $\boldsymbol{v} \neq 0$ and any $t \in \mathbb{R}$ we have

$$t\boldsymbol{v} = (1-t)\mathbf{0} + t\boldsymbol{v} \in S.$$

Thus, any multiple of any vector in S is also a vector in S. Thus, if $\boldsymbol{u} = \boldsymbol{v} \in S$ we have $\boldsymbol{u} + \boldsymbol{v} = 2\boldsymbol{u} \in S$. On the other hand, since S is an affine subspace, if $\boldsymbol{u}, \boldsymbol{v} \in S$, $\boldsymbol{u} \neq \boldsymbol{v}$, the vector $\boldsymbol{w} = \frac{1}{2}\boldsymbol{u} + \frac{1}{2}\boldsymbol{v}$ belongs to S. Hence the multiple $2\boldsymbol{w}$ belongs to S and therefore $\boldsymbol{u} + \boldsymbol{v} = 2\boldsymbol{w} \in S$.

(ii) $\Rightarrow$ (i) Let $\boldsymbol{u} \in S$. Hence $\mathbf{0} = 0 \cdot \boldsymbol{u} \in S$. Next observe that if $\boldsymbol{u}, \boldsymbol{v} \in S$, $\boldsymbol{u} \neq \boldsymbol{v}$, and $t \in \mathbb{R}$, then

$$(1-t)\boldsymbol{u},\ t\boldsymbol{v} \in S \ \Rightarrow\ (1-t)\boldsymbol{u} + t\boldsymbol{v} \in S.$$

This proves that S is an affine subspace. □

Definition 11.19 (Linear operators). Fix $m, n \in \mathbb{N}$. A map $A : \mathbb{R}^n \to \mathbb{R}^m$ is called *linear* or a *linear operator* if it satisfies the following two properties.

(i) (Additivity.) For any $\boldsymbol{x}, \boldsymbol{y} \in \mathbb{R}^n$ we have $A(\boldsymbol{x} + \boldsymbol{y}) = A(\boldsymbol{x}) + A(\boldsymbol{y})$.
(ii) (Homogeneity.) For any $t \in \mathbb{R}$ and any $\boldsymbol{x} \in \mathbb{R}^n$ we have $A(t\boldsymbol{x}) = tA(\boldsymbol{x})$.

We denote by $\operatorname{Hom}(\mathbb{R}^n, \mathbb{R}^m)$ the set of linear operators $\mathbb{R}^n \to \mathbb{R}^m$. □

Note that $\operatorname{Hom}(\mathbb{R}^n, \mathbb{R})$ is none other than the dual of $\mathbb{R}^n$, i.e., the space $(\mathbb{R}^n)^*$ of linear functionals on $\mathbb{R}^n$. Let us mention a simplifying convention that has been universally adopted. If $A : \mathbb{R}^n \to \mathbb{R}^m$ is a linear operator and $\boldsymbol{x} \in \mathbb{R}^n$, then we will often use the simpler notation $A\boldsymbol{x}$ when referring to $A(\boldsymbol{x})$.

The linear operators $\mathbb{R}^n \to \mathbb{R}^m$ have a rather simple structure. Let $A : \mathbb{R}^n \to \mathbb{R}^m$ be a linear operator. Denote by $\boldsymbol{e}_1, \dots, \boldsymbol{e}_n$ the canonical basis of $\mathbb{R}^n$ and by $x^1, \dots, x^n$ the canonical Cartesian coordinates. Similarly, we denote by $\boldsymbol{f}_1, \dots, \boldsymbol{f}_m$ the canonical basis of $\mathbb{R}^m$ and by $y^1, \dots, y^m$ the canonical Cartesian coordinates.

For any

$$\boldsymbol{x} = (x^1, \dots, x^n) = x^1\boldsymbol{e}_1 + \dots + x^n\boldsymbol{e}_n \in \mathbb{R}^n$$

we have

$$A\boldsymbol{x} = A(x^1\boldsymbol{e}_1 + \dots + x^n\boldsymbol{e}_n) = A(x^1\boldsymbol{e}_1) + \dots + A(x^n\boldsymbol{e}_n) = x^1A\boldsymbol{e}_1 + \dots + x^nA\boldsymbol{e}_n. \tag{11.12}$$

This shows that the operator A is completely determined by the m-dimensional vectors

$$A\boldsymbol{e}_1, \dots, A\boldsymbol{e}_n \in \mathbb{R}^m.$$

These m-dimensional vectors are described by columns of height m.

$$Ae_1 = \begin{bmatrix} A_1^1 \\ A_1^2 \\ \vdots \\ A_1^m \end{bmatrix}, \dots, Ae_j = \begin{bmatrix} A_j^1 \\ A_j^2 \\ \vdots \\ A_j^m \end{bmatrix}, \dots, Ae_n = \begin{bmatrix} A_n^1 \\ A_n^2 \\ \vdots \\ A_n^m \end{bmatrix}.$$

Arranging these columns one next to the other we obtain the rectangular array

$$\mathcal{M}_A = \begin{bmatrix} A_1^1 & \cdots & A_j^1 & \cdots & A_n^1 \\ A_1^2 & \cdots & A_j^2 & \cdots & A_n^2 \\ \vdots & \ddots & \vdots & \ddots & \vdots \\ \vdots & \ddots & \vdots & \ddots & \vdots \\ A_1^m & \cdots & A_j^m & \cdots & A_n^m \end{bmatrix}.$$

✍ We need to introduce some terminology and conventions.

- A rectangular array of numbers as above is called a *matrix*.
- The horizontal strings of numbers are called *rows*, and the vertical ones are called *columns*.
- We will denote by $\mathrm{Mat}_{m\times n}(\mathbb{R})$ the space of matrices with real entries, with m rows and n columns. The matrix $\mathcal{M}_A$ above is an $m \times n$ matrix.
- A *square matrix* is a matrix with an equal number of rows and columns. We will denote by $\mathrm{Mat}_n(\mathbb{R})$ the space of square matrices with n rows and columns.
- *The superscripts label the rows and the subscripts label the columns.* Thus, A_7^3 is the entry located at the intersection of the 3rd row with the 7th column of a matrix A.
- We denote by A_j the j-th column and by A^i the i-th row of a matrix A.

Note that a $1 \times k$ matrix is a length-k row

$$R = [r_1 \;\; r_2 \;\dots r_k],$$

while a $k \times 1$ matrix is a height-k column

$$C = \begin{bmatrix} c^1 \\ \vdots \\ c^k \end{bmatrix}.$$

The *pairing* between a row U and a column V of the same size k is defined to be the number

$$\boxed{R \bullet C := r_1c^1 + r_2c^2 + \cdots + r_kc^k}. \tag{11.13}$$

If we identify rows with linear functionals, then $U \bullet V$ is the real number that we get when we feed the vector V to the linear functional defined by U.

The above discussion shows that to any linear operator $A : \mathbb{R}^n \to \mathbb{R}^m$ we can canonically associate an $m \times n$ matrix called the *matrix associated to the linear operator.* This matrix has n columns $A_1, \dots, A_n$ that describe the coordinates of the vectors $Ae_1, \dots, Ae_n$.

Using (11.12) we deduce

$$Ax = x^1 Ae_1 + \cdots + x^n Ae_n$$

$$= x^1 \begin{bmatrix} A_1^1 \\ A_1^2 \\ \vdots \\ A_1^m \end{bmatrix} + x^2 \begin{bmatrix} A_2^1 \\ A_2^2 \\ \vdots \\ A_2^m \end{bmatrix} + \cdots + x^n \begin{bmatrix} A_n^1 \\ A_n^2 \\ \vdots \\ A_n^m \end{bmatrix}$$

$$= \begin{bmatrix} x^1 A_1^1 + x^2 A_2^1 + \cdots + x^n A_n^1 \\ x^1 A_1^2 + x^2 A_2^2 + \cdots + x^n A_n^2 \\ \vdots \\ x^1 A_1^m + x^2 A_2^m + \cdots + x^n A_n^m \end{bmatrix} = \begin{bmatrix} A_1^1 x^1 + A_2^1 x^2 + \cdots + A_n^1 x^n \\ A_1^2 x^1 + A_2^2 x^2 + \cdots + x^n A_n^2 x^n \\ \vdots \\ A_1^m x^1 + A_2^m x^2 + \cdots + A_n^m x^n \end{bmatrix}.$$

Let us analyze a bit the above sum equality. Note that the i-th coordinate of Ax is the quantity

$$\sum_{j=1}^n A_j^i x^j = A_1^i x^1 + A_2^i x^2 + \cdots + A_n^i x^n.$$

Note also that the above expression is obtained by pairing the i-th row $A^i = [A_1^i, \dots, A_n^i]$ of the matrix $\mathcal{M}_A$ with the column vector $x = [x^1, \dots, x^n]^\top$. Thus, the vector Ax in $\mathbb{R}^m$ is described by the column of height m

$$Ax = \begin{bmatrix} \sum_{j=1}^n A_j^1 x^j \\ \sum_{j=1}^n A_j^2 x^j \\ \vdots \\ \sum_{j=1}^n A_j^m x^j \end{bmatrix} = \begin{bmatrix} A^1 \bullet x \\ A^2 \bullet x \\ \vdots \\ A^m \bullet x \end{bmatrix}. \tag{11.14}$$

The above equality shows that each component of Ax is a linear functional in x.

Conversely, given an $m \times n$ matrix $\mathcal{A}$, its columns $A_1, \dots, A_n$ define vectors in $\mathbb{R}^m$ and we can use these vectors to define a linear operator $L = L_{\mathcal{A}} : \mathbb{R}^n \to \mathbb{R}^m$ via the formula

$$L_{\mathcal{A}}(x) = x^1 A_1 + \cdots + x^n A_n, \ \ x = (x^1, \ x^2, \dots, x^n).$$

In particular,

$$L_{\mathcal{A}} e_j = A_j,$$

so that the matrix associated to the operator $L_{\mathcal{A}}$ is the matrix $\mathcal{A}$ we started with. This proves the following very useful fact.

Theorem 11.20. *The correspondence that associates to a linear operator* $\mathbb{R}^n \to \mathbb{R}^m$ *its* $m \times n$ *matrix is a bijection between the set of linear operators* $\mathrm{Hom}(\mathbb{R}^n, \mathbb{R}^m)$ *and the set* $\mathrm{Mat}_{m\times n}(\mathbb{R})$ *of* $m \times n$ *matrices with real entries.* □

Because of the above bijective correspondence we will denote a linear operator and its associated matrix by the same symbol.

Proposition 11.21. *Let* $\ell, m, n \in \mathbb{N}$. *If* $A : \mathbb{R}^n \to \mathbb{R}^m$ *and* $B : \mathbb{R}^m \to \mathbb{R}^\ell$ *are linear operators, then so is their composition* $BA := B \circ A : \mathbb{R}^n \to \mathbb{R}^\ell$.

Proof. To prove the additivity of BA we choose $\boldsymbol{x}, \boldsymbol{y} \in \mathbb{R}^n$. Then

$$BA(\boldsymbol{x}+\boldsymbol{y}) = B\big(\,A(\boldsymbol{x}+\boldsymbol{y})\,\big)$$

(use the additivity of A)

$$= B\big(\,A\boldsymbol{x} + A\boldsymbol{y}\,\big)$$

(use the additivity of B)

$$= B(A\boldsymbol{x}) + B(A\boldsymbol{y}) = BA(\boldsymbol{x}) + BA(\boldsymbol{y}).$$

The homogeneity of BA is proved in a similar fashion. □

In Proposition 11.21 the operator A is represented by an $m \times n$ matrix $\mathcal{M}_A$ and the operator B by an $\ell \times m$ matrix $\mathcal{M}_B$

$$\mathcal{M}_A = \begin{bmatrix} A_1^1 & A_2^1 & \cdots & A_n^1 \\ & & & \\ A_1^2 & A_2^2 & \cdots & A_n^2 \\ \vdots & \vdots & \ddots & \vdots \\ A_1^m & A_2^m & \cdots & A_n^m \end{bmatrix}, \quad \mathcal{M}_B = \begin{bmatrix} B_1^1 & B_2^1 & \cdots & B_m^1 \\ & & & \\ B_1^2 & B_2^2 & \cdots & B_m^2 \\ \vdots & \vdots & \ddots & \vdots \\ B_1^\ell & B_2^\ell & \cdots & B_m^\ell \end{bmatrix}.$$

The operator $BA : \mathbb{R}^n \to \mathbb{R}^\ell$ is represented by an $\ell \times n$ matrix $\mathcal{M}_{BA}$ with entries $(BA)_j^i$ that we want to describe explicitly. Note that the columns of this matrix describe the coordinates of the vectors

$$B(A\boldsymbol{e}_1), \dots, B(A\boldsymbol{e}_n) \in \mathbb{R}^\ell.$$

Thus, for $i = 1, \dots, \ell$, the entry $(BA)_j^i$ denotes the i-th coordinate of the vector $B(A\boldsymbol{e}_j)$. The vector $A\boldsymbol{e}_j$ is described by the column

$$A\boldsymbol{e}_j = A_j = \begin{bmatrix} A_j^1 \\ \vdots \\ A_j^m \end{bmatrix}.$$

Since $(BA)^i_j$ is the i-th coordinate of $B(Ae_j)$, we deduce from (11.14) with $\boldsymbol{x} = Ae_j = A_j$ that

$$\boxed{(BA)^i_j = B^i \bullet A_j}. \tag{11.15}$$

More explicitly, given that $B^i = [B^i_1, \dots, B^i_m]$, we deduce from (11.13) with $U = B^i$ and $V = A_j$ that

$$(BA)^i_j = B^i_1 A^1_j + B^i_2 A^2_j + \cdots + B^i_m A^m_j.$$

Definition 11.22 (Matrix multiplication).[7] Given two matrices

$$A \in \mathrm{Mat}_{m\times n}(\mathbb{R}) \text{ and } B \in \mathrm{Mat}_{\ell\times m}(\mathbb{R})$$

(so that the number of columns of B is equal to the number of rows of A), their *product* is the $\ell \times n$ matrix $B \cdot A$ whose (i, j) entry is the pairing of the i-th row of B with the j-th column of A,

$$\boxed{(B \cdot A)^i_j = B^i \bullet A_j = B^i_1 A^1_j + B^i_2 A^2_j + \cdots + B^i_m A^m_j}. \qquad \square$$

The next result summarizes the above discussion.

Proposition 11.23. *The matrix associated to the composition of two linear operators*

$$A : \mathbb{R}^n \to \mathbb{R}^m, \quad B : \mathbb{R}^m \to \mathbb{R}^\ell$$

is the product of the matrices associated to these operators,

$$\mathcal{M}_{B\circ A} = \mathcal{M}_B \cdot \mathcal{M}_A. \qquad \square$$

Remark 11.24. According to Theorem 11.20, any matrix $A \in \mathrm{Mat}_{m\times n}(\mathbb{R})$ defines a linear operator $L_A : \mathbb{R}^n \to \mathbb{R}^m$. A vector $\boldsymbol{x} \in \mathbb{R}^n$ is represented by a column, i.e., by an $n \times 1$ matrix. The product of the matrices A and $\boldsymbol{x}$ is well defined and produces an $m \times 1$ matrix $A \cdot \boldsymbol{x}$ which can also be viewed as a vector in $\mathbb{R}^m$.

When we feed the vector $\boldsymbol{x}$ to the linear operator L_A defined by A we also obtain a vector in $\mathbb{R}^m$ given by (11.14)

$$L_A\boldsymbol{x} = \begin{bmatrix} A^1 \bullet \boldsymbol{x} \\ \\ A^2 \bullet \boldsymbol{x} \\ \vdots \\ A^m \bullet \boldsymbol{x} \end{bmatrix}.$$

The column on the right-hand side of the above equality is none other than the matrix multiplication $A \cdot \boldsymbol{x}$, i.e.,

$$L_A\boldsymbol{x} = A \cdot \boldsymbol{x}.$$

Thus, *when viewed as a linear operator, the action of a matrix on a vector coincides with the product of that matrix with the vector viewed as a matrix consisting of a single column.*

This remarkable coincidence is one of the main reasons we prefer to think of the vectors in $\mathbb{R}^n$ as *column* vectors. $\square$

[7]Check the site `http://matrixmultiplication.xyz/` that interactively shows you how to multiply matrices.

☞ **Important Convention** *In the sequel, to ease the notational burden, we will denote with the same symbol a linear operator and its associated matrix. With this convention, the equality $L_A\boldsymbol{x} = A \cdot \boldsymbol{x}$ above takes the simper form*

$$A\boldsymbol{x} = A \cdot \boldsymbol{x}. \tag{11.16}$$

Also, due to Proposition 11.23 we will use the simpler notation BA instead of $B \cdot A$ when referring to matrix multiplication.

Example 11.25. (a) A linear operator $\mathbb{R} \to \mathbb{R}$ corresponds to a 1×1 matrix which in turn can be identified with a number. If A is a real number, then the associated linear operator sends a real number x to the real number Ax. Thus, the real number A is the slope of the linear function $f(x) = Ax$. This simple example shows that the matrix associated to a linear operator is a sort of "generalized slope" of the linear operator.
(b) The identity operator $\mathbb{1} : \mathbb{R}^n \to \mathbb{R}^n$ is represented by the $n \times n$ diagonal matrix

$$\mathbb{1} = \mathbb{1}_n = \begin{bmatrix} 1\,0\,0 \cdots 0\,0 \\ 0\,1\,0 \cdots 0\,0 \\ \vdots\;\vdots\;\vdots \;\;\;\vdots\;\;\vdots\;\vdots \\ 0\,0\,0 \cdots 0\,1 \end{bmatrix}.$$

E.g.,

$$\mathbb{1}_2 = \begin{bmatrix} 1\,0 \\ 0\,1 \end{bmatrix}, \quad \mathbb{1}_3 = \begin{bmatrix} 1\,0\,0 \\ 0\,1\,0 \\ 0\,0\,1 \end{bmatrix}.$$

Note that the (i, j) entry of $\mathbb{1}_n$ is δ^i_j, where δ^i_j is the Kronecker symbol defined in (11.4). The identity operator (matrix) $\mathbb{1}_n$ has the property that

$$\mathbb{1}_n A = A\mathbb{1}_n = A, \quad \forall A \in \mathrm{Mat}_{n \times n}(\mathbb{R}).$$

We will denote by $\mathbf{0}$ a matrix whose entries are all equal to 0.
(c) The *diagonal* of a square $n \times n$ matrix A consists of the entries $A^1_1, A^2_2, \dots, A^n_n$. For example the diagonal of the 2×2 matrix

$$A = \begin{bmatrix} \boxed{1} & 2 \\ 3 & \boxed{4} \end{bmatrix}$$

consists of the boxed entries. An $n \times n$ *diagonal matrix* is a matrix of the form

$$\begin{bmatrix} c_1 & 0 & 0 & \cdots & 0 & 0 \\ 0 & c_2 & 0 & \cdots & 0 & 0 \\ \vdots & \vdots & \vdots & & \vdots & \vdots \\ 0 & 0 & 0 & \cdots & 0 & c_n \end{bmatrix}.$$

We will denote the above matrix by $\mathrm{Diag}(c_1, \dots, c_n)$.
(d) An $n \times n$ matrix A is called *symmetric* if $A^i_j = A^j_i$, $\forall i, j = 1, \dots, n$. For example, the matrix below is symmetric.

$$\begin{bmatrix} 1 & 2 & 3 \\ 2 & 4 & 5 \\ 3 & 5 & 6 \end{bmatrix}.$$

(e) We can add two matrices of the same dimensions. Thus

$$(A+B)^i_j = A^i_j + B^i_j,$$

i.e., the (i,j)-entry of $A+B$ is the sum of the (i,j)-entry of A with the (i,j)-entry of B. We can also multiply a matrix A by a scalar $c \in \mathbb{R}$. The new matrix is obtained by multiplying all entries of A by the constant c. □

Example 11.26. The multiplication of matrices resembles in some respects the multiplication of real numbers. For example, the multiplication of matrices is associative

$$(A \cdot B) \cdot C = A \cdot (B \cdot C)$$

for any matrices $A \in \mathrm{Mat}_{k\times\ell}(\mathbb{R})$, $B \in \mathrm{Mat}_{\ell\times m}(\mathbb{R})$, $C \in \mathrm{Mat}_{m\times n}(\mathbb{R})$. It is also distributive with respect to the addition of matrices

$$A \cdot (B+C) = AB + AC, \;\; \forall A \in \mathrm{Mat}_{\ell\times m}(\mathbb{R}), \;\; B, C \in \mathrm{Mat}_{m\times n}(\mathbb{R}).$$

However, there are some important differences. Consider for example the 2×2 matrices

$$A = \begin{bmatrix} 1 & 2 \\ 0 & 0 \end{bmatrix}, \;\; B = \begin{bmatrix} 0 & 3 \\ 0 & 4 \end{bmatrix}.$$

Observe that

$$A \cdot B = \begin{bmatrix} 0 & 3+8 \\ 0 & 0 \end{bmatrix} = \begin{bmatrix} 0 & 11 \\ 0 & 0 \end{bmatrix}, \;\; B \cdot A = \begin{bmatrix} 0 & 0 \\ 0 & 0 \end{bmatrix}.$$

□

This example shows two things.

- The multiplication of matrices is *not* commutative since obviously $AB \neq BA$ in the above example.
- The product of two matrices can be zero, although none of them is zero as in example $BA = 0$ above.

Definition 11.27. Suppose that $A : \mathbb{R}^n \to \mathbb{R}^m$ is a linear operator. The *kernel* of A, denoted by $\ker A$ is the set

$$\ker A := \left\{ \boldsymbol{x} \in \mathbb{R}^n;\;\; A\boldsymbol{x} = \boldsymbol{0} \right\} \subset \mathbb{R}^n. \qquad \square$$

We have the following useful result whose proof is left to you as an exercise.

Proposition 11.28. *Suppose that $A : \mathbb{R}^n \to \mathbb{R}^m$ is a linear operator and $S \subset \mathbb{R}^n$ is a vector subspace. Then its kernel* $\ker A$ *is a linear subspace of $\mathbb{R}^n$ and the image $A(S)$ of S is a vector subspace of $\mathbb{R}^m$. In particular, the range $\boldsymbol{R}(A) := A(\mathbb{R}^n)$ is a linear subspace of $\mathbb{R}^m$.* □

Example 11.29. Consider the 2×3 matrix

$$A = \begin{bmatrix} 1 & 2 & 3 \\ 4 & 5 & 6 \end{bmatrix}.$$

As such, it defines a linear operator $A : \mathbb{R}^3 \to \mathbb{R}^2$ described by

$$\mathbb{R}^3 \ni \boldsymbol{x} = \begin{bmatrix} x^1 \\ x^2 \\ x^3 \end{bmatrix} \mapsto \begin{bmatrix} 1 & 2 & 3 \\ 4 & 5 & 6 \end{bmatrix} \cdot \begin{bmatrix} x^1 \\ x^2 \\ x^3 \end{bmatrix} = \begin{bmatrix} x^1 + 2x^2 + 3x^3 \\ 4x^1 + 5x^2 + 6x^3 \end{bmatrix} \in \mathbb{R}^2.$$

If $\boldsymbol{e}_1, \boldsymbol{e}_2, \boldsymbol{e}_3$ is the natural basis, then $A\boldsymbol{e}_1, A\boldsymbol{e}_2, A\boldsymbol{e}_3$ are described respectively by the columns A_1, A_2, A_3 of A. E.g.,

$$A\boldsymbol{e}_1 = \begin{bmatrix} 1 \\ 4 \end{bmatrix} \in \mathbb{R}^2.$$

The kernel of this operator consists of vectors $\boldsymbol{x} = (x^1, x^2, x^3) \in \mathbb{R}^3$ satisfying $A\boldsymbol{x} = 0$, i.e., the system of linear equations

$$\begin{cases} x^1 + 2x^2 + 3x^3 = 0 \\ 4x^1 + 5x^2 + 6x^3 = 0. \end{cases}$$

If we multiply the first line above by 4 and then subtract it from the second line we deduce

$$\begin{cases} x^1 + 2x^2 + 3x^3 = 0 \\ -3x^2 - 6x^3 = 0 \end{cases} \iff \begin{cases} x^1 + 2x^2 + 3x^3 = 0 \\ x^2 + 2x^3 = 0. \end{cases}$$

We deduce that

$$x^2 = -2x^3, \;\; x^1 = -2x^2 - 3x^3 = x^3.$$

If we set $t := x^3$ we deduce that $(x^1, x^2, x^3) \in \ker A$ if and only if it has the form

$$x^1 = t, \;\; x^2 = -2t, \;\; x^3 = t, \;\; t \in \mathbb{R}.$$

Thus the kernel of A is the line through the origin with direction vector $\boldsymbol{v} = (1, -2, 1)$,

$$\ker A = \ell_{\boldsymbol{0}, \boldsymbol{v}}.$$

□

11.2. Basic Euclidean geometry

The space $\mathbb{R}^n$ has a considerably richer structure than the ones we have discussed in the previous section. The goal of the present section is to describe this additional structure and some of its consequences.

Definition 11.30 (Inner product). The *canonical inner product* in $\mathbb{R}^n$ is the map

$$\mathbb{R}^n \times \mathbb{R}^n \to \mathbb{R}$$

that associates to a pair of vectors $(\boldsymbol{x}, \boldsymbol{y}) \in \mathbb{R}^n \times \mathbb{R}^n$ and the real *number* $\langle \boldsymbol{x}, \boldsymbol{y} \rangle$ defined by

$$\langle \boldsymbol{x}, \boldsymbol{y} \rangle := \sum_{j=1}^{n} x^j y^j = x^1 y^1 + \cdots + x^n y^n.$$

□

Proposition 11.31. *The inner product* $\langle -,- \rangle : \mathbb{R}^n \times \mathbb{R}^n \to \mathbb{R}$ *satisfies the following properties.*

(i) For any $\boldsymbol{x}, \boldsymbol{y}, \boldsymbol{z} \in \mathbb{R}^n$ *we have*

$$\langle \boldsymbol{x}+\boldsymbol{y}, \boldsymbol{z} \rangle = \langle \boldsymbol{x}, \boldsymbol{z} \rangle + \langle \boldsymbol{y}, \boldsymbol{z} \rangle.$$

(ii) For any $\boldsymbol{x}, \boldsymbol{y} \in \mathbb{R}^n$ *and any* $t \in \mathbb{R}$ *we have*

$$\langle t\boldsymbol{x}, \boldsymbol{y} \rangle = \langle \boldsymbol{x}, t\boldsymbol{y} \rangle = t\langle \boldsymbol{x}, \boldsymbol{y} \rangle.$$

(iii) For any $\boldsymbol{x}, \boldsymbol{y} \in \mathbb{R}^n$ *we have*

$$\langle \boldsymbol{x}, \boldsymbol{y} \rangle = \langle \boldsymbol{y}, \boldsymbol{x} \rangle.$$

(iv) For any $\boldsymbol{x} \in \mathbb{R}^n$ *we have* $\langle \boldsymbol{x}, \boldsymbol{x} \rangle \geq 0$ *with equality if and only if* $\boldsymbol{x} = 0$.

Proof. (i) We have

$$\begin{aligned}\langle \boldsymbol{x}+\boldsymbol{y}, \boldsymbol{z} \rangle &= (x^1+y^1)z^1 + \cdots + (x^n+y^n)z^n \\ &= (x^1z^1 + \cdots + x^nz^n) + (y^1z^1 + \cdots + y^nz^n) = \langle \boldsymbol{x}, \boldsymbol{z} \rangle + \langle \boldsymbol{y}, \boldsymbol{z} \rangle.\end{aligned}$$

The properties (ii) and (iii) are obvious. As for (iv), note that

$$\langle \boldsymbol{x}, \boldsymbol{x} \rangle = (x^1)^2 + \cdots + (x^n)^2 \geq 0.$$

Clearly, we have equality if and only if $x^1 = \cdots = x^n = 0$. □

Definition 11.32. The *Euclidean norm* or *length* of a vector $\boldsymbol{x} = [x^1, \dots, x^n]^\top \in \mathbb{R}^n$ is the nonnegative real number $\|\boldsymbol{x}\|$ defined by

$$\|\boldsymbol{x}\| := \sqrt{\langle \boldsymbol{x}, \boldsymbol{x} \rangle} = \sqrt{(x^1)^2 + \cdots + (x^n)^2}.$$ □

Observe that

$$\|\boldsymbol{x}\|^2 = \langle \boldsymbol{x}, \boldsymbol{x} \rangle, \ \ \forall \boldsymbol{x} \in \mathbb{R}^n.$$

The Cauchy–Schwarz inequality (8.24) implies that for any

$$\boldsymbol{x} = [x^1, \dots, x^n]^\top, \ \boldsymbol{y} = [y^1, \dots, y^n]^\top$$

we have

$$\left| x^1y^1 + \cdots + x^ny^n \right| \leq \sqrt{(x^1)^2 + \cdots (x^n)^2} \cdot \sqrt{(y^1)^2 + \cdots + (y^n)^2}.$$

This can be rewritten in the more compact form

$$\boxed{\left| \langle \boldsymbol{x}, \boldsymbol{y} \rangle \right| \leq \|\boldsymbol{x}\| \cdot \|\boldsymbol{y}\|, \ \ \forall \boldsymbol{x}, \boldsymbol{y} \in \mathbb{R}^n.} \tag{11.17}$$

We will refer to (11.17) as the Cauchy–Schwarz inequality. Given the importance of this inequality we present below an alternate proof.

Alternate proof of the inequality (11.17). The inequality (11.17) obviously holds if $\boldsymbol{x} = \mathbf{0}$ or $\boldsymbol{y} = \mathbf{0}$ so it suffices to prove it in the case $\boldsymbol{x}, \boldsymbol{y} \neq 0$. Consider the function

$$f : \mathbb{R} \to \mathbb{R},\ \ f(t) = \langle t\boldsymbol{x} + \boldsymbol{y}, t\boldsymbol{x} + \boldsymbol{y} \rangle = \|t\boldsymbol{x} + \boldsymbol{y}\|^2.$$

Clearly, $f(t) \geq 0$ and $f(t_0) = 0$ for some $t_0 \in \mathbb{R}$ if and only if $\boldsymbol{x}, \boldsymbol{y}$ are collinear, $\boldsymbol{y} = -t_0\boldsymbol{x}$. Using Proposition 11.31 we deduce

$$\begin{aligned} f(t) &= \langle t\boldsymbol{x} + \boldsymbol{y}, t\boldsymbol{x} \rangle + \langle t\boldsymbol{x} + \boldsymbol{y}, \boldsymbol{y} \rangle = t\langle t\boldsymbol{x} + \boldsymbol{y}, \boldsymbol{x} \rangle + \langle t\boldsymbol{x} + \boldsymbol{y}, \boldsymbol{y} \rangle \\ &= t\big(\langle t\boldsymbol{x}, \boldsymbol{x} \rangle + \langle \boldsymbol{y}, \boldsymbol{x} \rangle \big) + t\langle \boldsymbol{x}, \boldsymbol{y} \rangle + \langle \boldsymbol{y}, \boldsymbol{y} \rangle \\ &= t^2\langle \boldsymbol{x}, \boldsymbol{x} \rangle + t\langle \boldsymbol{y}, \boldsymbol{x} \rangle + t\langle \boldsymbol{x}, \boldsymbol{y} \rangle + \langle \boldsymbol{y}, \boldsymbol{y} \rangle \\ &= \underbrace{\langle \boldsymbol{x}, \boldsymbol{x} \rangle}_{a}\, t^2 + \underbrace{2\langle \boldsymbol{x}, \boldsymbol{y} \rangle}_{b}\, t + \underbrace{\langle \boldsymbol{y}, \boldsymbol{y} \rangle}_{c} \\ &= at^2 + bt + c,\ \ a > 0. \end{aligned}$$

This shows that the quadratic polynomial $at^2 + bt + c$ with $a > 0$ is nonnegative for every $t \in \mathbb{R}$. From Exercise 3.10(a) we conclude that this is possible if and only if $b^2 - 4ac \leq 0$, i.e.,

$$4\big|\, \langle \boldsymbol{x}, \boldsymbol{y} \rangle \,\big|^2 - 4\|\boldsymbol{x}\|^2\|\boldsymbol{y}\|^2 \leq 0.$$

This implies $\big|\, \langle \boldsymbol{x}, \boldsymbol{y} \rangle \,\big| \leq \|\boldsymbol{x}\| \cdot \|\boldsymbol{y}\|$. □

Remark 11.33. The above argument shows that if

$$|\langle \boldsymbol{x}, \boldsymbol{y} \rangle| = \|\boldsymbol{x}\| \cdot \|\boldsymbol{y}\|,$$

then $b^2 - 4ac = 0$. In particular, this implies that there exists $t \in \mathbb{R}$ such that $t\boldsymbol{x} + \boldsymbol{y} = 0$, i.e., the vectors are collinear. Conversely, if the vectors $\boldsymbol{x}, \boldsymbol{y}$ are collinear, then clearly $|\langle \boldsymbol{x}, \boldsymbol{y} \rangle| = \|\boldsymbol{x}\| \cdot \|\boldsymbol{y}\|$. □

The Cauchy–Schwarz inequality implies that, for any nonzero vectors $\boldsymbol{x}, \boldsymbol{y} \in \mathbb{R}^n$, we have

$$\frac{\langle \boldsymbol{x}, \boldsymbol{y} \rangle}{\|\boldsymbol{x}\| \cdot \|\boldsymbol{y}\|} \in [-1, 1].$$

Thus, there exists a unique $\theta \in [0, \pi]$ such that

$$\cos\theta = \frac{\langle \boldsymbol{x}, \boldsymbol{y} \rangle}{\|\boldsymbol{x}\| \cdot \|\boldsymbol{y}\|}.$$

Definition 11.34. The *angle* between the *nonzero* vectors $\boldsymbol{x}, \boldsymbol{y} \in \mathbb{R}^n$, denoted by $\measuredangle(\boldsymbol{x}, \boldsymbol{y})$, is defined to be the unique number $\theta \in [0, \pi]$ such that

$$\cos\theta = \frac{\langle \boldsymbol{x}, \boldsymbol{y} \rangle}{\|\boldsymbol{x}\| \cdot \|\boldsymbol{y}\|}.$$

□

Thus, for any $\boldsymbol{x}, \boldsymbol{y} \in \mathbb{R}^n$, $\boldsymbol{x}, \boldsymbol{y} \neq \boldsymbol{0}$, we have

$$\boxed{\cos \measuredangle(\boldsymbol{x}, \boldsymbol{y}) = \frac{\langle \boldsymbol{x}, \boldsymbol{y} \rangle}{\|\boldsymbol{x}\| \cdot \|\boldsymbol{y}\|}} \text{ and } \boxed{\langle \boldsymbol{x}, \boldsymbol{y} \rangle = \|\boldsymbol{x}\| \cdot \|\boldsymbol{y}\| \cos \measuredangle(\boldsymbol{x}, \boldsymbol{y})}. \tag{11.18}$$

Classically, two nonzero vectors $\boldsymbol{x}, \boldsymbol{y}$ are orthogonal if $\measuredangle(\boldsymbol{x}, \boldsymbol{y}) = \frac{\pi}{2}$, i.e., $\cos \measuredangle(\boldsymbol{x}, \boldsymbol{y}) = 0$. The equality (11.18) shows that this happens iff $\langle \boldsymbol{x}, \boldsymbol{y} \rangle = 0$. This justifies our next definition.

Definition 11.35. We say that two vectors $\boldsymbol{x}, \boldsymbol{y} \in \mathbb{R}^n$ are *orthogonal*, and we write this $\boldsymbol{x} \perp \boldsymbol{y}$, if $\langle \boldsymbol{x}, \boldsymbol{y} \rangle = 0$. □

Example 11.36. If $\boldsymbol{e}_1, \dots, \boldsymbol{e}_n$ is the canonical basis of $\mathbb{R}^n$ (see (11.2)), then

$$\|\boldsymbol{e}_1\| = \cdots = \|\boldsymbol{e}_n\| = 1,$$

and

$$\boldsymbol{e}_i \perp \boldsymbol{e}_j, \ \ \forall i \neq j.$$

We can rewrite these facts in the more succinct form

$$\langle \boldsymbol{e}_i, \boldsymbol{e}_j \rangle = \delta_{ij} := \begin{cases} 1, & i = j, \\ 0, & i \neq j. \end{cases}$$

The collection (δ_{ij}) above is also called *Kronecker symbol.* Note that for any vector

$$\boldsymbol{x} = \begin{bmatrix} x^1 \\ \vdots \\ x^n \end{bmatrix} \in \mathbb{R}^n$$

we have

$$x^i = \langle \boldsymbol{x}, \boldsymbol{e}_i \rangle, \ \ \forall i = 1, 2, \dots, n,$$

and thus

$$\boldsymbol{x} = \langle \boldsymbol{x}, \boldsymbol{e}_1 \rangle \boldsymbol{e}_1 + \cdots + \langle \boldsymbol{x}, \boldsymbol{e}_n \rangle \boldsymbol{e}_n.$$

□

Theorem 11.37 (Pythagoras). *If $\boldsymbol{x}, \boldsymbol{y} \in \mathbb{R}^n$ and $\boldsymbol{x} \perp \boldsymbol{y}$, then*

$$\|\boldsymbol{x} + \boldsymbol{y}\|^2 = \|\boldsymbol{x}\|^2 + \|\boldsymbol{y}\|^2.$$

Proof. We have

$$\begin{aligned} \|\boldsymbol{x} + \boldsymbol{y}\|^2 &= \langle \boldsymbol{x} + \boldsymbol{y}, \boldsymbol{x} + \boldsymbol{y} \rangle = \langle \boldsymbol{x}, \boldsymbol{x} + \boldsymbol{y} \rangle + \langle \boldsymbol{y}, \boldsymbol{x} + \boldsymbol{y} \rangle \\ &= \langle \boldsymbol{x}, \boldsymbol{x} \rangle + \underbrace{\langle \boldsymbol{x}, \boldsymbol{y} \rangle + \langle \boldsymbol{y}, \boldsymbol{x} \rangle}_{=0} + \langle \boldsymbol{y}, \boldsymbol{y} \rangle = \langle \boldsymbol{x}, \boldsymbol{x} \rangle + \langle \boldsymbol{y}, \boldsymbol{y} \rangle = \|\boldsymbol{x}\|^2 + \|\boldsymbol{y}\|^2. \end{aligned}$$

□

Observe that any vector $\boldsymbol{x} \in \mathbb{R}^n$ defines a linear functional

$$\boldsymbol{x}^{\downarrow} : \mathbb{R}^n \to \mathbb{R}, \quad \boldsymbol{x}^{\downarrow}(\boldsymbol{y}) := \langle \boldsymbol{x}, \boldsymbol{y} \rangle.$$

We will refer to the functional $\boldsymbol{x}^{\downarrow}$ as the *dual* of $\boldsymbol{x}$. It is not hard to see that all the linear functionals on $\mathbb{R}^n$ are duals of vectors in $\mathbb{R}^n$.

Proposition 11.38. *Let $n \in \mathbb{N}$. Any linear functional $\boldsymbol{\xi} : \mathbb{R}^n \to \mathbb{R}$ is the dual of a unique vector in $\mathbb{R}^n$. This means that there exists a unique vector $\boldsymbol{z} \in \mathbb{R}^n$ such that $\boldsymbol{\xi} = \boldsymbol{z}^{\downarrow}$, i.e.,*

$$\boldsymbol{\xi}(\boldsymbol{x}) = \langle \boldsymbol{z}, \boldsymbol{x} \rangle, \quad \forall \boldsymbol{x} \in \mathbb{R}^n. \tag{11.19}$$

This unique vector $\boldsymbol{z}$ is called the dual *of $\boldsymbol{\xi}$ and it is denoted by $\boldsymbol{\xi}_{\uparrow}$.*

Proof. Let $\boldsymbol{e}_1, \dots, \boldsymbol{e}_n$ be the canonical basis of $\mathbb{R}^n$. Set

$$\xi_i := \boldsymbol{\xi}(\boldsymbol{e}_i), \quad i = 1, 2, \dots, n.$$

The vector $\boldsymbol{z} = (z^1, \dots, z^n)$ satisfies (11.19) if and only if

$$z^i = \langle \boldsymbol{z}, \boldsymbol{e}_i \rangle = \boldsymbol{\xi}(\boldsymbol{e}_i) = \xi_i, \quad i = 1, 2, \dots, n.$$

□

The above proof shows that, if the *linear form* $\boldsymbol{\xi}$ is described by the *row*

$$\boldsymbol{\xi} = [\xi_1, \dots, \xi_n],$$

then $\boldsymbol{\xi}_{\uparrow}$ is the *vector* described by the *column*

$$\boldsymbol{\xi}_{\uparrow} = \begin{bmatrix} \xi_1 \\ \vdots \\ \xi_n \end{bmatrix} \iff \boldsymbol{\xi}_{\uparrow}^i = \xi_i, \tag{11.20a}$$

$$\boldsymbol{\xi}(\boldsymbol{x}) = \boldsymbol{\xi} \bullet \boldsymbol{x} = \langle \boldsymbol{\xi}_{\uparrow}, \boldsymbol{x} \rangle, \quad \forall \boldsymbol{x} \in \mathbb{R}^n. \tag{11.20b}$$

Note that

$$(\boldsymbol{e}_i)^{\downarrow} = \boldsymbol{e}^i, \quad (\boldsymbol{e}^j)_{\uparrow} = \boldsymbol{e}_j, \quad \forall i, j = 1, \dots, n. \tag{11.21}$$

The duality operation defined above has a very simple intuitive description: it takes a row $\boldsymbol{\xi}$ and transforms into a column $\boldsymbol{\xi}_{\uparrow}$ with the same entries, and vice-versa, it takes a column $\boldsymbol{x}$ and transforms it into a row $\boldsymbol{x}^{\downarrow}$ with the same entries. E.g.,

$$[1, -2, 3]_{\uparrow} = \begin{bmatrix} 1 \\ -2 \\ 3 \end{bmatrix}, \quad \begin{bmatrix} 4 \\ 5 \\ 6 \end{bmatrix}^{\downarrow} = [4, 5, 6].$$

Proposition 11.39. *Let $n \in \mathbb{N}$ and $H \subset \mathbb{R}^n$. The following statements are equivalent.*

(i) The subset H is a hyperplane.
(ii) There exists a nonzero vector $\boldsymbol{N} \in \mathbb{R}^n$ and a constant $c \in \mathbb{R}$ such that $\boldsymbol{p} \in H$ if and only if $\langle \boldsymbol{N}, \boldsymbol{p} \rangle = c$.

Proof. (i) $\Rightarrow$ (ii) Since H is a hyperplane there exists a nonzero linear functional $\boldsymbol{\xi}:\mathbb{R}^n\to\mathbb{R}$ and a real number c such that

$$\boldsymbol{x}\in H\Longleftrightarrow\boldsymbol{\xi}(\boldsymbol{x})=c.$$

Let $\boldsymbol{N}:=\boldsymbol{\xi}_{\uparrow}$, i.e., $\langle\boldsymbol{N},\boldsymbol{x}\rangle=\boldsymbol{\xi}(\boldsymbol{x})$, $\forall\boldsymbol{x}\in\mathbb{R}^n$. Then, for any $\boldsymbol{p},\boldsymbol{q}\in H$, we have

$$\langle\boldsymbol{N},\boldsymbol{p}\rangle=\boldsymbol{\xi}(\boldsymbol{p})=c=\boldsymbol{\xi}(\boldsymbol{q})=\langle\boldsymbol{N},\boldsymbol{q}\rangle.$$

(ii) $\Rightarrow$ (i) Let $\boldsymbol{\xi}:=\boldsymbol{N}^{\downarrow}$. Then

$$\boldsymbol{p}\in H\Longleftrightarrow\langle\boldsymbol{N},\boldsymbol{p}\rangle=c\Longleftrightarrow\boldsymbol{\xi}(\boldsymbol{p})=c.$$

This shows that H is a hyperplane. □

Suppose that $H\subset\mathbb{R}^n$ is a hyperplane. Hence, there exist $\boldsymbol{N}\in\mathbb{R}^n\setminus\{\boldsymbol{0}\}$ and $c\in\mathbb{R}$ such that

$$\boldsymbol{x}\in H\Longleftrightarrow\langle\boldsymbol{N},\boldsymbol{x}\rangle=c.$$

If $\boldsymbol{p},\boldsymbol{q}\in H$ and $\boldsymbol{p}\neq\boldsymbol{q}$, then the direction of the line $\boldsymbol{pq}$ is given by the vector $\boldsymbol{q}-\boldsymbol{p}$. Now observe that

$$\langle\boldsymbol{N},\boldsymbol{q}-\boldsymbol{p}\rangle=\langle\boldsymbol{N},\boldsymbol{q}\rangle-\langle\boldsymbol{N},\boldsymbol{p}\rangle=0\Rightarrow\boldsymbol{N}\perp(\boldsymbol{q}-\boldsymbol{p}).$$

Thus, the defining vector $\boldsymbol{N}$ is *perpendicular to all the lines contained in* H. We say that $\boldsymbol{N}$ is orthogonal to H, we write this $\boldsymbol{N}\perp H$ and we will refer to $\boldsymbol{N}$ as *a normal vector* of H.

Example 11.40. (a) As we have mentioned earlier, any line in $\mathbb{R}^2$ is also an affine hyperplane. For example, the line given by the equation $2x+3y=5$ admits the vector $\boldsymbol{N}=(2,3)$ as normal vector.

(b) If $n\in\mathbb{N}$, then for any $\boldsymbol{p}\in\mathbb{R}^n$ and any $\boldsymbol{N}\in\mathbb{R}^n$, $\boldsymbol{N}\neq\boldsymbol{0}$, we denote by $H_{\boldsymbol{p},\boldsymbol{N}}$ the *hyperplane through* $\mathbf{p}$ *and normal* $\boldsymbol{N}$, i.e., the hyperplane

$$H_{\boldsymbol{p},\boldsymbol{N}}=\big\{\,\boldsymbol{x}\in\mathbb{R}^n;\;\;\langle\boldsymbol{N},\boldsymbol{x}\rangle=\langle\boldsymbol{N},\boldsymbol{p}\rangle\,\big\}.$$

Clearly $\boldsymbol{p}\in\boldsymbol{N}$. For example if $n=3$, $\boldsymbol{p}=(1,1,1)$ and $\boldsymbol{N}=(1,2,3)$ then $\langle\boldsymbol{N},\boldsymbol{p}\rangle=1+2+3=6$ and

$$H_{\boldsymbol{p},\boldsymbol{N}}=\big\{\,(x,y,z)\in\mathbb{R}^3;\;\;x+2y+3z=6\,\big\}.$$ □

Example 11.41 (The cross product in $\mathbb{R}^3$). The 3-dimensional Euclidean space $\mathbb{R}^3$ is equipped with another operation that is not available in any other dimensions. The *cross product* is the map

$$\times:\mathbb{R}^3\times\mathbb{R}^3\to\mathbb{R}^3,\;\;(\boldsymbol{u},\boldsymbol{v})\mapsto\boldsymbol{u}\times\boldsymbol{v}$$

uniquely characterized by the following conditions

(i) $\forall\boldsymbol{u},\boldsymbol{v},\boldsymbol{w}\in\mathbb{R}^3$

$$(\boldsymbol{u}+\boldsymbol{v})\times\boldsymbol{w}=(\boldsymbol{u}\times\boldsymbol{w})+(\boldsymbol{v}\times\boldsymbol{w}),$$
$$\boldsymbol{w}\times(\boldsymbol{u}+\boldsymbol{v})=(\boldsymbol{w}\times\boldsymbol{u})+(\boldsymbol{w}\times\boldsymbol{v}).$$

(ii)

$$(t\boldsymbol{u}) \times \boldsymbol{v} = \boldsymbol{u} \times (t\boldsymbol{v}) = t(\boldsymbol{u} \times \boldsymbol{v}), \quad \forall t \in \mathbb{R}, \quad \boldsymbol{u}, \boldsymbol{v} \in \mathbb{R}^3.$$

(iii)

$$\boldsymbol{u} \times \boldsymbol{v} = -(\boldsymbol{v} \times \boldsymbol{u}), \quad \forall \boldsymbol{u}, \boldsymbol{v} \in \mathbb{R}^3.$$

(iv)

$$\boldsymbol{e}_1 \times \boldsymbol{e}_2 = \boldsymbol{e}_3, \quad \boldsymbol{e}_2 \times \boldsymbol{e}_3 = \boldsymbol{e}_1, \quad \boldsymbol{e}_3 \times \boldsymbol{e}_1 = \boldsymbol{e}_2.$$

Note that (iii) implies that

$$\boldsymbol{u} \times \boldsymbol{u} = \boldsymbol{0}, \quad \forall \boldsymbol{u} \in \mathbb{R}^3.$$

Indeed

$$\boldsymbol{u} \times \boldsymbol{u} = -(\boldsymbol{u} \times \boldsymbol{u}) \Rightarrow 2(\boldsymbol{u} \times \boldsymbol{u}) = \boldsymbol{0} \Rightarrow \boldsymbol{u} \times \boldsymbol{u} = \boldsymbol{0}.$$

For example, if

$$\boldsymbol{u} = [1, 2, 3]^\top, \quad \boldsymbol{v} = [4, 5, 6]^\top,$$

then

$$\boldsymbol{u} \times \boldsymbol{v} = (\boldsymbol{e}_1 + 2\boldsymbol{e}_2 + 3\boldsymbol{e}_3) \times (4\boldsymbol{e}_1 + 5\boldsymbol{e}_2 + 6\boldsymbol{e}_3)$$

$$= \underbrace{\boldsymbol{e}_1 \times (4\boldsymbol{e}_1 + 5\boldsymbol{e}_2 + 6\boldsymbol{e}_3)}_{I} + \underbrace{2\boldsymbol{e}_2 \times (4\boldsymbol{e}_1 + 5\boldsymbol{e}_2 + 6\boldsymbol{e}_3)}_{II} + \underbrace{3\boldsymbol{e}_3 \times (4\boldsymbol{e}_1 + 5\boldsymbol{e}_2 + 6\boldsymbol{e}_3)}_{III}$$

$$= \underbrace{5\boldsymbol{e}_1 \times \boldsymbol{e}_2 + 6\boldsymbol{e}_1 \times \boldsymbol{e}_3}_{I} + \underbrace{8\boldsymbol{e}_2 \times \boldsymbol{e}_1 + 12\boldsymbol{e}_2 \times \boldsymbol{e}_3}_{II} + \underbrace{12\boldsymbol{e}_3 \times \boldsymbol{e}_1 + 15\boldsymbol{e}_3 \times \boldsymbol{e}_2}_{III}$$

$$= \underbrace{(5\boldsymbol{e}_3 - 6\boldsymbol{e}_2)}_{I} + \underbrace{(-8\boldsymbol{e}_3 + 12\boldsymbol{e}_1)}_{II} + \underbrace{(12\boldsymbol{e}_2 - 15\boldsymbol{e}_1)}_{III}$$

$$= -3\boldsymbol{e}_1 + 6\boldsymbol{e}_2 - 3\boldsymbol{e}_3 = [-3, 6, -3]^\top.$$

If we set $\boldsymbol{w} = \boldsymbol{u} \times \boldsymbol{v} = [-3, 6, -3]^\top$, then we observe that

$$\langle \boldsymbol{w}, \boldsymbol{u} \rangle = \langle \boldsymbol{w}, \boldsymbol{v} \rangle = 0.$$

We have

$$\|\boldsymbol{u}\| = \sqrt{1^2 + 2^2 + 3^2} = \sqrt{14}, \quad \|\boldsymbol{v}\| = \sqrt{4^2 + 5^2 + 6^2} = \sqrt{77},$$

$$\|\boldsymbol{u}\| \cdot \|\boldsymbol{v}\| = \sqrt{14 \cdot 77} = \sqrt{1078},$$

$$\langle \boldsymbol{u}, \boldsymbol{v} \rangle = 4 + 10 + 18 = 32.$$

If we denote by θ the angle between $\boldsymbol{u}$ and $\boldsymbol{v}$, then we deduce

$$\cos \theta = \frac{32}{\sqrt{1078}}.$$

Hence

$$\sin^2\theta = 1-\cos^2\theta = \frac{54}{1078}.$$

Note that

$$\|\boldsymbol{u}\times\boldsymbol{v}\| = \sqrt{3^2+6^2+3^2} = \sqrt{54}.$$

This proves that

$$\boldsymbol{u},\boldsymbol{v}\perp(\boldsymbol{u}\times\boldsymbol{v}),\ \ \|\boldsymbol{u}\times\boldsymbol{v}\| = \sqrt{54} = \sqrt{1078}\cdot\sqrt{\frac{54}{1078}} = \|\boldsymbol{u}\|\cdot\|\boldsymbol{v}\|\sin\theta.$$

Let us observe that the quantity $\|\boldsymbol{u}\|\cdot\|\boldsymbol{v}\|\sin\theta$ is the area of the parallelogram spanned by the vectors $\boldsymbol{u},\boldsymbol{v}$.

The above observations are manifestations of a more general phenomenon. Given any two vectors

$$\boldsymbol{u} = [u^1,u^2,u^3]^\top,\ \ \boldsymbol{v} = [v^1,v^2,v^3]^\top \in \mathbb{R}^3,$$

then the properties (i)–(iv) show that[8]

$$\boxed{\boldsymbol{u}\times\boldsymbol{v} = (u^2v^3-u^3v^2)\boldsymbol{e}_1 + (u^3v^1-u^1v^3)\boldsymbol{e}_2 + (u^1v^2-u^2v^1)\boldsymbol{e}_3}. \tag{11.22}$$

Using this equality one can show that $\boldsymbol{u}\times\boldsymbol{v}$ is a vector perpendicular to both $\boldsymbol{u}$ and $\boldsymbol{v}$ and its length is equal to the area of the parallelogram spanned by the vectors $\boldsymbol{u},\boldsymbol{v}$. These facts alone almost completely determine the vector $\boldsymbol{u}\times\boldsymbol{v}$. There are two vectors with these properties, and to determine which is the cross product we need to indicate the direction or orientation of this vector. This is achieved using the *right-hand rule*.

☛ Align your right-hand thumb with the vector $\boldsymbol{u}$ and your right-hand index with the vector $\boldsymbol{v}$. If you then move the right-hand middle finger so that it is perpendicular to your right-hand palm, then it will be aligned with $\boldsymbol{u}\times\boldsymbol{v}$. □

Definition 11.42. Suppose that $V\subset\mathbb{R}^n$ is a vector subspace. Its *orthogonal complement* is the subset

$$V^\perp := \big\{\,\boldsymbol{u}\in\mathbb{R}^n;\ \ \langle\boldsymbol{u},\boldsymbol{v}\rangle = 0,\ \ \forall\boldsymbol{v}\in V\,\big\}.$$ □

11.3. Basic Euclidean topology

The notions of convergence and continuity on the real axis have a multidimensional counterpart. The main reason why this happens is because the Euclidean norm $\|-\|$ behaves like the absolute value on $\mathbb{R}$. Observe first that

$$\|t\boldsymbol{x}\| = |t|\cdot\|\boldsymbol{x}\|,\ \ \forall\boldsymbol{x}\in\mathbb{R}^n,\ \ t\in\mathbb{R}, \tag{11.23a}$$

$$\|\boldsymbol{x}\|\geq 0,\ \ \|\boldsymbol{x}\| = 0 \Longleftrightarrow \boldsymbol{x} = \boldsymbol{0}. \tag{11.23b}$$

Additionally, and less trivially, we have the following key result.

[8]Do not try to memorize (11.22). Use (i)–(iv) whenever you want to compute a cross product.

Theorem 11.43 (Triangle inequality). *Let $n \in \mathbb{N}$. For any $\boldsymbol{x}, \boldsymbol{y} \in \mathbb{R}^n$ we have*

$$\|\boldsymbol{x} + \boldsymbol{y}\| \leq \|\boldsymbol{x}\| + \|\boldsymbol{y}\|. \tag{11.24a}$$

$$\Big|\, \|\boldsymbol{x}\| - \|\boldsymbol{y}\| \,\Big| \leq \|\boldsymbol{x} - \boldsymbol{y}\|. \tag{11.24b}$$

Proof. Observe that

$$\|\boldsymbol{x}+\boldsymbol{y}\|^2 = \langle \boldsymbol{x}+\boldsymbol{y}, \boldsymbol{x}+\boldsymbol{y}\rangle = \langle \boldsymbol{x}, \boldsymbol{x}\rangle + \langle \boldsymbol{x}, \boldsymbol{y}\rangle + \langle \boldsymbol{y}, \boldsymbol{x}\rangle + \langle \boldsymbol{y}, \boldsymbol{y}\rangle = \|\boldsymbol{x}\|^2 + 2\langle \boldsymbol{x}, \boldsymbol{y}\rangle + \|\boldsymbol{y}\|^2$$

(use the Cauchy–Schwarz inequality)

$$\leq \|\boldsymbol{x}\|^2 + 2\|\boldsymbol{x}\| \cdot \|\boldsymbol{y}\| + \|\boldsymbol{y}\|^2 = \big(\, \|\boldsymbol{x}\| + \|\boldsymbol{y}\| \,\big)^2.$$

Hence

$$\|\boldsymbol{x} + \boldsymbol{y}\|^2 \leq \big(\, \|\boldsymbol{x}\| + \|\boldsymbol{y}\| \,\big)^2.$$

This proves (11.24a).

Next, observe that (11.24a) implies

$$\|\boldsymbol{x}\| = \|\boldsymbol{y} + (\boldsymbol{x} - \boldsymbol{y})\| \leq \|\boldsymbol{y}\| + \|\boldsymbol{x} - \boldsymbol{y}\| \Rightarrow \|\boldsymbol{x}\| - \|\boldsymbol{y}\| \leq \|\boldsymbol{x} - \boldsymbol{y}\|.$$

Similarly

$$\begin{aligned} \|\boldsymbol{y}\| = \|\boldsymbol{x} + (\boldsymbol{y} - \boldsymbol{x})\| \leq \|\boldsymbol{x}\| + \|(\boldsymbol{y} - \boldsymbol{x})\| = \|\boldsymbol{x}\| + \|\boldsymbol{x} - \boldsymbol{y}\| \\ \Rightarrow \|\boldsymbol{y}\| - \|\boldsymbol{x}\| \leq \|\boldsymbol{x} - \boldsymbol{y}\|. \end{aligned}$$

Hence

$$\pm\big(\, \|\boldsymbol{x}\| - \|\boldsymbol{y}\| \,\big) \leq \|\boldsymbol{x} - \boldsymbol{y}\|.$$

This is clearly equivalent to (11.24b). □

Definition 11.44 (Euclidean distance). Let $n \in \mathbb{N}$ and $\boldsymbol{x}, \boldsymbol{y} \in \mathbb{R}^n$. The *Euclidean distance* between the points $\boldsymbol{x}, \boldsymbol{y}$ is the nonnegative real number

$$\operatorname{dist}(\boldsymbol{x}, \boldsymbol{y}) := \|\boldsymbol{x} - \boldsymbol{y}\|.$$ □

Example 11.45. (a) If $n = 1$, then for any $x, y \in \mathbb{R}$ we have $\operatorname{dist}(x, y) = |x - y|$.
(b) For any $n \in \mathbb{N}$ and any $\boldsymbol{x} \in \mathbb{R}^n$ we have $\|\boldsymbol{x}\| = \operatorname{dist}(\boldsymbol{x}, \boldsymbol{0})$. □

Proposition 11.46. *Let $n \in \mathbb{N}$. For any $\boldsymbol{x}, \boldsymbol{y}, \boldsymbol{z} \in \mathbb{R}^n$ the following hold.*

(i) $\operatorname{dist}(\boldsymbol{x}, \boldsymbol{y}) \geq 0$ *with equality if and only if* $\boldsymbol{x} = \boldsymbol{y}$.
(ii) $\operatorname{dist}(\boldsymbol{x}, \boldsymbol{y}) = \operatorname{dist}(\boldsymbol{y}, \boldsymbol{x})$.
(iii) (Triangle inequality) $\operatorname{dist}(\boldsymbol{x}, \boldsymbol{z}) \leq \operatorname{dist}(\boldsymbol{x}, \boldsymbol{y}) + \operatorname{dist}(\boldsymbol{y}, \boldsymbol{z})$.

Proof. We have

$$\operatorname{dist}(\boldsymbol{x},\boldsymbol{y}) = \|\boldsymbol{x}-\boldsymbol{y}\| = \big\|-(\boldsymbol{x}-\boldsymbol{y})\big\| = \|\boldsymbol{y}-\boldsymbol{x}\| = \operatorname{dist}(\boldsymbol{y},\boldsymbol{x}) \geq 0.$$

Clearly

$$\operatorname{dist}(\boldsymbol{x},\boldsymbol{y}) = 0 \Longleftrightarrow \|\boldsymbol{x}-\boldsymbol{y}\| = 0 \Longleftrightarrow \boldsymbol{x}=\boldsymbol{y}.$$

To prove (iii) note that

$$\begin{aligned}\operatorname{dist}(\boldsymbol{x},\boldsymbol{z}) &= \|\boldsymbol{x}-\boldsymbol{z}\| = \|(\boldsymbol{x}-\boldsymbol{y})+(\boldsymbol{y}-\boldsymbol{z})\|\\ &\overset{(11.24a)}{\leq} \|\boldsymbol{x}-\boldsymbol{y}\| + \|\boldsymbol{y}-\boldsymbol{z}\| = \operatorname{dist}(\boldsymbol{x},\boldsymbol{y}) + \operatorname{dist}(\boldsymbol{y},\boldsymbol{z}).\end{aligned}$$

□

Definition 11.47 (Open sets). Let $n \in \mathbb{N}$.

(i) For $r > 0$ and $\boldsymbol{p} \in \mathbb{R}^n$ we define the *open (Euclidean) ball of radius r and center $\boldsymbol{p}$* to be the set

$$B_r(\boldsymbol{p}) := \{\boldsymbol{x} \in \mathbb{R}^n;\ \operatorname{dist}(\boldsymbol{x},\boldsymbol{p}) < r\} = \{\boldsymbol{x} \in \mathbb{R}^n;\ \|\boldsymbol{x}-\boldsymbol{p}\| < r\}. \tag{11.25}$$

Sometimes, when we want to emphasize the ambient space $\mathbb{R}^n$ we will use the more precise notation $B^n_r(\boldsymbol{p})$ when referring to the open ball in $\mathbb{R}^n$ of radius r and center $\boldsymbol{p}$.

(ii) A set $U \subset \mathbb{R}^n$ is called *open* (in $\mathbb{R}^n$) if, for any $\boldsymbol{p} \in U$, there exists $r > 0$ such that $B_r(\boldsymbol{p}) \subset U$.

(iii) An *open neighborhood* of $\boldsymbol{x}_0$ in $\mathbb{R}^n$ is defined to be an open subset of $\mathbb{R}^n$ that contains $\boldsymbol{x}_0$.

□

Example 11.48. For any real numbers $a < b$ the intervals (a,b), $(-\infty,a)$ and (a,∞) are open subsets of $\mathbb{R}$. □

Proposition 11.49. *Let $n \in \mathbb{N}$. Then, for any $\boldsymbol{p} \in \mathbb{R}^n$ and any $r > 0$, the open ball $B_r(\boldsymbol{p})$ is an open subset of $\mathbb{R}^n$.*

Proof. Let $r > 0$ and $\boldsymbol{p} \in \mathbb{R}^n$. Given $\boldsymbol{q} \in B_r(\boldsymbol{p})$ let $\rho := \operatorname{dist}(\boldsymbol{p},\boldsymbol{q})$. Note that $\rho < r$. We claim that $B_{r-\rho}(\boldsymbol{q}) \subset B_r(\boldsymbol{p})$. Indeed, if $\boldsymbol{x} \in B_{r-\rho}(\boldsymbol{q})$, then $\operatorname{dist}(\boldsymbol{q},\boldsymbol{x}) < r-\rho$. Using the triangle inequality we deduce

$$\operatorname{dist}(\boldsymbol{p},\boldsymbol{x}) \leq \operatorname{dist}(\boldsymbol{p},\boldsymbol{q}) + \operatorname{dist}(\boldsymbol{q},\boldsymbol{x}) < \rho + (r-\rho) = r.$$

This proves that $\boldsymbol{x} \in B_r(\boldsymbol{p})$. □

Proposition 11.50. *Let $n \in \mathbb{N}$. Then the following hold.*

(i) The empty set and the whole space $\mathbb{R}^n$ are open subsets of $\mathbb{R}^n$.

(ii) The intersection of two open subsets of $\mathbb{R}^n$ is also an open subset of $\mathbb{R}^n$.

(iii) The union of a (possibly infinite) collection of open subsets of $\mathbb{R}^n$ is also an open subset of $\mathbb{R}^n$.

Proof. The statement (i) is obvious. To prove (ii) consider two open subsets $U_1, U_2 \subset \mathbb{R}^n$. We have to show that $U_1 \cap U_2$ is open, i.e., for any $\boldsymbol{p} \in U_1 \cap U_2$ there exists $r > 0$ such that $B_r(\boldsymbol{p}) \subset U_1 \cap U_2$.
Since U_1 is open, there exists $r_1 > 0$ such that $B_{r_1}(\boldsymbol{p}) \subset U_1$. Similarly, there exists $r_2 > 0$ such that $B_{r_2}(\boldsymbol{p}) \subset U_2$. If $r = \min(r_1, r_2)$, then

$$B_r(\boldsymbol{p}) = B_{r_1}(\boldsymbol{p}) \cap B_{r_2}(\boldsymbol{p}) \subset U_1 \cap U_2.$$

(iii) Suppose that $(U_i)_{i \in I}$ is a collection of open subsets of $\mathbb{R}^n$. Denote by U their union. If $\boldsymbol{p} \in U$, then there exists a set U_{i_0} of this collection that contains $\boldsymbol{p}$. Since U_{i_0} is open, there exists $r_0 > 0$ such that

$$B_{r_0}(\boldsymbol{p}) \subset U_{i_0} \subset U.$$

This proves that U is open. □

Definition 11.51. Let $n \in \mathbb{N}$. For any $\boldsymbol{x} = [x^1, \dots, x^n]^\top \in \mathbb{R}^n$ we set

$$\|\boldsymbol{x}\|_\infty := \max\{\, |x^1|, \dots, |x^n| \,\}.$$

We will refer to $\|\boldsymbol{x}\|_\infty$ as the *sup-norm* of $\boldsymbol{x}$. □

Example 11.52. If $\boldsymbol{x} = [3, 1, -7, 5]^\top \in \mathbb{R}^4$, then

$$\|\boldsymbol{x}\|_\infty = 7 \ \text{ and } \ \|\boldsymbol{x}\| = \sqrt{9 + 1 + 49 + 25} = \sqrt{84}.$$ □

The proof of the following result is left to you as an exercise.

Proposition 11.53. *Let $n \in \mathbb{N}$. Then*

$$\|\boldsymbol{x} + \boldsymbol{y}\|_\infty \le \|\boldsymbol{x}\|_\infty + \|\boldsymbol{y}\|_\infty, \ \ \forall \boldsymbol{x}, \boldsymbol{y} \in \mathbb{R}^n, \tag{11.26a}$$

$$\Big| \|\boldsymbol{x}\|_\infty - \|\boldsymbol{y}\|_\infty \Big| \le \|\boldsymbol{x} - \boldsymbol{y}\|_\infty, \ \ \forall \boldsymbol{x}, \boldsymbol{y} \in \mathbb{R}^n, \tag{11.26b}$$

and

$$\|\boldsymbol{x}\|_\infty \le \|\boldsymbol{x}\| \le \sqrt{n}\|\boldsymbol{x}\|_\infty, \ \ \forall \boldsymbol{x} \in \mathbb{R}^n. \tag{11.27}$$

□

Definition 11.54. Let $n \in \mathbb{N}$. For any $\boldsymbol{p} \in \mathbb{R}^n$ and $r > 0$ we define the *open cube* of center $\boldsymbol{p}$ and radius r to be the set

$$C_r(\boldsymbol{p}) := \big\{\, \boldsymbol{x} \in \mathbb{R}^n;\ \ \|\boldsymbol{x} - \boldsymbol{p}\|_\infty < r \,\big\}.$$ □

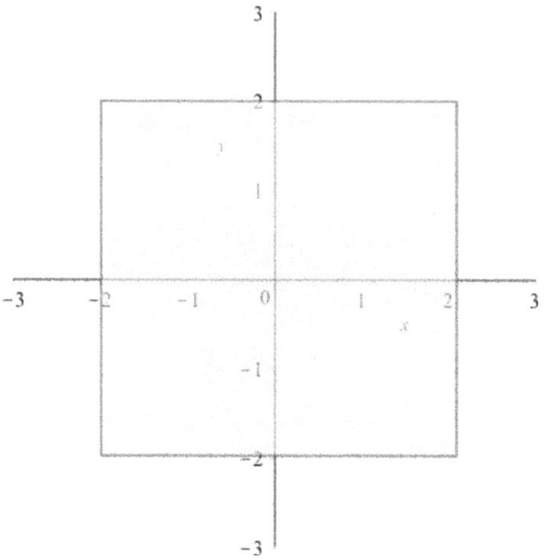

Fig. 11.9 *The open cube $C_2(\mathbf{0})$ of radius 2 and center $\mathbf{0} \in \mathbb{R}^2$.*

Observe that if $\boldsymbol{p} = [p^1, \dots, p^n]^\top \in \mathbb{R}^n$ and $r > 0$ then

$$\begin{aligned} \boldsymbol{x} \in C_r(\boldsymbol{p}) &\Longleftrightarrow |x^i - p^i| < r, \ \ \forall i = 1, 2, \dots, n \\ &\Longleftrightarrow x^i \in (p^i - r, p^i + r), \ \ \forall i = 1, 2, \dots, n \\ &\Longleftrightarrow \boldsymbol{x} \in (p^1 - r, p^1 + r) \times (p^2 - r, p^2 + r) \times \cdots \times (p^n - r, p^n + r). \end{aligned}$$

Note that the inequality (11.27) implies that

$$\forall \boldsymbol{p} \in \mathbb{R}^n, \ \ \forall r > 0 : \ \ C_{r/\sqrt{n}}(\boldsymbol{p}) \subset B_r(\boldsymbol{p}) \subset C_r(\boldsymbol{p}). \tag{11.28}$$

Proposition 11.55. *For any $n \in \mathbb{N}$, $\mathbf{p} \in \mathbb{R}^n$ and $r > 0$ the open cube $C_r(\mathbf{p})$ is an* open *subset of $\mathbb{R}^n$.* □

The proof is left to you as an exercise.

Proposition 11.56. *Let $n \in \mathbb{N}$ and $U \subset \mathbb{R}^n$. The following statements are equivalent.*

(i) The set U is open.
(ii) For all $\mathbf{p} \in U$, $\exists r > 0$ such that $C_r(\mathbf{p}) \subset U$.

□

Definition 11.57 (Closed sets). Let $n \in \mathbb{N}$. A subset $C \subset \mathbb{R}^n$ is called *closed* (in $\mathbb{R}^n$) if its complement $\mathbb{R}^n \setminus C$ is open in $\mathbb{R}^n$. More explicitly, this means that

$$\forall \boldsymbol{p} \in \mathbb{R}^n \setminus C \ \ \exists r > 0 : \ \ B_r(\boldsymbol{p}) \subset \mathbb{R}^n \setminus C.$$ □

Example 11.58. (a) For any real numbers $a < b$, the intervals $[a, b]$, $(-\infty, b]$, $[b, \infty)$ are closed subsets of $\mathbb{R}$.

(b) For $p \in \mathbb{R}^n$ and $r > 0$ we set

$$\overline{B_r(p)} := \{x \in \mathbb{R}^n;\ \|x - p\| \le r\}.$$

Then $\overline{B_r(p)}$ is a closed subset of $\mathbb{R}^n$, i.e., $\mathbb{R}^n \setminus \overline{B_r(p)}$ is open.

Indeed, let $q \in \mathbb{R}^n \setminus \overline{B_r(p)}$. Thus $\|q - p\| > r$. Set $R = \|q - p\|$. We claim that

$$B_{R-r}(q) \subset \mathbb{R}^n \setminus \overline{B_r(p)}.$$

Let $y \in B_{R-r}(q)$. We have

$$\begin{aligned} R = \|p - q\| &\le \|p - y\| + \|y - q\| < \|p - y\| + R - r \Rightarrow r < \|p - y\| \\ &\Rightarrow y \in \mathbb{R}^n \setminus \overline{B_r(p)}. \end{aligned}$$

(c) For $p \in \mathbb{R}^n$ and $r > 0$ we set

$$\overline{C_r(p)} := \{x \in \mathbb{R}^n;\ \|x - p\|_\infty \le r\}.$$

Then $\overline{C_r(p)}$ is a closed subset of $\mathbb{R}^n$. To prove this fact, imitate the argument in (b) with the Euclidean norm $\|-\|$ replaced by the sup-norm $\|-\|_\infty$ and then invoke Proposition 11.56. □

Definition 11.59. The sets $\overline{B_r(p)}$ and $\overline{C_r(p)}$ are called the *closed ball and respectively closed cube* of center p and radius r. □

According to the De Morgan Law (Proposition 1.12) the complement of a union of sets is the intersection of the complements of the sets, and the complement of an intersection of sets is the union of the complements of the sets. Invoking Proposition 11.50 we deduce the following result.

Proposition 11.60. *Let $n \in \mathbb{N}$. The following hold.*

(i) The empty set and the whole space $\mathbb{R}^n$ are closed subsets of $\mathbb{R}^n$.
(ii) The union of two closed subsets of $\mathbb{R}^n$ is also a closed subset of $\mathbb{R}^n$.
(iii) The intersection of a (possibly infinite) collection of closed subsets of $\mathbb{R}^n$ is also a closed subset of $\mathbb{R}^n$.

□

11.4. Convergence

The concept of convergence of sequences of real numbers has a multidimensional counterpart. In fact, the concept of convergence of a sequence of points in a Euclidean space $\mathbb{R}^n$ can be expressed in terms of the concept of convergence of sequences of real numbers.

Definition 11.61 (Convergent sequences). Let $n \in \mathbb{N}$. A sequence $(\boldsymbol{p}_\nu)_{\nu \geq 1}$ of points in $\mathbb{R}^n$ is said to be *convergent* if there exists $\boldsymbol{p}_\infty$ such that the sequence of real numbers $\big(\operatorname{dist}(\boldsymbol{p}_\nu, \boldsymbol{p}_\infty)\big)_{\nu \geq 1}$ converges to 0,

$$\lim_{\nu \to \infty} \operatorname{dist}(\boldsymbol{p}_\nu, \boldsymbol{p}_\infty) = 0.$$

More precisely, this means that $\forall \varepsilon > 0$, $\exists N = N(\varepsilon) > 0$ such that $\forall \nu > N(\varepsilon)$ we have $\|\boldsymbol{p}_\nu - \boldsymbol{p}_\infty\| < \varepsilon$. The point $\boldsymbol{p}_\infty$ is called the *limit* of the sequence $(\boldsymbol{p}_\nu)$ and we write this

$$\boldsymbol{p}_\infty = \lim_{\nu \to \infty} \boldsymbol{p}_\nu.$$

□

Note that

$$\boldsymbol{p}_\infty = \lim_{\nu \to \infty} \boldsymbol{p}_\nu \Longleftrightarrow \lim_{\nu \to \infty} \|\boldsymbol{p}_\nu - \boldsymbol{p}_\infty\| = 0 \Longleftrightarrow \lim_{\nu \to \infty} \operatorname{dist}(\boldsymbol{p}_\nu, \boldsymbol{p}_\infty) = 0. \tag{11.29}$$

The notion of convergence can be expressed in terms of open balls because the statement "$\operatorname{dist}(\boldsymbol{x}, \boldsymbol{p}) < \varepsilon$" is equivalent to the statement: "the point $\boldsymbol{x}$ belongs to the open ball of center $\boldsymbol{p}$ and radius ε". More precisely, we have the following result.

Proposition 11.62. *Let $n \in \mathbb{N}$ and $(\boldsymbol{p}_\nu)$ a sequence of points in $\mathbb{R}^n$. The following statements are equivalent.*

(i)

$$\boldsymbol{p}_\infty = \lim_{\nu \to \infty} \boldsymbol{p}_\nu.$$

(ii) For any $\varepsilon > 0$ there exists $N = N(\varepsilon) > 0$ such that, $\forall \nu > N(\varepsilon)$ we have $\boldsymbol{p}_\nu \in B_\varepsilon(\boldsymbol{p}_\infty)$.

□

The proof of the next result is left to you as an exercise.

Proposition 11.63. *Let $n \in \mathbb{N}$. Consider a sequence of points in $\mathbb{R}^n$*

$$\boldsymbol{p}_\nu = \begin{bmatrix} p_\nu^1 \\ \vdots \\ p_\nu^n \end{bmatrix}, \quad \nu = 1, 2, \dots,$$

and

$$\boldsymbol{p}_\infty = \begin{bmatrix} p_\infty^1 \\ \vdots \\ p_\infty^n \end{bmatrix} \in \mathbb{R}^n.$$

The following statements are equivalent.

(i)

$$\lim_{\nu \to \infty} \boldsymbol{p}_\nu = \boldsymbol{p}_\infty.$$

(ii)

$$\lim_{\nu\to\infty} \|\boldsymbol{p}_\nu - \boldsymbol{p}_\infty\|_\infty = 0.$$

(iii) For any $i = 1, 2, \dots, n$, *the* i*-th coordinate of* $\mathbf{p}_\nu$ *converges to the* i*-th coordinate of* $\mathbf{p}_\infty$, *i.e.,*

$$\lim_{\nu\to\infty} p_\nu^i = p_\infty^i, \quad \forall i = 1, 2, \dots, n.$$

□

Example 11.64. The sequence of points

$$\boldsymbol{p}_\nu = \begin{bmatrix} \frac{1}{\nu} \\ \\ \frac{\nu+1}{\nu^2} \\ \\ \frac{\nu}{\nu+1} \end{bmatrix} \in \mathbb{R}^3, \quad \nu \in \mathbb{N},$$

converges as $\nu \to \infty$ to the point

$$\boldsymbol{p}_\infty = \begin{bmatrix} 0 \\ 0 \\ 1 \end{bmatrix}$$

since

$$\lim_{\nu\to\infty} \frac{1}{\nu} = \lim_{\nu\to\infty} \frac{\nu+1}{\nu^2} = 0, \quad \lim_{\nu\to\infty} \frac{\nu}{\nu+1} = 1.$$

□

The following property of convergent sequences is an immediate generalization of its 1-dimensional cousin Proposition 4.9.

Proposition 11.65. *If the sequence* $(\mathbf{p}_\nu)$ *of points in* $\mathbb{R}^n$ *converges to a point* $\mathbf{p}$*, then any subsequence of* $(\mathbf{p}_\nu)$ *converges to the same point* $\mathbf{p}$. □

Definition 11.66. A sequence $(\boldsymbol{p}_\nu)_{\nu\geq 1}$ in $\mathbb{R}^n$ is called *bounded* if there exists $R > 0$ such that

$$\|\boldsymbol{p}_\nu\| < R, \quad \forall \nu \geq 1.$$

□

Proposition 11.67. *A convergent sequence of points on* $\mathbb{R}^n$ *is also bounded.*

Proof. Suppose that the sequence

$$\boldsymbol{p}_\nu = \begin{bmatrix} p_\nu^1 \\ \vdots \\ p_\nu^n \end{bmatrix}, \quad \nu = 1, 2, \dots$$

is convergent. According to Proposition 11.63, for each $i = 1, 2, \dots, n$ the sequence of coordinates (p_ν^i) is a convergent sequence of real numbers and thus, according to Proposition 4.14, it is bounded. Hence, there exists $C_i > 0$ such that

$$|p_\nu^i| < C_i, \quad \forall \nu = 1, 2, \dots .$$

Set

$$C := \max(C_1, \ldots, C_n).$$

Hence

$$\|\boldsymbol{p}_\nu\|_\infty = \max\big(|p_\nu^i|, \ldots, |p_\nu^i| \big) < C, \;\; \forall \nu \geq 1.$$

Using (11.27) we deduce

$$\|\boldsymbol{p}_\nu\| \leq \sqrt{n}\|\boldsymbol{p}_\nu\|_\infty < C\sqrt{n}, \;\; \forall \nu \geq 1.$$

This proves that the sequence $(\boldsymbol{p}_\nu)$ is bounded. □

Proposition 11.68. *Let $n \in \mathbb{N}$. Suppose that $(\boldsymbol{p}_\nu)_{\nu \geq 1}$ and $(\boldsymbol{q}_\nu)_{\nu \geq 1}$ are convergent sequences of points in $\mathbb{R}^n$. Denote by $\boldsymbol{p}_\infty$ and respectively $\boldsymbol{q}_\infty$ their limits. Then the following hold.*

(i)

$$\lim_{\nu \to \infty} (\boldsymbol{p}_\nu + \boldsymbol{q}_\nu) = \boldsymbol{p}_\infty + \boldsymbol{q}_\infty.$$

(ii) If $(t_\nu)_{\nu \geq 1}$ is a convergent sequence of real numbers with limit t_∞, then

$$\lim_{\nu \to \infty} t_\nu \boldsymbol{p}_\nu = t_\infty \boldsymbol{p}_\infty.$$

(iii)

$$\lim_{\nu \to \infty} \langle \boldsymbol{p}_\nu, \boldsymbol{q}_\nu \rangle = \langle \boldsymbol{p}_\infty, \boldsymbol{q}_\infty \rangle.$$

Proof. (i) We have

$$\begin{aligned} \operatorname{dist}\big(\boldsymbol{p}_\nu + \boldsymbol{q}_\nu, \boldsymbol{p}_\infty + \boldsymbol{q}_\infty\big) &= \|\,(\boldsymbol{p}_\nu + \boldsymbol{q}_\nu) - (\boldsymbol{p}_\infty + \boldsymbol{q}_\infty)\,\| = \|(\boldsymbol{p}_\nu - \boldsymbol{p}_\infty) + (\boldsymbol{q}_\nu - \boldsymbol{q}_\infty)\| \\ &\leq \|\boldsymbol{p}_\nu - \boldsymbol{p}_\infty\| + \|\boldsymbol{q}_\nu - \boldsymbol{q}_\infty\| = \operatorname{dist}(\boldsymbol{p}_\nu, \boldsymbol{p}_\infty) + \operatorname{dist}(\boldsymbol{q}_\nu, \boldsymbol{q}_\infty) \to 0 \;\text{ as } \nu \to \infty. \end{aligned}$$

The claim now follows from the Squeezing Principle.

(ii) Since the sequences (t_ν) and $(\boldsymbol{p}_\nu)$ are convergent, they are also bounded and thus there exists $C > 0$ such that

$$|t_\nu|, \;\; \|\boldsymbol{p}_\nu\| < C, \;\; \forall \nu \geq 1.$$

We have

$$\begin{aligned} \operatorname{dist}(t_\nu \boldsymbol{p}_\nu, t_\infty \boldsymbol{p}_\infty) &= \|t_\nu \boldsymbol{p}_\nu - t_\infty \boldsymbol{p}_\infty\| = \|t_\nu \boldsymbol{p}_\nu - t_\infty \boldsymbol{p}_\nu + t_\infty \boldsymbol{p}_\nu - t_\infty \boldsymbol{p}_\infty\| \\ &\leq \|t_\nu \boldsymbol{p}_\nu - t_\infty \boldsymbol{p}_\nu\| + \|t_\infty \boldsymbol{p}_\nu - t_\infty \boldsymbol{p}_\infty\| \\ &= \|(t_\nu - t_\infty)\boldsymbol{p}_\nu\| + \|t_\infty(\boldsymbol{p}_\nu - \boldsymbol{p}_\infty)\| \\ &= |t_\nu - t_\infty| \cdot \|\boldsymbol{p}_\nu\| + |t_\infty| \cdot \|\boldsymbol{p}_\nu - \boldsymbol{p}_\infty\| \\ &\leq C|t_\nu - t_\infty| + |t_\infty| \operatorname{dist}(\boldsymbol{p}_\nu, \boldsymbol{p}_\infty) \to 0 \;\text{ as } \nu \to \infty. \end{aligned}$$

(iii) Since the sequences $(\boldsymbol{p}_\nu)$ and $(\boldsymbol{q}_\nu)$ are convergent, they are also bounded and thus there exists $C > 0$ such that

$$\|\boldsymbol{p}_\nu\|, \;\; \|\boldsymbol{q}_\nu\| < C, \;\; \forall \nu \geq 1.$$

We have

$$
\begin{aligned}
\big|\, \langle \boldsymbol{p}_\nu, \boldsymbol{q}_\nu \rangle - \langle \boldsymbol{p}_\infty, \boldsymbol{q}_\infty \rangle \,\big| &= \big|\, \langle \boldsymbol{p}_\nu, \boldsymbol{q}_\nu \rangle - \langle \boldsymbol{p}_\infty, \boldsymbol{q}_\nu \rangle + \langle \boldsymbol{p}_\infty, \boldsymbol{q}_\nu \rangle - \langle \boldsymbol{p}_\infty, \boldsymbol{q}_\infty \rangle \,\big| \\
&\leq \big|\, \langle \boldsymbol{p}_\nu, \boldsymbol{q}_\nu \rangle - \langle \boldsymbol{p}_\infty, \boldsymbol{q}_\nu \rangle \,\big| + \big|\, \langle \boldsymbol{p}_\infty, \boldsymbol{q}_\nu \rangle - \langle \boldsymbol{p}_\infty, \boldsymbol{q}_\infty \rangle \,\big| \\
&= \big|\, \langle \boldsymbol{p}_\nu - \boldsymbol{p}_\infty, \boldsymbol{q}_\nu \rangle \,\big| + \big|\, \langle \boldsymbol{p}_\infty, \boldsymbol{q}_\nu - \boldsymbol{q}_\infty \rangle \,\big|
\end{aligned}
$$

(use the Cauchy–Schwarz inequality)

$$
\begin{aligned}
&\leq \|\boldsymbol{p}_\nu - \boldsymbol{p}_\infty\| \cdot \|\boldsymbol{q}_\nu\| + \|\boldsymbol{p}_\infty\| \cdot \|\boldsymbol{q}_\nu - \boldsymbol{q}_\infty\| \\
&\leq C \operatorname{dist}(\boldsymbol{p}_\nu, \boldsymbol{p}_\infty) + \|\boldsymbol{p}_\infty\| \operatorname{dist}(\boldsymbol{q}_\nu, \boldsymbol{q}_\infty) \to 0 \quad \text{as } \nu \to \infty.
\end{aligned}
$$

□

Definition 11.69. Let $n \in \mathbb{N}$. A sequence $(\boldsymbol{p}_\nu)_{\nu \geq 1}$ of points in $\mathbb{R}^n$ is called *Cauchy* or *fundamental* if $\forall \varepsilon > 0$, $\exists N = N(\varepsilon) > 0$ such that

$$
\forall \nu, \mu > N(\varepsilon): \quad \operatorname{dist}(\boldsymbol{p}_\mu, \boldsymbol{p}_\nu) = \|\boldsymbol{p}_\mu - \boldsymbol{p}_\nu\| < \varepsilon.
$$

□

Theorem 11.70 (Cauchy sequences). *Let $n \in \mathbb{N}$ and consider a sequence $(\boldsymbol{p}_\nu)_{\nu \geq 1}$ of points in $\mathbb{R}^n$. The following statements are equivalent.*

(i) The sequence $(\boldsymbol{p}_\nu)_{\nu \geq 1}$ is Cauchy.
(ii) The sequence $(\boldsymbol{p}_\nu)_{\nu \geq 1}$ converges to a point $\boldsymbol{p}_\infty \in \mathbb{R}^n$.

Proof. (i) ⇒ (ii) Assume

$$
\boldsymbol{p}_\nu = \begin{bmatrix} p_\nu^1 \\ \vdots \\ p_\nu^n \end{bmatrix}.
$$

For each $i = 1, \dots, n$ and any $\mu, \nu \in \mathbb{N}$ we have

$$
\big|\, p_\mu^i - p_\nu^i \,\big| = \sqrt{\left(p_\mu^i - p_\nu^i\right)^2} \leq \sqrt{\left(p_\mu^1 - p_\nu^1\right)^2 + \cdots + \left(p_\mu^n - p_\nu^n\right)^2} \leq \|\boldsymbol{p}_\mu - \boldsymbol{p}_\nu\|.
$$

The above inequality shows that, for each $i = 1, \dots, n$, the sequence of *real numbers* $(p_\nu^i)_{\nu \geq 1}$ is Cauchy. Invoking Cauchy's Theorem 4.33 we deduce that, for each $i = 1, \dots, n$, the sequence $(p_\nu^i)_{\nu \geq 1}$ is convergent. Hence, for every $i = 1, \dots, n$, there exists $p_\infty^i \in \mathbb{R}$ such that

$$
\lim_{\nu \to \infty} p_\nu^i = p_\infty^i.
$$

From Proposition 11.63 we now deduce that

$$
\lim_{\nu \to \infty} \begin{bmatrix} p_\nu^1 \\ \vdots \\ p_\nu^n \end{bmatrix} = \begin{bmatrix} p_\infty^1 \\ \vdots \\ p_\infty^n \end{bmatrix} =: \boldsymbol{p}_\infty.
$$

(ii) ⇒ (i) Suppose that

$$
\boldsymbol{p}_\infty = \lim_{\nu \to \infty} \boldsymbol{p}_\nu.
$$

Then, $\forall \varepsilon > 0$, $\exists N = N(\varepsilon) > 0$ such that $\forall \nu > N(\varepsilon)$ we have

$$\operatorname{dist}(\boldsymbol{p}_\nu, \boldsymbol{p}_\infty) < \frac{\varepsilon}{2}.$$

Then, for any $\mu, \nu > N(\varepsilon)$ we have

$$\operatorname{dist}(\boldsymbol{p}_\mu, \boldsymbol{p}_\nu) \leq \operatorname{dist}(\boldsymbol{p}_\mu, \boldsymbol{p}_\infty) + \operatorname{dist}(\boldsymbol{p}_\infty, \boldsymbol{p}_\nu) < \varepsilon.$$

□

Proposition 11.71. *Let $n \in \mathbb{N}$ and $C \subset \mathbb{R}^n$. Then the following statements are equivalent.*

(i) The set $C \subset \mathbb{R}^n$ is closed in $\mathbb{R}^n$.
(ii) For any convergent *sequence of points in C, its limit is also a point in C.*

Proof. (i) $\Rightarrow$ (ii). We know that $\mathbb{R}^n \setminus C$ is open and we have to show that if $(\boldsymbol{p}_\nu)_{\nu \geq 1}$ is a convergent sequence of points in C, then its limit $\boldsymbol{p}_\infty$ belongs to C. We argue by contradiction. Suppose that $\boldsymbol{p}_\infty \in \mathbb{R}^n \setminus C$. Since $\mathbb{R}^n \setminus C$ is open, there exists $r > 0$ such that $B_r(\boldsymbol{p}_\infty) \subset \mathbb{R}^n \setminus C$, i.e.,

$$B_r(\boldsymbol{p}_\infty) \cap C = \emptyset.$$

This proves that, $\forall \nu \geq 1$, $\boldsymbol{p}_\nu \notin B_r(\boldsymbol{p}_\infty)$, i.e.,

$$\operatorname{dist}(\boldsymbol{p}_\nu, \boldsymbol{p}_\infty) \geq r, \quad \forall \nu \geq 1.$$

This contradicts the fact that $\lim_{\nu \to \infty} \operatorname{dist}(\boldsymbol{p}_\nu, \boldsymbol{p}_\infty) = 0$.

(ii) $\Rightarrow$ (i). We have to show that $\mathbb{R}^n \setminus C$ is open. We argue by contradiction. Assume that there exists $\boldsymbol{p}_* \in \mathbb{R}^n \setminus C$ such that, $\forall r > 0$, the ball $B_r(\boldsymbol{p}_*)$ is not contained in $\mathbb{R}^n \setminus C$. Thus, for any $r > 0$ there exist $\boldsymbol{p}(r) \in B_r(\boldsymbol{p}_*) \cap C$, i.e., $\boldsymbol{p}(r) \in C$, $\operatorname{dist}(\boldsymbol{p}(r), \boldsymbol{p}_*) < r$. Thus, for any $\nu \in \mathbb{N}$, there exists $\boldsymbol{p}_\nu \in C$ such that

$$\operatorname{dist}(\boldsymbol{p}_\nu, \boldsymbol{p}_*) < \frac{1}{\nu}, \quad \forall \nu \in \mathbb{N}.$$

This shows that the sequence of points $(\boldsymbol{p}_\nu)$ in C converges to the point $\boldsymbol{p}_*$ *that is not in* C. This contradicts (ii). □

Example 11.72. Any affine line in $\mathbb{R}^n$ is a closed subset. We will prove this in two different ways. Consider the line $\ell_{\boldsymbol{p},\boldsymbol{v}} \subset \mathbb{R}^n$ passing through the point $\boldsymbol{p}$ in the direction $\boldsymbol{v} \neq \boldsymbol{0}$.

1st Method. Suppose that $(\boldsymbol{q}_\nu)$ is a convergent sequence of points on this line. We denote by $\boldsymbol{q}_\infty$ its limit. We want to prove that $\boldsymbol{q}_\infty$ also lies on the line $\ell_{\boldsymbol{p},\boldsymbol{v}}$.

To see this note first that since $\boldsymbol{q}_\nu \in \ell_{\boldsymbol{p},\boldsymbol{v}}$, there exists $t_\nu \in \mathbb{R}$ such that

$$\boldsymbol{q}_\nu = \boldsymbol{p} + t_\nu \boldsymbol{v}.$$

We deduce that for any $\mu, \nu \geq 1$ we have

$$\operatorname{dist}(\boldsymbol{q}_\mu, \boldsymbol{q}_\nu) = \|\boldsymbol{q}_\mu - \boldsymbol{q}_\nu\| = |t_\mu - t_\nu| \cdot \|\boldsymbol{v}\|. \Rightarrow |t_\mu - t_\nu| = \frac{1}{\|\boldsymbol{v}\|} \operatorname{dist}(\boldsymbol{q}_\mu, \boldsymbol{q}_\nu).$$

Since the sequence $(\boldsymbol{q}_\nu)$ is convergent, it is also Cauchy, and the above equality shows that the sequence (t_ν) is Cauchy as well. Hence the sequence (t_ν) is convergent in $\mathbb{R}$. If t_∞ is its limit, then Proposition 11.68 implies that

$$\boldsymbol{q}_\infty = \lim_{\nu\to\infty}(\boldsymbol{p}+t_\nu\boldsymbol{v}) = \boldsymbol{p}+t_\infty\boldsymbol{v} \in \ell_{\boldsymbol{p},\boldsymbol{v}}.$$

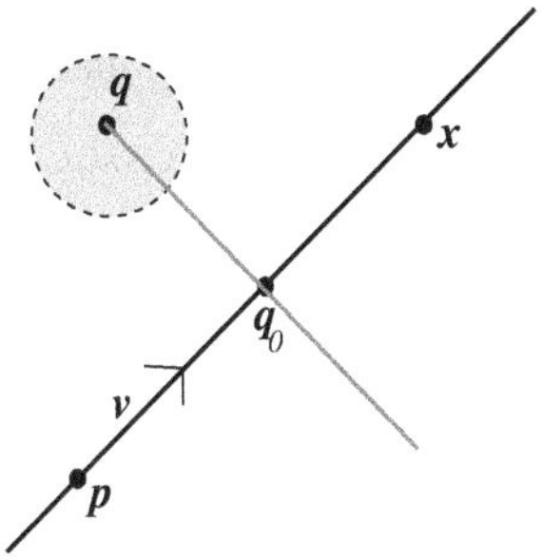

Fig. 11.10 $\operatorname{dist}(\boldsymbol{q},\boldsymbol{q}_0) \leq \operatorname{dist}(\boldsymbol{q},\boldsymbol{x}), \forall \boldsymbol{x}\in\ell_{\boldsymbol{p},\boldsymbol{v}}$.

2nd Method. We will prove that the complement of the line is open, i.e., if $\boldsymbol{q}$ is a point outside the line $\ell_{\boldsymbol{p},\boldsymbol{v}}$, then there exists an open ball centered at $\boldsymbol{q}$ that does not intersect the line; see Figure 11.10.

To do so, we will find the point $\boldsymbol{q}_0$ on the line closest to $\boldsymbol{q}$. Usual Euclidean geometry suggests that if $\boldsymbol{q}_0$ is such a point, then the line $\boldsymbol{q}\boldsymbol{q}_0$ should be perpendicular to $\ell_{\boldsymbol{p},\boldsymbol{v}}$; see Figure 11.10. So, instead of looking for a point on the line closest to $\boldsymbol{q}$, we will look for a point $\boldsymbol{q}_0$ such that $(\boldsymbol{q}-\boldsymbol{q}_0)\perp\boldsymbol{v}$. As we will see, such a $\boldsymbol{q}_0$ will indeed be the point on the line closest to $\boldsymbol{q}$. Observe that

$$(\boldsymbol{q}-\boldsymbol{q}_0)\perp\boldsymbol{v} \Longleftrightarrow \langle\boldsymbol{q}-\boldsymbol{q}_0,\boldsymbol{v}\rangle \Longleftrightarrow \langle\boldsymbol{q},\boldsymbol{v}\rangle = \langle\boldsymbol{q}_0,\boldsymbol{v}\rangle.$$

Since $\boldsymbol{q}_0$ is on the line $\ell_{\boldsymbol{p},\boldsymbol{v}}$ it has the form $\boldsymbol{q}=\boldsymbol{p}+t_0\boldsymbol{v}$ for some real number t_0. Using this in the above equality we deduce

$$\langle\boldsymbol{q},\boldsymbol{v}\rangle = \langle\boldsymbol{p}+t_0\boldsymbol{v},\boldsymbol{v}\rangle = \langle\boldsymbol{p},\boldsymbol{v}\rangle + t_0\langle\boldsymbol{v},\boldsymbol{v}\rangle = \langle\boldsymbol{p},\boldsymbol{v}\rangle + t_0\|\boldsymbol{v}\|^2$$

$$\Rightarrow t_0\|\boldsymbol{v}\|^2 = \langle\boldsymbol{q}-\boldsymbol{p},\boldsymbol{v}\rangle \Rightarrow t_0 = \frac{\langle\boldsymbol{q}-\boldsymbol{p},\boldsymbol{v}\rangle}{\|\boldsymbol{v}\|^2}.$$

Note that if $\boldsymbol{x}\in\ell_{\boldsymbol{p},\boldsymbol{q}}$, then $\boldsymbol{x}-\boldsymbol{q}_0$ is a multiple of $\boldsymbol{v}$ so $(\boldsymbol{x}-\boldsymbol{q}_0)\perp(\boldsymbol{q}-\boldsymbol{q}_0)$; see Exercise 11.2(b). Pythagoras' Theorem then implies that (Figure 11.10)

$$\operatorname{dist}(\boldsymbol{q},\boldsymbol{x})^2 = \operatorname{dist}(\boldsymbol{q},\boldsymbol{q}_0)^2 + \operatorname{dist}(\boldsymbol{q}_0,\boldsymbol{x})^2 \geq \operatorname{dist}(\boldsymbol{q},\boldsymbol{q}_0)^2.$$

Hence, if we set $r := \operatorname{dist}(\boldsymbol{q},\boldsymbol{q}_0)$, then we deduce that $r>0$ and $r\geq\operatorname{dist}(\boldsymbol{q},\boldsymbol{x})$, $\forall\boldsymbol{x}\in\ell_{\boldsymbol{p},\boldsymbol{v}}$. In particular this shows that the ball $B_{r/2}(\boldsymbol{q})$ of radius $r/2$ and centered at $\boldsymbol{q}$ does not intersect the line $\ell_{\boldsymbol{p},\boldsymbol{v}}$. □

Definition 11.73. Let $n\in\mathbb{N}$ and $X\subset\mathbb{R}^n$.

(i) A point $\boldsymbol{p}\in\mathbb{R}^n$ is a *cluster point* of X if, for any $\varepsilon>0$, the ball $B_\varepsilon(\boldsymbol{p})$ contains a point in X not equal to $\boldsymbol{p}$.

(ii) A subset $S\subset X$ is called *dense* in X if, for any $\boldsymbol{x}\in X$ and any $\varepsilon>0$, the ball $B_\varepsilon(\boldsymbol{x})$ contains a point in S.

□

Example 11.74. Proposition 3.33 shows that the set $\mathbb{Q}$ of rational numbers is dense in $\mathbb{R}$. More generally, the set $\mathbb{Q}^n$ is dense in $\mathbb{R}^n$. □

Proposition 11.75. *Let $n \in \mathbb{N}$, $X \subset \mathbb{R}^n$ and $\boldsymbol{p} \in \mathbb{R}^n$. The following statements are equivalent.*

(i) The point $\boldsymbol{p}$ is a cluster point of X.
(ii) There exists a sequence of points $(\boldsymbol{p}_\nu)$ in $X \setminus \{\boldsymbol{p}\}$ that converges to $\boldsymbol{p}$.

Proof. (i) $\Rightarrow$ (ii). Since $\boldsymbol{p}$ is a cluster point of X we deduce that, for any $\nu \in \mathbb{N}$, the ball $B_{1/\nu}(\boldsymbol{p})$ contains a point $\boldsymbol{p}_\nu \in X \setminus \{\boldsymbol{p}\}$. Observing that $\operatorname{dist}(\boldsymbol{p}_\nu, \boldsymbol{p}) < \frac{1}{\nu}$ we deduce that

$$\lim_{\nu \to \infty} \operatorname{dist}(\boldsymbol{p}_\nu, \boldsymbol{p}) = 0,$$

i.e., $(\boldsymbol{p}_\nu)$ is a sequence in $X \setminus \{\boldsymbol{p}\}$ that converges to $\boldsymbol{p}$.

(ii) $\Rightarrow$ (i). We know that there exists a sequence $(\boldsymbol{p}_\nu)$ in $X \setminus \{\boldsymbol{p}\}$ that converges to $\boldsymbol{p}$. Let $\varepsilon > 0$. There exists $N = N(\varepsilon) > 0$ such that $\operatorname{dist}(\boldsymbol{p}_\nu, \boldsymbol{p}) < \varepsilon$, $\forall \nu > N(\varepsilon)$. Thus the ball $B_\varepsilon(\boldsymbol{p})$ contains all the points $\boldsymbol{p}_\nu$, $\nu > N(\varepsilon)$ and none of these points is equal to $\boldsymbol{p}$. □

Definition 11.76. Let $n \in \mathbb{N}$ and $X \subset \mathbb{R}^n$.

(i) A subset $S \subset X$ is called *open in* X if there exists an open subset $U \subset \mathbb{R}^n$ such that $S = X \cap U$.
(ii) A subset $S \subset X$ is called *closed in* X if there exists a closed subset $C \subset \mathbb{R}^n$ such that $S = X \cap C$.

□

Example 11.77. (a) Let $X \subset \mathbb{R}$ denote the union of the intervals $[0, 1] \cup [3, 4]$. Clearly the set $S = [0, 1]$ is closed both in $\mathbb{R}$ and in X. On the other hand, S is also open in X since S is equal to the intersection of X with the open subset $(-\infty, 2) \subset \mathbb{R}$.□

The next result provides alternate characterizations of the sets that are open (closed) in a given set X. Its proof is left to you as an exercise.

Proposition 11.78. *Let $n \in \mathbb{N}$ and $X \subset \mathbb{R}^n$. Then the following hold.*

(i) The sets $\emptyset$ and X are simultaneously closed and open in X.
(ii) The subset $S \subset X$ is open in X if and only if the complement $X \setminus S$ is closed in X.
(iii) The subset $S \subset X$ is open in X if for any $\boldsymbol{p} \in S$ there exists $r > 0$ such that

$$\forall \boldsymbol{x} \in X: \;\; \operatorname{dist}(\boldsymbol{x}, \boldsymbol{p}) < r \Rightarrow \boldsymbol{x} \in S.$$

(iv) The subset $S \subset X$ is closed in X if and only if for any sequence $(\boldsymbol{p}_\nu)$ of points in S that converges to $\boldsymbol{p}_\infty \in X$, then $\boldsymbol{p}_\infty \in S$.

□

11.5. Normed vector spaces

To give the readers a broader perspective on the concepts we have presented so far, we have decided to include a brief introduction to some more abstract notions. This discussion will come in handy in later chapters

Definition 11.79 (Normed spaces). A (real) *normed vector space* is a pair $(\boldsymbol{X}, \|-\|)$, where $\boldsymbol{X}$ is a real vector space and $\|-\|$ is a *norm* on $\boldsymbol{X}$, i.e., a *nonnegative* function $\|-\| : \boldsymbol{X} \to [0, \infty)$ satisfying the following conditions.

(i) (Nondegeneracy.) $\forall \boldsymbol{x} \in \boldsymbol{X}$, $\boldsymbol{x} = \boldsymbol{0} \Longleftrightarrow \|\boldsymbol{x}\| = 0$.
(ii) (Positive homogeneity.) $\forall \boldsymbol{x} \in \boldsymbol{X}$, $\forall t \in \mathbb{R}$, $\|t\boldsymbol{x}\| = |t| \cdot \|\boldsymbol{x}\|$.
(iii) (Triangle Inequality.)

$$\|\boldsymbol{x} + \boldsymbol{y}\| \leq \|\boldsymbol{x}\| + \|\boldsymbol{y}\|, \;\; \forall \boldsymbol{x}, \boldsymbol{y} \in \boldsymbol{X}.$$

□

It turns out that there is a very large supply of normed spaces.

Example 11.80. (a) We have already discussed two examples: the Euclidean norm $\| - \|$ and the sup-norm $\| - \|_\infty$ on $\mathbb{R}^n$. There are many other examples of norms on $\mathbb{R}^n$. For $p \in [1, \infty)$ and $\boldsymbol{x} \in \mathbb{R}^n$ we define

$$\|\boldsymbol{x}\|_p = \left(\sum_{j=1}^{n} |x^j|^p\right)^{\frac{1}{p}}.$$

Minkowski's inequality (8.25) shows that $\| - \|_p$ is indeed a norm on $\mathbb{R}^n$. Note that,

$$\|\boldsymbol{x}\|_\infty = \lim_{p\to\infty} \|\boldsymbol{x}\|_p, \ \ \forall \boldsymbol{x} \in \mathbb{R}^n. \tag{11.30}$$

(b) Any finite dimensional vector space $\boldsymbol{X}$ admits norms. To see this *fix* a basis

$$\underline{\boldsymbol{b}} := \{\boldsymbol{b}_1, \dots, \boldsymbol{b}_n\}, \ \ n = \dim \boldsymbol{X}.$$

Any vector $\boldsymbol{x} \in \boldsymbol{X}$ admits a unique decomposition

$$\boldsymbol{x} = x^1\boldsymbol{b}_1 + \cdots + x^n\boldsymbol{b}_n,$$

where we recall that the real numbers $x^1, \dots, x^n$ are called *the coordinates of $\boldsymbol{x}$ in the basis $\underline{\boldsymbol{b}}$*. If we set

$$\|\boldsymbol{x}\| = \|\boldsymbol{x}\|_{\underline{\boldsymbol{b}}} = \max(|x^1|, \dots, |x^n|),$$

then it is easy to show that $\| - \|_{\underline{\boldsymbol{b}}}$ is indeed a norm on $\boldsymbol{X}$.

(c) We denote by $\mathcal{R}[0, 1]$ the space of Riemann integrable functions $f : [0, 1] \to \mathbb{R}$. This is a vector space. For $p \in [1, \infty)$ and $f \in \mathcal{R}[0, 1]$ we set

$$\|f\|_p := \left(\int_0^1 |f(x)|^p\right)^{\frac{1}{p}}.$$

The Minkowski type inequality in Exercise 9.8(c) shows that the correspondence $f \mapsto \|f\|_p$ defines a norm on $\mathcal{R}[0, 1]$ for each $p \in [1, \infty)$.

(d) Fix $a, b \in \mathbb{R}$, $a < b$. Denote by $C([a, b])$ the space of continuous functions $f : [a, b] \to \mathbb{R}$. For $f \in C([a, b])$ we set

$$\|f\| := \sup_{x\in[a,b]} |f(x)|.$$

The correspondence $f \mapsto \|f\|$ is a norm on $C[a, b]$ called the *sup-norm*. Note that $\mathcal{R}[a, b]$ and $C([a, b])$ *are not finite dimensional spaces.* □

Suppose that $(\boldsymbol{X}, \| - \|)$ is a normed spaced. Using the previous sections as inspiration we can define the concepts of open and closed subsets in $\boldsymbol{X}$. Thus, a set $U \subset \boldsymbol{X}$ is called *open in $\boldsymbol{X}$, with respect to the norm* $\| - \|$, if for any $\boldsymbol{u} \in U$, there exists $\varepsilon > 0$ such that

$$\forall \boldsymbol{x} \in \boldsymbol{X} : \ \ \|\boldsymbol{x} - \boldsymbol{u}\| < \varepsilon \Rightarrow \boldsymbol{x} \in U.$$

A subset $C \subset \boldsymbol{X}$ is called *closed in $\boldsymbol{X}$, with respect to the norm* $\| - \|$ if the complement $\boldsymbol{X} \setminus C$ is open in $\boldsymbol{X}$.

The concept of openness can be more visually described in terms of open balls. The *open ball of center $\boldsymbol{p} \in \boldsymbol{X}$ and radius* $r > 0$ is the set (compare with (11.25))

$$B_r(\boldsymbol{p}) := \{\boldsymbol{x} \in \boldsymbol{X};\ \|\boldsymbol{x} - \boldsymbol{p}\| < r\}.$$

We want to emphasize that the shape and size of the open ball $B_r(\boldsymbol{0})$ *depends on the norm* $\| - \|$. We see that set $U \subset \boldsymbol{X}$ is open in $\boldsymbol{X}$ with respect to the norm $\| - \|$ if for any point $\boldsymbol{p} \in U$ there exists $\varepsilon > 0$ such that $B_\varepsilon(\boldsymbol{p}) \subset U$.

Definition 11.81. Let $(\boldsymbol{X}, \| - \|)$ be a normed space and $S \subset \boldsymbol{X}$. A point $\boldsymbol{p}$ is a *cluster point* of S if for any $\varepsilon > 0$ the ball $B_\varepsilon(\boldsymbol{p})$ contains a point in $S \setminus \{\boldsymbol{p}\}$. □

Definition 11.82. Suppose that $(\boldsymbol{X}, \| - \|)$ is a normed space.

(i) A sequence $(\boldsymbol{x}_\nu)_{\nu\geq 1}$ of points in $\boldsymbol{X}$ is said to *converge* to the point $\boldsymbol{x}_*$ with respect to the norm $\| - \|$ if

$$\lim_{\nu\to\infty} \|\boldsymbol{x}_\nu - \boldsymbol{x}_\infty\| = 0.$$

(ii) A sequence $(\boldsymbol{x}_\nu)_{\nu\geq 1}$ of points in $\boldsymbol{X}$ is called *Cauchy* or *fundamental* if

$$\forall \varepsilon > 0,\ \exists N = N(\varepsilon) > 0\ \forall \nu, \mu \geq N(\varepsilon) : \ \ \|\boldsymbol{x}_\nu - \boldsymbol{x}_\mu\| < \varepsilon.$$

□

The proof of the implication (i) ⇒ (ii) in Theorem 11.70 extends word-for-word to the case of normed spaces and yields the following result.

Proposition 11.83. *In a normed space any convergent sequence is Cauchy.* □

Definition 11.84. A normed space $(\boldsymbol{X}, \| - \|)$ is called *complete* or *Banach* if any Cauchy sequence is convergent. □

Remark 11.85. One can show that not all normed spaces are Banach. However, one can prove that any *finite dimensional* normed space is Banach. □

11.6. Exercises

Exercise 11.1. Let $\boldsymbol{u}, \boldsymbol{v} \in \mathbb{R}^n \setminus \{\boldsymbol{0}\}$. Show that the following statements are equivalent.

(i) The vectors $\boldsymbol{u}, \boldsymbol{v}$ are collinear.
(ii) For any $\boldsymbol{p} \in \mathbb{R}^n$ the lines $\ell_{\boldsymbol{p},\boldsymbol{u}}, \ell_{\boldsymbol{p},\boldsymbol{v}}$ coincide, i.e., $\ell_{\boldsymbol{p},\boldsymbol{u}} = \ell_{\boldsymbol{p},\boldsymbol{v}}, \forall \boldsymbol{p} \in \mathbb{R}^n$.
(iii) The lines $\ell_{\boldsymbol{0},\boldsymbol{u}}, \ell_{\boldsymbol{0},\boldsymbol{v}}$ coincide, i.e., $\ell_{\boldsymbol{0},\boldsymbol{u}} = \ell_{\boldsymbol{0},\boldsymbol{v}}$.

□

Exercise 11.2. (a) Let $\boldsymbol{p}, \boldsymbol{v} \in \mathbb{R}^n$, $\boldsymbol{v} \neq \boldsymbol{0}$. Prove that if $\boldsymbol{q} \in \ell_{\boldsymbol{p},\boldsymbol{v}}$, then $\ell_{\boldsymbol{p},\boldsymbol{v}} = \ell_{\boldsymbol{q},\boldsymbol{v}}$.
(b) Let $\boldsymbol{p}, \boldsymbol{v} \in \mathbb{R}^n$, $\boldsymbol{v} \neq \boldsymbol{0}$. Prove that if $\boldsymbol{p}_1, \boldsymbol{p}_2 \in \ell_{\boldsymbol{p},\boldsymbol{v}}$ and $\boldsymbol{p}_1 \neq \boldsymbol{p}_2$, then the vectors $\boldsymbol{v}$ and $\boldsymbol{u} := \boldsymbol{p}_2 - \boldsymbol{p}_1$ are collinear and $\ell_{\boldsymbol{p},\boldsymbol{v}} = \ell_{\boldsymbol{p},\boldsymbol{u}} = \ell_{\boldsymbol{p}_1,\boldsymbol{u}} = \ell_{\boldsymbol{p}_2,\boldsymbol{u}}$.
(c) Let $\boldsymbol{p}, \boldsymbol{q}, \boldsymbol{u}, \boldsymbol{v} \in \mathbb{R}^n$, $\boldsymbol{u}, \boldsymbol{v} \neq \boldsymbol{0}$. Show that if the lines $\ell_{\boldsymbol{p},\boldsymbol{u}}$ and $\ell_{\boldsymbol{q},\boldsymbol{v}}$ have two distinct points in common, then they coincide. □

Exercise 11.3. Consider the points in $\mathbb{R}^2$

$$\boldsymbol{p}_0 = (0,0), \;\; \boldsymbol{q}_0 = (1,1), \;\; \boldsymbol{p}_1 = (1,0), \;\; \boldsymbol{q}_1 = (0,1).$$

(a) Depict these points and the lines $\ell_0 = \boldsymbol{p}_0\boldsymbol{q}_0$, $\ell_1 = \boldsymbol{p}_1\boldsymbol{q}_1$ on the same planar coordinate system of the type depicted in Figure 11.2.
(b) Find the coordinates of the point where the lines ℓ_0, ℓ_1 intersect. □

Exercise 11.4. Prove Proposition 11.11. □

Exercise 11.5. Let $n \in \mathbb{N}$ and $\boldsymbol{p}, \boldsymbol{q} \in \mathbb{R}^n$. Prove that the following statements are equivalent.

(i) $\boldsymbol{p} \neq \boldsymbol{q}$.
(ii) There exists a linear form $\boldsymbol{\xi} : \mathbb{R}^n \to \mathbb{R}$ such that $\boldsymbol{\xi}(\boldsymbol{p}) \neq \boldsymbol{\xi}(\boldsymbol{q})$.

□

Exercise 11.6. Find a parametric equation (see (11.6)) for the line in $\mathbb{R}^2$ described by the equation

$$x^1 + 2x^2 = 3.$$

Hint: Use the equality $x^1 = 3 - 2x^2$ to find two distinct points on this line. □

Exercise 11.7. Let $\boldsymbol{p} = (1,2,3) \in \mathbb{R}^3$ and $\boldsymbol{v} = (1,1,1) \in \mathbb{R}^3$. Find the coordinates of the point of intersection of the line $\ell_{\boldsymbol{p},\boldsymbol{v}}$ with the hyperplane

$$3x^1 + 4x^2 + 5x^3 = 6.$$ □

Exercise 11.8. Prove that the lines and the hyperplanes in $\mathbb{R}^n$ are affine subspaces. □

Exercise 11.9. Let S be a subset of the Euclidean space $\mathbb{R}^n$, $n \in \mathbb{N}$. Prove that the following statements are equivalent.

(i) The set S is an affine subspace.
(ii) For any $k \in \mathbb{N}$, any points $\boldsymbol{p}_0, \boldsymbol{p}_1, \ldots, \boldsymbol{p}_k \in S$ and any real numbers $t_0, t_1, \ldots, t_k$ such that $t_0 + t_1 + \cdots + t_k = 1$ we have

$$t_0\boldsymbol{p}_0 + t_1\boldsymbol{p}_1 + \cdots + t_k\boldsymbol{p}_k \in S.$$

Hint. The implication (ii) $\Rightarrow$ (i) is immediate. To prove the opposite implication (i) $\Rightarrow$ (ii) argue by induction on k. Observe that at least one of the numbers $t_0, t_1, \ldots, t_k$ is not equal to 1, say $t_k \neq 1$. Then $1 - t_k \neq 0$ and

$$t_0\boldsymbol{p}_0 + t_1\boldsymbol{p}_1 + \cdots + t_k\boldsymbol{p}_k = (1 - t_k)\underbrace{\left(\frac{t_0}{1-t_k}\boldsymbol{p}_1 + \cdots + \frac{t_{k-1}}{1-t_k}\boldsymbol{p}_{k-1}\right)}_{\boldsymbol{q}} + t_k\boldsymbol{p}_k.$$

Use the induction assumption to argue that $\boldsymbol{q} \in S$. Conclude using (i). □

Exercise 11.10. Prove Proposition 11.28. □

Exercise 11.11. Consider the linear operator $A : \mathbb{R}^3 \to \mathbb{R}^3$ characterized by the equalities

$$A\boldsymbol{e}_1 = \boldsymbol{e}_1 + 2\boldsymbol{e}_2 + 3\boldsymbol{e}_3, \quad A\boldsymbol{e}_2 = 4\boldsymbol{e}_1 + 5\boldsymbol{e}_2 + 5\boldsymbol{e}_3, \quad A\boldsymbol{e}_3 = 7\boldsymbol{e}_1 + 8\boldsymbol{e}_2 + 9\boldsymbol{e}_3,$$

where $\{\boldsymbol{e}_1, \boldsymbol{e}_2, \boldsymbol{e}_3\}$ is the canonical basis of $\mathbb{R}^3$.

(i) Find the 3×3 matrix associated to this linear operator.
(ii) Find the vector

$$A\begin{bmatrix} 1 \\ 1 \\ 1 \end{bmatrix}.$$

□

Exercise 11.12. Consider the linear operator $A : \mathbb{R}^3 \to \mathbb{R}^2$ given by the matrix

$$A = \begin{bmatrix} 1 & -2 & 3 \\ 0 & 1 & -4 \end{bmatrix}.$$

Show that there exists a nonzero vector $\boldsymbol{v} \in \mathbb{R}^3$ such that $\ker A$ is equal to the line $\ell_{\boldsymbol{0},\boldsymbol{v}}$. □

Exercise 11.13. Suppose that $A : \mathbb{R}^n \to \mathbb{R}^m$ is a linear operator. Prove that the following statements are equivalent.

(i) A is injective.
(ii) $\ker A = \{\boldsymbol{0}\}$.

□

Exercise 11.14. (a) An *automorphism* of $\mathbb{R}^k$ is a *bijective* linear operator $T : \mathbb{R}^k \to \mathbb{R}^k$. Prove that if T is an automorphism of $\mathbb{R}^k$ then its inverse is also an automorphism of $\mathbb{R}^k$.

(b) A $k \times k$ matrix A is called *invertible* if and only if there exists a $k \times k$ matrix A' such that $AA' = A'A = \mathbb{1}_k$. Prove that if A is invertible, then there exists *a unique* matrix A' with these properties. This unique matrix is called the *inverse* of A and it is denoted by A^{-1}.

(c) Show that T is an automorphism of $\mathbb{R}^k$ if and only if the $k \times k$ matrix representing T is invertible. □

Exercise 11.15. Let $m, n \in \mathbb{N}$, $B \in \mathrm{Mat}_m(\mathbb{R})$, $C \in \mathrm{Mat}_n(\mathbb{R})$, $D \in \mathrm{Mat}_{m\times n}(\mathbb{R})$ and $E \in \mathrm{Mat}_{n\times m}(\mathbb{R})$. Consider the square matrices $S, T \in \mathrm{Mat}_{m+n}(\mathbb{R})$ with block decompositions

$$S = \begin{bmatrix} B & D \\ \mathbf{0}_{n\times m} & C \end{bmatrix}, \quad T = \begin{bmatrix} B & \mathbf{0}_{m\times n} \\ E & C \end{bmatrix},$$

and $\mathbf{0}_{k\times \ell}$ denotes the $k \times \ell$ matrix with all entries 0. Show that if B, C are invertible, then so are S and T, and

$$S^{-1} = \begin{bmatrix} B^{-1} & -B^{-1}DC^{-1} \\ \mathbf{0}_{n\times m} & C^{-1} \end{bmatrix}, \quad T^{-1} = \begin{bmatrix} B^{-1} & \mathbf{0}_{m\times n} \\ -C^{-1}EB^{-1} & C^{-1} \end{bmatrix}.$$ □

Exercise 11.16. We say that a matrix $R \in \mathrm{Mat}_{k\times k}(\mathbb{R})$ is *nilpotent* if there exists $n \in \mathbb{N}$ such that $R^n = \mathbf{0}$. Show that if R is a $k \times k$ nilpotent matrix, then the matrix $\mathbb{1}_k - R$ is invertible.

Hint: Prove first that if $X \in \mathrm{Mat}_{k\times k}(\mathbb{R})$, then

$$\mathbb{1}_k - X^n = (\mathbb{1}_k - X)(\mathbb{1}_k + X + \cdots + X^{n-1}), \quad \forall n \in \mathbb{N}.$$ □

Exercise 11.17. Show that the space $\mathrm{Hom}(\mathbb{R}^n, \mathbb{R}^m)$ of linear operators $\mathbb{R}^n \to \mathbb{R}^m$ is a real vector space. □

Exercise 11.18. Consider the matrices

$$A = \begin{bmatrix} 1 & -2 & 3 \\ 0 & 1 & -4 \end{bmatrix}, \quad B = \begin{bmatrix} 1 & 0 \\ -2 & 1 \\ 3 & -4 \end{bmatrix}.$$

(i) Compute the products AB and BA.
(ii) Show that for any vectors $\boldsymbol{x} \in \mathbb{R}^2$, $\boldsymbol{y} \in \mathbb{R}^3$ we have

$$\langle \boldsymbol{x}, A\boldsymbol{y} \rangle = \langle B\boldsymbol{x}, \boldsymbol{y} \rangle.$$

□

Exercise 11.19. Let $m \in \mathbb{N}$, $m \geq 2$ and consider the $m \times m$ matrix

$$N = \begin{bmatrix} 0\,1\,0\,0 \cdots 0\,0 \\ 0\,0\,1\,0 \cdots 0\,0 \\ 0\,0\,0\,1 \cdots 0\,0 \\ \vdots\,\vdots\,\vdots\,\vdots \;\; \vdots \;\; \vdots\,\vdots \\ 0\,0\,0\,0 \cdots 0\,1 \\ 0\,0\,0\,0 \cdots 0\,0 \end{bmatrix}.$$

Compute the powers N^k, $k \in \mathbb{N}$.

Hint: Regard N as a linear operator $\mathbb{R}^m \to \mathbb{R}^m$ and observe that

$$N\boldsymbol{e}_1 = 0, \;\; N\boldsymbol{e}_2 = \boldsymbol{e}_1, \;\; N\boldsymbol{e}_3 = \boldsymbol{e}_2, \;\; \dots, \;\; N\boldsymbol{e}_m = \boldsymbol{e}_{m-1},$$

where $\boldsymbol{e}_1, \dots, \boldsymbol{e}_m$ is the natural basis of $\mathbb{R}^m$. Then use the fact that the composition of two linear operators corresponds to the multiplication of the corresponding matrices. □

Exercise 11.20. For every $\alpha \in [0, 2\pi]$ we denote by $R_\alpha : \mathbb{R}^2 \to \mathbb{R}^2$ the *counterclockwise rotation* of angle α about the origin **0**.

(i) Express the coordinates y^1, y^2 of $\boldsymbol{y} = R_\alpha \boldsymbol{x}$ in terms of the coordinates of $\boldsymbol{x} = [x^1, x^2]^\top$.

(ii) Show that R_α is a linear operator and compute its associated matrix. Continue to denote by R_α the associated matrix.

(iii) Given $\alpha, \beta \in [0, 2\pi]$ compute the product $R_\alpha \cdot R_\beta$.

Hint. (i) Set $r := \|\boldsymbol{x}\|$, and denote by θ the angle the vector $\boldsymbol{x}$ makes with the x^1-axis, measured counterclockwisely starting at the positive x^1-axis. Then $x^1 = r\cos\theta$, $x^2 = r\sin\theta$. Next, set $\boldsymbol{y} := R_\alpha \boldsymbol{x}$ and show that $y^1 = r\cos(\theta + \alpha)$, $y^2 = r\sin(\theta + \alpha)$. Conclude using the trig formulæ (5.33a). □

Exercise 11.21. The *trace* of an $n \times n$ matrix A is the scalar denoted by $\operatorname{tr} A$ and defined as the sum of the diagonal entries of A,

$$\operatorname{tr} A := A^1_1 + \cdots + A^n_n.$$

(i) Show that if $A, B \in \operatorname{Mat}_{n\times n}(\mathbb{R})$, $c \in \mathbb{R}$, then

$$\operatorname{tr}(A + B) = \operatorname{tr} A + \operatorname{tr} B, \;\; \operatorname{tr}(cA) = c \operatorname{tr} A.$$

(ii) Show that if $A \in \operatorname{Mat}_{m\times n}(\mathbb{R})$ and $B \in \operatorname{Mat}_{n\times m}(\mathbb{R})$, then

$$\operatorname{tr}(AB) = \operatorname{tr}(BA).$$

Hint: Use (11.15).

(iii) Show that there do not exist matrices $A, B \in \operatorname{Mat}_{n\times n}(\mathbb{R})$ such that $AB - BA = \mathbb{1}_n$.

□

Exercise 11.22. Prove (11.21). □

Exercise 11.23. Suppose that $A \in \mathrm{Mat}_{m\times n}(\mathbb{R})$. Denote by $(\boldsymbol{e}_j)_{1\leq j\leq n}$ the canonical basis of $\mathbb{R}^n$ and by $(\boldsymbol{f}_i)_{1\leq i\leq m}$ the canonical basis of $\mathbb{R}^m$. Prove that

$$A^i_j = \langle \boldsymbol{f}_i, A\boldsymbol{e}_j\rangle, \;\; \forall i = 1,\dots,m, \;\; j = 1,\dots,n.$$ □

Exercise 11.24. Suppose that $A \in \mathrm{Mat}_{m\times n}(\mathbb{R})$. The *transpose* of A is the $n \times m$ matrix $A^\top$ defined by the requirement

$$(A^\top)^j_i = A^i_j, \;\; \forall i = 1,\dots,m, \;\; j = 1,\dots,n.$$

In other words, the rows of $A^\top$ coincide with the columns of A. (For example, the transpose of the matrix A in Exercise 11.18 is the matrix B in the same exercise.)

(i) Suppose that $B \in \mathrm{Mat}_{p\times m}(\mathbb{R})$. Prove that

$$(B\cdot A)^\top = (A^\top)\cdot(B^\top).$$

Hint: Check this first in the special case when $p = m = 1$, i.e., B is a matrix consisting one row of size m, and A is a matrix consisting of one column of size m. Use this special case and the equality (11.15) to deduce the general case.

(ii) Prove that, for any $\boldsymbol{x} \in \mathbb{R}^m$ and $\boldsymbol{y} \in \mathbb{R}^n$, we have (identifying 1×1 matrices with numbers)

$$\langle \boldsymbol{x}, A\boldsymbol{y}\rangle = \boldsymbol{x}^\top \cdot A \cdot \boldsymbol{y} = \langle A^\top \boldsymbol{x}, \boldsymbol{y}\rangle.$$

(iii) Prove that for any $y \in \mathbb{R}^n$ we have

$$\langle A^\top A\boldsymbol{y}, \boldsymbol{y}\rangle \geq 0.$$

(iv) Prove that an $n \times n$ matrix A is symmetric if and only if

$$\langle A\boldsymbol{x}, \boldsymbol{y}\rangle = \langle \boldsymbol{x}, A\boldsymbol{y}\rangle, \;\; \forall \boldsymbol{x}, \boldsymbol{y} \in \mathbb{R}^n.$$

Hint: Use Exercise 11.23 and part (i) of this exercise. □

Exercise 11.25. Let $n \in \mathbb{N}$ and suppose that $A : \mathbb{R}^n \to \mathbb{R}^n$ is a linear operator. As usual we will continue to denote by A the associated matrix. Prove that the following statements are equivalent.

(i) $\langle A\boldsymbol{x}, A\boldsymbol{y}\rangle = \langle \boldsymbol{x}, \boldsymbol{y}\rangle, \forall \boldsymbol{x}, \boldsymbol{y} \in \mathbb{R}^n$.
(ii) $\|A\boldsymbol{x}\| = \|\boldsymbol{x}\|, \forall \boldsymbol{x} \in \mathbb{R}^n$.
(iii) $A^\top \cdot A = \mathbb{1}_n$.

An operator or matrix with any of the above three equivalent properties is called *orthogonal*. □

Exercise 11.26. Suppose that $\boldsymbol{\xi} : \mathbb{R}^n \to \mathbb{R}$ is a linear functional. Show that the graph of $\boldsymbol{\xi}$, defined as

$$G_{\boldsymbol{\xi}} = \big\{ (\boldsymbol{x}, y) \in \mathbb{R}^n \times \mathbb{R};\;\; y = \boldsymbol{\xi}(\boldsymbol{x}) \big\},$$

is a hyperplane in $\mathbb{R}^n \times \mathbb{R} = \mathbb{R}^{n+1}$ and then find a normal vector to this hyperplane. □

Exercise 11.27. Prove Proposition 11.53. □

Exercise 11.28. Prove (11.28). □

Exercise 11.29. Prove Proposition 11.55.

Hint: Use (11.28). □

Exercise 11.30. Prove Proposition 11.56. □

Exercise 11.31. Prove that if $U \subset \mathbb{R}^m$ is open in $\mathbb{R}^m$ and $V \subset \mathbb{R}^n$ is open in $\mathbb{R}^n$, then $U \times V$ is open in $\mathbb{R}^m \times \mathbb{R}^n = \mathbb{R}^{m+n}$.

Hint: Use Proposition 11.56 and observe several things. First, if $\boldsymbol{p} \in \mathbb{R}^m$ and $\boldsymbol{q} \in \mathbb{R}^n$ then the pair $(\boldsymbol{p}, \boldsymbol{q}) \in \mathbb{R}^m \times \mathbb{R}^n$ and the Cartesian product can be identified with $\mathbb{R}^{m+n}$. Next observe that the Cartesian product $C_r(\boldsymbol{p}) \times C_r(\boldsymbol{q}) \subset \mathbb{R}^m \times \mathbb{R}^n$ can be identified with $C_r\big((\boldsymbol{p}, \boldsymbol{q}) \big)$, the cube of radius r with center $(\boldsymbol{p}, \boldsymbol{q}) \in \mathbb{R}^{m+n}$. □

Exercise 11.32. Complete the proof of the claim in Example 11.58(c). □

Exercise 11.33. Prove Proposition 11.63.

Hint: Use (11.27). □

Exercise 11.34. (a) Prove that any finite subset of $\mathbb{R}^n$ is closed.
(b) Prove that any affine hyperplane in $\mathbb{R}^n$ is a closed subset. □

Exercise 11.35. Prove that any open subset $U \subset \mathbb{R}^n$ is the union of a (possibly infinite) family of open cubes. □

Exercise 11.36. Let $n \in \mathbb{N}$. Prove that for any $\boldsymbol{p} \in \mathbb{R}^n$ and any $r > 0$ the open Euclidean ball $B_r(\boldsymbol{p})$ and the closed Euclidean ball $\overline{B_r(\boldsymbol{p})}$ are convex sets. □

Exercise 11.37. Let $n \in \mathbb{N}$.
(a) Suppose that $(\boldsymbol{p}_\nu)$ is a sequence in $\mathbb{R}^n$ that converges to $\boldsymbol{p} \in \mathbb{R}^n$. Prove that

$$\lim_{\nu\to\infty} \|\boldsymbol{p}_\nu\| = \|\boldsymbol{p}\| \text{ and } \lim_{\nu\to\infty} \|\boldsymbol{p}_\nu\|_\infty = \|\boldsymbol{p}\|_\infty.$$

(b) Let $r > 0$. Prove that any point $\boldsymbol{x} \in \mathbb{R}^n$ such that $\|\boldsymbol{x}\| = r$ is a cluster point of the open ball $B_r(\boldsymbol{0})$.

Hint: (a) Use (11.24b) and (11.26b). □

Exercise 11.38. Let $n \in \mathbb{N}$ and $X \subset \mathbb{R}^n$. Prove that the following statements are equivalent.

(i) The set X is closed.
(ii) The set X contains all its cluster points.

□

Exercise 11.39. Prove that the set $\mathbb{Q}^n$ is dense in $\mathbb{R}^n$.

Hint: Use Proposition 3.33 and Proposition 11.63. □

Exercise 11.40. Let $n \in \mathbb{N}$. Consider a sequence of vectors $(\boldsymbol{x}_\nu)_{\nu\in\mathbb{N}}$ in $\mathbb{R}^n$. The *series*

$$\sum_{\nu\in\mathbb{N}} \boldsymbol{x}_\nu$$

associated to this sequence is the new sequence $(S_N)_{N\in\mathbb{N}}$ of vectors in $\mathbb{R}^n$ described by the *partial sums*

$$S_N = \boldsymbol{x}_1 + \cdots + \boldsymbol{x}_N, \quad N \in \mathbb{N}.$$

The series $\sum_{\nu\in\mathbb{N}} \boldsymbol{x}_\nu$ is called *convergent* if the sequence of partial sums S_N is convergent.

Prove that if the *series* of real numbers $\sum_{\nu\in\mathbb{N}} \|\boldsymbol{x}_\nu\|$ is convergent, then the *series* of vectors $\sum_{\nu\in\mathbb{N}} \boldsymbol{x}_\nu$ is also convergent.

Hint. It suffices to show that the sequence (S_N) is Cauchy. Define

$$S_N^* := \|\boldsymbol{x}_1\| + \cdots + \|\boldsymbol{x}_N\|, \quad N \in \mathbb{N}.$$

The *series* of real numbers $\sum_{\nu\in\mathbb{N}} \|\boldsymbol{x}_\nu\|$ is convergent and thus sequence (S_N^*) is convergent, hence Cauchy. Prove that this implies that the sequence S_N is Cauchy by imitating the proof of Absolute Convergence Theorem 4.46. □

Exercise 11.41 (Banach's fixed point theorem). Suppose that $X \subset \mathbb{R}^n$ is a *closed* subset and $F : X \to \mathbb{R}^n$ is a map satisfying the following conditions:

$$F(\boldsymbol{x}) \in X, \ \forall \boldsymbol{x} \in X. \tag{C_1}$$

$$\exists r \in (0,1) \ \text{such that} \ \forall \boldsymbol{x}_1, \boldsymbol{x}_2 \in X : \ \|F(\boldsymbol{x}_1) - F(\boldsymbol{x}_2)\| \le r\|\boldsymbol{x}_1 - \boldsymbol{x}_2\|. \tag{C_2}$$

Fix $\boldsymbol{x}_0 \in X$ and define inductively the sequence of points in X,

$$\boldsymbol{x}_1 = F(\boldsymbol{x}_0), \ \boldsymbol{x}_2 = F(\boldsymbol{x}_1), \dots, \boldsymbol{x}_\nu = F(\boldsymbol{x}_{\nu-1}), \ \forall \nu \in \mathbb{N}.$$

Prove that the following hold.

(i) For any $\nu \in \mathbb{N}$,

$$\|\boldsymbol{x}_{\nu+1} - \boldsymbol{x}_\nu\| \le r^\nu \|\boldsymbol{x}_1 - \boldsymbol{x}_0\|.$$

(ii) For any $\mu, \nu \in \mathbb{N}$, $\mu < \nu$

$$\|\boldsymbol{x}_\nu - \boldsymbol{x}_\mu\| \le \frac{r^\mu(1 - r^{\nu-\mu})}{1-r}\|\boldsymbol{x}_1 - \boldsymbol{x}_0\| \le \frac{r^\mu}{1-r}\|\boldsymbol{x}_1 - \boldsymbol{x}_0\|.$$

(iii) The sequence $(\boldsymbol{x}_\nu)_{\nu\ge 0}$ is Cauchy.

(iv) If $\boldsymbol{x}_*$ is the limit of the sequence $(\boldsymbol{x}_\nu)_{\nu\ge 0}$, then $F(\boldsymbol{x}_*) = \boldsymbol{x}_*$.

(v) Show that if $\boldsymbol{p} \in X$ is a *fixed point of* F, i.e., it satisfies $F(\boldsymbol{p}) = \boldsymbol{p}$, then $\boldsymbol{p}$ must be equal to the point $\boldsymbol{x}_*$ defined above.

□

Exercise 11.42. Suppose that $S \subset \mathbb{R}^{17}$ consists of $1,234,567,890$ points and $T : S \to S$ is a map such that

$$\|T\boldsymbol{s}_1 - T\boldsymbol{s}_2\| < \|\boldsymbol{s}_1 - \boldsymbol{s}_2\|, \;\; \forall \boldsymbol{s}_1, \boldsymbol{s}_2 \in S, \;\; \boldsymbol{s}_1 \neq \boldsymbol{s}_2.$$

Prove that there exists $\boldsymbol{s}_* \in S$ such that $T(\boldsymbol{s}_*) = \boldsymbol{s}_*$.

Hint: Use the result in the previous exercise. □

Exercise 11.43. Prove Proposition 11.78. □

Exercise 11.44. Consider the space $\mathbb{R}^2$ equipped with the norm $\| - \|_1$ described in Example 11.80(a). Draw a picture of the open ball $B_1(\boldsymbol{0})$ defined by this norm. □

11.7. More challenging problems

Problem 11.1. (a) Prove that if S is an affine subspace of $\mathbb{R}^2$, then S is either a point, or a line, or the whole $\mathbb{R}^2$.

(b) Prove that if S is an affine subspace of $\mathbb{R}^3$, then S is either a point, or a line, or a plane, or the whole $\mathbb{R}^3$. □

Problem 11.2. Suppose that $n \in \mathbb{N}$ and $A \in \mathrm{Mat}_{n\times n}(\mathbb{R})$. Prove that the following statements are equivalent.

(i) The matrix A is invertible in the sense defined in Exercise 11.14.
(ii) There exists $B \in \mathrm{Mat}_{n\times n}(\mathbb{R})$ such that $BA = \mathbb{1}_n$.
(iii) There exists $C \in \mathrm{Mat}_{n\times n}(\mathbb{R})$ such that $AC = \mathbb{1}_n$.
(iv) The linear operator $\mathbb{R}^n \to \mathbb{R}^n$ defined by A is bijective.
(v) The linear operator $\mathbb{R}^n \to \mathbb{R}^n$ defined by A is injective.
(vi) The linear operator $\mathbb{R}^n \to \mathbb{R}^n$ defined by A is surjective.

Hint: You need to use the fact that $\mathbb{R}^n$ is a *finite dimensional* vector space. □

Chapter 12

Continuity

A function $\boldsymbol{F} : \mathbb{R}^n \to \mathbb{R}^m$ can be viewed as transporting in some fashion the Euclidean space $\mathbb{R}^n$ into the Euclidean space $\mathbb{R}^m$. The space $\mathbb{R}^m$ is often called the *target space*. For example, a map $\boldsymbol{F} : \mathbb{R} \to \mathbb{R}^2$ "transports" the real axis $\mathbb{R}$ into a region of $\mathbb{R}^2$ that typically looks like a curve; see Figure 12.1. For this reason functions $\boldsymbol{F} : \mathbb{R}^n \to \mathbb{R}^m$ are often called *transformations*, *operators*, or *maps*.

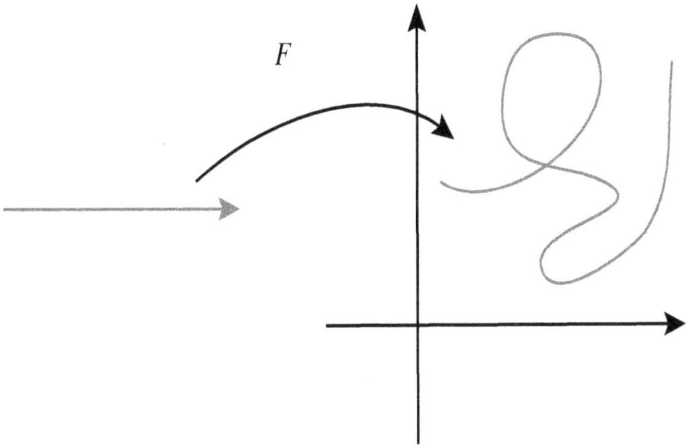

Fig. 12.1 *A map* $F : \mathbb{R} \to \mathbb{R}^2$.

Suppose that $\boldsymbol{F} : \mathbb{R}^n \to \mathbb{R}^m$ is a map. For any $\boldsymbol{x} \in \mathbb{R}^n$, its image $\boldsymbol{y} = \boldsymbol{F}(\boldsymbol{x})$ is a point in $\mathbb{R}^m$ and thus it is determined by a column vector

$$\boldsymbol{y} = \begin{bmatrix} y^1 \\ \vdots \\ y^m \end{bmatrix}.$$

The coordinates $y^1, \dots, y^m$ depend on the point $\boldsymbol{x}$ and thus they are described by functions

$$F^i : \mathbb{R}^n \to \mathbb{R}, \;\; y^i = F^i(x^1, \dots, x^n), \;\; i = 1, \dots, m.$$

We can turn this argument on its head, and think of a collection of functions

$$F^1, \dots, F^m : \mathbb{R}^n \to \mathbb{R}$$

as defining a map $\boldsymbol{F} : \mathbb{R}^n \to \mathbb{R}^m$. Often, when working with a map $\mathbb{R}^n \to \mathbb{R}^m$ and no confusion is possible, we will dispense of the extra symbol $\boldsymbol{F}$ and describe the map in a simpler way as a collection of functions

$$y^1 = y^1(x^1, \dots, x^n), \dots, y^m = y^m(x^1, \dots, x^n).$$

Example 12.1. When predicting the weather (on the surface of the Earth) we need to describe several quantities: temperature (T), pressure (P) and wind velocity $V = (V^1, V^2)$. These quantities depend on the location (determined by two coordinates x^1, x^2) and the time t. We thus have a collection of 4 functions P, T, V^1, V^2 depending on 3 variables x^1, x^2, t,

$$P = P(x^1, x^2, t), \;\; V^1 = V^1(x^1, x^2, t), \text{ etc.},$$

and thus we are dealing with a map $\mathbb{R}^3 \to \mathbb{R}^4$. □

Definition 12.2. Let $m, n \in \mathbb{N}$ and $X \subset \mathbb{R}^n$. The *graph* of a map $\boldsymbol{F} : X \to \mathbb{R}^m$ is the set

$$G_{\boldsymbol{F}} := \big\{ \big(\boldsymbol{x}, \boldsymbol{y} \big) \in X \times \mathbb{R}^m; \;\; \boldsymbol{y} = \boldsymbol{F}(\boldsymbol{x}) \big\} \subset X \times \mathbb{R}^m.$$ □

As we know, the graph of a function $f : \mathbb{R} \to \mathbb{R}$ can be visualized as a curve in $\mathbb{R}^2$. Similarly, the graph of a function $f : \mathbb{R}^2 \to \mathbb{R}$ can be visualized as surface in $\mathbb{R}^3$. If we denote by x, y, z the Euclidean coordinates in $\mathbb{R}^3$, then the graph of a function of two variables $f(x, y)$ is described by the equation $z = f(x, y)$. You can think of the graph as describing a form of relief on Earth, where the altitude z at the point with coordinates (x, y) is $f(x, y)$; see Figure 12.2.

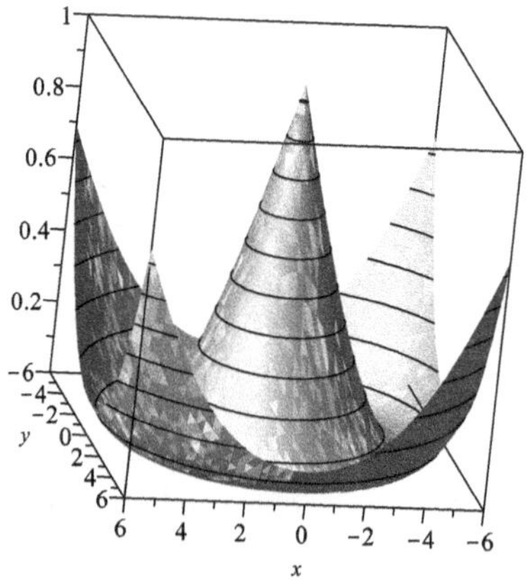

Fig. 12.2 *The graph of the function* $f : [-6, 6] \times [-6, 6] \to \mathbb{R}$, $f(x, y) = 1 - \sin \frac{\sqrt{x^2+y^2}}{3}$.

12.1. Limits and continuity

Definition 12.3. Let $m, n \in \mathbb{N}$, $X \subset \mathbb{R}^n$. Suppose we are given a map $\boldsymbol{F} : X \to \mathbb{R}^m$ and a cluster point $\boldsymbol{x}_0$ of X. (*The point $\boldsymbol{x}_0$ need not belong to X.*)

We say that *the limit of $\boldsymbol{F}(\boldsymbol{x})$ when $\boldsymbol{x}$ approaches $\boldsymbol{x}_0$ is the point $\boldsymbol{y}_0$* (in the target space $\mathbb{R}^m$) if

$$\forall \varepsilon > 0 \ \exists \delta = \delta(\varepsilon) > 0 \qquad \text{such that,} \tag{12.1}$$

$$\forall \boldsymbol{x} \in X \setminus \{\boldsymbol{x}_0\} : \ \|\boldsymbol{x} - \boldsymbol{x}_0\| < \delta \Rightarrow \|\boldsymbol{F}(\boldsymbol{x}) - \boldsymbol{y}_0\| < \varepsilon. \tag{12.2}$$

We will indicate this using the notation

$$\boldsymbol{y}_0 = \lim_{\boldsymbol{x} \to \boldsymbol{x}_0} \boldsymbol{F}(\boldsymbol{x}).$$ □

We have the following multidimensional counterpart of Theorem 5.4.

Proposition 12.4. *Let $m, n \in \mathbb{N}$, $X \subset \mathbb{R}^n$. Suppose we are given a map $\boldsymbol{F} : X \to \mathbb{R}^m$ and a cluster point $\boldsymbol{x}_0$ of X. The following statements are equivalent.*

(i)

$$\lim_{\boldsymbol{x} \to \boldsymbol{x}_0} \boldsymbol{F}(\boldsymbol{x}) = \boldsymbol{y}_0 \in \mathbb{R}^m.$$

(ii) For any sequence $(\boldsymbol{x}_\nu)$ in $X \setminus \{\boldsymbol{x}_0\}$ that converges to $\boldsymbol{x}_0$ we have

$$\lim_{\nu \to \infty} \boldsymbol{F}(\boldsymbol{x}_\nu) = \boldsymbol{y}_0.$$

Proof. (i) $\Rightarrow$ (ii) Suppose that $(\boldsymbol{x}_\nu)$ is a sequence in $X \setminus \{\boldsymbol{x}_0\}$ that converges to $\boldsymbol{x}_0$. We have to show that, given the condition (12.1), the sequence $F(\boldsymbol{x}_\nu)$ converges to $\boldsymbol{y}_0$.

Let $\varepsilon > 0$. Choose $\delta(\varepsilon) > 0$ determined by (12.1). Since $\boldsymbol{x}_\nu \to \boldsymbol{x}_0$, there exists $N = N(\varepsilon)$ such that, for all $\nu > N(\varepsilon)$ we have $\|\boldsymbol{x}_\nu - \boldsymbol{x}_0\| < \delta(\varepsilon)$. Invoking (12.1) we deduce that for all $\nu > N(\varepsilon)$ we have $\|\boldsymbol{F}(\boldsymbol{x}_\nu) - \boldsymbol{y}_0\| < \varepsilon$. This proves that

$$\lim_{\nu \to \infty} \boldsymbol{F}(\boldsymbol{x}_\nu) = \boldsymbol{y}_0.$$

(ii) $\Rightarrow$ (i) We argue by contradiction. Assume that (12.1) is false so that

$$\exists \varepsilon_0 > 0 : \ \forall \delta > 0, \ \exists \boldsymbol{x}_\delta \in X \setminus \{\boldsymbol{x}_0\} : \ \|\boldsymbol{x}_\delta - \boldsymbol{x}_0\| < \delta \ \text{ and } \ \|\boldsymbol{F}(\boldsymbol{x}_\delta) - \boldsymbol{y}_0\| \geq \varepsilon_0.$$

Thus, if we choose δ of the form $\delta = \frac{1}{\nu}$, $\nu \in \mathbb{N}$, we deduce that for any $\nu \in \mathbb{N}$ there exists $\boldsymbol{x}_\nu \in X \setminus \{\boldsymbol{x}_0\}$ such that

$$\|\boldsymbol{x}_\nu - \boldsymbol{x}_0\| < \frac{1}{\nu} \ \text{ and } \ \|F(\boldsymbol{x}_\nu) - \boldsymbol{y}_0\| \geq \varepsilon_0.$$

This shows that the sequence $(\boldsymbol{x}_\nu)$ in $X \setminus \{\boldsymbol{x}_0\}$ converges to $\boldsymbol{x}_0$, but the sequence $\boldsymbol{F}(\boldsymbol{x}_\nu)$ does not converge to $\boldsymbol{y}_0$. This contradicts (ii). □

Definition 12.5 (Continuity). Let $m, n \in \mathbb{N}$, $X \subset \mathbb{R}^n$.

(i) A map $\boldsymbol{F} : X \to \mathbb{R}^m$ is said to be *continuous at* $\boldsymbol{x}_0 \in X$ if

$$\boxed{\forall \varepsilon > 0\ \exists \delta = \delta(\varepsilon) > 0 \text{ such that } \forall \boldsymbol{x} \in X :\ \|\boldsymbol{x} - \boldsymbol{x}_0\| < \delta \Rightarrow \|\boldsymbol{F}(\boldsymbol{x}) - \boldsymbol{F}(\boldsymbol{x}_0)\| < \varepsilon}. \tag{12.3}$$

(ii) A map $\boldsymbol{F} : X \to \mathbb{R}^m$ is said to be *continuous on* X if it is continuous at every point $\boldsymbol{x}_0 \in X$.

□

Proposition 12.6. *Let $m, n \in \mathbb{N}$, $X \subset \mathbb{R}^n$. Consider a map*

$$\boldsymbol{F} : X \to \mathbb{R}^m, \quad \boldsymbol{F}(\boldsymbol{x}) = \begin{bmatrix} F^1(\boldsymbol{x}) \\ \vdots \\ F^m(\boldsymbol{x}) \end{bmatrix}.$$

The following statements are equivalent.

(i) The map $\boldsymbol{F}$ is continuous at $\boldsymbol{x}_0$.
(ii) For any sequence $(\boldsymbol{x}_\nu)$ in X that converges to $\boldsymbol{x}_0$ we have

$$\lim_{\nu\to\infty} \boldsymbol{F}(\boldsymbol{x}_\nu) = \boldsymbol{F}(\boldsymbol{x}_0).$$

(iii) The components $F^1, \dots, F^m : X \to \mathbb{R}$ are continuous at $\boldsymbol{x}_0$.

Proof. The proof of the equivalence (i) $\Longleftrightarrow$ (ii) is identical to the proof of Proposition 12.4 and the details are left to the reader. The proof of the equivalence (ii) $\Longleftrightarrow$ (iii) relies on the equivalence (i) $\Longleftrightarrow$ (ii).

(ii) $\Longleftrightarrow$ (iii) According to the equivalence (i) $\Longleftrightarrow$ (ii) applied to each component F^i individually, the functions $F^1, \dots, F^m$ are continuous at $\boldsymbol{x}_0$ *if and only if*, for any sequence $(\boldsymbol{x}_\nu)$ in X that converges to $\boldsymbol{x}_0$ we have

$$\lim_{\nu\to\infty} F^i(\boldsymbol{x}_\nu) = F^i(\boldsymbol{x}_0), \quad i = 1, 2, \dots, m.$$

Proposition 11.63 shows that these conditions are equivalent to

$$\lim_{\nu\to\infty} \boldsymbol{F}(\boldsymbol{x}_\nu) = \boldsymbol{F}(\boldsymbol{x}_0).$$

In turn, this is equivalent to the continuity of $\boldsymbol{F}$ at $\boldsymbol{x}_0$. □

Example 12.7. The *multiplication function* $\mu : \mathbb{R}^2 \to \mathbb{R}$ given by $\mu(x, y) = xy$ is continuous. We will prove this using Proposition 12.6. Consider a point $\boldsymbol{p}_0 = (x_0, y_0) \in \mathbb{R}^2$.

If $\boldsymbol{p}_\nu = (x_\nu, y_\nu) \in \mathbb{R}^2$ is a sequence of points converging to $\boldsymbol{p}_0$, then $x_\nu \to x_0$ and $y_\nu \to y_0$ as $\nu \to \infty$. Hence

$$\lim_{\nu\to\infty} \mu(\boldsymbol{p}_\nu) = \lim_{\nu\to\infty} (x_\nu y_\nu) = x_0 y_0 = \mu(\boldsymbol{p}_0).$$ □

Definition 12.8 (Paths). Let $n \in \mathbb{N}$. A *continuous path* in $\mathbb{R}^n$ is a continuous map

$$\gamma : I \to \mathbb{R}^n,$$

where $I \subset \mathbb{R}$ is an interval. □

A path $\gamma : I \to \mathbb{R}^n$ is completely determined by its components

$$\gamma^1, \ldots, \gamma^n : I \to \mathbb{R}$$

which are continuous functions. It is convenient to think of the interval I as a *time* interval so the components γ^i are functions of time, $\gamma^i = \gamma^i(t)$. As time goes by, the point

$$\gamma(t) = \begin{bmatrix} \gamma^1(t) \\ \vdots \\ \gamma^n(t) \end{bmatrix} \in \mathbb{R}^n$$

moves in space. Thus we can think of a path as describing the motion of a point in space during a given interval of time I. The image of a path $\boldsymbol{F} : I \to \mathbb{R}^n$ is the trajectory of this motion and it typically looks like a curve. Traditionally, a path is indicated by a system of equations

$$x^i = \gamma^i(t), \quad i = 1, \ldots, n,$$

meaning that the coordinates $x^1, \ldots, x^n$ of the moving point at time t are given by the functions $\gamma^1(t), \ldots, \gamma^n(t)$.

Example 12.9. The trajectory of the path

$$\gamma : [0, 4\pi] \to \mathbb{R}^2, \quad \gamma(t) = \begin{bmatrix} (t+1)\cos(2t) \\ (t+1)\sin(2t) \end{bmatrix} \in \mathbb{R}^2$$

is the *helix* depicted in Figure 12.3. □

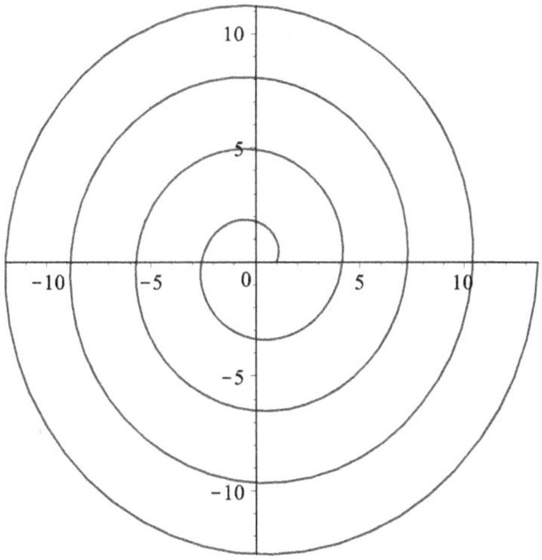

Fig. 12.3 *A linear spiral* $x = (1+t)\cos 2t$, $y = (1+t)\sin 2t$, $t \in [0, 4\pi]$.

Definition 12.10. Let $m, n \in \mathbb{N}$ and $X \subset \mathbb{R}^n$. A map $\boldsymbol{F} : X \to \mathbb{R}^m$ is called *Lipschitz* if it admits a *Lipschitz constant*, i.e., a constant $L > 0$ such that

$$\|\boldsymbol{F}(\boldsymbol{x}) - \boldsymbol{F}(\boldsymbol{y})\| \leq L\|\boldsymbol{x} - \boldsymbol{y}\|, \quad \forall \boldsymbol{x}, \boldsymbol{y} \in X. \tag{12.4}$$

□

Proposition 12.11. *Let $m, n \in \mathbb{N}$ and $X \subset \mathbb{R}^n$. Then a Lipschitz map $\boldsymbol{F} : X \to \mathbb{R}^m$ is continuous.*

Proof. Fix a Lipschitz constant $L > 0$ as in the Lipschitz condition (12.4). Let $\boldsymbol{x}_0 \in X$ be an arbitrary point in X. To prove that F is continuous at $\boldsymbol{x}_0$ we use Proposition 12.6(ii). Suppose that $(\boldsymbol{x}_\nu)$ is a sequence of points in X such that

$$\lim_{\nu\to\infty} \boldsymbol{x}_\nu = \boldsymbol{x}_0.$$

From the Lipschitz condition we deduce

$$\|\boldsymbol{F}(\boldsymbol{x}_\nu) - \boldsymbol{F}(\boldsymbol{x}_0)\| \leq L\|\boldsymbol{x}_\nu - \boldsymbol{x}_0\|.$$

Invoking the Squeezing Principle Proposition 4.10 we conclude that

$$\lim_{\nu\to\infty} \|\boldsymbol{F}(\boldsymbol{x}_\nu) - \boldsymbol{F}(\boldsymbol{x}_0)\| = 0 \Rightarrow \lim_{\nu\to\infty} \boldsymbol{F}(\boldsymbol{x}_\nu) = \boldsymbol{F}(\boldsymbol{x}_0).$$

This proves that $\boldsymbol{F}$ is continuous at $\boldsymbol{x}_0$. □

Proposition 12.12. *Let $m, n \in \mathbb{N}$. The following hold.*

(i) The norm functions

$$\mathbb{R}^n \ni \boldsymbol{x} \mapsto \|\boldsymbol{x}\| \in \mathbb{R}, \quad \mathbb{R}^n \ni \boldsymbol{x} \mapsto \|\boldsymbol{x}\|_\infty$$

are Lipschitz.

(ii) Any linear form $\boldsymbol{\xi} : \mathbb{R}^n \to \mathbb{R}$ is Lipschitz.

(iii) Any linear operator $A : \mathbb{R}^n \to \mathbb{R}^m$ is Lipschitz.

In particular, all the maps above are continuous.

Proof. (i) Using (11.24b) and (11.26b) we deduce

$$\big|\, \|\boldsymbol{x}\| - \|\boldsymbol{y}\| \,\big| \leq \|\boldsymbol{x} - \boldsymbol{y}\|, \quad \big|\, \|\boldsymbol{x}\|_\infty - \|\boldsymbol{y}\|_\infty \,\big| \leq \|\boldsymbol{x} - \boldsymbol{y}\|_\infty$$

which shows that the constant 1 is a Lipschitz constant of both functions $f(\boldsymbol{x}) = \|\boldsymbol{x}\|$ and $g(\boldsymbol{x}) = \|\boldsymbol{x}\|_\infty$.

(ii) Let $\boldsymbol{\xi}_\uparrow$ be the dual of $\boldsymbol{\xi}$ defined in Proposition 11.38. We recall that this means that $\boldsymbol{\xi}_\uparrow$ is the unique vector in $\mathbb{R}^n$ such that

$$\boldsymbol{\xi}(\boldsymbol{x}) = \langle \boldsymbol{\xi}_\uparrow, \boldsymbol{x} \rangle.$$

If $\boldsymbol{x}, \boldsymbol{y} \in \mathbb{R}^n$, then

$$\big|\, \boldsymbol{\xi}(\boldsymbol{x}) - \boldsymbol{\xi}(\boldsymbol{y}) \,\big| = \big|\, \boldsymbol{\xi}(\boldsymbol{x} - \boldsymbol{y}) \,\big| = \big|\, \langle \boldsymbol{\xi}_\uparrow, \boldsymbol{x} - \boldsymbol{y} \rangle \,\big|$$

(use the Cauchy–Schwarz inequality)

$$\leq \|\boldsymbol{\xi}_\uparrow\| \cdot \|\boldsymbol{x} - \boldsymbol{y}\|.$$

This proves that $\boldsymbol{\xi}$ is Lipschitz, and the norm of $\|\boldsymbol{\xi}_\uparrow\|$ is a Lipschitz constant of $\boldsymbol{\xi}$. In particular,

$$\big|\, \boldsymbol{\xi}(\boldsymbol{z}) \,\big| = \big|\, \boldsymbol{\xi} \bullet \boldsymbol{z} \,\big| \leq \|\boldsymbol{\xi}_\uparrow\| \cdot \|\boldsymbol{z}\|, \quad \forall \boldsymbol{z} \in \mathbb{R}^n. \tag{12.5}$$

(iii) As we have seen earlier, the components of $A\boldsymbol{x}$ are linear functionals in $\boldsymbol{x}$

$$A\boldsymbol{x} = \begin{bmatrix} A^1 \bullet \boldsymbol{x} \\ \vdots \\ A^m \bullet \boldsymbol{x} \end{bmatrix},$$

where $A^1, \dots, A^m$ are the rows of the $m \times n$ matrix associated to the operator A. From (12.5) we deduce

$$\left| A^i \bullet \boldsymbol{z} \right| \leq \left\| (A^i)_\uparrow \right\| \cdot \|\boldsymbol{z}\|, \;\; \forall \boldsymbol{z} \in \mathbb{R}^n, \;\; i = 1, \dots, m.$$

Given $\boldsymbol{x}, \boldsymbol{y} \in \mathbb{R}^n$, we set $\boldsymbol{z} := \boldsymbol{x} - \boldsymbol{y}$ and we have

$$A(\boldsymbol{x} - \boldsymbol{y}) = A\boldsymbol{z} = \begin{bmatrix} A^1 \bullet \boldsymbol{z} \\ \vdots \\ A^m \bullet \boldsymbol{z} \end{bmatrix}$$

so that

$$\begin{aligned} \|A(\boldsymbol{x} - \boldsymbol{y})\|^2 &= |A^1 \bullet \boldsymbol{z}|^2 + \cdots + |A^m \bullet \boldsymbol{z}|^2 \\ &\leq \|(A^1)_\uparrow\|^2 \cdot \|\boldsymbol{z}\|^2 + \cdots + \|(A^m)_\uparrow\|^2 \cdot \|\boldsymbol{z}\|^2 \\ &= \left(\|(A^1)_\uparrow\|^2 + \cdots + \|(A^m)_\uparrow\|^2 \right) \|\boldsymbol{z}\|^2 \\ &= \left(\|(A^1)_\uparrow\|^2 + \cdots + \|(A^m)_\uparrow\|^2 \right) \|\boldsymbol{x} - \boldsymbol{y}\|^2. \end{aligned}$$

□

Remark 12.13. (a) If $\boldsymbol{\xi}$ is a linear functional on $\mathbb{R}^n$ described by the *row* vector

$$[\xi_1, \dots, \xi_n],$$

then $\boldsymbol{\xi}_\uparrow$ is the *column* vector

$$\boldsymbol{\xi}_\uparrow = \begin{bmatrix} \xi_1 \\ \vdots \\ \xi_n \end{bmatrix}$$

and

$$\|\boldsymbol{\xi}_\uparrow\| = \sqrt{\xi_1^2 + \cdots + \xi_n^2} = \sqrt{\sum_{j=1}^n \xi_j^2}.$$

(b) Suppose that A is an $m \times n$ matrix with real entries. As usual, we denote by A^i the i-th row of A and by A_j the j-th column of A. The quantity

$$\sqrt{\sum_{i=1}^m \|(A^i)_\uparrow\|^2} = \sqrt{\|(A^1)_\uparrow\|^2 + \cdots + \|(A^m)_\uparrow\|^2}$$

that appears in the proof of Proposition 12.12(iii) is denoted by $\|A\|_{HS}$ and it is called the *Frobenius norm* or *Hilbert–Schmidt norm* of A. It can be given an alternate and more suggestive description.

Observe first that for any $i = 1, \dots, m$, the quantity $\|(A^i)_\uparrow\|^2$ is the sum of the squares of all the entries of A located on the i-th row. We deduce

$$\|A\|_{HS}^2 = \|(A^1)_\uparrow\|^2 + \cdots + \|(A^m)_\uparrow\|^2 = \text{the sum of the squares of all the entries of } A.$$

An $m \times n$ matrix A is a collection of mn real numbers and, as such, it can be viewed as an element of the Euclidean vector space $\mathbb{R}^{mn}$. We see that the Hilbert–Schmidt norm of A is none other than the Euclidean norm of A viewed as an element of $\mathbb{R}^{mn}$. In particular, if $A, B \in \mathrm{Mat}_{m\times n}(\mathbb{R})$ then

$$\|A + B\|_{HS} \leq \|A\|_{HS} + \|B\|_{HS}. \tag{12.6}$$

We can also speak of convergent sequences of matrices.

Definition 12.14. *A sequence (A_ν) of $m \times n$ matrices is said to converge to the $m \times n$ matrix A if*

$$\lim_{\nu\to\infty} \|A_\nu - A\|_{HS} = 0.$$

□

The proof of Proposition 12.12(iii) shows that we have the following important inequality

$$\|A \cdot \boldsymbol{x}\| \leq \|A\|_{HS} \cdot \|\boldsymbol{x}\|, \;\; \forall A \in \mathrm{Mat}_{m\times n}(\mathbb{R}), \;\; \boldsymbol{x} \in \mathbb{R}^n. \tag{12.7}$$

□

Corollary 12.15. *The* addition function $\alpha : \mathbb{R}^2 \to \mathbb{R}$, $\alpha(x, y) = (x + y)$ *is continuous.*

Proof. As shown in Example 11.10(b), the function α is linear and thus continuous according to Proposition 12.12(ii). □

Proposition 12.16. *Let $\ell, m, n \in \mathbb{N}$, $X \subset \mathbb{R}^\ell$ and $Y \subset \mathbb{R}^m$. If $\boldsymbol{F} : X \to \mathbb{R}^m$ and $\boldsymbol{G} : Y \to \mathbb{R}^n$ are continuous maps such that*

$$\boldsymbol{F}(X) \subset Y,$$

then the composition $\boldsymbol{G} \circ \boldsymbol{F} : X \to \mathbb{R}^n$ is also a continuous map.

Proof. Let $\boldsymbol{x}_0 \in X$ and set $\boldsymbol{y}_0 := \boldsymbol{F}(\boldsymbol{x}_0) \in Y$. We have to prove that if $(\boldsymbol{x}_\nu)$ is a sequence in X such that $\boldsymbol{x}_\nu \to \boldsymbol{x}_0$ as $\nu \to \infty$, then

$$\lim_{\nu\to\infty} \boldsymbol{G}(\boldsymbol{F}(\boldsymbol{x}_\nu)) = \boldsymbol{G}(\boldsymbol{F}(\boldsymbol{x}_0)) = \boldsymbol{G}(\boldsymbol{y}_0).$$

We set $\boldsymbol{y}_\nu := \boldsymbol{F}(\boldsymbol{x}_\nu)$. Then $\boldsymbol{y}_\nu \in Y$ and

$$\lim_{\nu\to\infty} \boldsymbol{y}_\nu = \lim_{\nu\to\infty} \boldsymbol{F}(\boldsymbol{x}_\nu) = \boldsymbol{F}(\boldsymbol{x}_0) = \boldsymbol{y}_0,$$

since F is continuous at $\boldsymbol{x}_0$. On the other hand, since G is continuous at $\boldsymbol{y}_0$ we have

$$\lim_{\nu\to\infty} \boldsymbol{G}(\boldsymbol{F}(\boldsymbol{x}_\nu)) = \lim_{\nu\to\infty} \boldsymbol{G}(\boldsymbol{y}_\nu) = \boldsymbol{G}(\boldsymbol{y}_0).$$

□

Corollary 12.17. *Suppose that $I \subset \mathbb{R}$ is an interval, $\gamma : I \to \mathbb{R}^m$ is a continuous path and $\boldsymbol{F} : \mathbb{R}^m \to \mathbb{R}^n$ is a continuous map. Then the composition $\boldsymbol{F} \circ \gamma : I \to \mathbb{R}^n$ is also a continuous path.* □

Definition 12.18. Let $n \in \mathbb{N}$. For any $X \subset \mathbb{R}^n$ we denote by $C(X)$ the space of continuous functions $f : X \to \mathbb{R}$. □

Corollary 12.19. *Let $n \in \mathbb{N}$ and $X \subset \mathbb{R}^n$. Then, for any $f, g \in C(X)$ and any $t \in \mathbb{R}$ the functions $f + g$, $t \cdot f$ and fg are continuous.*

Proof. Consider the maps

$$P : X \to \mathbb{R}^2, \;\; P(\boldsymbol{x}) = \begin{bmatrix} f(\boldsymbol{x}) \\ g(\boldsymbol{x}) \end{bmatrix}$$

$\mu_t : \mathbb{R} \to \mathbb{R}$, $\mu_t(u) = tu$, and $\alpha, \mu : \mathbb{R}^2 \to \mathbb{R}$, $\alpha(u, v) = u + v$, $\mu(u, v) = uv$. Each of these maps is continuous and we have

$$\alpha \circ P(\boldsymbol{x}) = f(\boldsymbol{x}) + g(\boldsymbol{x}), \;\; \mu \circ P(\boldsymbol{x}) = f(\boldsymbol{x})g(\boldsymbol{x}), \;\; \mu_t \circ f(\boldsymbol{x}) = tf(\boldsymbol{x}).$$

The desired conclusion follows by invoking Proposition 12.16. □

Remark 12.20. Let $n \in \mathbb{N}$ and $X \subset \mathbb{R}^n$. The set $C(X)$ is nonempty since obviously the constant functions belong to $C(X)$. However, if X consists of more than one point, then X also contains nonconstant functions. For example, given $\boldsymbol{x}_0 \in X$, the function

$$d_{\boldsymbol{x}_0} : X \to \mathbb{R}, \;\; d_{\boldsymbol{x}_0}(\boldsymbol{x}) = \|\boldsymbol{x} - \boldsymbol{x}_0\|$$

is continuous and nonconstant since $d_{\boldsymbol{x}_0}(\boldsymbol{x}_0) = 0$ and $d_{\boldsymbol{x}_0}(\boldsymbol{x}) > 0$, $\forall \boldsymbol{x} \in X$. □

Definition 12.21. Let $m, n \in \mathbb{N}$ and suppose that $X \subset \mathbb{R}^n$.

(i) The sequence of maps $\boldsymbol{F}_\nu : X \to \mathbb{R}^m$, $\nu \in \mathbb{N}$ is said to *converge pointwisely* to the map $\boldsymbol{F} : X \to \mathbb{R}^m$ if

$$\forall \boldsymbol{x} \in X \;\; \lim_{\nu \to \infty} \boldsymbol{F}_\nu(\boldsymbol{x}) = \boldsymbol{F}(\boldsymbol{x}),$$

i.e.,

$$\forall \boldsymbol{x} \in X, \;\; \forall \varepsilon > 0, \;\; \exists N = N(\varepsilon, \boldsymbol{x}) > 0 : \;\; \forall \nu > N \;\; \|\boldsymbol{F}_\nu(\boldsymbol{x}) - \boldsymbol{F}(\boldsymbol{x})\| < \varepsilon.$$

(ii) The sequence of maps $\boldsymbol{F}_\nu : X \to \mathbb{R}^m$, $\nu \in \mathbb{N}$ is said to *converge uniformly* to the map $\boldsymbol{F} : X \to \mathbb{R}^m$ if

$$\forall \varepsilon > 0, \;\; \exists N = N(\varepsilon) > 0 \;\text{ such that }\; \forall \boldsymbol{x} \in X, \; \forall \nu > N : \;\; \|\boldsymbol{F}_\nu(\boldsymbol{x}) - \boldsymbol{F}(\boldsymbol{x})\| < \varepsilon.$$

□

Theorem 12.22. *Let $m, n \in \mathbb{N}$ and $X \subset \mathbb{R}^n$. Suppose that the sequence of continuous maps $\boldsymbol{F}_\nu : X \to \mathbb{R}^m$ converges uniformly to the map $\boldsymbol{F} : X \to \mathbb{R}^m$. Then the following hold.*

(i) The sequence $(\boldsymbol{F}_\nu)$ converges pointwisely to $\boldsymbol{F}$.
(ii) The map $\boldsymbol{F}$ is continuous.

□

The proof of this theorem is very similar to the proof of Theorem 6.10 and is left to you as an exercise.

12.2. Connectedness and compactness

In this section we discuss two very important concepts that have many applications.

12.2.1. *Connectedness*

Definition 12.23. Let $n \in \mathbb{N}$. A subset $X \subset \mathbb{R}^n$ is called *path connected* if any two points in X can be connected by a continuous path contained in X. More precisely, this means that for any $\boldsymbol{x}_0, \boldsymbol{x}_1 \in X$, there exists a continuous path $\gamma : [t_0, t_1] \to \mathbb{R}^n$ satisfying the following properties.

(i) $\gamma(t) \in X, \forall t \in [t_0, t_1]$.
(ii) $\gamma(t_0) = \boldsymbol{x}_0$, $\gamma(t_1) = \boldsymbol{x}_1$.

□

Remark 12.24. The above definition has some built-in flexibility. Note that if, for some $t_0 < t_1$, there exists a continuous path $\gamma : [t_0, t_1] \to X$ such that $\gamma(t_0) = \boldsymbol{x}_0$ and $\gamma(t_1) = \boldsymbol{x}_1$, then, for any $s_0 < s_1$, there exists a continuous path $\tilde{\gamma} : [s_0, s_1] \to X$ such that $\tilde{\gamma}(s_0) = \boldsymbol{x}_0$ and $\tilde{\gamma}(s_1) = \boldsymbol{x}_1$. To see this consider the linear function

$$\ell : [s_0, s_1] \to \mathbb{R}, \;\; \ell(s) = t_0 + \frac{t_1 - t_0}{s_1 - s_0}(s - s_0).$$

This function is increasing,

$$\ell(s_0) = t_0, \;\; \ell(s_1) = t_0 + \frac{t_1 - t_0}{s_1 - s_0}(s_1 - s_0) = t_0 + t_1 - t_0 = t_1.$$

Now define $\tilde{\gamma} : [s_0, s_1] \to X$ by setting $\tilde{\gamma}(s) = \gamma\big(\ell(s)\big)$. Clearly

$$\tilde{\gamma}(s_0) = \gamma\big(\ell(s_0)\big) = \gamma(t_0) = \boldsymbol{x}_0$$

and, similarly, $\tilde{\gamma}(s_1) = \boldsymbol{x}_1$. □

Proposition 12.25. *Let $n \in \mathbb{N}$. If $X \subset \mathbb{R}^n$ is convex, then X is path connected.*

Proof. This should be intuitively very clear because in a convex set X, any two points $\boldsymbol{x}_0, \boldsymbol{x}_1$ are connected by the line segment $[\boldsymbol{x}_0, \boldsymbol{x}_1]$ which, by definition, is contained in X. Formally, the argument goes as follows. Consider the continuous path

$$\gamma : [0,1] \to \mathbb{R}^n, \;\; \gamma(t) = (1-t)\boldsymbol{x}_0 + t\boldsymbol{x}_1, \;\; \forall t \in [0,1].$$

The image (or trajectory) of this continuous path is the line segment $[\boldsymbol{x}_0, \boldsymbol{x}_1]$ which is contained in X since X is assumed convex. □

Proposition 12.26. *Let $X \subset \mathbb{R}$. The following statements are equivalent.*

(i) X is path connected.
(ii) X is an interval.

Proof. The implication (ii) $\Rightarrow$ (i) is immediate. If X is an interval, then X is convex and thus path connected according to the previous proposition.

Assume now that X is path connected. To prove that it is an interval we have to show (see Exercise 12.12) that for any $x_0, x_1 \in X$, $x_0 < x_1$, the interval $[x_0, x_1]$ is contained in X.

Let $x_0, x_1 \in X$, $x_0 < x_1$. We have to show that if $x_0 \leq u \leq x_1$, then $u \in X$. Since X is path connected there exists a continuous path $\gamma : [t_0, t_1] \to X \subset \mathbb{R}$ such that $\gamma(t_0) = x_0$, $\gamma(t_1) = x_1$. Since $x_0 \leq u \leq x_1$ we deduce from the intermediate value property that there exists $\tau \in [t_0, t_1]$ such that $u = \gamma(\tau)$. Since $\gamma(\tau) \in X$ we deduce $u \in X$. □

12.2.2. *Compactness*

Definition 12.27. Let $n \in \mathbb{N}$. A subset $K \subset \mathbb{R}^n$ satisfies the *Bolzano–Weierstrass property* or BW for brevity, if any sequence $(\boldsymbol{p}_\nu)_{\nu \in \mathbb{N}}$ of points in K contains a subsequence that converges to a point $\boldsymbol{p}$, *also in K*. □

Example 12.28. The Bolzano–Weierstrass Theorem 4.29 shows that intervals in $\mathbb{R}$ of the form $[a, b]$ satisfy BW. □

Proposition 12.29. *Let $m, n \in \mathbb{N}$. Suppose that $K \subset \mathbb{R}^m$ and $L \subset \mathbb{R}^n$ satisfy BW. Then the Cartesian product $K \times L \subset \mathbb{R}^{m+n}$ also satisfies BW.*

Proof. Let $(\boldsymbol{p}_\nu, \boldsymbol{q}_\nu) \in K \times L$, $\nu \in N$ be a sequence of points in $K \times L$. Since K satisfies BW, the sequence $(\boldsymbol{p}_\nu)$ of points in K contains a subsequence

$$(\boldsymbol{p}_{\nu_i}) = \boldsymbol{p}_{\nu_1},\ \boldsymbol{p}_{\nu_2}, \dots$$

that converges to a point $\boldsymbol{p} \in K$. Since L satisfies BW, the subsequence $(\boldsymbol{q}_{\nu_i})$ of points in L contains a sub-subsequence $(\boldsymbol{q}_{\mu_j})$ that converges to a point $\boldsymbol{q} \in L$.

The sub-subsequence $(\boldsymbol{p}_{\mu_j})$ of the subsequence $(\boldsymbol{p}_{\nu_i})$ converges to the same limit $\boldsymbol{p}$. Thus, the subsequence $(\boldsymbol{p}_{\mu_j}, \boldsymbol{q}_{\mu_j})$ of $(\boldsymbol{p}_\nu, \boldsymbol{q}_\nu)$ converges to $(\boldsymbol{p}, \boldsymbol{q}) \in K \times L$. □

Definition 12.30. Let $n \in \mathbb{N}$. An n-dimensional *closed box* (or *closed rectangle*) is a subset of $\mathbb{R}^n$ of the form

$$[a_1, b_1] \times \cdots \times [a_n, b_n],\quad a_1 \leq b_1, \dots, a_n \leq b_n.$$

An *open box* in $\mathbb{R}^n$ is a set of the form $(a_1, b_1) \times \cdots \times (a_n, b_n)$.

Note that the closed cubes are special examples of closed boxes.

Corollary 12.31. *The closed boxes in $\mathbb{R}^n$ satisfy BW.*

Proof. We argue by induction on n. For $n = 1$ this follows from the Bolzano–Weierstrass Theorem 4.29. For the inductive step suppose that $B \subset \mathbb{R}^{n+1}$ is a box,

$$B = \underbrace{[a_1, b_1] \times \cdots \times [a_n, b_n]}_{=B'} \times [a_{n+1}, b_{n+1}].$$

From the induction assumption we deduce that $B' \subset \mathbb{R}^n$ satisfies BW. Proposition 12.29 now implies that $B = B' \times [a_{n+1}, b_{n+1}]$ satisfies BW. □

Definition 12.32. Let $n \in \mathbb{N}$. A set $X \subset \mathbb{R}^n$ is called *bounded* if it is contained in some box $B \subset \mathbb{R}^n$. □

Proposition 12.33. *Let $n \in \mathbb{N}$ and $X \subset \mathbb{R}^n$. The following statements are equivalent.*

(i) The set X is bounded.
(ii) There exists $R > 0$ such that

$$\|\boldsymbol{x}\| \leq R, \quad \forall \boldsymbol{x} \in X. \tag{12.8}$$

Proof. (i) ⇒ (ii). Suppose that X is contained in the box

$$B = [a_1, b_1] \times \cdots \times [a_n, b_n].$$

Observe that there exists $M > 0$ large enough so that

$$[a_1, b_1], \ldots, [a_n, b_n] \subset [-M, M].$$

Thus, for any $\boldsymbol{x} = (x^1, \ldots, x^n) \in B$, we have

$$|x^i| \leq M, \quad \forall i = 1, \ldots, n$$

so that

$$\|\boldsymbol{x}\|^2 = |x^1|^2 + \cdots + |x^n|^2 \leq nM^2.$$

Hence

$$\|\boldsymbol{x}\| \leq M\sqrt{n}, \quad \forall \boldsymbol{x} \in B.$$

In particular, this shows that X satisfies (12.8).

(ii) ⇒ (i). Suppose that X satisfies (12.8). Thus there exists $R > 0$ such that X is contained in the closed Euclidean ball $\overline{B_R(\mathbf{0})}$ which in turn is contained in the closed cube $\overline{C_R(\mathbf{0})}$. □

Theorem 12.34 (Bolzano–Weierstrass). *Let $n \in \mathbb{N}$ and $X \subset \mathbb{R}^n$. The following statements are equivalent.*

(i) The set X satisfies BW.
(ii) The set X is closed and bounded.

Proof. (i) $\Rightarrow$ (ii). Assume that X satisfies BW. We have to prove that X is bounded and closed. To prove that X is closed we have to show that if $(\boldsymbol{p}_\nu)$ is a sequence of points in X that converges to some point $\boldsymbol{p} \in \mathbb{R}^n$, then $\boldsymbol{p} \in X$.

Since X satisfies BW, the sequence $(\boldsymbol{p}_\nu)$ contains a subsequence that converges to a point $\boldsymbol{p}_* \in X$. Since the limit of any subsequence is equal to the limit of the whole sequence, we deduce $\boldsymbol{p} = \boldsymbol{p}_* \in X$.

To prove that X is bounded we argue by contradiction. Thus, the condition (12.8) is violated. Hence, for any $\nu \in \mathbb{N}$ there exists $\boldsymbol{x}_\nu \in X$ such that $\|\boldsymbol{x}_\nu\| > \nu$. Since X satisfies BW, the sequence $(\boldsymbol{x}_\nu)$ contains a subsequence $(\boldsymbol{x}_{\nu_i})_{i\in\mathbb{N}}$ converging to $\boldsymbol{x}_* \in X$. We deduce

$$\lim_{i\to\infty} \|\boldsymbol{x}_{\nu_i}\| = \|\boldsymbol{x}_*\| < \infty.$$

This is impossible since

$$\|\boldsymbol{x}_{\nu_i}\| > \nu_i, \quad \lim_{i\to\infty} \nu_i = \infty$$

and thus

$$\lim_{i\to\infty} \|\boldsymbol{x}_{\nu_i}\| = \infty.$$

(ii) $\Rightarrow$ (i). Suppose that X is closed and bounded. Since X is bounded, it is contained in a closed box B. Suppose now that $(\boldsymbol{p}_\nu)$ is a sequence of points in X. According to Corollary 12.31 the box B satisfies BW, so the sequence $(\boldsymbol{p}_\nu)$ contains a subsequence $(\boldsymbol{p}_{\nu_i})$ that converges to a point $\boldsymbol{p} \in B$. On the other hand the limit of any convergent sequence of points in X is a point in X. Thus the limit of the sequence $(\boldsymbol{p}_{\nu_i})_{i\in N}$ must belong to X. This shows that X satisfies BW. □

Corollary 12.35. *Let $n \in \mathbb{N}$. For any $R > 0$ and any $\boldsymbol{p} \in \mathbb{R}^n$ the closed ball $\overline{B_R(\boldsymbol{p})}$ and the closed cube $\overline{C_R(\boldsymbol{p})}$ satisfy BW.* □

Proof. Indeed, the closed ball $\overline{B_R(\boldsymbol{p})}$ and the closed cube $\overline{C_R(\boldsymbol{p})}$ are closed and bounded. □

Corollary 12.36 (Bolzano–Weierstrass). *Let $n \in \mathbb{N}$. If $(\boldsymbol{x}_\nu)$ is a bounded sequence of points in $\mathbb{R}^n$, i.e.,*

$$\exists R > 0 \text{ such that } \|\boldsymbol{x}_\nu\| \leq R, \ \forall \nu \in \mathbb{N},$$

then $(\boldsymbol{x}_\nu)$ contains a convergent subsequence.

Proof. The sequence $(\boldsymbol{x}_\nu)$ is contained in the closed ball $\overline{B_R(\boldsymbol{p})}$ which satisfies BW. □

Definition 12.37. Let $n \in \mathbb{N}$ and $X \subset \mathbb{R}^n$.

(i) A (possibly infinite) collection of subsets of $\mathbb{R}^n$ is said to *cover* X if their union contains X.

(ii) The set X is said to satisfy the *weak Heine–Borel*[1] *property* (or wHB for brevity) if any collection of open boxes that covers X contains a finite subcollection that covers X.

(iii) The set X is said to satisfy the *Heine–Borel property* (or HB for brevity) if any collection of open sets that covers X contains a finite subcollection that covers X.

□

Often we use the expression "$\mathcal{U}$ *is an open cover of* X" to indicate that $\mathcal{U}$ is a collection of open sets that covers X. Given an open cover $\mathcal{U}$ of X, we define *subcover* of $\mathcal{U}$ is a subfamily of $\mathcal{U}$ that still covers X.

Example 12.38. The interval $(0,1]$ does not satisfy the HB property. Indeed, the family of open sets

$$U_n := (1/n, 2), \quad n \geq 2$$

covers $(0,1]$, but no finite subfamily covers $(0,1]$. Indeed if $U_{n_1}, \dots, U_{n_k}$ is a finite subfamily, $n_1 < \cdots < n_k$, then

$$U_{n_1} \subset \cdots \subset U_{n_k}, \quad U_{n_1} \cup \cdots \cup U_{n_k} = U_{n_k}$$

and the interval U_{n_k} does not contain $(0,1]$. □

Lemma 12.39. *A set satisfies* wHB *if and only if it satisfies* HB.

Proof. Clearly $HB \Rightarrow wHB$ so it suffices to show only that $wHB \Rightarrow HB$. Suppose that the collection $\mathcal{U}$ of open sets covers X. Each open set U in the family $\mathcal{U}$ is the union of a collection $\mathscr{C}_U$ open cubes; see Exercise 11.35.

The family $\mathscr{C}$ of all the cubes in all the collections $\mathscr{C}_U$, $U \in \mathcal{U}$, covers X. Since X satisfies wHB, there exists a finite subfamily $\mathcal{F} \subset \mathscr{C}$ that covers X. Each cube $C \in \mathcal{F}$ is contained in some open set $U = U_C$ of the family $\mathcal{U}$. It follows that the finite subfamily

$$\{ U_C; \; C \in \mathcal{F} \} \subset \mathcal{U}$$

covers X. □

Theorem 12.40 (Heine–Borel). *For any* $a, b \in \mathbb{R}$, *the closed interval* $[a, b]$ *satisfies* HB.

Proof. It suffices to verify only the wHB property. Let us observe that the open boxes in $\mathbb{R}$ are the open intervals. Suppose that $\mathcal{I} := (I_\alpha)_{\alpha \in \mathcal{A}}$ is a collection of open intervals that covers $[a, b]$. We have to prove that there exists a *finite* subcollection of $\mathcal{I}$ that covers $[a, b]$. We define

$$X := \{ x \in [a, b]; \; [a, x] \text{ is covered by some finite subcollection of } \mathcal{I} \}.$$

[1] Émile Borel (1871–1956) was a French mathematician and politician. As a mathematician, he was known for his founding work in the areas of measure theory and probability.

Note first that $a \in X$ because a is contained in some interval of the family $\mathcal{I}$. Thus X is nonempty and bounded above, and therefore it admits a supremum $x^* := \sup X$. Note that $x^* \in [a, b]$. It suffices to prove that

$$x^* \in X, \tag{12.9a}$$

$$x^* = b. \tag{12.9b}$$

Proof of (12.9a). Observe that there exists an increasing sequence (x_n) of points in X such that

$$x^* = \lim_{n\to\infty} x_n.$$

Since $x^* \in [a, b]$ there exists an open interval I^* in the family $\mathcal{I}$ that contains x^*. Since the sequence (x_n) converges to x^* there exists $k \in \mathbb{N}$ such that $x_k \in I^*$. The interval $[a, x_k]$ is covered by finitely many intervals $I_1, \dots, I_N \in \mathcal{I}$. Clearly $[x_k, x^*] \subset I^*$. Hence the finite collection $I^*, I_1, \dots, I_N \in \mathcal{I}$ covers $[a, x^*]$, i.e., $x^* \in X$. □

Proof of (12.9b). We argue by contradiction. Suppose that $x^* \neq b$. Hence $x^* < b$. Since $x^* \in X$, the interval $[a, x^*]$ is covered by finitely many open intervals $I^*, I_1, \dots, I_N \in \mathcal{I}$, where $I^* \ni x^*$. Since I^* is open, there exists $\varepsilon > 0$, such that $\varepsilon < b - x^*$ and $[x^*, x^* + \varepsilon] \subset I^*$. This shows that the interval $[a, x^* + \varepsilon]$ is covered by the finite family $I^*, I_1, \dots, I_N$ and $x^* + \varepsilon \in [a, b]$. Hence $x^* + \varepsilon \in X$. The inequality $x^* + \varepsilon > x^*$ contradicts the fact that $x^* = \sup X$. □

The proof of Theorem 12.40 is now complete. □

Proposition 12.41. *Let $m, n \in \mathbb{N}$. Suppose that $K \subset \mathbb{R}^m$ and $L \subset \mathbb{R}^n$ satisfy HB. Then the Cartesian product $K \times L \subset \mathbb{R}^{m+n}$ also satisfies HB.*

Proof. Again it suffices to verify only the wHB property. Suppose that $\mathcal{B}$ is a collection of open boxes in $\mathbb{R}^{m\times n}$ that covers $K \times L$. Each box $B \in \mathcal{B}$ is a product $B = B' \times B''$ where B' is an open box in $\mathbb{R}^m$ and B'' is an open box in $\mathbb{R}^n$. To see this note that each open box $B \in \mathcal{B}$ is a product of $m + n$ intervals

$$B = I_1 \times \cdots \times I_m \times I_{m+1} \times \cdots \times I_{m+n}.$$

Then

$$B' = I_1 \times \cdots \times I_m, \quad B'' = I_{m+1} \times \cdots \times I_{m+n}.$$

If you think of B as a rectangle in the xy-plane, then B' would be its "shadow" on the x axis and B'' would be its "shadow" on the y-axis. For each $\boldsymbol{x} \in K$ we denote by $\mathcal{B}_{\boldsymbol{x}}$ the subfamily of $\mathcal{B}$ consisting of boxes that intersect $\{\boldsymbol{x}\} \times L$; see Figure 12.4.

Lemma 12.42. *Fix $\boldsymbol{x} \in K \subset \mathbb{R}^m$. There exists an open box $\tilde{B}_{\boldsymbol{x}}$ in $\mathbb{R}^m$ containing $\boldsymbol{x}$ and a finite subcollection $\mathcal{F}_{\boldsymbol{x}} \subset \mathcal{B}_x$ that covers $\tilde{B}_{\boldsymbol{x}} \times L$.*

Proof of Lemma 12.42. The collection $\mathcal{B}_{\boldsymbol{x}}$ covers $\{\boldsymbol{x}\} \times L$ and thus the collection of open n-*dimensional* boxes

$$\left\{ B'';\ B' \times B'' \in \mathcal{B}_{\boldsymbol{x}} \right\}$$

covers L. Since L satisfies HB, there exists a finite subfamily $\mathcal{F}_{\boldsymbol{x}} \subset \mathcal{B}_x$ such that the collection of n-dimensional boxes

$$\{B'';\ B \in \mathcal{F}_{\boldsymbol{x}}\}$$

covers L. For each $B \in \mathcal{F}_{\boldsymbol{x}}$, the m-dimensional box B' contains $\boldsymbol{x}$. The intersection of the family $\{B';\ B' \times B \in \mathcal{F}_{\boldsymbol{x}}\}$ is therefore a nonempty m-dimensional box $\tilde{B}_{\boldsymbol{x}}$ that contains $\boldsymbol{x}$. Since

$$\tilde{B}_{\boldsymbol{x}} \times B'' \subset B' \times B'' = B, \ \ \forall B \in \mathcal{F}_{\boldsymbol{x}},$$

we deduce that

$$\tilde{B}_{\boldsymbol{x}} \times L \subset \bigcup_{B\in\mathcal{F}_{\boldsymbol{x}}} B'_{\boldsymbol{x}} \times B'' \subset \bigcup_{B\in\mathcal{F}_{\boldsymbol{x}}} B.$$

□

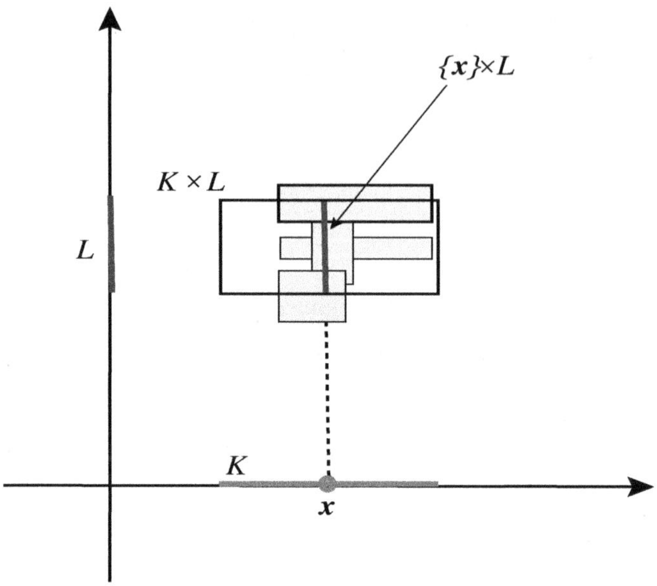

Fig. 12.4 *From the collection $\mathcal{B}$ of open boxes covering $K \times L$ we concentrate on the subcollection $\mathcal{B}_{\boldsymbol{x}}$ consisting of boxes that intersect the slice $\{\boldsymbol{x}\} \times L$.*

For any $\boldsymbol{x} \in K$ choose an open box $\tilde{B}_{\boldsymbol{x}} \subset \mathbb{R}^m$ as in Lemma 12.42. The collection of boxes

$$\left\{ \tilde{B}_{\boldsymbol{x}} \right\}_{\boldsymbol{x} \in K}$$

clearly covers K. Since K satisfies HB there exist finitely many points $\boldsymbol{x}_1, \ldots, \boldsymbol{x}_\nu \in K$ such that the finite subcollection

$$\left\{ \tilde{B}_{\boldsymbol{x}_j} \right\}_{1 \leq j \leq \nu}$$

covers K. Note that each finite subfamily $\mathcal{F}_{\boldsymbol{x}_j} \subset \mathcal{B}$ covers $\tilde{B}_{\boldsymbol{x}_j} \times L$ so the finite family

$$\mathcal{F} = \mathcal{F}_{\boldsymbol{x}_1} \cup \cdots \cup \mathcal{F}_{\boldsymbol{x}_\nu} \subset \mathcal{B}$$

covers $K \times L$. □

Corollary 12.43. *Any closed box in $\mathbb{R}^n$ satisfies HB.* □

We can now state and prove the following very important result.

Theorem 12.44. *Let $n \in \mathbb{N}$ and $X \subset \mathbb{R}^n$. Then the following statements are equivalent.*

(i) The set X is closed and bounded.
(ii) The set X satisfies BW.
(iii) The set X satisfies HB.

Proof. We already know that (i) $\iff$ (ii). Let us prove that (i) $\Rightarrow$ (iii). Thus we want to prove that if X is closed and bounded then X satisfies HB.

Observe first that since X is bounded X is contained in some *closed* cube C. Moreover, since X is closed, the set $U_0 = \mathbb{R}^n \setminus X$ is open. Suppose that $\mathfrak{U}$ is a family of open sets that covers X. The family $\mathfrak{U}_*$ of open sets obtained from $\mathfrak{U}$ by adding U_0 to the mix covers C. Indeed, $\mathfrak{U}$ covers X and U_0 covers the rest, $C \setminus X$. The closed cube C satisfies HB so there exists a finite subfamily $\mathcal{F}_*$ of $\mathfrak{U}_*$ that covers C. If $\mathcal{F}_*$ does not contain the set U_0 then clearly it is a finite subfamily of $\mathfrak{U}$ that covers C and, a fortiori, X. If U_0 belongs to $\mathcal{F}_*$, then the family $\mathcal{F}$ obtained from $\mathcal{F}_*$ by removing U_0 will cover X because U_0 does not cover any point on X.

(iii) $\Rightarrow$ (i). To prove that X is bounded choose a family $\mathscr{C}$ of open cubes that covers X. Since X satisfies HB, there exists a finite subfamily $\mathcal{F} \subset \mathscr{C}$ that covers X. The union of the cubes in the *finite* family $\mathcal{F}$ is contained in some large cube, hence X is contained in a large cube and it is therefore bounded.

To prove that X is closed we argue by contradiction. Suppose that $(\boldsymbol{x}_\nu)$ is a sequence of points in X that converges to a point $\boldsymbol{x}_*$ not in X. We set $r_\nu := \operatorname{dist}(\boldsymbol{x}_*, \boldsymbol{x}_\nu)$. Consider the family of open sets

$$U_\nu = \mathbb{R}^n \setminus \overline{B_{r_\nu}(\boldsymbol{x}_*)}, \quad \nu \in \mathbb{N}.$$

Since $r_\nu \to 0$ we have

$$\bigcup_{\nu \geq 1} U_\nu = \mathbb{R}^m \setminus \{\boldsymbol{x}_*\} \supset X.$$

However, no finite subfamily of this family covers X. Indeed the union of the open sets in such a finite family is the complement of a closed ball centered at $\boldsymbol{x}_*$ and such a ball contains infinitely many points in the sequence $(\boldsymbol{x}_\nu)$. □

Definition 12.45 (Compactness). Let $n \in \mathbb{N}$. A set $X \subset \mathbb{R}^n$ is called *compact* if it satisfies either one of the equivalent conditions (i), (ii) or (iii) in Theorem 12.44. □

Corollary 12.46. *Suppose that $S \subset \mathbb{R}$ is a nonempty compact subset of the real axis. Then there exist $s_*, s^* \in S$ such that $s_* \leq s \leq s^*$, $\forall s \in S$. In other words,*

$$\boxed{\inf S \in S, \ \ \sup S \in S}.$$

Proof. We set

$$s_* := \inf S, \ \ s^* := \sup S.$$

Since S is compact, it is bounded so that $-\infty < s_* \leq s^* < \infty$. We want to prove that $s_*, s^* \in S$.

Now choose a sequence of points (s_ν) in S such that $s_\nu \to s^*$. (The existence of such a sequence is guaranteed by Lemma 6.13.)

Since S is compact, it is closed, so the limit of any convergent sequence of points in S is also a point in S. Thus $s^* \in S$. A similar argument shows that $s_* \in S$. □

12.3. Topological properties of continuous maps

The continuous maps enjoy many useful properties not satisfied by many other types of maps. The first property we want to discuss generalizes the intermediate value property of continuous functions of one variable.

Theorem 12.47. *Let $m, n \in \mathbb{N}$ and suppose that $X \subset \mathbb{R}^n$ is path connected. If $\boldsymbol{F} : X \to \mathbb{R}^m$ is a continuous map, then its image $\boldsymbol{F}(X)$ is path connected.*

Proof. We have to show that for any $\boldsymbol{y}_0, \boldsymbol{y}_1 \in \boldsymbol{F}(X)$ there exists a continuous path in $\boldsymbol{F}(X)$ connecting $\boldsymbol{y}_0$ to $\boldsymbol{y}_1$.

Since $\boldsymbol{y}_0, \boldsymbol{y}_1 \in \boldsymbol{F}(X)$ there exist $\boldsymbol{x}_0, \boldsymbol{x}_1 \in X$ such that $\boldsymbol{F}(\boldsymbol{x}_0) = \boldsymbol{y}_0$ and $\boldsymbol{F}(\boldsymbol{x}_1) = \boldsymbol{y}_1$. Since X is path connected, there exists a continuous path $\gamma : [t_0, t_1] \to X$ such that $\gamma(t_0) = \boldsymbol{x}_0$ and $\gamma(t_1) = \boldsymbol{y}_1$. We obtain a continuous path

$$\boldsymbol{F} \circ \gamma : [t_0, t_1] \to \mathbb{R}^n$$

whose image is contained in the image $\boldsymbol{F}(X)$ of $\boldsymbol{F}$ and satisfying

$$\boldsymbol{F} \circ \gamma(t_i) = \boldsymbol{F}(\gamma(t_i)) = \boldsymbol{F}(\boldsymbol{x}_i) = y_i, \quad i = 0, 1.$$

Thus the continuous path $\boldsymbol{F} \circ \gamma$ in $\boldsymbol{F}(X)$ connects $\boldsymbol{y}_0$ to $\boldsymbol{y}_1$. □

Corollary 12.48. *Let $m, n \in \mathbb{N}$ and suppose that $X \subset \mathbb{R}^n$. If $\boldsymbol{F} : X \to \mathbb{R}^m$ is a continuous map, and the image $\boldsymbol{F}(X)$ is not path connected, then X is not path connected.* □

Corollary 12.49 (Multidimensional intermediate value theorem). *Let $n \in \mathbb{N}$ and suppose that $X \subset \mathbb{R}^n$ is a path connected subset. If $f : X \to \mathbb{R}$ is a continuous function, then its image $f(X) \subset \mathbb{R}$ is an interval. In particular, if $\boldsymbol{x}_0, \boldsymbol{x}_1 \in X$ and $c \in \mathbb{R}$ are such that $f(\boldsymbol{x}_0) < c < f(\boldsymbol{x}_1)$, then there exists $\boldsymbol{x} \in X$ such that $f(\boldsymbol{x}) = c$.*

Proof. Theorem 12.47 shows that $f(X) \subset \mathbb{R}$ is path connected while Proposition 12.26 shows that $f(X)$ must be an interval. In particular, for any points $\boldsymbol{x}_0, \boldsymbol{x}_1 \in X$ such that $f(\boldsymbol{x}_0) < f(\boldsymbol{x}_1)$, the interval $[f(\boldsymbol{x}_0), f(\boldsymbol{x}_1)] \subset \mathbb{R}$ is contained in the range $f(X)$ of f. □

Theorem 12.50. *Let $m, n \in \mathbb{N}$ and $X \subset \mathbb{R}^n$. If $\boldsymbol{F} : X \to \mathbb{R}^m$ is continuous and $K \subset X$ is compact, then $F(K)$ is compact.*

Proof. It suffices to prove that the set $\boldsymbol{F}(K)$ satisfies BW. Suppose that $(\boldsymbol{y}_\nu)_{\nu \in \mathbb{N}}$ is a sequence in $\boldsymbol{F}(K)$. We have to show that it admits a subsequence that converges to a point in $\boldsymbol{F}(K)$.

Since $\boldsymbol{y}_\nu \in \boldsymbol{F}(K)$, there exists $\boldsymbol{x}_\nu \in K$ such that $\boldsymbol{y}_\nu = \boldsymbol{F}(\boldsymbol{x}_\nu)$. On the other hand, K satisfies BW so the sequence $(\boldsymbol{x}_\nu)$ admits a subsequence $(\boldsymbol{x}_{\nu_i})$ that converges to a point $\boldsymbol{x}_* \in X$. Since $\boldsymbol{F}$ is continuous we deduce

$$\lim_{i \to \infty} \boldsymbol{y}_{\nu_i} = \lim_{i \to \infty} \boldsymbol{F}(\boldsymbol{x}_{\nu_i}) = \boldsymbol{F}(\boldsymbol{x}_*) \in \boldsymbol{F}(K).$$

□

Corollary 12.51 (Weierstrass). *Let $n \in \mathbb{N}$ and suppose that $K \subset \mathbb{R}^n$ is a compact set. If $f : K \to \mathbb{R}$ is continuous, then there exist $\boldsymbol{x}_*$ and $\boldsymbol{x}^*$ in K such that*

$$f(\boldsymbol{x}_*) \leq f(\boldsymbol{x}) \leq f(\boldsymbol{x}^*), \;\; \forall \boldsymbol{x} \in K.$$

Proof. According to Theorem 12.50 the set $f(K) \subset \mathbb{R}$ is compact. Corollary 12.46 implies that there exist $s_*, s^* \in f(K)$ such that $s_* = \inf f(K)$, $s^* = \sup f(K)$. In particular,

$$s_* \leq f(\boldsymbol{x}) \leq s^*, \;\; \forall \boldsymbol{x} \in K.$$

Since $s_*, s^* \in f(K)$, there exists $\boldsymbol{x}_*, \boldsymbol{x}^* \in K$ such that $s_* = f(\boldsymbol{x}_*)$, $s^* = f(\boldsymbol{x}^*)$. □

Definition 12.52. Let $m, n \in \mathbb{N}$ and $X \subset \mathbb{R}^n$. A map $\boldsymbol{F} : X \to \mathbb{R}^m$ is called *bounded* if its range $\boldsymbol{F}(X)$ is a bounded subset of $\mathbb{R}^m$. □

Corollary 12.53. *Let $m, n \in \mathbb{N}$. Suppose that $K \subset \mathbb{R}^n$ is a compact set and $\boldsymbol{F} : K \to \mathbb{R}^m$ is a continuous map. Then $\boldsymbol{F}$ is a bounded map.*

Proof. The range $\boldsymbol{F}(K)$ is compact, hence bounded. □

Definition 12.54. Let $n \in \mathbb{N}$. The *diameter* of a nonempty subset $S \subset \mathbb{R}^n$ is the quantity

$$\operatorname{diam}(S) = \sup_{\boldsymbol{x},\boldsymbol{y} \in S} \|\boldsymbol{x} - \boldsymbol{y}\| \in [0, \infty]. \qquad \square$$

We list below a few simple properties of the diameter. Their proofs are left to the reader as an exercise.

Proposition 12.55. *Let $n \in \mathbb{N}$. Then the following hold.*

(i) The set $S \subset \mathbb{R}^n$ is bounded if and only if $\operatorname{diam}(S) < \infty$.
(ii) If $S_1 \subset S_2 \subset \mathbb{R}^n$, then $\operatorname{diam}(S_1) \leq \operatorname{diam}(S_2)$.
(iii) For any $r > 0$

$$\operatorname{diam}(B_r) = 2r, \;\; \operatorname{diam}(C_r) = 2r\sqrt{n},$$

where $B_r, C_r \subset \mathbb{R}^n$ are the open ball and respectively the open cube of radius r centered at $0 \in \mathbb{R}^n$.

□

Definition 12.56. Let $n, m \in \mathbb{N}$, and $X \subset \mathbb{R}^n$. The *oscillation* of a function $\boldsymbol{F} : X \to \mathbb{R}^m$ on a subset S is the quantity

$$\operatorname{osc}(\boldsymbol{F}, S) = \sup_{\boldsymbol{x},\boldsymbol{y} \in S} \|\boldsymbol{F}(\boldsymbol{x}) - \boldsymbol{F}(\boldsymbol{y})\|. \qquad \square$$

The next result describes alternate characterizations of the oscillation of a *scalar* valued function. Its proof is left to you as an exercise.

Proposition 12.57. *Let $n \in \mathbb{N}$, $X \subset \mathbb{R}^n$. For any function $f : X \to \mathbb{R}$ and any subset $S \subset X$ we have the equalities*

$$\operatorname{osc}(f, S) = \sup_{\boldsymbol{x} \in S} f(\boldsymbol{x}) - \inf_{\boldsymbol{y} \in S} f(\boldsymbol{y}) = \sup_{\boldsymbol{x},\boldsymbol{y} \in S} \big(f(\boldsymbol{x}) - f(\boldsymbol{y})\big) = \operatorname{diam} f(S). \qquad \square$$

Definition 12.58. Let $n \in \mathbb{N}$, $X \subset \mathbb{R}^n$. A function $f : X \to \mathbb{R}$ is said to be *uniformly continuous* on the subset $Y \subset X$ if

$$\forall \varepsilon > 0 \ \exists \delta = \delta(\varepsilon) > 0 \ \text{ such that } \forall S \subset Y: \ \operatorname{diam}(S) \leq \delta \Rightarrow \operatorname{osc}(f, S) < \varepsilon. \qquad \square$$

Observe that the above uniform continuity condition can be rephrased in the following equivalent way.

$$\forall \varepsilon > 0 \ \exists \delta = \delta(\varepsilon) > 0 \ \text{ such that } \forall y_1, y_2 \in Y: \ \|\boldsymbol{y}_1 - \boldsymbol{y}_2\| \leq \delta \Rightarrow |f(\boldsymbol{y}_1) - f(\boldsymbol{y}_2)| < \varepsilon.$$

Theorem 12.59 (Weierstrass). *Let $n \in \mathbb{N}$, $X \subset \mathbb{R}^n$. Suppose that $f : X \to \mathbb{R}$ is continuous. Then f is* uniformly *continuous on any compact set $K \subset X$.*

Proof. Let K be a compact subset of X. We argue by contradiction so we assume that f is not uniformly continuous on K. Hence, there exists $\varepsilon_0 > 0$ such that for any $\nu \in \mathbb{N}$ there exists a subset $S_\nu \subset K$ such that

$$\operatorname{diam}(S_\nu) \leq \frac{1}{\nu} \ \text{ and } \ \operatorname{osc}(f, S_\nu) \geq \varepsilon_0.$$

Thus, for any $\nu \in \mathbb{N}$, there exist $\boldsymbol{x}_\nu, \boldsymbol{y}_\nu \in S_\nu$ such that

$$\big| f(\boldsymbol{x}_\nu) - f(\boldsymbol{y}_\nu) \big| \geq \frac{\varepsilon_0}{2}. \tag{12.10}$$

Note that because $\boldsymbol{x}_\nu, \boldsymbol{y}_\nu \in S_\nu$ and $\operatorname{diam}(S_\nu) < \frac{1}{\nu}$ we have

$$\operatorname{dist}(\boldsymbol{x}_\nu, \boldsymbol{y}_\nu) < \frac{1}{\nu} \to 0 \ \text{ as } \ \nu \to \infty.$$

Since K is compact, the sequence of points $(\boldsymbol{x}_\nu)$ in K has a convergent subsequence $(\boldsymbol{x}_{\nu_j})$

$$\lim_{j\to\infty} \boldsymbol{x}_{\nu_j} = \boldsymbol{x} \in K.$$

Observe that

$$\operatorname{dist}(\boldsymbol{y}_{\nu_j}, \boldsymbol{x}) \leq \underbrace{\operatorname{dist}(\boldsymbol{y}_{\nu_j}, \boldsymbol{x}_{\nu_j})}_{< \frac{1}{\nu_j}} + \operatorname{dist}(\boldsymbol{x}_{\nu_j}, \boldsymbol{x}) \to 0 \ \text{ as } j \to \infty.$$

Thus the subsequence $(\boldsymbol{y}_{\nu_j})$ also converges to $\boldsymbol{x}$. Since f is continuous at $\boldsymbol{x}$ we have

$$\lim_{j\to\infty} f(\boldsymbol{x}_{\nu_j}) = \lim_{j\to\infty} f(\boldsymbol{y}_{\nu_j}) = f(\boldsymbol{x})$$

so that

$$\lim_{j\to\infty} \big(f(\boldsymbol{x}_{\nu_j}) - f(\boldsymbol{y}_{\nu_j}) \big) = 0.$$

This contradicts (12.10). $\square$

Definition 12.60. Let $m, n \in \mathbb{N}$ and suppose that $X \subset \mathbb{R}^m$, $Y \subset \mathbb{R}^n$.

(i) A map $\boldsymbol{F} : X \to Y$ is called a *homeomorphism* if it is continuous, bijective and the inverse $\boldsymbol{F}^{-1} : Y \to X$ is also continuous.

(ii) The sets X, Y are called *homeomorphic* if there exists a homeomorphism $\boldsymbol{F} : X \to Y$.

□

Corollary 12.61. *Let $m, n \in \mathbb{N}$. Suppose that $X \subset \mathbb{R}^m$ and $Y \subset \mathbb{R}^n$ are homeomorphic sets. Then the following hold.*

(i) The set X is compact if and only if Y is.
(ii) The set X is path connected if and only if Y is.

Proof. Fix a homeomorphism $\boldsymbol{F} : X \to Y$. Then both $\boldsymbol{F}$ and $\boldsymbol{F}^{-1}$ are continuous and

$$Y = \boldsymbol{F}(X), \quad X = \boldsymbol{F}^{-1}(Y).$$

The desired conclusions now follow from Theorem 12.47 and 12.50. □

12.4. Continuous partitions of unity

We conclude this chapter by discussing a technical but very versatile result that will come in handy later. First, we need to discuss a few more topological concepts.

Definition 12.62. Let $n \in \mathbb{N}$ and suppose that $X \subset \mathbb{R}^n$.

(i) The *closure of* X, denoted by $\boldsymbol{cl}(X)$, is the intersection of all the closed subsets of $\mathbb{R}^n$ that contain X.
(ii) The *interior of* X, denoted by $\boldsymbol{int}(X)$, is the union of all the open sets contained in X.
(iii) The *boundary of* X, denoted ∂X, is the difference $\boldsymbol{cl}(X) \setminus \boldsymbol{int}(X)$.

□

In other words, the closure of a set X is the *smallest* closed subset containing X and its interior is the *largest* open set contained in X. The proof of the following result is left as an exercise.

Proposition 12.63. *Let $n \in \mathbb{N}$ and suppose $X \subset \mathbb{R}^n$. Then the following hold.*

(i) A point $\boldsymbol{x} \in \mathbb{R}^n$ belongs to the closure of X if and only if there exists a sequence of points in X that converges to $\boldsymbol{x}$.
(ii) A point $x \in \mathbb{R}^n$ belongs to the interior of X if and only if $\exists r > 0$ such that $B_r(\boldsymbol{x}) \subset X$.
(iii) $\partial X = \boldsymbol{cl}(X) \cap \boldsymbol{cl}(\mathbb{R}^n \setminus X)$.

□

Example 12.64. Using the above proposition it is not hard to see that for any $r > 0$, the closure of the open ball $B_r(\mathbf{0}) \subset \mathbb{R}^n$ is the closed ball $\overline{B_r(\mathbf{0})}$. Moreover

$$\partial B_r(\mathbf{0}) = \partial \overline{B_r(\mathbf{0})} := \Sigma_r(\mathbf{0}) = \left\{ \boldsymbol{x} \in \mathbb{R}^n;\ \ \|\boldsymbol{x}\| = r \right\}. \qquad \square$$

Definition 12.65. Let $n \in \mathbb{N}$. The *support* of function $f : \mathbb{R}^n \to \mathbb{R}$ is the subset $\operatorname{supp}(f) \subset \mathbb{R}^n$ defined as the closure of the set of points where f is not zero,

$$\operatorname{supp}(f) := \boldsymbol{cl}\Big(\left\{ \boldsymbol{x} \in \mathbb{R}^n;\ \ f(\boldsymbol{x}) \neq 0 \right\} \Big).$$

We denote by $C_{\text{cpt}}(\mathbb{R}^n)$ the set of *continuous* functions on $\mathbb{R}^n$ with compact support. □

Clearly, the function identically equal to zero has compact support: its support is empty. The function which is equal to 1 at the origin and zero elsewhere has compact support, but it is not continuous. It turns out that there are plenty of continuous functions with compact support. The next result describes a simple recipe for producing many examples of continuous functions $\mathbb{R}^n \to \mathbb{R}$.

Proposition 12.66. *Let $n \in \mathbb{N}$.*

(i) Suppose that $C, C' \subset \mathbb{R}^n$ are two closed subsets such that $C \cap C' = \emptyset$. Then there exists a continuous function $f : \mathbb{R}^n \to [0, 1]$ such that

$$C = f^{-1}(1),\ \ C' = f^{-1}(0).$$

(ii) For any positive real numbers $r < R$ and any $x_0 \in \mathbb{R}^n$ there exists a continuous function $f : \mathbb{R}^n \to [0, 1]$ such that

$$f(\boldsymbol{y}) = \begin{cases} 1, & \boldsymbol{y} \in \overline{B_r(\boldsymbol{x}_0)}, \\ \\ 0, & \boldsymbol{y} \in \mathbb{R}^n \setminus B_R(\boldsymbol{x}_0). \end{cases}$$

Proof. (i) We have (see Exercise 12.22)

$$\boldsymbol{x} \in C \Longleftrightarrow \operatorname{dist}(\boldsymbol{x}, C) = 0,\ \ \boldsymbol{x} \in C' \Longleftrightarrow \operatorname{dist}(\boldsymbol{x}, C') = 0.$$

Since C and C' are disjoint we deduce

$$\operatorname{dist}(\boldsymbol{x}, C) + \operatorname{dist}(\boldsymbol{x}, C') > 0,\ \ \forall \boldsymbol{x} \in \mathbb{R}^n.$$

Now define

$$f : \mathbb{R}^n \to \mathbb{R},\ \ f(\boldsymbol{x}) = \frac{\operatorname{dist}(\boldsymbol{x}, C')}{\operatorname{dist}(\boldsymbol{x}, C) + \operatorname{dist}(\boldsymbol{x}, C')}.$$

The function f is continuous (see Exercise 12.22) and $f(\boldsymbol{x}) \in [0, 1]$, $\forall \boldsymbol{x} \in [0, 1]$. Note that

$$f(\boldsymbol{x}) = 0 \Longleftrightarrow \operatorname{dist}(\boldsymbol{x}, C') = 0 \Longleftrightarrow \boldsymbol{x} \in C',$$

$$f(\boldsymbol{x}) = 1 \Longleftrightarrow \operatorname{dist}(\boldsymbol{x}, C) = 0 \Longleftrightarrow \boldsymbol{x} \in C.$$

(ii) This is a special case of (i) corresponding to $C = \overline{B_r(\boldsymbol{x}_0)}$ and $C' = \mathbb{R}^n \setminus B_R(\boldsymbol{x}_0)$. □

Definition 12.67. Let $n \in \mathbb{N}$, $X \subset \mathbb{R}^n$ and suppose that $\mathfrak{U}$ is an open cover of X. A *continuous partition of unity* on X, *subordinated to the open cover* $\mathfrak{U}$ is a finite collection of continuous functions $\chi_1, \dots, \chi_\ell : \mathbb{R}^n \to \mathbb{R}$ with the following properties.

(i) For any $i = 1, \dots, \ell$, there exists an open subset U_i in the collection $\mathfrak{U}$ such that $\operatorname{supp} \chi_i \subset U_i$.
(ii) $\chi_1(\boldsymbol{x}) + \cdots + \chi_\ell(\boldsymbol{x}) = 1, \forall \boldsymbol{x} \in X$.

The partition of unity $\chi_1, \dots, \chi_\ell$ is called *compactly supported* if, additionally,

$$\chi_1, \dots, \chi_\ell \in C_{\rm cpt}(\mathbb{R}^n). \qquad \square$$

Theorem 12.68 (Continuous partitions of unity). *Let $n \in \mathbb{N}$ and suppose that $K \subset \mathbb{R}^n$ is a compact subset. Then, for any open cover $\mathfrak{U}$ of K, there exists a compactly supported partition of unity on K subordinated to $\mathfrak{U}$.*

Proof. Since the collection $\mathfrak{U}$ covers K we deduce that for any $\boldsymbol{x} \in K$ there exists an open set $U_{\boldsymbol{x}}$ in the collection $\mathfrak{U}$ such that $\boldsymbol{x} \in U_{\boldsymbol{x}}$. For any $\boldsymbol{x} \in K$ choose $r(\boldsymbol{x}), R(\boldsymbol{x}) > 0$ such that $R(\boldsymbol{x}) > r(\boldsymbol{x})$ and $B_{R(\boldsymbol{x})}(\boldsymbol{x}) \subset U_{\boldsymbol{x}}$.

The family of open balls $\big(B_{r(\boldsymbol{x})}(\boldsymbol{x}) \big)_{\boldsymbol{x} \in K}$ obviously covers K and, since K is compact, we can find finitely many points $\boldsymbol{x}_1, \dots, \boldsymbol{x}_\ell$ such that the collection of open balls

$$B_{r(\boldsymbol{x}_1)}(\boldsymbol{x}_1), \dots, B_{r(\boldsymbol{x}_\ell)}(\boldsymbol{x}_\ell)$$

covers K. Using Proposition 12.66(ii) we deduce that, for any $i = 1, \dots, \ell$ there exists a continuous function $f_i : \mathbb{R}^n \to [0, 1]$ such that

$$f_i(\boldsymbol{y}) = \begin{cases} 1, & \boldsymbol{y} \in \overline{B_{r(\boldsymbol{x}_i)}(\boldsymbol{x}_i)}, \\ \\ 0, & \boldsymbol{y} \in \mathbb{R}^n \setminus B_{R(\boldsymbol{x}_i)}(\boldsymbol{x}_i). \end{cases}$$

Now define

$$\chi_1 := f_1, \;\; \chi_2 := (1 - f_1) f_2, \;\; \chi_3 := (1 - f_1)(1 - f_2) f_3,$$

$$\chi_j := (1 - f_1) \cdots (1 - f_{j-1}) f_j, \;\; \forall j = 2, \dots, \ell.$$

Note that

$$f_i(\boldsymbol{y}) = 0, \;\; \forall \boldsymbol{y} \in \mathbb{R}^n \setminus B_{R(\boldsymbol{x}_i)}(\boldsymbol{x}_i) \Rightarrow \chi_i(\boldsymbol{y}) = 0, \;\; \forall \boldsymbol{y} \in \mathbb{R}^n \setminus B_{R(\boldsymbol{x}_i)}(\boldsymbol{x}_i).$$

In particular, the function χ_j has compact support contained in $\overline{B_{R(\boldsymbol{x}_j)}(\boldsymbol{x}_j)}$. Since χ_j is the product of functions with values in $[0, 1]$, the function χ_j is also valued in $[0, 1]$.

Now observe that

$$\chi_1 + \chi_2 = 1 - (1 - f_1) + (1 - f_1) f_2 = 1 - (1 - f_1)(1 - f_2),$$

$$\chi_1 + \chi_2 + \chi_3 = 1 - (1 - f_1)(1 - f_2) + (1 - f_1)(1 - f_2) f_3$$

$$= 1 - \Big((1 - f_1)(1 - f_2) - (1 - f_1)(1 - f_2) f_3 \Big) = 1 - (1 - f_1)(1 - f_2)(1 - f_3).$$

We obtain inductively that

$$\chi_1 + \chi_2 + \cdots + \chi_\ell = 1 - (1 - f_1)(1 - f_2) \cdots (1 - f_\ell).$$

Finally note that

$$\boldsymbol{x} \in \bigcup_{j=1}^{\ell} B_{r(\boldsymbol{x}_j)}(\boldsymbol{x}_j) \Rightarrow \exists i : \ \boldsymbol{x} \in B_{r(\boldsymbol{x}_i)}(\boldsymbol{x}_i)$$

$$\Rightarrow \exists i : \ f_i(\boldsymbol{x}) = 1 \Rightarrow \prod_{j=1}^{\ell} \big(1 - f_j(\boldsymbol{x})\big) = 0 \Rightarrow \sum_{j=1}^{\ell} \chi_j(\boldsymbol{x}) = 1.$$

$\square$

12.5. Exercises

Exercise 12.1. Consider the function

$$f : \mathbb{R}^2 \setminus \{\mathbf{0}\} \to \mathbb{R}, \ f(x, y) = \frac{xy}{x^2 + y^2}.$$

Decide whether the limit

$$\lim_{\boldsymbol{p} \to 0} f(\boldsymbol{p})$$

exists. Justify your answer.

Hint. Analyze the behavior of f along the sequences

$$\boldsymbol{p}_\nu = (1/\nu, 1/\nu) \text{ and } \boldsymbol{q}_\nu = (1/\nu, 2/\nu).$$

□

Exercise 12.2. Let $n \in \mathbb{N}$. Prove that the function $f : \mathbb{R}^n \to \mathbb{R}$, $f(\boldsymbol{x}) = \|\boldsymbol{x}\|^2$ is not Lipschitz. □

Exercise 12.3. Let $n \in \mathbb{N}$ and suppose that

$$\boldsymbol{\alpha} : [a, b] \to \mathbb{R}^n, \ \boldsymbol{\beta} : [b, c] \to \mathbb{R}^n$$

are two continuous paths such that $\boldsymbol{\alpha}(b) = \boldsymbol{\beta}(b)$, i.e., $\boldsymbol{\alpha}$ ends where $\boldsymbol{\beta}$ begins. Define

$$\boldsymbol{\alpha} * \boldsymbol{\beta} : [a, c] \to \mathbb{R}^n, \ \boldsymbol{\alpha} * \boldsymbol{\beta}(t) = \begin{cases} \boldsymbol{\alpha}(t), & t \in [a, b], \\ \boldsymbol{\beta}(t), & t \in (b, c]. \end{cases}$$

Prove that $\boldsymbol{\alpha} * \boldsymbol{\beta}$ is a *continuous* path. □

Exercise 12.4. (a) Consider a map $\boldsymbol{F} : \mathbb{R}^n \to \mathbb{R}^m$. Show that the following statements are equivalent.

(i) The map $\boldsymbol{F}$ is continuous.
(ii) For any open set $U \subset \mathbb{R}^m$, the preimage $\boldsymbol{F}^{-1}(U)$ is open.
(iii) For any closed set $C \subset \mathbb{R}^m$, the preimage $\boldsymbol{F}^{-1}(C)$ is closed.

(b) Suppose that D is an open subset of $\mathbb{R}^n$ and $\boldsymbol{F} : D \to \mathbb{R}^m$ is a map. Show that the following statements are equivalent.

(i) The map $\boldsymbol{F}$ is continuous.
(ii) For any open set $U \subset \mathbb{R}^m$, the preimage $\boldsymbol{F}^{-1}(U)$ is open.

Hint. (a) You need to understand very well the definition of preimage (1.3). □

Exercise 12.5. Suppose that $f : \mathbb{R}^n \to \mathbb{R}$ is continuous and $c \in \mathbb{R}$.

(i) Prove that the set $E_1 = \{\, \boldsymbol{x} \in \mathbb{R}^n; \ f(\boldsymbol{x}) < c \,\}$ is open.
(ii) Prove that the set $E_2 = \{\, \boldsymbol{x} \in \mathbb{R}^n; \ f(\boldsymbol{x}) \leq c \,\}$ is closed.
(iii) Prove that the set $E_3 = \{\, \boldsymbol{x} \in \mathbb{R}^n; \ f(\boldsymbol{x}) = c \,\}$ is closed.

(iv) Find an example of a function $f : \mathbb{R} \to \mathbb{R}$ that is not continuous yet, for any $c \in \mathbb{R}$, the set $\{ x \in \mathbb{R};\ f(x) \leq c \}$ is closed.

Hint. (i)–(iii) Use the previous exercise and Example 11.48. □

Exercise 12.6. (a) Suppose that $A \in \mathrm{Mat}_{m\times n}(\mathbb{R})$ and $B \in \mathrm{Mat}_{n\times p}(\mathbb{R})$. Prove that

$$\|A\|_{HS}^2 = \mathrm{tr}(A^\top A) = \mathrm{tr}(AA^\top),$$

and

$$\|A \cdot B\|_{HS} \leq \|A\|_{HS} \cdot \|B\|_{HS}, \tag{12.11}$$

where $\|-\|_{HS}$ denotes the Hilbert–Schmidt norm defined in Remark 12.13 and "tr" denotes the trace of a square matrix defined in Exercise 11.21.

(b) Compute $\|A\|_{HS}$, where A is the 2×2 matrix

$$A = \begin{bmatrix} 1 & 2 \\ 3 & 4 \end{bmatrix}.$$

(c) Show that if (A_ν), (B_ν) are two sequences in $\mathrm{Mat}_n(\mathbb{R})$ that converge (see Definition 12.14) to the matrices A and respectively B, then $A_\nu B_\nu$ converges to AB.

Hint. (a) Denote by $(A \cdot B)^i_j$ the (i, j)-entry of the product matrix $A \cdot B$. Use (11.15) to prove that

$$|(A \cdot B)^i_j| \leq \|(A^i)_\dagger\| \cdot \|B_j\|.$$

(c) Use the same strategy as in the proof of Proposition 11.68. □

Exercise 12.7. Suppose that $(A_\nu)_{\nu\geq 1}$ is a sequence of $n \times n$ matrices and $A \in \mathrm{Mat}_n(\mathbb{R})$. Prove that the following statements are equivalent.

(i)

$$\lim_{\nu\to\infty} \|A_\nu - A\|_{HS} = 0.$$

(ii) For any $\boldsymbol{x} \in \mathbb{R}^n$

$$\lim_{\nu\to\infty} A_\nu \boldsymbol{x} = A\boldsymbol{x}.$$

(iii) For any $\boldsymbol{x}, \boldsymbol{y} \in \mathbb{R}^n$

$$\lim_{\nu\to\infty} \langle A_\nu \boldsymbol{x}, \boldsymbol{y} \rangle = \langle A\boldsymbol{x}, \boldsymbol{y} \rangle.$$

(iv) If the entries of A_ν are $A^i_j(\nu)$, $1 \leq i, j \leq n$, and the entries of A are A^i_j, then

$$\lim_{\nu\to\infty} A^i_j(\nu) = A^i_j,\ \ \forall 1 \leq i, j \leq n.$$

Hint. (i) ⇒ (ii) Use (12.11). (ii) ⇒ (iii) Use Cauchy–Schwarz. (iii) ⇒ (iv) Use Exercise 11.23. (iv) ⇒ (i) Use the definition of the Hilbert–Schmidt norm. □

Exercise 12.8. To a matrix $R \in \mathrm{Mat}_{n\times n}(\mathbb{R})$ we associate the series of matrices

$$\mathbb{1} + R + R^2 + \cdots$$

with partial sums

$$S_0 = \mathbb{1},\ \ S_1 = \mathbb{1} + R,\ \ S_2 = \mathbb{1} + R + R^2, \cdots.$$

(i) Show that if $\|R\|_{HS} < 1$, then the sequence (S_N) is convergent to a matrix S satisfying $S(\mathbb{1} - R) = (\mathbb{1} - R)S = \mathbb{1}$, i.e., $\mathbb{1} - R$ is invertible and its inverse is S.

(ii) Prove that the matrix S above satisfies

$$\|S - \mathbb{1}\|_{HS} \leq \frac{\|R\|_{HS}}{1 - \|R\|_{HS}}.$$

Hint: (i)–(ii) Use the results in Exercises 11.40 and 12.6. □

Exercise 12.9. Suppose that A is an invertible $n \times n$ matrix. Prove that there exists $\varepsilon > 0$ such that if B is an $n \times n$ matrix satisfying $\|A - B\|_{HS} < \varepsilon$, then B is also invertible.

Hint. Write $C = A - B$ so that $B = A - C = A(\mathbb{1} - A^{-1}C)$. Thus, to prove that B is invertible it suffices to show that $\mathbb{1} - A^{-1}C$ is invertible. Prove that if $\|C\|_{HS} < \frac{1}{\|A^{-1}\|_{HS}}$, then $\|A^{-1}C\|_{HS} < 1$. To conclude invoke Exercise 12.8. □

Exercise 12.10. Let $n \in \mathbb{N}$ and suppose that (A_ν) is a sequence of invertible $n \times n$ matrices that converges with respect to the Hilbert–Schmidt norm to an invertible matrix A. Prove that

$$\lim_{\nu \to \infty} \|A_\nu^{-1} - A^{-1}\|_{HS} = 0.$$

Hint: Write $C_\nu := A - A_\nu$, $R_\nu := A^{-1}C_\nu$. Observe that $C_\nu, R_\nu \to 0$, $A_\nu = A(\mathbb{1} - R_\nu)$ and for ν large

$$\begin{aligned} A_\nu^{-1} - A^{-1} &= (\mathbb{1} - R_\nu)^{-1}A^{-1} - A^{-1} = \left(\mathbb{1} + R_\nu + R_\nu^2 + \cdots\right)A^{-1} - A^{-1} \\ &= \left(R_\nu + R_\nu^2 + \cdots\right)A^{-1}. \end{aligned}$$

□

Exercise 12.11. Prove Theorem 12.22.

Hint. Mimic the proof of Theorem 6.10. □

Exercise 12.12. Suppose that X is a nonempty subset of the real axis $\mathbb{R}$. Prove that the following statements are equivalent.

(i) The set X is an interval, i.e., it has the form

$$(a, b),\ [a, b),\ (a, b],\ [a, b],\ (a, \infty),\ [a, \infty), (-\infty, b),\ \text{or } (-\infty, b],\ \text{or } (-\infty, \infty).$$

(ii) If $x_0, x_1 \in X$ and $x_0 < x_1$, then $[x_0, x_1] \subset X$.

(iii) The set X is convex.

Hint. Clearly (i) $\Rightarrow$ (ii) and (ii) $\Longleftrightarrow$ (iii). The tricky implication is (ii) $\Rightarrow$ (i). Set $m := \inf X$, $M := \sup X$. Show that (ii) $\Rightarrow (m, M) \subset X \subset [m, M]$. □

Exercise 12.13.

(i) Prove that the set $\mathbb{R} \setminus \{0\} \subset \mathbb{R}$ is not path connected.

(ii) Prove that if L is a line in $\mathbb{R}^n$ and $\boldsymbol{p} \in L$, then the set $L \setminus \{\boldsymbol{p}\}$ is not path connected.

(iii) Suppose that $\boldsymbol{\xi} : \mathbb{R}^n \to \mathbb{R}$ is a nonzero linear functional. Prove that the set

$$\{\, \boldsymbol{x} \in \mathbb{R}^n;\ \ \boldsymbol{\xi}(\boldsymbol{x}) \neq 0 \,\}$$

is not path connected.

Hint. For (ii) consider a point $\boldsymbol{q} \in L \setminus \{\boldsymbol{p}\}$ so that $L = \boldsymbol{pq}$. Define $f : \mathbb{R} \to L$, $f(t) = (1-t)\boldsymbol{p} + t\boldsymbol{q}$. Show that f is bijective, Lipschitz and $f^{-1} : L \to \mathbb{R}$ is also Lipschitz. Conclude using Corollary 12.61. For (iii) use (i) and Corollary 12.49. □

Exercise 12.14. Let $n \in \mathbb{N}$, $n > 1$.

(i) Show that the punctured space $\mathbb{R}^n \setminus \{\boldsymbol{0}\}$ is path connected.

(ii) Show that the *unit Euclidean sphere*

$$\Sigma_1(\boldsymbol{0}) := \{\boldsymbol{x} \in \mathbb{R}^n;\ \ \|\boldsymbol{x}\| = 1\},$$

is path connected.

(iii) Show that for any $r > 0$ and any $\boldsymbol{p} \in \mathbb{R}^n$ the *Euclidean sphere of center* $\boldsymbol{p}$ *and radius* r, i.e., the set

$$\Sigma_r(\boldsymbol{p}) := \{\, \boldsymbol{x} \in \mathbb{R}^n;\ \ \|\boldsymbol{x} - \boldsymbol{p}\| = r \,\}$$

is path connected.

(iv) Prove that for any positive numbers $r < R$ the *annulus*

$$A_{r,R} := \{\boldsymbol{x} \in \mathbb{R}^n;\ \ r < \|\boldsymbol{x}\| < R\}$$

is path connected but not convex.

Hint. (i) Let $\boldsymbol{p}, \boldsymbol{q} \in \mathbb{R}^n \setminus \{\boldsymbol{0}\}$. If the line $\boldsymbol{pq}$ does not contain $\boldsymbol{0}$ we are done since the segment $[\boldsymbol{p}, \boldsymbol{q}]$ will do the trick. If $\boldsymbol{0} \in \boldsymbol{pq}$, then choose a point $\boldsymbol{r} \in \mathbb{R}^n \setminus \{\boldsymbol{0}\}$ that does not belong to this line. (You need to use the assumption $n > 1$ to prove that such a point exists.) Then $\boldsymbol{0} \notin \boldsymbol{pr}$. Travel from $\boldsymbol{p}$ to $\boldsymbol{r}$ on $[\boldsymbol{p}, \boldsymbol{r}]$ and then from $\boldsymbol{r}$ to $\boldsymbol{q}$ on $[\boldsymbol{r}, \boldsymbol{q}]$. (Need to invoke Remark 12.24 and Exercise 12.3.) To prove (ii) use (i). To prove (iii) use (ii). To prove (iv) it helps to first visualize the region $A_{r,R}$ in the special case $n = 2$, $r = 1$, $R = 2$. Use (iii) to prove that this annulus is path connected. □

Exercise 12.15. Let $n \in \mathbb{N}$ and suppose that $S_1, S_2 \subset \mathbb{R}^n$ are two path connected subsets such that $S_1 \cap S_2 \neq \emptyset$. Prove that $S_1 \cup S_2$ is also path connected. □

Exercise 12.16. Prove Proposition 12.55. □

Exercise 12.17. Let $n \in \mathbb{N}$. Suppose that $(K_\nu)_{\nu \in \mathbb{N}}$ is a sequence of nonempty compact subsets of $\mathbb{R}^n$ such that

$$K_1 \supset K_2 \supset \cdots \supset K_\nu \supset \cdots.$$

Prove that

$$\bigcap_{\nu \in \mathbb{N}} K_\nu \neq \emptyset,$$

i.e.,

$$\exists \boldsymbol{p} \in \mathbb{R}^n \ \text{ such that } \boldsymbol{p} \in K_\nu,\ \ \forall \nu \in \mathbb{N}.$$

Hint. For any $\nu \in \mathbb{N}$ choose a point $\boldsymbol{p}_\nu \in K_\nu$. Show that a subsequence of $(\boldsymbol{p}_\nu)$ is convergent and then prove that its limit belongs to K_ν for any ν. □

Exercise 12.18. Let $n \in \mathbb{N}$ and suppose that $A, B \subset \mathbb{R}^n$ are nonempty. We regard $A \times B$ as a subset of $\mathbb{R}^n \times \mathbb{R}^n = \mathbb{R}^{2n}$ and we consider the function

$$f : A \times B \to \mathbb{R}, \quad f(\boldsymbol{a}, \boldsymbol{b}) = \|\boldsymbol{a} - \boldsymbol{b}\|.$$

Prove that f is continuous.

Hint. Use Proposition 12.6(ii). □

Exercise 12.19. Let $n \in \mathbb{N}$ and suppose that $K \subset \mathbb{R}^n$ is a nonempty compact subset. Recall (see Definition 12.54) that

$$\operatorname{diam}(K) := \sup_{\boldsymbol{x}, \boldsymbol{y} \in K} \|\boldsymbol{x} - \boldsymbol{y}\|.$$

Prove that there exist $\boldsymbol{x}_*, \boldsymbol{y}_* \in K$ such that

$$\operatorname{diam}(K) = \|\boldsymbol{x}_* - \boldsymbol{y}_*\|.$$

Hint. Use Exercise 12.18, Proposition 12.41, and Corollary 12.51. □

Exercise 12.20. Prove Proposition 12.57. □

Exercise 12.21. Let $X \subset \mathbb{R}^n$, and $f : X \to \mathbb{R}$ a Lipschitz function. Prove that f is uniformly continuous on X. □

Exercise 12.22. Let $n \in \mathbb{N}$. Suppose that $C \subset \mathbb{R}^n$ is a nonempty closed subset. For $\boldsymbol{x} \in \mathbb{R}^n$ we set

$$\operatorname{dist}(\boldsymbol{x}, C) := \inf_{\boldsymbol{p} \in C} \operatorname{dist}(\boldsymbol{x}, \boldsymbol{p}).$$

(i) Prove that for any $\boldsymbol{x} \in \mathbb{R}^n$ there exists $\boldsymbol{y} \in C$ such that

$$\|\boldsymbol{x} - \boldsymbol{y}\| = \operatorname{dist}(\boldsymbol{x}, C).$$

(ii) Prove that the function $f : \mathbb{R}^n \to \mathbb{R}$, $f(\boldsymbol{x}) = \operatorname{dist}(\boldsymbol{x}, C)$ is Lipschitz. More precisely

$$\big|\, f(\boldsymbol{x}) - f(\boldsymbol{y}) \,\big| \leq \big\|\, \boldsymbol{x} - \boldsymbol{y} \,\big\|, \quad \forall \boldsymbol{x}, \boldsymbol{y} \in \mathbb{R}^n.$$

(iii) Prove that

$$C = f^{-1}(0) = \big\{\, \boldsymbol{x} \in \mathbb{R}^n;\ \operatorname{dist}(\boldsymbol{x}, C) = 0 \,\big\}.$$

Hint. (i) Show that there exists a sequence $(\boldsymbol{y}_\nu)$ in C such that $\|\boldsymbol{x} - \boldsymbol{y}_\nu\| \to \operatorname{dist}(\boldsymbol{x}, C)$. Next prove that this sequence is bounded and thus it has a convergent subsequence. (ii) Use the triangle inequality, part (i) and the definition of $\operatorname{dist}(\boldsymbol{x}, C)$ to prove that $L = 1$ is a Lipschitz constant for $f(\boldsymbol{x})$. (iii) Use (i). □

Exercise 12.23. Let $n \in \mathbb{N}$ and suppose that $C \subset \mathbb{R}^n$ is a *closed, convex* subset and $\boldsymbol{x}_0 \in \mathbb{R}^n \setminus C$. Set

$$r := \operatorname{dist}(\boldsymbol{x}_0, C).$$

Prove that the sphere

$$\Sigma_r(\boldsymbol{x}_0) = \big\{\boldsymbol{x} \in \mathbb{R}^n;\ \|\boldsymbol{x} - \boldsymbol{x}_0\| = r\,\big\}$$

intersects the set C in *exactly one point*. This unique point of intersection is called the *projection of* $\boldsymbol{x}_0$ *on* C and it is denoted by $\operatorname{Proj}_C \boldsymbol{x}_0$.

Hint. You need to use Exercise 12.22. □

Exercise 12.24. Let $U \subset \mathbb{R}^n$ be an open set. Prove that the following statements are equivalent.

(i) The set U is path connected.

(ii) Any $\boldsymbol{p}, \boldsymbol{q} \in U$ can be joined by a broken line contained in U. More precisely, this means that for any $\boldsymbol{p}, \boldsymbol{q} \in U$ there exist points $\boldsymbol{p}_0, \boldsymbol{p}_1, \dots, \boldsymbol{p}_N \in U$ such that $\boldsymbol{p} = \boldsymbol{p}_0$, $\boldsymbol{q} = \boldsymbol{p}_N$ and all the line segments

$$[\boldsymbol{p}_0, \boldsymbol{p}_1], \ [\boldsymbol{p}_1, \boldsymbol{p}_2], \dots, [\boldsymbol{p}_{N-1}, \boldsymbol{p}_N]$$

are contained in U.

Hint. (i) $\Rightarrow$ (ii) Set $C = \mathbb{R}^n \setminus U$ and define $\rho : \mathbb{R}^n \to [0, \infty)$, $\rho(\boldsymbol{x}) = \operatorname{dist}(\boldsymbol{x}, C)$. Observe that ρ is Lipschitz and thus continuous. Consider a continuous path $\boldsymbol{\gamma} : [0, 1] \to U$ such that $\boldsymbol{\gamma}(0) = \boldsymbol{p}$ and $\boldsymbol{\gamma}(1) = \boldsymbol{q}$. Set $r_0 = \inf_{t \in [0,1]} \rho(\boldsymbol{\gamma}(t))$. Show that $r_0 > 0$ and $B_{r_0}(\boldsymbol{\gamma}(t)) \subset U$, $\forall t \in [0, 1]$. Use the uniform continuity of $\boldsymbol{\gamma} : [0, 1] \to U$ to show that, for N sufficiently large, we have

$$\| \boldsymbol{\gamma}(0) - \boldsymbol{\gamma}(1/N) \| < \frac{r_0}{2}, \ \dots, \| \boldsymbol{\gamma}((N-1)/N) - \boldsymbol{\gamma}(1) \| < \frac{r_0}{2},$$

and conclude that the broken line determined by the points

$$\boldsymbol{p}_0 = \boldsymbol{\gamma}(0), \ \boldsymbol{p}_1 = \boldsymbol{\gamma}(1/N), \ \boldsymbol{p}_i = \boldsymbol{\gamma}(i/N), \ i = 1, \dots, N,$$

is contained in U. □

Exercise 12.25. Suppose that $f : \mathbb{R}^n \to \mathbb{R}$ is a continuous function with the following property: there exist $A, B > 0$ such that

$$f(\boldsymbol{x}) \geq A\|\boldsymbol{x}\| - B, \ \forall \boldsymbol{x} \in \mathbb{R}^n.$$

(i) Prove that for any $R > 0$ the set

$$\{f \leq R\} := \{\, \boldsymbol{x} \in \mathbb{R}^n; \ f(\boldsymbol{x}) \leq R \,\}$$

is compact.

(ii) Prove that there exists $\boldsymbol{x}_* \in \mathbb{R}^n$ such that $f(\boldsymbol{x}_*) \leq f(\boldsymbol{x})$, $\forall \boldsymbol{x} \in \mathbb{R}^n$.

Hint. (i) Show that the set $\{f \leq R\}$ is bounded. (ii) Prove that there exists a sequence $(\boldsymbol{x}_\nu)$ in $\mathbb{R}^n$ such that

$$\lim_{\nu \to \infty} f(\boldsymbol{x}_\nu) = \inf_{\boldsymbol{x} \in \mathbb{R}^n} f(\boldsymbol{x}).$$

Deduce that the sequence $f(\boldsymbol{x}_\nu)$ is bounded above and then, using (i), prove that the sequence $(\boldsymbol{x}_\nu)$ is bounded and thus it has a convergent subsequence. □

Exercise 12.26. Let $n \in \mathbb{N}$, $d \in \mathbb{R}$. We say that a function $f : \mathbb{R}^n \to \mathbb{R}$ is *positively homogeneous of degree* d if

$$f(t\boldsymbol{x}) = t^d f(\boldsymbol{x}), \ \forall t > 0, \ \boldsymbol{x} \in \mathbb{R}^n \setminus \{0\}.$$

(i) Suppose that $f : \mathbb{R}^n \to \mathbb{R}$ is a nonconstant, continuous and positively homogeneous function of degree d. Prove that $d > 0$.

(ii) Given $d \in \mathbb{R}$ construct a nonconstant function $f : \mathbb{R}^n \to \mathbb{R}$ that is positively homogeneous of degree d and it is continuous at every point $\boldsymbol{x} \in \mathbb{R}^n \setminus \{\boldsymbol{0}\}$.

Hint. (i). Fix $\boldsymbol{x} \in \mathbb{R}^n \setminus \{\boldsymbol{0}\}$ and consider the sequence $f(\nu^{-1}\boldsymbol{x})$, $\nu \in \mathbb{N}$. □

Exercise 12.27. Let $n \in \mathbb{N}$ and suppose that $d > 0$ and $f : \mathbb{R}^n \to (0, \infty)$ is continuous, positively homogeneous of degree $d > 0$ and satisfies

$$f(\boldsymbol{x}) > 0, \ \ \forall \boldsymbol{x} \in \mathbb{R}^n \setminus \{0\}.$$

(i) Prove that there exists $c > 0$ such that

$$f(\boldsymbol{x}) \geq c\|\boldsymbol{x}\|^d, \ \ \forall \boldsymbol{x} \in \mathbb{R}^n.$$

(ii) Prove that for any $r > 0$ the *sublevel set*

$$\{f \leq r\} := \{\boldsymbol{x} \in \mathbb{R}^n; \ f(\boldsymbol{x}) \leq r\}$$

is compact.

Hint. (i) Consider the unit sphere

$$\Sigma_1 := \{\boldsymbol{x} \in \mathbb{R}^n; \ \|\boldsymbol{x}\| = 1\}.$$

Use Corollary 12.51 to show that the infimum of f on Σ_1 is *strictly positive*. (ii) Use (i). □

Exercise 12.28. For any linear operator $A : \mathbb{R}^n \to \mathbb{R}^m$ we set

$$\|A\| := \sup_{\|\boldsymbol{x}\|=1} \|A\boldsymbol{x}\|.$$

(i) Show that if $A : \mathbb{R}^n \to \mathbb{R}^m$ is a linear operator, then $\|A\| < \infty$ and

$$\|A\boldsymbol{x}\| \leq \|A\| \cdot \|\boldsymbol{x}\|, \ \ \forall \boldsymbol{x} \in \mathbb{R}^n.$$

(ii) Show that if $A : \mathbb{R}^n \to \mathbb{R}^m$ and $B : \mathbb{R}^m \to \mathbb{R}^\ell$ are linear operators, then

$$\|B \circ A\| \leq \|B\| \cdot \|A\|.$$

(iii) Show that the linear operator $A : \mathbb{R}^n \to \mathbb{R}^m$ is injective if and only if there exists $C > 0$ such that

$$\|A\boldsymbol{x}\| \geq C\|\boldsymbol{x}\|, \ \ \forall \boldsymbol{x} \in \mathbb{R}^n \setminus \{\boldsymbol{0}\}.$$

(iv) Prove that if $A, B : \mathbb{R}^n \to \mathbb{R}^m$ are linear operators and $t \in \mathbb{R}$ then

$$\|A + B\| \leq \|A\| + \|B\|, \ \ \|tA\| = |t| \cdot \|A\|.$$

Hint. (iii) Use Exercise 11.13 and Exercise 12.26 applied to the function $f(\boldsymbol{x}) = \|A\boldsymbol{x}\|$. □

Exercise 12.29. Prove Proposition 12.63. □

Exercise 12.30. Let $n \in \mathbb{N}$ and $X \subset \mathbb{R}^n$.

(i) Prove that the boundary ∂X is a closed subset of $\mathbb{R}^n$.
(ii) Show that if X is bounded, then ∂X is compact.

□

Exercise 12.31. Let $n \in \mathbb{N}$ and $X \subset \mathbb{R}^n$. Prove that the following statements are equivalent.

(i) $cl(X) = \mathbb{R}^n$.
(ii) The set X is dense in $\mathbb{R}^n$.

□

Exercise 12.32. Find the closures, the interiors and the boundaries of the following sets.

(i) $(0,1) \subset \mathbb{R}$.
(ii) $[0,1] \subset \mathbb{R}$.
(iii) $(0,1) \times \{0\} \subset \mathbb{R}^2$.
(iv) $\{\,(x,y) \in \mathbb{R}^2;\ \ 0 \leq x, y \leq 1\,\} \subset \mathbb{R}^2$.

□

12.6. More challenging problems

Problem 12.1. Let $n \in \mathbb{N}$ and suppose that $K \subset \mathbb{R}^n$ is a nonempty subset. Prove that the following statements are equivalent.

(i) The set K is compact
(ii) Any continuous function $f : K \to \mathbb{R}$ is bounded.

□

Problem 12.2. Suppose that $U \subset \mathbb{R}^n$ is an open set and $\boldsymbol{p}, \boldsymbol{q} \in U$ are points such that the line segment $[\boldsymbol{p}, \boldsymbol{q}]$ is contained in U. Prove that there exists an open *convex* set V such that

$$[\boldsymbol{p}, \boldsymbol{q}] \subset V \subset U.$$

□

Problem 12.3. Show that the set

$$S := \left\{ \frac{k}{2^m};\ \ k \in \mathbb{Z},\ \ m \in \mathbb{Z},\ \ m \geq 0 \right\}$$

is dense in $\mathbb{R}$. □

Problem 12.4. Let $n \in \mathbb{N}$ and suppose that $C \subset \mathbb{R}^n$ is a *closed, convex* subset and $\boldsymbol{x}_0 \in \mathbb{R}^n \setminus C$. Prove that there exists a linear functional $\boldsymbol{\xi} : \mathbb{R}^n \to \mathbb{R}$ and a real number c such that

$$\boldsymbol{\xi}(\boldsymbol{x}_0) > c > \boldsymbol{\xi}(\boldsymbol{x}),\ \ \forall \boldsymbol{x} \in C.$$

Hint. You may want to use the result in Exercise 12.23. □

Problem 12.5. Let $n \in \mathbb{N}$ and suppose that $f : \mathbb{R}^n \to (0, \infty)$ is continuous and positively homogeneous of degree $d > 0$. Prove that the following statements are equivalent.

(i) The function f is uniformly continuous on $\mathbb{R}^n$.
(ii) $d \leq 1$.

□

Problem 12.6. Let $n \in \mathbb{N}$ and suppose that $\| - \|_*$ is a norm on the vector space $\mathbb{R}^n$.

(i) Prove that there exists a constant $C > 0$ such that

$$\|\boldsymbol{x}\|_* \leq C\|\boldsymbol{x}\|, \quad \forall \boldsymbol{x} \in \mathbb{R}^n,$$

where $\|\boldsymbol{x}\|$ denotes the Euclidean norm of $\boldsymbol{x}$.

(ii) Prove that the function $f : \mathbb{R}^n \to \mathbb{R}$, $f(\boldsymbol{x}) = \|\boldsymbol{x}\|_*$ is continuous, i.e.,

$$\lim_{\nu\to\infty} \|\boldsymbol{x}_\nu - \boldsymbol{x}\| = 0 \Rightarrow \lim_{\nu\to\infty} \|\boldsymbol{x}_\nu\|_* = \|\boldsymbol{x}\|_*.$$

(iii) Prove that

$$\inf\{\, \|\boldsymbol{x}\|_*;\;\; \|\boldsymbol{x}\| = 1 \,\} \neq 0.$$

(iv) Prove that there exists $c > 0$ such that

$$\|\boldsymbol{x}\|_* \geq c\|\boldsymbol{x}\|, \quad \forall \boldsymbol{x} \in \mathbb{R}^n.$$

□

Problem 12.7. Suppose that $E \subset \mathbb{R}^n$ is an affine subspace.

(i) Prove that there exists $m \in \mathbb{N}$, a linear operator $A : \mathbb{R}^n \to \mathbb{R}^m$ and a vector $\boldsymbol{v} \in \mathbb{R}^m$ such that

$$E = \big\{ \boldsymbol{x} \in \mathbb{R}^n;\;\; A\boldsymbol{x} = \boldsymbol{v} \big\}.$$

(ii) Prove that E is a closed subset of $\mathbb{R}^n$.

□

Problem 12.8. Suppose that $T : \mathbb{R}^2 \to \mathbb{R}^2$ is a map satisfying the following conditions.

(i) T is continuous.
(ii) T is injective.
(iii) $T(\boldsymbol{0}) = \boldsymbol{0}, \;\; T(\boldsymbol{i}) = \boldsymbol{i}, \;\; T(\boldsymbol{j}) = \boldsymbol{j}$.
(iv) For any line $\ell \subset \mathbb{R}^2$, the image $T(\ell)$ is also a line in $\mathbb{R}^2$.

Prove that $T(\boldsymbol{v}) = \boldsymbol{v}$, $\forall \boldsymbol{v} \in \mathbb{R}^2$. □

Problem 12.9. Prove that $\mathbb{R}$ is *not* homeomorphic to $\mathbb{R}^2$. □

Problem 12.10. Suppose that $K \subset \mathbb{R}^n$ is a compact set and $\mathcal{U}$ is an open cover of K. Prove that there exists $r > 0$ such that for any $\boldsymbol{p} \in K$ there exists an open set U belonging to the cover $\mathcal{U}$ such that $U \supset B_r(\boldsymbol{p})$. □

Problem 12.11. Let $n \in \mathbb{N}$ and suppose that $C_1, \ldots, C_\nu, \ldots$ is a sequence of closed subsets of $\mathbb{R}^n$ such that

$$\mathbb{R}^n = \bigcup_{\nu=1}^{\infty} C_\nu.$$

Prove that there exists $\nu \in \mathbb{N}$ such that $\boldsymbol{int}\, C_\nu \neq \emptyset$. □

Chapter 13

Multi-variable Differential Calculus

The concept of differential of a one-variable function extends to functions of several variables. The several-variable situations add new complexity and subtleties, and the goal of the present chapter is to investigate them.

Recall that a function $f : \mathbb{R} \to \mathbb{R}$ is differentiable at a point $x_0 \in \mathbb{R}$ if and only if it admits a "best" linear approximation near x_0. More geometrically, the graph of f, which is a curve in $\mathbb{R}^2$, can be well approximated in a vicinity of the point $\boldsymbol{p}_0 = (x_0, y_0) \in \mathbb{R}^2$, $y_0 = f(x_0)$, by a straight line, the tangent line to the curve at the point $\boldsymbol{p}_0$. This tangent line is graph of a function of the form $L(x) = A(x - x_0) + y_0$. The slope A of this line is the derivative of f at x_0.

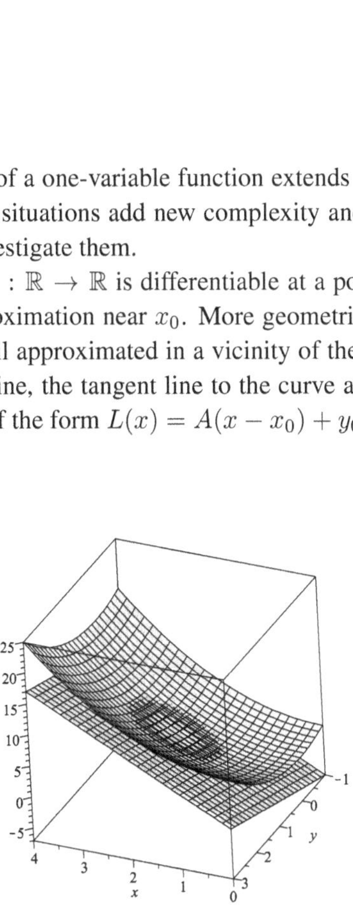

Fig. 13.1 *A "best" linear approximation of the function* $f(x, y) = x^2 + y^2$ *near the point* $(2, 1)$.

We want to extend this approach to maps $\boldsymbol{F} : \mathbb{R}^n \to \mathbb{R}^m$. The graph of such a map is an m-dimensional "curved" surface in $\mathbb{R}^n \times \mathbb{R}^m$. We seek to find a "best" approximation of this graph near $\boldsymbol{p}_0 = (\boldsymbol{x}_0, \boldsymbol{y}_0)$, $\boldsymbol{y}_0 = \boldsymbol{F}(\boldsymbol{x}_0)$, by a "straight" or "flat" m-dimensional surface; see Figure 13.1 where $n = 2$, $m = 1$. The "straight" surfaces in an Euclidean space are precisely the affine subspaces and we seek to approximate the graph of $\boldsymbol{F}$ near $\boldsymbol{p}_0$ by an affine subspace described as the graph of a map $L : \mathbb{R}^n \to \mathbb{R}^m$ of the form

$L(\boldsymbol{x}) = A(\boldsymbol{x} - \boldsymbol{x}_0) + \boldsymbol{y}_0$, where $A : \mathbb{R}^n \to \mathbb{R}^m$ is a linear operator. The concept of Fréchet derivative formalizes the above heuristics.

13.1. The differential of a map at a point

Suppose that $m, n \in \mathbb{N}$ and $U \subset \mathbb{R}^n$ is an open subset. Since U is open, we deduce that for any point $\boldsymbol{x}_0 \in U$ there exists $r = r(\boldsymbol{x}_0) > 0$ such that the open ball $B_r(\boldsymbol{x}_0)$ is contained in U. This means that (see Figure 13.2)

$$\boldsymbol{x}_0 + \boldsymbol{h} \in U, \ \ \forall \|\boldsymbol{h}\| < r.$$

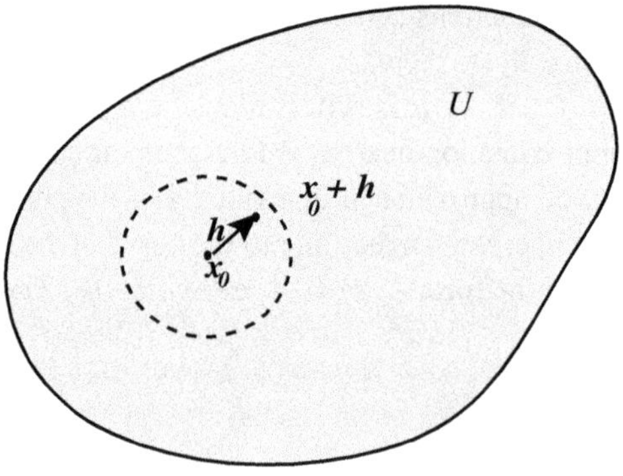

Fig. 13.2 *An open set.*

Definition 13.1. Suppose that $\boldsymbol{F} : U \to \mathbb{R}^m$ is a map and $\boldsymbol{x}_0 \in U$. We say that F is *Fréchet*[1] *differentiable at* $\boldsymbol{x}_0$ if there exists a linear operator $L : \mathbb{R}^n \to \mathbb{R}^m$ such that

$$\boxed{\lim_{\boldsymbol{h} \to \boldsymbol{0}} \frac{1}{\|\boldsymbol{h}\|}\Big(\boldsymbol{F}(\boldsymbol{x}_0 + \boldsymbol{h}) - \boldsymbol{F}(\boldsymbol{x}_0) - L\boldsymbol{h}\Big) = 0}. \tag{13.1}$$

□

Remark 13.2. Observe that if $\boldsymbol{F}$ is differentiable at $\boldsymbol{x}_0$, then there exists ***exactly one*** linear operator $L : \mathbb{R}^n \to \mathbb{R}^m$ satisfying the condition (13.1). More precisely, for any $\boldsymbol{h} \in \mathbb{R}^n \setminus 0$ we have

$$\boxed{L\boldsymbol{h} = \lim_{t \to 0} \frac{1}{t}\Big(F(\boldsymbol{x}_0 + t\boldsymbol{h}) - F(\boldsymbol{x}_0)\Big)}. \tag{13.2}$$

[1]Named after Maurice René Fréchet (1878–1973), a French mathematician. He made major contributions to point-set topology and introduced the concept of compactness; see Wikipedia.

Indeed, consider a sequence of real numbers $t_\nu \to 0$, $t_\nu \neq 0$. Set $\boldsymbol{h}_\nu = t_\nu \boldsymbol{h}$. Note that

$$\lim_{\nu\to\infty} \boldsymbol{h}_\nu = \boldsymbol{0}$$

so that $\boldsymbol{x}_0 + \boldsymbol{h}_\nu \in U$, for large ν. We have $L\boldsymbol{h}_\nu = t_\nu L\boldsymbol{h}$ and

$$\begin{aligned}
&\lim_{\nu\to\infty} \left\| \frac{1}{t_\nu}\Big(\boldsymbol{F}(\boldsymbol{x}_0 + t_\nu \boldsymbol{h}) - \boldsymbol{F}(\boldsymbol{x}_0) \Big) - L\boldsymbol{h} \right\| \\
&= \|\boldsymbol{h}\| \lim_{\nu\to\infty} \left\| \frac{1}{t_\nu \|\boldsymbol{h}\|}\Big(\boldsymbol{F}(\boldsymbol{x}_0 + t_\nu \boldsymbol{h}) - \boldsymbol{F}(\boldsymbol{x}_0) - t_\nu L\boldsymbol{h} \Big) \right\| \\
&= \|\boldsymbol{h}\| \lim_{\nu\to\infty} \frac{1}{|t_\nu| \cdot \|\boldsymbol{h}\|} \left\| \Big(\boldsymbol{F}(\boldsymbol{x}_0 + t_\nu \boldsymbol{h}) - \boldsymbol{F}(\boldsymbol{x}_0) - t_\nu L\boldsymbol{h} \Big) \right\|
\end{aligned}$$

$(|t_\nu| \cdot \|\boldsymbol{h}\| = \|t_\nu \boldsymbol{h}\| = \|\boldsymbol{h}_\nu\|)$

$$= \|\boldsymbol{h}\| \lim_{\nu\to\infty} \frac{1}{\|\boldsymbol{h}_\nu\|} \left\| \Big(\boldsymbol{F}(\boldsymbol{x}_0 + \boldsymbol{h}_\nu) - \boldsymbol{F}(\boldsymbol{x}_0) - L\boldsymbol{h}_\nu \Big) \right\| \overset{(13.1)}{=} 0. \qquad \square$$

Definition 13.3. The unique linear operator $L : \mathbb{R}^n \to \mathbb{R}^m$ such that the differentiability condition (13.1) or (13.6) is satisfied is called the *(Fréchet) differential* of $\boldsymbol{F}$ at $\boldsymbol{x}_0$ and it is denoted by $d\boldsymbol{F}(\boldsymbol{x}_0)$. □

The equality (13.2) shows that the Fréchet differential $d\boldsymbol{F}(\boldsymbol{x}_0)$ is determined uniquely by the equality

$$\boxed{d\boldsymbol{F}(\boldsymbol{x}_0)\boldsymbol{h} = \lim_{t\to 0} \frac{1}{t}\Big(\boldsymbol{F}(\boldsymbol{x}_0 + t\boldsymbol{h}) - \boldsymbol{F}(\boldsymbol{x}_0) \Big), \ \ \forall \boldsymbol{h} \in \mathbb{R}^n}. \tag{13.3}$$

Remark 13.4. (a) Suppose that $\boldsymbol{F} : U \to \mathbb{R}^m$ is differentiable at $\boldsymbol{x}_0$ and $L := d\boldsymbol{F}(\boldsymbol{x}_0)$. The main point of Definition 13.1 is that, for small $\boldsymbol{h}$, the variation

$$\Delta_{\boldsymbol{h}} \boldsymbol{F}(\boldsymbol{x}_0) = \boldsymbol{F}(\boldsymbol{x}_0 + \boldsymbol{h}) - \boldsymbol{F}(\boldsymbol{x}_0)$$

is very well approximated by the *linear* quantity $L\boldsymbol{h}$. For $\boldsymbol{h} \in \mathbb{R}^n$ the error of this approximation is

$$R(\boldsymbol{h}) := \boldsymbol{F}(\boldsymbol{x}_0 + \boldsymbol{h}) - \boldsymbol{F}(\boldsymbol{x}_0) - L\boldsymbol{h}.$$

The differentiability condition is equivalent to the fact that the error $R(\boldsymbol{h})$ is $o(\boldsymbol{h})$, where $o(\boldsymbol{h})$ stands for "a lot smaller" than $\boldsymbol{h}$ as $\boldsymbol{h} \to \boldsymbol{0}$. More precisely

$$\boxed{R(\boldsymbol{h}) = o(\boldsymbol{h}) \ \text{ as } \boldsymbol{h} \to \boldsymbol{0} \iff \lim_{\boldsymbol{h}\to 0} \frac{1}{\|\boldsymbol{h}\|} R(\boldsymbol{h}) = \boldsymbol{0}}. \tag{13.4}$$

One can prove (see Exercise 13.1) that the condition (13.4) is equivalent to the existence of a function

$$\varphi : [0, r) \to [0, \infty)$$

such that

$$\lim_{t\searrow 0} \varphi(t) = 0 \ \text{ and } \ \|R(\boldsymbol{h})\| \leq \varphi\big(\|\boldsymbol{h}\| \big) \| h\|, \ \ \forall \|\boldsymbol{h}\| < r. \tag{13.5}$$

The equality (13.4) can be rewritten as

$$\boldsymbol{F}(\boldsymbol{x}_0+\boldsymbol{h})-\boldsymbol{F}(\boldsymbol{x}_0)=L(\boldsymbol{h})+o(\boldsymbol{h}) \ \text{ as } \boldsymbol{h}\to\boldsymbol{0},$$

or, if we set $\boldsymbol{x}:=\boldsymbol{x}_0+\boldsymbol{h}$

$$\boldsymbol{F}(\boldsymbol{x})-\boldsymbol{F}(\boldsymbol{x}_0)=L(\boldsymbol{x}-\boldsymbol{x}_0)+o(\boldsymbol{x}-\boldsymbol{x}_0) \ \text{ as } \boldsymbol{x}\to\boldsymbol{x}_0. \tag{13.6}$$

This last equality can be taken as a definition of the Fréchet differential: *the linear operator $L:\mathbb{R}^n\to\mathbb{R}^m$ is the Fréchet differential of $\boldsymbol{F}$ at $\boldsymbol{x}_0$ iff it satisfies (13.6).*
(b) By definition, the differential $d\boldsymbol{F}(\boldsymbol{x}_0)$ is a linear operator $\mathbb{R}^n\to\mathbb{R}^m$ and, as such, it is represented by an $m\times n$ matrix sometimes called the *Jacobian matrix* of $\boldsymbol{F}$ at $\boldsymbol{x}_0$ denoted by

$$\boxed{J_{\boldsymbol{F}}(\boldsymbol{x}_0) \ \text{ or } \ \frac{\partial \boldsymbol{F}}{\partial \boldsymbol{x}}(\boldsymbol{x}_0)}.$$

The n columns of the matrix $J_{\boldsymbol{F}}(\boldsymbol{x}_0)$ consist of the vectors

$$d\boldsymbol{F}(\boldsymbol{x}_0)\boldsymbol{e}_1,\dots,d\boldsymbol{F}(\boldsymbol{x}_0)\boldsymbol{e}_n\in\mathbb{R}^m,$$

where $\{\boldsymbol{e}_1,\dots,\boldsymbol{e}_n\}$ is the natural basis of $\mathbb{R}^n$ and

$$d\boldsymbol{F}(\boldsymbol{x}_0)\boldsymbol{e}_j=\lim_{t\to 0}\frac{1}{t}\big(\boldsymbol{F}(\boldsymbol{x}_0+t\boldsymbol{e}_j)-\boldsymbol{F}(\boldsymbol{x}_0)\big), \ \ \forall j=1,\dots,n. \tag{13.7}$$

□

Definition 13.5. Let $m,n\in\mathbb{N}$, assume that $U\subset\mathbb{R}^n$ is an open set. If $\boldsymbol{F}:U\to\mathbb{R}^m$ is Fréchet differentiable at $\boldsymbol{x}_0$ and $L:\mathbb{R}^n\to\mathbb{R}^m$ is its Fréchet derivative, then the function $\mathscr{L}=\mathscr{L}_{\boldsymbol{F},\boldsymbol{x}_0}:\mathbb{R}^n\to\mathbb{R}^m$ defined by

$$\mathscr{L}(\boldsymbol{x})=\boldsymbol{F}(\boldsymbol{x}_0)+L(\boldsymbol{x}-\boldsymbol{x}_0) \tag{13.8}$$

is called the *linearization* or the *linear approximation* of $\boldsymbol{F}$ at $\boldsymbol{x}_0$. □

Note that the equality (13.4) where $\boldsymbol{h}=\boldsymbol{x}-\boldsymbol{x}_0$ (equivalently, $\boldsymbol{x}=\boldsymbol{x}_0+\boldsymbol{h}$), implies that

$$\boldsymbol{F}(\boldsymbol{x})-\mathscr{L}(\boldsymbol{x})=o\big(\boldsymbol{x}-\boldsymbol{x}_0\big) \ \text{ as } \boldsymbol{x}\to\boldsymbol{x}_0$$

i.e.,

$$\lim_{\boldsymbol{x}\to\boldsymbol{x}_0}\frac{\|\boldsymbol{F}(\boldsymbol{x})-\mathscr{L}(\boldsymbol{x})\|}{\|\boldsymbol{x}-\boldsymbol{x}_0\|}=0.$$

This shows that, when $\boldsymbol{x}\to\boldsymbol{x}_0$, the difference $\boldsymbol{F}(\boldsymbol{x})-\mathscr{L}(\boldsymbol{x})$ is a lot smaller than the very small quantity $\|\boldsymbol{x}-\boldsymbol{x}_0\|$.

The equality (13.4) implies the following result.

Proposition 13.6. *If $U\subset\mathbb{R}^n$ is open, $\boldsymbol{x}_0\in U$ and the map $\boldsymbol{F}:U\to\mathbb{R}^m$ is Fréchet differentiable at $\boldsymbol{x}_0$, then it is continuous at $\boldsymbol{x}_0$.*

Proof. Using the notation from Remark 13.2 we can write

$$\boldsymbol{F}(\boldsymbol{x}_0 + \boldsymbol{h}) = \boldsymbol{F}(\boldsymbol{x}_0) + L\boldsymbol{h} + R(\boldsymbol{h}).$$

Since L is a linear operator, it is a continuous map and thus

$$\lim_{\boldsymbol{h}\to\boldsymbol{0}} L\boldsymbol{h} = \boldsymbol{0}.$$

On the other hand, (13.4) shows that

$$\lim_{\boldsymbol{h}\to\boldsymbol{0}} R(\boldsymbol{h}) = \boldsymbol{0}.$$

Hence

$$\lim_{\boldsymbol{h}\to\boldsymbol{0}} \boldsymbol{F}(\boldsymbol{x}_0 + \boldsymbol{h}) = \lim_{\boldsymbol{h}\to\boldsymbol{0}} \big(\boldsymbol{F}(\boldsymbol{x}_0) + L\boldsymbol{h} + R(\boldsymbol{h})\big) = \boldsymbol{F}(\boldsymbol{x}_0).$$

□

Example 13.7. Before we proceed with the general theory let us look at a few special cases

(a) Suppose that $m = n = 1$ and $U \subset \mathbb{R}$ is an interval. In this case $F : U \to \mathbb{R}$ is a function of one real variable, $F = F(x)$. If F is differentiable at x_0, then the differential $dF(x_0)$ is a linear operator $\mathbb{R}^1 \to \mathbb{R}^1$ and, as such, it is described by a 1×1 matrix, i.e., a real number.

We see that F is differentiable at x_0 if and only if there exists a real number m such that

$$\lim_{h\to 0} \frac{1}{|h|}\Big(F(x_0 + h) - F(x_0) - mh\Big).$$

This happens if and only if F is differentiable at x_0 in the sense of Definition 7.2 and m is the derivative of F at x_0, $m = F'(x_0)$.

(b) Suppose that $m > 1$, $n = 1$ and U is an interval so that $\boldsymbol{F} : U \to \mathbb{R}^m$ is a vector valued function depending on a single real variable $x \in U \subset \mathbb{R}$

$$\boldsymbol{F}(x) = \begin{bmatrix} F^1(x) \\ \vdots \\ F^m(x) \end{bmatrix}.$$

If $\boldsymbol{F}$ is differentiable at $x_0 \in U$, then the differential of $d\boldsymbol{F}(x_0)$ is described by an $m \times 1$ matrix, i.e., a matrix consists of one column of height m. We have

$$\boldsymbol{F}(x + h) - \boldsymbol{F}(x) = \begin{bmatrix} F^1(x_0 + h) - F^1(x_0) \\ \vdots \\ F^m(x_0 + h) - F^m(x_0) \end{bmatrix}.$$

We deduce that $\boldsymbol{F}$ is differentiable at x_0 if and only if the functions $F^1, \dots, F^m$ are differentiable at x_0 and

$$d\boldsymbol{F}(x_0) = \begin{bmatrix} \frac{dF^1}{dx}(x_0) \\ \vdots \\ \frac{dF^m}{dx}(x_0) \end{bmatrix}.$$

(c) If $L : \mathbb{R}^n \to \mathbb{R}^m$ is a linear map, then L is Fréchet differentiable at any $\boldsymbol{x}_0 \in \mathbb{R}^n$. Moreover, the differential at $\boldsymbol{x}_0$ is the operator L itself. □

Deciding when a function or a map is Fréchet differentiable at a point $\boldsymbol{x}_0$ takes a bit of work. We will describe in the following sections some simple ways of recognizing Fréchet differentiable maps.

13.2. Partial derivatives and Fréchet differentials

Suppose that $U \subset \mathbb{R}^n$ is an open set and $\boldsymbol{F} : U \to \mathbb{R}^m$. The limits in the right-hand side of (13.3) play a very important role in differential calculus and for this reason they were given a special name.

Definition 13.8. Let $\boldsymbol{x}_0 \in U$ and $\boldsymbol{v} \in \mathbb{R}^n \setminus \{\boldsymbol{0}\}$. We say that $\boldsymbol{F}$ is *differentiable along the vector $\boldsymbol{v}$ at $\boldsymbol{x}_0$* if the limit

$$\boxed{\partial_{\boldsymbol{v}} \boldsymbol{F}(\boldsymbol{x}_0) = \frac{\partial \boldsymbol{F}(\boldsymbol{x}_0)}{\partial \boldsymbol{v}} := \lim_{t\to 0} \frac{1}{t}\Big(\boldsymbol{F}(\boldsymbol{x}_0 + t\boldsymbol{v}) - \boldsymbol{F}(\boldsymbol{x}_0)\Big)} \tag{13.9}$$

exists. This limit is called the *derivative of $\boldsymbol{F}$ along the vector $\boldsymbol{v}$ at the point $\boldsymbol{x}_0$*.

If $\boldsymbol{e}_1, \dots, \boldsymbol{e}_n$ is the natural basis of $\mathbb{R}^n$, then the derivatives of $\boldsymbol{F}$ along $\boldsymbol{e}_1, \dots, \boldsymbol{e}_n$ (when they exist) are called the *first order partial derivatives* of $\boldsymbol{F}$ at $\boldsymbol{x}_0$ and are denoted by

$$\boxed{\partial_{x^1} \boldsymbol{F}(\boldsymbol{x}_0) = \frac{\partial \boldsymbol{F}(\boldsymbol{x}_0)}{\partial x^1} := \frac{\partial \boldsymbol{F}(\boldsymbol{x}_0)}{\partial \boldsymbol{e}_1}, \dots, \partial_{x^n} \boldsymbol{F}(\boldsymbol{x}_0) = \frac{\partial F(\boldsymbol{x}_0)}{\partial x^n} := \frac{\partial F(\boldsymbol{x}_0)}{\partial \boldsymbol{e}_n}}.$$

We will refer to $\partial_{x^i} \boldsymbol{F}$ as the *partial derivative of the map $\boldsymbol{F}$ with respect to the variable x^i*. Often we will use the alternate notation

$$\boxed{\boldsymbol{F}'_{x^i} := \frac{\partial \boldsymbol{F}}{\partial x^i}}.$$

□

Remark 13.9. Suppose that $F : U \to \mathbb{R}$ is a real valued map depending on n real variables, $F = F(x^1, \dots, x^n)$. You should think of F as measuring some physical quantity at the point $\boldsymbol{x}$ such as temperature or pressure.

In this case the partial derivatives of F at $\boldsymbol{x}_0$ are real numbers. They can be computed as follows. Assume that $\boldsymbol{x}_0 = (x_0^1, \dots, x_0^n)$. Then, for any $t \in \mathbb{R}$ sufficiently small we have

$$\boldsymbol{x}_0 + t\boldsymbol{e}_k = \big(x_0^1, \dots, x_0^{k-1}, x_0^k + t, x_0^{k+1}, \dots, x_0^n \big)$$

and

$$\frac{F(\boldsymbol{x}_0 + t\boldsymbol{e}_k) - F(\boldsymbol{x}_0)}{t}$$
$$= \frac{F(x_0^1, \dots, x_0^{k-1}, x_0^k + t, x_0^{k+1}, \dots, \dots x_0^n) - F(x_0^1, \dots, x_0^k, \dots, x_0^n)}{t}.$$

Thus, when computing the partial derivative $\frac{\partial F}{\partial x^k}$ *we treat the variables x^i, $i \neq k$, as constants*, we regard F as a function of a single variable x^k and we derivate as such.

Equivalently, consider the function $g_k(t) = F(\boldsymbol{x}_0 + t\boldsymbol{e}_k)$, $|t|$ sufficiently small. Then

$$F'_{x^k}(\boldsymbol{x}_0) = g'_k(0).$$

In other words, if we think of F as measuring say the temperature at a point $\boldsymbol{x}$, then $F'_{x^k}(\boldsymbol{x}_0)$ is the rate of change in the temperature as we travel through the point $\boldsymbol{x}_0$, at unit speed, in the direction of the k-th axis of $\mathbb{R}^n$.

More generally, for any vector $\boldsymbol{v} \neq \boldsymbol{0}$, the image of the path $\gamma : \mathbb{R} \to \mathbb{R}^n$, $\gamma(t) = \boldsymbol{x}_0 + t\boldsymbol{v}$, is the line $\ell_{\boldsymbol{x}_0,\boldsymbol{v}}$ through $\boldsymbol{x}_0$ in the direction $\boldsymbol{v}$. Think of γ as describing the motion of a particle in $\mathbb{R}^n$ traveling with constant velocity $\boldsymbol{v}$. Next, think of a map $\boldsymbol{F} : U \to \mathbb{R}^m$ as associating m different physical quantities (e.g., pressure, temperature, external forces, etc.) to each point in U. These quantities can be measured by various sensors attached to the moving particle.

The derivative $\partial_{\boldsymbol{v}}\boldsymbol{F}(\boldsymbol{x}_0)$ measures the "infinitesimal rate of change" in the quantities aggregated in $\boldsymbol{F}$ as the moving particle travels through $\boldsymbol{x}_0$. As an object $\partial_{\boldsymbol{v}}\boldsymbol{F}(\boldsymbol{x}_0)$ is an m-dimensional vector. □

Proposition 13.10. *If $\boldsymbol{F} : U \to \mathbb{R}^m$ is Fréchet differentiable at $\boldsymbol{x}_0$, then $\boldsymbol{F}$ is differentiable along any direction $\boldsymbol{v}$ and*

$$\boxed{\partial_{\boldsymbol{v}}\boldsymbol{F}(\boldsymbol{x}_0) = d\boldsymbol{F}(\boldsymbol{x}_0)\boldsymbol{v}}. \tag{13.10}$$

In particular,

$$\boldsymbol{F}'_{x^j}(\boldsymbol{x}_0) = \frac{\partial \boldsymbol{F}}{\partial x^j}(\boldsymbol{x}_0) = \partial_{\boldsymbol{e}_j}\boldsymbol{F}(\boldsymbol{x}_0) = d\boldsymbol{F}(\boldsymbol{x}_0)\boldsymbol{e}_j, \quad \forall j = 1, \ldots, n,$$

and, if $\boldsymbol{v} = [v^1, \ldots, v^n]^\top$, then

$$\boxed{\partial_{\boldsymbol{v}}\boldsymbol{F}(\boldsymbol{x}_0) = v^1\frac{\partial \boldsymbol{F}(\boldsymbol{x}_0)}{\partial x^1} + \cdots + v^n\frac{\partial \boldsymbol{F}(\boldsymbol{x}_0)}{\partial x^n}}. \tag{13.11}$$

Proof. The equality (13.10) is in fact the equality (13.3) in disguise. To prove (13.11) observe first that the equality $\boldsymbol{v} = [v^1, \ldots, v^n]^\top$ signifies that

$$\boldsymbol{v} = v^1\boldsymbol{e}_1 + \cdots + v^n\boldsymbol{e}_n.$$

From (13.10) and the linearity of $d\boldsymbol{F}(\boldsymbol{x}_0)$ we deduce

$$\begin{aligned}\partial_{\boldsymbol{v}}\boldsymbol{F}(\boldsymbol{x}_0) &= d\boldsymbol{F}(\boldsymbol{x}_0)(v^1\boldsymbol{e}_1 + \cdots + v^n\boldsymbol{e}_n) = v^1 d\boldsymbol{F}(\boldsymbol{x}_0)\boldsymbol{e}_1 + \cdots + v^n d\boldsymbol{F}(\boldsymbol{x}_0)\boldsymbol{e}_n \\ &= v^1\frac{\partial \boldsymbol{F}(\boldsymbol{x}_0)}{\partial x^1} + \cdots + v^n\frac{\partial \boldsymbol{F}(\boldsymbol{x}_0)}{\partial x^n}.\end{aligned}$$

□

Remark 13.11. If $F : U \to \mathbb{R}$ is differentiable, then its differential is represented by a $1 \times n$ matrix, i.e., a matrix consisting of a single row of length n. Its entries are the real numbers

$$dF(\boldsymbol{x}_0)\boldsymbol{e}_1 = \frac{\partial F}{\partial x^1}(\boldsymbol{x}_0), \ldots, dF(\boldsymbol{x}_0)\boldsymbol{e}_n = \frac{\partial F}{\partial x^n}(\boldsymbol{x}_0).$$

In other words, the differential $dF(\boldsymbol{x}_0)$ is described by the row vector

$$\boxed{dF(\boldsymbol{x}_0) = \left[\frac{\partial F}{\partial x^1}(\boldsymbol{x}_0), \dots, \frac{\partial F}{\partial x^n}(\boldsymbol{x}_0)\right].} \tag{13.12}$$

Viewed as a linear form $\mathbb{R}^n \to \mathbb{R}$, the differential $dF(\boldsymbol{x}_0)$ admits the alternate description

$$dF(\boldsymbol{x}_0) = \frac{\partial F}{\partial x^1}(\boldsymbol{x}_0)\boldsymbol{e}^1 + \cdots + \frac{\partial F}{\partial x^n}(\boldsymbol{x}_0)\boldsymbol{e}^n = \sum_{j=1}^n \frac{\partial F}{\partial x^j}(\boldsymbol{x}_0)\boldsymbol{e}^j, \tag{13.13}$$

where we recall that $\boldsymbol{e}^j$ denotes the linear functional $\mathbb{R}^n \to \mathbb{R}$ given by

$$\boldsymbol{e}^j(\boldsymbol{x}) = x^j.$$

In terms of row vectors we have

$$\boldsymbol{e}^1 = [1, 0, 0, \dots 0], \;\; \boldsymbol{e}^2 = [0, 1, 0, \dots, 0], \dots . \qquad \square$$

Example 13.12. For example, if $n = 3$,

$$x := x^1, \;\; y := x^2, \;\; z := x^3, \;\; \boldsymbol{x}_0 = (x_0, y_0, z_0)$$

and

$$F : \mathbb{R}^3 \to \mathbb{R}, \;\; F(x, y, z) = e^{3x+4y+5z},$$

then

$$\frac{\partial F}{\partial x}(\boldsymbol{x}_0) = 3e^{3x_0+4y_0+5z_0}, \;\; \frac{\partial F}{\partial y}(\boldsymbol{x}_0) = 4e^{3x_0+4y_0+5z_0}, \;\; \frac{\partial F}{\partial z}(\boldsymbol{x}_0) = 5e^{3x_0+4y_0+5z_0}.$$

If $\boldsymbol{x}_0 = \boldsymbol{0} = (0, 0, 0)$, then the differential $dF(\boldsymbol{0})$, *if it exists,*[2] *must* be the *single row matrix*

$$dF(\boldsymbol{0}) = [3, 4, 5] = 3\boldsymbol{e}^1 + 4\boldsymbol{e}^2 + 5\boldsymbol{e}^3.$$

In particular, for any vector $\boldsymbol{v} = (v^1, v^3, v^3) \in \mathbb{R}^3 \setminus \{\boldsymbol{0}\}$, we have

$$\partial_{\boldsymbol{v}} f(\boldsymbol{0}) \overset{(13.11)}{=} 3v^1 + 4v^2 + 5v^3. \qquad \square$$

We saw that the differentiability of a map at a point $\boldsymbol{x}_0$ guarantees the existence of derivatives at $\boldsymbol{x}_0$ in any direction. We want to investigate the extent to which a converse is true. To do this we first need to clarify a bit the concept of differentiability.

Proposition 13.13. *Let $m, n \in \mathbb{N}$ and suppose that $U \subset \mathbb{R}^n$ is an open set. Consider a map*

$$\boldsymbol{F} : U \to \mathbb{R}^m, \;\; \boldsymbol{F}(\boldsymbol{x}) = \begin{bmatrix} F^1(\boldsymbol{x}) \\ \vdots \\ F^m(\boldsymbol{x}) \end{bmatrix}, \;\; \boldsymbol{x} \in U.$$

Then the following statements are equivalent.

[2] We will see a bit later in Example 13.18 that the differential does indeed exist.

(i) The map $\boldsymbol{F}$ is Fréchet differentiable at $\boldsymbol{x}_0 \in U$.
(ii) Each of the scalar valued functions $F^1, \dots, F^m : U \to \mathbb{R}$ is Fréchet differentiable at $\boldsymbol{x}_0$.

Proof. (i) $\Rightarrow$ (ii) Suppose that $\boldsymbol{F}$ is differentiable at $\boldsymbol{x}_0$. We denote by L its differential. We identify L with an $m \times n$ matrix. For $i = 1, \dots, m$ we denote by L^i the i-th *row* of L and we view L^i as a linear map $L^i : \mathbb{R}^n \to \mathbb{R}$. We will show that L^i is the differential of F^i at $\boldsymbol{x}_0$. For $\boldsymbol{h} \in \mathbb{R}^n$ sufficiently small we have

$$\frac{1}{\|\boldsymbol{h}\|}\Big(\boldsymbol{F}(\boldsymbol{x}_0+\boldsymbol{h})-\boldsymbol{F}(\boldsymbol{x}_0)-L\boldsymbol{h}\Big) = \frac{1}{\|\boldsymbol{h}\|}\begin{bmatrix} F^1(\boldsymbol{x}_0+\boldsymbol{h})-F^1(\boldsymbol{x}_0)-L^1\boldsymbol{h} \\ \vdots \\ F^m(\boldsymbol{x}_0+\boldsymbol{h})-F^m(\boldsymbol{x}_0)-L^m\boldsymbol{h} \end{bmatrix}. \tag{13.14}$$

We deduce

$$\begin{gathered} \lim_{\boldsymbol{h}\to 0}\frac{1}{\|\boldsymbol{h}\|}\Big(\boldsymbol{F}(\boldsymbol{x}_0+\boldsymbol{h})-\boldsymbol{F}(\boldsymbol{x}_0)-L\boldsymbol{h}\Big) = \boldsymbol{0} \\ \iff \lim_{\boldsymbol{h}\to 0}\frac{1}{\|\boldsymbol{h}\|}\Big(F^i(\boldsymbol{x}_0+\boldsymbol{h})-F^i(\boldsymbol{x}_0)-L^i\boldsymbol{h}\Big) = \boldsymbol{0},\ \ \forall i = 1,\dots,m. \end{gathered} \tag{13.15}$$

The top line of this equivalence states the differentiability of $\boldsymbol{F}$ at $\boldsymbol{x}_0$ and the bottom line of this equivalence amounts to the differentiability at $\boldsymbol{x}_0$ of each of the components $F^1, \dots, F^m$.

(ii) $\Rightarrow$ (i) Suppose that each of the functions F^i is differentiable at $\boldsymbol{x}_0$. We denote by L^i the differential of F^i at $\boldsymbol{x}_0$. This is a linear map $\mathbb{R}^n \to \mathbb{R}$ which we identify with a row of length n. Denote by L the $m \times n$ matrix with i-th row is L^i, $\forall i = 1, \dots, m$.

The matrix L satisfies the equality (13.14) and the equivalence (13.15) holds as well. This proves (i). □

Example 13.14. Suppose that $m, n \in \mathbb{N}$ and $U \subset \mathbb{R}^n$ is an open set. If the map is differentiable at $\boldsymbol{x}_0 \in U$, then the differential $d\boldsymbol{F}(\boldsymbol{x}_0)$ is represented by the $m \times n$ Jacobian matrix $J_{\boldsymbol{F}}(\boldsymbol{x}_0)$ with columns

$$d\boldsymbol{F}(\boldsymbol{x}_0)e_1 = \frac{\partial \boldsymbol{F}(\boldsymbol{x}_0)}{\partial x^1}, \dots, d\boldsymbol{F}(\boldsymbol{x}_0)e_n = \frac{\partial \boldsymbol{F}(\boldsymbol{x}_0)}{\partial x^n}.$$

Let $F^1, \dots, F^m$ be the components of $\boldsymbol{F}$, so that

$$\boldsymbol{F}(\boldsymbol{x}) = \begin{bmatrix} F^1(\boldsymbol{x}) \\ \vdots \\ F^m(\boldsymbol{x}) \end{bmatrix}.$$

Each component F^j, viewed as a function $F^j : U \to \mathbb{R}$ is differentiable at $\boldsymbol{x}_0$. For any $j = 1, \dots, n$ we have

$$\frac{\partial \boldsymbol{F}(\boldsymbol{x}_0)}{\partial x^j} = \lim_{t\to 0}\frac{1}{t}\big(\,\boldsymbol{F}(\boldsymbol{x}_0+te_j)-\boldsymbol{F}(\boldsymbol{x}_0)\big)$$

$$= \lim_{t\to 0} \frac{1}{t} \begin{bmatrix} F^1(\boldsymbol{x}_0 + t\boldsymbol{e}_j) - F^1(\boldsymbol{x}_0) \\ \vdots \\ F^m(\boldsymbol{x}_0 + t\boldsymbol{e}_j) - F^m(\boldsymbol{x}_0) \end{bmatrix} = \begin{bmatrix} \frac{\partial F^1(\boldsymbol{x}_0)}{\partial x^j} \\ \vdots \\ \frac{\partial F^m(\boldsymbol{x}_0)}{\partial x^j} \end{bmatrix}.$$

Hence, the Jacobian matrix of $\boldsymbol{F}$ at $\boldsymbol{x}_0$ is

$$\frac{\partial \boldsymbol{F}}{\partial \boldsymbol{x}}(\boldsymbol{x}_0) = J_{\boldsymbol{F}}(\boldsymbol{x}_0) = \begin{bmatrix} \frac{\partial F^1(\boldsymbol{x}_0)}{\partial x^1} & \cdots & \frac{\partial F^1(\boldsymbol{x}_0)}{\partial x^n} \\ \vdots & \vdots & \vdots \\ \frac{\partial F^m(\boldsymbol{x}_0)}{\partial x^1} & \cdots & \frac{\partial F^m(\boldsymbol{x}_0)}{\partial x^n} \end{bmatrix}.$$

Using the equality (13.12) we see that the first row of $J_{\boldsymbol{F}}(\boldsymbol{x}_0)$ is the differential of F^1 at $\boldsymbol{x}_0$, the second row of $J_{\boldsymbol{F}}(\boldsymbol{x}_0)$ is the differential of F^2 at $\boldsymbol{x}_0$, etc. Thus we can describe the Jacobian $J_{\boldsymbol{F}}$ in the simplified form

$$J_{\boldsymbol{F}} = \begin{bmatrix} dF^1 \\ dF^2 \\ \vdots \\ dF^m \end{bmatrix}.$$ □

Proposition 13.10 shows that *the maps $\boldsymbol{F} : U \to \mathbb{R}^m$ that are differentiable at a point $\boldsymbol{x}_0$ admit partial derivatives at $\boldsymbol{x}_0$.* This is a rather special property. However, the existence of partial derivatives at $\boldsymbol{x}_0$ is not enough to guarantee the Fréchet differentiability at $\boldsymbol{x}_0$. The next result describes one very simple and useful condition guaranteeing Fréchet differentiability.

Theorem 13.15. *Let $m, n \in \mathbb{N}$ and $U \subset \mathbb{R}^n$ be an open set. Suppose $F : U \to \mathbb{R}^m$ is a map and $\boldsymbol{x}_0$ is a point in U satisfying the following conditions.*

(i) There exists $r > 0$ such that $B_r(\boldsymbol{x}_0) \subset U$ and the map F admits partial derivatives at any point $\boldsymbol{x} \in B_r(\boldsymbol{x}_0)$.
(ii) For any $j = 1, \ldots, n$

$$\frac{\partial F(\boldsymbol{x}_0)}{\partial x^j} = \lim_{\boldsymbol{x}\to\boldsymbol{x}_0} \frac{\partial F(\boldsymbol{x})}{\partial x^j}.$$

Then the map F is Fréchet differentiable at $\boldsymbol{x}_0$.

Proof. According to Proposition 13.13 it suffices to consider only the case $m = 1$, i.e., F is a real valued function, $F : U \to \mathbb{R}$. Denote by L the linear map

$$L : \mathbb{R}^n \to \mathbb{R}, \quad L\boldsymbol{h} = \sum_{j=1}^{n} \frac{\partial F(\boldsymbol{x}_0)}{\partial x^j} h^j.$$

We want to prove that L is the Fréchet differential of F at $\boldsymbol{x}_0$, i.e.,

$$\lim_{\boldsymbol{h}\to \boldsymbol{0}} \frac{1}{\|\boldsymbol{h}\|} \left| F(\boldsymbol{x}_0 + \boldsymbol{h}) - F(\boldsymbol{x}_0) - L\boldsymbol{h} \right| = 0. \tag{13.16}$$

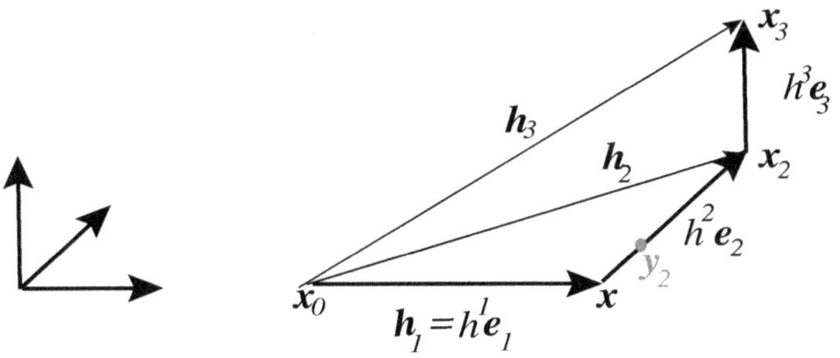

Fig. 13.3 *Zig-zagging from* $\boldsymbol{x}_0$ *to* $\boldsymbol{x}_n = \boldsymbol{x} + \boldsymbol{h}$, $n = 3$.

Given $\boldsymbol{h} = h^1\boldsymbol{e}_1 + \cdots + h^n e^n$, $\|\boldsymbol{h}\| < \frac{r}{2}$, we set (see Figure 13.3)

$$\boldsymbol{h}_1 := h^1\boldsymbol{e}_1,\ \ \boldsymbol{h}_2 = h^1\boldsymbol{e}_1 + h^2\boldsymbol{e}_2,\ \ \boldsymbol{h}_j := h^1\boldsymbol{e}_1 + \cdots + h^j\boldsymbol{e}_j, \dots, j = 1, \dots, n,$$

$$\boldsymbol{x}_j = \boldsymbol{x}_0 + \boldsymbol{h}_j,\ \ j = 1, \dots, n.$$

We have

$$F(\boldsymbol{x}_0 + \boldsymbol{h}) - F(\boldsymbol{x}_0) = F(\boldsymbol{x}_n) - F(\boldsymbol{x}_{n-1}) + F(\boldsymbol{x}_{n-1}) - F(\boldsymbol{x}_{n-2}) + \cdots + F(\boldsymbol{x}_1) - F(\boldsymbol{x}_0).$$

For each $j = 1, \dots, n$ define[3]

$$g_j : (-r/2, r/2) \to \mathbb{R},\ \ g_j(t) = F(\boldsymbol{x}_{j-1} + t\boldsymbol{e}_j).$$

Note that,

$$\boldsymbol{x}_j = \boldsymbol{x}_{j-1} + h^j\boldsymbol{e}_j,\ \ F(\boldsymbol{x}_{j-1}) = g_j(0),\ \ F(\boldsymbol{x}_j) = g_j(h^j).$$

Since F admits partial derivatives at every $\boldsymbol{x} \in B_r(\boldsymbol{x}_0)$ we deduce that the function g_j is differentiable and

$$g_j'(t) = \frac{\partial F(\boldsymbol{x}_{j-1} + t\boldsymbol{e}_j)}{\partial x^j}. \tag{13.17}$$

The Lagrange Mean Value Theorem implies that there exists τ_j in the interval $[0, h^j]$ such that

$$F(\boldsymbol{x}_j) - F(\boldsymbol{x}_{j-1}) = g_j(h^j) - g_j(0) = g_j'(\tau_j)h^j.$$

We set $\boldsymbol{y}_j = \boldsymbol{y}_j(\boldsymbol{h}) = \boldsymbol{x}_{j-1} + \tau_j\boldsymbol{e}_k$. Note that $\boldsymbol{y}_j$ is situated on the line segment connecting $\boldsymbol{x}_{j-1}$ to $\boldsymbol{x}_j$. From (13.17) we deduce

$$F(\boldsymbol{x}_j) - F(\boldsymbol{x}_{j-1}) = \frac{\partial F(\boldsymbol{y}_j)}{\partial x^k}h^j.$$

Let us observe that

$$\|\boldsymbol{h}_j\| \leq \|\boldsymbol{h}\|,\ \ \forall j = 1, \dots, n$$

proving that

$$\operatorname{dist}(\boldsymbol{x}_j, \boldsymbol{x}_0) \leq \|\boldsymbol{h}\|,\ \ \forall j = 1, \dots, n.$$

Thus all the points $\boldsymbol{x}_0, \boldsymbol{x}_1, \dots, \boldsymbol{x}_n = \boldsymbol{x}_0 + \boldsymbol{h}$ lie in $\overline{B_{\|\boldsymbol{h}\|}(\boldsymbol{x}_0)}$, the closed Euclidean ball of center $\boldsymbol{x}_0$ and radius $\|\boldsymbol{h}\|$. This is a convex subset, and since $\boldsymbol{y}_j$ is situated on the line segment $[\boldsymbol{x}_{j-1}, \boldsymbol{x}_j]$, it is also contained $\overline{B_{\|\boldsymbol{h}\|}(\boldsymbol{x}_0)}$. Hence

$$\lim_{\boldsymbol{h}\to\boldsymbol{0}} y_j(\boldsymbol{h}) = \boldsymbol{x}_0,\ \ \forall j = 1, \dots, n. \tag{13.18}$$

We can now put together all the facts above. We have

$$F(\boldsymbol{x}_0 + \boldsymbol{h}) - F(\boldsymbol{x}_0) = \sum_{j=1}^{n} \frac{\partial F(\boldsymbol{y}_j)}{\partial x^k}h^j$$

[3]Observe that $\boldsymbol{x}_{j-1} + t\boldsymbol{e}_j \in U, \forall |t| < r/2$.

$$F(\boldsymbol{x}_0+\boldsymbol{h})-F(\boldsymbol{x}_0)-L\boldsymbol{h}=\sum_{j=1}^{n}\left(\frac{\partial F(\boldsymbol{y}_j)}{\partial x^k}-\frac{\partial F(\boldsymbol{x}_0)}{\partial x^j}\right)h^j,$$

so that

$$\left|F(\boldsymbol{x}_0+\boldsymbol{h})-F(\boldsymbol{x}_0)-L\boldsymbol{h}\right|\leq\sum_{j=1}^{n}\left|\frac{\partial F(\boldsymbol{y}_j)}{\partial x^k}-\frac{\partial F(\boldsymbol{x}_0)}{\partial x^j}\right|\cdot|h^j|$$

(use the Cauchy–Schwarz inequality)

$$\leq\sqrt{\left|\frac{\partial F(\boldsymbol{y}_j)}{\partial x^k}-\frac{\partial F(\boldsymbol{x}_0)}{\partial x^j}\right|^2}\cdot\|\boldsymbol{h}\|.$$

Hence

$$\frac{1}{\|\boldsymbol{h}\|}\left|F(\boldsymbol{x}_0+\boldsymbol{h})-F(\boldsymbol{x}_0)-L\boldsymbol{h}\right|\leq\sqrt{\left|\frac{\partial F(\boldsymbol{y}_j)}{\partial x^k}-\frac{\partial F(\boldsymbol{x}_0)}{\partial x^j}\right|^2}.$$

If we let $\boldsymbol{h}\to 0$, and take (13.18) into account, we obtain the desired conclusion, (13.14). □

Definition 13.16. Let $m,n\in\mathbb{N}$, $U\subset\mathbb{R}^n$ an open set, and $\boldsymbol{F}:U\to\mathbb{R}^m$ a map.

(i) We say that the map $\boldsymbol{F}:U\to\mathbb{R}^m$ is *Fréchet differentiable on* U if it is Fréchet differentiable at every point $\boldsymbol{x}\in U$.

(ii) We say that $\boldsymbol{F}$ is *continuously differentiable*, or C^1, on U, if it admits first order partial derivatives at any $\boldsymbol{x}\in U$ and, for any $j=1,\ldots,n$, the function

$$U\ni\boldsymbol{x}\mapsto\frac{\partial\boldsymbol{F}(\boldsymbol{x})}{\partial x^j}\in\mathbb{R}^m$$

is continuous.

□

✍ *We will denote by $C^1(U,\mathbb{R}^m)$ the set of C^1-maps $\boldsymbol{F}:U\to\mathbb{R}^m$. For simplicity we will write $C^1(U)$ instead of $C^1(U,\mathbb{R})$.*

From Theorem 13.15 we obtain the following very useful result.

Corollary 13.17. *Let $m,n\in\mathbb{N}$ and $U\subset\mathbb{R}^n$. If the map $\boldsymbol{F}:U\to\mathbb{R}^m$ is C^1 on U, then it is Fréchet differentiable on U. Moreover, if $\boldsymbol{e}_1,\ldots,\boldsymbol{e}_n$ is the canonical basis of $\mathbb{R}^n$, then*

$$d\boldsymbol{F}(\boldsymbol{x})\boldsymbol{e}_j=\frac{\partial\boldsymbol{F}(\boldsymbol{x})}{\partial x^j},\quad\forall\boldsymbol{x}\in U,\quad\forall j=1,\ldots,n.$$ □

Example 13.18. (a) Consider a linear functional

$$\boldsymbol{\xi}:\mathbb{R}^n\to\mathbb{R},\ \ \boldsymbol{\xi}(\boldsymbol{x})=\sum_{j=1}^{n}\xi_j x^j.$$

We deduce that, for any $\boldsymbol{x}\in\mathbb{R}^n$, and any $j=1,\ldots,n$,

$$\frac{\partial\xi(\boldsymbol{x})}{\partial x^j}=\xi_j.$$

Thus the functions

$$\mathbb{R}^n\ni\boldsymbol{x}\mapsto\frac{\partial\xi(\boldsymbol{x})}{\partial x^j}\in\mathbb{R}$$

are constant and, in particular, continuous. Corollary 13.17 implies that the linear function $\boldsymbol{\xi}$ is differentiable on $\mathbb{R}^n$, and its differential at a point $\boldsymbol{x}$ is represented by the row vector

$$[\xi_1, \dots, \xi_n].$$

This is the same row vector that represents $\boldsymbol{\xi}$. Thus we have the equality of linear functions

$$d\boldsymbol{\xi}(\boldsymbol{x}) = \boldsymbol{\xi}, \ \ \forall \boldsymbol{x} \in \mathbb{R}^n. \tag{13.19}$$

At this point it is worth mentioning a classical convention that we will use frequently in the sequel.

Note that for any $j = 1, \dots, n$, the linear functional $\boldsymbol{e}^j$ associates to the vector $\boldsymbol{x}$ its j-th coordinate x^j. We can rephrase this by saying that $\boldsymbol{e}^j$ is the function x^j, i.e., $\boldsymbol{e}^j(\boldsymbol{x}) = x^j$. We write this in the less precise fashion $\boldsymbol{e}^j = x^j$. The equality (13.19) applied to the linear functional $\boldsymbol{e}^j$ yields the classical convention

$$\boxed{dx^j = d\boldsymbol{e}^j = \boldsymbol{e}^j}. \tag{13.20}$$

If now $f : \mathbb{R}^n \to \mathbb{R}$ is a C^1-function, then it is differentiable everywhere and, according to (13.13), its differential at $\boldsymbol{x} \in \mathbb{R}^n$ is the linear functional

$$df(\boldsymbol{x}) = \sum_{j=1}^n \frac{\partial f(\boldsymbol{x})}{\partial x^j}\boldsymbol{e}^j = \frac{\partial f(\boldsymbol{x})}{\partial x^1}\boldsymbol{e}^1 + \cdots + \frac{\partial f(\boldsymbol{x})}{\partial x^n}\boldsymbol{e}^n.$$

Using the convention (13.20) we obtain another frequently used convention/notation

$$\boxed{df = \sum_{j=1}^n \frac{\partial f}{\partial x^j}dx^j = \frac{\partial f}{\partial x^1}dx^1 + \cdots + \frac{\partial f}{\partial x^n}dx^n.} \tag{13.21}$$

The right-hand side of the above equality is classically referred to as the *total differential* of the function f. Moreover, in the above equality we interpret both sides as *functions* on $\mathbb{R}^n$ with values in the space $\operatorname{Hom}(\mathbb{R}^n, \mathbb{R})$ of linear functionals on $\mathbb{R}^n$.

For example, if $n = 1$, so f is a function of a single real variable, then the above equality takes the known form (7.6)

$$df = \frac{df}{dx}dx = f'(x)dx.$$

(b) Consider the function $r : \mathbb{R}^2 \setminus \{\boldsymbol{0}\} \to \mathbb{R}$, $r(x,y) = \sqrt{x^2+y^2}$. For fixed y the function $x \mapsto \sqrt{x^2+y^2}$ is differentiable as long as $x^2 + y^2 \neq 0$. Its derivative is

$$\frac{\partial r}{\partial x} = \frac{x}{\sqrt{x^2+y^2}} = \frac{x}{r}.$$

A similar argument shows that $\frac{\partial r}{\partial y}$ exists as long as $x^2 + y^2 \neq 0$ and we have

$$\frac{\partial r}{\partial y} = \frac{y}{\sqrt{x^2+y^2}} = \frac{y}{r}.$$

Thus the function r is differentiable at every point in $\mathbb{R}^2 \setminus \{\boldsymbol{0}\}$ and

$$dr = \frac{x}{\sqrt{x^2+y^2}}dx + \frac{y}{\sqrt{x^2+y^2}}dy.$$

The associated Jacobian matrix is the single row matrix

$$J_r = \begin{bmatrix} \dfrac{x}{\sqrt{x^2+y^2}} & \dfrac{y}{\sqrt{x^2+y^2}} \end{bmatrix}.$$

The differential of r at the point $(x_0, y_0) = (3, 4)$ is therefore represented by the row vector

$$\left[\frac{3}{5}, \frac{4}{5}\right].$$

(c) Consider again the function

$$F : \mathbb{R}^3 \to \mathbb{R}, \quad F(x,y,z) = e^{3x+4y+5z}$$

we discussed in Remark 13.11(b). The function F admits partial derivatives

$$\frac{\partial F}{\partial x} = 3e^{3x+4y+5z}, \quad \frac{\partial F}{\partial y} = 4e^{3x+4y+5z}, \quad \frac{\partial F}{\partial z} = 5e^{3x+4y+5z}$$

which are continuous functions. Thus $F \in C^1(\mathbb{R}^3)$ and, in particular, it is Fréchet differentiable on $\mathbb{R}^3$. Moreover

$$dF = 3e^{3x+4y+5z}\,dx + 4e^{3x+4y+5z}\,dy + 5e^{3x+4y+5z}\,dz.$$

Again, we interpret both sides of the above equality as functions $\mathbb{R}^3 \to \operatorname{Hom}(\mathbb{R}^3, \mathbb{R})$.

(d) Consider the map $\boldsymbol{F} : \mathbb{R}^2 \to \mathbb{R}^2$ defined by

$$\mathbb{R}^2 \ni \begin{bmatrix} r \\ \theta \end{bmatrix} \overset{\boldsymbol{F}}{\mapsto} \begin{bmatrix} x \\ y \end{bmatrix} = \begin{bmatrix} r\cos\theta \\ r\sin\theta \end{bmatrix} \in \mathbb{R}^2.$$

You should read the above as follows: the components of the map $\boldsymbol{F}$ are two functions called x and y depending on two variables (r, θ) and

$$x(r,\theta) = r\cos\theta, \quad y(r,\theta) = r\sin\theta.$$

Clearly the functions x, y are C^1 on their domains and the Jacobian matrix of $\boldsymbol{F}$ is

$$J_{\boldsymbol{F}} = \begin{bmatrix} \frac{\partial x}{\partial r} & \frac{\partial x}{\partial \theta} \\ \frac{\partial y}{\partial r} & \frac{\partial y}{\partial \theta} \end{bmatrix} = \begin{bmatrix} \cos\theta & -r\sin\theta \\ \sin\theta & r\cos\theta \end{bmatrix}.$$

In particular for $(r, \theta) = (1, \pi/2)$ we have

$$J_{\boldsymbol{F}}(1, \pi/2) = \begin{bmatrix} 0 & -1 \\ 1 & 0 \end{bmatrix}.$$

If $\boldsymbol{v} = [3, 4]^\top$, then

$$\partial_{\boldsymbol{v}} \boldsymbol{F}(1, \pi/2) = \begin{bmatrix} 0 & -1 \\ 1 & 0 \end{bmatrix} \cdot \begin{bmatrix} 3 \\ 4 \end{bmatrix} = \begin{bmatrix} -4 \\ 3 \end{bmatrix}. \qquad \square$$

Example 13.19 (Linearizations). Suppose that $n \in \mathbb{N}$, $U \subset \mathbb{R}^n$ is an open set and $f : U \to \mathbb{R}$ is a C^1-function. Then, according to Definition 13.5, the linearization (or linear approximation) of f at $\boldsymbol{x}_0$ is the affine function

$$\mathcal{L} : \mathbb{R}^n \to \mathbb{R}, \;\; \mathcal{L}(\boldsymbol{x}) = f(\boldsymbol{x}_0) + df(\boldsymbol{x}_0)(\boldsymbol{x} - \boldsymbol{x}_0).$$

To see how this looks concretely, consider the function $f : \mathbb{R}^2 \to \mathbb{R}$, $f(x, y) = x^2 + y^2$. We have

$$\partial_x f = 2x, \;\; \partial_y f = 2y.$$

Let us find the linear approximation of this function at the point $(x_0, y_0) = (2, 1)$. We have

$$f(x_0, y_0) = 2^2 + 1^2 = 5, \;\; \partial_x f(x_0, y_0) = 4, \;\; \partial_y f(x_0, y_0) = 2.$$

The differential $df(x_0, y_0)$ is thus described by the row vector $[4, 2]$. The linearization of f at $(2, 1)$ is the affine function

$$\begin{aligned}\mathcal{L}(x, y) &= f(x_0, y_0) + \partial_x f(x_0, y_0)(x - x_0) + \partial_y f(x_0, y_0)(y - y_0) \\ &= 5 + 4(x - 2) + 2(y - 1) = 4x + 2y - 5.\end{aligned}$$

The surface Figure 13.1 is the graph of f, while the plane in the same figure is the graph of $\mathcal{L}$. □

Definition 13.20 (Gradient). Let $n \in \mathbb{N}$. Suppose that $U \subset \mathbb{R}^n$ is an open set and $f : U \to \mathbb{R}$ is a function differentiable at $\boldsymbol{x}_0$. The *gradient* of f at $\boldsymbol{x}_0$ is the *vector* $df(\boldsymbol{x}_0)_\uparrow$ dual to the differential of f at $\boldsymbol{x}_0$. We denote by $\nabla f(\boldsymbol{x}_0)$ the gradient of f at $\boldsymbol{x}_0$. The symbol ∇ is pronounced *nabla*.[4] □

The above definition is rather dense. Let us unpack it. The differential $df(\boldsymbol{x}_0)$ of f at $\boldsymbol{x}_0$ is a linear form $\mathbb{R}^n \to \mathbb{R}$ (or covector) represented by the single row matrix

$$\left[\frac{\partial f(\boldsymbol{x}_0)}{\partial x^1}, \dots, \frac{\partial f(\boldsymbol{x}_0)}{\partial x^n}\right].$$

As explained in (11.20a) the dual of this covector is the (column) vector

$$\begin{bmatrix}\frac{\partial f(\boldsymbol{x}_0)}{\partial x^1} \\ \vdots \\ \frac{\partial f(\boldsymbol{x}_0)}{\partial x^n}\end{bmatrix} =: \nabla f(\boldsymbol{x}_0).$$

Using (13.11) we deduce that, for any $\boldsymbol{v} \in \mathbb{R}^n \setminus \{\boldsymbol{0}\}$ we have

$$\partial_{\boldsymbol{v}} f(\boldsymbol{x}_0) = \partial_{x^1} f(\boldsymbol{x}_0) v^1 + \cdots + \partial_{x^n} f(\boldsymbol{x}_0) v^n,$$

i.e.,

$$\boxed{\partial_{\boldsymbol{v}} f(\boldsymbol{x}_0) = df(\boldsymbol{x}_0)\boldsymbol{v} = \langle \nabla f(\boldsymbol{x}_0), \boldsymbol{v} \rangle, \;\; \forall \boldsymbol{v} \in \mathbb{R}^n}. \tag{13.22}$$

[4]The name nabla originates from an ancient stringed musical instrument shaped as a harp.

The construction of the gradient might appear to the uninitiated as "much ado about nothing" because all we have done was to take a row and then transform it into a column. Temporarily it is difficult to justify this algebraic contortion. For now, please take it as an article of faith that there is a method to this "madness."

Example 13.21. Consider the function $f : \mathbb{R}^2 \to \mathbb{R}$, $f(x, y) = x^2 + y^2$. Then

$$df(x, y) = 2xdx + 2ydy, \;\; \nabla f(x, y) = \begin{bmatrix} 2x \\ 2y \end{bmatrix}.$$

The correspondence $\mathbb{R}^2 \ni (x, y) \mapsto \nabla f(x, y) \in \mathbb{R}^2$ is often viewed as a vector field on $\mathbb{R}^2$ in that it assigns an "arrow" (or vector) to each point of $\mathbb{R}^2$. □

Example 13.22. We define a *direction* in $\mathbb{R}^n$ to be a unit length vector $\boldsymbol{n}$, $\|\boldsymbol{n}\| = 1$. Observe that any nonzero vector $\boldsymbol{v} \in \mathbb{R}^n$ determines a direction

$$\boldsymbol{n} = \boldsymbol{n}(\boldsymbol{v}) = \frac{1}{\|\boldsymbol{v}\|}\boldsymbol{v}.$$

A point $\boldsymbol{x}_0 \in \mathbb{R}^n$ and a direction $\boldsymbol{n}$ canonically determine a path

$$\gamma_{\boldsymbol{x}_0,\boldsymbol{n}} : \mathbb{R} \to \mathbb{R}^n, \;\; \gamma_{\boldsymbol{x}_0,\boldsymbol{n}}(t) = \boldsymbol{x}_0 + t\boldsymbol{n}$$

whose image is the line through $\boldsymbol{x}_0$ in the direction $\boldsymbol{n}$.

Given an open set $U \subset \mathbb{R}^n$, a C^1-function $f : U \to \mathbb{R}$, a point $\boldsymbol{x}_0$ and a direction $\boldsymbol{n}$, we define the derivative of f in the direction $\boldsymbol{n}$ at $\boldsymbol{x}_0$ to be the derivative of f along the vector $\boldsymbol{n}$. From (13.22) we deduce that

$$\partial_{\boldsymbol{n}} f(\boldsymbol{x}_0) = \langle \nabla f(\boldsymbol{x}_0), \boldsymbol{n} \rangle.$$

Suppose that $\nabla f(\boldsymbol{x}_0) \neq 0$ and let $\theta \in [0, \pi]$ be the angle between the vectors $\nabla f(\boldsymbol{x}_0)$ and $\boldsymbol{n}$. We have

$$\partial_{\boldsymbol{n}} f(\boldsymbol{x}_0) = \langle \nabla f(\boldsymbol{x}_0), \boldsymbol{n} \rangle = \|\nabla f(\boldsymbol{x}_0)\| \cos\theta \leq \|\nabla f(\boldsymbol{x}_0)\|.$$

Above, we have equality if and only if $\theta = 0$. Thus $\partial_{\boldsymbol{n}} f(\boldsymbol{x}_0)$ takes its highest possible value if and only if $\boldsymbol{n}$ points in the same direction as $\nabla f(\boldsymbol{x}_0)$ or, equivalently, $\boldsymbol{n}$ is the direction determined by the gradient vector $\nabla f(\boldsymbol{x}_0)$. This shows that *the direction determined by the gradient of a function at a point is the direction of fastest growth* of the function at that given point. □

13.3. The chain rule

We can now state and prove a key result in several variable calculus.

Theorem 13.23 (Chain rule). *Let $\ell, m, n \in \mathbb{N}$. Suppose that we are given open sets $U \subset \mathbb{R}^n$ and $V \subset \mathbb{R}^m$, maps $\boldsymbol{F} : U \to \mathbb{R}^m$, $\boldsymbol{G} : V \to \mathbb{R}^\ell$, and a point $\boldsymbol{u}_0 \in U$ satisfying the following conditions.*

(i) $\boldsymbol{F}(U) \subset V$.

(ii) $\boldsymbol{F}$ *is differentiable at* $\boldsymbol{u}_0$ *and* $\boldsymbol{G}$ *is differentiable at* $\boldsymbol{v}_0 := \boldsymbol{F}(\boldsymbol{u}_0)$.

Then the composition $\boldsymbol{G} \circ \boldsymbol{F} : U \to \mathbb{R}^\ell$ *is differentiable at* $\boldsymbol{u}_0$ *and*

$$d(\boldsymbol{G} \circ \boldsymbol{F})(\boldsymbol{u}_0) = d\boldsymbol{G}(\boldsymbol{v}_0) \circ d\boldsymbol{F}(\boldsymbol{u}_0). \tag{13.23}$$

Idea of proof. Set $A := d\boldsymbol{F}(\boldsymbol{u}_0)$, $B := d\boldsymbol{G}(\boldsymbol{v}_0)$ so that $A \in \operatorname{Hom}(\mathbb{R}^n, \mathbb{R}^m)$, $B \in \operatorname{Hom}(\mathbb{R}^m, \mathbb{R}^\ell)$. Form the definition of Fréchet differential we deduce

$$\boldsymbol{G}\big(\boldsymbol{F}(\boldsymbol{u})\big) - \boldsymbol{G}\big(\boldsymbol{F}(\boldsymbol{u}_0)\big) \approx B\big(\boldsymbol{F}(\boldsymbol{u}) - \boldsymbol{F}(\boldsymbol{u}_0)\big),$$
$$\boldsymbol{F}(\boldsymbol{u}) - \boldsymbol{F}(\boldsymbol{u}_0) \approx A(\boldsymbol{u} - \boldsymbol{u}_0).$$

Hence

$$\boldsymbol{G}\big(\boldsymbol{F}(\boldsymbol{u})\big) - \boldsymbol{G}\big(\boldsymbol{F}(\boldsymbol{u}_0)\big) \approx B \circ A\big(\boldsymbol{u} - \boldsymbol{u}_0\big).$$

This shows that $B \circ A$ is the Fréchet differential of $\boldsymbol{G} \circ \boldsymbol{F}$ at $\boldsymbol{u}_0$. □

The above argument is an almost complete proof capturing the essence of the main idea. We present the missing details below.

Proof. We set $A := d\boldsymbol{F}(\boldsymbol{u}_0)$ and $B := d\boldsymbol{G}(\boldsymbol{v}_0)$. We have to prove that

$$\lim_{\boldsymbol{h} \to \boldsymbol{0}} \frac{1}{\|\boldsymbol{h}\|} \big\| \boldsymbol{G}\big(\boldsymbol{F}(\boldsymbol{u}_0 + \boldsymbol{h})\big) - \boldsymbol{G}(\boldsymbol{v}_0) - B(A\boldsymbol{h}) \big\| = 0. \tag{13.24}$$

We set

$$T\boldsymbol{h} := \boldsymbol{F}(\boldsymbol{u}_0 + \boldsymbol{h}) - \boldsymbol{F}(\boldsymbol{u}_0) = \boldsymbol{F}(\boldsymbol{u}_0 + \boldsymbol{h}) - \boldsymbol{v}_0$$

and we deduce

$$\begin{aligned} \boldsymbol{G}\big(\boldsymbol{F}(\boldsymbol{u}_0 + \boldsymbol{h})\big) - \boldsymbol{G}(\boldsymbol{v}_0) - B(A\boldsymbol{h}) &= \boldsymbol{G}(\boldsymbol{v}_0 + T\boldsymbol{h}) - G(\boldsymbol{v}_0) - B(A\boldsymbol{h}) \\ &= \boldsymbol{G}(\boldsymbol{v}_0 + T\boldsymbol{h}) - G(\boldsymbol{v}_0) - B(T\boldsymbol{h}) + B(T\boldsymbol{h} - A\boldsymbol{h}). \end{aligned}$$

Set

$$R_F(\boldsymbol{h}) := \boldsymbol{F}(\boldsymbol{u}_0 + \boldsymbol{h}) - \boldsymbol{F}(\boldsymbol{u}_0) - A\boldsymbol{h}, \quad R_G(\boldsymbol{k}) := \boldsymbol{G}(\boldsymbol{v}_0 + \boldsymbol{k}) - \boldsymbol{G}(\boldsymbol{v}_0) - B\boldsymbol{k}.$$

Since $\boldsymbol{F}$ is differentiable at $\boldsymbol{u}_0$ and $\boldsymbol{G}$ is differentiable at $\boldsymbol{v}_0$ we deduce from (13.5) that there exist $r > 0$ and functions $\varphi_F, \varphi_G : [0, r) \to \mathbb{R}$ such that

$$0 = \varphi_F(0) = \lim_{t \searrow 0} \varphi_F(t), \quad 0 = \varphi_G(0) = \lim_{t \searrow 0} \varphi_G(t), \tag{13.25a}$$

$$\|R_F(\boldsymbol{h})\| \le \varphi_F(\|\boldsymbol{h}\|)\|\boldsymbol{h}\|, \quad \|R_G(\boldsymbol{k})\| \le \varphi_G(\|\boldsymbol{k}\|)\|\boldsymbol{k}\|, \quad \forall \|\boldsymbol{h}\|, \ \|\boldsymbol{k}\| < r. \tag{13.25b}$$

Note that

$$T\boldsymbol{h} - A\boldsymbol{h} = \boldsymbol{F}(\boldsymbol{u}_0 + \boldsymbol{h}) - \boldsymbol{F}(\boldsymbol{u}_0) - A\boldsymbol{h} = R_F(\boldsymbol{h}),$$
$$\boldsymbol{G}(\boldsymbol{v}_0 + T\boldsymbol{h}) - G(\boldsymbol{v}_0) - B(T\boldsymbol{h}) = R_G(T\boldsymbol{h}),$$

and

$$\begin{aligned} \boldsymbol{G}\big(\boldsymbol{F}(\boldsymbol{u}_0 + \boldsymbol{h})\big) - \boldsymbol{G}(\boldsymbol{v}_0) - B(A\boldsymbol{h}) &= R_G(T\boldsymbol{h}) + B\big(R_F(\boldsymbol{h})\big) \\ &= R_G\big(A\boldsymbol{h} + R_F(\boldsymbol{h})\big) + B\big(R_F(\boldsymbol{h})\big). \end{aligned}$$

Hence

$$\begin{aligned} \big\| \boldsymbol{G}\big(\boldsymbol{F}(\boldsymbol{u}_0 + \boldsymbol{h})\big) - \boldsymbol{G}(\boldsymbol{v}_0) - B(A\boldsymbol{h}) \big\| &\le \big\| R_G\big(A\boldsymbol{h} + R_F(\boldsymbol{h})\big) \big\| + \|B\big(R_F(\boldsymbol{h})\big)\| \\ &\le \big\| R_G\big(A\boldsymbol{h} + R_F(\boldsymbol{h})\big) \big\| + \|B\|_{HS} \cdot \big\| R_F(\boldsymbol{h})\big) \big\| \\ &\overset{(13.25b)}{\le} \varphi_G\big(A\boldsymbol{h} + R_F(\boldsymbol{h})\big)\|A\boldsymbol{h} + R_F(\boldsymbol{h})\| + \varphi_F(\|\boldsymbol{h}\|)\|B\|_{HS} \cdot \|\boldsymbol{h}\| \\ &\overset{(13.25b)}{\le} \varphi_G\big(A\boldsymbol{h} + R_F(\boldsymbol{h})\big)\Big(\|A\|_{HS} + \varphi_F(\|\boldsymbol{h}\|)\Big)\|\boldsymbol{h}\| + \varphi_F(\|\boldsymbol{h}\|)\|B\|_{HS} \cdot \|\boldsymbol{h}\|, \end{aligned}$$

and thus

$$\begin{aligned} &\frac{1}{\|\boldsymbol{h}\|} \big\| \boldsymbol{G}(\boldsymbol{F}(\boldsymbol{u}_0 + \boldsymbol{h})) - \boldsymbol{G}(\boldsymbol{v}_0) - B(A\boldsymbol{h}) \big\| \\ &\le \varphi_G\big(A\boldsymbol{h} + R_F(\boldsymbol{h})\big)\Big(\|A\|_{HS} + \varphi_F(\|\boldsymbol{h}\|)\Big) + \varphi_F(\|\boldsymbol{h}\|)\|B\|_{HS}. \end{aligned}$$

The conclusion (13.24) is obtained by letting $\boldsymbol{h} \to \boldsymbol{0}$ in the above inequality and invoking (13.25a). □

Let us rewrite the chain rule (13.23) in a less precise, but more intuitive manner.

We denote by $(u^i)_{1\leq i\leq n}$ the Euclidean coordinates on $\mathbb{R}^n$, by $(v^j)_{1\leq j\leq m}$ the Euclidean coordinates on $\mathbb{R}^m$ and by $(x^k)_{1\leq k\leq \ell}$ the Euclidean coordinates in $\mathbb{R}^\ell$. The map $\boldsymbol{F}$ is described by m functions depending on the variables (u^i)

$$v^j = F^j(u^1,\dots,u^n),\;\; j=1,\dots,m,$$

while the map $\boldsymbol{G}$ is described by ℓ functions depending on the variables (v^j)

$$x^k = G^k(v^1,\dots,v^m),\;\; k=1,\dots,\ell.$$

The differential of $\boldsymbol{F}$ at $\boldsymbol{u}_0$ is described by the $m\times n$ Jacobian matrix $J_{\boldsymbol{F}}$ with entries

$$(J_{\boldsymbol{F}})^j_i = \frac{\partial F^j}{\partial u^i} = \frac{\partial v^j}{\partial u^i}.$$

The differential of $\boldsymbol{G}$ at $\boldsymbol{v}_0 = \boldsymbol{F}(\boldsymbol{u}_0)$ is described by the $m\times n$ Jacobian matrix $J_{\boldsymbol{G}}$ with entries

$$(J_{\boldsymbol{G}})^k_j = \frac{\partial G^k}{\partial v^j} = \frac{\partial x^k}{\partial v^j}.$$

The composition $\boldsymbol{G}\circ\boldsymbol{F}$ is described by ℓ functions depending on the variables (u^i)

$$x^k = G^k\big(F^1(u^1,\dots,u^n),\dots,F^m(u^1,\dots,u^n)\big),\;\; k=1,\dots,\ell.$$

The differential of $\boldsymbol{G}\circ\boldsymbol{F}$ at $\boldsymbol{u}_0$ is described by the $\ell\times n$ matrix $J_{\boldsymbol{G}\circ\boldsymbol{F}}$ with entries

$$(J_{\boldsymbol{G}\circ\boldsymbol{F}})^k_i = \frac{\partial x^k}{\partial u^i}.$$

The chain rule (13.23) states that

$$J_{\boldsymbol{G}\circ\boldsymbol{F}}(\boldsymbol{u}_0) = J_{\boldsymbol{G}}(\boldsymbol{v}_0)J_{\boldsymbol{F}}(\boldsymbol{u}_0) = J_{\boldsymbol{G}}\big(\boldsymbol{F}(\boldsymbol{u}_0)\big)J_{\boldsymbol{F}}(\boldsymbol{u}_0), \tag{13.26}$$

or, equivalently,

$$\boxed{\frac{\partial x^k}{\partial u^i} = \sum_{j=1}^m \frac{\partial x^k}{\partial v^j}\cdot\frac{\partial v^j}{\partial u^i} = \frac{\partial x^k}{\partial v^1}\cdot\frac{\partial v^1}{\partial u^i} + \cdots + \frac{\partial x^k}{\partial v^m}\cdot\frac{\partial v^m}{\partial u^i}}. \tag{13.27}$$

Example 13.24. Consider the function

$$f:\mathbb{R}^2\to\mathbb{R},\;\; f(x,y) = (x^2+y^2+1)^{\sin(xy)}.$$

This is the composition of two C^1-maps

$$(x,y)\mapsto (u,v) = \big(1+x^2+y^2,\, \sin(xy)\big),\;\; (u,v)\mapsto f = u^v.$$

Then

$$\begin{aligned}\frac{\partial f}{\partial x} &= \frac{\partial f}{\partial u}\cdot\frac{\partial u}{\partial x} + \frac{\partial f}{\partial v}\cdot\frac{\partial v}{\partial x}\\ &= vu^{v-1}\cdot(2x) + u^v\ln u\cdot y\cos(xy) = u^v\cdot\left(\frac{v}{u}\cdot(2x) + (\ln u)\cdot y\cos(xy)\right)\\ &= (x^2+y^2+1)^{\sin(xy)}\left(\frac{2x\sin(xy)}{x^2+y^2+1} + y\cos(xy)\ln(x^2+y^2+1)\right).\end{aligned}$$

□

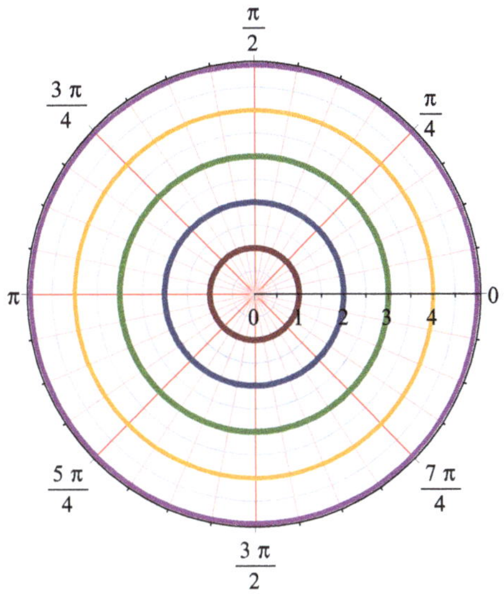

Fig. 13.4 *Polar grid.*

Example 13.25. Suppose that $f : \mathbb{R}^2 \to \mathbb{R}$ is a differentiable function depending on two variables $f = f(x, y)$. Suppose additionally that x, y are themselves functions of two variables

$$x = x(r, \theta) = r\cos\theta, \quad y = y(r, \theta) = r\sin\theta. \tag{13.28}$$

We want to compute the partial derivatives $\frac{\partial f}{\partial r}$ and $\frac{\partial f}{\partial \theta}$. First, let us give a geometric interpretation to the functions (13.28).

If we fix r, say $r = 4$, then we get a path

$$\theta \mapsto \big(\, 4\cos\theta, 4\sin\theta \,\big) \in \mathbb{R}^2.$$

This describes the motion of a point in the plane with constant angular velocity along the circle of radius 4 centered at the origin; see the thick orange circle in Figure 13.4. If we keep θ fixed, $\theta = \theta_0$, then the resulting path

$$r \mapsto \big(\, r\cos\theta_0, r\sin\theta_0 \,\big)$$

describes the motion with speed 1 along a ray emanating at the origin that makes angle θ_0 with the x-axis.

We get two families of curves in the plane: the family of curves obtained by fixing r (circles centered at the origin) and the family of curves obtained by fixing θ (rays). These two families form a *curvilinear grid* in the plane (see Figure 13.4) known as the *polar grid.*

The function f depends on the variables x, y, which themselves depend on the quantities r, θ. $\frac{\partial f}{\partial r}$ measures how fast is f changing when we travel at unit speed along a ray, while $\frac{\partial f}{\partial \theta}$ measures how fast is f changing when we travel along a circle at constant angular velocity $1\text{rad}/\text{sec}$. The chain rule shows that

$$\frac{\partial f}{\partial r} = \frac{\partial f}{\partial x}\frac{\partial x}{\partial r} + \frac{\partial f}{\partial y}\frac{\partial y}{\partial r} = \frac{\partial f}{\partial x}\cos\theta + \frac{\partial f}{\partial y}\sin\theta \tag{13.29a}$$

$$\frac{\partial f}{\partial \theta} = -\frac{\partial f}{\partial x} r \sin\theta + \frac{\partial f}{\partial y} r \cos\theta. \tag{13.29b}$$

Suppose for example that $f(x,y) = x^2 + y^2$. Note that if x, y depend on r, θ as in (13.28), then $x^2 + y^2 = r^2$ and thus

$$\frac{\partial f}{\partial r} = 2r.$$

On the other hand (13.29a) implies that

$$\frac{\partial f}{\partial r} = 2x\cos\theta + 2y\sin\theta \stackrel{(13.28)}{=} 2r\cos^2\theta + 2r\sin^2\theta = 2r. \qquad \square$$

Remark 13.26 (The naturality of the differential). We want to describe a remarkable "accident" which is extremely important in differential geometry and theoretical physics.

Suppose that f is a differentiable function depending on the n variables $x^1, \dots, x^n$. Using the convention (13.21) we have

$$\boxed{df = \frac{\partial f}{\partial x^1} dx^1 + \cdots + \frac{\partial f}{\partial x^n} dx^n}. \tag{13.30}$$

Suppose that the quantities $x^1, \dots, x^n$ themselves depend differentiably on a number of variables

$$x^i = x^i(u^1, \dots, u^m), \;\; i = 1, \dots, n. \tag{13.31}$$

Through this new dependence we can view the quantity f as a function of the variables $u^1, \dots, u^m$ and, as such, we have

$$df = \frac{\partial f}{\partial u^1} du^1 + \cdots + \frac{\partial f}{\partial u^m} du^m. \tag{13.32}$$

! *How do we reconcile (13.30) with (13.32)?*

The chain rule comes to the rescue. To see that (13.30) and (13.32) are compatible (noncontradictory) regard the quantities $dx^1, \dots, dx^n$ as the differentials of the functions in (13.31), i.e.,

$$dx^i = \frac{\partial x^i}{\partial u^1} du^1 + \cdots + \frac{\partial x^i}{\partial u^m} du^m, i = 1, \dots, n.$$

The equality (13.30) becomes

$$\begin{aligned} df &= \frac{\partial f}{\partial x^1}\left(\frac{\partial x^1}{\partial u^1} du^1 + \cdots + \frac{\partial x^1}{\partial u^m} du^m\right) + \cdots + \frac{\partial f}{\partial x^n}\left(\frac{\partial x^n}{\partial u^1} du^1 + \cdots + \frac{\partial x^n}{\partial u^m} du^m\right) \\ &= \underbrace{\left(\frac{\partial f}{\partial x^1}\frac{\partial x^1}{\partial u^1} + \cdots + \frac{\partial f}{\partial x^n}\frac{\partial x^n}{\partial u^1}\right)}_{=:q_1} du^1 + \cdots + \underbrace{\left(\frac{\partial f}{\partial x^1}\frac{\partial x^1}{\partial u^m} + \cdots + \frac{\partial f}{\partial x^n}\frac{\partial x^n}{\partial u^m}\right)}_{=:q_m} du^m. \end{aligned}$$

Hence

$$df = q_1 du^1 + \cdots + q_m du^m. \tag{13.33}$$

The chain rule (13.27) shows that

$$q_1 = \frac{\partial f}{\partial u^1}, \dots, q_m = \frac{\partial f}{\partial u^m}$$

so the equality (13.33) is none other than (13.32) in disguise. $\square$

Let us discuss a few simple but useful applications of the chain rule.

Definition 13.27 (Differentiable paths). Let $n \in \mathbb{N}$. A *differentiable path* in $\mathbb{R}^n$ is a differentiable map $\gamma : I \to \mathbb{R}^n$, where $I \subset \mathbb{R}$ is an interval. □

A differentiable path $\gamma : (a, b) \to \mathbb{R}^n$ is described by n differentiable functions

$$x^i : (a, b) \to \mathbb{R}, \quad i = 1, \dots, n,$$

such that

$$\gamma(t) = \begin{bmatrix} x^1(t) \\ x^2(t) \\ \vdots \\ x^n(t) \end{bmatrix}.$$

The differential of the map γ is an $n \times 1$ matrix, i.e., a matrix consisting of a single column of height n. This matrix is

$$\frac{d}{dt}\gamma(t) = \begin{bmatrix} \frac{dx^1(t)}{dt} \\ \frac{dx^2(t)}{dt} \\ \vdots \\ \frac{dx^n(t)}{dt} \end{bmatrix}.$$

We will adopt a convention frequently used by physicists and will denote by an upper dot "˙" the *time derivatives*. With this convention we can rewrite the above equality as

$$\dot{\gamma}(t) = \begin{bmatrix} \dot{x}^1(t) \\ \dot{x}^2(t) \\ \vdots \\ \dot{x}^n(t) \end{bmatrix}.$$

If we think of γ as describing the motion of a point in $\mathbb{R}^n$, then the vector $\dot{\gamma}(t)$ is the *velocity* of that moving point at the moment of time t.

Proposition 13.28 (Derivatives along paths). *Let $n \in \mathbb{N}$. Assume that $U \subset \mathbb{R}^n$ is an open set, $f : U \to \mathbb{R}$ is a Fréchet differentiable function and $\gamma : (a, b) \to U$ a differentiable path. Then*

$$\boxed{\frac{d}{dt} f\big(\gamma(t)\big) = \big\langle \nabla f\big(\gamma(t)\big), \dot{\gamma}(t) \big\rangle, \ \ \forall t \in (a, b)}, \tag{13.34}$$

where we recall that $\nabla f(\boldsymbol{x})$ denotes the gradient of f at $\boldsymbol{x}$. The quantity $\langle \nabla f(\gamma), \dot{\gamma} \rangle$ is called the derivative of f along the path γ.

Proof. As explained above, the path γ is described by n differentiable functions

$$\gamma(t) = \big(\, x^1(t), \dots, x^n(t)\,\big).$$

We have

$$f\big(\,\gamma(t)\,\big) = f\big(\, x^1(t), \dots, x^n(t)\,\big).$$

Using the chain rule (13.27) we deduce

$$\begin{aligned}\frac{d}{dt} f\big(\,\gamma(t)\,\big) &= \frac{\partial f(\gamma(t))}{\partial x^1}\frac{dx^1(t)}{dt} + \cdots + \frac{\partial f(\gamma(t))}{\partial x^n}\frac{dx^n(t)}{dt} \\ &= \frac{\partial f(\gamma)}{\partial x^1}\dot{x}^1 + \cdots + \frac{\partial f(\gamma)}{\partial x^n}\dot{x}^n = \langle \nabla f(\gamma), \dot{\gamma}\rangle.\end{aligned}$$

□

If we think of the function $f : U \to \mathbb{R}$ as a physical quantity associated to each point in U (say temperature) and of the path γ as describing the motion of a point in U, then the derivative of f along the path is the rate of change of f (per unit of time) during the motion.

Example 13.29 (Euler identity). Suppose that $f : \mathbb{R}^n \to \mathbb{R}$ is positively homogeneous of degree k, i.e.,

$$f(t\boldsymbol{x}) = t^k f(\boldsymbol{x}), \;\; \forall t > 0, \;\; \forall \boldsymbol{x} \in \mathbb{R}^n \setminus \{\boldsymbol{0}\}.$$

If f is differentiable on $\mathbb{R}^n \setminus \{\boldsymbol{0}\}$, then f satisfies *Euler's identity*

$$\big\langle\, \boldsymbol{x}, \nabla f(\boldsymbol{x})\,\big\rangle = kf(\boldsymbol{x}), \;\; \forall \boldsymbol{x} \in \mathbb{R}^n \setminus \{\boldsymbol{0}\}. \tag{13.35}$$

To prove the above identity, fix $\boldsymbol{x} \in \mathbb{R}^n \setminus \{\boldsymbol{0}\}$ and consider the path

$$\gamma_{\boldsymbol{x}} : (0, \infty) \to \mathbb{R}^n, \;\; \gamma_{\boldsymbol{x}}(t) = t\boldsymbol{x}, \;\; \forall t > 0.$$

Observe that

$$f\big(\,\gamma_{\boldsymbol{x}}(t)\,\big) = f(t\boldsymbol{x}) = t^k f(\boldsymbol{x}), \;\; \dot{\gamma}_{\boldsymbol{x}}(t) = \boldsymbol{x}, \;\; \forall t > 0.$$

Thus

$$\frac{d}{dt} f\big(\,\gamma_{\boldsymbol{x}}(t)\,\big) = kt^{k-1} f(\boldsymbol{x}), \;\; \forall t > 0.$$

On the other hand, the derivative of f along $\gamma_{\boldsymbol{x}}(t)$ is given by (13.34)

$$\frac{d}{dt} f\big(\,\gamma_{\boldsymbol{x}}(t)\,\big) = \big\langle \dot{\gamma}_{\boldsymbol{x}}(t), \nabla f\big(\,\gamma_{\boldsymbol{x}}(t)\,\big)\,\big\rangle = \big\langle \boldsymbol{x}, \nabla f(t\boldsymbol{x})\,\big\rangle.$$

We deduce

$$\big\langle \boldsymbol{x}, \nabla f(t\boldsymbol{x})\,\big\rangle = kt^{k-1} f(\boldsymbol{x}), \;\; \forall t > 0.$$

If we set $t = 1$ in the above equality we obtain Euler's identity (13.35). □

Theorem 13.30 (Lagrange Mean Value Theorem). *Suppose that $U \subset \mathbb{R}^n$ is an open set and $f : U \to \mathbb{R}$ is a differentiable function. Then, for any $\boldsymbol{x}_0, \boldsymbol{x}_1 \in U$ such that $[\boldsymbol{x}_0, \boldsymbol{x}_1] \subset U$, there exists a point $\mathbf{p}$ on the line segment $[\boldsymbol{x}_0, \boldsymbol{x}_1]$ such that*

$$f(\boldsymbol{x}_1) - f(\boldsymbol{x}_0) = \langle \nabla f(\boldsymbol{p}), \boldsymbol{x}_1 - \boldsymbol{x}_0\rangle.$$

Proof. Consider the restriction of f to the line segment $[\boldsymbol{x}_0, \boldsymbol{x}_1]$, i.e., the function $g : [0,1] \to \mathbb{R}$

$$g(t) = f\big(\boldsymbol{x}_0 + t(\boldsymbol{x}_1 - \boldsymbol{x}_0)\big).$$

According to the 1-dimensional Lagrange Mean Value Theorem there exists $\tau \in (0,1)$ such that

$$f(\boldsymbol{x}_1) - f(\boldsymbol{x}_0) = g(1) - g(0) = g'(\tau).$$

On the other hand, the derivative of f along the path $t \mapsto \boldsymbol{x}_0 + t(\boldsymbol{x}_1 - \boldsymbol{x}_0)$ is

$$g'(t) = \big\langle\, \nabla f(\boldsymbol{x}_0 + t(\boldsymbol{x}_1 - \boldsymbol{x}_0)\,), \boldsymbol{x}_1 - \boldsymbol{x}_0 \,\big\rangle.$$

This yields the desired conclusion with $\boldsymbol{p} = \boldsymbol{x}_0 + \tau(\boldsymbol{x}_1 - \boldsymbol{x}_0)$. □

Corollary 13.31. *Suppose that $U \subset \mathbb{R}^n$ is an open convex set and $f : U \to \mathbb{R}$ is a differentiable function. If there exists $C > 0$ such that $\|\nabla f(\boldsymbol{x})\| \leq C$, $\forall \boldsymbol{x} \in U$, then*

$$|f(\boldsymbol{x}) - f(\boldsymbol{y})| \leq C\|\boldsymbol{x} - \boldsymbol{y}\|, \;\; \forall \boldsymbol{x}, \boldsymbol{y} \in U. \tag{13.36}$$

Proof. Let $\boldsymbol{x}, \boldsymbol{y} \in U$. The Mean Value Theorem shows that there exists a point $\boldsymbol{p}$ on the line segment $[\boldsymbol{x}, \boldsymbol{y}]$ such that

$$|f(\boldsymbol{x}) - f(\boldsymbol{y})| = \big|\langle \nabla f(\boldsymbol{p}), \boldsymbol{x} - \boldsymbol{y}\rangle\big|.$$

The desired conclusion now follows by invoking the Cauchy–Schwarz inequality

$$\big|\langle \nabla f(\boldsymbol{p}), \boldsymbol{x} - \boldsymbol{y}\rangle\big| \leq \|\nabla f(\boldsymbol{p})\| \cdot \|\boldsymbol{x} - \boldsymbol{y}\| \leq C\|\boldsymbol{x} - \boldsymbol{y}\|.$$

□

Corollary 13.32. *Suppose that $U \subset \mathbb{R}^n$ is an open and path connected set and $f : U \to \mathbb{R}$ is a differentiable function such that $\nabla f(\boldsymbol{x}) = \boldsymbol{0}$, $\forall \boldsymbol{x} \in U$. Then the function f is constant.*

Proof. Fix $\boldsymbol{p}_0 \in U$. Let $\boldsymbol{q}$ be an arbitrary point in U. Since U is path connected, Exercise 12.24 shows that there exist points $\boldsymbol{p}_1, \dots, \boldsymbol{p}_N$ such that $\boldsymbol{q} = \boldsymbol{p}_N$ and the line segments $[\boldsymbol{p}_{i-1}, \boldsymbol{p}_i]$, $i = 1, \dots, N$ are contained in U. Corollary 13.31 then implies

$$f(\boldsymbol{p}_0) = f(\boldsymbol{p}_1) = f(\boldsymbol{p}_2) = \cdots = f(\boldsymbol{p}_{N-1}) = f(\boldsymbol{p}_N) = f(\boldsymbol{q}).$$

We have thus proved that $f(\boldsymbol{q}) = f(\boldsymbol{p}_0)$, $\forall \boldsymbol{q} \in U$, i.e., f is constant. □

Corollary 13.33. *Suppose that $U \subset \mathbb{R}^n$ is an open convex set and $\boldsymbol{F} : U \to \mathbb{R}^m$ is a C^1-map. Suppose that there exists a constant $C > 0$ such that $\|J_{\boldsymbol{F}}(\boldsymbol{x})\|_{HS} < C$, $\forall \boldsymbol{x} \in U$, where $\| - \|_{HS}$ denotes the Frobenius norm of an $m \times n$ matrix; see Remark 12.13. Then*

$$\|\boldsymbol{F}(\boldsymbol{x}) - \boldsymbol{F}(\boldsymbol{y})\| \leq C\sqrt{m}\|\boldsymbol{x} - \boldsymbol{y}\|, \;\; \forall \boldsymbol{x}, \boldsymbol{y} \in U. \tag{13.37}$$

Proof. Denote by $F^1, \dots, F^m$ the components of $\boldsymbol{F}$. Note that

$$\|J_{\boldsymbol{F}}(\boldsymbol{x})\|_{HS}^2 = \sum_{i=1}^m \|\nabla F^i(\boldsymbol{x})\|^2, \quad \forall \boldsymbol{x} \in \mathbb{R}^n.$$

Hence

$$\|\nabla F^i(\boldsymbol{x})\| \leq \|J_{\boldsymbol{F}}(\boldsymbol{x})\|_{HS}, \quad \forall \boldsymbol{x} \in U, \quad i = 1, \dots, m.$$

Then, for any $\boldsymbol{x}, \boldsymbol{y} \in U$ we have

$$\|\boldsymbol{F}(\boldsymbol{x}) - \boldsymbol{F}(\boldsymbol{y})\|^2 = \sum_{i=1}^m \|F^i(\boldsymbol{x}) - F^i(\boldsymbol{y})\|^2$$
$$\stackrel{(13.36)}{\leq} \sum_{i=1}^m C^2\|\boldsymbol{x} - \boldsymbol{y}\|^2 = C^2 m\|\boldsymbol{x} - \boldsymbol{y}\|^2.$$

□

Definition 13.34 (Vector fields). Let $n \in \mathbb{N}$.

(i) A *vector field* on a set $S \subset \mathbb{R}^n$ is a map

$$\boldsymbol{V} : S \to \mathbb{R}^n, \quad S \ni \boldsymbol{x} \mapsto \boldsymbol{V}(\boldsymbol{x}).$$

(ii) An *integral curve* or *flow line* of a vector field $\boldsymbol{V}$ on a set $S \subset \mathbb{R}^n$ is a differentiable path $\gamma : (a, b) \to \mathbb{R}^n$ such that

$$\gamma(t) \in S, \quad \dot{\gamma}(t) = \boldsymbol{V}\big(\gamma(t)\big), \quad \forall t \in (a, b). \tag{13.38}$$

□

Let us emphasize one aspect in the definition of a vector field that you may overlook. The domain of the vector field, i.e., the set S, lives inside the space $\mathbb{R}^n$ and $\boldsymbol{V}$ takes values in <u>*the same*</u> vector space $\mathbb{R}^n$. Intuitively, a vector field $\boldsymbol{V}$ on a set $S \subset \mathbb{R}^n$ associates to each point $\boldsymbol{x} \in S$ a vector $\boldsymbol{V}(\boldsymbol{x}) \in \mathbb{R}^n$ that should be visualized as an arrow $V(\boldsymbol{x})$ originating at $\boldsymbol{x}$. The result is a "hairy" region S, with one "hair" $\boldsymbol{V}(\boldsymbol{x})$ at each location $\boldsymbol{x} \in S$; see Figure 13.5.

An integral curve of the vector field $\boldsymbol{V}$ is then a path $\gamma(t)$ in S whose velocity $\dot{\gamma}(t)$ each point $\gamma(t)$ is equal to $\boldsymbol{V}\big(\gamma(t)\big)$: this is precisely the arrow the vector field associates to $\gamma(t)$. In particular, this arrow is tangent to the path at this point; see the curves in Figure 13.5.

A vector field $\boldsymbol{V}$ on S is determined by n functions on S

$$\boldsymbol{V}(\boldsymbol{x}) = \begin{bmatrix} V^1(\boldsymbol{x}) \\ \vdots \\ V^n(\boldsymbol{x}) \end{bmatrix}, \forall \boldsymbol{x} \in S, \quad V^i : S \to \mathbb{R}, \quad i = 1, \dots, n.$$

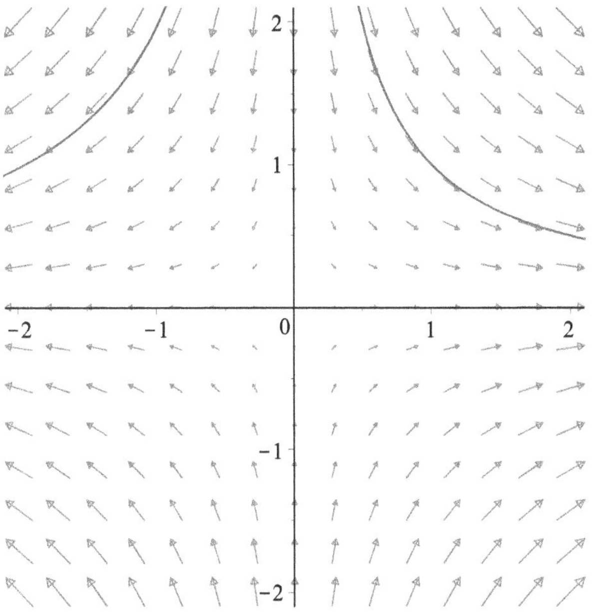

Fig. 13.5 *The vector field $V(x,y) = (2x, -2y)$ on the square $S = [-2,2] \times [-2,2] \subset \mathbb{R}^2$ and two integral curves of this vector field.*

An integral curve $\gamma : (a,b) \to S \subset \mathbb{R}^n$ of $\boldsymbol{V}$ is then given by n functions $x^1(t), \dots, x^n(t)$, $t \in (a,b)$ satisfying the system of differential equations

$$\begin{cases} \dot{x}^1(t) = V^1\big(x^1(t), \dots, x^n(t)\big) \\ \vdots \quad \vdots \qquad \vdots \\ \dot{x}^n(t) = V^n\big(x^1(t), \dots, x^n(t)\big) \end{cases}. \tag{13.39}$$

Example 13.35 (Gradient vector fields). Suppose that $U \subset \mathbb{R}^n$ and $f : U \to \mathbb{R}$ is a smooth function. The gradient of f defines a vector field on U,

$$U \ni \boldsymbol{x} \mapsto \nabla f(\boldsymbol{x}) \in \mathbb{R}^n.$$

This vector field is called the *gradient vector field* of (or associated to) the function f. Such a function f is called a *potential* of the gradient vector field.

The vector field depicted in Figure 13.5 is the gradient vector field of the function $f(x,y) = x^2 - y^2$,

$$\nabla f(x,y) = (2x, -2y).$$

The integral curves of this vector field are differentiable maps

$$\mathbb{R} \ni t \mapsto \begin{bmatrix} x(t) \\ y(t) \end{bmatrix} \in \mathbb{R}^2$$

satisfying the system of differential equations

$$\begin{cases} \dot{x} = 2x \\ \dot{y} = -2y \end{cases}.$$

Arguing as in Example 8.51 we deduce that the solutions of the first equations have the form $x(t) = ae^{2t}$, where a is constant, while the solutions of the second equation are $y(t) = be^{-2t}$, where b is another constant.

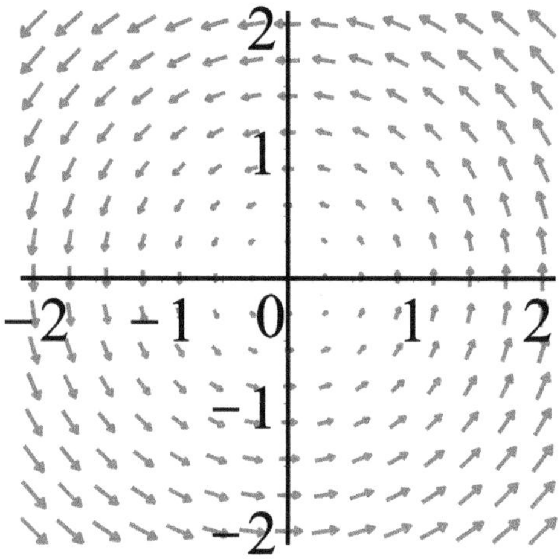

Fig. 13.6 *A non-gradient vector field on the square* $S = [-2,2] \times [-2,2] \subset \mathbb{R}^2$.

The vector field

$$\mathbb{R}^2 \ni (x,y) \mapsto \boldsymbol{V}(x,y) = \begin{bmatrix} -y \\ x \end{bmatrix} \in \mathbb{R}^2 \tag{13.40}$$

depicted in Figure 13.6 *is not* the gradient of any function. Exercise 13.19 asks you to prove this. □

Definition 13.36. Suppose that $\boldsymbol{V}$ is a vector field on the set $S \subset \mathbb{R}^n$. A *prime integral* or *conservation law* of $\boldsymbol{V}$ is a continuous function $f : S \to \mathbb{R}$ that is constant along the flow lines of $\boldsymbol{V}$, i.e., for any integral curve $\gamma : (a,b) \to S$ of $\boldsymbol{V}$, the function

$$(a,b) \ni t \mapsto f(\gamma(t)) \in \mathbb{R}$$

is constant. □

Proposition 13.37. *Suppose that* $\boldsymbol{V}$ *is a vector field on the open set* $U \subset \mathbb{R}^n$ *and* $f : U \to \mathbb{R}$ *is a differentiable function such that*

$$\big\langle \nabla f(\boldsymbol{x}), \boldsymbol{V}(\boldsymbol{x}) \big\rangle = 0, \;\; \forall \boldsymbol{x} \in U.$$

Then f *is a prime integral of* $\boldsymbol{V}$.

Proof. Let $\gamma : (a,b) \to U$ be an integral curve of $\boldsymbol{V}$. Then

$$\dot{\gamma}(t) = \boldsymbol{V}(\gamma(t)),$$

and

$$\frac{d}{dt} f(\gamma(t)) = \langle \nabla f(\gamma(t)), \dot{\gamma}(t) \rangle = \big\langle \nabla f(\gamma(t)), \boldsymbol{V}(\gamma(t)) \big\rangle = 0.$$

□

13.4. Higher order partial derivatives

Let $U \subset \mathbb{R}^n$ be an open set and $f : U \to \mathbb{R}$ be a function such that the partial derivatives $\partial_{x^1} f(\boldsymbol{x}), \dots, \partial_{x^n} f(\boldsymbol{x})$ exist at every point $\boldsymbol{x} \in U$. We obtain n new functions

$$\partial_{x^1} f, \dots, \partial_{x^n} f : U \to \mathbb{R}. \tag{13.41}$$

We say that f admits *second order* partial derivatives on U if each of the functions (13.41) admit partial derivatives on U. We say that f admits *third order* partial derivatives on U if each of the functions (13.41) admits second order partial derivatives on U. Inductively, if $k \in \mathbb{N}$, we say that f admits *partial derivatives of order* k on U if each of the functions (13.41) admits partial derivatives of order $k-1$ on U.

Recall that f is said to be C^1 on U if the functions (13.41) are continuous on U. We say that f is C^2 on U if the functions (13.41) are C^1 on U. We say that f is C^3 on U if the functions (13.41) are C^2 on U. Inductively, if $k \in \mathbb{N}$, we say that f is C^k on U if the functions (13.41) are C^{k-1} on U. We will write $f \in C^k(U)$ to indicate that f is C^k on U. We say that the function f is *smooth* or C^∞ on U, and we denote this $f \in C^\infty(U)$ if

$$f \in C^k(U), \quad \forall k \in \mathbb{N}.$$

Note that $C^k(U)$ stands for the collection of all functions $f : U \to \mathbb{R}$ that are C^k on U. This collection is a vector space.

Suppose that $f : U \to \mathbb{R}$ is a function that admits second order derivatives on U. Thus, each of the first order derivatives $\partial_{x^j} f$, $j = 1, \dots, n$ admits first order derivatives. We denote

$$\partial^2_{x^k x^j} f \quad \text{or} \quad \frac{\partial^2 f}{\partial x^k \partial x^j}$$

the partial derivative of the function $\partial_{x^j} f$ with respect to the variable x^k, i.e.,

$$\boxed{\partial^2_{x^k x^j} f := \partial_{x^k} \big(\partial_{x^j} f \big)}.$$

More generally, if f is a C^k-function, then for any $i_1, \dots, i_k \in \{1, \dots, n\}$ we define inductively

$$\partial^k_{x^{i_k} \cdots x^{i_1}} f := \partial_{x^{i_k}} \big(\partial^{k-1}_{x^{i_{k-1}} \cdots x^{i_1}} f = \partial_{x^{i_k}} \big).$$

We have the following important result.

Theorem 13.38 (Partial derivatives commute). *Let $n \in \mathbb{N}$ and suppose that $U \subset \mathbb{R}^n$ is an open set. Then for any function $f \in C^2(U)$ we have*

$$\partial^2_{x^k x^j} f(\boldsymbol{x}) = \partial^2_{x^j x^k} f(\boldsymbol{x}), \ \ \forall \boldsymbol{x} \in U, \ \ \forall j, k = 1, \dots, n.$$

Proof. The result is obviously true when $j = k$ so it suffices to consider the case $j \neq k$, say $j < k$. Fix a point $\boldsymbol{a} \in U$. We have to prove that

$$\partial^2_{x^k x^j} f(\boldsymbol{a}) = \partial^2_{x^j x^k} f(\boldsymbol{a}). \tag{13.42}$$

We have

$$\partial_{x^j} f(\boldsymbol{x}) = \lim_{s \to 0} \frac{f(\boldsymbol{x} + s\boldsymbol{e}_j) - f(\boldsymbol{x})}{s}, \ \ \forall \boldsymbol{x} \in U \tag{13.43a}$$

$$\partial_{x^k} f(\boldsymbol{x}) = \lim_{t \to 0} \frac{f(\boldsymbol{x} + t\boldsymbol{e}_k) - f(\boldsymbol{x})}{t}, \ \ \forall \boldsymbol{x} \in U \tag{13.43b}$$

$$\partial^2_{x^k x^j} f(\boldsymbol{a}) = \lim_{t \to 0} \frac{\partial_{x^j} f(\boldsymbol{a} + t\boldsymbol{e}_k) - \partial_{x^j} f(\boldsymbol{a})}{t}$$

$$\stackrel{(13.43a)}{=} \lim_{t \to 0} \left(\lim_{s \to 0} \frac{1}{t} \cdot \frac{f(\boldsymbol{a} + t\boldsymbol{e}_k + s\boldsymbol{e}_j) - f(\boldsymbol{a} + t\boldsymbol{e}_k) - f(\boldsymbol{a} + s\boldsymbol{e}_j) + f(\boldsymbol{a})}{s} \right).$$

Similarly

$$\partial^2_{x^j x^k} f(\boldsymbol{a}) = \lim_{s \to 0} \frac{\partial_{x^k} f(\boldsymbol{a} + s\boldsymbol{e}_j) - \partial_{x^k} f(\boldsymbol{a})}{s}$$

$$\stackrel{(13.43a)}{=} \lim_{s \to 0} \left(\lim_{t \to 0} \frac{1}{s} \cdot \frac{f(\boldsymbol{a} + t\boldsymbol{e}_k + s\boldsymbol{e}_j) - f(\boldsymbol{a} + s\boldsymbol{e}_j) - f(\boldsymbol{a} + t\boldsymbol{e}_k) + f(\boldsymbol{a})}{t} \right).$$

If we denote by $Q(s,t)$, $s, t \neq 0$ the quantity

$$Q(s,t) := f(\boldsymbol{a} + t\boldsymbol{e}_k + s\boldsymbol{e}_j) - f(\boldsymbol{a} + t\boldsymbol{e}_k) - f(\boldsymbol{a} + s\boldsymbol{e}_j) + f(\boldsymbol{a}),$$

then we see that (13.42) is equivalent with the equality

$$\lim_{s \to 0} \left(\lim_{t \to 0} \frac{Q(s,t)}{st} \right) = \lim_{t \to 0} \left(\lim_{s \to 0} \frac{Q(s,t)}{st} \right). \tag{13.44}$$

The two sides above are examples of *iterated limits*, and they differ only in the order we take the limits. It suggests that Theorem 13.38 is at least plausible. However, the complete proof of (13.44) is not trivial and requires a bit of sweat.

For simplicity, for $s, t \in \mathbb{R}$ we set (see Figure 13.7)

$$\boldsymbol{a}_{s,t} := \boldsymbol{a} + s\boldsymbol{e}_j + t\boldsymbol{e}_k.$$

Denote by $\mathcal{R}_{s,t}$ the rectangle with vertices $\boldsymbol{a}, \boldsymbol{a}_{s,0}, \boldsymbol{a}_{0,t}, \boldsymbol{a}_{s,t}$. Note also that

$$Q(s,t) = f(\boldsymbol{a}_{s,t}) - f(\boldsymbol{a}_{0,t}) - f(\boldsymbol{a}_{s,0}) + f(\boldsymbol{a}),$$

and that st is the area of this rectangle.

Applying the Lagrange Mean Value Theorem to the function $\lambda \mapsto g_t(\lambda)$ we deduce that, for any s, t small, there exists $\bar{\lambda} = \lambda_{s,t} \in (0, s)$ such that

$$\frac{Q(s,t)}{s} = \frac{g_t(s) - g_t(0)}{s}$$
$$= g_t'(\lambda) = \partial_{x^j} f(\boldsymbol{a} + \lambda\boldsymbol{e}_j + t\boldsymbol{e}_k) - \partial_{x^j} f(\boldsymbol{a} + \lambda\boldsymbol{e}_j) = \partial_{x^j} f(\boldsymbol{a}_{\lambda,t}) - \partial_{x^j} f(\boldsymbol{a}_{\lambda,0}).$$

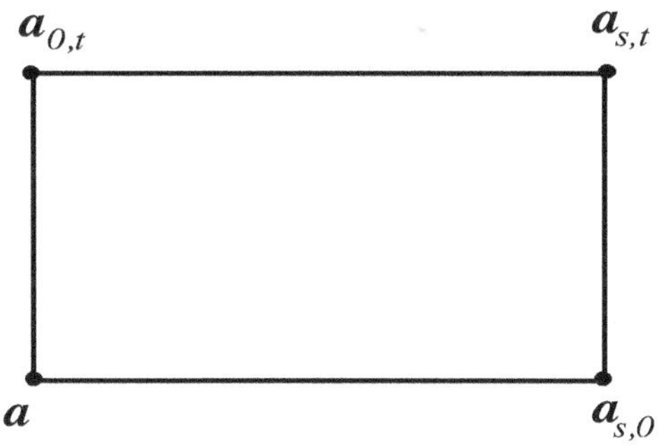

Fig. 13.7 *The rectangle $\mathcal{R}_{s,t}$ at a spanned by the vectors se_j and te_k.*

Applying the Lagrange Mean Value Theorem to the function

$$h(\mu) = \partial_{x^j} f(\boldsymbol{a} + \mu \boldsymbol{e}_k + \lambda_{s,t}\boldsymbol{e}_j)$$

we deduce that there exists $\bar{\mu} = \mu_{s,t}$ in the interval $(0, t)$ such that

$$\begin{aligned}\frac{Q(s,t)}{st} &= \frac{\partial_{x^j} f(\boldsymbol{a} + t\boldsymbol{e}_k + \lambda_{s,t}\boldsymbol{e}_j) - \partial_{x^j} f(\boldsymbol{a} + \lambda_{s,t}\boldsymbol{e}_j)}{t} = \frac{h(t) - h(0)}{t} \\ &= h'(\mu_{s,t}) = \partial^2_{x^k x^j} f(\boldsymbol{a} + \mu_{s,t}\boldsymbol{e}_k + \lambda_{s,t}\boldsymbol{e}_j) = \partial^2_{x^k x^j} f(\boldsymbol{a}_{\bar{\lambda},\bar{\mu}}).\end{aligned}$$

We denote by $\boldsymbol{p}_{s,t}$ the point $\boldsymbol{a} + \mu_{s,t}\boldsymbol{e}_k + \lambda_{s,t}\boldsymbol{e}_j$. Thus

$$\frac{Q(s,t)}{st} = \partial^2_{x^k x^j} f(\boldsymbol{p}_{s,t}). \tag{13.45}$$

Applying the Lagrange Mean Value Theorem to the function $\beta \mapsto u_s(\beta) = Q(s, \beta)$ we deduce that for every s, t sufficiently small there exists $\bar{\beta} = \beta_{s,t}$ in $(0, t)$ such that

$$\frac{Q(s,t)}{t} = \frac{u_s(t) - u_s(0)}{t} = u_s'(\beta) = \partial_{x^k} f(\boldsymbol{a} + s\boldsymbol{e}_j + \bar{\beta}\boldsymbol{e}_k) - \partial_{x^k} f(\boldsymbol{a} + \bar{\beta}\boldsymbol{e}_k).$$

Applying the Lagrange Mean Value Theorem to the function

$$\alpha \mapsto v(\alpha) = \partial_{x^k} f(\boldsymbol{a} + \alpha\boldsymbol{e}_j + \bar{\beta}\boldsymbol{e}_k)'$$

we deduce that there exists $\bar{\alpha} = \alpha_{s,t}$ in the interval $(0, s)$ such that

$$\begin{aligned}\frac{Q(s,t)}{st} &= \frac{u_s(t) - u_s(0)}{st} = \frac{\partial_{x^k} f(\boldsymbol{a} + s\boldsymbol{e}_j + \bar{\beta}\boldsymbol{e}_k) - \partial_{x^k} f(\boldsymbol{a} + \bar{\beta}\boldsymbol{e}_k)}{s} \\ &= \frac{v(s) - v(0)}{s} = v'(\bar{\alpha}) = \partial^2_{x^j x^k} f(\boldsymbol{a} + \bar{\alpha}\boldsymbol{e}_j + \bar{\beta}\boldsymbol{e}_k).\end{aligned}$$

We denote by $\boldsymbol{q}_{s,t}$ the point $\boldsymbol{a} + \alpha_{s,t}\boldsymbol{e}_j + \beta_{s,t}\boldsymbol{e}_k$. Thus

$$\frac{Q(s,t)}{st} = \partial^2_{x^j x^k} f(\boldsymbol{q}_{s,t}). \tag{13.46}$$

In particular, we deduce

$$\partial^2_{x^k x^j} f(\boldsymbol{p}_{s,t}) = \frac{Q(s,t)}{st} = \partial^2_{x^j x^k} f(\boldsymbol{q}_{s,t}). \tag{13.47}$$

Note that since $\alpha, \lambda \in (0, s)$, $\beta, \mu \in (0, t)$ we have

$$\operatorname{dist}(\boldsymbol{a}, \boldsymbol{p}_{s,t}) = \sqrt{\bar{\lambda}^2 + \bar{\mu}^2} \leq \sqrt{s^2 + t^2}, \tag{13.48a}$$

$$\operatorname{dist}(\boldsymbol{a}, \boldsymbol{q}_{s,t}) = \sqrt{\bar{\alpha}^2 + \bar{\beta}^2} \leq \sqrt{s^2 + t^2}. \tag{13.48b}$$

Fix $r > 0$ sufficiently small such that the closed ball $\overline{B_r(\boldsymbol{a})}$ is contained in U. Since the functions

$$\partial^2_{x^k x^j} f, \quad \partial^2_{x^j x^k} f : U \to \mathbb{R}$$

are continuous, they are continuous at $\boldsymbol{a}$. Hence, for any $\varepsilon > 0$, there exists $\delta = \delta(\varepsilon) > 0$ such that

$$\forall \boldsymbol{x} \in U,\ \ \operatorname{dist}(\boldsymbol{a}, \boldsymbol{x}) < \delta(\varepsilon) \Rightarrow \big|\,\partial^2_{x^k x^j} f(\boldsymbol{a}) - \partial^2_{x^k x^j}(\boldsymbol{x})\,\big| < \frac{\varepsilon}{2}, \quad \big|\,\partial^2_{x^j x^k} f(\boldsymbol{a}) - \partial^2_{x^j x^k}(\boldsymbol{x})\,\big| < \frac{\varepsilon}{2}. \tag{13.49}$$

Fix $\varepsilon > 0$. Choose $s, t > 0$ small enough such that $\sqrt{s^2 + t^2} < \delta(\varepsilon)$ and $\mathcal{R}_{s,t} \subset B_r(\boldsymbol{a})$. The points $\boldsymbol{p}_{s,t}$ and $\boldsymbol{q}_{s,t}$ belong to the rectangle $\mathcal{R}_{s,t}$ and thus, also to U. We deduce

$$\big|\,\partial^2_{x^k x^j} f(\boldsymbol{a}) - \partial^2 f_{x^j x^k}(\boldsymbol{a})\,\big|$$

$$\leq \big|\,\partial^2_{x^k x^j} f(\boldsymbol{a}) - \partial^2_{x^k x^j} f(\boldsymbol{p}_{s,t})\,\big| + \boxed{\big|\,\partial^2_{x^k x^j} f(\boldsymbol{p}_{s,t}) - \partial^2_{x^j x^k} f(\boldsymbol{q}_{st})\,\big|} + \big|\,\partial^2_{x^j x^k} f(\boldsymbol{q}_{st}) - \partial^2 f_{x^j x^k}(\boldsymbol{a})\,\big|$$

$$\stackrel{(13.47)}{=} \big|\,\partial^2_{x^k x^j} f(\boldsymbol{a}) - \partial^2_{x^k x^j} f(\boldsymbol{p}_{s,t})\,\big| + \big|\,\partial^2_{x^j x^k} f(\boldsymbol{q}_{st}) - \partial^2 f_{x^j x^k}(\boldsymbol{a})\,\big|$$

(use (13.48a), (13.48b), (13.49))

$$< \frac{\varepsilon}{2} + \frac{\varepsilon}{2} = \varepsilon.$$

This shows that

$$\big|\,\partial^2_{x^k x^j} f(\boldsymbol{a}) - \partial^2 f_{x^j x^k}(\boldsymbol{a})\,\big| < \varepsilon, \ \ \forall \varepsilon > 0.$$

This proves (13.42). □

Example 13.39 (Gradient vector fields again). Suppose that $U \subset \mathbb{R}^n$ is an open set and $\boldsymbol{V} : U \to \mathbb{R}^n$ is a C^1-vector field

$$\boldsymbol{V}(x^1, \dots, x^n) = \begin{bmatrix} V^1(x^1, \dots, x^n) \\ \vdots \\ V^n(x^1, \dots, x^n) \end{bmatrix}.$$

Let us show that

$$\boxed{\boldsymbol{V} \text{ is a gradient vector field } \Rightarrow \partial_{x^j} V^i = \partial_{x^i} V^j, \ \ \forall \boldsymbol{x} \in U, \ \ i \neq j}. \tag{13.50}$$

Indeed, if $\boldsymbol{V}$ is the gradient of some function $f : U \to \mathbb{R}$, then

$$V^i = \partial_{x^i} f, \ \ \forall i.$$

In particular this shows that the function f is C^2 since its partial derivatives are C^1. We deduce

$$\partial_{x^j} V^i = \partial_{x^j}(\partial_{x^i} f) = \partial_{x^i}(\partial_{x^j} f) = \partial_{x^i} V^j.$$

For example, if $\boldsymbol{V}(x, y)$ is a *gradient* vector field on $\mathbb{R}^2$,

$$\boldsymbol{V}(x, y) = \begin{bmatrix} P(x, y) \\ Q(x, y) \end{bmatrix} = P(x, y)\boldsymbol{i} + Q(x, y)\boldsymbol{j},$$

then

$$\frac{\partial P}{\partial y} = \frac{\partial Q}{\partial x}.$$

Similarly, if $\boldsymbol{V}(x, y, z)$ is a *gradient* vector field on $\mathbb{R}^3$,

$$\boldsymbol{V}(x, y, z) = \begin{bmatrix} P(x, y, z) \\ Q(x, y, z) \\ R(x, y, z) \end{bmatrix} = P(x, y)\boldsymbol{i} + Q(x, y)\boldsymbol{j} + R(x, y, z)\boldsymbol{k},$$

then

$$\frac{\partial P}{\partial y} = \frac{\partial Q}{\partial x}, \quad \frac{\partial R}{\partial y} = \frac{\partial Q}{\partial z}, \quad \frac{\partial P}{\partial z} = \frac{\partial R}{\partial x}.$$

The converse of (13.50) is not true. More precisely, there exist open sets $U \subset \mathbb{R}^n$ and C^1 vector fields $\boldsymbol{V} : U \to \mathbb{R}^n$ satisfying (13.50) yet they are not gradient vector fields.

A famous example is the vector field

$$\Theta : \mathbb{R}^2 \setminus \{0\} \to \mathbb{R}^2, \quad \Theta(x, y) = \begin{bmatrix} P(x,y) \\ Q(x,y) \end{bmatrix} := \begin{bmatrix} -\frac{y}{x^2+y^2} \\ \\ \frac{x}{x^2+y^2} \end{bmatrix}.$$

Indeed

$$\partial_y P = -\frac{1}{x^2+y^2} + \frac{2y^2}{(x^2+y^2)^2} = \frac{-(x^2+y^2)+2y^2}{(x^2+y^2)^2} = \frac{y^2-x^2}{(x^2+y^2)^2}.$$

$$\partial_x Q = \frac{1}{x^2+y^2} - \frac{2x^2}{(x^2+y^2)^2} = \frac{(x^2+y^2)-2x^2}{(x^2+y^2)^2} = \frac{y^2-x^2}{(x^2+y^2)^2}.$$

The reason why Θ *is not* a gradient vector field is rather subtle and can be properly explained once we introduce the concept of integration along paths. What is more surprising, H. Poincaré proved that if $U \subset \mathbb{R}^n$ is an open *convex* set and $\boldsymbol{V} \to \mathbb{R}^n$ is a C^1 vector field satisfying (13.50), then $\boldsymbol{V}$ *is* a gradient vector field. Thus, for any open convex subset $C \subset \mathbb{R}^2 \setminus \{\boldsymbol{0}\}$ there exists a C^2 function $f_C : C \to \mathbb{R}$ such that

$$\Theta(\boldsymbol{x}) = \nabla f_C(\boldsymbol{x}), \quad \forall \boldsymbol{x} \in C.$$

However, we cannot find a function $f : \mathbb{R}^2 \setminus \{\boldsymbol{0}\} \to \mathbb{R}$ such that $\Theta = \nabla f$! □

Example 13.40. Suppose that $f : \mathbb{R}^2 \to \mathbb{R}$ is a C^3 function of two variables x, y. Then $\partial_y f$ is a C^2 function and we have

$$\boxed{\partial^3_{xyy} f} = \partial^2_{xy}\big(\partial_y f\big) = \partial^2_{yx}\big(\partial_y f\big) = \boxed{\partial^3_{yxy} f} = \partial_y\big(\partial^2_{xy} f\big) = \partial_y\big(\partial^2_{yx} f\big) = \boxed{\partial^3_{yyx} f}.$$

If additionally f is C^4, then a similar argument shows

$$\partial^4_{xxyy} f = \partial^4_{xyxy} f = \partial^4_{yxxy} f = \partial^4_{yxyx} f = \partial^4_{yyxx} f = \partial^4_{xyyx} f.$$

It is now time to introduce a more convenient notation. Fix $n \in \mathbb{N}$. A *multi-index* of dimension n is an n-tuple

$$\alpha = (\alpha_1, \dots, \alpha_n), \quad \alpha_1, \dots, \alpha_n \in \mathbb{Z}_{\geq 0}.$$

The *size* of the multi-index α is the nonnegative integer

$$|\alpha| := \alpha_1 + \cdots + \alpha_n.$$

Suppose now that $m \in \mathbb{N}$, $U \subset \mathbb{R}^n$ is an open set and $f : U \to \mathbb{R}$ is a C^m-function. Given $k \leq m$ we define

$$\partial^k_{x^1} f := \partial^k_{x^1 \cdots x^1} f, \dots, \partial^k_{x^n} f := \partial^k_{x^n \cdots x^n} f.$$

Thus, instead of $\partial^2_{x^1 x^1}$ we will write $\partial^2_{x^1} f$. We define $\partial^0_{x^1} f := f$.

For any multi-index α of dimension n and size $|\alpha| \leq m$ we set

$$\partial^\alpha_x f := \partial^{\alpha_1}_{x^1} \partial^{\alpha_2}_{x^2} \cdots \partial^{\alpha_n}_{x^n} f.$$

13.5. Exercises

Exercise 13.1. Let $m, n \in \mathbb{N}$, $U \subset \mathbb{R}^n$, $\boldsymbol{x}_0 \in U$ and $\boldsymbol{F} : U \to \mathbb{R}^m$ a map. Assume U is open. Prove that the following statements are equivalent.

(i) The map $\boldsymbol{F}$ is Fréchet differentiable at $\boldsymbol{x}_0$.
(ii) There exists a linear operator $L : \mathbb{R}^n \to \mathbb{R}^m$ with the following property:

$$\forall \varepsilon > 0, \exists \delta = \delta(\varepsilon) > 0 : \forall \boldsymbol{h} \in \mathbb{R}^n, \|\boldsymbol{h}\| < \delta \Rightarrow \big\|\boldsymbol{F}(\boldsymbol{x}_0 + \boldsymbol{h}) - \boldsymbol{F}(\boldsymbol{x}_0) - L\boldsymbol{h}\big\| \leq \varepsilon\|\boldsymbol{h}\|.$$

(iii) There exists a linear operator $L : \mathbb{R}^n \to \mathbb{R}^m$, a number $r > 0$ such that $B_r(\boldsymbol{x}_0) \subset U$ and a function $\varphi : [0, r) \to [0, \infty)$ with the following properties

$$\big\|\boldsymbol{F}(\boldsymbol{x}_0 + \boldsymbol{h}) - \boldsymbol{F}(\boldsymbol{x}_0) - L\boldsymbol{h}\big\| \leq \varphi\big(\|\boldsymbol{h}\|\big)\|\boldsymbol{h}\|.$$

$$\lim_{t \searrow 0} \varphi(t) = 0 = \varphi(0).$$

Hint. For (ii) $\Rightarrow$ (iii) use

$$\varphi(t) := \sup_{\|\boldsymbol{h}\|=t} \frac{1}{\|\boldsymbol{h}\|}\big\|\boldsymbol{F}(\boldsymbol{x}_0 + \boldsymbol{h}) - \boldsymbol{F}(\boldsymbol{x}_0) - L\boldsymbol{h}\big\|, \quad t > 0.$$

□

Exercise 13.2. Consider the map $\boldsymbol{F} : \mathbb{R}^3 \to \mathbb{R}^2$ given by

$$\boldsymbol{F}(x, y, z) = \begin{bmatrix} x^3 + y^3 + z^3 \\ xyz \end{bmatrix}.$$

(i) Show that $\boldsymbol{F}$ is differentiable at any point $(x_0, y_0, z_0) \in \mathbb{R}^3$.
(ii) Find the Jacobian matrix of $\boldsymbol{F}$ at the point $(x_0, y_0, z_0) = (1, 1, 1)$.

□

Exercise 13.3. Compute the Jacobian matrices of the maps $\boldsymbol{F} : \mathbb{R}^2 \to \mathbb{R}^2$, $\boldsymbol{G}, \boldsymbol{H} : \mathbb{R}^3 \to \mathbb{R}^3$ defined by

$$\begin{bmatrix} r \\ \theta \end{bmatrix} \overset{\boldsymbol{F}}{\mapsto} \begin{bmatrix} x \\ y \end{bmatrix} = \begin{bmatrix} r\cos\theta \\ r\sin\theta \end{bmatrix},$$

$$\begin{bmatrix} \rho \\ \theta \\ \varphi \end{bmatrix} \overset{\boldsymbol{G}}{\mapsto} \begin{bmatrix} x \\ y \\ z \end{bmatrix} = \begin{bmatrix} \rho\sin\varphi\cos\theta \\ \rho\sin\varphi\sin\theta \\ \rho\cos\varphi \end{bmatrix}, \quad \begin{bmatrix} r \\ \theta \\ z \end{bmatrix} \overset{\boldsymbol{H}}{\mapsto} \begin{bmatrix} x \\ y \\ z \end{bmatrix} = \begin{bmatrix} r\cos\theta \\ r\sin\theta \\ z \end{bmatrix}.$$ □

Exercise 13.4. Show that the function $f : \mathbb{R}^2 \to \mathbb{R}$, $f(x, y) = 2^{xy}$ is C^1 and then find its linear approximation at the point $(x_0, y_0) = (1, 1)$.

Hint. Use Example 13.19 as inspiration.

□

Exercise 13.5. Let $n \in \mathbb{N}$ and suppose that A is a symmetric $n \times n$ matrix. Define

$$q_A : \mathbb{R}^n \to \mathbb{R}, \ \ q_A(\boldsymbol{x}) = \frac{1}{2}\langle A\boldsymbol{x}, \boldsymbol{x}\rangle, \ \ \forall \boldsymbol{x} \in \mathbb{R}^n.$$

Prove that

$$\nabla q_A(\boldsymbol{x}) = A\boldsymbol{x}, \ \ \forall \boldsymbol{x} \in \mathbb{R}^n.$$

Hint. You need to use the results in Exercise 11.24. □

Exercise 13.6. A function $f : \mathbb{R}^n \to \mathbb{R}$ is called *homogeneous of degree* 1, if

$$f(t\boldsymbol{x}) = tf(\boldsymbol{x}), \ \ \forall t \in \mathbb{R}, \ \ \boldsymbol{x} \in \mathbb{R}^n.$$

(i) Prove that if $f : \mathbb{R}^n \to \mathbb{R}$ is homogeneous of degree 1, then for any $\boldsymbol{v} \in \mathbb{R}^n \setminus \{\boldsymbol{0}\}$, the function f is differentiable along $\boldsymbol{v}$ at $\boldsymbol{0}$.

(ii) Show that the function

$$f : \mathbb{R}^2 \to \mathbb{R}, \ \ f(x,y) = \begin{cases} \frac{x^3 - y^3}{x^2 + y^2}, & (x,y) \neq (0,0), \\ 0, & (x,y) = (0,0) \end{cases}$$

is homogeneous of degree 1, and it is continuous at $\boldsymbol{0}$, but it is not Fréchet differentiable at $\boldsymbol{0}$.

(iii) Prove that if $f : \mathbb{R}^n \to \mathbb{R}$ is homogeneous of degree 1 and Fréchet differentiable at $\boldsymbol{0}$, then f is linear.

□

Exercise 13.7. Let $m, n \in \mathbb{N}$. Suppose that U is an open subset of $\mathbb{R}^n$, $I \subset \mathbb{R}$ is an open interval, $\gamma : I \to U$ is a C^1-path and $\boldsymbol{F} : U \to \mathbb{R}^m$ is a C^1-map. Let $\boldsymbol{\omega} : I \to \mathbb{R}^m$ denote the C^1-path $\boldsymbol{\omega}(t) = \boldsymbol{F}(\gamma(t))$. Prove that

$$\dot{\boldsymbol{\omega}}(t) = d\boldsymbol{F}(\gamma(t))\dot{\gamma}(t), \ \ \forall t \in I.$$ □

Exercise 13.8. Let $a, b \in \mathbb{R}$, $a < b$ and suppose that $\boldsymbol{\alpha}, \boldsymbol{\beta} : (a,b) \to \mathbb{R}^3$ are two C^1-paths. Prove that

$$\frac{d}{dt}\Big(\boldsymbol{\alpha}(t) \times \boldsymbol{\beta}(t)\Big) = \dot{\boldsymbol{\alpha}}(t) \times \boldsymbol{\beta}(t) + \boldsymbol{\alpha}(t) \times \dot{\boldsymbol{\beta}}(t).$$

Hint. Use (11.22). □

Exercise 13.9. Let $f : \mathbb{R}^n \to \mathbb{R}$, $f(\boldsymbol{x}) = \|\boldsymbol{x}\|^2$ and suppose that $\boldsymbol{\alpha}, \boldsymbol{\beta} : (-1,1) \to \mathbb{R}^n$ are two differentiable paths.

(i) Show that

$$\frac{d}{dt}\langle \boldsymbol{\alpha}(t), \boldsymbol{\beta}(t) \rangle = \langle \dot{\boldsymbol{\alpha}}(t), \boldsymbol{\beta}(t) \rangle + \langle \boldsymbol{\alpha}(t), \dot{\boldsymbol{\beta}}(t) \rangle, \ \ \forall t \in (-1,1).$$

(ii) Compute the gradient ∇f. **Hint.** Compare with Exercise 13.5.

(iii) Compute $\frac{d}{dt}\|\boldsymbol{\alpha}(t)\|^2$.

(iv) Show that the function $t \mapsto \|\boldsymbol{\alpha}(t)\|$ is constant if and only if $\boldsymbol{\alpha}(t) \perp \dot{\boldsymbol{\alpha}}(t)$, $\forall t \in (-1, 1)$.

(v) What can you say about the motion described by the path $\boldsymbol{\alpha}(t)$ when $\boldsymbol{\alpha}(t) \perp \dot{\boldsymbol{\alpha}}(t)$, $\forall t$?

□

Exercise 13.10. Let $f : \mathbb{R}^n \setminus \{\mathbf{0}\} \to \mathbb{R}$ be a function that is positively homogeneous of degree k, i.e.,

$$f(t\boldsymbol{x}) = t^k f(\boldsymbol{x}), \quad \forall t > 0, \quad \boldsymbol{x} \in \mathbb{R}^n \setminus \{\mathbf{0}\}.$$

Show that if f is differentiable, then the partial derivatives $\frac{\partial f}{\partial x^i}$, $i = 1, \dots, n$, are positively homogeneous of degree $k - 1$. □

Exercise 13.11. Suppose that the path

$$\gamma : \mathbb{R} \to \mathbb{R}^2, \quad \gamma(t) = \begin{bmatrix} x(t) \\ y(t) \end{bmatrix} \in \mathbb{R}^2$$

is an integral curve of the vector field $\boldsymbol{V}$ defined in (13.40).

(i) Prove that $\|\gamma(t)\| = \|\gamma(0)\|$, $\forall t$.

(ii) Deduce from the above that $\dot{x}(t)^2 + x(t)^2 = \dot{x}(0)^2 + x(0)^2$, $\forall t$.

(iii) Prove that $\ddot{x} = -x$, where $\ddot{f}$ denotes the second order time derivative of a function f.

(iv) Given that $\gamma(0) = (1, 0)$ determine $\gamma(t)$.

Hint. For (i)–(iii) use the differential equations (13.39). (iv) Compare with Exercise 7.16. □

Exercise 13.12. Prove that the function $r : \mathbb{R}^n \setminus \{\mathbf{0}\} \to \mathbb{R}$, $r(\boldsymbol{x}) = \|\boldsymbol{x}\|$ is C^1 and then describe its differential. □

Exercise 13.13. Let $n \in \mathbb{N}$. Fix a C^1-function $U : \mathbb{R}^n \setminus \{\mathbf{0}\} \to \mathbb{R}$. Suppose that I is an open interval of the real axis and

$$\gamma : I \to \mathbb{R}^n \setminus \{\mathbf{0}\}, \quad t \mapsto \gamma(t) = \big(x^1(t), \dots, x^n(t) \big)$$

is a C^2-path satisfying Newton's (2nd Law of Dynamics) differential equations

$$\ddot{\gamma}(t) = -\nabla U\big(\gamma(t)\big), \quad \forall t \in I.$$

(i) (*Conservation of energy*) Prove that the function $E : I \to \mathbb{R}$

$$E(t) = \frac{1}{2}\|\dot{\gamma}(t)\|^2 + U\big(\gamma(t)\big),$$

is constant.

(ii) (*Conservation of momenta*) Suppose that there exists a C^1-function $f : (0,\infty) \to \mathbb{R}$ such that

$$U(\boldsymbol{x}) = f(\|\boldsymbol{x}\|), \ \ \forall \boldsymbol{x} \in \mathbb{R}^n \setminus \{\boldsymbol{0}\}.$$

Prove that, for any $1 \leq k < \ell \leq n$, the function $P^{k\ell} : I \to \mathbb{R}$

$$P^{k\ell}(t) = \dot{x}^k(t)x^\ell(t) - \dot{x}^\ell(t)x^k(t)$$

is constant.

□

Exercise 13.14.

(i) Let $k, n \in \mathbb{N}$. Prove that if $f : \mathbb{R}^n \to \mathbb{R}$ and $u : \mathbb{R} \to \mathbb{R}$ are C^k-functions, then so is their composition $u \circ f : \mathbb{R}^n \to \mathbb{R}$.

(ii) Let $n \in \mathbb{N}$ and $r > 0$. Prove that for any $0 < r < R$ there exists a nonzero function $f \in C^\infty(\mathbb{R}^n)$ such that $f(\boldsymbol{x}) = 1$ if $\|\boldsymbol{x}\| \leq r$ and $f(\boldsymbol{x}) = 0$, if $\|\boldsymbol{x}\| \geq R$.

(iii) Let $n \in \mathbb{N}$. Suppose that $K \subset \mathbb{R}^n$ is a compact set and $\mathcal{U}$ is an open cover of K. Prove that there exist compactly supported smooth functions $\chi_1, \dots, \chi_\ell : \mathbb{R}^n \to [0,\infty)$ with the following properties.

- For any $i = 1, \dots, \ell$ there exists an open set $U = U_i$ in the family $\mathcal{U}$ such that $\operatorname{supp} \chi_i \subset U_i$.
- $\chi_1(\boldsymbol{x}) + \cdots + \chi_\ell(\boldsymbol{x}) = 1, \forall \boldsymbol{x} \in K$.

Hint. (i) Argue by induction on k. (ii) Use the result proved in Exercise 7.8 to construct a smooth function $u : \mathbb{R} \to [0,\infty)$ such that $u(s) = 1$ if $s \leq 0$ and $u(s) = 0$ if $s \geq 1$. Then, for $a < b$, define

$$u_{a,b} : \mathbb{R} \to [0,\infty), \ \ u_{a,b}(t) = u\left(\frac{t-a}{b-a}\right)$$

and show that $u_{a,b}$ is smooth and satisfies $u_{a,b}(t) = 1$ if $t \leq a$ and $u_{a,b}(t) = 0$ if $t > b$. Finally, set $f(\boldsymbol{x}) = u_{a,b}(\|\boldsymbol{x}\|^2)$ with $a = r^2$, $b = R^2$ and then show that f will do the trick. (iii) Use (ii) and imitate the proof of Theorem 12.68. □

Exercise 13.15. Let $n \in \mathbb{N}$. For any open set $\mathcal{O} \subset \mathbb{R}^n$ we define the *Laplacian* to be the map

$$\Delta : C^2(\mathcal{O}) \to C(\mathcal{O}), \ \ (\Delta f)(\boldsymbol{x}) = \sum_{k=1}^{n} \partial^2_{x^k} f(\boldsymbol{x}). \tag{13.51}$$

(i) Show that, $\forall f, g \in C^2(\mathcal{O})$, we have

$$\Delta(f+g) = \Delta f + \Delta g,$$

$$\Delta(fg) = f\Delta g + 2\langle \nabla f, \nabla g\rangle + g\Delta f.$$

(ii) Show

$$\Delta\|\boldsymbol{x}\|^p = p(p+n-2)\|\boldsymbol{x}\|^{p-2}, \ \ \forall \boldsymbol{x} \in \mathbb{R}^n \setminus \{\boldsymbol{0}\}, \ \ p \in \mathbb{R}.$$

□

Exercise 13.16. Let $n \in \mathbb{N}$, $n \geq 2$. Consider the function $U : \mathbb{R}^n \setminus \{\mathbf{0}\} \to \mathbb{R}$

$$U(\boldsymbol{x}) = \begin{cases} \log \|\boldsymbol{x}\|, & n = 2, \\ \frac{1}{\|\boldsymbol{x}\|^{n-2}}, & n > 2. \end{cases}$$

Compute $\Delta U(\boldsymbol{x})$, where Δ is the Laplacian defined as in (13.51). □

Exercise 13.17. Let $n \in \mathbb{N}$ and consider the function $K : (0, \infty) \times \mathbb{R}^n \to \mathbb{R}$ given by

$$K(t, \boldsymbol{x}) = t^{-n/2} e^{-\frac{\|\boldsymbol{x}\|^2}{4t}}.$$

Compute

$$\partial_t K - \Delta_{\boldsymbol{x}} K = \partial_t K - \left(\partial^2_{x^1} K + \cdots + \partial^2_{x^n} K \right).$$

□

Exercise 13.18. Suppose that $f, g : \mathbb{R} \to \mathbb{R}$ are C^2 functions. Define

$$w : \mathbb{R}^2 \to \mathbb{R}, \;\; w(t, x) = f(x + t) + g(x - t).$$

Compute

$$\partial^2_t w - \partial^2_x w.$$ □

Exercise 13.19. Show that the vector field

$$\boldsymbol{V} : \mathbb{R}^2 \to \mathbb{R}^2, \;\; \boldsymbol{V}(x, y) = \begin{bmatrix} -y \\ x \end{bmatrix}$$

is not a gradient vector field.

Hint. Have a look at Example 13.39. □

13.6. More challenging problems

Problem 13.1. Let $n \in \mathbb{N}$ and denote by $\mathrm{Mat}^*_n(\mathbb{R})$ the set of invertible $n \times n$ matrices.

(i) Prove that $\mathrm{Mat}^*_n(\mathbb{R})$ is open (in $\mathrm{Mat}_n(\mathbb{R})$).
(ii) Prove that the map $\boldsymbol{F} : \mathrm{Mat}^*_n(\mathbb{R}) \to \mathrm{Mat}_n(\mathbb{R})$ given by

$$\boldsymbol{F}(A) = A^{-1}$$

is differentiable and then compute its differential at $A_0 \in \mathrm{Mat}^*_n(\mathbb{R})$.

□

Problem 13.2. Let $k, n \in \mathbb{N}$ and $U \subset \mathbb{R}^n$ be an open subset.

(i) Prove that the collection $C^k(U)$ of functions that are C^k on U is a real vector space.

(ii) Let $m \in \mathbb{N}$ and suppose that $\boldsymbol{p}_1, \dots, \boldsymbol{p}_m$ are m distinct points in U and $c_1, \dots, c_m \in \mathbb{R}$. Prove that there exists $f \in C^k(U)$ such that

$$f(\boldsymbol{p}_j) = j, \ \ \forall j = 1, \dots, m.$$

Conclude that the vector space $C^k(U)$ is *infinite dimensional*.

□

Problem 13.3. Let $n \in \mathbb{N}$ and suppose that A is an $n \times n$ matrix.

(i) Prove that for any $\boldsymbol{x} \in \mathbb{R}^n$ the series

$$\sum_{k=0}^{\infty} \frac{1}{k!} A^k \boldsymbol{x} = \boldsymbol{x} + A\boldsymbol{x} + \frac{1}{2!} A^2 \boldsymbol{x} + \frac{1}{3!} A^3 \boldsymbol{x} + \cdots$$

is absolutely convergent.

(ii) Prove that for any $\boldsymbol{x} \in \mathbb{R}^n$ the path

$$\gamma : \mathbb{R} \to \mathbb{R}^n, \ \ \gamma(t) = \sum_{k=0}^{\infty} \frac{1}{k!} (tA)^k \boldsymbol{x}$$

is differentiable and

$$\dot{\gamma}(t) = A\gamma(t), \ \ \forall t \in \mathbb{R}.$$

(iii) Compute $\gamma(t)$ when $n = 2$,

$$\boldsymbol{x} = \begin{bmatrix} x^1 \\ x^2 \end{bmatrix} \text{ and } A = \begin{bmatrix} \lambda_1 & 0 \\ 0 & \lambda_2 \end{bmatrix}.$$

Chapter 14

Applications of Multi-variable Differential Calculus

We present below a few of the most frequently encountered applications of multi-dimensional differential calculus.

14.1. Taylor formula

Just like functions of one real variable, the differentiable functions of several variables can be well approximated by certain explicit polynomials. We present in this section two such approximation formulæ that are used frequently in applications. We refer to [5, §2.8] for more general results.

Fix $n \in \mathbb{N}$, an open set $U \subset \mathbb{R}^n$ and a point $\boldsymbol{x}_0 \in U$. Then there exists $r_0 > 0$ such that $\overline{B_{r_0}(\boldsymbol{x}_0)} \subset U$. Suppose that $f : U \to \mathbb{R}$ is a differentiable function.

Theorem 14.1 (Multidimensional Taylor formula). *(a) If the function f is C^2, then for any $\boldsymbol{h} = (h^1, \dots, h^n) \in \mathbb{R}^n$ such that $\|\boldsymbol{h}\| < r_0$ we have*

$$f(\boldsymbol{x}_0 + \boldsymbol{h}) = f(\boldsymbol{x}_0) + \sum_{i=1}^{n} \partial_{x^i} f(\boldsymbol{x}_0) h^i + R_1(\boldsymbol{x}_0, \boldsymbol{h}), \tag{14.1}$$

where the remainder $R_1(\boldsymbol{x}_0, \boldsymbol{h})$ is described by the integral formula

$$R_1(\boldsymbol{x}_0, \boldsymbol{h}) = \int_0^1 (1-t)\rho_2(\boldsymbol{x}_0, \boldsymbol{h}, t)dt, \;\; \rho_2(\boldsymbol{x}_0, \boldsymbol{h}, t) = \sum_{i,j=1}^{n} \partial^2_{x^i x^j} f(\boldsymbol{x}_0 + t\boldsymbol{h}) h^i h^j. \tag{14.2}$$

Moreover, there exists a constant $C > 0$, $\boxed{\text{independent of } \boldsymbol{h}}$, *such that*

$$\big|\, R_1(\boldsymbol{x}_0, \boldsymbol{h}) \,\big| \leq C\|\boldsymbol{h}\|^2, \;\; \forall \|\boldsymbol{h}\| < r_0. \tag{14.3}$$

(b) If the function f is C^3, then for any $\boldsymbol{h} = (h^1, \dots, h^n) \in \mathbb{R}^n$ such that $\|\boldsymbol{h}\| < r_0$ we have

$$f(\boldsymbol{x}_0 + \boldsymbol{h}) = f(\boldsymbol{x}_0) + \sum_{i=1}^{n} \partial_{x^i} f(\boldsymbol{x}_0) h^i + \frac{1}{2} \sum_{i,j=1}^{n} \partial^2_{x^i x^j} f(\boldsymbol{x}_0) h^i h^j + R_2(\boldsymbol{x}_0, \boldsymbol{h}), \tag{14.4}$$

where the remainder $R_2(\boldsymbol{x}_0, \boldsymbol{h})$ is described by the integral formula

$$R_2(\boldsymbol{x}_0, \boldsymbol{h}) = \frac{1}{2!}\int_0^1 (1-t)^2 \rho_3(\boldsymbol{x}_0, \boldsymbol{h}, t)dt,$$
$$\rho_3(\boldsymbol{x}_0, \boldsymbol{h}, t) = \sum_{i,j,k=1}^{n} \partial^3_{x^i x^j x^k} f(\boldsymbol{x}_0 + t\boldsymbol{h})h^i h^j h^k. \tag{14.5}$$

Moreover, there exists a constant $C > 0$, $\boxed{\text{independent of } \boldsymbol{h}}$*, such that*

$$\big| R_2(\boldsymbol{x}_0, \boldsymbol{h}) \big| \leq C\|\boldsymbol{h}\|^3, \quad \forall \|\boldsymbol{h}\| < r_0. \tag{14.6}$$

Proof. We prove only (b). The case (a) is similar and involves simpler computations. Fix $\boldsymbol{h} \in \mathbb{R}^n$ such that $\|\boldsymbol{h}\| < r_0$. Consider the C^3-function

$$g : [-1, 1] \to \mathbb{R}, \ \ g(t) = f(\boldsymbol{x}_0 + t\boldsymbol{h}).$$

Using the 1-dimensional Taylor formula with integral remainder, Proposition 9.51, we deduce

$$g(1) = g(0) + g'(0) + \frac{1}{2}g''(0) + R_2, \ \ R_2 = \frac{1}{2!}\int_0^1 g^{(3)}(t)(1-t)^2 dt. \tag{14.7}$$

Using the chain rule (13.34) repeatedly we deduce

$$g'(t) = \sum_{i=1}^{n} \partial_{x^i} f(\boldsymbol{x}_0 + t\boldsymbol{h})h^i, \ \ g''(t) = \sum_{i,j=1}^{n} \partial^2_{x^i x^j} f(\boldsymbol{x}_0 + t\boldsymbol{h})h^i h^j,$$

$$g^{(3)}(t) = \sum_{i,j,k=1}^{n} \partial^3_{x^i x^j x^k} f(\boldsymbol{x}_0 + t\boldsymbol{h})h^i h^j h^k.$$

Using these equalities in (14.7) we obtain (14.4) and (14.5). It remains to prove (14.6).

Observe that $|h^i| \leq \|\boldsymbol{h}\|$ for any $i = 1, \dots, n$. Hence

$$|\rho_3(\boldsymbol{x}_0, \boldsymbol{h}, t)| \leq \sum_{i,j,k=1}^{n} \big| \partial^3_{x^i x^j x^k} f(\boldsymbol{x}_0 + t\boldsymbol{h})h^i h^j h^k \big| \leq \|\boldsymbol{h}\|^3 \sum_{i,j,k=1}^{n} \big| \partial^3_{x^i x^j x^k} f(\boldsymbol{x}_0 + t\boldsymbol{h}) \big|.$$

For each $i, j, k = 1, \dots, n$ we set

$$M_{i,j,k} = \sup_{\boldsymbol{x} \in \overline{B_{r_0}(\boldsymbol{x}_0)}} \big| \partial^3_{x^i x^j x^k} f(\boldsymbol{x}) \big|.$$

The quantity $M_{i,j,k}$ is finite since the function $\partial^3_{x^i x^j x^k} f(\boldsymbol{x})$ is continuous on the *compact* set $\overline{B_{r_0}(\boldsymbol{x}_0)}$. We set

$$M := \max_{i,j,k} M_{i,j,k}.$$

Clearly, the number M is independent of $\boldsymbol{h}$. We deduce that for any $\|\boldsymbol{h}\| < r_0$ we have

$$|\rho_3(\boldsymbol{x}_0, \boldsymbol{h}, t)| \leq \|\boldsymbol{h}\|^3 \sum_{i,jk=1}^{n} M_{i,j,k} \leq \|\boldsymbol{h}\|^3 \sum_{i,j,k=1}^{n} M = Mn^3\|\boldsymbol{h}\|^3.$$

Hence

$$|R_2(\boldsymbol{x}_0, \boldsymbol{h})| \leq \frac{1}{2}\int_0^1 (1-t)^2 \, |\rho_3(\boldsymbol{x}_0, \boldsymbol{h}, t)| dt \leq \frac{Mn^3}{2}\|\boldsymbol{h}\|^3.$$

□

Definition 14.2. The $n \times n$ matrix $\boldsymbol{H}(f, \boldsymbol{x}_0)$ with entries

$$\boldsymbol{H}(f, \boldsymbol{x}_0)_{ij} = \partial^2_{x^i x^j} f(\boldsymbol{x}_0), \quad 1 \leq i, j \leq n,$$

is called the *Hessian* of f at $\boldsymbol{x}_0$.[1] □

Since partial derivatives commute, we see that the Hessian is a *symmetric* matrix, i.e.,

$$\partial^2_{x^i x^j} f(\boldsymbol{x}_0) = \partial^2_{x^j x^i} f(\boldsymbol{x}_0).$$

The matrix $\boldsymbol{H}(f, \boldsymbol{x}_0)$ defines a linear operator $\mathbb{R}^n \to \mathbb{R}^n$ given by

$$\begin{aligned} \boldsymbol{H}(f, \boldsymbol{x}_0)\boldsymbol{h} &= \Big(\sum_{j=1}^n \boldsymbol{H}(f, \boldsymbol{x}_0)_{1j} h^j\Big)\boldsymbol{e}_1 + \cdots + \Big(\sum_{j=1}^n \boldsymbol{H}(f, \boldsymbol{x}_0)_{nj} h^j\Big)\boldsymbol{e}_n \\ &= \sum_{i=1}^n \Big(\sum_{j=1}^n \boldsymbol{H}(f, \boldsymbol{x}_0)_{ij} h^j\Big)\boldsymbol{e}_i, \quad \forall \boldsymbol{h} \in \mathbb{R}^n. \end{aligned}$$

We deduce that

$$\sum_{i,j=1}^n \partial^2_{x^i x^j} f(\boldsymbol{x}_0) h^i h^j = \langle\, \boldsymbol{h}, \boldsymbol{H}(f, \boldsymbol{x}_0)\boldsymbol{h} \,\rangle. \tag{14.8}$$

We can rewrite the equality (14.4) in the more compact form

$$\boxed{f(\boldsymbol{x}_0 + \boldsymbol{h}) = f(\boldsymbol{x}_0) + \langle\, \nabla f(\boldsymbol{x}_0), \boldsymbol{h} \,\rangle + \frac{1}{2}\langle\, \boldsymbol{h}, \boldsymbol{H}(f, \boldsymbol{x}_0)\boldsymbol{h} \,\rangle + R_2(\boldsymbol{x}_0, \boldsymbol{h})}. \tag{14.9}$$

Example 14.3. Consider the function

$$f : \mathbb{R}^2 \to \mathbb{R}, \quad f(x, y) = 3x^2 + 4xy + 5y^2.$$

Then

$$\partial^2_x f = 6, \quad \partial^2_{xy} f = 4, \quad \partial^2_y f = 10.$$

The Hessian of f at $\boldsymbol{0}$ is the symmetric 2×2 matrix

$$\boldsymbol{H}(f, \boldsymbol{0}) = \begin{bmatrix} 6 & 4 \\ 4 & 10 \end{bmatrix}.$$

□

14.2. Extrema of functions of several variables

Fix a natural number n.

Definition 14.4. Let $X \subset \mathbb{R}^n$ and $\boldsymbol{x}_0 \in X$. Fix a function $f : X \to \mathbb{R}$.

(i) The point $\boldsymbol{x}_0$ is said to be a *local minimum* of the function if there exists $r > 0$ with the following property

$$\forall \boldsymbol{x} \in X \ \ \operatorname{dist}(\boldsymbol{x}, \boldsymbol{x}_0) < r \Rightarrow f(\boldsymbol{x}_0) \leq f(\boldsymbol{x}).$$

[1]Note that when describing the Hessian matrix both indices are *subscripts*. This differs from the way we described the matrix associated to an operator where one index is a superscript, the other is a subscript. This discrepancy is a reflection of the fact that the Hessian of a function is intrinsically a different beast than a linear operator.

(ii) The point $\boldsymbol{x}_0$ is said to be a *local maximum* of the function f if there exists $r > 0$ with the following property

$$\forall \boldsymbol{x} \in X \quad \operatorname{dist}(\boldsymbol{x}, \boldsymbol{x}_0) < r \Rightarrow f(\boldsymbol{x}_0) \geq f(\boldsymbol{x}).$$

(iii) The point $\boldsymbol{x}_0$ is said to be a *local extremum* of the function f if it is either a local minimum or a local maximum of f.

□

The 1-dimensional Fermat Principle (Theorem 7.24) has the following multi-dimensional counterpart.

Theorem 14.5 (Multidimensional Fermat Principle). *Suppose that $U \subset \mathbb{R}^n$ is an open set and $f : U \to \mathbb{R}$ is a C^1-function. If $\boldsymbol{x}_0 \in U$ is a local extremum of f, then $df(\boldsymbol{x}_0) = \boldsymbol{0}$, i.e.,*

$$\partial_{x^1} f(\boldsymbol{x}_0) = \cdots = \partial_{x^n} f(\boldsymbol{x}_0) = 0.$$

Proof. Assume for simplicity that $\boldsymbol{x}_0$ is a local minimum of f. (When $\boldsymbol{x}_0$ is a local maximum of f, then it is a local minimum of $-f$.) Fix $r > 0$ sufficiently small with the following properties.

- $B_r(\boldsymbol{x}_0) \subset U$.
- $f(\boldsymbol{x}_0) \leq f(\boldsymbol{x}), \forall \boldsymbol{x} \in B_r(\boldsymbol{x}_0)$.

Fix a vector $\boldsymbol{h} \in \mathbb{R}^n$. For $\varepsilon > 0$ sufficiently small, the line segment $[\boldsymbol{x}_0 - \varepsilon\boldsymbol{h}, \boldsymbol{x}_0 + \varepsilon\boldsymbol{h}]$ is contained in the ball $B_r(\boldsymbol{x}_0)$. Consider now the function

$$g : [-\varepsilon, \varepsilon] \to \mathbb{R}, \quad g(t) = f(\boldsymbol{x}_0 + t\boldsymbol{h}).$$

We can identify g with the restriction of f to the line segment $[\boldsymbol{x}_0 - \varepsilon\boldsymbol{h}, \boldsymbol{x}_0 + \varepsilon\boldsymbol{h}]$. Note that $g(0) = f(\boldsymbol{x}_0) \leq f(\boldsymbol{x}_0 + t\boldsymbol{h}) = g(t)$, $\forall t \in [-\varepsilon, \varepsilon]$. Thus 0 is a minimum point of g and the 1-dimensional Fermat Principle implies that $g'(0) = 0$. The chain rule (13.34) now implies

$$\langle \nabla f(\boldsymbol{x}_0), \boldsymbol{h} \rangle = g'(0) = 0.$$

We have thus shown that $\langle \nabla f(\boldsymbol{x}_0), \boldsymbol{h} \rangle = 0$, for any vector $\boldsymbol{h} \in \mathbb{R}^n$. If we choose $\boldsymbol{h} = \nabla f(\boldsymbol{x}_0)$, then we deduce

$$\|\nabla f(\boldsymbol{x}_0)\|^2 = \langle \nabla f(\boldsymbol{x}_0), \nabla f(\boldsymbol{x}_0) \rangle = 0.$$

□

Definition 14.6. Let $U \subset \mathbb{R}^n$ be an open set. A *critical point* of a differentiable function $f : U \to \mathbb{R}$ is a point $\boldsymbol{x}_0 \in U$ such that $df(\boldsymbol{x}_0) = 0$. □

We can rephrase Theorem 14.5 as follows.

If $U \subset \mathbb{R}^n$ is an open set, and $\boldsymbol{x}_0 \in U$ is a local extremum of a C^1-function $f : U \to \mathbb{R}$, then $\boldsymbol{x}_0$ must be a critical point of f.

We know now that the local extrema of a C^1-function $f : U \to \mathbb{R}$, if any, are located among the critical points of f. We want to address a sort of converse. Suppose that $\boldsymbol{x}_0 \in U$ is a critical point. Is there any way of deciding whether $\boldsymbol{x}_0$ is a local min, max or neither?

To answer this question we need to introduce some more terminology.

Definition 14.7. Suppose that A is a *symmetric* $n \times n$ matrix A. We denote by a_{ij} the entry located on the i-th row and j-th column.

(i) The *quadratic function* associated to A is the function $Q_A : \mathbb{R}^n \to \mathbb{R}$ given by

$$Q_A(\boldsymbol{h}) = \langle \boldsymbol{h}, A\boldsymbol{h} \rangle = \sum_{i,j=1}^{n} a_{ij} h^i h^j.$$

(ii) The matrix A is called *positive definite* if

$$Q_A(\boldsymbol{h}) > 0, \ \ \forall \boldsymbol{h} \in \mathbb{R}^n \setminus \{\boldsymbol{0}\}.$$

(iii) The matrix A is called *negative definite* if

$$Q_A(\boldsymbol{h}) < 0, \ \ \forall \boldsymbol{h} \in \mathbb{R}^n \setminus \{\boldsymbol{0}\}.$$

(iv) The matrix A is called *indefinite* if there exist $\boldsymbol{h}_0, \boldsymbol{h}_1 \in \mathbb{R}^n \setminus \{\boldsymbol{0}\}$ such that

$$Q_A(\boldsymbol{h}_0) < 0 < Q_A(\boldsymbol{h}_1).$$

□

Let us observe that the quadratic function associated to a symmetric $n \times n$ matrix A is homogeneous of degree 2, i.e.,

$$Q_A(t\boldsymbol{h}) = t^2 Q_A(\boldsymbol{h}), \ \ \forall t \in \mathbb{R}, \ \ \boldsymbol{h} \in \mathbb{R}^n. \tag{14.10}$$

Example 14.8. Suppose that $n = 2$,

$$A = \begin{bmatrix} a & b \\ b & c \end{bmatrix}, \ \ \boldsymbol{h} = \begin{bmatrix} x \\ y \end{bmatrix}.$$

Then

$$A\boldsymbol{h} = \begin{bmatrix} ax + by \\ bx + cy \end{bmatrix}, \ \ Q_A(\boldsymbol{h}) = \langle \boldsymbol{h}, A\boldsymbol{h} \rangle = x(ax + by) + y(bx + cy) = ax^2 + 2bxy + cy^2.$$

□

Theorem 14.9. *Let $U \subset \mathbb{R}$ be an open set and $f : U \to \mathbb{R}$ a C^3-function. Suppose that $\boldsymbol{x}_0$ is a critical point of f. Denote by A the Hessian of f at $\boldsymbol{x}_0$, $A := \boldsymbol{H}(f, \boldsymbol{x}_0)$. Then the following hold.*

(i) If A is positive definite, then $\boldsymbol{x}_0$ is a local minimum of f.

(ii) If A is negative definite, then $\boldsymbol{x}_0$ is a local maximum of f.
(iii) If A is indefinite, then $\boldsymbol{x}_0$ is not a local extremum of f.

Proof. The above claims are immediate consequences of Taylor's formula (14.4). Fix $r > 0$ sufficiently small such that $\overline{B_r(\boldsymbol{x}_0)} \subset U$. According to (14.4) for any $\boldsymbol{h}$ such that $\|\boldsymbol{h}\| < r$ we have

$$f(\boldsymbol{x}_0 + \boldsymbol{h}) = f(\boldsymbol{x}_0) + \frac{1}{2}Q_A(\boldsymbol{h}) + R_2(\boldsymbol{x}_0, \boldsymbol{h}). \tag{14.11}$$

Moreover, there exists $C > 0$ such that

$$|R_2(\boldsymbol{x}_0, \boldsymbol{h})| \leq C\|\boldsymbol{h}\|^3, \ \ \forall \|\boldsymbol{h}\| < r. \tag{14.12}$$

To prove (i) we need to use the following very useful technical fact whose proof is outlined in Exercise 14.5.

Lemma 14.10. *Suppose that A is a symmetric, positive definite matrix. Then there exists $m > 0$ such that*

$$Q_A(\boldsymbol{h}) \geq m\|\boldsymbol{h}\|^2, \ \ \forall \boldsymbol{h} \in \mathbb{R}^n. \qquad \square$$

Suppose now that $A = \boldsymbol{H}(f, \boldsymbol{x}_0)$ is positive definite. Choose a number $m > 0$ as in Lemma 14.10. From (14.11) and (14.12) we deduce

$$f(\boldsymbol{x}_0 + \boldsymbol{h}) \geq f(\boldsymbol{x}_0) + \frac{m}{2}\|\boldsymbol{h}\|^2 - C\|\boldsymbol{h}\|^3 = f(\boldsymbol{x}_0) + \|\boldsymbol{h}\|^2\Big(\frac{m}{2} - C\|\boldsymbol{h}\|\Big).$$

Choose $\varepsilon > 0$ smaller than both r and $\frac{m}{2C}$. Then, for any $\boldsymbol{h}$ such that $\|\boldsymbol{h}\| < \varepsilon$ we have

$$\boldsymbol{x}_0 + \boldsymbol{h} \in B_\varepsilon(\boldsymbol{x}_0), \ \ \frac{m}{2} - C\|\boldsymbol{h}\| > 0.$$

Thus for any $\boldsymbol{h}$ such that $\|\boldsymbol{h}\| < \varepsilon$ we have $f(\boldsymbol{x}_0 + \boldsymbol{h}) > f(\boldsymbol{x}_0)$. This proves that $\boldsymbol{x}_0$ is a local minimum of f.

The statement (ii) reduces to (i) by observing that the Hessian of $-f$ at $\boldsymbol{x}_0$ is $-A$ and it is positive definite. Thus $\boldsymbol{x}_0$ is a local minimum of $-f$, therefore a local maximum of f.

To prove (iii) choose vectors $\boldsymbol{h}_0, \boldsymbol{h}_1$ such that

$$Q_A(\boldsymbol{h}_0) < 0 < Q_A(\boldsymbol{h}_1).$$

For $t > 0$ sufficiently small we have $\boldsymbol{x}_0 + t\boldsymbol{h}_0, \boldsymbol{x}_0 + t\boldsymbol{h}_1 \in B_r(\boldsymbol{x}_0)$ and

$$\begin{aligned} f(\boldsymbol{x}_0 + t\boldsymbol{h}_0) &= f(\boldsymbol{x}_0) + \frac{1}{2}Q_A(t\boldsymbol{h}_0) + R_2(\boldsymbol{x}_0, t\boldsymbol{h}_0) \overset{(14.10)}{=} f(\boldsymbol{x}_0) + \frac{t^2}{2}Q_A(\boldsymbol{h}_0) + R_2(\boldsymbol{x}_0, t\boldsymbol{h}_0) \\ &\leq f(\boldsymbol{x}_0) + \frac{t^2}{2}Q_A(\boldsymbol{h}_0) + Ct^3\|\boldsymbol{h}_0\|^3 = f(\boldsymbol{x}_0) + \frac{t^2}{2}\underbrace{\Big(Q_A(\boldsymbol{h}_0) + 2tC\|\boldsymbol{h}_0\|\Big)}_{=:u(t)}. \end{aligned}$$

Observe that

$$\lim_{t\to 0} u(t) = Q_A(\boldsymbol{h}_0) < 0$$

so $u(t)$ is negative for t sufficiently small. Thus, for all t sufficiently small we have

$$f(\boldsymbol{x}_0 + t\boldsymbol{h}_0) < f(\boldsymbol{x}_0),$$

so $\boldsymbol{x}_0$ cannot be a local minimum.

Similarly

$$f(\boldsymbol{x}_0 + t\boldsymbol{h}_1) = f(\boldsymbol{x}_0) + \frac{1}{2}Q_A(t\boldsymbol{h}_1) + R_2(\boldsymbol{x}_0, t\boldsymbol{h}_1) = f(\boldsymbol{x}_0) + \frac{t^2}{2}Q_A(\boldsymbol{h}_1) + R_2(\boldsymbol{x}_0, t\boldsymbol{h}_1)$$

$$\geq f(\boldsymbol{x}_0) + \frac{t^2}{2}Q_A(\boldsymbol{h}_1) - Ct^3\|\boldsymbol{h}_1\| \stackrel{(14.10)}{=} f(\boldsymbol{x}_0) + \frac{t^2}{2}\underbrace{\Big(Q_A(\boldsymbol{h}_1) - 2Ct\|\boldsymbol{h}_1\|\Big)}_{=:v(t)}.$$

Observe that

$$\lim_{t\to 0} v(t) = Q_A(\boldsymbol{h}_1) > 0$$

so $v(t) > 0$ for all t sufficiently small. Hence, for all t sufficiently small we have

$$f(\boldsymbol{x}_0 + t\boldsymbol{h}_1) > f(\boldsymbol{x}_0)$$

so $\boldsymbol{x}_0$ cannot be a local maximum either. □

Remark 14.11. Deciding when a symmetric matrix A is positive/negative definite or indefinite is a nontrivial task. All the known techniques rely on more linear algebra than we are prepared to assume at this point. It is known that all the eigenvalues of a real symmetric matrix are real. The matrix A is positive/negative definite if all its eigenvalues are positive/negative. The matrix A is indefinite if it admits both positive and negative eigenvalues.

If the dimension of the matrix A is small, one can conceive faster ad-hoc methods of deciding if S is positive/negative definite. In Exercise 14.6 we describe a simple method of deciding when a 2×2 symmetric matrix is positive/negative definite. This is a special case of a theorem of J. J. Sylvester[2] [21, Chap. 7]. □

Example 14.12. Consider the function

$$f : (0,\infty) \times (0,\infty) \to \mathbb{R}, \quad f(x,y) = x^3y^2(6 - x - y).$$

The critical points of f are found solving the system of equations $\partial_x f = \partial_y f = 0$, i.e.,

$$\begin{cases} 3x^2y^2(6-x-y) - x^3y^2 = 0 \\ 2x^3y(6-x-y) - x^3y^2 \ = 0. \end{cases} \tag{14.13}$$

The first equality in (14.13) can be rewritten as

$$x^2y^2\Big(3(6-x-y) - x\Big) = 0.$$

Since $x, y > 0$ we deduce

$$18 - 3x - 3y - x = 0 \Rightarrow 4x + 3y = 18.$$

The second equality in (14.13) can be rewritten as

$$x^3y\Big(2(6-x-y) - y\Big) = 0$$

[2] J. J. Sylvester was an English mathematician. He made fundamental contributions to matrix theory, invariant theory, number theory, partition theory, and combinatorics. He played a leadership role in American mathematics in the later half of the 19th century as a professor at the Johns Hopkins University and as founder of the American Journal of Mathematics. `https://en.wikipedia.org/wiki/James_Joseph_Sylvester`

and we conclude as above that

$$2x + 3y = 12.$$

Hence

$$2x = (4x + 3y) - (2x + 3y) = 18 - 12 = 6 \Rightarrow x = 3.$$

Using this information in the equality $2x + 3y = 12$ we deduce $3y = 6$ so $y = 2$. Thus, the only critical point of f is $(3, 2)$. Let us find the Hessian at this point. We have

$$\partial_x f = x^2y^2\big(3(6 - x - y) - x\big) = x^2y^2\big(18 - 4x - 3y\big),$$

$$\partial_y f = x^3y\Big(2(6 - x - y) - y\Big) = x^3y\big(12 - 2x - 3y\big),$$

$$\partial^2_{xx} f = 2xy^2(18 - 4x - 3y) - 4x^2y^2, \;\; \partial^2_{xy} f = 2x^2y(18 - 4x - 3y) - 3x^2y^2,$$

$$\partial^2_{yy} f = 3x^2(12 - 2x - 3y) - 3x^3y.$$

Hence

$$\partial^2_{xx} f(3, 2) = -4 \cdot 3^2 \cdot 2^2 = -144, \;\; \partial^2_{yy}(3, 2) = -3 \cdot 3^3 \cdot 2 = -162,$$

$$\partial^2_{xy} f(3, 2) = -3 \cdot 3^2 \cdot 2^2 = -108.$$

Hence, the Hessian of f at $(3, 2)$ is

$$A := \begin{bmatrix} -144 & -108 \\ -108 & -162 \end{bmatrix}.$$

To decide whether the matrix A is positive/negative definite we use the criterion in Exercise 14.6. Note that $-144 < 0$ and

$$\det A = (-144)(-162) - (-108)^2 = 144 \cdot 162 - (108)^2 = 11644 > 0.$$

Hence A is negative definite and thus the stationary point $(3, 2)$ is a local maximum. □

14.3. Diffeomorphisms and the inverse function theorem

We can now discuss a classical theorem that plays a key role in modern differential geometry/topology. The remainder of this chapter assumes familiarity with basic linear algebra concepts such as linear combinations, linear independence, rank and determinant of a matrix.

We begin by introducing a key concept.

Definition 14.13 (Diffeomorphisms). Let $n \in \mathbb{N}$, $k \in \mathbb{N} \cup \{\infty\}$ and suppose that $U \subset \mathbb{R}^n$ is an open set. A map $\boldsymbol{F} : U \to \mathbb{R}^n$ is called a C^k-*diffeomorphism* if the following hold.

- The map $\boldsymbol{F}$ is injective and its range $\boldsymbol{F}(U)$ is also an open subset of $\mathbb{R}^n$.
- The inverse map $F^{-1} : \boldsymbol{F}(U) \to U$ is also C^k.

□

Example 14.14. (a) Any invertible linear map $L : \mathbb{R}^n \to \mathbb{R}^n$ is a diffeomorphism.

(b) The bijective C^1-map $f : \mathbb{R} \to \mathbb{R}$, $f(x) = x^3$ is *not* a diffeomorphism because its inverse is *not* differentiable at 0. □

Example 14.15. The map

$$\boldsymbol{F} : (0, \infty) \times (0, 2\pi) \to \mathbb{R}^2, \quad \boldsymbol{F}(r, \theta) = \begin{bmatrix} x \\ y \end{bmatrix} = \begin{bmatrix} r\cos\theta \\ r\sin\theta \end{bmatrix}$$

is a C^1-diffeomorphism. Indeed, it is a C^1 map. To see that it is injective observe that if

$$x = r\cos\theta, \quad y = r\sin\theta$$

then

$$x^2 + y^2 = r^2 \Rightarrow r = \sqrt{x^2 + y^2}.$$

Thus, $r > 0$ is uniquely determined by (x, y). Note that $(x/r, y/r)$ is a point on the unit circle, it is not equal to $(1, 0)$ and uniquely determines the angle θ; recall the trigonometric circle in Section 5.6.

This proves that $\boldsymbol{F}$ is injective and the range is the plane $\mathbb{R}^2$ with the nonnegative x-semiaxis removed. Hence the range is open. One can show directly that $\boldsymbol{F}^{-1}$ is C^1, but this is a rather tedious job. Fortunately there is a faster alternate approach that relies on the main theorem of this section, namely, the inverse function theorem. We will present this approach after we discuss this very important theorem. □

We have the following useful consequence of the chain rule. Its proof is left to the reader as an exercise.

Proposition 14.16. *Let $n \in \mathbb{N}$. Suppose that $U \subset \mathbb{R}^n$ is an open set and $\boldsymbol{F} : U \to \mathbb{R}^n$ is a C^1-diffeomorphism. If $\boldsymbol{x}_0 \in U$ and $\boldsymbol{y}_0 = \boldsymbol{F}(\boldsymbol{x}_0)$, then the differential $d\boldsymbol{F}(\boldsymbol{x}_0)$ of $\boldsymbol{F}$ at $\boldsymbol{x}_0$ is invertible and*

$$d\boldsymbol{F}^{-1}(\boldsymbol{y}_0) = d\boldsymbol{F}(\boldsymbol{x}_0)^{-1}.$$ □

The above result gives a necessary condition for a map to be a diffeomorphism, namely its differential has to be invertible. The next result is a very versatile criterion for recognizing diffeomorphisms. Roughly speaking, it states that maps with invertible differentials are very close to being diffeomorphisms.

Theorem 14.17 (Inverse function theorem). *Let $n \in \mathbb{N}$, $k \in \mathbb{N} \cup \{\infty\}$. Suppose that $U \subset \mathbb{R}^n$ is an open set and $F : U \to \mathbb{R}^n$ is a C^k-map. If $\boldsymbol{x}_0 \in U$ is such that the differential $d\boldsymbol{F}(\boldsymbol{x}_0) : \mathbb{R}^n \to \mathbb{R}^n$ is invertible, then there exists an open neighborhood V of $\boldsymbol{x}_0$ with the following properties.*

(i) $V \subset U$.
(ii) The restriction of $\boldsymbol{F}$ to V defines a C^k-diffeomorphism $\boldsymbol{F} : V \to \mathbb{R}^n$.

Proof. For simplicity we consider only the case $k = 1$. The case $k > 1$ follows inductively from this special case [5, Prop. 3.2.9]. We follow closely the approach in the proof of [18, Thm. 2.11]. Denote by L the differential of $\boldsymbol{F}$ at $\boldsymbol{x}_0$ and set $\boldsymbol{y}_0 := \boldsymbol{F}(\boldsymbol{x}_0)$. We begin with an apparently very special case.

A. *The differential L is the identity operator $\mathbb{R}^n \to \mathbb{R}^n$.* We complete the proof in several steps.

Step 1. *We prove that there exists $r > 0$ such that the closed ball $\overline{B_r(\boldsymbol{x}_0)}$ is contained in U and the restriction of $\boldsymbol{F}$ to this closed ball is injective.*

Using the definition of the differential we can write

$$\boldsymbol{F}(\boldsymbol{x}_0 + \boldsymbol{h}) = \boldsymbol{F}(\boldsymbol{x}_0) + \boldsymbol{h} + R(\boldsymbol{h}) = \boldsymbol{y}_0 + \boldsymbol{h} + R(\boldsymbol{h}), \tag{14.14}$$

where

$$\lim_{\|\boldsymbol{h}\| \to 0} \frac{1}{\|\boldsymbol{h}\|} R(h) = 0. \tag{14.15}$$

Observe that

$$\boldsymbol{F}(\boldsymbol{x}_0 + \boldsymbol{h}_1) = \boldsymbol{F}(\boldsymbol{x}_0 + \boldsymbol{h}_2) \overset{(14.14)}{\iff} R(\boldsymbol{h}_1) - R(\boldsymbol{h}_2) = -(\boldsymbol{h}_1 - \boldsymbol{h}_2).$$

We will show that the last equality above cannot happen if $\boldsymbol{h}_1, \boldsymbol{h}_2$ are sufficiently small and $\boldsymbol{h}_1 \neq \boldsymbol{h}_2$.

The correspondence $\boldsymbol{x} \mapsto J_{\boldsymbol{F}}(\boldsymbol{x})$ is continuous and the Jacobian matrix $J_{\boldsymbol{F}}(\boldsymbol{x}_0)$ is invertible. Thus, for $\boldsymbol{x}$ close to $\boldsymbol{x}_0$ the Jacobian $J_{\boldsymbol{F}}(\boldsymbol{x})$ is also invertible; see Exercise 12.9. Fix a radius $r_0 > 0$ such that $\overline{B_{2r_0}(\boldsymbol{x}_0)} \subset U$ and

$$J_{\boldsymbol{F}}(\boldsymbol{x}) \text{ is invertible } \forall \boldsymbol{x} \in B_{2r_0}(\boldsymbol{x}_0). \tag{14.16}$$

Observe that for $\|\boldsymbol{h}\| \leq 2r_0$ we have $R(\boldsymbol{h}) = \boldsymbol{F}(\boldsymbol{x}_0 + \boldsymbol{h}) - \boldsymbol{h} - \boldsymbol{y}_0$. This proves that the map $R : B_{2r_0}(\boldsymbol{0}) \to \mathbb{R}^n$ is differentiable and

$$J_R(\boldsymbol{h}) = J_{\boldsymbol{F}}(\boldsymbol{x}_0 + \boldsymbol{h}) - \mathbb{1} = J_{\boldsymbol{F}}(\boldsymbol{x}_0 + \boldsymbol{h}) - J_{\boldsymbol{F}}(\boldsymbol{x}_0).$$

Since the map $\boldsymbol{F}$ is C^1 we have

$$\lim_{\boldsymbol{h} \to \boldsymbol{0}} \|J_{\boldsymbol{F}}(\boldsymbol{x}_0 + \boldsymbol{h}) - J_{\boldsymbol{F}}(\boldsymbol{x}_0)\|_{HS} = 0,$$

where $\| - \|_{HS}$ denotes the Frobenius norm of a matrix described in Remark 12.13. Fix a very small positive constant $\hbar$,

$$\hbar < \frac{1}{10n}. \tag{14.17}$$

There exists $r < r_0$ sufficiently small such that

$$\|J_R(\boldsymbol{h})\|_{HS} = \|J_{\boldsymbol{F}}(\boldsymbol{x}_0 + \boldsymbol{h}) - J_{\boldsymbol{F}}(\boldsymbol{x}_0)\|_{HS} < \hbar, \ \forall \|\boldsymbol{h}\| < 2r. \tag{14.18}$$

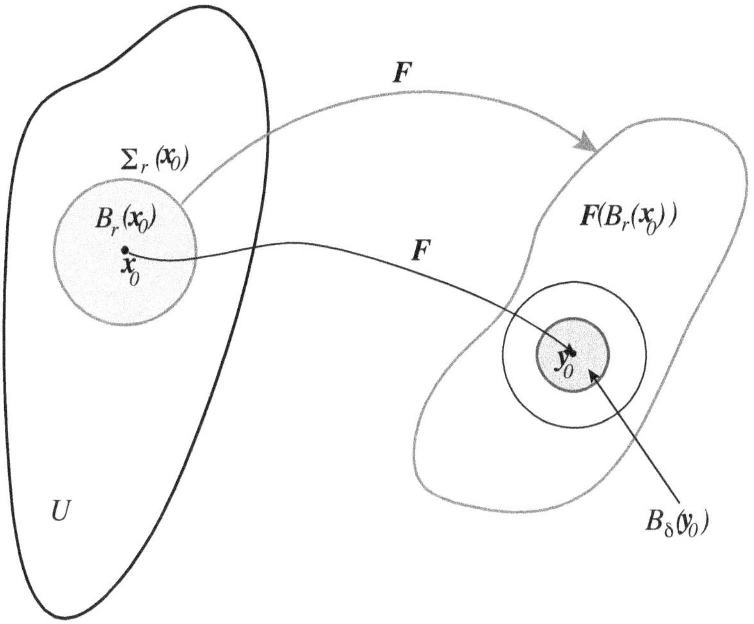

Fig. 14.1 *The map $\boldsymbol{F}$ is injective on $B_r(\boldsymbol{x}_0)$, and the image of this ball contains a small ball $B_\delta(\boldsymbol{y}_0)$.*

Corollary 13.33 implies that

$$\begin{aligned} &\|\boldsymbol{F}(\boldsymbol{x}_0+\boldsymbol{h}_1)-\boldsymbol{F}(\boldsymbol{x}_0+\boldsymbol{h}_2)-(\boldsymbol{h}_1-\boldsymbol{h}_2)\| = \|R(\boldsymbol{h}_1)-R(\boldsymbol{h}_2)\| \\ &\overset{(14.18)}{\leq} \hbar\sqrt{n}\|\boldsymbol{h}_1-\boldsymbol{h}_2\| \overset{(14.17)}{<} \|\boldsymbol{h}_1-\boldsymbol{h}_2\|, \ \ \forall \boldsymbol{h}_1,\boldsymbol{h}_2 \in B_{2r}(\boldsymbol{0}), \ \ \boldsymbol{h}_1 \neq \boldsymbol{h}_2. \end{aligned} \tag{14.19}$$

This proves that if $\|\boldsymbol{h}_1\|, \|\boldsymbol{h}_2\| < 2r$ and $\boldsymbol{h}_1 \neq \boldsymbol{h}_2$, then

$$\boldsymbol{F}(\boldsymbol{x}_0+\boldsymbol{h}_1)-\boldsymbol{F}(\boldsymbol{x}_0+\boldsymbol{h}_2)-(\boldsymbol{h}_1-\boldsymbol{h}_2) = R(\boldsymbol{h}_1)-R(\boldsymbol{h}_2) \neq -(\boldsymbol{h}_1-\boldsymbol{h}_2).$$

Hence

$$\boldsymbol{F}(\boldsymbol{x}_0+\boldsymbol{h}_1)-\boldsymbol{F}(\boldsymbol{x}_0+\boldsymbol{h}_2) \neq \boldsymbol{0}.$$

In particular, this shows that the restriction of $\boldsymbol{F}$ on $\overline{B_r(\boldsymbol{x}_0)} \subset B_{2r}(\boldsymbol{x}_0) \subset U$ is injective.

The sphere

$$\Sigma_r(\boldsymbol{x}_0) = \{\, \boldsymbol{x} \in \mathbb{R}^n; \ \ \|\boldsymbol{x}-\boldsymbol{x}_0\| = r \,\}$$

is compact, and thus its image $\boldsymbol{F}\big(\Sigma_r(\boldsymbol{x}_0)\big)$ is also compact; see Figure 14.1. Because of the injectivity of $\boldsymbol{F}$ on $\overline{B_r(\boldsymbol{x}_0)}$, the point $\boldsymbol{y}_0 = \boldsymbol{F}(\boldsymbol{x}_0)$ does not belong to the image $\boldsymbol{F}\big(\Sigma_r(\boldsymbol{x}_0)\big)$ of this sphere. Hence,

$$\operatorname{dist}\Big(\boldsymbol{y}_0, \boldsymbol{F}\big(\Sigma_r(\boldsymbol{x}_0)\big)\Big) > 0,$$

so there exists $\delta > 0$ such that

$$\|\boldsymbol{y}_0 - \boldsymbol{F}(\boldsymbol{x})\| > 2\delta, \ \ \forall \boldsymbol{x} \in \Sigma_r(\boldsymbol{x}_0). \tag{14.20}$$

Step 2. *We will prove that* $B_\delta(\boldsymbol{y}_0) \subset \boldsymbol{F}\big(B_r(\boldsymbol{x}_0)\big)$, i.e.,

$$\forall \boldsymbol{y} \in B_\delta(\boldsymbol{y}_0),\ \ \exists \boldsymbol{h} \in \mathbb{R}^n \ \text{ such that } \ \|\boldsymbol{h}\| < r \ \text{ and } \ \boldsymbol{y} = \boldsymbol{F}(\boldsymbol{x}_0 + \boldsymbol{h}). \tag{14.21}$$

To do this, let $\boldsymbol{y} \in B_\delta(\boldsymbol{y}_0)$ and consider the function

$$g_{\boldsymbol{y}} : \overline{B_r(\boldsymbol{x}_0)} \to \mathbb{R},\ \ g_{\boldsymbol{y}}(\boldsymbol{x}) = \|\boldsymbol{y} - \boldsymbol{F}(\boldsymbol{x})\|^2.$$

The function $g_{\boldsymbol{y}}$ is continuous and the closed ball $\overline{B_r(\boldsymbol{x}_0)}$ is compact and thus $g_{\boldsymbol{y}}$ admits a global minimum

$$\boldsymbol{z} \in \overline{B_r(\boldsymbol{x}_0)},\ \ g_{\boldsymbol{y}}(\boldsymbol{z}) \leq g_{\boldsymbol{y}}(\boldsymbol{x}),\ \ \forall \boldsymbol{x} \in \overline{B_r(\boldsymbol{x}_0)}.$$

Let us first observe that $\boldsymbol{z} \in B_r(\boldsymbol{x}_0)$. We argue by contradiction. If $\boldsymbol{z} \in \Sigma_r(\boldsymbol{x}_0)$, then

$$\|\boldsymbol{y} - \boldsymbol{F}(\boldsymbol{z})\| \geq \|\boldsymbol{y}_0 - \boldsymbol{F}(\boldsymbol{z})\| - \|\boldsymbol{y}_0 - \boldsymbol{y}\| \stackrel{(14.20)}{>} 2\delta - \underbrace{\|\boldsymbol{y}_0 - \boldsymbol{y}\|}_{<\delta} > \delta > \|\boldsymbol{y} - \boldsymbol{F}(\boldsymbol{x}_0)\|.$$

Hence

$$g_{\boldsymbol{y}}(\boldsymbol{z}) > g_{\boldsymbol{y}}(\boldsymbol{x}_0),\ \ \forall \boldsymbol{z} \in \Sigma_r(\boldsymbol{x}_0)$$

proving that the absolute minimum $\boldsymbol{z}$ of $g_{\boldsymbol{y}}$ is achieved somewhere inside the *open* ball $B_r(\boldsymbol{x}_0)$. The Multidimensional Fermat Principle then implies

$$\nabla g_{\boldsymbol{y}}(\boldsymbol{z}) = \boldsymbol{0} \Longleftrightarrow J_{\boldsymbol{F}}(\boldsymbol{z})\big(\boldsymbol{y} - \boldsymbol{F}(\boldsymbol{z})\big) = \boldsymbol{0}.$$

On the other hand, according to (14.16), the differential $d\boldsymbol{F}(\boldsymbol{z})$ is invertible. We deduce from the above equality that $\boldsymbol{y} = \boldsymbol{F}(\boldsymbol{z})$ for some $\boldsymbol{z} \in B_r(\boldsymbol{x}_0)$.

Since $\boldsymbol{F} : U \to \mathbb{R}^n$ is continuous the preimage $\boldsymbol{F}^{-1}\big(B_\delta(\boldsymbol{y}_0)\big)$ is open (see Exercise 12.4(b)) and so is the set

$$V := \boldsymbol{F}^{-1}\big(B_\delta(\boldsymbol{y}_0)\big) \cap B_r(\boldsymbol{x}_0).$$

The above discussion shows that the resulting map $\boldsymbol{F} : V \to B_\delta(\boldsymbol{y}_0)$ is bijective.

Step 3. We prove that the inverse $\boldsymbol{G} := \boldsymbol{F}^{-1} : B_\delta(\boldsymbol{y}_0) \to V$ is Lipschitz continuous.

Let $\boldsymbol{y}_*, \boldsymbol{y} \in B_\delta(\boldsymbol{y}_0)$. We set $\boldsymbol{x} := \boldsymbol{G}(\boldsymbol{y})$, $\boldsymbol{x}_* = \boldsymbol{G}(\boldsymbol{y}_*)$. Then $\boldsymbol{x} = \boldsymbol{x}_0 + \boldsymbol{h}$, $\boldsymbol{x}_* = \boldsymbol{x}_0 + \boldsymbol{h}_*$. From (14.19) we deduce

$$\|\boldsymbol{x} - \boldsymbol{x}_*\| - \|\boldsymbol{y} - \boldsymbol{y}_*\| \leq \|\boldsymbol{y} - \boldsymbol{y}_* - (\boldsymbol{x} - \boldsymbol{x}_*)\| = \|R(\boldsymbol{h}) - R(\boldsymbol{h}_*)\| \stackrel{(14.19)}{\leq} \hbar\sqrt{n}\|\boldsymbol{h} - \boldsymbol{h}_*\|$$
$$\stackrel{(14.17)}{\leq} \frac{1}{10\sqrt{n}}\|\boldsymbol{h} - \boldsymbol{h}_*\| = \frac{1}{10\sqrt{n}}\|\boldsymbol{x} - \boldsymbol{x}_*\| \leq \frac{1}{10}\|\boldsymbol{x} - \boldsymbol{x}_*\|.$$

We deduce that

$$\|\boldsymbol{G}(\boldsymbol{y}) - \boldsymbol{G}(\boldsymbol{y}_*)\| = \|\boldsymbol{x} - \boldsymbol{x}_*\| \leq \frac{10}{9}\|\boldsymbol{y} - \boldsymbol{y}_*\|.$$

Step 4. We prove that the inverse $\boldsymbol{G} := \boldsymbol{F}^{-1} : B_\delta(\boldsymbol{y}_0) \to V$ is differentiable.

Fix $\boldsymbol{y}_* \in B_\delta(\boldsymbol{y}_0)$. There exists $\boldsymbol{x}_* \in V$ such that $\boldsymbol{F}(\boldsymbol{x}_*) = \boldsymbol{y}_*$. Proposition 14.16 suggests that the differential of $\boldsymbol{G}$ at $\boldsymbol{y}_*$ should be the inverse of the differential of $\boldsymbol{F}$ at $\boldsymbol{x}_0$. For $\boldsymbol{y} \in B_\delta(\boldsymbol{y}_0)$ we set

$$R(\boldsymbol{y}, \boldsymbol{y}_*) := \Big(\boldsymbol{G}(\boldsymbol{y}) - \boldsymbol{G}(\boldsymbol{y}_*) - d\boldsymbol{F}(\boldsymbol{x}_*)^{-1}(\boldsymbol{y} - \boldsymbol{y}_*)\Big).$$

We have to prove that

$$\lim_{y \to y_*} \frac{\|R(\boldsymbol{y}, \boldsymbol{y}_*)\|}{\|\boldsymbol{y} - \boldsymbol{y}_*\|} = 0.$$

Observe first that

$$d\boldsymbol{F}(\boldsymbol{x}_*)R(\boldsymbol{y}, \boldsymbol{y}_*) = d\boldsymbol{F}(\boldsymbol{x}_*)\Big(\boldsymbol{G}(\boldsymbol{y}) - \boldsymbol{G}(\boldsymbol{y}_*)\Big) - (\boldsymbol{y} - \boldsymbol{y}_*)$$

$$= \underbrace{d\boldsymbol{F}(\boldsymbol{x}_*)\Big(\boldsymbol{G}(\boldsymbol{y}) - \boldsymbol{G}(\boldsymbol{y}_*)\Big) - \Big(\boldsymbol{F}(\boldsymbol{G}(\boldsymbol{y})) - \boldsymbol{F}(\boldsymbol{G}(\boldsymbol{y}_*))\Big)}_{=:Q(\boldsymbol{y}, \boldsymbol{y}_*)}.$$

Since $\boldsymbol{F}$ is differentiable at $\boldsymbol{x}_*$ we have

$$\lim_{y \to y_*} \frac{\|Q(\boldsymbol{y}, \boldsymbol{y}_*)\|}{\|\boldsymbol{G}(\boldsymbol{y}) - \boldsymbol{G}(\boldsymbol{y}_*)\|} = 0. \tag{14.22}$$

On the other hand,

$$\|\boldsymbol{G}(\boldsymbol{y}) - \boldsymbol{G}(\boldsymbol{y}_*)\| \leq \frac{10}{9}\|\boldsymbol{y} - \boldsymbol{y}_*\|$$

so that

$$\frac{10}{9}\frac{\|Q(\boldsymbol{y}, \boldsymbol{y}_*)\|}{\|\boldsymbol{G}(\boldsymbol{y}) - \boldsymbol{G}(\boldsymbol{y}_*)\|} \geq \frac{\|Q(\boldsymbol{y}, \boldsymbol{y}_*)\|}{\|\boldsymbol{y} - \boldsymbol{y}_*\|}$$

We deduce that

$$\frac{\|d\boldsymbol{F}(\boldsymbol{x}_*)R(\boldsymbol{y}, \boldsymbol{y}_*)\|}{\|\boldsymbol{y} - \boldsymbol{y}_*\|} \leq \frac{10}{9}\frac{\|Q(\boldsymbol{y}, \boldsymbol{y}_*)\|}{\|\boldsymbol{G}(\boldsymbol{y}) - \boldsymbol{G}(\boldsymbol{y}_*)\|}.$$

On the other hand, since $d\boldsymbol{F}(\boldsymbol{x}_*)$ is invertible, we deduce from Exercise 12.28(iii) that there exists a constant $C > 0$ such that

$$C\|\boldsymbol{h}\| \leq \|d\boldsymbol{F}(\boldsymbol{x}_*)\boldsymbol{h}\|, \quad \forall \boldsymbol{h} \in \mathbb{R}^n.$$

We conclude that

$$C\frac{\|R(\boldsymbol{y}, \boldsymbol{y}_*)\|}{\|\boldsymbol{y} - \boldsymbol{y}_*\|} \leq \frac{10}{9}\frac{\|Q(\boldsymbol{y}, \boldsymbol{y}_*)\|}{\|\boldsymbol{G}(\boldsymbol{y}) - \boldsymbol{G}(\boldsymbol{y}_*)\|}.$$

Invoking the Squeezing Principle and (14.22) we deduce from the above

$$\lim_{y \to y_*} \frac{\|R(\boldsymbol{y}, \boldsymbol{y}_*)\|}{\|\boldsymbol{y} - \boldsymbol{y}_*\|} = 0.$$

This proves the differentiability of $\boldsymbol{G}$ at $\boldsymbol{y}_*$.

Step 5. We finally prove that map $\boldsymbol{G}$ is C^1. We have to show that the map $\boldsymbol{y} \mapsto J_{\boldsymbol{G}}(\boldsymbol{y})$ is continuous, i.e., the map

$$\boldsymbol{y} \mapsto J_{\boldsymbol{F}}\big(\boldsymbol{G}(\boldsymbol{y})\big)^{-1}$$

is continuous. This follows from Exercise 12.10.

B. We now discuss the general case when we do not assume that $d\boldsymbol{F}(\boldsymbol{x}_0) = \mathbb{1}$. Set $L = d\boldsymbol{F}(\boldsymbol{x}_0)$. Define

$$\Phi : U \to \mathbb{R}^n, \quad \Phi = L^{-1} \circ \boldsymbol{F}.$$

The chain rule implies that

$$d\Phi = dL^{-1} \circ d\boldsymbol{F} = L^{-1} \circ d\boldsymbol{F} = \mathbb{1}.$$

From Case **A** we deduce that there exists an open neighborhood V of $\boldsymbol{x}_0$ contained in U such that the restriction of Φ to V is a diffeomorphism. From the equality $F = L \circ \Phi$ we deduce that the restriction of F to V is also a diffeomorphism. □

Remark 14.18. (a) The assumption that $d\boldsymbol{F}(\boldsymbol{x}_0)$ is invertible is equivalent with the condition

$$\det J_{\boldsymbol{F}}(\boldsymbol{x}_0) \neq 0.$$

This is easier to verify especially when n is not too large.

(b) If $V \subset \mathbb{R}^n$ is an open neighborhood of $\boldsymbol{x}_0$ satisfying the conditions (i) and (ii) in Theorem 14.17, then any smaller open neighborhood $W \subset V$ of $\boldsymbol{x}_0$ satisfies these conditions. □

We have the following useful consequence of the inverse function theorem. Its proof is left to you as an exercise.

Corollary 14.19. *Let $n \in \mathbb{N}$. Suppose that U is an open subset of $\mathbb{R}^n$ and $\boldsymbol{F} : U \to \mathbb{R}^n$ is a C^1-map satisfying the following conditions.*

(i) The map $\boldsymbol{F}$ is injective.
(ii) For any $\boldsymbol{x} \in U$, the differential $d\boldsymbol{F}(\boldsymbol{x}) : \mathbb{R}^n \to \mathbb{R}^n$ is bijective.

Then the map $\boldsymbol{F}$ is a C^1-diffeomorphism. □

Remark 14.20. The condition (ii) in the above corollary is equivalent with the condition

$$\det J_{\boldsymbol{F}}(\boldsymbol{x}) \neq 0, \;\; \forall \boldsymbol{x} \in U.$$

This is easier to verify especially when n is not too large. □

Example 14.21. Consider again the map

$$\boldsymbol{F} : (0, \infty) \times (0, 2\pi) \to \mathbb{R}^2, \;\; \boldsymbol{F}(r, \theta) = \begin{bmatrix} x \\ y \end{bmatrix} = \begin{bmatrix} r\cos\theta \\ r\sin\theta \end{bmatrix}$$

in Example 14.15. We have seen there that it is injective. According to Corollary 14.19, to prove that it is a diffeomorphism it suffices to show that for any $(r, \theta) \in (0, \infty) \times (0, 2\pi)$ the Jacobian matrix $J_{\boldsymbol{F}}(r, \theta)$ is invertible. We have

$$J_{\boldsymbol{F}}(r, \theta) = \begin{bmatrix} \frac{\partial x}{\partial r} & \frac{\partial x}{\partial \theta} \\ \\ \frac{\partial y}{\partial r} & \frac{\partial y}{\partial \theta} \end{bmatrix} = \begin{bmatrix} \cos\theta & -r\sin\theta \\ \sin\theta & r\cos\theta \end{bmatrix}.$$

The determinant of the above matrix is

$$\det J_{\boldsymbol{F}} = (\cos\theta) \cdot (r\cos\theta) - (-r\sin\theta) \cdot (\sin\theta) = r\cos^2\theta + r\sin^2\theta = r > 0.$$

Thus the matrix $J_{\boldsymbol{F}}(r, \theta)$ is invertible for any $(r, \theta) \in (0, \infty) \times (0, 2\pi)$. □

Example 14.22. The transformation $\boldsymbol{F}$ in Example 14.21 is often referred to as the *change to polar coordinates*. A function u depending on the Cartesian coordinates (x, y) can be transformed to a function depending on the coordinates (r, θ), $u(x, y) = u(r\cos\theta, r\sin\theta)$.

Often in physics and geometry one is faced with the problem of transforming various quantities expressed in the (x, y)-coordinates to quantities expressed in the polar coordinates (r, θ). We discuss below one such important example.

Suppose $u = u(x, y)$ is a C^2-function. We are deliberately vague about the domain of definition of u since this detail is irrelevant to the computations we are about to perform. Its *Laplacian* is the function

$$\Delta u = \frac{\partial^2 u}{\partial x^2} + \frac{\partial^2 u}{\partial y^2}.$$

We want to express the Laplacian in polar coordinates. The chain rule, cleverly deployed, will do the trick.

Note first that for any function (or quantity) q depending on the variables (x, y), $q = q(x, y)$, we have

$$\frac{\partial q}{\partial x} = \frac{\partial q}{\partial r}\frac{\partial r}{\partial x} + \frac{\partial q}{\partial \theta}\frac{\partial \theta}{\partial x}, \tag{14.23a}$$

$$\frac{\partial q}{\partial y} = \frac{\partial q}{\partial r}\frac{\partial r}{\partial y} + \frac{\partial q}{\partial \theta}\frac{\partial \theta}{\partial y}. \tag{14.23b}$$

Let us concentrate first on the x-derivative. We rewrite (14.23a) in the form,

$$\frac{\partial q}{\partial x} = \frac{\partial r}{\partial x}\frac{\partial q}{\partial r} + \frac{\partial \theta}{\partial x}\frac{\partial q}{\partial \theta}.$$

Since the exact nature of the quantity q is not important in the sequel, we will drop the letter q from our notations. Hence, the above equality becomes

$$\frac{\partial}{\partial x} = \frac{\partial r}{\partial x}\frac{\partial}{\partial r} + \frac{\partial \theta}{\partial x}\frac{\partial}{\partial \theta}. \tag{14.24}$$

From the equalities $r^2 = x^2 + y^2$, $x = r\cos\theta$ and $y = r\sin\theta$ we deduce

$$\partial_x r = \frac{x}{r} = \cos\theta, \;\; \partial_y r = \frac{y}{r} = \sin\theta. \tag{14.25}$$

Derivating the equality $y = r\sin\theta$ with respect to x we deduce

$$0 = \partial_x r(\sin\theta) + r(\cos\theta)\partial_x\theta = \cos\theta\sin\theta + (r\cos\theta)\partial_x\theta = \cos\theta\big(\sin\theta + r\partial_x\theta\big)$$

$$\Rightarrow \partial_x\theta = -\frac{\sin\theta}{r}.$$

Hence (14.24) becomes

$$\boxed{\frac{\partial}{\partial x} = \cos\theta\frac{\partial}{\partial r} - \frac{\sin\theta}{r}\frac{\partial}{\partial \theta}}. \tag{14.26}$$

Then

$$\begin{aligned}
\frac{\partial^2 u}{\partial x^2} &= \frac{\partial}{\partial x}\frac{\partial u}{\partial x} = \Big(\cos\theta\frac{\partial}{\partial r} - \frac{\sin\theta}{r}\frac{\partial}{\partial\theta}\Big)\Big(\cos\theta\frac{\partial u}{\partial r} - \frac{\sin\theta}{r}\frac{\partial u}{\partial\theta}\Big)\\
&= \cos\theta\frac{\partial}{\partial r}\Big(\cos\theta\frac{\partial u}{\partial r} - \frac{\sin\theta}{r}\frac{\partial u}{\partial\theta}\Big) - \frac{\sin\theta}{r}\frac{\partial}{\partial\theta}\Big(\cos\theta\frac{\partial u}{\partial r} - \frac{\sin\theta}{r}\frac{\partial u}{\partial\theta}\Big)\\
&= \cos\theta\Big(\cos\theta\frac{\partial^2 u}{\partial r^2} + \frac{\sin\theta}{r^2}\frac{\partial u}{\partial\theta} - \frac{\sin\theta}{r}\frac{\partial^2 u}{\partial r\partial\theta}\Big)\\
&\quad - \frac{\sin\theta}{r}\Big(-\sin\theta\frac{\partial u}{\partial r} + \cos\theta\frac{\partial^2 u}{\partial\theta\partial r} - \frac{\sin\theta}{r}\frac{\partial^2 u}{\partial\theta^2}\Big)\\
&= \cos^2\theta\,\partial_r^2 u + \frac{\sin\theta\cos\theta}{r^2}\partial_\theta u - 2\frac{\sin\theta\cos\theta}{r}\partial^2_{r\theta}u + \frac{\sin^2\theta}{r}\partial_r u + \frac{\sin^2\theta}{r^2}\partial^2_\theta u.
\end{aligned}$$

Arguing in a similar fashion we have

$$\frac{\partial}{\partial y}=\frac{\partial r}{\partial y}\frac{\partial}{\partial r}+\frac{\partial \theta}{\partial y}\frac{\partial}{\partial \theta}.$$

Derivating with respect to y the equality $x = r\cos\theta$ we deduce in similar fashion that $\partial_y\theta=\frac{\cos\theta}{r}$ so that

$$\boxed{\frac{\partial}{\partial y}=\sin\theta\frac{\partial}{\partial r}+\frac{\cos\theta}{r}\frac{\partial}{\partial \theta}},$$

$$\begin{aligned}\frac{\partial^2 u}{\partial y^2}&=\Big(\sin\theta\frac{\partial}{\partial r}+\frac{\cos\theta}{r}\frac{\partial}{\partial\theta}\Big)\Big(\sin\theta\frac{\partial u}{\partial r}+\frac{\cos\theta}{r}\frac{\partial u}{\partial\theta}\Big)\\
&=\sin\theta\frac{\partial}{\partial r}\Big(\sin\theta\frac{\partial u}{\partial r}+\frac{\cos\theta}{r}\frac{\partial u}{\partial\theta}\Big)+\frac{\cos\theta}{r}\frac{\partial}{\partial\theta}\Big(\sin\theta\frac{\partial u}{\partial r}+\frac{\cos\theta}{r}\frac{\partial u}{\partial\theta}\Big)\\
&=\sin\theta\Big(\sin\theta\frac{\partial^2 u}{\partial r^2}-\frac{\cos\theta}{r^2}\frac{\partial u}{\partial\theta}+\frac{\cos\theta}{r}\frac{\partial^2 u}{\partial r\partial\theta}\Big)\\
&\quad+\frac{\cos\theta}{r}\Big(\cos\theta\frac{\partial u}{\partial r}+\sin\theta\frac{\partial^2 u}{\partial\theta\partial r}-\frac{\sin\theta}{r}\frac{\partial u}{\partial\theta}+\frac{\cos\theta}{r}\frac{\partial^2 u}{\partial\theta^2}\Big)\\
&=\sin^2\theta\partial_r^2u-\frac{\sin\theta\cos\theta}{r^2}\partial_\theta u+\frac{2\sin\theta\cos\theta}{r}\partial^2_{r\theta}u+\frac{\cos^2\theta}{r}\partial_r u+\frac{\cos^2\theta}{r^2}\partial^2_\theta u.\end{aligned}$$

Putting together all of the above we deduce

$$\boxed{\Delta u=\partial_r^2u+\frac{1}{r}\partial_r u+\frac{1}{r^2}\partial_\theta^2 u=\frac{1}{r}\frac{\partial}{\partial r}\Big(r\frac{\partial u}{\partial r}\Big)+\frac{1}{r^2}\frac{\partial u}{\partial\theta^2}}. \tag{14.27}$$

To see how this works in practice, consider the special case $u = (x^2+y^2)^{\frac{p}{2}}$. Since $x^2+y^2=r^2$ we deduce $u=r^p$ and

$$\Delta u=\partial_r^2(r^p)+\frac{1}{r}\partial_r(r^p)=p(p-1)r^{p-2}+pr^{p-2}=p^2r^{p-2}.$$ □

14.4. The implicit function theorem

To understand the meaning of the implicit function theorem it is useful to start with a simple example.

Example 14.23. Consider the function $f:\mathbb{R}^2\to\mathbb{R}$, $f(x,y)=x^2+y^2-1$. The level set

$$f^{-1}(0)=\big\{(x,y)\in\mathbb{R}^2;\;\; f(x,y)=1\big\}$$

is the circle C_1 of radius 1 centered at the origin of $\mathbb{R}^2$. This curve cannot be the graph of any function, but portions of it are graphs. For example, the part of C_1 above the x-axis

$$\big\{\,(x,y)\in C_1;\;\; y>0\,\big\}$$

is the graph of a function. To see this, we solve for y the equality $x^2+y^2=1$, and since $y>0$, we obtain the unique solution

$$y=\sqrt{1-x^2}.$$

We say that the function $\sqrt{1-x^2}$ is a function defined *implicitly* by the equality $f(x,y)$.

This is not an isolated phenomenon. The implicit function theorem states that for many equations of the type $f(x,y)=const$ the solution set is locally the graph of a function g, although we cannot describe g as explicitly as in the above simple example. □

Theorem 14.24 (Implicit function theorem. Version 1). *Let $m, n \in \mathbb{N}$. Suppose that*

$$\mathcal{O} \subset \mathbb{R}^n \times \mathbb{R}^m$$

is an open set, $\boldsymbol{F} = \boldsymbol{F}(\boldsymbol{u}, \boldsymbol{v}) : \mathcal{O} \to \mathbb{R}^m$ is a C^1 map and $(\boldsymbol{u}_0, \boldsymbol{v}_0) \in \mathcal{O}$ is a point satisfying the following properties.

(i) $\boldsymbol{F}(\boldsymbol{u}_0, \boldsymbol{v}_0) = \boldsymbol{0}$.
(ii) The restriction of the differential $d\boldsymbol{F}(0, \boldsymbol{v}_0)$ to the subspace

$$\boldsymbol{0} \times \mathbb{R}^m \subset \mathbb{R}^n \times \mathbb{R}^m$$

induces an invertible linear map $\boldsymbol{0} \times \mathbb{R}^m \to \mathbb{R}^m$.

Then there exists an open neighborhood U of $\boldsymbol{u}_0 \in \mathbb{R}^n$, an open neighborhood V of $\boldsymbol{v}_0 \in \mathbb{R}^m$ and a C^1-map $\boldsymbol{G} : U \to V$ with the following properties

- $U \times V \subset \mathcal{O}$.
- *If $(\boldsymbol{u}, \boldsymbol{v}) \in U \times V$, then $\boldsymbol{F}(\boldsymbol{u}, \boldsymbol{v}) = \boldsymbol{0}$ if and only if $\boldsymbol{v} = \boldsymbol{G}(\boldsymbol{u})$.*

In other words, for any $\boldsymbol{u} \in U$, the equation $\boldsymbol{F}(\boldsymbol{u}, \boldsymbol{v}) = \boldsymbol{0}$ has a unique solution $\boldsymbol{v} \in V$. This unique solution is denoted by $\boldsymbol{G}(\boldsymbol{u})$. We say that $\boldsymbol{G}$ is the function implicitly defined by the equation $\boldsymbol{F}(\boldsymbol{u}, \boldsymbol{v}) = \boldsymbol{0}$.

Proof. If we represent $L := d\boldsymbol{F}(\boldsymbol{u}_0, \boldsymbol{v}_0)$ as an $m \times (n+m)$ matrix, then it has a block decomposition

$$L = \left[\frac{\partial \boldsymbol{F}}{\partial \boldsymbol{u}}, \frac{\partial \boldsymbol{F}}{\partial \boldsymbol{v}} \right] = [A\ B],$$

where A is an $m \times n$ matrix and B is a $m \times m$ matrix. The matrix $B = \frac{\partial \boldsymbol{F}}{\partial \boldsymbol{v}}$ describes the restriction of L to the subspace $\boldsymbol{0} \times \mathbb{R}^m$ and assumption (ii) implies that B is invertible.

Consider the new map $\boldsymbol{H} : \mathcal{O} \to \mathbb{R}^n \times \mathbb{R}^m$,

$$\boldsymbol{H}(\boldsymbol{u}, \boldsymbol{v}) = \big(\boldsymbol{u}, \boldsymbol{F}(\boldsymbol{u}, \boldsymbol{v}) \big).$$

The differential of $\boldsymbol{H}$ at $(\boldsymbol{u}_0, \boldsymbol{v}_0)$ is a linear map $T : \mathbb{R}^m \times \mathbb{R}^n \to \mathbb{R}^m \times \mathbb{R}^n$ described by an $(m+n) \times (m+n)$ matrix with block decomposition

$$T = \begin{bmatrix} \mathbb{1}_n & \boldsymbol{0} \\ & \\ \frac{\partial \boldsymbol{F}}{\partial \boldsymbol{u}} & \frac{\partial \boldsymbol{F}}{\partial \boldsymbol{v}} \end{bmatrix} = \begin{bmatrix} \mathbb{1}_n & \boldsymbol{0} \\ A & B \end{bmatrix}.$$

Since B is invertible, we deduce that T is also invertible since $\det T = \det B \neq 0$. Note that $\boldsymbol{H}(\boldsymbol{u}_0, \boldsymbol{v}_0) = (\boldsymbol{u}_0, \boldsymbol{0})$.

From the inverse function theorem we deduce that there exists an open neighborhood W of $(\boldsymbol{u}_0, \boldsymbol{v}_0)$ contained in $\mathcal{O}$ such that the restriction of $\boldsymbol{H}$ to W is a diffeomorphism. By making W smaller as in Remark 14.18, we can assume that W has the form $W = U \times V$, where $U \subset \mathbb{R}^n$ is an open neighborhood of $\boldsymbol{u}_0$ in $\mathbb{R}^n$ and $V \subset \mathbb{R}^m$ is an open neighborhood of $\boldsymbol{v}_0$ in $\mathbb{R}^m$.

We denote by $\mathcal{W}$ the image of $U \times V$ via $\boldsymbol{H}$, $\mathcal{W} := \boldsymbol{H}(U \times V)$. Let $\Phi : \mathcal{W} \to U \times V$ denote the inverse of $\boldsymbol{H} : U \times V \to \mathcal{W}$. The diffeomorphism Φ has the form

$$\Phi(\boldsymbol{x}, \boldsymbol{y}) = (\boldsymbol{u}, \boldsymbol{v}) = \big(\, \Psi(\boldsymbol{x}, \boldsymbol{y}), \Xi(\boldsymbol{x}, \boldsymbol{y}) \,\big) \in U \times V \subset \mathbb{R}^n \times \mathbb{R}^m,$$

where

$$\Psi : \mathcal{W} \to \mathbb{R}^n, \quad \Xi : \mathcal{W} \to \mathbb{R}^m$$

are C^1-maps. Note that if $(\boldsymbol{x}, \boldsymbol{y}) \in \mathcal{W}$ and

$$(\boldsymbol{u}, \boldsymbol{v}) = \Phi(\boldsymbol{x}, \boldsymbol{y}) = \big(\, \Psi(\boldsymbol{x}, \boldsymbol{y}), \Xi(\boldsymbol{x}, \boldsymbol{y}) \,\big),$$

then

$$\begin{aligned}(\boldsymbol{x}, \boldsymbol{y}) &= \boldsymbol{H}(\boldsymbol{u}, \boldsymbol{v}) = \big(\, \boldsymbol{u}, \boldsymbol{F}(\boldsymbol{u}, \boldsymbol{v}) \,\big) \\ &= \Big(\, \Psi(\boldsymbol{x}, \boldsymbol{y}), \boldsymbol{F}\big(\Phi(\boldsymbol{x}, \boldsymbol{y}), \Xi(\boldsymbol{x}, \boldsymbol{y}) \big) \,\Big).\end{aligned}$$

We deduce that $\boldsymbol{u} = \boldsymbol{x}$, i.e., $\Xi(\boldsymbol{x}, \boldsymbol{y}) = \boldsymbol{v}$. Hence the inverse Φ has the form

$$\Phi(\boldsymbol{x}, \boldsymbol{y}) = (\boldsymbol{u}, \boldsymbol{v}) = \big(\, \boldsymbol{x}, \Xi(\boldsymbol{x}, \boldsymbol{y}) \,\big),$$

where

$$\boldsymbol{u} = \boldsymbol{x}, \quad \boldsymbol{y} = \boldsymbol{F}(\boldsymbol{u}, \boldsymbol{v}).$$

Note that

$$\begin{aligned}\boldsymbol{F}(\boldsymbol{u}, \boldsymbol{v}) = 0 &\Longleftrightarrow (\boldsymbol{x}, \boldsymbol{y}) = \boldsymbol{H}(\boldsymbol{u}, \boldsymbol{v}) = (\boldsymbol{u}, \boldsymbol{0}) \\ &\Longleftrightarrow (\boldsymbol{u}, \boldsymbol{v}) = \Phi(\boldsymbol{u}, \boldsymbol{0}) = \big(\, \boldsymbol{u}, \Xi(\boldsymbol{u}, \boldsymbol{0}) \,\big).\end{aligned}$$

The sought out map $\boldsymbol{G}$ is then

$$\boldsymbol{G}(\boldsymbol{u}) = \Xi(\boldsymbol{u}, \boldsymbol{0}).$$

□

Remark 14.25. (a) The above proof shows that there exist

- an open set $\mathcal{W} \subset \mathbb{R}^n \times \mathbb{R}^m$ containing $(\boldsymbol{u}_0, \boldsymbol{0})$,
- an open neighborhood V of $\boldsymbol{v}_0$ in $\mathbb{R}^m$,
- an open neighborhood U of $\boldsymbol{u}_0$ in $\mathbb{R}^n$, and
- a diffeomorphism $\Phi : \mathcal{W} \to \mathbb{R}^n \times \mathbb{R}^m$,

with the following properties:

(i) $\Phi(\boldsymbol{u}_0, \boldsymbol{0}) = (\boldsymbol{u}_0, \boldsymbol{v}_0)$, $\Phi(\mathcal{W}) = U \times V$.

(ii) The diffeomorphism Φ maps the part of the plane $\mathbb{R}^n \times \boldsymbol{0}$ contained in $\mathcal{W}$ bijectively to the part of the set $\boldsymbol{F} = \boldsymbol{0}$ contained in $U \times V$.

(b) The assumption (ii) in the statement of the implicit function theorem can be rephrased in a more convenient way. In the proof this was used to conclude that the $(n+m) \times m$ matrix $J_{\boldsymbol{F}}$ representing $d\boldsymbol{F}(\boldsymbol{x}_0, \boldsymbol{y}_0)$ has the property that the matrix B determined by the columns and the *last* m rows is invertible. The condition (ii) is then equivalent with the condition

$$\det B \neq 0.$$

Note that if $F^1, \dots, F^m$ are the components of $\boldsymbol{F}$ and $v^1, \dots, v^m$ are the components of $\boldsymbol{v}$, then B is the $m \times m$ matrix with entries

$$B^i_j = \frac{\partial F^i}{\partial v^j}, \quad 1 \leq i, j \leq m.$$

The condition implies that $d\boldsymbol{F}(\boldsymbol{u}_0, \boldsymbol{v}_0)$ is surjective.

If we assume *only* that the differential $d\boldsymbol{F}(\boldsymbol{u}_0, \boldsymbol{v}_0) : \mathbb{R}^{n+m} \to \mathbb{R}^m$ is *surjective*, then the $m \times (n+m)$ matrix representing this linear operator has maximal rank m and thus, there exist m columns so that the matrix determined by these columns and all the m rows is invertible; see [21, Thm. 6.1]. If we reorder the components of a vector in $\mathbb{R}^{n+m}$ we can then assume that these m columns are the last m-columns.

(c) Note that the surjectivity of the linear operator $d\boldsymbol{F}(\boldsymbol{u}_0, \boldsymbol{v}_0) : \mathbb{R}^{n+m} \to \mathbb{R}^m$ is equivalent with the linear independence of the m rows of $J_{\boldsymbol{F}}(\boldsymbol{u}_0, \boldsymbol{v}_0)$. If $F^1, \dots, F^m$ are the components of F, then the rows of $J_{\boldsymbol{F}}$ describe the differentials $dF^1, \dots, dF^n$ and we see that the rows are linearly independent if and only if the gradients $\nabla F^1, \dots, \nabla F^m$ are linearly independent. □

In view of last remark, we can give an equivalent but more flexible formulation of Theorem 14.24. First, let us introduce some convenient terminology. A *codimension m coordinate subspace* of an Euclidean space $\mathbb{R}^n$ is a linear subspace of $\mathbb{R}^n$ described by the vanishing of a given group of m coordinates.

For example, the subspace of $\mathbb{R}^5$ of the form

$$S = \{ (x^1, 0, x^3, x^4, 0); \;\; x^1, x^3, x^4 \in \mathbb{R} \}$$

is a codimension 2 coordinate subspace described by the vanishing of the coordinates x^2, x^5. It is naturally isomorphic to $\mathbb{R}^3 = \mathbb{R}^{5-2}$. The codimension 3 subspace

$$\{ (0, x^2, 0, 0, x^5); \;\; x^2, x^5 \in \mathbb{R} \}$$

described by the vanishing of the coordinates x^1, x^3, x^4 is none other than $S^{\perp}$, the orthogonal complement of S. It is naturally isomorphic to $\mathbb{R}^2$. Note that we have a natural decomposition

$$(x^1, x^2, x^3, x^4, x^5) = \underbrace{(x^1, 0, x^3, x^4, 0)}_{\in S} + \underbrace{(0, x^2, 0, 0, x^5)}_{\in S^{\perp}}.$$

In general, a codimension m coordinate subspace of $\mathbb{R}^N$ is naturally isomorphic to $\mathbb{R}^{N-m}$ and thus has dimension $N - m$. The orthogonal complement $S^{\perp}$ is another coordinate subspace of codimension $N - m$. Moreover any $\boldsymbol{z} \in \mathbb{R}^N$ admits a *unique decomposition* of the form

$$\boldsymbol{z} = \boldsymbol{u} + \boldsymbol{v}, \;\; \boldsymbol{u} \in S, \;\; \boldsymbol{v} \in S^{\perp}.$$

The vectors $\boldsymbol{u}, \boldsymbol{v}$ are called the *projections* of $\boldsymbol{z}$ on S and respectively $S^{\perp}$.

Theorem 14.26 (Implicit function theorem.Version 2). *Let $m, n \in \mathbb{N}$ and set $N := n + m$. Suppose that $\mathcal{O} \subset \mathbb{R}^N$ is an open set, $\boldsymbol{F} = \boldsymbol{F}(\boldsymbol{x}) : \mathcal{O} \to \mathbb{R}^m$ is a C^1 map and $\boldsymbol{p}_0 \in \mathcal{O}$ is a point satisfying the following properties.*

(i) $\boldsymbol{F}(\boldsymbol{p}_0) = \boldsymbol{0}$.
(ii) The differential $L = d\boldsymbol{F}(\boldsymbol{p}_0) : \mathbb{R}^N \to \mathbb{R}^m$ is surjective.

Set $n = N - m$. Label the coordinates $(x^i)_{1 \leq i \leq N}$ of $\boldsymbol{x} \in \mathbb{R}^N$ so that

$$\det \left[\frac{\partial F^i}{\partial x^{n+j}}(\boldsymbol{p}_0) \right]_{1 \leq i,j \leq m} \neq 0.$$

Denote by S the codimension m coordinate plane defined by the equations

$$x^{n+1} = \cdots = x^{n+m} = 0.$$

Then there exist

- *an open ball $U \subset S$ centered at $\boldsymbol{u}_0$, the projection of $\boldsymbol{p}_0$ on S,*
- *an open ball $V \subset S^\perp$ centered at $\boldsymbol{v}_0$, the projection of $\boldsymbol{p}_0$ on $S^\perp$ and a C^1-map $\boldsymbol{G} : U \to V$*

with the following properties:

- *$V \times U \subset \mathcal{O}$;*
- *If $(\boldsymbol{v}, \boldsymbol{u}) \in U \times V$, then $\boldsymbol{F}(\boldsymbol{u}, \boldsymbol{v}) = \boldsymbol{0}$ if and only if $\boldsymbol{v} = \boldsymbol{G}(\boldsymbol{u})$.*

□

Remark 14.27. Under the assumptions of the above theorem, we say that we can solve for $x^{n+1}, \dots, x^{n+m}$ in terms of the remaining variables $x^1, \dots, x^n$. The map $\boldsymbol{G}$ is then described by m functions $g^1, \dots, g^m$, depending on the "free" variables $x^1, \dots, x^n$, such that

$$x^{m+1} = g^1(x^1, \dots, x^n), \dots, x^m = g^m(x^1, \dots, x^n),$$

if and only if

$$F^1(x^1, \dots, x^m, x^{m+1}, \dots, x^{m+n}) = \cdots = F^m(x^1, \dots, x^m, x^{m+1}, \dots, x^{m+n}) = 0,$$

where $F^1(\boldsymbol{x}), \dots, F^m(\boldsymbol{x})$ are the components of $\boldsymbol{F}(\boldsymbol{x}) \in \mathbb{R}$. In the notation of the above theorem we have

$$\boldsymbol{v} = (x^{n+1}, \dots, x^{n+m}), \quad \boldsymbol{u} = (x^1, \dots, x^n).$$ □

Example 14.28. Consider a function $f : \mathbb{R}^{n+1} \to \mathbb{R}$. We denote by $(x^0, x^1, \dots, x^n)$ the Cartesian coordinates in $\mathbb{R}^{n+1}$. Consider the zero set of f,

$$Z_f := \{ (x^0, x^1, \dots, x^n) \in \mathbb{R}^{n+1};\ f(x^0, x^1, \dots, x^n) = 0 \}.$$

Assume for some $\boldsymbol{x}_0 = (x_0^0, x_0^1, \dots, x_0^n)$ we have $\boldsymbol{x}_0 \in Z_f$ and the differential of f at $\boldsymbol{x}_0$ is surjective as a linear map $\mathbb{R}^{n+1} \to \mathbb{R}$.

The differential of f at $\boldsymbol{x}_0$ is the $1 \times (n+1)$ matrix

$$\big[\, \partial_{x^0} f(\boldsymbol{x}_0), \partial_{x^1} f(\boldsymbol{x}_0), \dots, \partial_{x^n} f(\boldsymbol{x}_0) \,\big].$$

The differential of $df(\mathbf{0})$ is surjective if and only if it is nonzero, i.e., one of the partial derivatives

$$\partial_{x^0} f(\boldsymbol{x}_0), \partial_{x^1} f(\boldsymbol{x}_0), \dots, \partial_{x^n} f(\boldsymbol{x}_0)$$

is nonzero. Without loss of generality we can assume that $\partial_{x^0} f(\boldsymbol{x}_0) \neq 0$.

Consider the coordinate subspace S described by $x^0 = 0$. Explicitly,

$$S = \big\{\, (0, x^1, \dots, x^n);\;\; x^1, \dots, x^n \in \mathbb{R} \,\big\}.$$

Loosely speaking, the implicit function theorem says that, in a neighborhood of $\boldsymbol{x}_0$, we can solve the equation

$$f(x^0, x^1, \dots, x^n) = 0$$

uniquely for x^0 in terms of $x^1, \dots, x^n$.

More precisely, the implicit function theorem states that there exists an open (n-dimensional) ball B^n in $S \cong \mathbb{R}^n$ centered at $\boldsymbol{u}_0 = (\boldsymbol{x}_0^1, \dots, \boldsymbol{x}_0^n)$, an open interval $I \subset \mathbb{R}$ containing $v_0 = x_0^0$, and a C^1-function $g : B \to I$ such that

$$(x^0, x^1, \dots, x^n) \in I \times B^n \text{ and } f(x^0, x^1, \dots, x^n) = 0 \Longleftrightarrow x^0 = g(x^1, \dots, x^n).$$

The function g is only locally defined, and it is called the *implicit function* determined by the equation

$$f(x^0, x^1, \dots, x^n) = 0,$$

i.e.,

$$\begin{aligned} f(x^0, x^1, \dots, x^n) = 0 &\Longleftrightarrow x^0 = g(x^1, \dots, x^n) \\ &\Longleftrightarrow f\big(\, g(x^1, \dots, x^n),\; x^1, \dots, x^n \,\big) = 0. \end{aligned} \tag{14.28}$$

We often express this by saying that, in an open neighborhood of $\boldsymbol{x}_0$, along the zero set Z_f the coordinate x^0 is a function of the remaining coordinates

$$x^0 = x^0(x^1, \dots, x^n)$$

and thus, locally, Z is the graph of a C^1-function depending on the n variables $(x^1, \dots, x^n)$.

From the equality $f(x^0, x^1, \dots, x^n) = 0$ we can determine the partial derivatives of x^0 at $\boldsymbol{u}_0$, when x^0 is viewed as a function of $(x^1, \dots, x^n)$. Derivating the equality

$$f(x^0, x^1, \dots, x^n) = 0$$

with respect to x^i, $i = 1, \dots, n$, while keeping in mind that x^0 is really a function of the remaining variables $x^1, \dots, x^n$, we deduce from the chain rule that

$$f'_{x^0} \frac{\partial x^0}{\partial x^i} + f'_{x_i} = 0 \Rightarrow \frac{\partial x^0}{\partial x^i} = -\frac{f'_{x^i}}{f'_{x^0}}.$$

Hence

$$\frac{\partial x^0}{\partial x^i}(\boldsymbol{u}_0) = g'_{x^i}(\boldsymbol{u}_0) = -\frac{f'_{x^i}\big(\, x_0^0, \boldsymbol{u}_0 \big)}{f'_{x^0}\big(\, x_0^0, \boldsymbol{u}_0 \,\big)}. \tag{14.29}$$

□

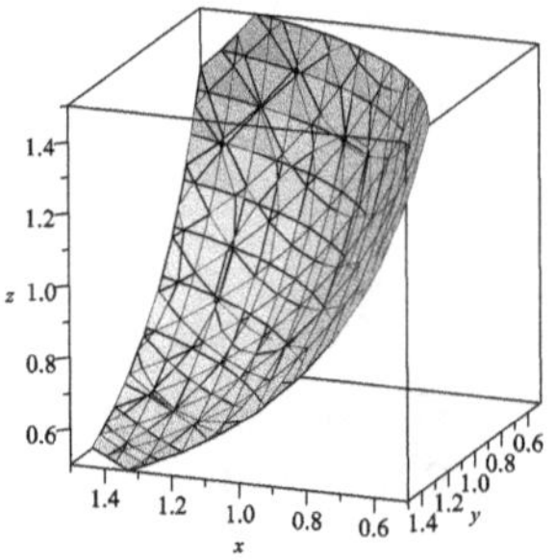

Fig. 14.2 *The surface in* $\mathbb{R}^3$ *described by the equation* $2^{xyz} = 2$.

Example 14.29. Consider subset

$$Z = \{(x, y, z) \in \mathbb{R}^3;\ \ 2^{xyz} = 2\}.$$

A portion of this set is depicted in Figure 14.2.

Note that $Z \neq \emptyset$ since $(1, 1, 1) \in Z$. Equivalently, Z is the zero set of the function $f : \mathbb{R}^3 \to \mathbb{R}$, $f(x, y, z) = 2^{xyz} - 2$. Note that

$$\frac{\partial f}{\partial z} = xy2^{xyz} \ln 2, \ \ \frac{\partial f}{\partial z}(1, 1, 1) = 2 \ln 2.$$

The implicit function theorem shows that there exists a small open ball B in $\mathbb{R}^2$ centered at $(1, 1)$, an open interval $I \subset \mathbb{R}$ centered at 1 and a C^1-function $g : B \to I$ such that

$$(x, y, z) \in Z \cap \Big(B \times I \Big) \Longleftrightarrow z = g(x, y).$$

In other words, in an open neighborhood of $(1, 1, 1)$, the set Z is the graph of a C^1-function $z = z(x, y)$. Let us compute the partial derivatives of $z(x, y)$ at $(1, 1)$.

Derivating with respect to x the equality $2^{xyz} = 2$ in which we treat z as a function of the variables (x, y), we deduce

$$(yz + x\partial_x z)2^{xyz} \ln 2 = 0 \Rightarrow x\frac{\partial z}{\partial x} + zy = 0 \Rightarrow \frac{\partial z}{\partial x} = -\frac{zy}{x}.$$

When $(x, y) = (1, 1)$, we have $z = 1$ and we deduce

$$\frac{\partial z}{\partial x}(1, 1) = -1.$$

We can give an alternate verification of this equality. Namely, observe that we can solve for z *explicitly* the equality $2^{xyz} = 2$. More precisely, we have

$$\log_2 \Big(2^{xyz}\Big) = \log_2(2) \Rightarrow xyz = 1 \Rightarrow z = \frac{1}{xy} \Rightarrow \frac{\partial z}{\partial x} = -\frac{1}{x^2 y} \Rightarrow \frac{\partial z}{\partial x}(1, 1) = -1. \ \ \square$$

Example 14.30. Consider the map

$$\boldsymbol{F} : \mathbb{R}^3 \to \mathbb{R}^2, \ \ \boldsymbol{F}(x, y, z) = \begin{bmatrix} u \\ v \end{bmatrix} = \begin{bmatrix} xyz - 2 \\ x + y + z - 4 \end{bmatrix}.$$

The zero set Z of $\boldsymbol{F}$ consists of the points $(x, y, z) \in \mathbb{R}^3$ satisfying the equations

$$xyz = 2, \quad x + y + z = 4. \tag{14.30}$$

Note that $(1, 1, 2) \in Z$. The Jacobian of the map $\boldsymbol{F}$ at a point $(x, y, z) \in \mathbb{R}^3$ is the 2×3 matrix

$$J = J(x, y, z) = \begin{bmatrix} \frac{\partial u}{\partial x} & \frac{\partial u}{\partial y} & \frac{\partial u}{\partial z} \\ & & \\ \frac{\partial v}{\partial x} & \frac{\partial v}{\partial y} & \frac{\partial v}{\partial z} \end{bmatrix} = \begin{bmatrix} yz & xz & xy \\ 1 & 1 & 1 \end{bmatrix}.$$

Consider the minor of the above matrix determined by the y and z columns,

$$\det \begin{bmatrix} xz & xy \\ & \\ 1 & 1 \end{bmatrix} = xz - xy = x(y - z).$$

Note that this minor is nonzero at the point $(x, y, z) = (1, 1, 2)$. The implicit function theorem then implies that, near $(1, 1, 2)$ we can solve (14.30) for y, z in terms of x. In other words, there exists a tiny (open) box B centered at $(1, 1, 2)$ such that the intersection of Z with B coincides with the graph of a C^1-map

$$I \ni x \mapsto \big(y(x), z(x)\big) \in \mathbb{R}^2,$$

where I is some open interval on the x-axis centered at $x = 1$. To find the derivatives of $y(x)$ and $z(x)$ at $x = 1$ we derivate (14.30) with respect to x, keeping in mind that y and z depend on x. We deduce

$$yz + xzy' + xyz' = 0, \quad 1 + y' + z' = 0.$$

At the point $(z, y, z) = (1, 1, 2)$ we have $yz = xz = 2$, $xy = 1$, and the above equations become

$$\begin{cases} 2y' + z' = -2 \\ \;\; y' + z' = -1. \end{cases}$$

Note that the matrix of this linear system is the sub-matrix of $J(1, 1, 2)$ corresponding to the y, z columns. This matrix is nondegenerate so we can solve the above system uniquely. In fact, if we subtract the second equation from the first we deduce $y' = -1$. Using this information in the second equation we deduce $z' = 0$. Hence

$$y'(x)\big|_{x=1} = -1, \quad z'(x)\big|_{x=1} = 0. \qquad \square$$

14.5. Submanifolds of $\mathbb{R}^n$

The implicit function theorem discussed in the previous section leads to a very important concept that clarifies and generalizes our intuitive concepts of curves and surfaces.

14.5.1. *Definition and basic examples*

A submanifold of dimension m in the n-dimensional Euclidean space $\mathbb{R}^n$ is a set that locally "feels" like an m-dimensional vector subspace of $\mathbb{R}^n$. This is not very precise and we will address this lack of precision in Definition 14.31. Before we do this we want to build some intuition. Let us consider a controversy that plagued humanity for centuries.

We now know that the surface of the Earth is spherical, but initially, this was not what people believed. Anybody that walked in a wide open field could see clearly that the Earth is "obviously flat" as far as the eyes can see. The problem is that "*as far as the eyes can see*" is not far enough when compared to the size of the Earth. Our eyesight can only reach as far as the horizon: this is where the Earth's surface begins "to bend".

This phenomenon is not restricted to spheres. Take for example a surface S in $\mathbb{R}^3$, say the one depicted in Figure 14.3. Any tiny region on this surface is nearly flat, and it can appear to be so to an observer living there.

Another way to express it is to say that *any* tiny region on S, together with a tiny region around it in space, can be straightened so it now looks like a tiny region of the vector subspace $\mathbb{R}^2$ sitting in $\mathbb{R}^3$.

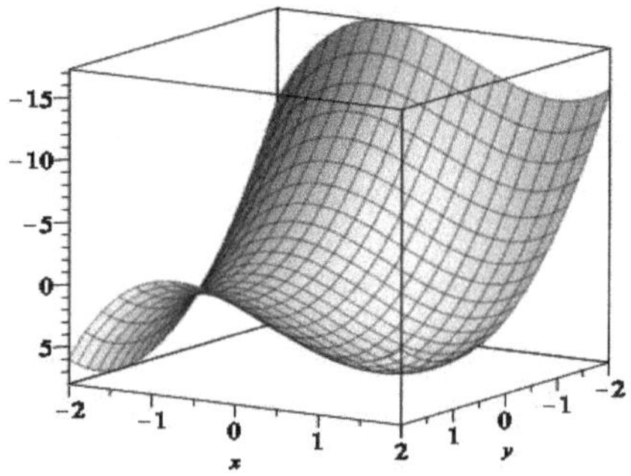

Fig. 14.3 *The surface* $S := \{z = -x^3 + x^2 - 2y^2 + 3x - 4y, |x|, |y| < 2\}$.

For example, the origin $(0, 0, 0)$ lives on the surface S. In Figure 14.4 we depicted the image of the tiny region $|x|, |y| < 0.2$ of S, magnified by a factor of 10. In fact, if we push the magnification factor to ∞, then this tiny region will approach a 2-dimensional vector subspace of $\mathbb{R}^3$ that is intimately related to the surface, namely, the plane tangent to S at the origin.

The local straightening property is indeed the defining feature of a surface in $\mathbb{R}^3$. The next definition is mouthful but it describes in precise terms the essential features of a surface and its higher dimensional cousins, the submanifolds of Euclidean spaces. Let $n \in \mathbb{N}$, $m \in \mathbb{N}_0$, $m \leq n$ and $k \in \mathbb{N} \cup \{\infty\}$.

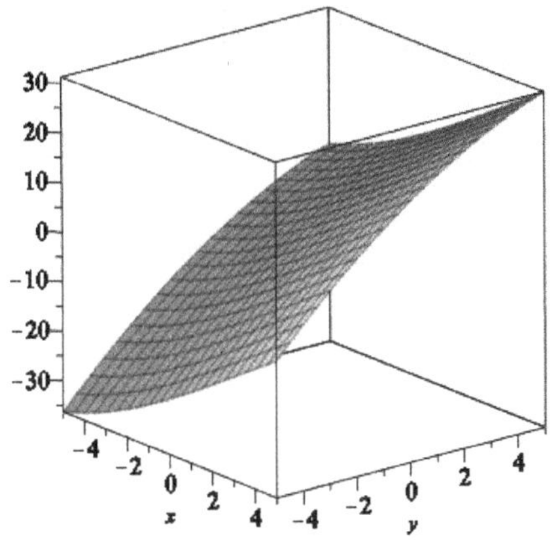

Fig. 14.4 *The tiny region of the surface* $z = -x^3 + x^2 - 2y^2 + 3x - 4y$ *corresponding to* $|x|, |y| < 0.2$ *could seem flat under magnification.*

Definition 14.31 (Submanifolds). An *m-dimensional C^k-submanifold* of $\mathbb{R}^n$ is a subset $X \subset \mathbb{R}^n$ such that, for any $\boldsymbol{p}_0 \in X$, there exists a pair $(\mathcal{U}, \Psi)$ with the following properties.

(i) $\mathcal{U}$ is an open neighborhood of $\boldsymbol{p}_0$ in $\mathbb{R}^n$,
(ii) $\Psi : \mathcal{U} \to \mathbb{R}^n$ is a C^k-diffeomorphism. We set $\boldsymbol{q}_0 := \Psi(\boldsymbol{p}_0), U := \Psi(\mathcal{U})$.
(iii) If $\mathbb{R}^m \times \mathbf{0}$ denotes the coordinate subspace

$$\mathbb{R}^m \times \mathbf{0} = \Big\{ \big(x^1, \dots, x^m, x^{m+1}, \dots, x^n \big) \in \mathbb{R}^n;\ \ x^{m+1} = \dots = x^n = 0 \Big\},$$

then $\boldsymbol{q}_0 \in \mathbb{R}^m \times \mathbf{0}$ and $\Psi\big(X \cap \mathcal{U} \big) = \big(\mathbb{R}^m \times \mathbf{0} \big) \cap U$.

An open set $\mathcal{U}$ as above is called a *coordinate neighborhood* of p_0 *adapted to* X. The pair $(\mathcal{U}, \Psi)$ is called a *straightening diffeomorphism* or *straightening chart* near $\boldsymbol{p}_0$. The induced map $\Psi : X \cap \mathcal{U} \to \mathbb{R}^m$ is called a *local coordinate chart*of X at $\boldsymbol{p}_0$. The inverse map $\Psi^{-1} : \big(\mathbb{R}^m \times \mathbf{0}\big) \cap U \to X \cap \mathcal{U}$ is called a *local parametrization* of X near $\boldsymbol{p}_0$. □

Remark 14.32. (a) A local chart maps a piece of the m-dimensional submanifold X bijectively onto an open subset of the "flat" m-dimensional space $\mathbb{R}^m$. A local parametrization of X "deforms" an open subset of the "flat" m-dimenisonal space $\mathbb{R}^m$ bijectively onto a piece of the m-dimensional submanifold.

Intuitively, a 1-dimensional submanifold of $\mathbb{R}^3$ is a curve, while a 2-dimensional submanifold of $\mathbb{R}^3$ is a surface.

(b) If $X \subset \mathbb{R}^n$ is an m-dimensional submanifold, $\boldsymbol{p}_0 \in X$ and $\mathcal{U}$ is a coordinate neighborhood of $\boldsymbol{p}_0$ adapted to X, then any open neighborhood $\mathcal{V}$ of $\boldsymbol{p}_0$ in $\mathbb{R}^n$ such that $\mathcal{V} \subset \mathcal{U}$ is also a coordinate neighborhood of $\boldsymbol{p}_0$ adapted to X. □

Example 14.33. (a) A point in $\mathbb{R}^n$ is a 0-dimensional submanifold of $\mathbb{R}^n$. An open subset of $\mathbb{R}^n$ is an n-dimensional submanifold of $\mathbb{R}^n$.[3] □

Our next result is a direct consequence of the inverse function theorem and describes an alternate characterization of submanifolds.

Proposition 14.34 (Parametric description of a submanifold). *Let $m, n \in \mathbb{N}$, $m < n$. Suppose that $U \subset \mathbb{R}^m$ is an open set and*

$$\Phi : U \to \mathbb{R}^n, \quad \Phi(\boldsymbol{u}) = \begin{bmatrix} \Phi^1(\boldsymbol{u}) \\ \vdots \\ \Phi^n(\boldsymbol{u}) \end{bmatrix}$$

*is a C^k-*parametrization, *i.e., a C^k-map satisfying the following properties.*

(i) The map Φ is injective.
(ii) The map Φ is an immersion, *i.e., for any $\boldsymbol{u} \in U$ the $n \times m$ Jacobian matrix*

$$J_\Phi := \Big(\partial_{u^j} \Phi^i(\boldsymbol{u}) \Big)_{\substack{1 \le i \le n \\ 1 \le j \le m}}$$

has maximal rank m, i.e., it is injective when viewed as a linear operator $\mathbb{R}^m \to \mathbb{R}^n$.
(iii) The inverse $\Phi^{-1} : \Phi(U) \to U$ is continuous.

Then the following hold.

(A) The set $X = \Phi(U)$ is an m-dimensional C^k-submanifold of $\mathbb{R}^n$. (The map Φ is referred to as a parametrization *of X.)*
(B) If $\ell \in \mathbb{N}$, $V \subset \mathbb{R}^\ell$ is an open set and $\boldsymbol{G} : V \to \mathbb{R}^n$ is a C^k-map such that $\boldsymbol{G}(V) \subset X$, then the map $\Phi^{-1} \circ \boldsymbol{G} : V \to U$ is C^k.

Proof. Fix $\boldsymbol{u}_0 \in U$ and set $\boldsymbol{x}_0 := \Phi(\boldsymbol{u}_0)$. The Jacobian matrix $J_\Phi(\boldsymbol{u}_0)$ is an $n \times m$ matrix with (maximal) rank m. Thus (see [21, Thm. 6.1]) there exist m rows such that the matrix determined by these rows and all the m columns of $J_\Phi(\boldsymbol{x}_0)$ is invertible. Without loss of generality we can assume that these m rows are the first m rows. We denote by $J_\Phi^m(\boldsymbol{u}_0)$ this $m \times m$ matrix. Define

$$\boldsymbol{F} : U \times \mathbb{R}^{n-m} \to \mathbb{R}^n, \quad \boldsymbol{F}(\boldsymbol{u}, \boldsymbol{v}) = \begin{bmatrix} \Phi^1(\boldsymbol{u}) \\ \vdots \\ \Phi^m(\boldsymbol{u}) \\ \Phi^{m+1}(\boldsymbol{u}) + v^1 \\ \vdots \\ \Phi^n(\boldsymbol{u}) + v^n \end{bmatrix}.$$

Note that $\boldsymbol{F}(\boldsymbol{u}_0, \boldsymbol{0}) = \Phi(\boldsymbol{u}_0) = \boldsymbol{p}_0$. The Jacobian matrix of $\boldsymbol{F}$ and $(\boldsymbol{u}_0, \boldsymbol{0})$ is the $n \times n$ matrix with the block decomposition

$$J_{\boldsymbol{F}}(\boldsymbol{u}_0, \boldsymbol{0}) = \begin{bmatrix} J_\Phi^m(\boldsymbol{u}_0) & \boldsymbol{0}_{m \times (n-m)} \\ A_{(n-m) \times m} & \mathbb{1}_{n-m} \end{bmatrix},$$

where $\boldsymbol{0}_{m \times (n-m)}$ denotes the $m \times (n-m)$ matrix with all entries equal to 0 and $A_{(n-m) \times m}$ is an $(n-m) \times m$ matrix whose explicit description is irrelevant for our argument.

Since $\det J_\Phi^m(\boldsymbol{u}_0) \neq 0$, we deduce that $\det J_{\boldsymbol{F}}(\boldsymbol{u}_0, \boldsymbol{0}) \neq 0$ so $J_{\boldsymbol{F}}(\boldsymbol{u}_0, \boldsymbol{0})$ is invertible. We can then apply the inverse function theorem to conclude that there exists $\rho > 0$ sufficiently small such that the restriction of $\boldsymbol{F}$ to the open set

[3] Can you verify this claim?

$B^m_\rho(\boldsymbol{u}_0) \times B^{n-m}_\rho(\mathbf{0}) \subset \mathbb{R}^m \times \mathbb{R}^{n-m}$ is a diffeomorphism. Since $\boldsymbol{F}^{-1}$ is continuous, there is an open neighborhood $\mathcal{O}$ of $\boldsymbol{p}_0$ in $\mathbb{R}^n$ such that

$$\mathcal{O} \cap \Phi(U) = \Phi\big(B^m_\rho(\boldsymbol{u}_0)\big).$$

Now choose $r < \rho$ sufficiently small so that, if $W_r = B^m_r(\boldsymbol{u}_0) \times B^{n-m}_r(\mathbf{0})$, then $\mathcal{W}_r := \boldsymbol{F}(W_r) \subset \mathcal{O}$.

The inverse $\Psi : \mathcal{W}_r \to W_r \subset \mathbb{R}^m \times \mathbb{R}^{n-m} = \mathbb{R}^n$ is a local straightening of $X = \Phi(U)$ around $\Phi(\boldsymbol{u}_0)$. It sends $\mathcal{W}_r \cap X$ to the m-dimensional ball $B^m_r(\boldsymbol{u}_0) \times \mathbf{0}_{n-m} \subset \mathbb{R}^m \times \mathbb{R}^{n-m}$.

Suppose $\boldsymbol{G}$ is as in the statement of the proposition. Fix $\boldsymbol{v}_0 \in \mathbb{V}$. Then there exists $\boldsymbol{u}_0 \in U$ such that $\Phi(\boldsymbol{u}_0) = \boldsymbol{G}(\boldsymbol{v}_0)$. Choose a local straightening Ψ of X around $\Phi(\boldsymbol{u}_0)$. Now observe that the restriction $\Phi^{-1} \circ \boldsymbol{G}$ to the *open* neighborhood $\boldsymbol{G}^{-1}(\mathcal{W})$ of $\boldsymbol{v}_0$ is equal to $\Psi \circ \boldsymbol{G}$. □

Remark 14.35. A C^1-map $\Phi : I \to \mathbb{R}^n$, $I \subset \mathbb{R}$ interval, is an immersion if and only if the derivative $\Phi'(t)$ is nonzero for any $t \in I$. □

Example 14.36. (a) Let $\boldsymbol{p} \in \mathbb{R}^n$, $\boldsymbol{v} \in \mathbb{R}^n \setminus \{\mathbf{0}\}$. The line $\ell_{\boldsymbol{p},\boldsymbol{v}}$ is a 1-dimensional submanifold. To see this consider the map

$$\gamma : \mathbb{R} \to \mathbb{R}^n, \;\; \gamma(t) = \boldsymbol{p} + t\boldsymbol{v}.$$

This is an immersion since $\dot{\gamma}(t) = \boldsymbol{v} \neq \mathbf{0}$, $\forall t \in \mathbb{R}$. It is also an injection since $\boldsymbol{v} \neq \mathbf{0}$. According to (11.5), the image of γ is the line $\ell_{\boldsymbol{p},\boldsymbol{v}}$. The inverse

$$\gamma^{-1} : \ell_{\boldsymbol{p},\boldsymbol{v}} \to \mathbb{R}$$

is given by

$$\gamma^{-1}(q) = \frac{1}{\|\boldsymbol{v}\|}\langle q - p, \boldsymbol{v}\rangle.$$

The above map is clearly continuous.

(b) Consider the map $\Phi : (0, \pi) \to \mathbb{R}^2$,

$$\Phi(\theta) = (\cos\theta, \sin\theta).$$

This map is injective (why?) and it is an immersion since

$$\Phi'(\theta) = (-\sin\theta, \cos\theta), \;\; \|\Phi'(\theta)\|^2 = 1 \neq 0.$$

Its image is the half circle centered at the origin and contained in the upper half space $\{y > 0\}$. Its inverse associates to a point (x, y) on this circle the angle $\theta = \arccos x \in (0, \pi)$. The map $(x, y) \mapsto \arccos x$ is obviously continuous.

(c) A helix H in $\mathbb{R}^3$ is a curve described by the parametrization (see Figure 14.5)

$$\boldsymbol{\alpha} : (0, 1) \to \mathbb{R}^3, \;\; \boldsymbol{\alpha}(t) = \big(r\cos(at), r\sin(at), bt\big),$$

where r, a, b are fixed *nonzero* real numbers and $r > 0$. Note that the above map is an immersion since

$$\|\dot{\boldsymbol{\alpha}}(t)\|^2 = a^2r^2\sin^2 t + a^2r^2\cos^2 t + b^2 = a^2r^2 + b^2 \neq 0.$$

The map $\boldsymbol{\alpha}$ is clearly injective since its third component bt is such. Its inverse associates to a point (x, y, z) on the helix, the real number $t = z/b$. The map $(x, y, z) \mapsto z/b$ is obviously continuous. In Figure 14.5 we have depicted an example of helix with $r = 1$, $a = 4\pi$, $b = 1$.

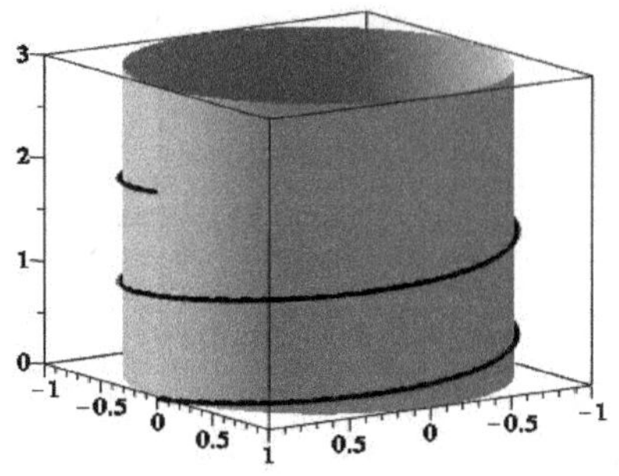

Fig. 14.5 *The helix described by the parametrization* $\big(\cos(4\pi t), \sin(4\pi t), 2t\big)$ *is winding up a cylinder of radius* $r = 1$. *During one second, it winds twice around the cylinder while climbing up* 2 *units of distance.*

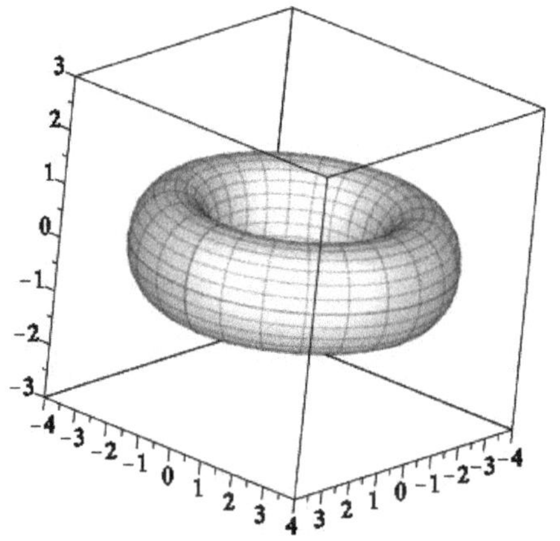

Fig. 14.6 *A 2-dimensional torus in* $\mathbb{R}^3$.

(d) Consider the map $\Phi : (-\pi, \pi) \times (-\pi, \pi) \to \mathbb{R}^3$ given by

$$\Phi(\theta, \varphi) = \begin{bmatrix} (3 + \cos\varphi)\cos\theta \\ (3 + \cos\varphi)\sin\theta \\ \sin\varphi \end{bmatrix}.$$

This is an injective immersion; see Exercise 14.14. Its image is the torus in Figure 14.6. □

Corollary 14.37 (Graphical description of a submanifold). *Let* $m, k \in \mathbb{N}$. *Suppose that* $U \subset \mathbb{R}^m$ *is an open set and* $\boldsymbol{F} : \mathbb{R}^m \to \mathbb{R}^k$ *is a* C^1*-map. Then the graph of* $\boldsymbol{F}$,

$$\Gamma_{\boldsymbol{F}} := \Big\{ \big(\boldsymbol{x}, \boldsymbol{F}(\boldsymbol{x})\big) \in \mathbb{R}^m \times \mathbb{R}^k;\ \ \boldsymbol{x} \in U \Big\} \subset \mathbb{R}^m \times \mathbb{R}^k \cong \mathbb{R}^{m+k},$$

is an m*-dimensional* C^1*-submanifold of* $\mathbb{R}^m \times \mathbb{R}^k$.

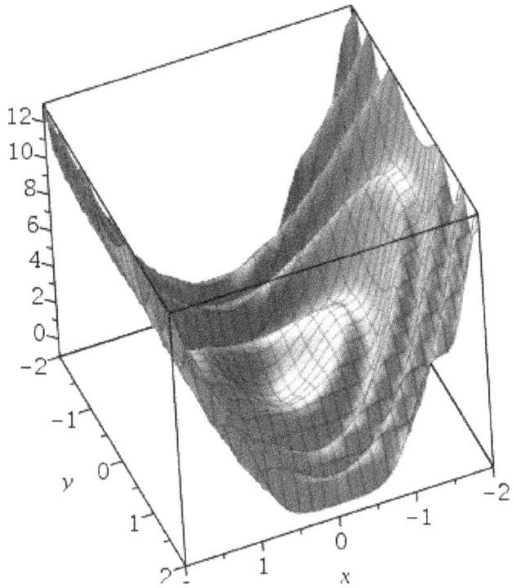

Fig. 14.7 *The graph of the map* $f : (-2,2)\times(-2,2)\to\mathbb{R}$, $f(x,y) = 3x^2+\sin(3x^2+3y^2)$ *is a 2-dimensional submanifold of* $\mathbb{R}^3$.

Proof.[4] Observe that the map $\Phi:\mathbb{R}^m\to\mathbb{R}^{m+k}$, $\Phi(\boldsymbol{x}) = \big(\boldsymbol{x}, \boldsymbol{F}(\boldsymbol{x})\big)$ is a parametrization. The conclusion now follows from Proposition 14.34. □

Remark 14.38. The condition (iii) in Proposition 14.34 is difficult to verify in concrete situations. However, the parametrizations play an important role in integration problems and it would be desirable to have a simple way of recognizing them. We mention below, without proof, one such method.

Suppose that $X\subset\mathbb{R}^n$ is an m-dimensional submanifold, $U\subset\mathbb{R}^m$ is an open set and $\Phi:U\to\mathbb{R}^n$ is an injective immersion such that $\Phi(U)\subset X$. Then Φ is a parametrization, i.e., it satisfies assumption (iii) in Proposition 14.34. □

Remark 14.25 coupled with Corollary 14.37 imply immediately the following important result. We let the reader supply the proof.

Proposition 14.39 (Implicit description of a submanifold). *Let* $k,m,n\in\mathbb{N}$, $m<n$. *Suppose that U is an open subset of* $\mathbb{R}^n$ *and* $\boldsymbol{F}:U\to\mathbb{R}^m$ *is a* C^k*-map satisfying*

$$\forall\boldsymbol{x}\in U:\ \ \boldsymbol{F}(\boldsymbol{x})=0\Rightarrow \textit{the differential } d\boldsymbol{F}(\boldsymbol{x}):\mathbb{R}^n\to\mathbb{R}^m \textit{ is surjective.} \tag{14.31}$$

Then the set

$$\{\boldsymbol{F}=\boldsymbol{0}\} := \big\{\,\boldsymbol{x}\in U\subset\mathbb{R}^n;\boldsymbol{F}(\boldsymbol{x})=\boldsymbol{0}\,\big\}$$

is an $(n-m)$*-dimensional* C^k*-submanifold of* $\mathbb{R}^n$. □

[4] Remember the simple trick used in this proof. It will come in handy later.

Remark 14.40. Let us rephrase the above result in, hopefully, more intuitive terms.

Recall that an $m \times n$ matrix $m < n$ has maximal rank if and only if its rows are linearly independent. The differential $d\boldsymbol{F}(\boldsymbol{x}) : \mathbb{R}^n \to \mathbb{R}^m$ is surjective if and only if it has maximal rank m. Denote by $F^1, \dots, F^m$ the components map $\boldsymbol{F} : U \to \mathbb{R}^m$. Then the rows of $J_{\boldsymbol{F}}(\boldsymbol{x})$ correspond to the gradients of the components F^j.

The zero set $\{\boldsymbol{F} = \boldsymbol{0}\}$ is a subset $U \subset \mathbb{R}^n$ described by m equations in n unknowns

$$F^1(x^1, \dots, x^n) = 0, \cdots, F^m(x^1, \dots, x^n) = 0.$$

The condition (14.31) is equivalent with the following *transversality* property.

If $\boldsymbol{x} \in U$ and $F^1(\boldsymbol{x}) = \cdots = F^m(\boldsymbol{x}) = 0$, then the gradients $\nabla F^1(\boldsymbol{x}), \dots, \nabla F^m(\boldsymbol{x})$ are linearly independent.

The above result shows that, if the transversality condition is satisfied, then the common zero locus

$$\mathcal{Z}(F^1, \dots, F^m) := \big\{ \boldsymbol{x} \in U;\ F^1(\boldsymbol{x}) = \cdots = F^m(\boldsymbol{x}) = 0 \big\}$$

is a C^k-submanifold of $\mathbb{R}^n$ of dimension $n - m$. In this case we say that the submanifold $\mathcal{Z}(F^1, \dots, F^m)$ $\mathcal{Z}$ *is cut out transversally* by the equations $F^j(\boldsymbol{x}) = 0$, $j = 1, \dots, m$.

Note that the transversality is automatically satisfied if $\boldsymbol{F}(\boldsymbol{x})$ is a *submersion*, i.e., *for any* $\boldsymbol{x} \in U$, the differential $d\boldsymbol{F} : \mathbb{R}^n \to \mathbb{R}^m$ is surjective.

As an illustration, consider the situation in Example 14.30. There we proved that the equations

$$xyz = 2, \quad x + y + z = 4$$

satisfy the transversality conditions in a small open neighborhood U of the point $(1, 1, 2)$. Thus in this neighborhood these equations describe a submanifold of dimension $3 - 2 = 1$, i.e., a curve. □

From Propositions 14.34 and 14.39 we deduce the following useful characterization of submanifolds.

Theorem 14.41. *Let $m, n \in \mathbb{N}$, $m < n$, and $X \subset \mathbb{R}^n$. The following statements are equivalent.*

(i) The set X is an m-dimensional submanifold of $\mathbb{R}^n$.

(ii) For any $\boldsymbol{x}_0 \in X$ there exists an open neighborhood V of $\boldsymbol{x}_0$ and a submersion $\boldsymbol{F} : V \to \mathbb{R}^{n-m}$ such that

$$V \cap X = \big\{ \boldsymbol{x} \in V;\ \boldsymbol{F}(\boldsymbol{x}) = \boldsymbol{0} \big\}.$$

(iii) For any $\boldsymbol{x}_0 \in X$ there exists an open neighborhood V of $\boldsymbol{x}_0$, an open neighborhood U of $\boldsymbol{0}$ in $\mathbb{R}^m$ and a parametrization $\Phi : U \to \mathbb{R}^n$ such that

$$V \cap X = \Phi(U).$$

□

Outline of the proof. Proposition 14.34 shows that (iii) $\Rightarrow$ (i) while Proposition 14.39 shows that (ii) $\Rightarrow$ (i). The opposite implications (i) $\Rightarrow$ (ii) and (i) $\Rightarrow$ (iii) follow from the definition of a submanifold. □

Example 14.42 (The unit circle). The unit circle is the closed subset of $\mathbb{R}^2$ defined by

$$\mathbb{S}^1 := \{\, (x,y) \in \mathbb{R}^2;\ x^2 + y^2 = 1 \,\}.$$

Then $\mathbb{S}^1$ is a curve in $\mathbb{R}^2$, i.e., a 1-dimensional submanifold of $\mathbb{R}^2$. To see this consider the smooth function

$$f : \mathbb{R}^2 \to \mathbb{R},\ \ f(x,y) = x^2 + y^2 - 1.$$

Then $\mathbb{S}^1$ can be identified with the level set $\{f = 0\}$. Note that $df = 2xdx + 2ydy$. Hence if $(x_0, y_0) \in \mathbb{S}^1$, then at least one of the coordinates x_0, y_0 is nonzero, so that $df(x_0, y_0) \neq \mathbf{0}$, proving that the differential $df(x_0, y_0) : \mathbb{R}^2 \to \mathbb{R}$ is surjective. The implicit function theorem then implies that $\mathbb{S}^1$ is a 1-dimensional submanifold of $\mathbb{R}^2$. There are several ways of constructing useful local coordinates.

For example, in the region $y > 0$ the correspondence

$$\mathbb{S}^1 \cap \{y > 0\} \to \mathbb{R},\ \ (x,y) \mapsto x$$

is a local coordinate chart. The corresponding parametrization is the map

$$(-1,1) \to \mathbb{S}^1 \cap \{y > 0\},\ \ x \mapsto \big(x, \sqrt{1-x^2}\,\big).$$

This follows from the fact that the portion $\mathbb{S}^1 \cap \{y > 0\}$ is the graph of the smooth map

$$F : (-1,1) \to \mathbb{R},\ \ F(x) = \sqrt{1-x^2}.$$

Another very convenient choice is that of *polar coordinates*.

The location of a point $\boldsymbol{p} = (x,y)$ in the Cartesian plane $\mathbb{R}^2$, other than the origin $\mathbf{0}$, is uniquely determined by two parameters: the distance to the origin $r = \|\boldsymbol{p}\| = \sqrt{x^2+y^2}$, and the angle θ the vector $\boldsymbol{p}$ makes with the x-axis, measured *counterclockwisely*; see Figure 14.8.

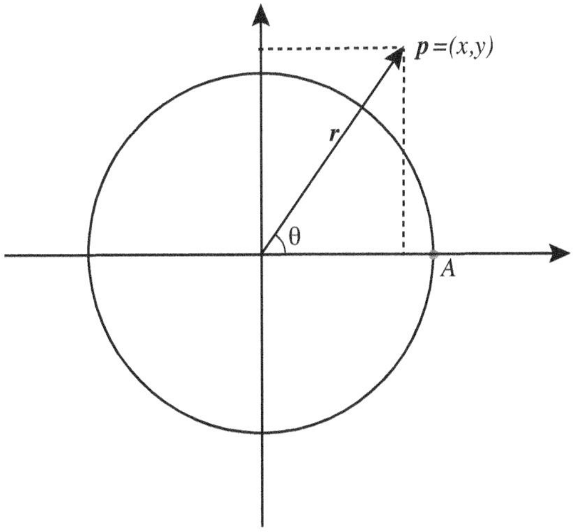

Fig. 14.8 *Constructing the polar coordinates.*

The Cartesian coordinates x, y are related to the parameters r, θ via the equalities

$$\begin{cases} x = r\cos\theta \\ y = r\sin\theta. \end{cases} \tag{14.32}$$

Exercise 14.12 shows that the map

$$\Psi : (0,\infty)\times(0,2\pi)\to\mathbb{R}^2,\ \ (r,\theta)\mapsto(r\cos\theta, r\sin\theta)$$

is a diffeomorphism whose image is the region $\mathbb{R}^2_*$, the plane $\mathbb{R}^2$ with the nonnegative x-semiaxis removed. The parameters (r,θ) are called *polar coordinates*. Denote by $\mathbb{S}^1_*$ the circle $\mathbb{S}^1$ with the point $A = (1,0)$ removed; see Figure 14.8. The inverse of the diffeomorphism Ψ maps $\mathbb{S}^1$ to a portion line $r = 1$ in the (r,θ) plane. The correspondence that associates to a point $\boldsymbol{p}\in\mathbb{S}^1_*$ the angle θ is a local coordinate chart. □

Example 14.43 (The unit sphere). The unit sphere is the closed subset $\mathbb{S}^2$ of $\mathbb{R}^3$ defined by

$$\mathbb{S}^2 := \big\{\,(x,y,z)\in\mathbb{R}^3;\ \ x^2+y^2+z^2=1\,\big\}.$$

Arguing exactly as in the case of the unit circle, we can invoke the implicit function theorem to deduce that $\mathbb{S}^2$ is a surface in $\mathbb{R}^3$, i.e., 2-dimensional submanifold of $\mathbb{R}^3$.

Besides the Cartesian coordinates in $\mathbb{R}^3$ there are two other particularly useful choices of coordinates. To describe them pick a point $\boldsymbol{p} = (x,y,z)\in\mathbb{R}^3$ not situated on the z-axis. Note that the location of $\boldsymbol{p}$ is completely known if we know the altitude z of $\boldsymbol{p}$ and the location of the projection of $\boldsymbol{p}$ on the (x,y)-plane. We denote by $\boldsymbol{q}$ this projection so that $\boldsymbol{q} = (x,y)\in\mathbb{R}^2$; see Figure 14.9.

The location of $\boldsymbol{q}$ is completely determined by its polar coordinates (r,θ) so that the location of $\boldsymbol{p}$ is completely determined if we know the parameters r,θ,z. These parameters are called the *cylindrical coordinates* in $\mathbb{R}^3$. The Cartesian coordinates are related to the cylindrical coordinates via the equalities

$$\begin{cases} x = r\cos\theta \\ y = r\sin\theta \\ z = \ \ z. \end{cases} \tag{14.33}$$

Observe that if we know the distance ρ of $\boldsymbol{p}$ to the origin, $\rho = \|\boldsymbol{p}\| = \sqrt{x^2+y^2+z^2}$, and the angle $\varphi\in(0,\pi)$ the vector $\boldsymbol{p}$ makes with the z-axis, then we can determine the altitude z via the equality $z = \rho\cos\varphi$ and the parameter r via the equality $r = \rho\sin\varphi$; see Figure 14.9. This shows that the parameters r,θ,φ uniquely determine the location of $\boldsymbol{p}$. These parameters are called the *spherical coordinates* in $\mathbb{R}^3$.

The Cartesian coordinates are related to the spherical coordinates via the equalities

$$\begin{cases} x = \rho\sin\varphi\cos\theta \\ y = \rho\sin\varphi\sin\theta \\ z = \quad \rho\cos\varphi. \end{cases} \tag{14.34}$$

Exercise 14.12 shows that the equalities (14.33) and (14.34) describe diffeomorphisms defined on certain open subsets of $\mathbb{R}^3$. Note that in spherical coordinates the unit sphere $\mathbb{S}^2$ is described by the very simple equation $\rho = 1$. The position of a point on $\mathbb{S}^2$ not situated at the poles is completely determined by the two angles φ and θ. Intuitively, φ gives the Latitude of the point, while θ determines the Longitude. □

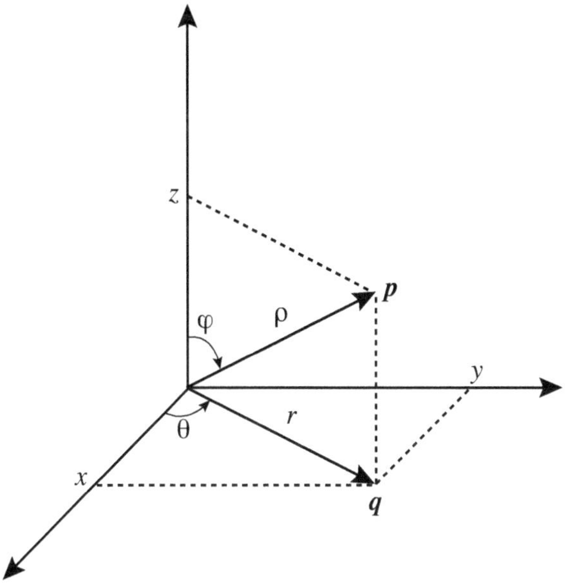

Fig. 14.9 *Constructing the cylindrical and spherical coordinates.*

14.5.2. *Tangent spaces*

Definition 14.44. Let $m, n \in \mathbb{N}$, $m \leq n$, suppose that $X \subset \mathbb{R}^n$ is an m-dimensional submanifold of $\mathbb{R}^n$ and $\boldsymbol{x}_0 \in X$.

(i) A *path in X through $\boldsymbol{x}_0$* is a C^1-path $\gamma : I \to \mathbb{R}^n$, $I \subset \mathbb{R}$ open interval, such that $0 \in I$, $\gamma(0) = \boldsymbol{x}_0$ and $\gamma(I) \subset X$; see Figure 14.10.

(ii) A vector $\boldsymbol{v} \in \mathbb{R}^n$ is said to be *tangent to X at $\boldsymbol{x}_0$* if there exists a path $\gamma : I \to \mathbb{R}^n$ in X through $\boldsymbol{x}_0$ such that $\dot{\gamma}(0) = \boldsymbol{v}$. We denote by $T_{\boldsymbol{x}_0}X$ the set of vectors tangent to X at $\boldsymbol{x}_0$. We will refer to $T_{\boldsymbol{x}_0}X$ as the *tangent space to X at $\boldsymbol{x}_0$*.

□

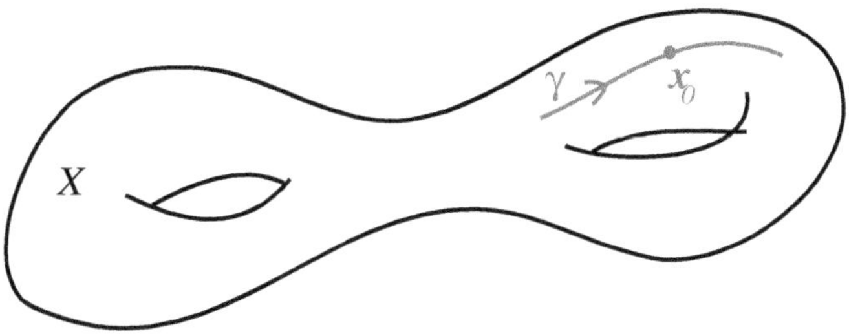

Fig. 14.10 *A path γ in the surface $X \subset \mathbb{R}^3$ through a point $\boldsymbol{x}_0 \in X$. The velocity of γ at $\boldsymbol{x}_0$ is, by definition, a vector tangent to X at $\boldsymbol{x}_0$.*

Example 14.45. Let $m, n \in \mathbb{N}$, $m < n$. Denote by $\mathbb{R}^m \times \mathbf{0}$ the subspace of $\mathbb{R}^n$ defined by the equations $x^{m+1} = \cdots = x^n = 0$. Fix an open set $U \subset \mathbb{R}^n$ and denote by Y the intersection $Y := U \cap (\mathbb{R}^m \times \mathbf{0})$. Then Y is an m-dimensional submanifold of $\mathbb{R}^n$. We want to prove that

$$T_{\boldsymbol{y}_0} Y = \mathbb{R}^m \times \mathbf{0}, \;\; \forall \boldsymbol{y}_0 \in Y. \tag{14.35}$$

In particular $T_{\boldsymbol{y}_0} Y$ is a vector subspace of $\mathbb{R}^n$.

Clearly $T_{\boldsymbol{y}_0} Y \subset \mathbb{R}^m \times \mathbf{0}$. To see this observe that if $\gamma : I \to \mathbb{R}^n$ is a path in Y through $\boldsymbol{y}_0$, then it has the form

$$\gamma(t) = \begin{bmatrix} \gamma^1(t) \\ \vdots \\ \gamma^m(t) \\ 0 \\ \vdots \\ 0 \end{bmatrix} \in \mathbb{R}^n.$$

In particular, $\dot{\gamma}(0) \in \mathbb{R}^m \times \mathbf{0}$.

To prove that $\mathbb{R}^m \times \mathbf{0} \subset T_{\boldsymbol{y}_0} Y$ consider a vector

$$\boldsymbol{v} = \begin{bmatrix} v^1 \\ \vdots \\ v^m \\ \\ 0 \\ \vdots \\ 0 \end{bmatrix} \in \mathbb{R}^m \times \mathbf{0}.$$

Then, there exists $\varepsilon > 0$ sufficiently small such that $\boldsymbol{x}_0 + t\boldsymbol{v} \in U$, $\forall t \in (-\varepsilon, \varepsilon)$. The path

$$\gamma : (-\varepsilon, \varepsilon) \to \mathbb{R}^n, \;\; \gamma(t) = \boldsymbol{x}_0 + t\boldsymbol{v}$$

is in Y through $\boldsymbol{y}_0$ and $\dot{\gamma}(0) = \boldsymbol{v}$, i.e., $\boldsymbol{v} \in T_{\boldsymbol{y}_0} Y$. □

The above example is a manifestation of a more general phenomenon.

Proposition 14.46. *Let $m, n \in \mathbb{N}$ and suppose that $X \subset \mathbb{R}^n$ is an m-dimensional C^1-submanifold. Then for any $\boldsymbol{x}_0 \in X$ the tangent space $T_{\boldsymbol{x}_0} X$ is an m-dimensional vector subspace of $\mathbb{R}^n$.*

Proof. Let $\boldsymbol{x}_0 \in X$. Fix a straightening diffeomorphism $\Psi : \mathcal{U} \to \mathbb{R}^n$ of X at $\boldsymbol{x}_0$ and set $U := \Psi(\mathcal{U})$. Denote by Φ the inverse $\Psi^{-1} : U \to \mathbb{R}^n$ and by L the differential of Φ at $(\boldsymbol{u}_0, \mathbf{0})$. Then $\Psi(\boldsymbol{x}_0) = (\boldsymbol{u}_0, \mathbf{0}) \in \mathbb{R}^m \times \mathbf{0} \subset \mathbb{R}^n$. Note that ω is a path in X through $\boldsymbol{x}_0$ if and only if $\gamma := \Psi \circ \omega$ is a path in $\mathbb{R}^m \times \mathbf{0}$ through $(\boldsymbol{u}_0, \mathbf{0})$. Moreover, $\omega = \Phi \circ \gamma$. Thus any path ω in X through $\boldsymbol{x}_0$ has the form $\omega = \Phi \circ \gamma$ for some path γ in $\mathbb{R}^m \times \mathbf{0}$ through $(\boldsymbol{u}_0, \mathbf{0})$ and (see Exercise 13.7)

$$\dot{\omega}(0) = L\dot{\gamma}(0).$$

This proves that

$$T_{\boldsymbol{x}_0}X = L\big(T_{\boldsymbol{x}_0,\mathbf{0}}\mathbb{R}^m \times \mathbf{0}\big) \stackrel{(14.35)}{=} L\big(\mathbb{R}^m \times \mathbf{0}\big).$$

Thus $T_{\boldsymbol{x}_0}X$ is the image of the m-dimensional vector subspace $\mathbb{R}^m \times \mathbf{0}$ via the linear map L. Since L is injective, the image also has dimension m. □

The above proof leads to the following useful consequence.

Corollary 14.47. *Let $m, n \in \mathbb{N}$, $m < n$. Suppose that $U \subset \mathbb{R}^m$ is open and $\Phi : U \to \mathbb{R}^n$ is a parametrization (see Proposition 14.34) with image $X = \Phi(U)$. If $\boldsymbol{u}_0 \in U$ and $\boldsymbol{x}_0 = \Phi(\boldsymbol{u}_0)$, then the tangent space $T_{\boldsymbol{x}_0}X$ is equal to the range of the differential $d\Phi(\boldsymbol{u}_0)$, i.e.,*

$$T_{\boldsymbol{x}_0}X = d\Phi(\boldsymbol{u}_0)\mathbb{R}^m.$$

In particular, the vectors

$$\partial_{u^1}\Phi(\boldsymbol{u}_0),\ \partial_{u^2}\Phi(\boldsymbol{u}_0), \dots, \partial_{u^m}\Phi(\boldsymbol{u}_0)$$

form a basis of $T_{\boldsymbol{x}_0}X$. □

Proposition 14.48. *Let $m, n \in \mathbb{N}$, $m < n$. Suppose that $V \subset \mathbb{R}^n$ is open and $\boldsymbol{F} : V \to \mathbb{R}^m$ is a C^1-map such that, for any $\boldsymbol{x} \in X = \boldsymbol{F}^{-1}(\mathbf{0})$, the differential $d\boldsymbol{F} : \mathbb{R}^n \to \mathbb{R}^m$ is onto, so the Jacobian matrix $J_{\boldsymbol{F}}(\boldsymbol{x})$ has rank m. Then X is a smooth submanifold of $\mathbb{R}^n$ of dimension $n - m$ and*

$$\forall \boldsymbol{x}_0 \in X,\ \ T_{\boldsymbol{x}_0}X = \ker d\boldsymbol{F}(\boldsymbol{x}_0) = \big\{\, \boldsymbol{v} \in \mathbb{R}^n;\ \ d\boldsymbol{F}(\boldsymbol{x}_0)\boldsymbol{v} = 0 \,\big\}.$$

Proof. The fact that X is a smooth submanifold of dimension $n - m$ follows from the implicit function theorem; see Proposition 14.39. Let $\boldsymbol{x}_0 \in X$. The range $\boldsymbol{R}\big(\, d\boldsymbol{F}(\boldsymbol{x}_0)\big)$ of the linear operator $d\boldsymbol{F}(\boldsymbol{x}_0) : \mathbb{R}^n \to \mathbb{R}^m$ has dimension m since this linear operator is surjective. We deduce

$$\dim \ker d\boldsymbol{F}(\boldsymbol{x}_0) = n - \dim \boldsymbol{R}\big(\, d\boldsymbol{F}(\boldsymbol{x}_0)\big) = n - m = \dim T_{\boldsymbol{x}_0}X.$$

Hence it suffices to show that $T_{\boldsymbol{x}_0}X \subset \ker d\boldsymbol{F}(\boldsymbol{x}_0)$.

Let $\boldsymbol{v} \in T_{\boldsymbol{x}_0}X$. Thus, there exists a C^1-path $\gamma : (-\varepsilon, \varepsilon) \to \mathbb{R}^n$ such that $\gamma(t) \in X$, $\forall t \in (-\varepsilon, \varepsilon)$, $\gamma(0) = \boldsymbol{x}_0$, $\dot{\gamma}(0) = \boldsymbol{v}$ and consequently

$$\boldsymbol{F}\big(\gamma(t)\big) = \mathbf{0},\ \ \forall t \in (-\varepsilon, \varepsilon).$$

Derivating the last equality at $t = 0$ using the chain rule we deduce

$$\mathbf{0} = \frac{d}{dt}\Big|_{t=0} \boldsymbol{F}\big(\gamma(t)\big) = d\boldsymbol{F}\big(\gamma(0)\big)\dot{\gamma}(0) = d\boldsymbol{F}(\boldsymbol{x}_0)\boldsymbol{v} \Rightarrow \boldsymbol{v} \in \ker d\boldsymbol{F}(\boldsymbol{x}_0).$$

□

Remark 14.49. The last result has a more geometric equivalent reformulation. The map $\boldsymbol{F}$ in Proposition 14.48 has m components,

$$\boldsymbol{F}(\boldsymbol{x}) = \begin{bmatrix} F^1(\boldsymbol{x}) \\ \vdots \\ F^m(\boldsymbol{x}) \end{bmatrix}.$$

The differential of $\boldsymbol{F}$ is represented by the $m \times n$ matrix

$$d\boldsymbol{F}(\boldsymbol{x}) = \begin{bmatrix} dF^1(\boldsymbol{x}) \\ \vdots \\ dF^m(\boldsymbol{x}) \end{bmatrix},$$

where the i-th row describes the differential of F^i. Note that

$$\boldsymbol{v} \in \ker d\boldsymbol{F}(\boldsymbol{x}) \Longleftrightarrow dF^i(\boldsymbol{x})(\boldsymbol{v}) = 0, \ \ \forall i = 1, \dots, m$$

$$\Longleftrightarrow \langle \nabla F^i(\boldsymbol{x}), \boldsymbol{v} \rangle = 0, \ \ \forall i = 1, \dots, n-m \Longleftrightarrow \boldsymbol{v} \perp \nabla F^i(\boldsymbol{x}), \ \ \forall i = 1, \dots, m$$

$$\Longleftrightarrow v^1 \partial_{x^1} F^i(\boldsymbol{x}_0) + v^2 \partial_{x^2} F^i(\boldsymbol{x}_0) + \cdots + v^n \partial_{x^n} F^i(\boldsymbol{x}_0) = 0, \ \ \forall i = 1, \dots, m. \qquad \square$$

To put the above remark in its proper geometric perspective we need to survey a few linear algebra facts. For a given vector subspace $V \subset \mathbb{R}^n$ we denote by $V^\perp$ the set of vectors $\boldsymbol{x} \in \mathbb{R}^n$ such that $\boldsymbol{x} \perp \boldsymbol{v}$, $\forall \boldsymbol{v} \in V$. The set $V^\perp$ is called the *orthogonal complement* of V in $\mathbb{R}^n$. Often we will use the notation $\boldsymbol{x} \perp V$ to indicate $\boldsymbol{x} \in V^\perp$. The orthogonal complement enjoys several useful properties.

Proposition 14.50. *Let $n \in \mathbb{N}$ and suppose that V is a vector subspace of $\mathbb{R}^n$. Then the following hold.*

(i) The orthogonal complement $V^\perp$ is also a vector subspace of $\mathbb{R}^n$. Moreover, if the vectors $\boldsymbol{v}_1, \dots, \boldsymbol{v}_m$ span V, then

$$\boldsymbol{x} \in V^\perp \Longleftrightarrow \boldsymbol{x} \perp \boldsymbol{v}_i, \ \ \forall i = 1, \dots, m.$$

(ii) For any $\boldsymbol{x} \in \mathbb{R}^n$ there exists a unique $\boldsymbol{v} = \boldsymbol{v}(\boldsymbol{x}) \in V$ such that $\boldsymbol{x} - \boldsymbol{v} \in V^\perp$. This vector is called the orthogonal projection *of $\boldsymbol{x}$ on V.*

(iii) $\dim V + \dim V^\perp = n = \dim \mathbb{R}^n$.

(iv) $(V^\perp)^\perp = V$.

$\square$

For a proof of the above proposition and additional information we refer to [21, Sec. 5.3].

Corollary 14.51. *Let $k, n \in \mathbb{N}$, $k < n$. Suppose that $U \subset \mathbb{R}^n$ is an open set and*

$$F^1, \dots, F^k : U \to \mathbb{R}$$

are C^1-functions. Set

$$X := \{ \boldsymbol{x} \in \mathbb{R}^n;\ \ F^1(\boldsymbol{x}) = \cdots = F^k(\boldsymbol{x}) = 0 \}.$$

Assume that

$$\boxed{\text{for any } \boldsymbol{x} \in X, \text{ the vectors } \nabla F^1(\boldsymbol{x}), \dots, \nabla F^k(\boldsymbol{x}) \text{ are linearly independent}}. \tag{14.36}$$

Then the following hold.

(i) The subset X is a C^1-submanifold of dimension $m = n - k$.
(ii) For any $\boldsymbol{x} \in X$ we have

$$T_{\boldsymbol{x}}X = \operatorname{span}\{ \nabla F^1(\boldsymbol{x}), \dots, \nabla F^k(\boldsymbol{x}) \}^{\perp}.$$

(iii) For any $\boldsymbol{x} \in X$, $\boldsymbol{v} \in \mathbb{R}^n$ we have

$$\boxed{\boldsymbol{v} \perp T_{\boldsymbol{x}}X \Longleftrightarrow \boldsymbol{v} \in \{\nabla F^1(\boldsymbol{x}), \dots, \nabla F^k(\boldsymbol{x}) \}}. \tag{14.37}$$

Proof. Consider the map $\boldsymbol{F} : U \to \mathbb{R}^k$,

$$\boldsymbol{F}(\boldsymbol{x}) = \begin{bmatrix} F^1(\boldsymbol{x}) \\ \vdots \\ F^k(\boldsymbol{x}) \end{bmatrix}.$$

The Jacobian matrix $J_{\boldsymbol{F}}(\boldsymbol{x})$ that represents $d\boldsymbol{F}(\boldsymbol{x})$ is a $k \times n$ matrix and its rows are described by the differentials $dF^1(\boldsymbol{x}), \dots, dF^k(\boldsymbol{x})$; see Example 13.14. The assumption (14.36) shows that for $\boldsymbol{x} \in X$ the (*row*) rank of the matrix $J_{\boldsymbol{F}}(\boldsymbol{x})$ is k. This implies that, for any $\boldsymbol{x} \in X$, the operator $J_{\boldsymbol{F}}(\boldsymbol{x})$ is onto; see [21, Sec. 2.7]. Proposition 14.48 now implies that X is a C^1-submanifold of dimension $m = n - k$.

From Remark 14.49 and Proposition 14.50(i) we deduce

$$\boxed{T_{\boldsymbol{x}}X = \big(\operatorname{span}\{ \nabla F^1(\boldsymbol{x}), \dots, \nabla F^k(\boldsymbol{x}) \} \big)^{\perp}}.$$

The equivalence (14.37) is now a consequence of Proposition 14.50(iv). □

Example 14.52 (Hypersurfaces). A *hypersurface* in $\mathbb{R}^n$ is a C^1-submanifold of dimension $n - 1$. We can use Corollary 14.51 to produce hypersurfaces as follows.

Suppose that $U \subset \mathbb{R}^n$ is an open subset and $f : U \to \mathbb{R}$ is a C^1-function such that

$$\boxed{\forall \boldsymbol{x} \in U,\ \ f(\boldsymbol{x}) = 0 \Rightarrow \nabla f(\boldsymbol{x}) \neq \boldsymbol{0}}.$$

Then the zero set of f,

$$X := \{ \boldsymbol{u} \in U;\ \ f(\boldsymbol{u}) = 0 \},$$

is a hypersurface in $\mathbb{R}^n$. Moreover, for all $\boldsymbol{x}_0 \in X$, the tangent space of X at $\boldsymbol{x}_0$ is a hyperplane (through $\boldsymbol{0}$) and $\nabla f(\boldsymbol{x}_0)$ is a normal vector of this hyperplane. In particular, it is described by the equation

$$T_{\boldsymbol{x}_0}X = \{ \boldsymbol{v} \in \mathbb{R}^n;\ \ \langle \nabla f(\boldsymbol{x}_0), \boldsymbol{v} \rangle = 0 \}.$$

As an example, consider a C^1-function $h:\mathbb{R}^2\to\mathbb{R}$. As we know, its graph

$$\Gamma_h := \{\, (x,y,z)\in\mathbb{R}^3;\ \ z=h(x,y)\,\}$$

is a hypersurface in $\mathbb{R}^3$. If we define $f:\mathbb{R}^3\to\mathbb{R}$, $f(x,y,z)=h(x,y)-z$, we see that we can alternatively characterize Γ_h as the zero set of f.

Note that for any $(x_0,y_0,z_0)\in\mathbb{R}^3$ we have

$$\nabla f(x_0,y_0,z_0) = \begin{bmatrix}\partial_x h(x_0,y_0)\\ \partial_y h(x_0,y_0)\\ -1\end{bmatrix}\neq \mathbf{0}.$$

Thus the tangent space to Γ_h at $\boldsymbol{p}_0=(x_0,y_0,z_0)$, $z_0=h(x_0,y_0)$ consists of the vectors

$$\dot{\boldsymbol{r}} = \begin{bmatrix}\dot{x}\\ \dot{y}\\ \dot{z}\end{bmatrix}\in\mathbb{R}^3$$

such that $\langle\nabla f(\boldsymbol{p}_0),\dot{\boldsymbol{r}}\rangle=0$, i.e.,

$$\partial_x h(x_0,y_0)\dot{x}+\partial_y h(x_0,y_0)\dot{y}-\dot{z}\Longleftrightarrow \dot{z}=\partial_x h(x_0,y_0)\dot{x}+\partial_y h(x_0,y_0)\dot{y}.$$

We see that $T_{\boldsymbol{p}_0}\Gamma_h$ is the graph of the differential

$$dh(x_0,y_0):\mathbb{R}^2\to\mathbb{R},\ \ dh(x_0,y_0)(\dot{x},\dot{y})=\partial_x h(x_0,y_0)\dot{x}+\partial_y h(x_0,y_0)\dot{y}.$$

As an even more concrete example consider the sphere

$$\Sigma := \{\, (x,y,z)\in\mathbb{R}^3;\ \ x^2+y^2+z^2=3\,\}.$$

It is the zero set of the function $f(x,y,z)=x^2+y^2+z^2-3$. Note that

$$\nabla f(x,y,z) = \begin{bmatrix}2x\\ 2y\\ 2z\end{bmatrix}\neq\mathbf{0},\ \ \forall (x,y,z)\neq\mathbf{0}.$$

This shows that Σ is a hypersurface in $\mathbb{R}^3$. The point $\boldsymbol{p}_0=(1,1,1)$ lives on this sphere and

$$T_{\boldsymbol{p}_0}\Sigma = \{\dot{\boldsymbol{r}}=(\dot{x},\dot{y},\dot{z})\in\mathbb{R}^3;\ \ \dot{x}+\dot{y}+\dot{z}=0\,\}.$$

We treat the equality $\dot{x}+\dot{y}+\dot{z}=0$ as a homogeneous linear system consisting of one equation in the three unknowns $\dot{x},\dot{y},\dot{z}$. We see that the solutions of this system satisfy $\dot{x}=-\dot{y}-\dot{z}$ so that

$$\begin{bmatrix}\dot{x}\\ \dot{y}\\ \dot{z}\end{bmatrix} = \begin{bmatrix}-\dot{y}-\dot{z}\\ \dot{y}\\ \dot{z}\end{bmatrix} = \dot{y}\begin{bmatrix}-1\\ 1\\ 0\end{bmatrix}+\dot{z}\begin{bmatrix}-1\\ 0\\ 1\end{bmatrix}$$

where $\dot{y},\dot{z}$ are arbitrary. This shows that the vectors

$$\begin{bmatrix}-1\\ 1\\ 0\end{bmatrix},\ \begin{bmatrix}-1\\ 0\\ 1\end{bmatrix}$$

form a basis of $T_{\boldsymbol{p}_0}\Sigma$. □

Example 14.53. Consider the map

$$\boldsymbol{F}:\mathbb{R}^4\to\mathbb{R}^2,\ \ \boldsymbol{F}(\boldsymbol{x})=\begin{bmatrix}F^1(\boldsymbol{x})\\F^2(\boldsymbol{x})\end{bmatrix}=\begin{bmatrix}\|\boldsymbol{x}\|^2-1\\ \\ x^1+x^2+x^3+x^4-1\end{bmatrix}$$

and the set

$$S:=\big\{\,\boldsymbol{x}\in\mathbb{R}^4;\ \ \boldsymbol{F}(\boldsymbol{x})=\boldsymbol{0}\,\big\}.$$

In more concrete terms, S is the locus of points $\boldsymbol{x}\in\mathbb{R}^4$ satisfying the equations

$$\begin{cases}\|\boldsymbol{x}\|^2=1\\ x^1+x^2+x^3+x^4=1.\end{cases}$$

Let us observe first that $S\neq\emptyset$ since the basic vectors $\boldsymbol{e}_1,\dots,\boldsymbol{e}_4\in\mathbb{R}^4$ belong to S. We will prove that S is a 2-dimensional submanifold. In view of Corollary 14.51 it suffices to verify that for any $\boldsymbol{x}\in S$ the gradients $\nabla F^1(\boldsymbol{x})$ and $\nabla F^2(\boldsymbol{x})$ are linearly independent, i.e., they are not collinear.

Observe that

$$\nabla F^1(\boldsymbol{x})=2\boldsymbol{x},\ \ \nabla F^2(\boldsymbol{x})=\begin{bmatrix}1\\1\\1\\1\end{bmatrix}.$$

Note that if $\nabla F^1(\boldsymbol{x})$ and $\nabla F^2(\boldsymbol{x})$ were collinear, then

$$x^1=x^2=x^3=x^4=c.$$

Since $\boldsymbol{x}\in S$ we deduce $4c^2=1$ and $4c=1$. This is obviously impossible. Hence S is a 2-dimensional submanifold. To find the tangent space of S at $\boldsymbol{e}_1=(1,0,0,0)$ observe first that

$$\nabla F^1(\boldsymbol{e}_1)=2\boldsymbol{e}_1=(2,0,0,0),\ \ \nabla F^2(\boldsymbol{e}_1)=(1,1,1,1),$$

and we deduce that $T_{\boldsymbol{e}_1}S$ consists of vectors $(\dot{x}^1,\dot{x}^2,\dot{x}^3,\dot{x}^4)$ satisfying the homogeneous linear system

$$\begin{aligned}\dot{x}^1\qquad\qquad\quad&=0\\ \dot{x}^1+\dot{x}^2+\dot{x}^3+\dot{x}^4&=0.\end{aligned}$$

This system is equivalent to the system in upper echelon form

$$\begin{aligned}2\dot{x}^1\qquad\qquad\quad&=0\\ \dot{x}^2+\dot{x}^3+\dot{x}^4&=0.\end{aligned}$$

The general solution of the last system is

$$\begin{bmatrix}\dot{x}^1\\\dot{x}^2\\\dot{x}^3\\\dot{x}^4\end{bmatrix}=\dot{x}^3\begin{bmatrix}0\\-1\\1\\0\end{bmatrix}+\dot{x}^4\begin{bmatrix}0\\-1\\0\\1\end{bmatrix}.$$

This shows that the vectors

$$\begin{bmatrix}0\\-1\\1\\0\end{bmatrix},\ \begin{bmatrix}0\\-1\\0\\1\end{bmatrix}$$

form a basis of the tangent space $T_{\boldsymbol{e}_1}S$. □

14.5.3. *Lagrange multipliers*

To understand the significance of the Lagrange Multipliers Theorem we consider simple question that can be addressed by it.

Example 14.54. Find the minimum of the *cost function*

$$h : \mathbb{R}^3 \to \mathbb{R}, \quad h(x, y, z) = x + y + z,$$

subject to the *constraint*

$$x^2 + y^2 + z^2 = 3.$$

Note that the above constraint equation defines a submanifold S in $\mathbb{R}^3$, more precisely, the sphere of radius $\sqrt{3}$ centered at the origin. The question can now be rephrased as asking to find the minimum value of the restriction of h to S.

If you think of h as describing, e.g., the temperature in $\mathbb{R}^3$ at a given moment, then the question asks to find the coldest point on the sphere S. The Lagrange Multipliers Theorem describes a simple criterion for recognizing (local) minima or maxima of functions defined on a submanifold of $\mathbb{R}^n$. □

Theorem 14.55. *Let $n \in \mathbb{N}$. Suppose that S is a submanifold of $\mathbb{R}^n$, $\mathcal{O} \subset \mathbb{R}^n$ is an open subset containing S and $h : \mathcal{O} \to \mathbb{R}$ is a C^1 function. If $\boldsymbol{x}_0$ is a local minimum (or maximum) of the restriction of h to S, then*

$$\nabla h(\boldsymbol{x}_0) \perp T_{\boldsymbol{x}_0} S, \;\; i.e., \; \big\langle \nabla h(\boldsymbol{x}_0), \boldsymbol{v} \big\rangle = 0, \;\; \forall \boldsymbol{v} \in T_{\boldsymbol{x}_0} S.$$

Proof. Suppose that $\boldsymbol{x}_0$ is a local minimum of the restriction of h to S. (The local maximum follows from this case applied to the function $-h$.) Let $\boldsymbol{v} \in T_{\boldsymbol{x}_0} S$. We deduce that there exists an open interval $I \subset \mathbb{R}$ containing 0 and a C^1 path $\gamma : I \to \mathbb{R}^n$ such that $\gamma(t) \in S$, $\forall t \in I$ and $\dot{\gamma}(0) = \boldsymbol{v}$.

Since $\boldsymbol{x}_0$ is a local minimum of h on S, there exists $r > 0$ such that

$$h(\boldsymbol{x}_0) \leq h(\boldsymbol{x}), \;\; \forall \boldsymbol{x} \in B_r(\boldsymbol{x}_0) \cap S.$$

On the other hand, since γ is continuous and $\gamma(0) = \boldsymbol{x}_0$, there exists $\varepsilon > 0$ sufficiently small such that $(-\varepsilon, \varepsilon) \subset I$ and $\gamma(t) \in B_r(\boldsymbol{x}_0)$, $\forall t \in (-\varepsilon, \varepsilon)$. Hence

$$h(\gamma(0)) = h(\boldsymbol{x}_0) \leq h(\gamma(t)), \;\; \forall t \in (-\varepsilon, \varepsilon).$$

In other words, $0 \in I$ is a local minimum of the function $h \circ \gamma : I \to \mathbb{R}$. Fermat's theorem then implies

$$0 = \frac{d}{dt}\Big|_{t=0} h(\gamma(t)) \overset{(13.34)}{=} \big\langle \nabla h(\,\gamma(0)\,), \dot{\gamma}(0) \,\big\rangle = \langle \nabla h(\boldsymbol{x}_0), \boldsymbol{v} \rangle.$$

□

Corollary 14.56 (Lagrange Multipliers Theorem). *Let $k, n \in \mathbb{N}$, $k < n$. Suppose that $\mathcal{O} \subset \mathbb{R}^n$ is an open set, and we are given C^1-functions $h, F^1, \dots, F^k : \mathcal{O} \to \mathbb{R}$ with the property*

$$F^1(\boldsymbol{x}) = \cdots = F^k(\boldsymbol{x}) = 0 \;\; \Rightarrow \textit{the vectors } \nabla F^1(\boldsymbol{x}), \dots, \nabla F^k(\boldsymbol{x}) \textit{ are linearly independent.} \tag{14.38}$$

If $\boldsymbol{x}_0 \in \mathcal{O}$ minimizes h subject to the constraints

$$F^1(\boldsymbol{x}) = \cdots = F^k(\boldsymbol{x}) = 0,$$

then there exist real numbers $\lambda_1, \dots, \lambda_k$ such that

$$\nabla h(\boldsymbol{x}_0) = \lambda_1 \nabla F^1(\boldsymbol{x}_0) + \cdots + \lambda_k \nabla F^k(\boldsymbol{x}_0).$$

The real numbers $\lambda_1, \dots, \lambda_k$ are called Lagrange multipliers.[5]

Proof. In view of Corollary 14.51, the assumption (14.38) implies that the constrained set

$$S := \{\, \boldsymbol{x} \in \mathbb{R}^n;\ F^1(\boldsymbol{x}) = \cdots = F^k(\boldsymbol{x}) = 0 \,\}$$

is a submanifold of $\mathbb{R}^n$ of dimension $n - k$.

If $\boldsymbol{x}_0$ is a minimum of the restriction of h on S, then Theorem 14.55 shows that

$$\nabla h(\boldsymbol{x}_0) \in (T_{\boldsymbol{x}_0} S)^{\perp}.$$

Corollary 14.51 implies that

$$\nabla h(\boldsymbol{x}_0) \in \operatorname{span}\{\, \nabla F^1(\boldsymbol{x}_0), \dots, \nabla F^k(\boldsymbol{x}_0) \,\} = (T_{\boldsymbol{x}_0} S)^{\perp}.$$

This implies the existence of numbers $\lambda_1, \dots, \lambda_k \in \mathbb{R}$ such that

$$\nabla h(\boldsymbol{x}_0) = \lambda_1 \nabla F^1(\boldsymbol{x}_0) + \cdots + \lambda_k \nabla F^k(\boldsymbol{x}_0).$$

□

Example 14.57. We want to find the minimum of the function

$$h : \mathbb{R}^3 \to \mathbb{R},\ \ h(x, y, z) = x + y + z,$$

subject to the constraint

$$x^2 + y^2 + z^2 = 1.$$

The set S consisting of the points satisfying this constraint is the unit sphere in $\mathbb{R}^3$ centered at the origin. This is compact and, since h is continuous, its restriction to S has an absolute minimum attained at some point $\boldsymbol{p}_0 = (x_0, y_0, z_0)$.

Set $f(x, y, z) := x^2 + y^2 + z^2 - 1$ so the constraint is described by the equation $f(x, y, z) = 0$. Since

$$\nabla f(x, y, z) = 2 \begin{bmatrix} x \\ y \\ z \end{bmatrix},$$

we deduce that if $x^2 + y^2 + z^2 = 1$, then $\nabla f(x, y, z) \neq \mathbf{0}$ so (14.38) is satisfied.

Corollary 14.56 implies that there exists a Lagrange multiplier $\lambda \in \mathbb{R}$ such that

$$\nabla h(\boldsymbol{p}_0) = \lambda \nabla f(\boldsymbol{p}_0) \Longleftrightarrow (1, 1, 1) = \lambda(2x_0, 2y_0, 2z_0).$$

[5]Observe that the number of Lagrange multipliers is equal to the number of constraints $F^1(\boldsymbol{x}) = 0, \dots, F^k(\boldsymbol{x}) = 0$.

We obtain the system of 4 equations

$$\begin{cases} x_0^2 + y_0^2 + z_0^2 = 1 \\ 1 = 2\lambda x_0 \\ 1 = 2\lambda y_0 \\ 1 = 2\lambda z_0 \end{cases}$$

in 4 unknowns, x_0, y_0, z_0, λ. From the last 3 equations we deduce

$$x_0 = y_0 = z_0 = \frac{1}{2\lambda}$$

and

$$3 = (2\lambda)^2(x_0^2 + y_0^2 + z_0^2) \Rightarrow 3 = 4\lambda^2 \Rightarrow \lambda^2 = \frac{3}{4} \Rightarrow \lambda = \pm\frac{\sqrt{3}}{2}.$$

Thus $\boldsymbol{p}_0$ can only be one of the two points

$$\boldsymbol{p}_0^{\pm} = \pm\frac{1}{\sqrt{3}}\begin{bmatrix} 1 \\ 1 \\ 1 \end{bmatrix}.$$

Since $h(\boldsymbol{p}_0^-) < 0 < h(\boldsymbol{p}_0^+)$ we deduce that the minimum of h subject to the constraint $x^2 + y^2 + z^2 = 1$ is $-\sqrt{3}$ and it is attained at the point $\boldsymbol{p}_0^-$. □

14.6. Exercises

Exercise 14.1. Let $n \in \mathbb{N}$, $r > 0$ and suppose that $f : \mathbb{R}^n \to \mathbb{R}$ is a C^1-function such that $f(\boldsymbol{x}) = 0$, $\forall \boldsymbol{x} \in \mathbb{R}^n$, $\|\boldsymbol{x}\| = r$. Show that there exists $\boldsymbol{x}_0 \in \mathbb{R}^n$ such that

$$\|\boldsymbol{x}_0\| < r \text{ and } \nabla f(\boldsymbol{x}_0) = \boldsymbol{0}.$$

Hint. Use the proof of Rolle's Theorem 7.27 as inspiration. □

Exercise 14.2. Consider the symmetric 3×3 matrix

$$A = \begin{bmatrix} 1 & 2 & 3 \\ 2 & 4 & 5 \\ 3 & 5 & 6 \end{bmatrix}.$$

Show that the associated quadratic function $Q_A : \mathbb{R}^3 \to \mathbb{R}$ satisfies

$$Q_A(x, y, z) = x^2 + 4y^2 + 6z^2 + 4xy + 6xz + 10yz, \quad \forall x, y, z \in \mathbb{R}.$$ □

Exercise 14.3. Suppose that A is a symmetric $n \times n$ matrix with associated quadratic function Q_A. Prove that

$$\nabla Q_A(\boldsymbol{x}) = 2A\boldsymbol{x}, \quad \boldsymbol{H}(Q_A, \boldsymbol{x}) = 2A, \quad \forall \boldsymbol{x} \in \mathbb{R}^n.$$ □

Exercise 14.4. Suppose that $f : \mathbb{R}^n \to \mathbb{R}$ is a C^2-function and $\boldsymbol{p}_0 \in \mathbb{R}^n$. Show that the Hessian of f at $\boldsymbol{p}_0$ is equal to the Jacobian of the map $\nabla f : \mathbb{R}^n \to \mathbb{R}^n$ at the same point. □

Exercise 14.5. Suppose that A is a symmetric, positive definite $n \times n$ matrix. Prove that there exists $m > 0$ such that

$$Q_A(\boldsymbol{h}) \geq m\|\boldsymbol{h}\|^2, \quad \forall \boldsymbol{h} \in \mathbb{R}^n.$$

Hint. Set

$$\Sigma_1 := \{\boldsymbol{h} \in \mathbb{R}^n;\ \|\boldsymbol{h}\| = 1\}, \quad m := \inf_{\boldsymbol{h} \in \Sigma_1} Q_A(\boldsymbol{h}).$$

Show that Σ_1 is compact and deduce that $m > 0$. Next, use (14.10) to prove that $Q_A(\boldsymbol{h}) \geq m\|\boldsymbol{h}\|^2$, $\forall \boldsymbol{h}$. □

Exercise 14.6. Consider the symmetric 2×2 matrix

$$A = \begin{bmatrix} a & b \\ b & c \end{bmatrix}.$$

(i) Prove that A is positive definite if and only if $a > 0$ and $ac - b^2 > 0$.
(ii) Prove that A is negative definite if and only if $a < 0$ and $ac - b^2 > 0$.
(iii) Prove that A is indefinite if and only if $ac - b^2 < 0$.

Hint: (i) Investigate when $Q_A(x, 1) > 0$ for any $x \in \mathbb{R}$. (ii) Investigate when $Q_A(x, 1) < 0$ for any $x \in \mathbb{R}$. □

Exercise 14.7. Let $n \in \mathbb{N}$ and suppose that $U \subset \mathbb{R}^n$ is an open and convex set. A function $f : U \to \mathbb{R}$ is called *convex* if

$$f(t\boldsymbol{p} + (1-t)\boldsymbol{q}) \leq tf(\boldsymbol{p}) + (1-t)f(\boldsymbol{q}), \quad \forall t \in [0, 1], \quad \boldsymbol{p}, \boldsymbol{q} \in U.$$

(i) Prove that if the C^1 function $f : U \to \mathbb{R}$ is convex, then

$$f(\boldsymbol{q}) \geq f(\boldsymbol{p}) + \langle \nabla f(\boldsymbol{p}), \boldsymbol{q} - \boldsymbol{p} \rangle, \ \ \forall \boldsymbol{p}, \boldsymbol{q} \in U.$$

(ii) Prove that a C^1 function $f : U \to \mathbb{R}$ is convex if and only if

$$\langle \nabla f(\boldsymbol{p}) - \nabla f(\boldsymbol{q}), \boldsymbol{p} - \boldsymbol{q} \rangle \geq 0, \ \ \forall \boldsymbol{p}, \boldsymbol{q} \in U.$$

(iii) Prove that a C^2 function $f : U \to \mathbb{R}$ is convex if and only if

$$\langle \boldsymbol{H}(f, \boldsymbol{p})\boldsymbol{v}, \boldsymbol{v} \rangle \geq 0, \ \ \forall \boldsymbol{p} \in U, \ \ \boldsymbol{v} \in \mathbb{R}^n.$$

Hint: Have a look at Section 8.3. □

Exercise 14.8. Consider the smooth function

$$f : (0, \infty) \times (0, \infty) \to \mathbb{R}, \ \ f(x, y) = xy + \frac{1}{x} + \frac{1}{y}.$$

(i) Show that the point $\boldsymbol{p}_0 = (1, 1)$ is the only critical point of f.
(ii) Show that the point $\boldsymbol{p}_0 = (1, 1)$ is a local minimum of f.
(iii) Prove that

$$f(x, y) \geq 2\sqrt{x + y}, \ \ \forall x, y > 0.$$

Hint: Use the inequality $a^2 + b^2 \geq 2ab$, $\forall a, b \in \mathbb{R}$.

(iv) Prove that the point $\boldsymbol{p}_0$ in (i) is *the global minimum point of* f, i.e.,

$$f(\boldsymbol{p}_0) < f(\boldsymbol{p}), \ \ \forall \boldsymbol{p} \in (0, \infty) \times (0, \infty), \ \ \boldsymbol{p} \neq \boldsymbol{p}_0.$$

Hint. Set $\mu := \inf_{x,y>0} f(x, y) \geq 0$. Choose a sequence (x_n, y_n) such that

$$x_n, y_n > 0, \ \ \mu \leq f(x_n, y_n) < \mu + \frac{1}{n}, \ \ \forall n \geq 1.$$

Prove that (x_n, y_n) is bounded. Conclude using Bolzano–Weierstrass.

□

Exercise 14.9. Let $n \in \mathbb{N}$ and suppose that $U, V \subset \mathbb{R}^n$ are open sets.

(i) Prove that if $\boldsymbol{G} : V \to \mathbb{R}^n$ is a C^1 diffeomorphism and $W \subset V$ is an open set, then $\boldsymbol{G}(W)$ is also an open set.
(ii) Prove that if $\boldsymbol{F} : U \to \mathbb{R}^n$ and $\boldsymbol{G} : V \to \mathbb{R}^n$ are C^1-diffeomorphisms and $\boldsymbol{F}(U) \subset V$, then the composition $\boldsymbol{G} \circ \boldsymbol{F} : U \to \mathbb{R}^n$ is also a C^1-diffeomorphism.

□

Exercise 14.10. Let $m, n \in \mathbb{N}$, $m \leq n$. Suppose that $U \subset \mathbb{R}^n$ is an open set and $\boldsymbol{F} : \mathcal{U} \to \mathbb{R}^m$ is a C^1-map. Prove that the set of points $\boldsymbol{p} \in U$, such that the differential $d\boldsymbol{F}(\boldsymbol{p}) : \mathbb{R}^n \to \mathbb{R}^m$ is surjective, is an *open* subset of U.

Hint. Use the characterization of the rank of a matrix in terms of its minors. □

Exercise 14.11. Prove Corollary 14.19. □

Exercise 14.12. Prove that the following maps are C^1 diffeomorphisms,[6] and then find their ranges and inverses.

$$\boldsymbol{F} : (0,\infty) \times (0, 2\pi) \to \mathbb{R}^2, \ \ \boldsymbol{F}(r, \theta) = [r\cos\theta, r\sin\theta]^\top,$$

$$\boldsymbol{G} : (0,\infty) \times (0, 2\pi) \times \mathbb{R} \to \mathbb{R}^3, \ \ \boldsymbol{G}(r, \theta, z) = [r\cos\theta, r\sin\theta, z]^\top,$$

$$\boldsymbol{H} : (0,\infty) \times (0, 2\pi) \times (0, \pi) \to \mathbb{R}^3, \ \ \boldsymbol{H}(\rho, \theta, \varphi) = [\rho\cos\varphi\cos\theta, \rho\cos\varphi\sin\theta, \rho\cos\varphi]^\top.$$

Hint. Prove that each of the above maps is an immersion and then show that Corollary 14.19 applies in each of these cases. □

Exercise 14.13. Let $m, n \in \mathbb{N}$, $m < n$. Prove that if S_1, S_2 are two codimension m coordinate subspaces of $\mathbb{R}^n$, then there exists a bijective linear map $T : \mathbb{R}^n \to \mathbb{R}^n$ such that $T(S_1) = S_2$. □

Exercise 14.14. Consider the map $\Phi : (-\pi, \pi) \times (-\pi, \pi) \to \mathbb{R}^3$ given by

$$\Phi(\theta, \varphi) = \begin{bmatrix} (2+\cos\varphi)\cos\theta \\ (2+\cos\varphi)\sin\theta \\ \sin\varphi \end{bmatrix}.$$

Prove that Φ is an injective immersion. □

Exercise 14.15. Prove Proposition 14.16. □

Exercise 14.16. Consider the function $f : \mathbb{R}^2 \to \mathbb{R}$, $f(x, y) = e^{\sin(xy)} - 1$.

(i) Show that $f(1, 0) = 0$.
(ii) Show that there exist open intervals I centered at 1 and J centered at 0 and a C^1-function $g : I \to \mathbb{R}$ such that the intersection of the level set $\{f = 0\}$ with the rectangle $I \times J \subset \mathbb{R}^2$ coincides with the graph of g.
(iii) Compute $g'(1)$.

□

Exercise 14.17. Show that the equation

$$xy - z\log y + e^{xz} = 1$$

can be solved uniquely in the form $y = g(x, z)$ in an open neighborhood of $(0, 1, 1)$. □

Exercise 14.18. Show that the system of equations

$$\begin{cases} u^2 + v^2 - x^2 - y = 0 \\ u + v - x^2 + y = 0, \end{cases}$$

can be solved uniquely for (u, v) in terms of (x, y) in an open neighborhood of

$$(u_0, v_0, x_0, y_0) = (1, 2, 2, 1).$$ □

[6] Exercise 13.3 asks you to compute the Jacobian matrices of these maps.

Exercise 14.19. Consider the map

$$\boldsymbol{F} : (0, 2\pi) \times (0, \pi/2) \to \mathbb{R}^3, \quad \boldsymbol{F}(\theta, \varphi) = \begin{bmatrix} \sin\varphi\cos\theta \\ \sin\varphi\sin\theta \\ \cos\varphi \end{bmatrix}.$$

Show that $\boldsymbol{F}$ is an injective immersion and then find its image. □

Exercise 14.20. (a) Suppose that $f : \mathbb{R} \to (0, \infty)$ is a C^1-function. Its graph is the curve

$$\Gamma_f := \{ (x, y) \in \mathbb{R}^2;\ y = f(x) \}.$$

Denote by S_f the region in the space $\mathbb{R}^3$ swept when rotating Γ_f about the x axis; see Figure 14.11. Show that S_f is 2-dimensional submanifold of $\mathbb{R}^3$.

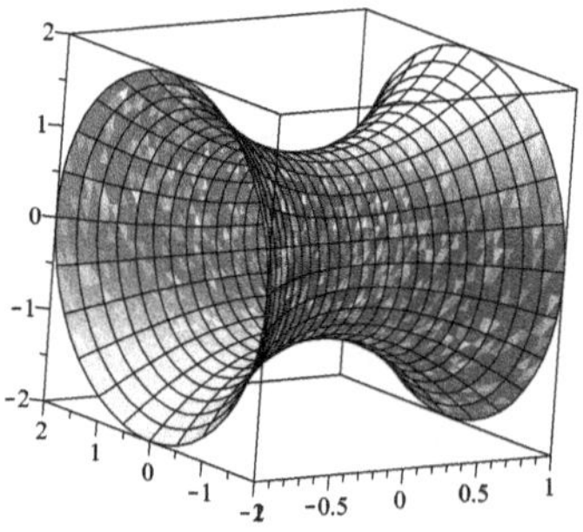

Fig. 14.11 *The surface of revolution S_f, $f(x) = x^2 + 1$.*

(b) Suppose that $0 < a < b$ and $h : (a, b) \to \mathbb{R}$ is a C^1-function. Denote by Σ_h the region in the space $\mathbb{R}^3$ swept when rotating Γ_h about the y-axis; see Figure 14.12 in the special case $h(x) = (x - 1)(x - 2)$. Show that Σ_h is a 2-dimensional submanifold of $\mathbb{R}^3$.

Hint. (a) Show that S_f is described by the equation $f(x)^2 = y^2 + z^2$ and then show that this equation satisfies the assumptions of Proposition 14.39. (b) Show that Σ_h is the graph of a function depending on x, z, i.e., it can be described by an equation of the form $y = F(x, z)$ for some C^1-function F. □

Exercise 14.21. Prove that the map $\boldsymbol{F} : (0, \infty) \times \mathbb{R} \to \mathbb{R}^3$ given by

$$F(r, t) = \begin{bmatrix} r\cos t \\ r\sin t \\ t \end{bmatrix}$$

satisfies all the conditions (i)–(iii) in Proposition 14.34. (Its image is a *helicoid* and it is depicted Figure 14.13.) □

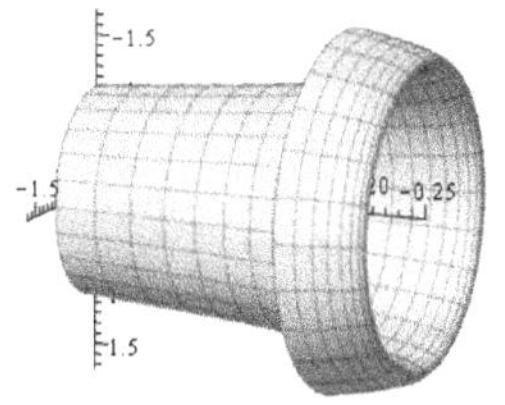

Fig. 14.12 *The surface of revolution* Σ_h, $h(x) = (x-1)(x-2)$, $x \in (1, 1.75)$.

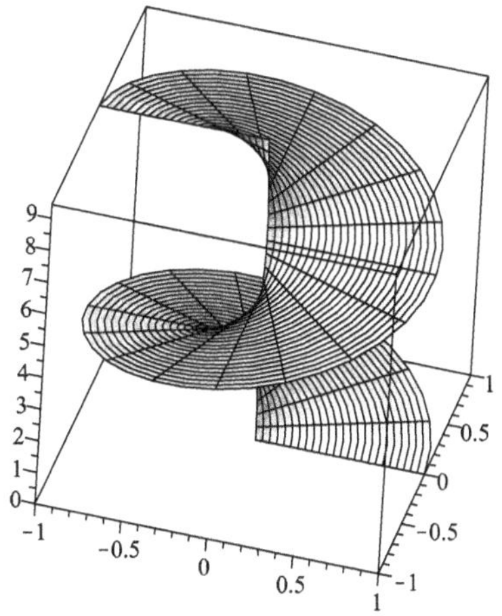

Fig. 14.13 *A helicoid.*

Exercise 14.22. Consider the function $f : \mathbb{R}^3 \to \mathbb{R}$, $f(x, y, z) = xy - z \log y + e^{xz} - 1$.

(i) Show that there exists an open neighborhood U of $\boldsymbol{p}_0 := (0, 1, 1)$ such that $\nabla f(\boldsymbol{p}) \neq 0 \; \forall \boldsymbol{p} \in U$.

(ii) Let U be as above. Show that the set

$$Z = \{ \boldsymbol{p} \in U; \; f(\boldsymbol{p}) = 0 \}$$

is a 2-dimensional submanifold of $\mathbb{R}^3$ containing $\boldsymbol{p}_0$.

(iii) Find a basis of the tangent space $T_{\boldsymbol{p}_0} Z$.

□

Exercise 14.23. Consider the map $\boldsymbol{F} : \mathbb{R}^4 \to \mathbb{R}^2$ given by

$$\boldsymbol{F}(u, v, x, y) = \begin{bmatrix} u^2 + v^2 - x^2 - y \\ u + v - x^2 + y \end{bmatrix}.$$

Set $\boldsymbol{p}_0 := (1, 2, 2, 1) \in \mathbb{R}^4$.

(i) Show that there exists an open neighborhood U of $\boldsymbol{p}_0$ in $\mathbb{R}^4$ such that, $\forall \boldsymbol{p} \in U$, the differential $d\boldsymbol{F}(\boldsymbol{p})$ is surjective as a linear map $\mathbb{R}^4 \to \mathbb{R}^2$, i.e., the Jacobian $J_{\boldsymbol{F}}(\boldsymbol{p})$ has rank 2 for any $\boldsymbol{p} \in U$.
(ii) Let U be as above. Show that the set

$$Z = \{\, \boldsymbol{p} \in U;\ \ \boldsymbol{F}(\boldsymbol{p}) = 0 \,\}$$

is a 2-dimensional submanifold of $\mathbb{R}^4$ containing $\boldsymbol{p}_0$.
(iii) Find a basis of the tangent space $T_{\boldsymbol{p}_0} Z$.

Hint. (i) Use the main theorem in Sec. 6, Chapter 3 of [21]. □

Exercise 14.24. Show that the set

$$S := \{\, (x, y, z) \in \mathbb{R}^3;\ \ x^2 + y^2 - z^2 = 1 \,\}$$

is a 2-dimensional submanifold of $\mathbb{R}^3$ and then describe a basis of the tangent space to S at the point $\boldsymbol{p}_0 = (1, 1, 1)$. □

Exercise 14.25. Let $n \in \mathbb{N}$, $U \subset \mathbb{R}^n$ be an open set and $S \subset U$ a C^1-submanifold of dimension k. Prove that if $\boldsymbol{F} : U \to \mathbb{R}^n$ is a C^1-diffeomorphism, then $\boldsymbol{F}(S)$ is also C^1-submanifold of $\mathbb{R}^n$ of dimension k.

Hint. Observe that $\boldsymbol{F}^{-1} : \boldsymbol{F}(U) \to \mathbb{R}^n$ is a diffeomorphism. Use Exercise 14.9 to show that if (V, Ψ) is a straightening diffeomorphism of S at $\boldsymbol{p}_0$, then $(\boldsymbol{F}(V), \Psi \circ \boldsymbol{F}^{-1})$ is a straightening diffeomorphism of $\boldsymbol{F}(S)$ at $\boldsymbol{F}(\boldsymbol{p}_0)$. □

Exercise 14.26. Let $n \in \mathbb{N}$.

(i) Prove that any vector subspace $U \subset \mathbb{R}^n$ is a submanifold of dimension $\dim U$.
(ii) Suppose that $X \subset \mathbb{R}^n$ is a nonempty affine subspace and $\boldsymbol{p}_0 \in X$. Define $T : \mathbb{R}^n \to \mathbb{R}^n$, $T(\boldsymbol{x}) = \boldsymbol{x} - \boldsymbol{p}_0$. Show that $U = T(X)$ is a vector subspace of $\mathbb{R}^n$.
(iii) Prove that the above map T is a diffeomorphism with image $\mathbb{R}^n$.
(iv) Deduce that X is a submanifold of $\mathbb{R}^n$ of dimension $\dim U$.

Hint. (i) requires linear algebra, namely that a basis of U can be extended to a basis of $\mathbb{R}^n$. (ii) and (iii) are easy. For (iv) use (i)–(iii) and Exercise 14.25. □

Exercise 14.27. Let $n \in \mathbb{N}$, $n \geq 2$.

(i) Show that a hyperplane H of $\mathbb{R}^n$ is a submanifold of dimension $n - 1$.
(ii) Let $f : \mathbb{R}^n \to \mathbb{R}$ be a C^1-function and set

$$Z_f := \{\boldsymbol{p} \in \mathbb{R}^n;\ \ f(\boldsymbol{p}) = 0\,\}.$$

Suppose that $\nabla f(\boldsymbol{p}) \neq \boldsymbol{0}$, $\forall \boldsymbol{p} \in Z_f$. According to Example 14.52 the zero set Z_f is a *hypersurface* of $\mathbb{R}^n$, i.e., a submanifold of dimension $n - 1$. Fix a point $\boldsymbol{p}_0 \in Z_f$ and set $\boldsymbol{v}_0 := \nabla f(\boldsymbol{p}_0)$. Describe explicitly in terms of $\boldsymbol{p}_0$ and $\boldsymbol{v}_0$ the hyperplane of $\mathbb{R}^n$ satisfying

$$\boldsymbol{p}_0 \in H \text{ and } T_{\boldsymbol{p}_0} H = T_{\boldsymbol{p}_0} Z_f.$$

This hyperplane is called the <u>affine</u> *tangent space* of Z_f at $\boldsymbol{p}_0$.

□

Exercise 14.28. Let $n \in \mathbb{N}$ and suppose that $U \subset \mathbb{R}^n$ is an open set containing the origin $\mathbf{0}$. Consider a C^1-function $f : U \to \mathbb{R}$. We set

$$c_0 = f(\mathbf{0}), \;\; c_i := \partial_{x^i} f(\mathbf{0}), \;\; i = 1, \dots, n.$$

The graph of f,

$$\Gamma_f := \big\{ (\boldsymbol{x}, y) \in U \times \mathbb{R};\;\; y = f(\boldsymbol{x}) \big\} \subset \mathbb{R}^{n+1}$$

is an n-dimensional submanifold of $\mathbb{R}^{n+1}$. Note that the point $\boldsymbol{p}_0 := (\, \mathbf{0}, f(\mathbf{0}) \,)$ belongs to the graph.

(i) Describe explicitly in terms of the constants $c_1, \dots, c_n$ a basis of the tangent space $T_{\boldsymbol{p}_0} \Gamma_f$.

(ii) Describe explicitly in terms of the constants $c_0, c_1, \dots, c_n$ a hyperplane $H \subset \mathbb{R}^{n+1}$ such that

$$\boldsymbol{p}_0 \in H \text{ and } T_{\boldsymbol{p}_0} H = T_{\boldsymbol{p}_0} \Gamma_f.$$

Hint. Observe that Γ_f can be described as the hypersurface of $\mathbb{R}^{n+1}$ described in the coordinates $(x^1, \dots, x^n, y)$ by the equation $f(x^1, \dots, x^n) - y = 0$. □

Exercise 14.29. Find the minimum of $h(x, y, z, t) = t$ subject to the constraints

$$x^2 + y^2 + z^2 + t^2 = x + y + z + t = 1.$$

Hint. Apply Corollary 14.56. You also need to use the results in Example 14.53. □

Exercise 14.30. From all rectangular parallelepipeds of given volume $V > 0$ find the ones that have the least surface area. □

Exercise 14.31. Determine the outer dimensions of a plastic box, with no lid, with walls of a given thickness δ and a given internal volume V that requires the least amount of plastic to produce. □

14.7. More challenging problems

Problem 14.1. Let $n \in \mathbb{N}$ and suppose that $f : \mathbb{R}^n \to \mathbb{R}$ is a convex C^1-function. Show that the map

$$\Phi : \mathbb{R}^n \to \mathbb{R}^n, \;\; \Phi(x) = \boldsymbol{x} + \nabla f(\boldsymbol{x})$$

is surjective.

Hint. Use Exercise 12.25 to prove that for any $\boldsymbol{y} \in \mathbb{R}$ the function $f_{\boldsymbol{y}} : \mathbb{R}^n \to \mathbb{R}$, $f_{\boldsymbol{y}}(\boldsymbol{x}) = \frac{1}{2} \|\boldsymbol{x}\|^2 + f(\boldsymbol{x}) - \langle \boldsymbol{y}, \boldsymbol{x} \rangle$ has at least one critical point. □

Problem 14.2. Prove Proposition 14.39.

Hint. Use Version 2 of the implicit function theorem. □

Problem 14.3 (Rayleigh–Ritz). Let $n \in \mathbb{N}$ and suppose that A is a symmetric $n \times n$ matrix. Define

$$q_A : \mathbb{R}^n \to \mathbb{R}, \;\; q_A(\boldsymbol{x}) = \langle A\boldsymbol{x}, \boldsymbol{x}\rangle, \;\; \forall \boldsymbol{x} \in \mathbb{R}^n.$$

Set

$$\mu := \inf_{\|\boldsymbol{x}\|=1} q_A(\boldsymbol{x}).$$

(i) Show that there exists $\boldsymbol{u} \in \mathbb{R}^n$ such that $\|\boldsymbol{u}\| = 1$ and $q_A(\boldsymbol{u}) = \mu$.
(ii) Use Lagrange multipliers to show that any vector $\boldsymbol{u}$ as above is an eigenvector of A corresponding to the eigenvalue μ, i.e., $A\boldsymbol{u} = \mu\boldsymbol{u}$.

Hint. You may want to have a look at Exercise 13.5. □

Problem 14.4. Suppose that $X \subset \mathbb{R}^n$ is a *compact, smooth* submanifold. Prove that there exists $r_0 > 0$ with the following property: for any $r \in (0, r_0)$ and for any $\boldsymbol{p}, \boldsymbol{q} \in X$ such that $\|\boldsymbol{p} - \boldsymbol{q}\| < r$, the intersection $B_r(\boldsymbol{p}) \cap B_r(\boldsymbol{q}) \cap X$ is a nonempty path connected set.
□

Chapter 15

Multidimensional Riemann Integration

In this chapter we want to extend the concept of Riemann integral to functions of several variables. While there are many similarities, the higher dimensional situation displays new phenomena and difficulties that do not have a 1-dimensional counterpart.

15.1. Riemann integrable functions of several variables

15.1.1. *The Riemann integral over a box*

In this chapter, for simplicity, we define a *box* in $\mathbb{R}^n$ to be a *closed* box in the sense of Definition 12.30, i.e., a closed set B of the form

$$B = [a_1, b_1] \times \cdots \times [a_n, b_n], \tag{15.1}$$

where a's and b's are real numbers satisfying $a_1 \leq b_1, \dots, a_n \leq b_n$. Note that we allow the possibility that $a_i = b_i$ for some i's. In particular, a set consisting of a single point is a very special case of box.

A *vertex* of the box in (15.1) is a point $\boldsymbol{x} = (x^1, \dots, x^n)$ such that

$$x^i = a_i \;\text{ or }\; x^i = b_i, \;\; \forall i = 1, \dots, n.$$

A *facet* of B is the set obtained by intersecting B with a coordinate hyperplane of the form $x^i = a_i$ or $x^i = b_i$.[1]

The n*-dimensional volume* of the box B in (15.1) is the nonnegative real number

$$\operatorname{vol}(B) = \operatorname{vol}_n(B) := (b_1 - a_1)(b_2 - a_2) \cdots (b_n - a_n). \tag{15.2}$$

The box B is called *nondegenerate* if $\operatorname{vol}_n(B) > 0$, i.e., $a_i < b_i$, $\forall i = 1, \dots, n$. Note that when $n = 1$, a box B in $\mathbb{R}$ is a compact interval and $\operatorname{vol}_1(B)$ is precisely the length of the interval B. In this case the facets of B are the endpoints of the interval B

Recall that the diameter of a set $S \subset \mathbb{R}^n$ is (see Definition 12.54)

$$\operatorname{diam}(S) = \sup_{\boldsymbol{x}, \boldsymbol{y} \in S} \operatorname{dist}(\boldsymbol{x}, \boldsymbol{y}).$$

It is not hard to see that if B is the box in (15.1), then

$$\operatorname{diam}(B) = \sqrt{(b_1 - a_1)^2 + \cdots + (b_n - a_n)^2}.$$

[1] In dimension 2 a box is a rectangle and the facets are the boundary edges of that rectangle.

Note that the intersection of two boxes B, B' is either empty, or another box. For example if

$$B = I_1 \times \cdots \times I_n, \;\; B' = I'_1 \times \cdots \times I'_n,$$

the $I_j, I'_k \subset \mathbb{R}$ are compact intervals, then $B \cap B'$ is the (possibly empty) box

$$(I_1 \cap I'_1) \times \cdots \times (I_n \cap I'_n).$$

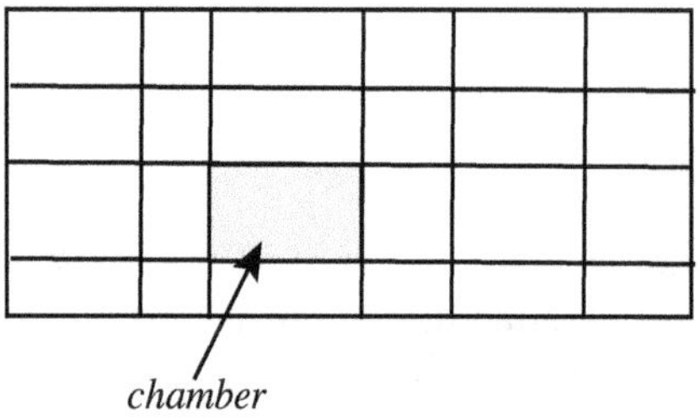

Fig. 15.1 *A partition of a 2-dimensional box. The sum of the areas of the chambers is equal to the area of the big box that contains them.*

Definition 15.1. Let $n \in \mathbb{N}$ and suppose that

$$B = [a_1, b_1] \times [a_2, b_2] \times \cdots \times [a_n, b_n] \subset \mathbb{R}^n$$

is a nondegenerate box.

(a) A *partition* of B is an n-tuple $\boldsymbol{P} = (\boldsymbol{P}_1, \ldots, \boldsymbol{P}_n)$, where, for each $j = 1, \ldots, n$, $\boldsymbol{P}_j$ is a partition of the interval $[a_j, b_j]$; see Definition 9.1.

(b) A *chamber* of $\boldsymbol{P}$ is a box of the form $I_1 \times \cdots \times I_n$, where $I_j \subset [a_j, b_j]$ is an interval of the partition $\boldsymbol{P}_j$. We denote by $\mathscr{C}(\boldsymbol{P})$ the set of chambers of a partition $\boldsymbol{P}$ and by $\mathcal{P}(B)$ the set of partitions of B.

(c) The *mesh size* or *mesh* of a partition $\boldsymbol{P}$ is the positive number

$$\|\boldsymbol{P}\| := \max_{C \in \mathscr{C}(\boldsymbol{P})} \operatorname{diam}(C).$$

(d) A partition $\boldsymbol{P}'$ of B is said to be *finer* than another partition $\boldsymbol{P}$, and we denote this by $\boldsymbol{P}' \succ \boldsymbol{P}$, if any chamber of $\boldsymbol{P}'$ is contained in some chamber of $\boldsymbol{P}$. □

The next result is immediate in dimension 1 and, although it is very intuitive, it takes a bit more work in higher dimensions. We leave its proof to you as an exercise.

Lemma 15.2. *Let $n \in \mathbb{N}$. Suppose that $B \subset \mathbb{R}^n$ is a nondegenerate box and $\boldsymbol{P}$ is a partition of B. Then (see Figure 15.1)*

$$\operatorname{vol}_n(B) = \sum_{C \in \mathscr{C}(\boldsymbol{P})} \operatorname{vol}_n(C). \tag{15.3}$$

□

✍ In the sequel, for the sake of readability, we introduce the following notation:

$$m_S(f) := \inf_{x\in S} f(x), \quad M_S(f) := \sup_{x\in S} f(x),$$

for any function $f : X \to \mathbb{R}$, and any $S \subset X$.

Definition 15.3. Let $n \in \mathbb{N}$. Suppose that $B \subset \mathbb{R}^n$ is a nondegenerate box and $f : B \to \mathbb{R}$ is a *bounded* function.

(a) For any partition $\boldsymbol{P}$ of B we define the *lower Darboux sum* of f over $\boldsymbol{P}$ to be

$$\boldsymbol{S}_*(f, \boldsymbol{P}) := \sum_{C\in\mathscr{C}(\boldsymbol{P})} m_C(f)\operatorname{vol}_n(C).$$

The *upper Darboux sum* of f over $\boldsymbol{P}$ is

$$\boldsymbol{S}^*(f, \boldsymbol{P}) := \sum_{C\in\mathscr{C}(\boldsymbol{P})} M_C(f)\operatorname{vol}_n(C).$$

(b) The *mean oscillation* of f over a partition $\boldsymbol{P}$ of B is the real number

$$\omega(f, \boldsymbol{P}) := \sum_{C\in\mathscr{C}(\boldsymbol{P})} \operatorname{osc}(f, C)\operatorname{vol}_n(C). \qquad \square$$

Arguing as in the 1-dimensional case (see Proposition 9.8) one can show that for any nondegenerate box $B \subset \mathbb{R}^n$, any bounded function $f : B \to \mathbb{R}$ and any partition $\boldsymbol{P}$ of B we have

$$m_B(f)\operatorname{vol}_n(B) \leq \boldsymbol{S}_*(f, \boldsymbol{P}) \leq \boldsymbol{S}^*(f, \boldsymbol{P}) \leq M_B(f)\operatorname{vol}_n(B), \tag{15.4a}$$

$$\omega(f, \boldsymbol{P}) = \boldsymbol{S}^*(f, \boldsymbol{P}) - \boldsymbol{S}_*(f, \boldsymbol{P}). \tag{15.4b}$$

Indeed

$$m_B(f)\operatorname{vol}_n(B) \overset{(15.3)}{=} \sum_{C\in\mathscr{C}(\boldsymbol{P})} m_B(f)\operatorname{vol}_n(C)$$

$(m_B(f) \leq m_C(f), \forall C \in \mathscr{C}(\boldsymbol{P}))$

$$\leq \sum_{C\in\mathscr{C}(\boldsymbol{P})} m_C(f)\operatorname{vol}_n(C) \leq \sum_{C\in\mathscr{C}(\boldsymbol{P})} M_C(f)\operatorname{vol}_n(C)$$

$(M_C(f) \leq M_B(f), \forall C \in \mathscr{C}(\boldsymbol{P}))$

$$\leq \sum_{C\in\mathscr{C}(\boldsymbol{P})} M_B(f)\operatorname{vol}_n(C) \overset{(15.3)}{=} M_B(f)\operatorname{vol}_n(B).$$

The next result is the higher dimensional counterpart of Proposition 9.12.

Lemma 15.4. *Let $n \in \mathbb{N}$. Assume that $B \subset \mathbb{R}^n$ is a nondegenerate box and $\boldsymbol{P}$, $\boldsymbol{P}'$ are partitions of B such that $\boldsymbol{P}'$ is finer than $\boldsymbol{P}$, $\boldsymbol{P}' \succ \boldsymbol{P}$. Then, for any bounded function $f : B \to \mathbb{R}$ we have*

$$\boldsymbol{S}_*(f, \boldsymbol{P}) \leq \boldsymbol{S}_*(f, \boldsymbol{P}') \leq \boldsymbol{S}^*(f, \boldsymbol{P}') \leq \boldsymbol{S}^*(f, \boldsymbol{P}), \tag{15.5a}$$

$$\omega(f, \boldsymbol{P}') \leq \omega(f, \boldsymbol{P}). \tag{15.5b}$$

Proof. We already know that $\boldsymbol{S}_*(f,\boldsymbol{P}')\leq \boldsymbol{S}^*(f,\boldsymbol{P}')$ so it suffices to prove

$$\boldsymbol{S}_*(f,\boldsymbol{P})\leq \boldsymbol{S}_*(f,\boldsymbol{P}') \text{ and } \boldsymbol{S}^*(f,\boldsymbol{P}')\leq \boldsymbol{S}^*(f,\boldsymbol{P}).$$

We will prove only the first one. The proof of the second inequality above is entirely similar, and it follows from the first inequality applied to the function $-f$. Suppose that $\boldsymbol{P}=(B_\alpha)_{\alpha\in\mathcal{A}}$.

For every chamber $C\in\mathscr{C}(\boldsymbol{P})$ we denote by $\boldsymbol{P}'_C$ the collection of chambers of the partition $\boldsymbol{P}'$ that are contained in C. Note that the collection $\boldsymbol{P}'_C$ is the collection of chambers of some partition of C. We have

$$\boldsymbol{S}^*(f,\boldsymbol{P}')=\sum_{C'\in\mathscr{C}(\boldsymbol{P}')}M_{C'}(f)\operatorname{vol}_n(C')=\sum_{C\in\mathscr{C}(\boldsymbol{P})}\left(\sum_{C'\in\boldsymbol{P}'_C}M_{C'}(f)\operatorname{vol}_n(C')\right)$$

$(M_{C'}(f)\leq M_C(f)$ when $C'\subset C)$

$$\leq \sum_{C\in\mathscr{C}(\boldsymbol{P})}\left(\sum_{C'\in\boldsymbol{P}'_C}M_C(f)\operatorname{vol}_n(C')\right)=\sum_{C\in\mathscr{C}(\boldsymbol{P})}M_C(f)\boxed{\left(\sum_{C'\in\boldsymbol{P}'_C}\operatorname{vol}_n(C')\right)}$$

$$\stackrel{(15.3)}{=}\sum_{C\in\mathscr{C}(\boldsymbol{P})}M_C(f)\boxed{\operatorname{vol}_n(C)}=\boldsymbol{S}^*(f,\boldsymbol{P}).$$

This proves (15.5a). To prove (15.5b) note that

$$\omega(f,\boldsymbol{P}')=\boldsymbol{S}^*(f,\boldsymbol{P}')-\boldsymbol{S}_*(f,\boldsymbol{P}')\stackrel{(15.5a)}{\leq}\boldsymbol{S}^*(f,\boldsymbol{P})-\boldsymbol{S}_*(f,\boldsymbol{P})\leq\omega(f,\boldsymbol{P}).$$

□

Consider a nondegenerate box

$$B=[a_1,b_1]\times\cdots\times[a_n,b_n]\subset\mathbb{R}^n$$

and two partitions $\boldsymbol{P}',\boldsymbol{P}''$ of it. The intersection of a chamber $C'\in\mathscr{C}(\boldsymbol{P}')$ with a chamber $C''\in\mathscr{C}(\boldsymbol{P}'')$ is either empty, a degenerate box or a nondegenerate box. The collection of all the possible nondegenerate boxes formed by such overlaps coincides with the collection of chambers of a new partition of B that we denote by $\boldsymbol{P}'\vee\boldsymbol{P}''$. Equivalently if $\boldsymbol{P}'$ and $\boldsymbol{P}''$ are described by n-tuples of partitions of the intervals $[a_j,b_j]$,

$$\boldsymbol{P}'=(\boldsymbol{P}'_1,\dots,\boldsymbol{P}'_n),\;\;\boldsymbol{P}'=(\boldsymbol{P}''_1,\dots,\boldsymbol{P}''_n),$$

then

$$\boldsymbol{P}'\vee\boldsymbol{P}''=(\boldsymbol{P}'_1\vee\boldsymbol{P}''_1,\dots,\boldsymbol{P}'_n\vee\boldsymbol{P}''_n),$$

where $\boldsymbol{P}'_j\vee\boldsymbol{P}''_j$ was defined on page 248.

By construction, any chamber of $\boldsymbol{P}'\vee\boldsymbol{P}''$ is contained both in a chamber of the partition $\boldsymbol{P}'$ and in a chamber of $\boldsymbol{P}''$. In other words,

$$\boldsymbol{P}'\vee\boldsymbol{P}''\succ\boldsymbol{P}',\;\;\boldsymbol{P}''.$$

Lemma 15.4 implies that, for any bounded function $f:B\to\mathbb{R}$ we have

$$\boldsymbol{S}_*(f,\boldsymbol{P}')\leq\boldsymbol{S}_*(f,\boldsymbol{P}'\vee\boldsymbol{P}'')\leq\boldsymbol{S}^*(f,\boldsymbol{P}'\vee\boldsymbol{P}'')\leq\boldsymbol{S}^*(f,\boldsymbol{P}'').$$

We have thus shown that

$$\boldsymbol{S}_*(f,\boldsymbol{P}')\leq\boldsymbol{S}^*(f,\boldsymbol{P}''),\;\;\forall\boldsymbol{P}',\boldsymbol{P}''\in\mathcal{P}(B).$$

Hence, the collection of lower Darboux sums of f is bounded above by any upper Darboux sum, and the collection of upper Darboux sums of f is bounded below by any lower Darboux sum. We set

$$\underline{\int}_B f(\boldsymbol{x})|d\boldsymbol{x}| := \sup_{\boldsymbol{P}\in\mathcal{P}(B)} \boldsymbol{S}_*(f,\boldsymbol{P}),\ \overline{\int}_B f(\boldsymbol{x})|d\boldsymbol{x}| := \inf_{\boldsymbol{P}\in\mathcal{P}(B)} \boldsymbol{S}^*(f,\boldsymbol{P}).$$

The number $\underline{\int}_B f(\boldsymbol{x})|d\boldsymbol{x}|$ is called the *lower Darboux integral* of f over B, and the number $\overline{\int}_B f(\boldsymbol{x})|d\boldsymbol{x}|$ is called the *upper Darboux integral* of f over B. Note that

$$\underline{\int}_B f(\boldsymbol{x})|d\boldsymbol{x}| \leq \overline{\int}_B f(\boldsymbol{x})|d\boldsymbol{x}|.$$

Definition 15.5. Let $n\in\mathbb{N}$. Suppose that $B\subset\mathbb{R}^n$ is a nondegenerate box and $f:B\to\mathbb{R}$ is a bounded function. The function f is called *Riemann integrable* over B if

$$\underline{\int}_B f(\boldsymbol{x})|d\boldsymbol{x}| = \overline{\int}_B f(\boldsymbol{x})|d\boldsymbol{x}|.$$

The common value of these numbers is called the *Riemann integral* of f over B and it is denoted by

$$\boxed{\int_B f(\boldsymbol{x})|d\boldsymbol{x}| \text{ or } \int_B f(x^1,\dots,x^n)|dx^1\cdots dx^n|.}$$

We denote by $\mathcal{R}(B)$ the set of Riemann integrable functions $f:B\to\mathbb{R}$. □

Remark 15.6. When $n=1$, and $B=[a,b]$, $a<b$, then

$$\int_{[a,b]} f(x)|dx| = \int_a^b f(x)dx = -\int_b^a f(x)dx,$$

where $\int_a^b f(x)dx$ is the usual 1-dimensional Riemann integral. For this reason we will set

$$\int_a^b f(x)|dx| = \int_b^a f(x)|dx| := \int_{[a,b]} f(\boldsymbol{x})|d\boldsymbol{x}|.$$ □

Example 15.7. Suppose $f:B\to\mathbb{R}$ is a constant function, $f(\boldsymbol{x})=c_0$, $\forall\boldsymbol{x}\in B$. Then, for any partition $\boldsymbol{P}$ of B, we have

$$\boldsymbol{S}_*(f,\boldsymbol{P}) = \sum_{C\in\mathscr{C}(\boldsymbol{P})} c_0\operatorname{vol}_n(C) = c_0\sum_{C\in\mathscr{C}(\boldsymbol{P})}\operatorname{vol}_n(C) \stackrel{(15.3)}{=} c_0\operatorname{vol}_n(B)$$

and, similarly,

$$\boldsymbol{S}^*(f,\boldsymbol{P}) = c_0\operatorname{vol}_n(B).$$

This proves that f is integrable and

$$\int_B c_0|d\boldsymbol{x}| = c_0\operatorname{vol}_n(B).$$ □

Definition 15.8. Let $n \in \mathbb{N}$ and suppose that $B \subset \mathbb{R}^n$ is a nondegenerate box and $f : B \to \mathbb{R}$ is a *bounded* function. We define a *sample* of a partition of $\boldsymbol{P}$ to be an assignment $\underline{\xi}$ that associates to each chamber C of $\boldsymbol{P}$ a point ξ_C located in the chamber C. The *Riemann sum of f* determined by the partition $\boldsymbol{P}$ and the sample $\underline{\xi}$ is the real number

$$\boldsymbol{S}(f, \boldsymbol{P}, \underline{\xi}) := \sum_{C \in \mathscr{C}(\boldsymbol{P})} f(\xi_C) \operatorname{vol}_n(C). \qquad \square$$

From the definition of Riemann and Darboux sums we deduce immediately that, for any partition $\boldsymbol{P}$ of B and any sample $\underline{\xi}$ of $\boldsymbol{P}$ we have

$$\boldsymbol{S}_*(f, \boldsymbol{P}) \leq \boldsymbol{S}(f, \boldsymbol{P}, \underline{\xi}) \leq \boldsymbol{S}^*(f, \boldsymbol{P}).$$

Note also that if $f, g : B \to \mathbb{R}$ are two bounded functions, $a, b \in \mathbb{R}$, $\boldsymbol{P}$ is a partition of B and $\underline{\xi}$ is a sample of $\boldsymbol{P}$, then

$$\boldsymbol{S}\big(af + bg, \boldsymbol{P}, \underline{\xi},\big) = a\boldsymbol{S}(f, \boldsymbol{P}, \underline{\xi}) + b\boldsymbol{S}(g, \boldsymbol{P}, \underline{\xi}). \tag{15.6}$$

Our next result suggests a method of approximation of Riemann integrals.

Proposition 15.9. *Let $n \in \mathbb{N}$. Suppose that $B \subset \mathbb{R}^n$ is a nondegenerate box and $f : B \to \mathbb{R}$ is a Riemann integrable function. Then, for any partition $\boldsymbol{P}$ of B and any sample $\underline{\xi}$ of $\boldsymbol{P}$, we have*

$$\left| \boldsymbol{S}(f, \boldsymbol{P}, \underline{\xi}) - \int_B f(\boldsymbol{x}) |d\boldsymbol{x}| \right| \leq \omega(f, \boldsymbol{P}), \tag{15.7a}$$

$$\left| \boldsymbol{S}_*(f, \boldsymbol{P}) - \int_B f(\boldsymbol{x})\, |d\boldsymbol{x}| \right|, \ \left| \boldsymbol{S}^*(f, \boldsymbol{P}) - \int_B f(\boldsymbol{x})\, |d\boldsymbol{x}| \right| \leq \omega(f, \boldsymbol{P}). \tag{15.7b}$$

Proof. We have

$$\boldsymbol{S}_*(f, \boldsymbol{P}) \leq \int_B f(\boldsymbol{x})\, |d\boldsymbol{x}| \leq \boldsymbol{S}^*(f, \boldsymbol{P})$$

and

$$\boldsymbol{S}_*(f, \boldsymbol{P}) \leq \boldsymbol{S}(f, \boldsymbol{P}, \underline{\xi}) \leq \boldsymbol{S}^*(f, \boldsymbol{P}).$$

The conclusion follows by observing that the two numbers

$$\boldsymbol{S}(f, \boldsymbol{P}, \underline{\xi}), \ \int_B f(\boldsymbol{x}) |d\boldsymbol{x}|$$

are both situated in the interval $\big[\, \boldsymbol{S}_*(f, \boldsymbol{P}), \boldsymbol{S}^*(f, \boldsymbol{P}) \,\big]$ of length $\omega(f, \boldsymbol{P})$. $\square$

Our next result is a higher dimensional version of the Riemann–Darboux Theorem 9.17.

Theorem 15.10 (Riemann–Darboux). *Let $n \in \mathbb{N}$. Suppose that $B \subset \mathbb{R}^n$ is a nondegenerate box and $f : B \to \mathbb{R}$ is a bounded function. Then the following statements are equivalent.*

(i) The function f is Riemann integrable over B.

(ii) For any $\varepsilon > 0$ there exists a partition $\boldsymbol{P}$ of B such that the mean oscillation of f over $\boldsymbol{P}$ is $< \varepsilon$, i.e.,

$$\omega(f, \boldsymbol{P}) < \varepsilon.$$

Proof. (i) $\Rightarrow$ (ii). We know that f is Riemann integrable. We set

$$I := \int_B f(\boldsymbol{x})|d\boldsymbol{x}|.$$

We have

$$\underline{\int}_B f(\boldsymbol{x})|d\boldsymbol{x}| = \sup_{\boldsymbol{P}\in\mathcal{P}(B)} \boldsymbol{S}_*(f, \boldsymbol{P}) = \overline{\int}_B f(\boldsymbol{x})|d\boldsymbol{x}| = \inf_{\boldsymbol{P}\in\mathcal{P}(B)} \boldsymbol{S}^*(f, \boldsymbol{P}) = I.$$

Thus, for any $\varepsilon > 0$ there exists partitions $\boldsymbol{P}', \boldsymbol{P}''$ of B such that

$$I - \frac{\varepsilon}{2} < \boldsymbol{S}_*(f, \boldsymbol{P}') \leq I \leq \boldsymbol{S}^*(f, \boldsymbol{P}'') < I + \frac{\varepsilon}{2}.$$

If we set $\boldsymbol{P} := \boldsymbol{P}' \vee \boldsymbol{P}''$, then we deduce

$$I - \frac{\varepsilon}{2} < \boldsymbol{S}_*(f, \boldsymbol{P}') \leq \boldsymbol{S}_*(f, \boldsymbol{P}) \leq \boldsymbol{S}^*(f, \boldsymbol{P}) \leq \boldsymbol{S}^*(f, \boldsymbol{P}'') < I + \frac{\varepsilon}{2}.$$

Hence

$$\omega(f, \boldsymbol{P}) = \boldsymbol{S}^*(f, \boldsymbol{P}) - \boldsymbol{S}_*(f, \boldsymbol{P}) < I + \frac{\varepsilon}{2} - \Big(I - \frac{\varepsilon}{2}\Big) = \varepsilon.$$

(ii) $\Rightarrow$ (i). Let $\varepsilon > 0$. There exists a partition $\boldsymbol{P}$ of B such that

$$\omega(f, \boldsymbol{P}) = \boldsymbol{S}^*(f, \boldsymbol{P}) - \boldsymbol{S}_*(f, \boldsymbol{P}) < \varepsilon.$$

On the other hand,

$$\boldsymbol{S}_*(f, \boldsymbol{P}) \leq \underline{\int}_B f(\boldsymbol{x})|d\boldsymbol{x}| \leq \overline{\int}_B f(\boldsymbol{x})|d\boldsymbol{x}| \leq \boldsymbol{S}^*(f, \boldsymbol{P})$$

so that

$$\overline{\int}_B f(\boldsymbol{x})|d\boldsymbol{x}| - \underline{\int}_B f(\boldsymbol{x})|d\boldsymbol{x}| \leq \boldsymbol{S}^*(f, \boldsymbol{P}) - \boldsymbol{S}_*(f, \boldsymbol{P}) < \varepsilon.$$

In other words

$$0 \leq \overline{\int}_B f(\boldsymbol{x})|d\boldsymbol{x}| - \underline{\int}_B f(\boldsymbol{x})|d\boldsymbol{x}| \leq \varepsilon, \ \ \forall \varepsilon > 0,$$

i.e.,

$$\overline{\int}_B f(\boldsymbol{x})|d\boldsymbol{x}| - \underline{\int}_B f(\boldsymbol{x})|d\boldsymbol{x}| = 0$$

and thus the function f is Riemann integrable. □

Corollary 15.11. *Suppose that $B \subset \mathbb{R}^n$ is a nondegenerate box and $f : B \to \mathbb{R}$ is a continuous function. Then f is Riemann integrable over B.*

Proof. The box B is compact and thus f is uniformly continuous. Thus, for any $\varepsilon > 0$ there exists $\delta(\varepsilon) > 0$ such that, for any set $S \subset B$ satisfying $\operatorname{diam}(S) < \delta(\varepsilon)$ we have

$$\operatorname{osc}(f,S) < \frac{\varepsilon}{\operatorname{vol}_n(B)}.$$

Choose a partition $\boldsymbol{P}$ of B such that $\|\boldsymbol{P}\| < \delta(\varepsilon)$. In particular, we deduce that

$$\operatorname{osc}(f,C) < \frac{\varepsilon}{\operatorname{vol}_n(B)},\quad \forall C \in \mathscr{C}(\boldsymbol{P}).$$

We have

$$\omega(f,\boldsymbol{P}) = \sum_{C\in\mathscr{C}(\boldsymbol{P})} \operatorname{osc}(f,C)\operatorname{vol}_n(C) < \frac{\varepsilon}{\operatorname{vol}_n(B)} \sum_{C\in\mathscr{C}(\boldsymbol{P})} \operatorname{vol}_n(C) = \varepsilon.$$

□

Theorem 15.12. *Suppose that $B \subset \mathbb{R}^n$ is a nondegenerate box and $f_1,\dots,f_N : B \to \mathbb{R}$ are Riemann integrable functions. Fix a positive constant R such that*

$$|f_i(\boldsymbol{x})| \le R,\ \ \forall i = 1,\dots,N,\ \ \boldsymbol{x} \in B.$$

If

$$H : \underbrace{[-R,R] \times \cdots \times [-R,R]}_{N} \to \mathbb{R}$$

is a Lipschitz function, then the function

$$f : B \to \mathbb{R},\ \ f(\boldsymbol{x}) = H\big(\,f_1(\boldsymbol{x}),\dots,f_N(\boldsymbol{x})\,\big)$$

is also Riemann integrable.

Proof. Fix a Lipschitz constant $L > 0$ of H, i.e.,

$$|H(\boldsymbol{y}_1) - H(\boldsymbol{y}_2)| \le L\|\boldsymbol{y}_1 - \boldsymbol{y}_2\|,\ \ \forall \boldsymbol{y} \in \underbrace{[-R,R] \times \cdots [-R,R]}_{N} = \overline{C_R(\boldsymbol{0})} \subset \mathbb{R}^N.$$

Define $\boldsymbol{F} : \mathbb{R}^n \to \overline{C_R(\boldsymbol{0})}$

$$\boldsymbol{F}(\boldsymbol{x}) = \begin{bmatrix} f_1(\boldsymbol{x}) \\ \vdots \\ f_N(\boldsymbol{x}) \end{bmatrix}.$$

We first prove that, for any subset $S \subset B$, we have

$$\operatorname{osc}(f,S) \le L\sqrt{N}\sum_{i=1}^{N} \operatorname{osc}(f_i,S). \tag{15.8}$$

Indeed, for any $\boldsymbol{x}_1, \boldsymbol{x}_2 \in S$, we have

$$\big|\,f(\boldsymbol{x}_1) - f(\boldsymbol{x}_2)\,\big| = \big|\,H(\,\boldsymbol{F}(\boldsymbol{x}_1)\,) - H(\,\boldsymbol{F}(\boldsymbol{x}_2)\,)\,\big|$$

$$\le L\big\|\,\boldsymbol{F}(\boldsymbol{x}_1) - \boldsymbol{F}(\boldsymbol{x}_2)\,\big\| \le L\sqrt{N}\big\|\,\boldsymbol{F}(\boldsymbol{x}_1) - \boldsymbol{F}(\boldsymbol{x}_2)\,\big\|_\infty$$

$$= L\sqrt{N} \max_{1\leq i\leq N} \big| f_i(\boldsymbol{x}_1) - f_i(\boldsymbol{x}_2) \big| \leq L\sqrt{N}\sum_{i=1}^{N} \operatorname{osc}(f_i, S).$$

Since the functions f_i are Riemann integrable, we deduce that, for any $i = 1, \dots, N$, and for any $\varepsilon > 0$, we can find a partition $\boldsymbol{P}_i$ of B such that

$$\omega(f_i, \boldsymbol{P}_i) < \frac{\varepsilon}{LN\sqrt{N}}. \tag{15.9}$$

Choose a partition $\boldsymbol{P}$ that is finer than all the partitions $\boldsymbol{P}_1, \dots, \boldsymbol{P}_N$. E.g., we can choose

$$\boldsymbol{P} = \boldsymbol{P}_1 \vee \boldsymbol{P}_2 \vee \cdots \vee \boldsymbol{P}_N.$$

We deduce from (15.5b) that

$$\omega(f_i, \boldsymbol{P}) < \frac{\varepsilon}{LN\sqrt{N}}, \quad \forall i = 1, \dots, N.$$

We have

$$\omega(f, \boldsymbol{P}) = \sum_{C\in\mathscr{C}(\boldsymbol{P})} \operatorname{osc}(f, C)\operatorname{vol}_n(C) \overset{(15.8)}{\leq} L\sqrt{N} \sum_{C\in\mathscr{C}(\boldsymbol{P})} \sum_{i=1}^{N} \operatorname{osc}(f_i, C)\operatorname{vol}_n(C)$$

$$= L\sqrt{N}\sum_{i=1}^{N}\left(\sum_{C\in\mathscr{C}(\boldsymbol{P})} \operatorname{osc}(f_i, C)\operatorname{vol}_n(C)\right) = L\sqrt{N}\sum_{i=1}^{N} \omega(f_i, \boldsymbol{P})$$

$$\overset{(15.9)}{<} L\sqrt{N}\sum_{i=1}^{N} \frac{\varepsilon}{LN\sqrt{N}} = \varepsilon.$$

This proves that $f(\boldsymbol{x})$ is Riemann integrable. □

Theorem 15.13. *Let $n \in \mathbb{N}$ and suppose that $B \subset \mathbb{R}^n$ is a nondegenerate box. Then the following hold.*

(i) If $f, g \in \mathcal{R}(B)$ and $s, t \in \mathbb{R}$, then $sf + tg \in \mathcal{R}(B)$ and

$$\int_B \big(sf(\boldsymbol{x}) + tg(\boldsymbol{x}) \big)|d\boldsymbol{x}| = s\int_B f(\boldsymbol{x})|d\boldsymbol{x}| + t\int_B g(\boldsymbol{x})\,|d\boldsymbol{x}|. \tag{15.10}$$

(ii) If $f, g \in \mathcal{R}(B)$, then $fg \in \mathcal{R}(B)$.
(iii) If $f, g \in \mathcal{R}(B)$ and $f(\boldsymbol{x}) \leq g(\boldsymbol{x})$, $\forall \boldsymbol{x} \in B$, then

$$\int_B f(\boldsymbol{x})|d\boldsymbol{x}| \leq \int_B g(\boldsymbol{x})|d\boldsymbol{x}|.$$

(iv) If $f \in \mathcal{R}(B)$, then $|f| \in \mathcal{R}(B)$ and

$$\left|\int_B f(\boldsymbol{x})|d\boldsymbol{x}|\right| \leq \int_B |f(\boldsymbol{x})|\,|d\boldsymbol{x}|.$$

Proof. (i) Let $H : \mathbb{R}^2 \to \mathbb{R}$ be the linear function $H(x, y) = sx + ty$. Then H is Lipschitz and

$$sf(\boldsymbol{x}) + tg(\boldsymbol{x}) = H\big(\, f(\boldsymbol{x}), g(\boldsymbol{x})\,\big), \;\; \forall \boldsymbol{x} \in B.$$

Theorem 15.12 now implies that $sf(\boldsymbol{x}) + tg(\boldsymbol{x})$ is Riemann integrable.

Arguing as in the proof of Theorem 15.12, we can find a sequence of partitions $\boldsymbol{P}_\nu$, $\nu \in \mathbb{N}$ of B such that

$$\omega(sf + tg, \boldsymbol{P}_\nu), \;\; \omega(f, \boldsymbol{P}_\nu), \;\; \omega(g, \boldsymbol{P}_\nu) < \frac{1}{\nu}, \;\; \forall \nu \in \mathbb{N}.$$

Next, choose a sample $\underline{\xi}_\nu$ of $\boldsymbol{P}_\nu$ for any $\nu \in \mathbb{N}$. Proposition 15.9 now implies that

$$\lim_{\nu\to\infty} \boldsymbol{S}(f, \boldsymbol{P}_\nu, \underline{\xi}_\nu) = \int_B f(\boldsymbol{x})|d\boldsymbol{x}|,$$

$$\lim_{\nu\to\infty} \boldsymbol{S}(g, \boldsymbol{P}_\nu, \underline{\xi}_\nu) = \int_B g(\boldsymbol{x})|d\boldsymbol{x}|,$$

$$\lim_{\nu\to\infty} \boldsymbol{S}(\, sf + tg, \boldsymbol{P}_\nu, \underline{\xi}_\nu\,) = \int_B \big(\, sf(\boldsymbol{x}) + tg(\boldsymbol{x})\,\big)|d\boldsymbol{x}|.$$

On the other hand

$$\boldsymbol{S}(\, sf + tg, \boldsymbol{P}_\nu, \underline{\xi}_\nu\,) = s\boldsymbol{S}(f, \boldsymbol{P}_\nu, \underline{\xi}_\nu) + t\boldsymbol{S}(g, \boldsymbol{P}_\nu, \underline{\xi}_\nu).$$

If we let $\nu \to \infty$ in the last equality we obtain (15.10).

(ii) We begin by proving that for any $u \in \mathcal{R}(B)$, its square u^2 is also Riemann integrable. To see this, fix $R > 0$ such that $|u(\boldsymbol{x})| \leq R$, $\forall \boldsymbol{x} \in B$. The function $H : [-R, R] \to \mathbb{R}$, $H(t) = t^2$ is Lipschitz because, for any $s, t \in [-R, R]$, we have

$$|H(s) - H(t)| = |s^2 - t^2| = |s + t| \cdot |s - t| \leq (|s| + |t|) \cdot |t - s| \leq 2R|s - t|.$$

Then $u(\boldsymbol{x})^2 = H(\, u(\boldsymbol{x})\,)$ is Riemann integrable.

To deal with the general case, note that, according to (i) $f + g, f - g \in \mathcal{R}(B)$. We deduce that $(f + g)^2, (f - g)^2 \in \mathcal{R}(B)$ and thus

$$fg = \frac{1}{4}\Big(\, (f + g)^2 - (f - g)^2 \,\Big) \in \mathcal{R}(B).$$

(iii) The function $g(\boldsymbol{x}) - f(\boldsymbol{x})$ is Riemann integrable and nonnegative. In particular, we deduce that, for any partition $\boldsymbol{P}$ of B we have

$$0 \leq \boldsymbol{S}_*(g - f, \boldsymbol{P}) \leq \int_B \big(\, g(\boldsymbol{x}) - f(\boldsymbol{x})\,\big)|d\boldsymbol{x}| = \int_B g(\boldsymbol{x})|d\boldsymbol{x}| - \int_B f(\boldsymbol{x})|d\boldsymbol{x}|.$$

(iv) The function $H : \mathbb{R} \to \mathbb{R}$, $H(x) = |x|$ is Lipschitz and Theorem 15.12 implies that $|f| \in \mathcal{R}(B)$ for any $f \in \mathcal{R}(B)$. Observe next that

$$-|f(\boldsymbol{x})| \leq f(\boldsymbol{x}) \leq |f(\boldsymbol{x})|.$$

Using (i) and (iii) we deduce

$$-\int_B |f(\boldsymbol{x})||d\boldsymbol{x}| \leq \int_B f(\boldsymbol{x})|d\boldsymbol{x}| \leq \int_B |f(\boldsymbol{x})||d\boldsymbol{x}| \Longleftrightarrow \left|\int_B f(\boldsymbol{x})|d\boldsymbol{x}|\right| \leq \int_B |f(\boldsymbol{x})||d\boldsymbol{x}|.$$

□

Proposition 15.14. *Fix $n \in \mathbb{N}$ and a nondegenerate box $B \subset \mathbb{R}^n$. If $f_\nu : B \to \mathbb{R}$, $\nu \in \mathbb{N}$ is a sequence of Riemann integrable functions that converges uniformly to the function $f : B \to \mathbb{R}$, then f is also Riemann integrable and*

$$\lim_{\nu\to\infty} \int_B f_\nu(\boldsymbol{x})|d\boldsymbol{x}| = \int_B f(\boldsymbol{x})|d\boldsymbol{x}|. \tag{15.11}$$

Proof. Let $\hbar > 0$. Since f_ν converges uniformly to f, there exists $N = N(\hbar)$ such that

$$\forall \nu \geq N(\hbar),\ \ \forall \boldsymbol{x} \in B : f_\nu(\boldsymbol{x}) - \hbar < f(\boldsymbol{x}) < f_\nu(\boldsymbol{x}) + \hbar. \tag{15.12}$$

We deduce from the above inequality that for any box $C \subset B$ and $\nu \geq N(\hbar)$ we have

$$m_C(f_\nu) - \hbar \leq m_C(f) \leq M_C(f) \leq M_C(f_\nu) + \hbar.$$

Hence, for any partition $\boldsymbol{P}$ of B we have

$$\boldsymbol{S}_*(f_\nu, \boldsymbol{P}) - \hbar \operatorname{vol}_n(B) \leq \boldsymbol{S}_*(f, \boldsymbol{P}) \leq \boldsymbol{S}^*(f, \boldsymbol{P}) \leq \boldsymbol{S}^*(f_\nu, \boldsymbol{P}) + \hbar \operatorname{vol}_n(B). \tag{15.13}$$

In particular, for any partition $\boldsymbol{P}$ of B, we have

$$\omega(f, \boldsymbol{P}) \leq \omega(f_\nu, \boldsymbol{P}) + 2\hbar \operatorname{vol}_n(B),\ \ \forall \nu \geq N(\hbar). \tag{15.14}$$

Now let $\varepsilon > 0$. Choose $\hbar = \hbar(\varepsilon)$ such that

$$2\hbar \operatorname{vol}_n(B) < \frac{\varepsilon}{2}.$$

Now fix a natural number $\nu > N(\hbar(\varepsilon))$, where $N(\hbar)$ is as in (15.12). Since f_ν is Riemann integrable we can find a partition $\boldsymbol{P}_\varepsilon$ of B such that

$$\omega(f_\nu, \boldsymbol{P}_\varepsilon) < \frac{\varepsilon}{2}.$$

We deduce from (15.14) that

$$\omega(f, \boldsymbol{P}_\varepsilon) < \varepsilon$$

proving that f is Riemann integrable. From the inequalities (15.13) we can now conclude that

$$\int_B f_\nu(\boldsymbol{x})|d\boldsymbol{x}| - \hbar \operatorname{vol}_n(B) \leq \int_B f(\boldsymbol{x})|d\boldsymbol{x}| \leq \int_B f_\nu(\boldsymbol{x})|d\boldsymbol{x}| + \hbar \operatorname{vol}_n(B),\ \ \forall \nu \geq N(\hbar)$$

i.e.,

$$\left| \int_B f_\nu(\boldsymbol{x})|d\boldsymbol{x}| - \int_B f(\boldsymbol{x})|d\boldsymbol{x}| \right| \leq \hbar \operatorname{vol}_n(B),\ \ \forall \nu \geq N(\hbar).$$

This last inequality proves (15.11). □

Let us interrupt the flow of arguments to take stock of what we have achieved so far.

- We have defined concepts of Riemann integrability/integral associated to functions of several variables defined on a box B.
- We showed that the set $\mathcal{R}(B)$ of functions that are Riemann integrable on B is quite large: it contains all the continuous functions, and it is closed with respect to the algebraic operations of addition and multiplication of functions.
- The uniform limits of Riemann integrable functions are Riemann integrable.

If we compare the current state of affairs with the 1-dimensional situation we realize that we have several glaring gaps in our developing story. First, our supply of Riemann integrable functions is still "meagre" since, unlike the 1-dimensional case, we have not yet produced any example of a Riemann integrable function that is not continuous. Second, we have not yet indicated any concrete and practical way of computing Riemann integrals of functions of several variables.

Exercise 15.3 describes a simple method of producing discontinuous Riemann integrable functions of several variables. A remarkable result of *Henri Lebesgue*[2] states that a function is Riemann integrable if and only if it is "nearly continuous". To state it precisely we need to define the concept of *negligible subset of* $\mathbb{R}^n$.

Definition 15.15. A subset $S \subset \mathbb{R}^n$ is called *negligible* if, for any $\varepsilon > 0$, there exists a countable family $(B_\nu)_{\nu\in\mathbb{N}}$ of *closed* boxes in $\mathbb{R}^n$ that covers S and such that

$$\sum_{\nu\in\mathbb{N}} \mathrm{vol}_n(B_\nu) < \varepsilon. \qquad \square$$

Example 15.16. Suppose that $f : [a,b] \to \mathbb{R}$ is a Riemann integrable function. Then its graph

$$\Gamma_f := \{ (x, f(x)) \in \mathbb{R}^2;\ \ x \in [a,b] \}$$

is a negligible subset of $\mathbb{R}^2$.

To see this fix $\varepsilon > 0$ and choose a partition $\boldsymbol{P}$ of $[a,b]$ such that $\omega(f,\boldsymbol{P}) < \varepsilon$. Suppose that

$$\boldsymbol{P} = a = x_0 < x_1 < x_2 < \cdots < x_{n-1} < x_n = b.$$

Set

$$m_i := \inf_{x\in[x_{i-1},x_i]} f(x),\ \ M_i := \sup_{x\in[x_{i-1},x_i]} f(x),\ \ i = 1,\dots,n.$$

For $i = 1,\dots,n$ we denote by B_i the box $[x_{i-1},x_i] \times [m_i, M_i]$. From the definition of m_i and M_i we deduce that

$$\Gamma_f \subset \bigcup_{i=1}^n B_i,$$

$$\sum_{i=1}^n \mathrm{vol}_2(B_i) = \sum_{i=1}^n (M_i - m_i)(x_i - x_{i-1}) = \omega(f,\boldsymbol{P}) < \varepsilon.$$

This shows that, for any $\varepsilon > 0$ we can find a fine collection of rectangles that covers the graph and such that the sum of their areas is $< \varepsilon$. $\square$

In Exercise 15.8 we describe several examples of negligible sets, and some elementary properties of such sets. We have the following result that vastly generalizes Corollary 15.11.

[2]Henri Lebesgue (1875–1941) was a French mathematician famous for his theory of integration; see Wikipedia for more details on his life and work.

Theorem 15.17 (Lebesgue). *Let $n \in \mathbb{N}$ and suppose that $f : \mathbb{R}^n \to \mathbb{R}$ is a bounded function that is identically zero outside some box $B \subset \mathbb{R}^n$. Then the following statements are equivalent.*

(i) The function f is Riemann integrable.
(ii) The set of points of discontinuity of f is negligible.

□

For a proof we refer to [24, §11.1.2].

15.1.2. *A conditional Fubini theorem*

In this subsection we take a stab at the second problem and we describe a very versatile result showing that, under certain conditions, one can reduce the computation of an integral of a function of n-variables to the computation of Riemann integrals of functions with *fewer than n variables.*

Theorem 15.18 (Fubini). *Let $m, n \in \mathbb{N}$. Suppose that*

$$B^m = [a_1, b_1] \times \cdots \times [a_m, b_m]$$

is a nondegenerate box in $\mathbb{R}^m$ and

$$B^n = [a_{m+1}, b_{m+1}] \times \cdots \times [a_{m+n}, b_{m+n}]$$

is a nondegenerate box in $\mathbb{R}^n$. Suppose that

- *the function $f : B^m \times B^n \to \mathbb{R}$ is Riemann integrable on the box*
$$B = B^m \times B^n \subset \mathbb{R}^{m+n}$$
and,
- *for any $\boldsymbol{x} \in B^m$, the function*
$$B^n \ni \boldsymbol{y} \mapsto f_{\boldsymbol{x}}(\boldsymbol{y}) := f(\boldsymbol{x}, \boldsymbol{y}) \in \mathbb{R}$$
is Riemann integrable.

Then, the marginal *(function)*

$$B^m \ni \boldsymbol{x} \mapsto M_f^1(\boldsymbol{x}) := \int_{B^n} f(\boldsymbol{x}, \boldsymbol{y})|d\boldsymbol{y}| \in \mathbb{R}$$

is Riemann integrable and

$$\int_{B^m \times B^n} f(\boldsymbol{x}, \boldsymbol{y})|d\boldsymbol{x}d\boldsymbol{y}| = \int_{B^m} M_f^1(\boldsymbol{x})|d\boldsymbol{x}| = \int_{B^m} \left(\int_{B^n} f(\boldsymbol{x}, \boldsymbol{y})|d\boldsymbol{y}| \right) |d\boldsymbol{x}|. \tag{15.15}$$

The last term in the above equality is called a repeated *or* iterated integral. *Often, for mnemonic purposes, we use the alternate notation*

$$\int_{B^m} |d\boldsymbol{x}| \left(\int_{B^n} f(\boldsymbol{x}, \boldsymbol{y})|d\boldsymbol{y}| \right) := \int_{B^m} \left(\int_{B^n} f(\boldsymbol{x}, \boldsymbol{y})|d\boldsymbol{y}| \right) |d\boldsymbol{x}|.$$

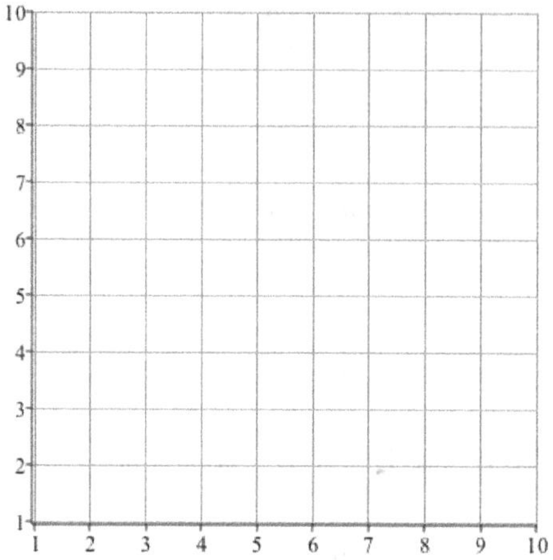

Fig. 15.2 *Adding numbers on a grid.*

Main idea behind Fubini Before we embark in the proof we want to explain the simple principle behind this result. Suppose that we want to add all the numbers situated at the nodes of a grid such as the one in Figure 15.2.

Fubini says that one could proceed as follows: for each number $k = 1, \dots, 10$ on the horizontal (or the first) margin there is a tower of grid nodes above it. Add the numbers above it to obtain the value M_k^1 (the 1st marginal at k). Next add all these marginal values $M_1^1 + \dots + M_{10}^1$ to recover the sum of all the numbers situated at nodes. One can proceed in a similar fashion using the vertical margin, obtaining first a marginal M^2.

Proof. We follow the approach in [18, Thm. 3.10]. Consider a partition

$$\boldsymbol{P}_\varepsilon = (\boldsymbol{P}_1, \dots, \boldsymbol{P}_m, \boldsymbol{P}_{m+1}, \dots, \dots, \boldsymbol{P}_{m+n})$$

of B. We denote by $\boldsymbol{P}^m$ the partition

$$(\boldsymbol{P}_1, \dots, \boldsymbol{P}_m)$$

of B^m, and by $\boldsymbol{P}^n$ the partition

$$(\boldsymbol{P}_{m+1}, \dots, \dots, \boldsymbol{P}_{m+n})$$

of B^n. Every chamber C of $\boldsymbol{P}$ is the Cartesian product of a chamber C^m of $\boldsymbol{P}^m$ and a chamber C^n of $\boldsymbol{P}^n$. Moreover

$$\operatorname{vol}_{m+n}(C) = \operatorname{vol}_m(C^m)\operatorname{vol}_n(C^n).$$

For simplicity we set $\mathscr{C}^m := \mathscr{C}(\boldsymbol{P}^m)$ and $\mathscr{C}^n := \mathscr{C}(\boldsymbol{P}^n)$. To simplify the exposition we will continue to use the notation

$$m_U(g) := \inf_{u \in U} g(u), \quad M_U(g) := \sup_{u \in U} g(u),$$

for any set U and for any bounded real valued function g defined on a set containing U. We have

$$\begin{aligned}\boldsymbol{S}_*(f,\boldsymbol{P}) &= \sum_{\substack{C^m\in\mathscr{C}^m,\\ C^n\in\mathscr{C}^n}} m_{C^m\times C^n}(f)\cdot \operatorname{vol}_m(C^m)\operatorname{vol}_n(C^n)\\ &= \sum_{C^m\in\mathscr{C}^m}\left(\sum_{C^n\in\mathscr{C}^n} m_{C^m\times C^n}(f)\cdot\operatorname{vol}_n(C^n)\right)\operatorname{vol}_m(C^m).\end{aligned}$$

Now observe that for any $C^m\in\mathscr{C}^m$, any $C^n\in\mathscr{C}^n$, and any $\boldsymbol{x}\in C^m$ we have

$$m_{C^m\times C^n}(f)\le m_{C^n}(f_{\boldsymbol{x}}).$$

Hence, for any $\boldsymbol{x}\in\mathscr{C}^m$ we have

$$\begin{aligned}\sum_{C^n\in\mathscr{C}^n} m_{C^m\times C^n}(f)\cdot\operatorname{vol}_n(C^n) &\le \sum_{C^n\in\mathscr{C}^n} m_{C^n}(f_{\boldsymbol{x}})\cdot\operatorname{vol}_n(C^n)=\boldsymbol{S}_*(f_{\boldsymbol{x}},\boldsymbol{P}^n)\\ &\le \int_{B^n} f_{\boldsymbol{x}}(\boldsymbol{y})|d\boldsymbol{y}| = M_f^1(\boldsymbol{x}).\end{aligned}$$

Thus

$$\sum_{C^n\in\mathscr{C}^n} m_{C^m\times C^n}(f)\cdot\operatorname{vol}_n(C^n)\le \inf_{\boldsymbol{x}\in C^m} M_f^1(\boldsymbol{x})$$

so that

$$\begin{aligned}\boldsymbol{S}_*(f,\boldsymbol{P}) &= \sum_{C^m\in\mathscr{C}^m}\left(\sum_{C^n\in\mathscr{C}^n} m_{C^m\times C^n}(f)\cdot\operatorname{vol}_n(C^n)\right)\operatorname{vol}_m(C^m)\\ &\le \sum_{C^m\in\mathscr{C}^m}\inf_{\boldsymbol{x}\in C^m} M_f^1(\boldsymbol{x})\cdot\operatorname{vol}_m(C^m)=\boldsymbol{S}_*\big(M_f^1,\boldsymbol{P}^m\big).\end{aligned}$$

A similar argument shows that

$$\boldsymbol{S}^*\big(M_f^1,\boldsymbol{P}^m\big)\le\boldsymbol{S}^*(f,\boldsymbol{P}).$$

Hence, for any partition $\boldsymbol{P}$ of $B^m\times B^n$ we have

$$\boldsymbol{S}_*(f,\boldsymbol{P})\le\boldsymbol{S}_*\big(M_f^1,\boldsymbol{P}^m\big)\le\boldsymbol{S}^*\big(M_f^1,\boldsymbol{P}^m\big)\le\boldsymbol{S}^*(f,\boldsymbol{P}).$$

We deduce from the above that

$$\begin{aligned}\underline{\int}_{B^m\times B^n} f(\boldsymbol{x},\boldsymbol{y})|d\boldsymbol{x}||d\boldsymbol{y}| &\le \underline{\int}_{B^m} M_f^1(\boldsymbol{x})|d\boldsymbol{x}|\\ &\le \overline{\int}_{B^m} M_f^1(\boldsymbol{x})|d\boldsymbol{x}|\le\overline{\int}_{B^m\times B^n} f(\boldsymbol{x},\boldsymbol{y})|d\boldsymbol{x}||d\boldsymbol{y}|.\end{aligned}$$

Since f is Riemann integrable,

$$\underline{\int}_{B^m\times B^n} f(\boldsymbol{x},\boldsymbol{y})|d\boldsymbol{x}||d\boldsymbol{y}| = \overline{\int}_{B^m\times B^n} f(\boldsymbol{x},\boldsymbol{y})|d\boldsymbol{x}||d\boldsymbol{y}|$$

so all the above inequalities are in fact equalities. □

Remark 15.19. (a) Note that when $f : B^m \times B^n \to \mathbb{R}$ is continuous, all the assumptions in Theorem 15.18 are automatically satisfied. In particular, if $f : [a_1, b_1] \times \cdots \times [a_n, b_n] \to \mathbb{R}$ is continuous, then

$$\begin{aligned}
&\int_{[a_1,b_1]\times\cdots\times[a_n,b_n]} f(x^1,\dots,x^n)|dx^1\cdots dx^n| \\
&= \int_{[a_1,b_1]} |dx^1| \int_{[a_2,b_2]\times\cdots\times[a_n,b_n]} f(x^1,x^2,\dots,x^n)|dx^2\cdots dx^n| \\
&= \int_{[a_1,b_1]} |dx^1| \int_{[a_2,b_2]} |dx^2| \int_{[a_3,b_3]\times\cdots\times[a_n,b_n]} f(x^1,x^2,x^3,\dots,x^n)|dx^2\cdots dx^n| \\
&= \int_{[a_1,b_1]} |dx^1| \int_{[a_2,b_2]} |dx^2| \cdots \int_{[a_n,b_n]} f(x^1,x^2,\dots,x^n)|dx^n| \\
&= \int_{a_1}^{b_1} dx^1 \int_{a_2}^{b_2} dx^2 \cdots \int_{a_n}^{b_n} f(x^1,x^2,\dots,x^n)dx^n.
\end{aligned}$$

For example

$$\begin{aligned}
&\int_{[0,\pi/2]\times[0,\pi]} \sin(x+y)|dxdy| = \int_0^{\frac{\pi}{2}} dx \int_0^{\pi} \sin(x+y)\,dy \\
&= \int_0^{\frac{\pi}{2}} \Big(-\cos(x+y)\Big|_{y=0}^{y=\pi} \Big) dx = \int_0^{\frac{\pi}{2}} \Big(\cos x - \cos(\pi+x) \Big) dx \\
&= \int_0^{\frac{\pi}{2}} \cos x\,dx - \int_0^{\frac{\pi}{2}} \cos(x+\pi)dx \\
&= \sin x\Big|_{x=0}^{x=\frac{\pi}{2}} - \sin(x+\pi)\Big|_{x=0}^{x=\frac{\pi}{2}} = \sin\frac{\pi}{2} - \sin\frac{3\pi}{2} + \sin\pi = 2.
\end{aligned}$$

(b) A completely similar argument shows that when $f : B^m \times B^n \to \mathbb{R}$ is Riemann integrable and, for any $\boldsymbol{y} \in B^n$, the function

$$f_{\boldsymbol{y}} : B^m \to \mathbb{R},\ \ f_{\boldsymbol{y}}(\boldsymbol{x}) = f(\boldsymbol{x}, \boldsymbol{y})$$

is Riemann integrable, then the second *marginal* function

$$M_f^2 : B^n \to \mathbb{R},\ \ M_f^2(\boldsymbol{y}) = \int_{B^n} f_{\boldsymbol{y}}(\boldsymbol{x})|d\boldsymbol{x}|$$

is Riemann integrable and we have

$$\int_{B^m\times B^n} f(\boldsymbol{x},\boldsymbol{y})|d\boldsymbol{x}|\,|d\boldsymbol{y}| = \int_{B^n} M_f^2(\boldsymbol{y})|d\boldsymbol{y}| = \int_{B^n} \left(\int_{B^m} f(\boldsymbol{x},\boldsymbol{y})|d\boldsymbol{x}| \right) |d\boldsymbol{y}|. \tag{15.16}$$

□

Remark 15.20 (Changing the order of integration). Suppose $B = [a, b] \times [c, d] \subset\in \mathbb{R}^2$ is a nondegenerate box and $f : B \to \mathbb{R}$ is a Riemann integrable function such that for any $x \in [a, b]$ the function

$$[c, d] \ni y \mapsto f(x, y)$$

is Riemann integrable and, for any $y \in [c, d]$, the function

$$[a, b] \ni x \mapsto f(x, y)$$

is Riemann integrable. We obtain in this fashion *two* marginals

$$M_f^1 : [a, b] \to \mathbb{R}, \quad M_f^1(x) = \int_c^d f(x, y)dy$$

and

$$M_f^2 : [c, d] \to \mathbb{R}, \quad M_f^2(y) = \int_a^b f(x, y)dx.$$

Fubini's theorem then implies that

$$\int_a^b M_f^1(x)dx = \int_{[a,b]\times[c,d]} f(x, y)|dxdy| = \int_c^d M_f^2(y)dy.$$

Using the concrete definitions of the marginals we obtain the equality

$$\boxed{\int_a^b dx \int_c^d f(x, y)dy = \int_c^d dy \int_a^b f(x, y)dx}. \tag{15.17}$$

This equality often leads to surprising conclusions and its commonly known as the *changing-the-order-of-integration trick.* □

15.1.3. *Functions Riemann integrable over* $\mathbb{R}^n$

Let $n \in \mathbb{N}$ and suppose that $X \subset \mathbb{R}^n$ is an arbitrary set. For any function $f : X \to \mathbb{R}$ we denote by f^0 its *extension by zero outside* X, i.e., f^0 is defined on the entire space $\mathbb{R}^n$, and

$$f^0(\boldsymbol{x}) = \begin{cases} f(\boldsymbol{x}), & \boldsymbol{x} \in X, \\ 0, & \boldsymbol{x} \in \mathbb{R}^n \setminus X. \end{cases}$$

Proposition 15.21. *Suppose that $B \subset \mathbb{R}^n$ is a nondegenerate box and $f : B \to \mathbb{R}$ is a Riemann integrable function. Then, for any box B' that contains B, the restriction $f^0_{B'}$ of f^0 to B' is Riemann integrable and*

$$\int_{B'} f^0_{B'}(\boldsymbol{x})\,|d\boldsymbol{x}| = \int_B f(\boldsymbol{x})|d\boldsymbol{x}|.$$

Proof. It is important to have a heuristic explanation why the above result is plausible. We know that the Riemann integral can be very well approximated by appropriate Riemann sums. Start with a partition $\boldsymbol{P}$ of the inside box B. Extend it in some way to a partition $\boldsymbol{P}'$ of the surrounding box B'. Observe that a "typical" Riemann sum of $f^0_{B'}$ associated to $\boldsymbol{P}'$ is equal to a Riemann sum of f associated to $\boldsymbol{P}$ since the value of f^0 outside B is 0. Thus the Riemann integral of f over B ought to be arbitrarily close to the Riemann integral of f^0 over B'. Here are the details.

The set D_f of points of discontinuity of f is negligible since f is Riemann integrable. The set of points of discontinuity of $f^0_{B'}$ is contained in the union $D_f \cup \partial B$. Since each of the faces of B is contained in a coordinate hyperplane, we deduce from Exercise 15.8 that each facet is negligible. The boundary ∂B is the union of facets and thus it is negligible; see Exercise 15.8. Hence $f^0_{B'}$ is Riemann integrable.

Fix $\varepsilon > 0$. Now choose a partition $\boldsymbol{P}$ of B such that

$$\omega(f, \boldsymbol{P}) < \frac{\varepsilon}{2}.$$

Now, extend $\boldsymbol{P}$ to a partition $\boldsymbol{P}'$ of B'. Since $f^0_{B'}$ is Riemann integrable we can find a partition $\boldsymbol{Q}' \succ \boldsymbol{P}'$ of B' such that

$$\omega\big(f^0_{B'}\, \boldsymbol{Q}' \big) < \frac{\varepsilon}{2}. \tag{15.18}$$

The partition $\boldsymbol{Q}'$ induces a partition $\boldsymbol{Q}$ of B that is finer than $\boldsymbol{P}$. Thus

$$\omega(f, \boldsymbol{Q}) \leq \omega(f, \boldsymbol{P}) < \frac{\varepsilon}{2}. \tag{15.19}$$

Now choose a sample $\underline{\xi}$ of $\boldsymbol{Q}'$ such that, for each chamber C of $\boldsymbol{Q}'$ not contained in B, the corresponding sample $\xi(C)$ is contained in the interior of C. In particular, this shows that $f^0\big(\xi(C) \big) = 0$ for such a chamber and sample point. We deduce

$$\begin{aligned} \boldsymbol{S}(f^0_{B'}, \boldsymbol{Q}', \underline{\xi}) &= \sum_{C \in \mathscr{C}(\boldsymbol{Q}')} f^0\big(\xi(C) \big) \operatorname{vol}_n(C) = \sum_{\substack{C \in \mathscr{C}(\boldsymbol{Q}') \\ C \subset B}} f\big(\xi(C) \big) \operatorname{vol}_n(C) \\ &= \sum_{C \in \mathscr{C}(\boldsymbol{Q})} f\big(\xi(C) \big) \operatorname{vol}_n(C) = \boldsymbol{S}(f, \boldsymbol{Q}, \underline{\xi}). \end{aligned}$$

On the other hand, Proposition 15.9 coupled with (15.18) and (15.19) imply that

$$\left| \int_{B'} f^0_{B'}(\boldsymbol{x}) |d\boldsymbol{x}| - \boldsymbol{S}(f^0_{B'}, \boldsymbol{Q}', \underline{\xi}) \right| < \omega\big(f^0_{B'}\, \boldsymbol{Q}' \big) < \frac{\varepsilon}{2},$$

$$\left| \int_{B} f(\boldsymbol{x}) |d\boldsymbol{x}| - \boldsymbol{S}(f, \boldsymbol{Q}, \underline{\xi}) \right| < \omega\big(f^0_{B'}\, \boldsymbol{Q}' \big) < \frac{\varepsilon}{2}.$$

Hence

$$\left| \int_{B'} f^0_{B'}(\boldsymbol{x}) |d\boldsymbol{x}| - \int_{B} f(\boldsymbol{x}) |d\boldsymbol{x}| \right| < \varepsilon, \ \ \forall \varepsilon > 0,$$

and thus

$$\int_{B'} f^0_{B'}(\boldsymbol{x}) |d\boldsymbol{x}| = \int_{B} f(\boldsymbol{x}) |d\boldsymbol{x}|.$$

□

Let us introduce an important concept.

Definition 15.22. Let $n \in \mathbb{N}$. The *indicator function* of a subset $S \subset \mathbb{R}^n$ is the function

$$I_S : \mathbb{R}^n \to \mathbb{R}, \ \ I_S(\boldsymbol{x}) = \begin{cases} 1, & \boldsymbol{x} \in S, \\ 0, & \boldsymbol{x} \in \mathbb{R}^n \setminus S. \end{cases}$$

In other words, I_S is the extension by 0 of the function on S equal to the constant 1. □

If B, B' are nondegenerate boxes such that $B \subset B'$, then Proposition 15.21 shows that the restriction of I_B to B' is Riemann integrable and it is discontinuous if $B \neq B'$. Moreover

$$\int_{B'} I_B|_{B'}(\boldsymbol{x}) |d\boldsymbol{x}| = \int_{B} |d\boldsymbol{x}| = \operatorname{vol}_n(B).$$

Definition 15.23. We say that a function $f : \mathbb{R}^n \to \mathbb{R}$ is *Riemann integrable* (over $\mathbb{R}^n$) if it satisfies the following conditions.

(i) There exists a nondegenerate box $B \subset \mathbb{R}^n$ such that $f(\boldsymbol{x}) = 0, \forall \boldsymbol{x} \in \mathbb{R}^n \setminus B$.

(ii) The restriction of $f|_B$ of f to B is Riemann integrable.

We set

$$\int_{\mathbb{R}^n} f(\boldsymbol{x})|d\boldsymbol{x}| := \int_B f|_B(\boldsymbol{x})|d\boldsymbol{x}|,$$

where B is a box satisfying the conditions (i) and (ii) above. We denote by $\mathcal{R}_n$ the set of Riemann integrable functions $f : \mathbb{R}^n \to \mathbb{R}$. □

Remark 15.24. (a) From the definition we deduce that if $f : \mathbb{R}^n \to \mathbb{R}$ is Riemann integrable, then it must have compact support.
(b) We need to verify the consistency of the above definition of the integral. More precisely, we need to verify that, if $f : \mathbb{R}^n \to \mathbb{R}^n$ is Riemann integrable and B_1, B_2 are two boxes satisfying the conditions (i)–(ii) in the Definition 15.23, then

$$\int_{B_1} f|_{B_1}(\boldsymbol{x})|d\boldsymbol{x}| = \int_{B_2} f|_{B_2}(\boldsymbol{x})|d\boldsymbol{x}|.$$

To check this choose a box B such that $B \supset B_1 \cup B_2$. Observe that the function f can be identified with the extension by 0 of either function $f|_{B_1}$ and $f|_{B_2}$. From Proposition 15.21 we now deduce that $f|_B$ is Riemann integrable (on B) and

$$\int_{B_1} f|_{B_1}(\boldsymbol{x})|d\boldsymbol{x}| = \int_B f|_B(\boldsymbol{x})|d\boldsymbol{x}| = \int_{B_2} f|_{B_2}(\boldsymbol{x})|d\boldsymbol{x}|.$$

We see that the indicator function of a nondegenerate (closed) box $B \subset \mathbb{R}^n$ is Riemann integrable and

$$\int_{\mathbb{R}^n} I_B(\boldsymbol{x})|d\boldsymbol{x}| = \operatorname{vol}_n(B).$$

(c) Theorem 15.13 shows that if $f, g \in \mathcal{R}_n$ and $s, t \in \mathbb{R}$, then

$$sf + tg,\ fg \in \mathcal{R}_n$$

and

$$\int_{\mathbb{R}^n} \big(sf(\boldsymbol{x}) + tg(\boldsymbol{x}) \big) d\boldsymbol{s} = s\int_{\mathbb{R}^n} f(\boldsymbol{x})|d\boldsymbol{x}| + t\int_{\mathbb{R}^n} g(\boldsymbol{x})|d\boldsymbol{x}|.$$

If additionally $f(\boldsymbol{x}) \leq g(\boldsymbol{x})$, $\forall \boldsymbol{x} \in \mathbb{R}^n$, then

$$\int_{\mathbb{R}^n} f(\boldsymbol{x})|d\boldsymbol{x}| \leq \int_{\mathbb{R}^n} g(\boldsymbol{x})|d\boldsymbol{x}|.$$ □

Recall that $C_{\mathrm{cpt}}(\mathbb{R}^n)$ denotes the set of continuous functions $\mathbb{R}^n \to \mathbb{R}$ with compact support.

Corollary 15.25. *Let $n \in \mathbb{N}$. Then $C_{\mathrm{cpt}}(\mathbb{R}^n) \subset \mathcal{R}_n$, i.e., any compactly supported function $f : \mathbb{R}^n \to \mathbb{R}$ is Riemann integrable.*

Proof. Since the support of f is compact, there exists a (closed) box $B \supset \operatorname{supp}(f)$. Thus B contains all the points where f is nonzero so that f is identically zero outside B. Moreover, since f is continuous, it is integrable on B. □

The compactly supported continuous functions play an important role in the theory of Riemann integration due to the following approximation result.

Theorem 15.26. *Let $n \in \mathbb{N}$ and suppose that $f : \mathbb{R}^n \to \mathbb{R}$ is a bounded function and U is an open set containing the support of f. Then the following statements are equivalent.*

(i) The function f is Riemann integrable (on $\mathbb{R}^n$).
(ii) For any $\varepsilon > 0$ there exist functions $g, G \in C_{\mathrm{cpt}}(\mathbb{R}^n)$ such that

$$\operatorname{supp}(g),\ \operatorname{supp}(G) \subset U,$$

$$g(\boldsymbol{x}) \leq f(\boldsymbol{x}) \leq G(\boldsymbol{x}),\ \ \forall \boldsymbol{x} \in \mathbb{R}^n \ \ \textit{and} \ \ 0 \leq \int_{\mathbb{R}^n} \big(G(\boldsymbol{x}) - g(\boldsymbol{x}) \big) |d\boldsymbol{x}| < \varepsilon.$$

□

The proof of this theorem is not extremely demanding but it would distract us from the main "story". The curious reader can find the details in [6, §6.9].

15.1.4. *Volume and Jordan measurability*

The concept of Riemann integral can be used to define the notion of n-dimensional volume. Intuitively, the n-dimensional volume of subset $S \subset \mathbb{R}^n$ ought to be a nonnegative number $\operatorname{vol}_n(S)$ that satisfies the Inclusion-Exclusion Principle

$$\operatorname{vol}_n(S_1 \cup S_2) = \operatorname{vol}_n(S_1) + \operatorname{vol}_n(S_2) - \operatorname{vol}_n(S_1 \cap S_2),$$

it "depends continuously" on S and, when S is a box, this notion of volume should coincide with our original definition (15.2). Additionally, we would like this volume to stay unchanged when we rigidly move S around $\mathbb{R}^n$. (Typical examples of rigid transformations are translations and rotations about an "axis".)

The famous Banach–Tarski "paradox"[3] shows that we cannot associate a notion of n-dimensional volume with the above properties to *all* subsets of $\mathbb{R}^n$. However, we can associate a notion of volume with these properties to many subsets of $\mathbb{R}^n$.

Definition 15.27. (a) A bounded set $S \subset \mathbb{R}^n$ is called *Jordan*[4] *measurable* if the indicator function I_S is Riemann integrable. We denote by $\mathfrak{J}(\mathbb{R}^n)$ the collection of Jordan measurable subsets of $\mathbb{R}^n$.

(b) The *n-dimensional volume* of a Jordan measurable set $S \subset \mathbb{R}^n$ is the nonnegative number

$$\operatorname{vol}_n(S) := \int_{\mathbb{R}^n} I_S(\boldsymbol{x}) |d\boldsymbol{x}|.$$

□

Remark 15.28. If B is a (closed) box, then the volume of B as defined in the above definition, coincides with the volume of B as defined in (15.2). This follows from Example 15.7 and Proposition 15.21. □

[3] Search Wikipedia for more details about this famous result.

[4] Named after the French mathematician Camille Jordan (1838–1922). See Wikipedia for more on Jordan.

We mention several useful consequences of the above result.

Corollary 15.29. *A bounded subset $S \subset \mathbb{R}^n$ is Jordan measurable if and only if its boundary ∂S is negligible.*

Proof. Note that S is Jordan measurable if and only if its indicator function

$$I_S : \mathbb{R}^n \to \mathbb{R}, \quad I_S(\boldsymbol{x}) = \begin{cases} 1, & \boldsymbol{x} \in S, \\ 0, & \boldsymbol{x} \in \mathbb{R}^n \setminus S, \end{cases}$$

is Riemann integrable. According to Lebesgue's theorem this happens if and only if the set of points of discontinuity of I_S is negligible. Now observe that the boundary of S is precisely the set of discontinuities of I_S. □

Proposition 15.30. *Let $n \in \mathbb{N}$.*

(i) (Inclusion-Exclusion Principle) *If $S_1, S_2 \in \mathfrak{J}(\mathbb{R}^n)$, then*

$$S_1 \cup S_2, \quad S_1 \cap S_2 \in \mathfrak{J}(\mathbb{R}^n)$$

and

$$\operatorname{vol}_n(S_1 \cup S_2) = \operatorname{vol}_n(S_1) + \operatorname{vol}_n(S_2) - \operatorname{vol}_n(S_1 \cap S_2). \tag{15.20}$$

(ii) (Monotonicity) *If $S_1, S_2 \in \mathfrak{J}(\mathbb{R}^n)$ and $S_1 \subset S_2$, then*

$$\operatorname{vol}_n(S_1) \leq \operatorname{vol}_n(S_2).$$

Proof. (i) Since S_1, S_2 are bounded, there exists a box B that contains both S_1 and S_2. We deduce that the restrictions to B of both functions I_{S_1} and I_{S_2} are Riemann integrable. Thus the restrictions to B of the functions $I_{S_1} + I_{S_2}$ and $I_{S_1} I_{S_2}$ are Riemann integrable. Observing that

$$I_{S_1 \cap S_2} = I_{S_1} I_{S_2} \text{ and } I_{S_1 \cup S_2} = I_{S_1} + I_{S_2} - I_{S_1 \cap S_2}$$

we deduce that $S_1 \cap S_2$ and $S_1 \cup S_2$ are Jordan measurable and

$$\begin{aligned} \operatorname{vol}_n(S_1 \cup S_2) &= \int_{\mathbb{R}^n} \big(I_{S_1}(\boldsymbol{x}) + I_{S_2}(\boldsymbol{x}) - I_{S_1 \cap S_2}(\boldsymbol{x}) \big) |d\boldsymbol{x}| \\ &= \operatorname{vol}_n(S_1) + \operatorname{vol}_n(S_2) - \operatorname{vol}_n(S_1 \cap S_2). \end{aligned}$$

(i) Note that

$$S_1 \subset S_2 \Rightarrow I_{S_1}(\boldsymbol{x}) \leq I_{S_2}(\boldsymbol{x}), \ \forall \boldsymbol{x} \in \mathbb{R}^n \Rightarrow \int_{\mathbb{R}^n} I_{S_1}(\boldsymbol{x}) |d\boldsymbol{x}| \leq \int_{\mathbb{R}^n} I_{S_2}(\boldsymbol{x}) |d\boldsymbol{x}|.$$

□

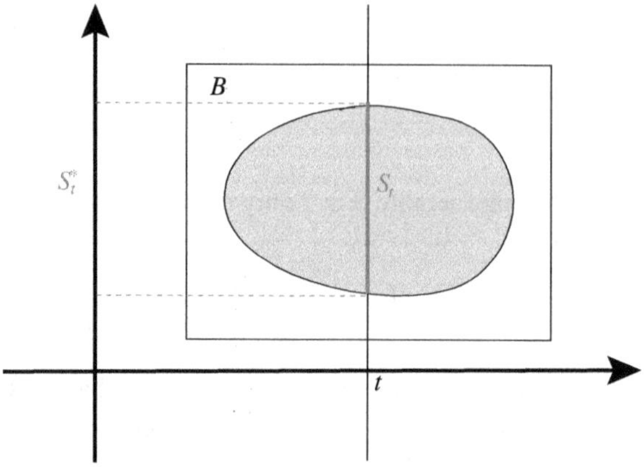

Fig. 15.3 *Slicing a 2-dimensional region by vertical lines.*

Example 15.31 (Cavalieri's Principle). Let $n \in \mathbb{N}$. Consider a Jordan measurable set $S \subset \mathbb{R} \times \mathbb{R}^n = \mathbb{R}^{1+n}$. We denote by $x^0, \dots, x^n$ the coordinates in $\mathbb{R}^{1+n}$. For any $t \in \mathbb{R}$ we denote by S_t the intersection of S with the hyperplane $\{x^0 = t\}$. We will refer to S_t as the *slice* of S over t; see Figure 15.3.

Denote by S_t^* the projection of S_t on the coordinate subspace $\{0\} \times \mathbb{R}^n$ (pictured as the vertical axis in Figure 15.3). More precisely

$$S_t^* := \left\{ (x^1, \dots, x^n) \in \mathbb{R}^n;\ \ (t, x^1, \dots, x^n) \in S_t \right\}.$$

Cavalieri's Principle states that if the all the (projected) slices $S_t^* \subset \mathbb{R}^n$ are Jordan measurable then

$$\operatorname{vol}_{n+1}(S) = \int_{\mathbb{R}} \operatorname{vol}_n(S_t^*)|dt|. \tag{15.21}$$

This is an immediate consequence of the Fubini Theorem 15.18. Since S is bounded, there exists a box

$$B = [a_0, b_0] \times \underbrace{[a_1, b_1] \times \cdots \times [a_n, b_n]}_{B'}.$$

Let $f : \mathbb{R}^{1+n} \to \mathbb{R}$ be the indicator function of S, $f = I_S$. Note that f is zero outside the box B. Since S is Jordan measurable, the restriction of f to B is Riemann integrable. For any $t \in \mathbb{R}$ the function

$$\mathbb{R} \ni (x^1, \dots, x^n) \mapsto f_t(x^1, \dots, x^n) \in \mathbb{R}$$

is the indicator function of S_t^*,

$$f_t(x^1, \dots, x^n) = I_{S_t^*}(x^1, \dots, x^n),$$

and thus it is Riemann integrable. Note that $f_t = 0$ if $t \notin [a_0, b_0]$. The marginal function is

$$M_f(t) = \int_{B'} f_t(x^1, \dots, x^n) dx^1 \cdots dx^n = \operatorname{vol}_n(S_t^*).$$

Fubini's Theorem now implies

$$\operatorname{vol}_{n+1}(S) = \int_B f(x^0, x^1, \dots, x^n)|dx^0 dx^1 \cdots dx^n| = \int_{a_0}^{b_0} \operatorname{vol}_n(S_t^*)|dt|. \qquad \square$$

15.1.5. *The Riemann integral over arbitrary regions*

Definition 15.32. Let $S \subset \mathbb{R}^n$. A bounded function $f : S \to \mathbb{R}$ is called *Riemann integrable over S* if f^0, its extension by 0 to $\mathbb{R}^n$, is Riemann integrable over $\mathbb{R}^n$. In this case we define the Riemann integral of f over the set S to be

$$\int_S f(\boldsymbol{x})|d\boldsymbol{x}| := \int_{\mathbb{R}^n} f^0(\boldsymbol{x})|d\boldsymbol{x}|.$$

We denote by $\mathcal{R}_n(S)$ the set of Riemann integrable functions on S.

A function $f : \mathbb{R}^n \to \mathbb{R}$ is said to be Riemann integrable over S if $f\boldsymbol{I}_S$ is integrable over $\mathbb{R}^n$. □

Remark 15.33. (a) Note that we can rephrase the above definition in a more concise way,

$$\boxed{f \in \mathcal{R}_n(S) \Longleftrightarrow f^0 I_S \in \mathcal{R}_n}.$$

(b) Proposition 15.21 shows that, if B is a box, then the set $\mathcal{R}(B)$ of functions $f : B \to \mathbb{R}$ Riemann integrable in the sense of Definition 15.5 coincides with the set $\mathcal{R}_n(B)$ in the above definition.

(c) If $f \in \mathcal{R}(S)$, then

$$\int_S f(\boldsymbol{x})\,|d\boldsymbol{x}| = \int_B f^0(\boldsymbol{x})\boldsymbol{I}_S(\boldsymbol{x})\,|d\boldsymbol{x}|,$$

where B is *any* box in $\mathbb{R}^n$ that contains S. □

Proposition 15.34. *Let $n \in \mathbb{N}$.*

(i) (Additivity of integrals with respect to domains) *Suppose that $S_1, S_2 \subset \mathbb{R}^n$ are Jordan measurable sets and $f : S_1 \cup S_2 \to \mathbb{R}$ is Riemann integrable. Then the restrictions of f to S_1 and S_2 are Riemann integrable and*

$$\int_{S_1\cup S_2} f(\boldsymbol{x})|d\boldsymbol{x}| = \int_{S_1} f(\boldsymbol{x})|d\boldsymbol{x}| + \int_{S_2} f(\boldsymbol{x})|d\boldsymbol{x}| - \int_{S_1\cap S_2} f(\boldsymbol{x})|d\boldsymbol{x}|. \tag{15.22}$$

(ii) (Monotonicity) *If S is Jordan measurable and $f, g : S \to \mathbb{R}$ are Riemann integrable functions such that $f(\boldsymbol{x}) \le g(\boldsymbol{x})$, $\forall \boldsymbol{x} \in S$, then*

$$\int_S f(\boldsymbol{x})|d\boldsymbol{x}| \le \int_S g(\boldsymbol{x})|d\boldsymbol{x}|. \tag{15.23}$$

In particular

$$\left|\int_S f(\boldsymbol{x})|d\boldsymbol{x}|\right| \le \Big(\sup_{\boldsymbol{x}\in S} |f(\boldsymbol{x})|\Big)\operatorname{vol}_n(S). \tag{15.24}$$

Proof. (i) Observe that $f^0, I_{S_1}, I_{S_2} \in \mathcal{R}_n$ so that

$$I_{S_1}f^0,\;\; I_{S_2}f^0 \in \mathcal{R}_n,$$

$$f^0 = I_{S_1\cup S_2}f^0 = I_{S_1}f^0 + I_{S_2}f^0 - I_{S_1\cap S_2}f^0.$$

The equality (15.22) follows by integrating over $\mathbb{R}^n$ the above equality.

(ii) The equality (15.23) follows from Remark 15.24(b) by observing that $f^0(\boldsymbol{x}) \leq g^0(\boldsymbol{x})$, $\forall \boldsymbol{x} \in \mathbb{R}^n$. The inequality (15.24) follows by integrating over S the inequality

$$-\Big(\sup_{\boldsymbol{x}\in S} |f(\boldsymbol{x})|\Big) \leq f(\boldsymbol{x}) \leq \Big(\sup_{\boldsymbol{x}\in S} |f(\boldsymbol{x})|\Big), \quad \forall \boldsymbol{x} \in S.$$

□

Corollary 15.35. *Let $n \in \mathbb{N}$. Suppose that $S_1, S_2 \subset \mathbb{R}^n$ are Jordan measurable sets and $f : S_1 \cup S_2 \to \mathbb{R}$ is Riemann integrable. If $\operatorname{vol}_n(S_1 \cap S_2) = 0$, then*

$$\int_{S_1\cup S_2} f(\boldsymbol{x})|d\boldsymbol{x}| = \int_{S_1} f(\boldsymbol{x})|d\boldsymbol{x}| + \int_{S_2} f(\boldsymbol{x})|d\boldsymbol{x}|. \tag{15.25}$$

Proof. From (15.24) we deduce that

$$\int_{S_1\cap S_2} f(\boldsymbol{x})|d\boldsymbol{x}| = 0.$$

We now see that (15.25) is a special case of (15.22). □

Proposition 15.36. *Suppose that $K \subset \mathbb{R}^n$ is a* compact *Jordan measurable set and $f : K \to \mathbb{R}$ is a continuous function. Then f is integrable over K.*

Proof. Fix a box B that contains K. As usual, denote by f^0 the extension by 0 of f. Observe that the set of points of discontinuity of f^0 is contained in the boundary ∂K which is negligible since K is Jordan measurable. Lebesgue's Theorem 15.17 then shows that f^0 is integrable. □

The next two results are also consequences of Lebesgue's Theorem. Their proofs are left to you as an exercise.

Corollary 15.37. *If $f : \mathbb{R}^n \to \mathbb{R}$ is Riemann integrable, then its support is a compact Jordan measurable subset of $\mathbb{R}^n$.* □

Corollary 15.38. *Suppose that $K \subset \mathbb{R}^n$ is a compact Jordan measurable set, $Z \subset \mathbb{R}^n$ is a negligible closed subset and $f : K \to \mathbb{R}$ is a bounded function. Then the following are equivalent.*

(i) The function f is Riemann integrable over K.
(ii) The function f is Riemann integrable over $K \setminus Z$.

If either (i) or (ii) holds, then

$$\int_{K\setminus Z} f(\boldsymbol{x})\,|d\boldsymbol{x}| = \int_K f(\boldsymbol{x})\,|d\boldsymbol{x}|.$$ □

15.2. Fubini theorem and iterated integrals

We now have at our disposal all the information we need to prove a version of the Fubini Theorem 15.18 that involves easily verifiable assumptions.

15.2.1. *An unconditional Fubini theorem*

Before we can state the version of the Fubini theorem most frequently used in applications we need to introduce a very versatile concept.

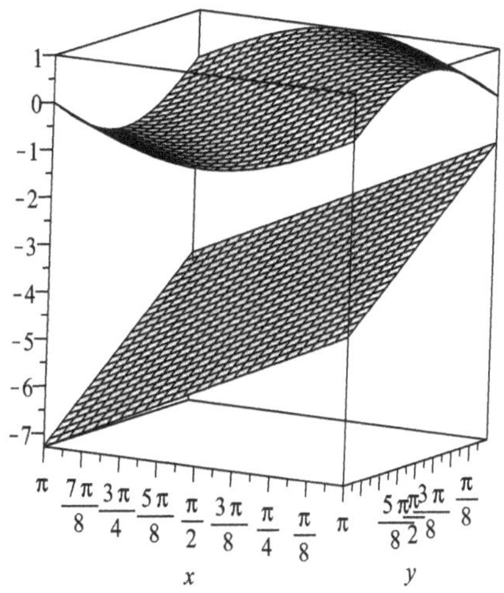

Fig. 15.4 *A simple type region in* $\mathbb{R}^3$ *with cross section* $K = [0, \pi] \times [0, \pi]$, *a flat bottom* $\beta(x, y) = -2 - x - y$ *and curved top* $\tau(x, y) = \sin(x + y)$.

Definition 15.39. Let $n \in \mathbb{N}$. A compact set $D \subset \mathbb{R}^{n+1}$ is called a *domain of simple type*if there exist a *compact Jordan measurable* set $K \subset \mathbb{R}^n$ and *continuous* functions

$$\beta, \tau : K \to \mathbb{R}$$

with the following properties

- $\beta(x^1, \ldots, x^n) \leq \tau(x^1, \ldots, x^n)$, $\forall (x^1, \ldots, x^n) \in K$.
- The point $(x^1, \ldots, x^n, y) \in \mathbb{R}^{n+1}$ belongs to D if and only if

$$(x^1, \ldots, x^n) \in K \ \text{ and } \ \beta(x^1, \ldots, x^n) \leq y \leq \tau(x^1, \ldots, x^n).$$

We will denote this domain by $D(K, \beta, \tau)$. The region K is called the *cross section* of D, the function β is called the *bottom* of D and the function τ is called the *top* of D; see Figure 15.4. □

Thus $D(K, \beta, \tau)$ is the region between the graphs of β and τ, where β sits at the bottom and τ at the top.

Proposition 15.40. *Let $n \in \mathbb{N}$. Suppose that $D = D(K, \beta, \tau) \subset \mathbb{R}^{n+1}$ is a simple type domain with cross section $K \subset \mathbb{R}^n$ and top/bottom functions $\tau, \beta : K \to \mathbb{R}$. Then D is compact and Jordan measurable.*

Proof. As usual we denote by $m_S(f)$ and $M_S(f)$ the infimum and respectively the supremum of a function $f : S \to \mathbb{R}$. Note that

$$D(K, \beta, \tau) \subset K \times [m_\beta, M_\tau]$$

so $D(K, \beta, \tau)$ is bounded. Since K is compact, we deduce that K is closed. The continuity of β, τ shows that $D(K, \beta, \tau)$ is closed and thus compact.

Fix a box $B \subset \mathbb{R}^n$ that contains K. Set

$$\widehat{B} = B \times I, \;\; I := [m_\beta, M_\tau], \;\; L = M_\tau - m_\beta.$$

Fix a partition $\boldsymbol{P} = (\boldsymbol{P}_1, \dots, \boldsymbol{P}_n)$ of B. The set of chambers $\mathscr{C}(\boldsymbol{P})$ decomposes as

$$\mathscr{C}(\boldsymbol{P}) = \mathscr{C}_e(\boldsymbol{P}) \cup \mathscr{C}_b(\boldsymbol{P}) \cup \mathscr{C}_i(\boldsymbol{P}),$$

where $\mathscr{C}_e(\boldsymbol{P})$ consists of chambers that do not intersect K, $\mathscr{C}_i(\boldsymbol{P})$ consists of chambers located in the interior of K and $\mathscr{C}_b(\boldsymbol{P})$ consists of chambers that intersect both K and its complement. For $\nu \in \mathbb{N}$ we denote by $\boldsymbol{U}_\nu$ the uniform partition of I into ν sub-intervals of equal length L/ν. We denote by $I_1, \dots, I_\nu$ the sub-intervals of the partition $\boldsymbol{U}_\nu$.

We obtain a partition $\widehat{\boldsymbol{P}}_\nu = (\boldsymbol{P}_1, \dots, \boldsymbol{P}_n, \boldsymbol{U}_\nu)$ of $\widehat{B} = B \times I$. The chambers of $\widehat{\boldsymbol{P}}_\nu$ have the form $C \times I_k$, $C \in \mathscr{C}(\boldsymbol{P})$, $k = 1, \dots, \nu$. Note that

$$\operatorname{osc}(I_D, C \times I_k) = 0, \;\; \forall C \in \mathscr{C}_e(\boldsymbol{P}), \; k = 1, \dots, \nu,$$

and

$$\operatorname{osc}(I_D, C \times I_k) \leq 1, \;\; \forall C \in \mathscr{C}_b(\boldsymbol{P}), \; k = 1, \dots, \nu.$$

In particular, we deduce that,

$$\forall C \in \mathscr{C}_b(\boldsymbol{P}), \; \sum_{k=1}^{\nu} \operatorname{osc}(I_D, C \times I_k) \operatorname{vol}_{n+1}(C \times I_k) \leq \sum_{k=1}^{\nu} \operatorname{vol}_n(C)\frac{L}{\nu} = L \operatorname{vol}_n(C).$$

If $C \in \mathscr{C}_i(\boldsymbol{P})$, and $I_k \subset \big(M_C(\beta), m_C(\tau) \big)$, then $\operatorname{osc}(I_D, C \times I_k\,) = 0$. We deduce that if $\operatorname{osc}(I_D, C \times I_k\,) = 1$, then

$$I_k \subset \mathfrak{I}_\nu(C) := \left[m_C(\beta) - \frac{L}{\nu}, M_C(\beta) + \frac{L}{\nu} \right] \cup \left[m_C(\tau) - \frac{L}{\nu}, M_C(\tau) + \frac{\nu}{\nu} \right].$$

Hence, $\forall C \in \mathscr{C}_i(\boldsymbol{P})$ we have

$$\begin{aligned} \sum_{k=1}^{\nu} \operatorname{osc}(I_D, C \times I_k) \operatorname{vol}_{n+1}(C \times I_k) &\leq \operatorname{vol}_n(C) \sum_{k \in \mathfrak{I}_\nu(C)} \operatorname{vol}_1(I_k) \\ &\leq \operatorname{vol}_n(C) \operatorname{vol}_1\big(\mathfrak{I}_\nu(C)\big) = \left(\operatorname{osc}(\beta, C) + \operatorname{osc}(\tau, C) + \frac{4L}{\nu} \right) \operatorname{vol}_n(C). \end{aligned}$$

Putting together all of the above we deduce

$$\begin{aligned} \omega(I_D, \widehat{\boldsymbol{P}}_\nu) = &\sum_{C \in \mathscr{C}_b(\boldsymbol{P})} \sum_{k=1}^{\nu} \operatorname{osc}(I_D, C \times I_k) \operatorname{vol}_{n+1}(C \times I_k) \\ &+ \sum_{C \in \mathscr{C}_i(\boldsymbol{P})} \sum_{k=1}^{\nu} \operatorname{osc}(I_D, C \times I_k) \operatorname{vol}_{n+1}(C \times I_k) \\ &\leq L \sum_{C \in \mathscr{C}_b(\boldsymbol{P})} \operatorname{vol}_n(C) + \frac{4L}{\nu} \underbrace{\sum_{C \in \mathscr{C}_i(\boldsymbol{P})} \operatorname{vol}_n(C)}_{\leq \operatorname{vol}_n(B)} + \sum_{C \in \mathscr{C}_i(\boldsymbol{P})} \big(\operatorname{osc}(\beta, C) + \operatorname{osc}(\tau, C) \big) \operatorname{vol}_n(C). \end{aligned}$$

Hence

$$\begin{aligned} \omega(I_D, \widehat{\boldsymbol{P}}_\nu) \leq L &\sum_{C \in \mathscr{C}_b(\boldsymbol{P})} \operatorname{vol}_n(C) + \frac{4L \operatorname{vol}_n(B)}{\nu} \\ &+ \sum_{C \in \mathscr{C}_i(\boldsymbol{P})} \big(\operatorname{osc}(\beta, C) + \operatorname{osc}(\tau, C) \big) \operatorname{vol}_n(C). \end{aligned} \tag{15.26}$$

Fix $\varepsilon > 0$. Since K is Jordan measurable, we can find a partition $\boldsymbol{Q}_\varepsilon$ of B such that

$$\sum_{C \in \mathscr{C}_b(\boldsymbol{Q}^\varepsilon)} \operatorname{vol}_n(C) = \omega(I_K, \boldsymbol{Q}_\varepsilon) < \frac{\varepsilon}{3}.$$

Choose $\nu = \nu(\varepsilon) > 0$ sufficiently large so that

$$\frac{4L}{\nu} \operatorname{vol}_n(B) < \frac{\varepsilon}{3}.$$

Since $\beta, \tau : K \to \mathbb{R}$ are uniformly continuous, we can find $\delta = \delta(\varepsilon) > 0$ such that, for any set $S \subset K$ with $\operatorname{diam}(S) < \delta$ we have

$$\operatorname{osc}(\beta, S) + \operatorname{osc}(\tau, C) < \frac{\varepsilon}{3 \operatorname{vol}_n(B)}.$$

Now choose a partition $\boldsymbol{P}^\varepsilon$ of $\boldsymbol{P}$ such that $\boldsymbol{P}^\varepsilon \succ \boldsymbol{Q}^\varepsilon$ and $\|\boldsymbol{P}^\varepsilon\| < \delta(\varepsilon)$. We deduce from (15.26) that for $\nu > \nu(\varepsilon)$ we have

$$\omega\big(I_D, \widehat{\boldsymbol{P}^\varepsilon_\nu}\big) < \varepsilon.$$

□

Theorem 15.41 (Fubini). *Let $n \in \mathbb{N}$ and suppose $D = D(K, \beta, \tau)$ is a simple type domain with cross section K and bottom/top functions $\beta, \tau : K \to \mathbb{R}$. We denote by $(\boldsymbol{x}, \boldsymbol{y})$ the coordinates in $\mathbb{R}^{n+1} = \mathbb{R}^n \times \mathbb{R}$. If $f : D \to \mathbb{R}$ is continuous, then f is integrable over D, the marginal function*

$$M_f : K \to \mathbb{R}, \ \ M_f(\boldsymbol{x}) = \int_{\beta(\boldsymbol{x})}^{\tau(\boldsymbol{x})} f(\boldsymbol{x}, y)|dy|$$

is Riemann integrable and

$$\int_D f(\boldsymbol{x}, y)|d\boldsymbol{x}dy| = \int_K M_f(\boldsymbol{x})|d\boldsymbol{x}| = \int_K \left(\int_{\beta(\boldsymbol{x})}^{\tau(\boldsymbol{x})} f(\boldsymbol{x}, y)\,|dy| \right) |d\boldsymbol{x}|. \tag{15.27}$$

Proof. According to Proposition 15.40 the region $D \subset \mathbb{R}^{n+1}$ is compact and Jordan measurable. Proposition 15.36 now implies that f is Riemann integrable on D.

Fix a box in $B \subset \mathbb{R}^n$ and a compact interval $I = [m, M] \subset \mathbb{R}$ such that

$$\beta(\boldsymbol{x}), \ \ \tau(\boldsymbol{x}) \in I, \ \ \forall \boldsymbol{x} \in K.$$

Then the box $B' = B \times [m, M] \subset \mathbb{R}^{n+1}$ contains D and the extension f^0 is Riemann integrable on B'. Next observe that for any $\boldsymbol{x} \in B$ the function

$$f^0_{\boldsymbol{x}} : I \to \mathbb{R}, \ \ f^0_{\boldsymbol{x}}(y) = f(\boldsymbol{x}, y)$$

is Riemann integrable. Indeed, if $\boldsymbol{x} \in B \setminus K$, then $f^0_{\boldsymbol{x}}$ is identically 0. On the other hand, if $\boldsymbol{x} \in K$, then

$$f^0_{\boldsymbol{x}}(y) = \begin{cases} f(\boldsymbol{x}, y), & y \in [\beta(\boldsymbol{x}), \tau(\boldsymbol{x})], \\ 0, & y \in [m, \beta(\boldsymbol{x})) \cup (\tau(\boldsymbol{x}), M]. \end{cases}$$

The continuity of f coupled with Corollary 9.27 imply the Riemann integrability of $f^0_{\boldsymbol{x}}$. We can now apply the conditional Fubini Theorem 15.18 to the function $f^0 : B' \to \mathbb{R}$. Note that the marginal function $M^1_{f^0}$ is Riemann integrable on B and coincides with the extension by 0 of the marginal function $M_f : K \to \mathbb{R}$, i.e., $M^1_{f^0} = (M_f)^0$. Thus M_f is Riemann integrable on K. We deduce

$$\int_{B'} f(\boldsymbol{x}, y)|d\boldsymbol{x}dy| = \int_{B \times I} f^0(\boldsymbol{x}, y)|d\boldsymbol{x}dy|$$

$$= \int_B M^1_{f^0}(\boldsymbol{x})|d\boldsymbol{x}| = \int_B (M^1_f)^0(\boldsymbol{x})|d\boldsymbol{x}| = \int_K M^1_f(\boldsymbol{x})|d\boldsymbol{x}|.$$

□

Remark 15.42. (a) The last integral in (15.27) is an example of *iterated* or *repeated* integral.

(b) In the definition of simple type domains in $\mathbb{R}^{n+1}$ the last coordinate x^{n+1} plays a distinguished role. We did this only to simplify the notation in the various proofs. The results we proved above hold for regions $D \subset \mathbb{R}^{n+1}$ of the type

$$D = \left\{ (x^1, \dots, x^{n+1}) \in \mathbb{R}^{n+1};\ \boldsymbol{x} \in K,\ \beta(\boldsymbol{x}) \leq x^j \leq \tau(\boldsymbol{x}) \right\}$$

where

$$\boldsymbol{x} := (x^1, \dots, x^{j-1}, x^{j+1}, \dots, x^{n+1}) \in \mathbb{R}^n.$$

We will refer to domains of this type as *domains of simple type with respect to the* x^j*-axis*. We want to point out that a given domain could be of simple type with respect to many axes.

Theorem 15.41 has an obvious extension to domains that are simple type with respect to an arbitrary axis x^j: replace y by x^j everywhere in the statement and the proof of this theorem. □

15.2.2. *Some applications*

Theorem 15.41 is best understood by witnessing it at work.

Example 15.43. Consider the triangle T in Figure 15.5. Its vertices have coordinates $(0,0), (1,1), (0,2)$. It is limited by the y-axis and the lines $y = x$, $y = 2 - x$. This triangle is a domain of simple type with respect to both x and y-axis.

Viewed as a simple type domain with respect to the y-axis it has description

$$T = \left\{ (x, y) \in \mathbb{R}^2;\ x \in [0, 1],\ \beta(x) \leq y \leq \tau(x) \right\},$$

where $\beta(x) = x$ and $\tau(x) = 2 - x$.

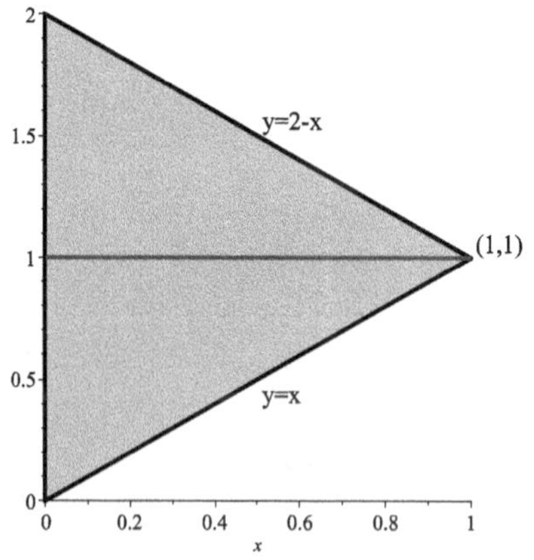

Fig. 15.5 *An isosceles triangle in the plane.*

Viewed as a simple type domain with respect to the x-axis it has description

$$T=\{\,(x,y)\in\mathbb{R}^2;\;\; y\in[0,2],\;\; \bar{\beta}(y)\le x\le\bar{\tau}(y)\,\},$$

where $\bar{\beta}(y)=0$ and

$$\bar{\tau}(y)=\begin{cases} y, & y\in[0,1],\\ 2-y, & y\in(1,2].\end{cases}$$

Consider a continuous function $f:T\to\mathbb{R}$. Using Fubini's theorem we deduce

$$\begin{aligned}\int_0^1\Big(\int_x^{2-x} f(x,y)|dy|\Big)|dx| &= \int_T f(x,y)|dxdy|\\ &=\int_0^1\Big(\int_0^y f(x,y)|dx|\Big)\,|dy|+\int_1^2\Big(\int_0^{2-y} f(x,y)|dx|\Big)\,|dy|.\end{aligned}$$

□

Our next application is another version of *Cavalieri's Principle.*

Proposition 15.44. *Let $n\in\mathbb{N}$. Suppose that $K\subset\mathbb{R}^n$ is a compact Jordan measurable set and $\beta,\tau:\mathbb{K}\to\mathbb{R}$ continuous functions such that*

$$\beta(\boldsymbol{x})\le\tau(\boldsymbol{x}),\;\;\forall\boldsymbol{x}\in K.$$

Then the $(n+1)$-dimensional volume of the region $D(K,\beta,\tau)\subset\mathbb{R}^{n+1}$ between the graphs of β and τ is

$$\operatorname{vol}_{n+1}\big(D(K,\beta,\tau)\big)=\int_K\big(\tau(\boldsymbol{x})-\beta(\boldsymbol{x})\big)|d\boldsymbol{x}|. \tag{15.28}$$

In particular, for any continuous function $h:K\to\mathbb{R}$, the $(n+1)$-dimensional volume of the graph Γ_h of h is 0

$$\operatorname{vol}_{n+1}(\Gamma_h)=0. \tag{15.29}$$

Proof. The equality (15.28) is the special case of (15.27) corresponding to $f=1$. The equality (15.29) follows from (15.28) by choosing $\beta=\tau=h$. □

Example 15.45. For every $n\in\mathbb{N}$ we denote by $\boldsymbol{T}_n$ the n-dimensional *simplex*[5] defined by the conditions

$$\boldsymbol{T}_n:=\big\{\,\boldsymbol{x}\in\mathbb{R}^n;\;\; x^i\ge 0,\;\;\forall i=1,\dots,n,\;\; x^1+\cdots+x^n\le 1\,\big\}.$$

The region $\boldsymbol{T}_n$ can be alternatively described as the region

$$\boldsymbol{T}_n:=\big\{\,(\boldsymbol{x}_*,x^n)\in\mathbb{R}^n;\;\;\boldsymbol{x}_*\in\boldsymbol{T}_{n-1},\;\; 0\le x^n\le 1-(x^1+\cdots+x^{n-1})\,\big\},$$

where $\boldsymbol{x}_*:=(x^1,\dots,x^{n-1})$.

[5] The concept of simplex is the higher dimensional generalization of the more familiar concepts of triangles and tetrahedra.

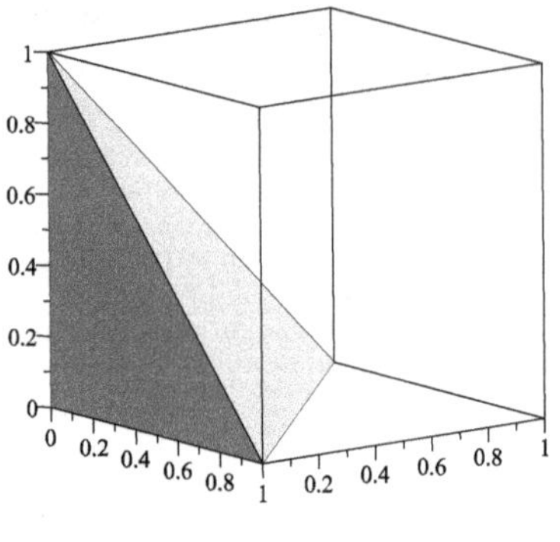

Fig. 15.6 *The tetrahedron $\boldsymbol{T}_3$.*

Observe that $\boldsymbol{T}_1 = [0,1]$, that $\boldsymbol{T}_2$ is a domain of simple type in $\mathbb{R}^2$ with respect to the x^2 axis, with bottom 0 and top $1 - x^1$ and cross section $\boldsymbol{T}_1$ and thus $\boldsymbol{T}_2$ is compact and Jordan measurable. We deduce inductively that $\boldsymbol{T}_n$ is simple type with respect to the x^n-axis, with bottom 0, top $1 - (x^1 + \cdots + x^{n-1})$ and cross section $\boldsymbol{T}_{n-1}$ and thus $\boldsymbol{T}_n$ is compact and Jordan measurable. We want to compute its volume.

To keep the insanity at bay we introduce the simplifying notation

$$s_k = s_k(x^1, \dots, x^k) := x^1 + \cdots + x^k.$$

Note that $s_k = s_{k-1} + x^k$ and

$$\boldsymbol{T}_k = \big\{ (x^1, \dots, x^k) \in \mathbb{R}^k;\ \ (x^1, \dots, x^{k-1}) \in \boldsymbol{T}_{k-1},\ 0 \le x^k \le 1 - s_{k-1} \big\}.$$

For any $k \in \mathbb{N}$ and $s \in \mathbb{R}$ we set

$$I_k(s) := \int_0^{1-s} (1 - s - x)^k dx.$$

Making the change in variables $u := 1 - s - x$ we deduce

$$I_k(s) = -\int_{1-s}^0 u^k du = \frac{1}{k}(1-s)^k. \tag{15.30}$$

Using Fubini's Theorem (15.27) and the equality (15.30) we deduce

$$
\begin{aligned}
\mathrm{vol}_n(\boldsymbol{T}_n) &\overset{(15.27)}{=} \int_{\boldsymbol{T}_{n-1}} (1 - s_{n-1})|dx^1 \cdots dx^{n-1}| \\
&\overset{(15.27)}{=} \int_{\boldsymbol{T}_{n-2}} \left(\int_0^{1-s_{n-2}} \big(1 - (s_{n-2} + x^{n-1})\big) dx^{n-1} \right) |dx^1 \cdots dx^{n-2}| \\
&= \int_{\boldsymbol{T}_{n-2}} \underbrace{\left(\int_0^{1-s_{n-2}} \big((1 - s_{n-2}) - x^{n-1}\big) dx^{n-1} \right)}_{I_1(s_{n-2})} |dx^1 \cdots dx^{n-2}| \\
&\overset{(15.30)}{=} \frac{1}{2} \int_{\boldsymbol{T}_{n-2}} (1 - s_{n-2})^2 |dx^1 \cdots dx^{n-2}| \\
&\overset{(15.27)}{=} \frac{1}{2} \int_{\boldsymbol{T}_{n-3}} \underbrace{\left(\int_0^{1-s_{n-3}} \big((1 - s_{n-3}) - x_{n-2}\big)^2 dx^{n-2} \right)}_{I_2(s_{n-3})} |dx^1 \cdots dx^{n-3}| \\
&\overset{(15.30)}{=} \frac{1}{2 \cdot 3} \int_{\boldsymbol{T}_{n-3}} (1 - s_{n-3})^3 |dx^1 \cdots dx^{n-3}|.
\end{aligned}
$$

Continuing in this fashion we deduce

$$
\mathrm{vol}_n(\boldsymbol{T}_n) = \frac{1}{k!} \int_{\boldsymbol{T}_{n-k}} (1 - s_{n-k})^k |dx^1 \cdots dx^{n-k}|, \quad \forall k = 1, \dots, n.
$$

In particular, if we let $k = n - 1$, we deduce

$$
\begin{aligned}
\mathrm{vol}_n(\boldsymbol{T}_n) &= \frac{1}{(n-1)!} \int_{\boldsymbol{T}_1} (1 - s_1)^{n-1} |dx^1| \\
&= \frac{1}{(n-1)!} \int_0^1 (1 - x^1)^{n-1} dx^1 = \frac{1}{n!}.
\end{aligned}
$$

□

15.3. Change in variables formula

In the last section of this chapter we discuss a fundamental result in the theory of integration of functions of several variables. The importance of change-in-variables formula goes beyond its applications to the computation of many concrete integrals. It will serve as a guiding principle when defining the integral over "curved" spaces, i.e., submanifolds.

15.3.1. *Formulation and some classical examples*

Let $n \in \mathbb{N}$ and suppose that $U \subset \mathbb{R}^n$ is open and $\Phi : U \to \mathbb{R}^n$ is a C^1-diffeomorphism

$$
U \ni \boldsymbol{x} \mapsto \boldsymbol{y} = \begin{bmatrix} y^1 \\ \vdots \\ y^n \end{bmatrix} = \begin{bmatrix} \Phi^1(\boldsymbol{x}) \\ \vdots \\ \Phi^n(\boldsymbol{x}) \end{bmatrix}.
$$

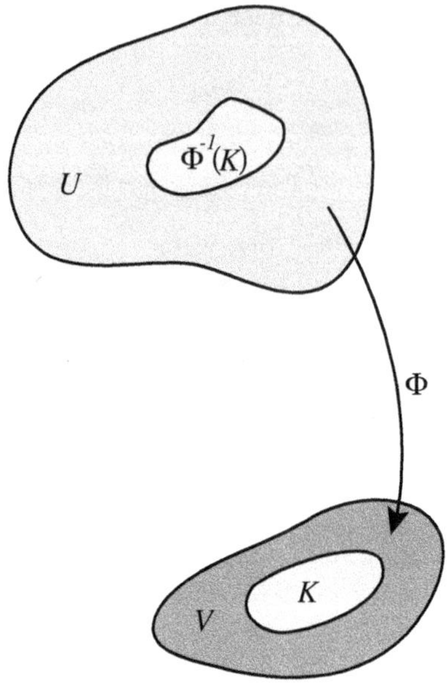

Fig. 15.7 *A diffeomorphism* $\Phi : U \to V$.

We denote by $J_\Phi(\boldsymbol{x})$ the Jacobian matrix of Φ at the point $\boldsymbol{x} \in U$. We set $V := \Phi(U)$ so V is an open subset of the target space $\mathbb{R}^n$.

Theorem 15.46. *Suppose that $f : V \to \mathbb{R}$ is a bounded function that vanishes outside a compact subset $K \subset V$. Then the following hold.*

(i) The function f is integrable if and only if the function

$$U \ni \boldsymbol{x} \mapsto f\big(\,\Phi(\boldsymbol{x})\,\big)|\det J_\Phi(\boldsymbol{x})| \in \mathbb{R}$$

is Riemann integrable.

(ii) We have the change in variables formula *(see Figure 15.7)*

$$\boxed{\int_V f(\boldsymbol{y})|d\boldsymbol{y}| = \int_U f\big(\,\Phi(\boldsymbol{x})\,\big)\,\big|\det J_\Phi(\boldsymbol{x})\big|\,|d\boldsymbol{x}|}\,. \tag{15.31}$$

□

Remark 15.47. (a) We say that Φ changes the "*old*" variables (or coordinates) $y^1, \dots, y^n$ on V to the "*new*" variables (or coordinates) $x^1, \dots, x^n$ on U. The change is described by the equations

$$y^k = \Phi^k(x^1, \dots, x^n), \;\; k = 1, \dots, n.$$

Thus the "*old*" variables $\boldsymbol{y}$ are expressed as functions of the "*new*" variables $\boldsymbol{x}$. We often use the slightly ambiguous but more suggestive notation

$$\boldsymbol{y} = \boldsymbol{y}(\boldsymbol{x}) = \Phi(\boldsymbol{x}),$$

to express the dependence of the "old" coordinates $\boldsymbol{y}$ on the "new" coordinates $\boldsymbol{x}$. The inverse transformation $\Phi^{-1}(\boldsymbol{y})$ is often replaced by the simpler and more intuitive notation

$$\boldsymbol{x} = \boldsymbol{x}(\boldsymbol{y}) = \Phi^{-1}(\boldsymbol{y}),$$

indicating that $\boldsymbol{x}$ depends on $\boldsymbol{y}$ via the unspecified transformation Φ^{-1}.

Frequently we will use the more intuitive notation

$$\left|\frac{\partial \boldsymbol{y}}{\partial \boldsymbol{x}}\right| := |\det J_\Phi|.$$

In concrete applications we are given a compact Jordan measurable subset $K \subset V$ and a Riemann integrable function $f : K \to \mathbb{R}$. The change of variables formula applied to the function $I_K(\boldsymbol{y})f(\boldsymbol{y})$ can then be rewritten in the more intuitive form (15.31)

$$\int_K f(\boldsymbol{y})|d\boldsymbol{y}| = \int_{\Phi^{-1}(K)} f\big(\boldsymbol{y}(\boldsymbol{x})\big)\left|\frac{\partial \boldsymbol{y}}{\partial \boldsymbol{x}}\right| |d\boldsymbol{x}|. \tag{15.32}$$

Implicit in the above equality is the conclusion that the compact $\Phi^{-1}(K)$ is also Jordan measurable.

(b) Let us observe that if in (15.31) we set $g(\boldsymbol{x}) := f\big(\Phi(x)\big)$, then we can rewrite this equality in the form

$$\int_V g\big(\boldsymbol{x}(\boldsymbol{y})\big)|d\boldsymbol{y}| = \int_U g(\boldsymbol{x})\left|\frac{\partial \boldsymbol{y}}{\partial \boldsymbol{x}}\right| |d\boldsymbol{x}|. \tag{15.33}$$

Clearly (15.33) is equivalent to (15.31). □

We will present an outline of the proof in the next subsection. The best way of understanding how it works is through concrete examples.

Example 15.48 (The volume of a parallelepiped). Let $n \in \mathbb{N}$. Suppose that we are given n *linearly independent* vectors $\boldsymbol{v}_1, \dots, \boldsymbol{v}_n \in \mathbb{R}^n$.

The *parallelepiped* spanned by $\boldsymbol{v}_1, \dots, \boldsymbol{v}_n$ is the set $P(\boldsymbol{v}_1, \dots, \boldsymbol{v}_n) \subset \mathbb{R}^n$ consisting of all the vectors $\boldsymbol{y}$ of the form

$$\boldsymbol{y} = x^1\boldsymbol{v}_1 + \cdots + x^n\boldsymbol{v}_n, \;\; x^1, \dots, x^n \in [0,1]. \tag{15.34}$$

We want to prove that $P(\boldsymbol{v}_1, \dots, \boldsymbol{v}_n)$ is Jordan measurable and then compute its volume. To this end consider the linear map $V : \mathbb{R}^n \to \mathbb{R}^n$ uniquely determined by the requirements

$$V\boldsymbol{e}_j = \boldsymbol{v}_j, \;\; \forall j = 1, \dots, n.$$

In other words, the columns of the matrix representing V are given by the (column) vectors $\boldsymbol{v}_1, \dots, \boldsymbol{v}_n$. Since the vectors $\boldsymbol{v}_1, \dots, \boldsymbol{v}_n$ are linearly independent the operator V is invertible and thus defines a diffeomorphism $\mathbb{R}^n \to \mathbb{R}^n$. Being linear, the operator V coincides with its differential at every $\boldsymbol{x} \in \mathbb{R}^n$ so

$$J_V = V \;\;\; \det J_V = \det V.$$

We denote by C the cube $C = [0,1]^n \subset \mathbb{R}^n$. The equation (15.34) can be written in the form

$$\boldsymbol{y} \in P(\boldsymbol{v}_1, \dots, \boldsymbol{v}_n) \Longleftrightarrow \boldsymbol{y} = x^1V\boldsymbol{e}_1 + \cdots + x^nV\boldsymbol{e}_n, \;\; \boldsymbol{x} = (x^1, \dots, x^n) \in [0,1]^n = C.$$

In other words, $P = V(C)$. In particular, this shows that P is compact. The equality $P = V(C)$ translates into an equality of indicator functions

$$I_P(V\boldsymbol{x}) = I_C(\boldsymbol{x}).$$

Theorem 15.46 implies that P is Jordan measurable and

$$\operatorname{vol}_n\big(P(\boldsymbol{v}_1,\dots,\boldsymbol{v}_n)\big) = \int_{\mathbb{R}^n} I_P(\boldsymbol{y})|d\boldsymbol{y}| = \int_{\mathbb{R}^n} I_C(\boldsymbol{x})|\det V||d\boldsymbol{x}| = |\det V|. \quad (15.35)$$

When $n = 2$, and $\boldsymbol{v}_1, \boldsymbol{v}_2 \in \mathbb{R}^n$ are not collinear, then $P(\boldsymbol{v}_1, \boldsymbol{v}_2)$ is the parallelogram spanned by $\boldsymbol{v}_1$ and $\boldsymbol{v}_2$. For example, if

$$\boldsymbol{v}_1 = \begin{bmatrix} 1 \\ 2 \end{bmatrix}, \quad \boldsymbol{v}_2 = \begin{bmatrix} 3 \\ 4 \end{bmatrix},$$

then

$$V = \begin{bmatrix} 1 & 3 \\ 2 & 4 \end{bmatrix}, \quad \det V = 1\cdot 4 - 2\cdot 3 = -2, \quad \operatorname{area}\big(P(\boldsymbol{v}_1, \boldsymbol{v}_2)\big) = |\det V| = 2. \qquad \square$$

Example 15.49 (Polar coordinates). Consider the map

$$\Phi : \mathbb{R}^2 \to \mathbb{R}^2, (r, \theta) \mapsto (x, y) = (r\cos\theta, r\sin\theta).$$

The geometric significance of this map was explained in Example 14.42 and can be seen in Figure 15.8.

The location of a point $\boldsymbol{p} \in \mathbb{R}^2 \setminus \{\boldsymbol{0}\}$ can be indicated using the "old" *Cartesian coordinates*, or the "new" *polar coordinates* (r, θ), where $r = \operatorname{dist}(\boldsymbol{p}, \boldsymbol{0})$ and θ is the angle between the line segment $[\boldsymbol{0}, \boldsymbol{p}]$ and the x-axis, measured counterclockwisely.

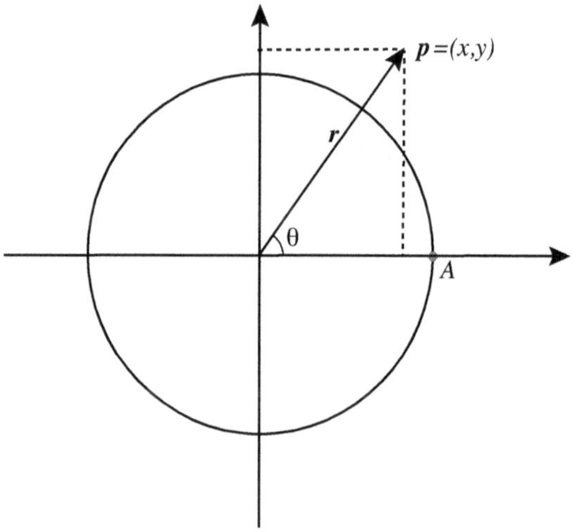

Fig. 15.8 *Constructing the polar coordinates.*

The Jacobian of this map is

$$J_\Phi = \begin{bmatrix} \frac{\partial x}{\partial r} & \frac{\partial x}{\partial \theta} \\ \\ \frac{\partial y}{\partial r} & \frac{\partial y}{\partial \theta} \end{bmatrix} = \begin{bmatrix} \cos\theta & -r\sin\theta \\ \sin\theta & r\cos\theta \end{bmatrix}$$

so

$$\det J_\Phi = r. \tag{15.36}$$

Let us observe that, for any $T \in \mathbb{R}$, the restriction of Φ to the region $(0,\infty) \times (T, T+2\pi)$ produces a bijection onto $\mathbb{R}^2_* :=$ the plane $\mathbb{R}^2$ with the nonnegative x-semiaxis removed. We will work exclusively with the restriction

$$\Phi_{\mathrm{polar}} := \Phi\big|_{[0,\infty)\times[0,2\pi]}.$$

Note two "problems" with Φ_{polar}.

- The domain $[0,\infty) \times [0, 2\pi]$ is *not* an open subset of $\mathbb{R}^2$.
- The map Φ_{polar} is *not* injective, but its restriction to the open subset $(0,\infty) \times (0, 2\pi)$ is injective.

For every Jordan measurable compact set $K \subset \mathbb{R}^2$ we set

$$K_{\mathrm{polar}} := \Phi^{-1}_{\mathrm{polar}}(K) = \big\{ (r,\theta) \in [0,\infty) \times [0,2\pi];\ \ (r\cos\theta, r\sin\theta) \in K \big\}. \tag{15.37}$$

Observe that K_{polar} is compact. If K *does not* intersect the nonnegative x-semiaxis, then Theorem 15.46 applies directly to this situation and shows that if $f = f(x,y) : K \to \mathbb{R}$ is Riemann integrable, then so is the function

$$f \circ \Phi_{\mathrm{polar}} : K_{\mathrm{polar}} \to \mathbb{R},\ \ f \circ \Phi_{\mathrm{polar}}(r,\theta) = f(r\cos\theta, r\sin\theta),$$

and we have

$$\int_{K_{\mathrm{polar}}} f(r\cos\theta, r\sin\theta) r |drd\theta| = \int_K f(x,y)|dxdy|. \tag{15.38}$$

When K does intersect this semi-axis, Theorem 15.46 does not apply directly because of the above two "problems". However the equality (15.38) continues to hold even in this case. This requires a separate argument.

Since we will be frequently confronted with such problems, we state below a general result that deals with these situations. We refer the curious reader to [16, Thm. XX.4.7] or [24, Sec. 11.5.7, Thm. 2] for a proof of this result.

Theorem 15.50. *Let $n \in \mathbb{N}$. We are given a compact Jordan measurable set $K \subset \mathbb{R}^n$, an open set $U \supset K$ and a C^1 map $\Phi : U \to \mathbb{R}^n$. Suppose that the restriction of Φ to the interior of K is a diffeomorphism. If $f : \Phi(K) \to \mathbb{R}$, is Riemann integrable, then the function*

$$K \ni \boldsymbol{x} \mapsto f\big(\Phi(\boldsymbol{x})\big)|\det J_\Phi(\boldsymbol{x})| \in \mathbb{R}$$

is Riemann integrable and

$$\int_{\Phi(K)} f(\boldsymbol{y})|d\boldsymbol{y}| = \int_K f\big(\Phi(\boldsymbol{x})\big)\big|\det J_\Phi(\boldsymbol{x})\big||d\boldsymbol{x}|. \tag{15.39}$$

□

Here is an immediate application of the above result. Suppose that $K = K_R$ is the rectangle (see Figure 15.9)

$$K = K_R := \{ (r,\theta) \in \mathbb{R}^2;\ \ r \in [0,R],\ \theta \in [0,2\pi] \}.$$

We have a differentiable map $\Phi : \mathbb{R}^2 \to \mathbb{R}^2$,

$$\Phi(r,\theta) = (r\cos\theta, r\sin\theta),$$

and $\Phi(K) = D = D_R$ is the closed disk (in $\mathbb{R}^2$) of radius R centered at the origin,

$$D_R = \{ (x,y) \in \mathbb{R}^2;\ \ x^2+y^2 \leq R^2 \}.$$

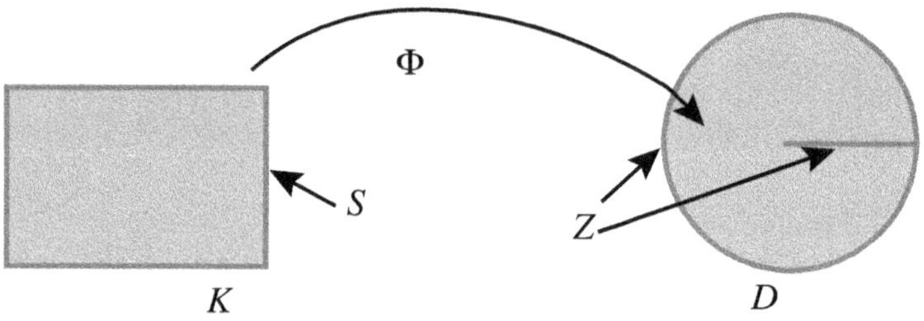

Fig. 15.9 *The polar coordinate change transformation sends a rectangle U to a disk V.*

Using the terminology in (15.37) we have $K = D_{\text{polar}}$. The boundary S of K is depicted on Figure 15.9. The interior of K is $K \setminus S$. The induced map $\Phi : K \setminus S \to \mathbb{R}^2$ is a diffeomorphism with image the complement of the set Z also depicted on Figure 15.9. Theorem 15.50 implies that for any Riemann integrable function $f : D \to \mathbb{R}$, the function

$$K \ni (r,\theta) \mapsto f(r\cos\theta, r\sin\theta) \in \mathbb{R}$$

is Riemann integrable and we have

$$\int_D f(x,y)|dxdy| = \int_K f(r\cos\theta, \sin\theta) r|drd\theta|. \tag{15.40}$$

More generally, suppose that S is a compact, Jordan measurable subset of the (x,y)-plane. Since S is bounded, it is contained in some closed disk D_R of radius R, centered at the origin. Form the associated closed rectangle K_R in the (r,θ)-plane,

$$K_R := \{ (r,\theta);\ \ r \in [0,R],\ \ \theta \in [0,2\pi] \}.$$

Suppose that $f : S \to \mathbb{R}$ is a Riemann integrable function. As usual, we denote by f^0 the extension by 0 of f and by I_{D_R} the indicator function of D_R. Note that $f^0 = f^0 I_{D_R}$. By definition

$$\int_S f|dxdy| = \int_{\mathbb{R}^2} I_{D_R} f^0|dxdy| = \int_{D_R} f^0|dxdy|.$$

Applying (15.40) to the function $f^0 : D_R \to \mathbb{R}$ we deduce

$$\boxed{\int_{S_{\text{polar}}} f(r\cos\theta, r\sin\theta) r |dr d\theta| = \int_S f(x,y)|dxdy|}, \tag{15.41}$$

where S_{polar} is defined as in (15.37).

To see how (15.41) works in practice, consider the continuous function

$$f : \mathbb{R}^2 \setminus \{\mathbf{0}\} \to \mathbb{R}, \;\; f(x,y) = \log(x^2+y^2),$$

where log denotes the natural logarithm. We want to compute the integral of this function over the annulus

$$A := \{\, \boldsymbol{p} \in \mathbb{R}^2; \;\; 1 \leq \operatorname{dist}(\boldsymbol{p}, \mathbf{0}) \leq 2 \,\}.$$

Then

$$A_{\text{polar}} = \{\, (r,\theta); \;\; r \in [1,2], \;\; \theta \in [0, 2\pi] \,\}$$

and

$$\log(x^2+y^2) = \log(r^2) = 2\log r.$$

We deduce

$$\int_A \log(x^2+y^2)|dxdy| = \int_{\substack{1\leq r\leq 2,\\ 0\leq\theta\leq 2\pi}} 2(\log r) r |dr d\theta|$$

(use Fubini)

$$= 2\int_1^2 \left(\int_0^{2\pi} |d\theta| \right) r\log r |dr| = 2\pi \int_1^2 2r\log r dr.$$

We have

$$\begin{aligned} \int_1^2 2r\log r dr &= \int_1^2 \log r d(r^2) = (r^2\log r)\Big|_{r=1}^{r=2} - \int_1^2 r^2 d(\log r) \\ &= 4\log 2 - \int_1^2 r dr = 4\log 2 - \frac{1}{2}\left(r^2\Big|_{r=1}^{r=2} \right) = 4\log 2 - \frac{3}{2}. \end{aligned}$$

Thus

$$\int_A \log(x^2+y^2) dxdy = \pi\big(8\log 2 - 3 \big).$$ □

Example 15.51 (Cylindrical and spherical coordinates). Denote by $\mathcal{O}$ the open subset of $\mathbb{R}^3$ obtained by removing the half-plane H contained in the xz-plane defined by

$$H := \{\, (x,y,z) \in \mathbb{R}^3; \;\; y = 0, \;\; x \geq 0 \,\}. \tag{15.42}$$

The location of a point $\boldsymbol{p} \in \mathcal{O}$ is determined either by its Cartesian coordinates (x,y,z), or by its altitude and the location of its projection $\boldsymbol{q}$ on the xy-plane. In turn, this projection is determined by its polar coordinates (r,θ); see Figure 15.10.

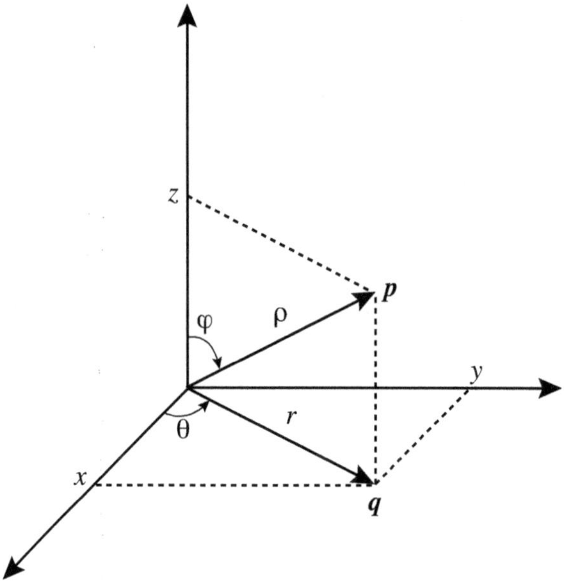

Fig. 15.10 *Constructing the cylindrical and spherical coordinates.*

The *cylindrical coordinates* (r,θ,z) are related to the Cartesian coordinates via the equalities

$$\begin{cases} x = r\cos\theta \\ y = r\sin\theta \quad r>0,\ \ \theta\in(0,2\pi),\ \ z\in\mathbb{R}. \\ z = z, \end{cases}$$

The above equalities define a transformation

$$\Phi_{\text{Cart}\leftarrow\text{cyl}} : [0,\infty)\times[0,2\pi]\times\mathbb{R}\to\mathbb{R}^3,\ \ (r,\theta,z)\mapsto\begin{bmatrix} x\\ y\\ z\end{bmatrix} = \begin{bmatrix} r\cos\theta\\ r\sin\theta\\ z\end{bmatrix},$$

whose restriction to the open set $(0,\infty)\times(0,2\pi)\times\mathbb{R}$ is a diffeomorphism with image $\mathcal{O} = \mathbb{R}^3\setminus H$, where H is the half-plane in (15.42). The Jacobian matrix of the transformation $\Phi_{\text{Cart}\leftarrow\text{cyl}}$ is

$$J_{\text{Cart}\leftarrow\text{cyl}} := \begin{bmatrix} \cos\theta & -r\sin\theta & 0\\ \sin\theta & r\cos\theta & 0\\ 0 & 0 & 1\end{bmatrix}.$$

We have

$$\det J_{\text{Cart}\leftarrow\text{cyl}} = \det\begin{bmatrix} \cos\theta & -r\sin\theta\\ \sin\theta & r\cos\theta\end{bmatrix} = r. \tag{15.43}$$

Alternatively, the location of a point $\boldsymbol{p}\in\mathcal{O}$ is determined if we know the distance ρ to the origin $\rho = \|\boldsymbol{p}\|$, the angle $\varphi\in(0,\pi)$ the line $\mathbf{0}\boldsymbol{p}$ makes with the z-axis, and the polar coordinate θ of the projection of $\boldsymbol{p}$ on the xy-plane; see Figure 15.10. The parameters ρ, φ, θ are called the *spherical coordinates* of $\boldsymbol{p}$. The spherical coordinates (ρ, φ, θ) determine the cylindrical coordinates (r, θ, z) via the equalities

$$r = \rho \sin\varphi, \quad \theta = \theta, \quad z = \rho\cos\varphi.$$

The Jacobian matrix of the transformation $\Phi_{\mathrm{cyl}\leftarrow\mathrm{sph}}(\rho, \varphi, \theta) = (r, \theta, z)$ is

$$J_{\mathrm{cyl}\leftarrow\mathrm{sph}} = \begin{bmatrix} \frac{\partial r}{\partial\rho} & \frac{\partial r}{\partial\varphi} & \frac{\partial r}{\partial\theta} \\ \\ \frac{\partial \theta}{\partial\rho} & \frac{\partial \theta}{\partial\varphi} & \frac{\partial \theta}{\partial\theta} \\ \\ \frac{\partial z}{\partial\rho} & \frac{\partial z}{\partial\varphi} & \frac{\partial z}{\partial\theta} \end{bmatrix} = \begin{bmatrix} \sin\varphi & \rho\cos\varphi & 0 \\ 0 & 0 & 1 \\ \cos\varphi & -\rho\sin\varphi & 0 \end{bmatrix}.$$

Expanding along the second row we deduce

$$\det J_{cyl\leftarrow sph} = -\det \begin{bmatrix} \sin\varphi & \rho\cos\varphi \\ \cos\varphi & -\rho\sin\varphi \end{bmatrix} = \rho. \tag{15.44}$$

We deduce that the Cartesian coordinates (x, y, z) are related to the spherical coordinates (ρ, φ, θ) via the equalities

$$\begin{cases} x = \rho\sin\varphi\cos\theta \\ y = \rho\sin\varphi\sin\theta \\ z = \rho\cos\varphi \end{cases}, \quad \rho > 0, \quad \theta \in (0, 2\pi), \quad \varphi \in (0, \pi).$$

The above equations define a transformation

$$\Phi_{\mathrm{Cart}\leftarrow\mathrm{cyl}} : [0, \infty) \times [0, 2\pi) \times \mathbb{R} \to \mathbb{R}^3, \quad (\rho, \varphi, \theta) \mapsto \begin{bmatrix} x \\ y \\ x \end{bmatrix} = \begin{bmatrix} \rho\sin\varphi\cos\theta \\ \rho\sin\varphi\sin\theta \\ \rho\cos\varphi \end{bmatrix}.$$

The restriction of $\Phi_{\mathrm{Cart}\leftarrow\mathrm{sph}}$ to the open set $(0, \infty) \times (0, \pi) \times (0, 2\pi)$ is a diffeomorphism with image $\mathcal{O}$. The Jacobian matrix of the above transformation is

$$J_{\mathrm{Cart}\leftarrow\mathrm{sph}} = \begin{bmatrix} \frac{\partial x}{\partial\rho} & \frac{\partial x}{\partial\varphi} & \frac{\partial x}{\partial\theta} \\ \\ \frac{\partial y}{\partial\rho} & \frac{\partial y}{\partial\varphi} & \frac{\partial y}{\partial\theta} \\ \\ \frac{\partial z}{\partial\rho} & \frac{\partial z}{\partial\varphi} & \frac{\partial z}{\partial\theta} \end{bmatrix} = \begin{bmatrix} \sin\varphi\cos\theta & \rho\cos\varphi\cos\theta & -\rho\sin\varphi\sin\theta \\ \\ \sin\varphi\sin\theta & \rho\cos\varphi\sin\theta & \rho\sin\varphi\cos\theta \\ \\ \cos\varphi & -\rho\sin\varphi & 0 \end{bmatrix}.$$

Since $\Phi_{\mathrm{Cart}\leftarrow\mathrm{sph}} = \Phi_{\mathrm{Cart}\leftarrow\mathrm{cyl}} \circ \Phi_{\mathrm{cyl}\leftarrow\mathrm{sph}}$ we deduce from the chain rule that

$$J_{\mathrm{Cart}\leftarrow\mathrm{sph}} = J_{\mathrm{Cart}\leftarrow\mathrm{cyl}} \cdot J_{\mathrm{cyl}\leftarrow\mathrm{sph}}.$$

Hence $\det J_{\mathrm{Cart}\leftarrow\mathrm{sph}} = \det J_{\mathrm{Cart}\leftarrow\mathrm{cyl}} \det J_{\mathrm{cyl}\leftarrow\mathrm{sph}}$. Using (15.43) and (15.44) we deduce

$$\det J_{\mathrm{Cart}\leftarrow\mathrm{sph}} = r\rho = \rho^2\sin\varphi. \tag{15.45}$$

Let us see how we can use these facts in concrete situations. Suppose that $K \subset \mathbb{R}^3$ is a compact measurable set contained in some large ball $\overline{B_R(\mathbf{0})} \subset \mathbb{R}^3$. We set

$$K_{\mathrm{cyl}} := \big\{ (r, \theta, z) \in [0, \infty) \times [0, 2\pi] \times \mathbb{R};\ \ \Phi_{\mathrm{Cart}\leftarrow\mathrm{cyl}}(r, \theta, z) \in K \big\},$$

$$K_{\text{sph}} := \left\{ (r, \varphi, z) \in [0, \pi] \times [0, 2\pi] \times \mathbb{R};\ \ \Phi_{\text{Cart}\leftarrow\text{sph}}(r, \varphi, \theta) \in K \right\}.$$

Suppose that $f : K \to \mathbb{R}$ is a Riemann integrable function. If K does not intersect the "forbidden" half-plane H, then Theorem 15.46 applies and yields

$$\int_K f(x, y, z) dx dy dz = \int_{K_{\text{cyl}}} f(r\cos\theta, r\sin\theta, z) r dr d\theta dz. \tag{15.46a}$$

$$\int_K f(x, y, z)\, |dx dy dz| = \int_{K_{\text{sph}}} f(\rho\sin\varphi\cos\theta, \rho\sin\varphi\sin\theta, \rho\cos\varphi)\rho^2 \sin\varphi\, |d\rho d\varphi d\theta|. \tag{15.46b}$$

If K does intersect the "forbidden" half-plane H, then using Theorem 15.50 we deduce as in Example 15.49 that (15.46a) and (15.46b) continue to hold in this case as well. Let us see how the above formulæ work in concrete situations.

Fix a number $\varphi_0 \in (0, \pi/2)$. The locus of points in $\mathbb{R}^3$ such that $\varphi = \varphi_0$ describes a circular cone; Figure 15.11. The z-axis is a symmetry axis of this cone.

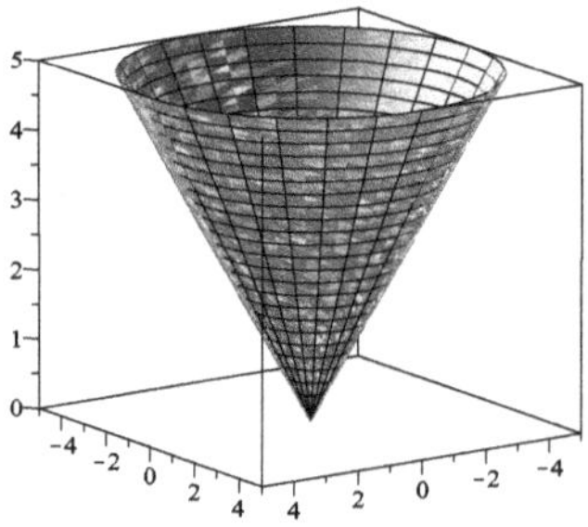

Fig. 15.11 *A cone.*

If we set $m_0 := \tan\varphi_0$, then we observe that, along the surface of this cone the cylindrical coordinates r, z satisfy

$$m_0 = \tan\varphi_0 = \frac{r}{z}\quad r = m_0 z.$$

The "inside part" of this cone is described by the inequality

$$\frac{r}{z} \le m_0 \Longleftrightarrow r \le m_0 z.$$

Consider two positive numbers $z_0 < z_1$ and denote by $R = R(m_0, z_0, z_1)$ the region inside this cone contained between the horizontal planes $z = z_0$ and $z = z_1$; see Figure 15.12.

We want to compute the volume of this truncated cone. In cylindrical coordinates it corresponds to the region R_{cyl} described by the inequalities

$$0 \le r \le m_0 z,\ \ z_0 \le z \le z_1,\ \ 0 \le \theta \le 2\pi.$$

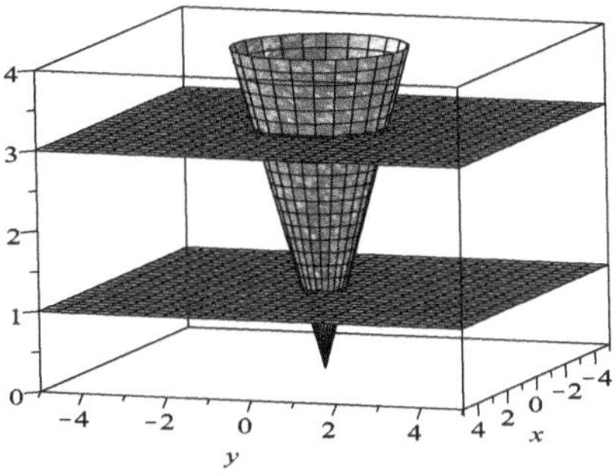

Fig. 15.12 *A truncated cone.*

Using the change in variables formula (15.46a)

$$\begin{aligned}
\mathrm{vol}_3(R) &= \int_R |dxdydz| = \int_{\substack{(\theta,z)\in[0,2\pi]\times[z_0,z_1]\\ 0\le r\le m_0 z}} r\,|drd\theta dz| \\
&= \int_0^{2\pi}\left(\int_{z_0}^{z_1}\left(\int_0^{m_0 z} rdr\right)dz\right)d\theta \\
&= \int_0^{2\pi}\left(\int_{z_0}^{z_1}\frac{1}{2}(m_0 z)^2 dz\right)d\theta \\
&= \frac{m_0^2}{2}\int_0^{2\pi}\left(\frac{z_1^3 - z_0^3}{3}d\theta\right) = \frac{\pi m_0^2}{3}\left(z_1^3 - z_0^3\right).
\end{aligned}$$

To see the spherical coordinates at work, it is useful to relate them to more familiar notions. Note that the surface $\rho = const$ is a sphere. The surface $\varphi = const$ is a cone while the surface $\theta = const$ is a half-plane with edge z-axis. Since $z = \rho\cos\varphi$ we deduce that in spherical coordinates the truncated cone R corresponds to the region R_{sph} described by the inequalities

$$0 \le \varphi \le \varphi_0,\ \ \theta \in [0, 2\pi],\ \ z_0 \le \rho\cos\varphi \le z_1.$$

We deduce

$$\begin{aligned}
\mathrm{vol}_3(R) &= \int_{R_{\mathrm{sph}}} \rho^2 \sin\varphi |d\rho d\varphi d\theta| \\
&= \int_0^{2\pi}\left(\int_0^{\varphi_0}\left(\int_{\frac{z_0}{\cos\varphi}}^{\frac{z_1}{\cos\varphi}} \rho^2 d\rho\right)\sin\varphi d\varphi\right)d\theta \\
&= \frac{z_1^3 - z_0^3}{3}\int_0^{2\pi}\left(\int_0^{\varphi_0}\frac{\sin\varphi}{\cos^3\varphi}d\varphi\right)d\theta
\end{aligned}$$

(make the change in variables $u = \cos\varphi$)

$$= \frac{z_1^3 - z_0^3}{3}\int_0^{2\pi}\left(\int_{\cos\varphi_0}^1 \frac{1}{u^3}du\right)d\theta$$
$$= \frac{z_1^3 - z_0^3}{6}\int_0^{2\pi}\underbrace{\left(\frac{1}{\cos^2\varphi_0} - 1\right)}_{=\tan^2\varphi_0 = m_0^2} d\theta = \frac{\pi m_0^2(z_1^3 - z_0^3)}{3}.$$

This is in perfect agreement with the computation using cylindrical coordinates.

To verify the validity of these computations, we present an alternate computation based on Cavalieri's Principle. The intersection of the region R with the horizontal plane $z = t$ is a disk R_t of radius $r_t = m_0 t$. We have

$$\mathrm{vol}_2(R_t) = \pi(m_0 t)^2.$$

Then

$$\mathrm{vol}_3(R) = \int_{z_0}^{z_1} \mathrm{vol}_2(R_t)dt = \pi m_0^2 \int_{z_0}^{z_1} t^2 dt = \frac{\pi m_0^2}{3}\left(z_1^3 - z_0^3\right).$$

Let us rewrite this volume formula in a more familiar form. The bottom of the truncated cone R is a disk of radius $r_0 = m_0 z_0$ and the top is a disk of radius $r_1 = m_0 z_1$. We denote by h the height of this truncated cone $h := z_1 - z_0$. Then

$$\mathrm{vol}_3(R) = \frac{\pi}{3}(z_1 - z_0)\left((m_0 z_1)^2 + (m_0 z_1)(m_0 z_0) + (m_0 z_0)^2\right) = \frac{\pi h}{3}\left(r_1^2 + r_1 r_0 + r_0^2\right).$$

□

Example 15.52 (Spherical coordinates in arbitrary dimensions). Let $n \in \mathbb{N}$, $n \geq 2$. We will construct inductively spherical coordinates on $\mathbb{R}^n$.

For $n = 2$, these are versions of the polar coordinates (r, θ). Define (ρ_2, θ_2) via the well known equalities

$$x^1 = \rho_2 \cos\theta_2, \;\; x^2 = \rho_2 \sin\theta_2, \;\; \rho_2 = \|\boldsymbol{x}\|, \;\; \theta_2 \in [0, 2\pi]. \tag{15.47}$$

Observe that the numbers ρ_2 and θ_2 completely determine the location of the point $\boldsymbol{x}$ in $\mathbb{R}^2$.

Suppose now that we have constructed the spherical coordinates $\theta_2, \dots, \theta_n, \rho_n$ on $\mathbb{R}^n$. In particular, the coordinates $x^1, \dots, x^n$ are functions depending on these coordinates,

$$x^1 = f^1(\theta_2, \dots, \theta_{n-1}, \rho_n), \dots, x^n = f^n(\theta_2, \dots, \theta_{n-1}, \rho_n), \tag{15.48a}$$

$$\rho_n(\boldsymbol{x}) = \|\boldsymbol{x}\|, \;\; \forall \boldsymbol{x} \in \mathbb{R}^n. \tag{15.48b}$$

We will construct spherical coordinates $\theta_2, \dots, \theta_n, \theta_{n+1}, \rho_{n+1}$ on $\mathbb{R}^n$.

For $\boldsymbol{x} = (x^1, \dots, x^n, x^{n+1}) \in \mathbb{R}^{n+1}$ we denote by $\bar{\boldsymbol{x}}$ the projection onto the coordinate hyperplane $\mathbb{R}^n \times \{0\} = \{x^{n+1} = 0\}$. We think of $\bar{\boldsymbol{x}}$ as a point in $\mathbb{R}^n$; see Figure 15.13. In other words, we have

$$\boldsymbol{x} = (\underbrace{x^1, \dots, x^n}_{\bar{\boldsymbol{x}}}, x^{n+1}) = (\bar{\boldsymbol{x}}, x^{n+1}).$$

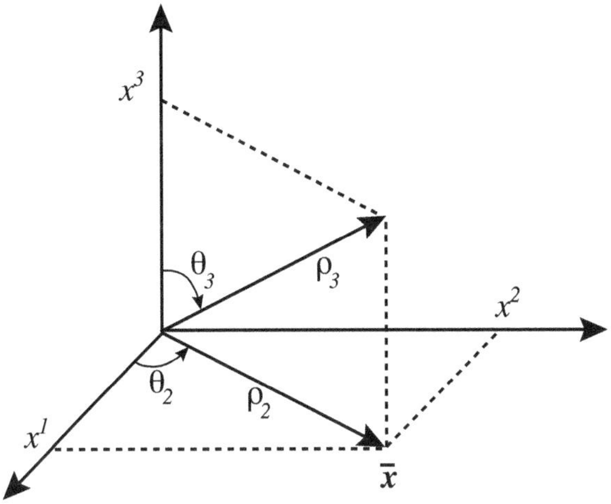

Fig. 15.13 *Constructing spherical coordinates in n dimensions.*

We denote by ρ_{n+1} the distance from $\boldsymbol{x}$ to the origin, i.e.,

$$\rho_{n+1} = \|\boldsymbol{x}\| = \sqrt{(x^0)^2 + (x^1)^2 + \cdots + (x^n)^2}.$$

Observe that the location of $\boldsymbol{x}$ is completely determined once we know x^0 and the location of $\bar{\boldsymbol{x}}$ in $\mathbb{R}^n$; see Figure 15.13. According to the induction assumption the location of $\bar{\boldsymbol{x}}$ is completely determined by the previously defined spherical coordinates $(\theta_2, \dots, \theta_n, \rho_n)$, $\rho_n = \|\bar{\boldsymbol{x}}\|$.

For $\boldsymbol{x} \in \mathbb{R}^{n+1} \setminus \{\boldsymbol{0}\}$ denote by $\theta_{n+1} = \theta_{n+1}(\boldsymbol{x}) \in [0, \pi]$ the angle that the vector $\boldsymbol{x}$ makes with the x^{n+1}-axis. More precisely, we have

$$x^{n+1} = \langle \boldsymbol{x}, \boldsymbol{e}_{n+1} \rangle = \|\boldsymbol{x}\| \cdot \|\boldsymbol{e}_{n+1}\| \cos\theta_{n+1} = \rho_{n+1} \cos\theta_{n+1}.$$

We have (see Figure 15.13)

$$\rho_{n+1}^2 = \|\boldsymbol{x}\|^2 = (x^1)^2 + \cdots + (x^n)^2 + (x^{n+1})^2 = \|\bar{\boldsymbol{x}}\|^2 + (x^{n+1})^2 = \rho_{n+1}^2 + \rho_{n+1}^2 \cos^2\theta_{n+1}.$$

We deduce

$$\rho_n^2 = \rho_{n+1}^2 - \rho_{n+1}^2 \cos^2\theta_{n+1} = \rho_{n+1}^2 \sin^2\theta_{n+1}.$$

Since $\theta_{n+1} \in [0, \pi]$, we have $\sin\theta_{n+1} \geq 0$ and thus

$$\boxed{\rho_n = \rho_{n+1} \sin\theta_{n+1}}.$$

This shows that the quantities $(\rho_{n+1}, \theta_{n+1})$ determine the coordinate x^0 and the spherical coordinate ρ_n of $\bar{\boldsymbol{x}}$. Thus, the quantities

$$\theta_2, \dots, \theta_n, \theta_{n+1}, \rho_{n+1} \sin\theta_{n+1}$$

completely determine the location of $\boldsymbol{x}$ in $\mathbb{R}^{n+1}$. More precisely, using (15.48a) we obtain equalities

$$\begin{aligned} x^1 &= f^1(\theta_2, \dots, \theta_{n-1}), \rho_{n+1} \sin \theta_{n+1}) \\ \vdots & \;\vdots\; \vdots \\ x^n &= f^n(\theta_2, \dots, \theta_{n-1}, \rho_{n+1} \sin \theta_{n+1}) \\ x^{n+1} &= \rho_{n+1} \cos \theta_{n+1}, \end{aligned} \tag{15.49}$$

where

$$\rho_{n+1} > 0, \;\; \theta_2 \in [0, 2\pi], \;\; \theta_3, \dots, \theta_{n+1} \in [0, \pi].$$

Let us see more explicitly the manner in which the spherical coordinates

$$\theta_2, \dots, \theta_{n+1}, \rho_{n+1}$$

determine the Cartesian coordinates $x^1, \dots, x^{n+1}$. Using (15.47) we deduce that, for $n + 1 = 3$ we have

$$x^1 = \rho_3 \sin \theta_3 \cos \theta_2, \;\; x^2 = \rho_3 \sin \theta_3 \sin \theta_2, \;\; x^3 = \rho_3 \cos \theta_3.$$

We recognize here an old "friend", the spherical coordinates in $\mathbb{R}^3$, $\rho = \rho_3, \theta = \theta_2, \varphi = \theta_3$; see Figure 15.13. Using these freshly obtained equalities and the inductive scheme (15.49) we deduce that for $n + 1 = 4$ we have

$$\begin{aligned} x^1 &= \rho_4 \sin \theta_4 \sin \theta_3 \cos \theta_2, \\ x^2 &= \rho_4 \sin \theta_4 \sin \theta_3 \sin \theta_2, \\ x^3 &= \rho_4 \sin \theta_4 \cos \theta_3, \\ x^4 &= \rho_4 \cos \theta_4. \end{aligned}$$

The general pattern should be clear:

$$\boxed{\begin{aligned} x^1 &= \rho_n \sin \theta_n \cdots \sin \theta_4 \sin \theta_3 \cos \theta_2, \\ x^2 &= \rho_n \sin \theta_n \cdots \sin \theta_4 \sin \theta_3 \sin \theta_2, \\ x^3 &= \rho_n \sin \theta_n \cdots \sin \theta_4 \cdots \cos \theta_3, \\ \vdots & \;\vdots\; \vdots \\ x^{n-1} &= \rho_n \sin \theta_n \cos \theta_{n-1}, \\ x^n &= \rho_n \cos \theta_n. \end{aligned}} \tag{15.50}$$

We interpret the above equalities as defining a map $\Phi_n = \Phi_n(\theta_2, \cdots \theta_n, \rho_n)$ from an open subset of a vector space with coordinates $(\theta_2, \dots, \theta_n, \rho_n)$ to another vector space with coordinates $(x^1, \dots, x^n)$. We want to compute $\delta_n := \det J_{\Phi_n}$.

We will achieve this inductively by observing that we can write Φ_{n+1} as a composition

$$\begin{bmatrix} \theta_2 \\ \vdots \\ \theta_{n+1} \\ \rho_{n+1} \end{bmatrix} \overset{\Psi_{n+1}}{\mapsto} \begin{bmatrix} \theta_2 \\ \vdots \\ \theta_n \\ \rho_n \\ x^{n+1} \end{bmatrix} = \begin{bmatrix} \theta_2 \\ \vdots \\ \theta_n \\ \rho_{n+1} \sin \theta_{n+1} \\ \rho_{n+1} \cos \theta_{n+1} \end{bmatrix},$$

$$\begin{bmatrix} \theta_2 \\ \vdots \\ \theta_n \\ \rho_n \\ x^{n+1} \end{bmatrix} \overset{\hat{\Phi}_n}{\mapsto} \begin{bmatrix} \overline{\boldsymbol{x}} \\ x^{n+1} \end{bmatrix} = \begin{bmatrix} \Phi_n(\theta_2, \dots, \theta_n, \rho_n) \\ x^{n+1} \end{bmatrix}.$$

From the equality

$$\Phi_{n+1} = \hat{\Phi}_n \circ \Psi_{n+1}$$

and the chain rule we deduce

$$\det J_{\Phi_{n+1}} = \det J_{\hat{\Phi}_n} \cdot \det J_{\Psi_{n+1}}.$$

A simple computation[6] shows that

$$\det J_{\hat{\Phi}_n} = \det J_{\Phi_n}, \quad \det J_{\Psi_{n+1}} = \rho_{n+1}. \tag{15.51}$$

Hence

$$\begin{aligned} \delta_{n+1} &= \rho_{n+1}\delta_n = \rho_{n+1}\rho_n\delta_{n-1} = \cdots \rho_{n+1}\rho_n \cdots \rho_3\delta_2 \\ &= \rho_{n+1}\rho_n \cdots \rho_3\rho_2. \end{aligned}$$

From the equalities

$$\rho_n = \rho_{n+1}\sin\theta_{n+1}, \quad \rho_2 = \rho_3 \sin\theta_3$$

we deduce

$$\delta_2 = \rho_2, \quad \delta_3 = \rho_3^2 \sin\theta_3, \quad \delta_4 = \rho_4\rho_3^2 \sin\theta_3 = \rho_4^3 (\sin\theta_4)^2 \sin\theta_3,$$

and, in general,

$$\boxed{\det J_{\Phi_{n+1}} = \delta_{n+1} = \rho_{n+1}^n (\sin\theta_{n+1})^{n-1} (\sin\theta_n)^{n-2} \cdots \sin\theta_3}. \tag{15.52}$$

Note that since $\theta_3, \dots, \theta_n \in (0, \pi)$ we deduce that $\det J_{\Phi_{n+1}} > 0$, so $\det J_{\Phi_{n+1}} = |\det J_{\Phi_{n+1}}|$. □

Example 15.53 (The volume of the unit n-dimensional ball). Denote by ω_n the volume of the closed unit n-dimensional ball

$$\overline{B_1^n(\mathbf{0})} := \{ \boldsymbol{x} \in \mathbb{R}^n; \ \|\boldsymbol{x}\| \leq 1 \}.$$

Consider the box

$$\mathcal{B}_n := \{ (\theta_2, \theta_n, \rho_n) \in \mathbb{R}^n, \ \theta_2 \in [0, 2\pi], \ \theta_3, \dots, \theta_n \in [0, \pi], \ \rho_n \in [0, 1] \}.$$

The transformation Φ_n sends this box to the closed unit n-dimensional ball $\overline{B_1^n(\mathbf{0})}$.

The equality (15.52) shows that the determinant of the Jacobian of Φ_n is bounded on $\mathcal{B}_n$. Applying Theorem 15.50 we deduce that

$$\omega_n = \operatorname{vol}_n\left(\overline{B_1^n(\mathbf{0})}\right) = \int_{\mathcal{B}_n} \rho_n^{n-1} (\sin\theta_n)^{n-2} (\sin\theta_{n-1})^{n-3} \cdots \sin\theta_3 |d\theta_2 d\theta_3 \cdots d\theta_n d\rho_n|$$

(set $\rho := \rho_n$ and use Fubini)

$$= \left(\int_0^1 \rho^{n-1} d\rho \right) \left(\int_0^{2\pi} d\theta_2 \right) \left(\int_0^{\pi} \sin\theta_3 d\theta_3 \right) \cdots \left(\int_0^{\pi} (\sin\theta_n)^{n-2} d\theta_n \right)$$

[6] You need to perform this simple computation.

$$= \frac{2\pi}{n}\left(\int_0^\pi \sin\theta d\theta\right)\cdots\left(\int_0^\pi (\sin\theta)^{n-2}d\theta\right).$$

If we set

$$J_k := \int_0^\pi (\sin\theta)^k d\theta,$$

then we deduce

$$\omega_n = \frac{2\pi}{n}J_1J_2\cdots J_{n-2}.$$

Using the equality

$$\sin(\pi-\theta) = \sin\theta,\ \ \forall\theta\in\mathbb{R}$$

we deduce

$$\int_0^\pi (\sin\theta)^k d\theta = \int_0^{\frac{\pi}{2}} (\sin\theta)^k d\theta + \int_{\frac{\pi}{2}}^\pi (\sin\theta)^k d\theta = 2\underbrace{\int_0^{\frac{\pi}{2}} (\sin\theta)^k d\theta}_{=:I_k}.$$

Hence

$$\omega_n = \frac{2^{n-1}\pi}{n}I_1I_2\cdots I_{n-2}. \tag{15.53}$$

We have computed the integrals I_k earlier in (9.47).

$$I_{2j} = \frac{\pi}{2}\frac{(2j-1)!!}{(2j)!!},\ \ I_{2j-1} = \frac{(2j-2)!!}{(2j-1)!!},$$

where the bi-factorial $n!!$ is defined in (9.46).

If $n = 2k$, then

$$I_1\cdots I_{n-2} = (I_1I_2)\cdots(I_{2k-3}I_{2k-2}) = \prod_{j=1}^{k-1} I_{2j-1}I_{2j} = \left(\frac{\pi}{2}\right)^{k-1}\prod_{j=1}^{k-1}\frac{(2j-2)!!}{(2j)!!}$$

$$= \left(\frac{\pi}{2}\right)^{k-1}\frac{1}{((2k-2)!!} = \frac{\pi^{k-1}}{2^{2k-2}(k-1)!}.$$

If $n = 2k+1$, then

$$I_1\cdots I_{n-2} = (I_1I_2)\cdots(I_{2k-3}I_{2k-2})I_{2k-1} = \left(\frac{\pi}{2}\right)^{k-1}\frac{1}{((2k-2)!!}\cdot\frac{(2k-2)!!}{(2k-1)!!}$$

$$= \left(\frac{\pi}{2}\right)^{k-1}\frac{1}{(2k-1)!!}.$$

Using (15.53) we deduce

$$\boxed{\omega_{2k} = \frac{\pi^k}{k!},\ \ \omega_{2k+1} = \frac{2^{k+1}\pi^k}{(2k+1)!!}}. \tag{15.54}$$

We list below the values of ω_n for small n.

n	0	1	2	3	4	5
ω_n	1	2	π	$\frac{4\pi}{3}$	$\frac{\pi^2}{2}$	$\frac{8\pi^2}{15}$

Let us mention one simple consequence of the above computations that will come in handy later.

$$n\omega_n = \left(\int_0^{2\pi} d\theta_2\right)\left(\int_0^\pi \sin\theta_3\, d\theta_3\right)\cdots\left(\int_0^\pi (\sin\theta_n)^{n-2}\, d\theta_n\right). \tag{15.55}$$

□

Example 15.54 (Integrals of radially symmetric functions). Here is another useful application of the n-dimensional spherical coordinates. For $0 \leq r < R$ define

$$A(r,R) = A^n(r,R) = \big\{ \boldsymbol{x} \in \mathbb{R}^n;\ \ r \leq \|\boldsymbol{x}\| \leq R \big\}.$$

Suppose that $f : A(r,R) \to \mathbb{R}$ is a continuous *radially symmetric* function. This means that there exists a continuous function $u : [r,R] \to \mathbb{R}$ such that

$$f(\boldsymbol{x}) = u\big(\|\boldsymbol{x}\| \big),\ \ \forall \boldsymbol{x} \in \mathbb{R}^n.$$

We want to show that

$$\int_{A^n(r,R)} u(\|\boldsymbol{x}\|)|d\boldsymbol{x}| = \int_{r \leq \|\boldsymbol{x}\| \leq R} u(\|\boldsymbol{x}\|)\,|d\boldsymbol{x}| = n\boldsymbol{\omega}_n \int_r^R u(\rho)\rho^{n-1} d\rho. \tag{15.56}$$

When $n = 2$ this formula reads

$$\int_{r \leq \sqrt{x^2+y^2} \leq R} u\big(\sqrt{x^2+y^2}\,\big)\,|dxdy| = 2\pi \int_r^R tu(t)dt,$$

and when $n = 3$ it reads

$$\int_{r \leq \sqrt{x^2+y^2+z^2} \leq R} u\big(\sqrt{x^2+y^2+z^2}\,\big)\,|dxdydz| = 4\pi \int_r^R t^2 u(t)dt.$$

To prove (15.56) we use the n-dimensional spherical coordinates

$$(\theta_2, \dots, \theta_n, \rho = \rho_n).$$

We deduce

$$\begin{aligned}
&\int_{A^n(r,R)} u(\|\boldsymbol{x}\|)|d\boldsymbol{x}| \\
&= \int_{\substack{r \leq \rho \leq R \\ \theta_2 \in [0,2\pi]\ \theta_3,\dots,\theta_n \in [0,\pi]}} u(\rho)\rho^{n-1} \left(\prod_{j=2}^n (\sin\theta_j)^{j-2} \right) |d\theta_2 d\theta_3 \cdots d\theta_n d\rho| \\
&= \left(\int_r^R u(\rho)\rho^{n-1} d\rho \right) \left(\int_0^{2\pi} d\theta_2 \right) \left(\int_0^{\pi} \sin\theta_3 d\theta_3 \right) \cdots \left(\int_0^{\pi} (\sin\theta_n)^{n-2} d\theta_n \right) \\
&\overset{(15.55)}{=} n\boldsymbol{\omega}_n \int_r^R u(\rho)\rho^{n-1} d\rho.
\end{aligned}$$

□

15.3.2. *Proof of the change of variables formula*

We will carry the proof of the change in variables formula (15.31) or, equivalently, (15.33), in several steps that we describe loosely below.

Step 1. *(De)composition.* If the change in variables formula is valid for the diffeomorphisms Φ_0, Φ_1, then it is valid for their composition $\Phi_1 \circ \Phi_0$, whenever this composition makes sense.

Step 2. *Localization.* Suppose that the diffeomorphism $\Phi : U \to \mathbb{R}^n$ has the property that, for any $\boldsymbol{p} \in U$, there exists an open neighborhood $\mathcal{O}_{\boldsymbol{p}}$ of $\boldsymbol{p}$ such that $\mathcal{O}_{\boldsymbol{p}} \subset U$ and the restriction of Φ to $\mathcal{O}_{\boldsymbol{p}}$ is a diffeomorphism satisfying Theorem 15.46. We will show that the entire diffeomorphism Φ satisfies this theorem. This step uses the partition of unity trick.

Step 3. *Elementary diffeomorphism.* We describe a class of so-called *elementary* diffeomorphisms for which the change in variables holds. In conjunction with Step 1 we deduce that the change in variable formula holds for *quasi-elementary* diffeomorphisms, i.e., diffeomorphisms that are compositions of several elementary ones. This step uses Fubini's theorem coupled with the 1-dimensional change in variables formula.

Step 4. *Everything is (quasi)elementary.* We show that for any diffeomorphism $\Phi : U \to \mathbb{R}^n$, ($U$ open subset in $\mathbb{R}^n$) and for any $\boldsymbol{x} \in U$, there exists a tiny open neighborhood $\mathcal{O}$ of $\boldsymbol{x}$ such that $\mathcal{O} \subset U$ and the restriction of Φ to $\mathcal{O}$ is a quasi-elementary diffeomorphism. This step relies on the inverse function theorem.

Clearly Steps 1–4 imply the validity of Theorem 15.46. We present the details below. For simplicity we consider only the case when the integrand f in (15.31) is a continuous function that vanishes outside a compact set $K \subset V$. The general case follows from this special case by using Theorem 15.26.

Step 1. Let $\Phi : U_0 \to \mathbb{R}^n$, $\Psi : U_1 \to \mathbb{R}^n$ be two diffeomorphisms satisfying Theorem 15.46 such that $\Phi(U_0) \subset U_1$. We want to prove that $\Psi \circ \Phi$ also satisfies this theorem. Set

$$V_1 := \Phi(U_0), \;\; V_2 := \Psi(V_1).$$

Suppose that $f : V_2 \to \mathbb{R}$ is continuous and vanishes outside a compact set $K_2 \subset V_2$. The function

$$g : V_1 \to \mathbb{R}, \;\; g(\boldsymbol{y}) = f\big(\,\Psi(\boldsymbol{y})\,\big)\big|\det J_\Psi(\boldsymbol{y})\,\big|$$

is continuous and vanishes outside the compact set $K_1 := \Psi^{-1}(K_2) \subset V_1$. Since Ψ satisfies Theorem 15.46 we deduce that

$$\int_{V_2} f(\boldsymbol{z})|d\boldsymbol{z}| = \int_{V_1} g(\boldsymbol{y})|d\boldsymbol{y}|. \tag{15.57}$$

Similarly, the function

$$h : U_0 \to \mathbb{R}, \;\; h(\boldsymbol{x}) = g\big(\,\Phi(\boldsymbol{x})\,\big)\big|\det J_\Phi(\boldsymbol{x})\,\big|$$

is continuous and vanishes outside the compact set $K_0 = \Phi^{-1}(K_1) \subset U_0$. Since Φ satisfies Theorem 15.46 we conclude that

$$\int_{V_1} g(\boldsymbol{y})|d\boldsymbol{y}| = \int_{U_0} h(\boldsymbol{x})|d\boldsymbol{x}|. \tag{15.58}$$

Now observe that

$$h(\boldsymbol{x}) = f\big(\,\Psi \circ \Phi(\boldsymbol{x})\,\big) \cdot \big|\det J_\Psi(\Phi(\boldsymbol{x}))\,\big| \cdot \big|\det J_\Phi(\boldsymbol{x})\,\big|$$

The chain rule (13.26) implies that

$$J_{\Psi\circ\Phi}(\boldsymbol{x}) = J_\Psi(\Phi(\boldsymbol{x}))J_\Phi(\boldsymbol{x})$$

so

$$\left|\det J_{\Psi\circ\Phi}(\boldsymbol{x})\right| = \left|\det J_{\Psi}(\Phi(\boldsymbol{x}))\right| \cdot \left|\det J_{\Phi}(\boldsymbol{x})\right|$$

and

$$h(\boldsymbol{x}) = f\big(\Psi\circ\Phi(\boldsymbol{x})\big)\cdot\left|\det J_{\Psi\circ\Phi}(\boldsymbol{x})\right|.$$

We deduce from (15.57) and (15.58) that

$$\int_{V_2} f(\boldsymbol{z})|d\boldsymbol{z}| = \int_{U_0} f\big(\Psi\circ\Phi(\boldsymbol{x})\big)\cdot\left|\det J_{\Psi\circ\Phi}(\boldsymbol{x})\right||d\boldsymbol{x}|.$$

This shows that $\Psi\circ\Phi$ satisfies Theorem 15.46.

Step 2. Suppose that $\Phi : U \to \mathbb{R}^n$ is a diffeomorphism such that, for any point $\boldsymbol{p}\in U$ there exists an open neighborhood $\mathcal{O}_{\boldsymbol{p}}$ of $\boldsymbol{p}$ with the following properties.

(i) $\mathcal{O}_{\boldsymbol{p}}\subset U$. Set $\hat{\mathcal{O}}_{\boldsymbol{p}} := \Phi(\mathcal{O}_{\boldsymbol{p}})$.
(ii) Any continuous function $g : V\to\mathbb{R}$ that vanishes outside a compact set $C_{\boldsymbol{p}}\subset\hat{\mathcal{O}}_{\boldsymbol{p}}$ satisfies the change in variables formula (15.33).

We will show that this condition implies that any continuous function $g : V\to\mathbb{R}$ that vanishes outside a compact set $C\subset V$ satisfies the change in variables formula (15.33).

The collection of open sets $\{\hat{\mathcal{O}}_{\boldsymbol{p}}\}_{\boldsymbol{p}\in\Phi^{-1}(C)}$ is an open cover of C. Fix a compactly supported partition of unity on K subordinated to this open cover. This consists of finitely many *compactly supported* continuous functions

$$\chi_1,\dots,\chi_\ell : \mathbb{R}^n \to [0,1]$$

satisfying the following properties.

- For any $j = 1,\dots,\ell$ there exists $\boldsymbol{p}_j\in C$ such that $\operatorname{supp}\chi_j\subset\hat{\mathcal{O}}_{\boldsymbol{p}_j}$.
- $\chi_1(\boldsymbol{y}) + \cdots + \chi_\ell(\boldsymbol{y}) = 1$, $\forall\boldsymbol{y}\in C$.

Set $g_j := \chi_j g$. Observe that since $g(\boldsymbol{y}) = 0$, $\forall\boldsymbol{y}\in V\setminus C$, we have

$$g_1(\boldsymbol{y}) + \cdots + g_\ell(\boldsymbol{y}) = \big(\chi_1(\boldsymbol{y}) + \cdots + \chi_\ell(\boldsymbol{y})\big)g(\boldsymbol{y}) = g(\boldsymbol{y}),\quad \forall\boldsymbol{y}\in V.$$

Clearly g_j is continuous and supported on a compact set contained in $\mathcal{O}_{\boldsymbol{p}_j}$. It satisfies the change-in-variables formula (15.33)

$$\int_V g_j(\boldsymbol{x})\left|\det J_\Phi(\boldsymbol{x})\right||d\boldsymbol{x}| = \int_{\Phi(\mathcal{O}_{\boldsymbol{p}_j})} g_j(\boldsymbol{y})|d\boldsymbol{x}|$$

$$= \int_{\mathcal{O}_{\boldsymbol{p}_j}} g_j\big(\Phi(\boldsymbol{x})\big)\left|\det J_\Phi(\boldsymbol{x})\right||d\boldsymbol{y}| = \int_U g_j\big(\Phi(\boldsymbol{x})\big)\left|\det J_\Phi(\boldsymbol{x})\right||d\boldsymbol{x}|.$$

Summing these equalities we deduce

$$\int_U g\big(\Phi(\boldsymbol{x})\big)\left|\det J_\Phi(\boldsymbol{x})\right||d\boldsymbol{x}| = \sum_{j=1}^{\ell}\int_U g_j\big(\Phi(\boldsymbol{x})\big)\left|\det J_\Phi(\boldsymbol{x})\right||d\boldsymbol{x}|$$

$$= \sum_{j=1}^{\ell}\int_V g_j(\boldsymbol{y})|d\boldsymbol{y}| = \int_V g(\boldsymbol{y})|d\boldsymbol{y}|.$$

This proves that the diffeomorphism Φ satisfies the change-in-variables formula (15.33).

Step 3. Let $j \in \{1, \dots n\}$. We say that a diffeomorphism

$$\Phi : U \to \mathbb{R}^n, \;\; U \ni \boldsymbol{x} \mapsto \boldsymbol{y} = \Phi(\boldsymbol{x}) = \begin{bmatrix} \Phi^1(\boldsymbol{x}) \\ \Phi^2(\boldsymbol{x}) \\ \vdots \\ \Phi^n(\boldsymbol{x}) \end{bmatrix}$$

is *j-elementary* if, for any $i \neq j$, we have

$$\Phi^i(x^1, \dots, x^n) = x^i.$$

Note that the inverse of a j-elementary diffeomorphism is also a j-elementary diffeomorphism. We say that Φ is elementary if it is j-elementary for some index $j = 1, \dots, n$. We will show that if $\Phi : U \to \mathbb{R}^n$ is elementary, then it satisfies Theorem 15.46. To complete this we rely on the localization trick discussed in Step 2. Assume for simplicity that Φ is 1-elementary, i.e.,

$$\Phi(\boldsymbol{x}) = \begin{bmatrix} \varphi(x^1, \dots, x^n) \\ x^2 \\ \vdots \\ x^n \end{bmatrix}$$

where $\varphi : U \to \mathbb{R}$ is a C^1 function. The Jacobian matrix of Φ is

$$J_\Phi(\boldsymbol{x}) = \begin{bmatrix} \varphi'_{x^1}(\boldsymbol{x}) & \varphi'_{x^2}(\boldsymbol{x}) & \varphi'_{x^3}(\boldsymbol{x}) & \varphi'_{x^4}(\boldsymbol{x}) & \cdots & \varphi'_{x^n}(\boldsymbol{x}) \\ 0 & 1 & 0 & 0 & \cdots & 0 \\ 0 & 0 & 1 & 0 & \cdots & 0 \\ \vdots & \vdots & \vdots & \ddots & \ddots & \vdots \\ \vdots & \vdots & \vdots & \vdots & \ddots & \vdots \\ 0 & 0 & 0 & 0 & \cdots & 1 \end{bmatrix}.$$

Observe that

$$\det J_\Phi(\boldsymbol{x}) = \varphi'_{x^1}(\boldsymbol{x}). \tag{15.59}$$

Since Φ is a diffeomorphism, we deduce $\det J_\Phi(\boldsymbol{x}) \neq 0$, $\forall \boldsymbol{x} \in U$, i.e.,

$$\varphi'_{x^1}(\boldsymbol{x}) \neq 0, \;\; \forall \boldsymbol{x} \in U.$$

Let $\boldsymbol{p} \in U$. We want to show that there exists an open neighborhood $\mathcal{O}_{\boldsymbol{p}}$ of $\boldsymbol{p}$ in U such that the restriction of Φ to that neighborhood satisfies Theorem 15.46.

Fix $r > 0$ small enough such that the open cube $C_{2r}(\boldsymbol{p})$ (see Definition 11.54) is contained in U. We will prove that the restriction of Φ to $C_r(\boldsymbol{p})$ satisfies the change-in-variables formula.

The cube $C_{2r}(\boldsymbol{p})$ is connected and the continuous function φ'_{x^1} does not vanish in this cube. Hence it must have constant sign. Assume for simplicity that[7]

$$\varphi'_{x^1}(\boldsymbol{x}) > 0, \ \ \forall \boldsymbol{x} \in C_{2r}(\boldsymbol{p}).$$

Note that

$$C_r(\boldsymbol{p}) = \big\{ \, \boldsymbol{x} \in \mathbb{R}^n; \ \ |x^i - p^i| < r, \ \ \forall i = 1, \dots, n \, \big\}.$$

For any $\boldsymbol{x} = (x^1, \dots, x^n)$ we set $\overline{\boldsymbol{x}} := (x^2, \dots, x^n)$ and denote by $C'_r(\overline{\boldsymbol{p}}) \subset \mathbb{R}^{n-1}$ the open $(n-1)$-dimensional cube of radius r centered at $\overline{\boldsymbol{p}}$. Using this notation we have

$$C_r(\boldsymbol{p}) = \big\{ \, (x^1, \overline{\boldsymbol{x}}) \in \mathbb{R} \times \mathbb{R}^{n-1}; \ \ x^1 \in (p^1 - r, p^1 + r), \ \ \overline{\boldsymbol{x}} \in C'_r(\overline{\boldsymbol{p}}) \, \big\}.$$

For each $\overline{\boldsymbol{x}} \in C'_r(\overline{\boldsymbol{p}})$ the function $x^1 \mapsto \varphi(x^1, \overline{\boldsymbol{x}})$ sends the interval $(p^1 - r, p^1 + r)$ to the interval

$$I_{\overline{\boldsymbol{x}}} := \big\{ \, y^1 \in \mathbb{R}; \ \ \varphi(p^1 - r, \overline{\boldsymbol{x}}) < y^1 < \varphi(p^1 + r, \overline{\boldsymbol{x}}) \, \big\}.$$

We deduce that the image of $C_r(\boldsymbol{p})$ via Φ is the simple type domain

$$\mathscr{C}(r, \boldsymbol{p}) = \big\{ \, (y^1, \overline{\boldsymbol{y}}) \in \mathbb{R} \times \mathbb{R}^{n-1}; \ \ \varphi(p^1 - r, \overline{\boldsymbol{y}}) < y^1 < \varphi(p^1 + r, \overline{\boldsymbol{y}}), \ \ \overline{\boldsymbol{y}} \in C'_r(\overline{\boldsymbol{p}}) \, \big\}.$$

Suppose that $f : \mathscr{C}(r, \boldsymbol{p}) \to \mathbb{R}$ is a continuous function that vanishes outside a compact set $K \subset \mathscr{C}(r, \boldsymbol{p})$. Using Fubini's theorem we deduce

$$\int_{\mathscr{C}(r,\boldsymbol{p})} f(\boldsymbol{y}) d\boldsymbol{y} = \int_{C'_r(\overline{\boldsymbol{p}})} \left(\int_{\varphi(p^1 - r, \overline{\boldsymbol{x}})}^{\varphi(p^1 + r, \overline{\boldsymbol{x}})} f(y^1, \overline{\boldsymbol{y}}) dy^1 \right) d\overline{\boldsymbol{y}}.$$

Fix $\overline{\boldsymbol{x}}$ and thus $\overline{\boldsymbol{y}} = \overline{\boldsymbol{x}}$. Using the 1-dimensional change-in-variables formula (9.52) we deduce

$$\int_{\varphi(p^1 - r, \overline{\boldsymbol{x}})}^{\varphi(p^1 + r, \overline{\boldsymbol{x}})} f(y^1, \overline{\boldsymbol{y}}) dy^1 = \int_{p^1 - r}^{p^1 + r} f\big(\varphi(x^1, \overline{\boldsymbol{x}}), \overline{\boldsymbol{y}} \big) \varphi'_{x^1}(x^1, \overline{\boldsymbol{y}}) dx^1.$$

Hence

$$\int_{\mathscr{C}(r,\boldsymbol{p})} f(\boldsymbol{y}) d\boldsymbol{y} = \int_{C'_r(\overline{\boldsymbol{p}})} \left(\int_{p^1 - r}^{p^1 + r} f\big(\varphi(x^1, \overline{\boldsymbol{x}}), \overline{\boldsymbol{y}} \big) \varphi'_{x^1}(x^1, \overline{\boldsymbol{y}}) dx^1 \right) d\overline{\boldsymbol{y}}$$

(rename by $\overline{\boldsymbol{x}}$ the variables $\overline{\boldsymbol{y}}$)

$$= \int_{C'_r(\overline{\boldsymbol{p}})} \left(\int_{p^1 - r}^{p^1 + r} f\big(\varphi(x^1, \overline{\boldsymbol{x}}), \overline{\boldsymbol{x}} \big) \varphi'_{x^1}(x^1, \overline{\boldsymbol{x}}) dx^1 \right) d\overline{\boldsymbol{x}}$$

(use Fubini again)

$$= \int_{C_r(\boldsymbol{p})} f\big(\varphi(x^1), \overline{\boldsymbol{x}} \big) \varphi'_{x^1}(x^1, \overline{\boldsymbol{x}}) |d\boldsymbol{x}| \overset{(15.59)}{=} \int_{C_r(\boldsymbol{p})} f\big(\Phi(\boldsymbol{x}) \big) \cdot \big| \det J_\Phi(\boldsymbol{x}) \big| |d\boldsymbol{x}|.$$

Step 4. To give you a taste of the main idea we consider first the special case $n = 2$. Then, $\Phi(\boldsymbol{x})$ has the form

$$\boldsymbol{x} = \begin{bmatrix} x^1 \\ x^2 \end{bmatrix} \mapsto \Phi(\boldsymbol{x}) = \begin{bmatrix} y^1 \\ y^2 \end{bmatrix} = \begin{bmatrix} \phi^1(x^1, x^2) \\ \phi^2(x^1, x^2) \end{bmatrix}.$$

[7] The case $\varphi'_{x^1} < 0$ is dealt with in a similar fashion.

The Jacobian matrix of Φ is

$$J_\Phi = \begin{bmatrix} \frac{\partial\phi^1}{\partial x^1} & \frac{\partial\phi^1}{\partial x^2} \\ \\ \frac{\partial\phi^2}{\partial x^1} & \frac{\partial\phi^2}{\partial x^2} \end{bmatrix}.$$

Since Φ is a diffeomorphism, $\det J_\Phi(\boldsymbol{p}) \neq 0$, so at least one of the entries $\frac{\partial\phi^i}{\partial x^j}$ must be nonzero at $\boldsymbol{p}$. After a possible relabeling of the variables $\boldsymbol{y}$ and/or $\boldsymbol{x}$ we can assume that $\frac{\partial\phi^1}{\partial x^1} \neq 0$. The implicit function theorem shows that in the equation

$$y^1 - \phi^1(x^1, x^2) = 0$$

we can locally solve for x^1 in terms of y^1 and x^2, i.e., we can regard x^1 as an implicitly defined C^1-function depending on the variables y^1, x^2,

$$x^1 = \psi^1(y^1, x^2) \Longleftrightarrow y^1 = \phi^1\big(\psi^1(y^1, x^2), x^2\big). \tag{15.60}$$

This shows that the map

$$\boldsymbol{x} = \begin{bmatrix} x^1 \\ x^2 \end{bmatrix} \mapsto \Phi_1(\boldsymbol{x}) = \begin{bmatrix} y^1 \\ x^2 \end{bmatrix} = \begin{bmatrix} \phi^1(x^1, x^2) \\ x^2 \end{bmatrix}$$

is a 1-elementary diffeomorphism defined on an open neighborhood U_1 and its inverse, denoted by Υ_1, is the 1-elementary diffeomorphism described explicitly by

$$\begin{bmatrix} y^1 \\ x^2 \end{bmatrix} \overset{\Upsilon_1}{\mapsto} \begin{bmatrix} x^1 \\ x^2 \end{bmatrix} = \begin{bmatrix} \psi^1(y^1, x^2) \\ x^2 \end{bmatrix}.$$

Now consider the composition $\Psi_2 = \Phi \circ \Upsilon_1$. More precisely,

$$\begin{bmatrix} y^1 \\ x^2 \end{bmatrix} \overset{\Upsilon_1}{\mapsto} \begin{bmatrix} x^1 \\ x^2 \end{bmatrix} \overset{\Phi}{\mapsto} \begin{bmatrix} y^1 \\ y^2 \end{bmatrix} \overset{(15.60)}{=} \begin{bmatrix} y^1 \\ \phi^2\big(\psi^1(y^1, x^2), x^2\big) \end{bmatrix}.$$

This shows that Ψ_2 is 2-elementary. The equality $\Psi_2 = \Phi \circ \Upsilon_1 = \Phi \circ \Phi_1^{-1}$ implies that

$$\Phi = \Psi_2 \circ \Phi_1,$$

i.e., locally, Φ is the composition of elementary diffeomorphisms. Here are the details for the general case $n \geq 2$.

We outline now how the above approach extends to arbitrary n, referring for details to [23, Sec. 8.6.4]. Given a diffeomorphism $\Phi : U \to \mathbb{R}^n$ and a point $\boldsymbol{p} \in U$ one shows, using the implicit function theorem, that there exist elementary diffeomorphisms $\Psi_n, \dots, \Psi_1$ such that Ψ_1 is defined on an open neighborhood $\mathcal{O}$ of $\boldsymbol{p}$ and

$$\Phi(\boldsymbol{x}) = \Psi_n \circ \Psi_{n-1} \circ \cdots \circ \Psi_1(\boldsymbol{x}), \quad \forall \boldsymbol{x} \in \mathcal{O}.$$

This process proceeds gradually. First, using the inverse function theorem one constructs an open neighborhood U_{n-1} of $\boldsymbol{p}$ in U and a diffeomorphism $\Phi_{n-1} : U_{n-1} \to \mathbb{R}^n$ with the following two properties.

(A) The n-th component of Φ_{n-1} has the special form $\Phi^n_{n-1}(\boldsymbol{x}) = x^n$, $\forall \boldsymbol{x} \in U_{n-1}$.
(B) The diffeomorphism $\Psi_n := \Phi \circ \Phi^{-1}_{n-1}$ is n-elementary.

Note that $\Phi = \Psi_n \circ \Phi_{n-1}$. Proceeding inductively, again relying on the implicit function theorem, one constructs open sets

$$U_n \supset U_{n-1} \supset U_{n-2} \supset \cdots \supset U_1 \ni \boldsymbol{p}$$

and, for any $k = 2, \dots, n$, diffeomorphisms $\Phi_k : U_k \to \mathbb{R}^n$, with the following properties.

- $\Phi_n = \Phi$.
- If Φ_k^j is the j-th component of Ψ_k, then

$$\Phi_k^j(\boldsymbol{x}) = x^j, \quad \forall j > k, \quad \boldsymbol{x} \in U_k. \tag{15.61}$$

- The diffeomorphism $\Psi_k := \Phi_k \circ \Phi_{k-1}^{-1}$ is k-elementary.

We deduce

$$\begin{aligned}\Phi(\boldsymbol{x}) = \Phi_n &= (\Phi_n \circ \Phi_{n-1}^{-1}) \circ (\Phi_{n-1} \circ \Phi_{n-2}^{-1}) \circ \cdots \circ (\Phi_2 \circ \Phi_1^{-1}) \circ \Phi_1 \\ &= \Psi_n \circ \cdots \circ \Psi_2 \circ \Phi_1(\boldsymbol{x}), \quad \forall \boldsymbol{x} \in U_1.\end{aligned}$$

By construction, the diffeomorphism Ψ_k is k-elementary, $\forall k \geq 2$, while (15.61) with $k = 1$, shows that Φ_1 is 1-elementary.

This completes our outline of the proof of the change-in-variables formula (15.31). For a different, more intuitive but more laborious approach we refer to [16, §XX.4].

15.4. Improper integrals

Concrete problems arising in mathematics and natural sciences force us to integrate functions that are not covered by the theory developed so far. For example, we might want to integrate an unbounded function, or we might want to integrate a function over an unbounded region. The goal of this section is to explain how to handle such issues. Let us first introduce a notation.

✍ For any set $A \subset \mathbb{R}^n$ we denote by $\mathfrak{J}(A)$ the collection of Jordan measurable subsets of A and by $\mathfrak{J}_c(A)$ the collection of compact Jordan measurable subsets of A.

15.4.1. *Locally integrable functions*

Let $n \in \mathbb{N}$ and suppose that $U \subset \mathbb{R}^n$ is an open set.

Definition 15.55. A function $f : U \to \mathbb{R}$ is called *locally integrable* if, for any $\boldsymbol{x} \in U$, there exists a closed box $B = B_{\boldsymbol{x}}$ such that $B_{\boldsymbol{x}} \subset U$, $\boldsymbol{x} \in \boldsymbol{int}(B_{\boldsymbol{x}})$ and the restriction of f to $B_{\boldsymbol{x}}$ is Riemann integrable. □

Example 15.56. (a) Any continuous function $f : U \to \mathbb{R}$ is locally integrable.

(b) If the open set $U \subset \mathbb{R}^n$ is Jordan measurable, then any Riemann integrable function $f : U \to \mathbb{R}$ is also locally integrable. □

Proposition 15.57. *For any function $f : U \to \mathbb{R}$ the following statements are equivalent.*

(i) The function f is locally integrable.
(ii) For any compact Jordan measurable set $K \subset U$, the restriction of f to K is Riemann integrable on K.

Proof. Clearly (ii) ⇒ (i) since any closed box is compact and Jordan measurable. Let us prove (i) ⇒ (ii). Suppose that $K \subset U$ is compact and Jordan measurable. For any $\boldsymbol{x} \in K$

choose a closed box $B_{\boldsymbol{x}}$ satisfying the conditions in Definition 15.55. Consider a partition of unity on K subordinated to the open cover

$$\Big\{ \boldsymbol{int}(B_{\boldsymbol{x}});\ \ \boldsymbol{x} \in K \Big\}_{\boldsymbol{x}\in K}.$$

Recall (see Definition 12.67) that this consists of a finite collection of compactly supported continuous functions

$$\chi_1, \dots, \chi_\ell : \mathbb{R}^n \to \mathbb{R}$$

with the following properties;

$$\chi_1(\boldsymbol{x}) + \cdots + \chi_\ell(\boldsymbol{x}) = 1,\ \ \forall \boldsymbol{x} \in K, \tag{15.62a}$$

$$\forall i = 1, \dots, \ell\ \ \exists \boldsymbol{x}_i \in K\ \ \operatorname{supp}\chi_i \subset \boldsymbol{int}(B_{\boldsymbol{x}_i}). \tag{15.62b}$$

Denote by f^0 the extension of f by 0. The functions $I_{B_{\boldsymbol{x}_i}} f^0$ are Riemann integrable and so are the functions

$$\chi_i I_{B_{\boldsymbol{x}_i}} f^0 \stackrel{(15.62a)}{=} \chi_i f^0.$$

Hence $\chi_1 f^0 + \cdots + \chi_\ell f^0$ is Riemann integrable. Since K is Jordan measurable, we deduce that the function

$$I_K\big(\chi_1 + \cdots + \chi_\ell\big) f^0 \stackrel{(15.62b)}{=} I_K f^0$$

is also Riemann integrable. □

Definition 15.58. A *compact exhaustion* of U is a sequence of compact sets $(K_\nu)_{\nu\in\mathbb{N}}$ with the following properties.

(i) $K_\nu \subset \boldsymbol{int}(K_{\nu+1})$, $\forall \nu \in \mathbb{N}$.

(ii)

$$U = \bigcup_{\nu\in\mathbb{N}} K_\nu.$$

The compact exhaustion $(K_\nu)_{\nu\in\mathbb{N}}$ is called *Jordan measurable* if all the compact sets K_ν are Jordan measurable. □

Observe that if $(K_\nu)_{\nu\in\mathbb{N}}$ is a compact exhaustion of the open set U, then the collection of interiors $(\boldsymbol{int}\, K_\nu)_{\nu\in\mathbb{N}}$ is *increasing* and covers U, i.e.,

$$\boldsymbol{int}\, K_1 \subset \boldsymbol{int}\, K_2 \subset \cdots \subset \boldsymbol{int}\, K_\nu \subset \boldsymbol{int}\, K_{\nu+1} \subset \cdots,\ \ \bigcup_{\nu\geq 1} \boldsymbol{int}\, K_\nu = U.$$

Example 15.59. The collection

$$K_\nu = \big\{ \boldsymbol{x} \in \mathbb{R}^n;\ \ \|\boldsymbol{x}\| \leq \nu \big\},\ \ \nu \in \mathbb{N}$$

is a Jordan measurable compact exhaustion of $\mathbb{R}^n$. The collection

$$K_\nu = \Big\{ \boldsymbol{x} \in \mathbb{R}^n;\ \ \frac{1}{\nu} \leq \|\boldsymbol{x}\| \leq 1 - \frac{1}{\nu} \Big\},\ \ \nu \in \mathbb{N}$$

is a Jordan measurable compact exhaustion of $B_1(\mathbf{0}) \setminus \{\mathbf{0}\}$. □

Proposition 15.60. *Any open set $U \subset \mathbb{R}^n$ admits Jordan measurable compact exhaustions.*

Proof. Denote by C the complement of U in $\mathbb{R}^n$, $C := \mathbb{R}^n \setminus U$. By definition, C is a closed set. For $\nu \in \mathbb{N}$ we set

$$K_\nu := \overline{B_\nu(\mathbf{0})} \cap \Big\{ \boldsymbol{x} \in \mathbb{R}^n;\ \operatorname{dist}(\boldsymbol{x}, C) \geq \frac{1}{\nu} \Big\}.$$

Clearly K_ν is closed as the intersection of two closed sets. It is bounded since it is contained in the closed ball of radius ν. Hence K_ν is compact. Note that K_ν is contained in U since $\boldsymbol{x} \in U = \mathbb{R}^n \setminus C$ if and only if $\operatorname{dist}(\boldsymbol{x}, C) > 0$. Obviously

$$U = \bigcup_{\nu \in \mathbb{N}} K_\nu.$$

Note that

$$K_\nu \subset B_{\nu+1}(\mathbf{0}) \cap \Big\{ \boldsymbol{x} \in \mathbb{R}^n;\ \operatorname{dist}(\boldsymbol{x}, C) > \frac{1}{\nu+1} \Big\} \subset \boldsymbol{int}(K_{\nu+1}).$$

Hence the collection $(K_\nu)_{\nu \in \mathbb{N}}$ is a compact exhaustion of U. However, it may not be Jordan measurable. We can modify it to a Jordan measurable one as follows.

Since K_ν is compact there exist finitely many closed boxes contained in $\boldsymbol{int}(K_{\nu+1})$ such that their interiors cover K_ν. Denote by $\tilde{K}_\nu$ the union of these finitely many closed boxes. Clearly $\tilde{K}_\nu$ is compact and Jordan measurable, and

$$K_\nu \subset \tilde{K}_\nu \subset K_{\nu+1} \subset \tilde{K}_{\nu+1}.$$

The collection $(\tilde{K}_\nu)_{\nu \in \mathbb{N}}$ is a Jordan measurable compact exhaustion of U. □

Proposition 15.61. *Suppose that $U \subset \mathbb{R}^n$ is a Jordan measurable open set and $f : U \to \mathbb{R}$ is a Riemann integrable function. Then, for any Jordan measurable compact exhaustion $(K_\nu)_{\nu \in \mathbb{N}}$ of U we have*

$$\int_U f(\boldsymbol{x})|d\boldsymbol{x}| = \lim_{\nu \to \infty} \int_{K_\nu} f(\boldsymbol{x})|d\boldsymbol{x}|. \tag{15.63}$$

Proof. Fix a Jordan measurable compact exhaustion $(K_\nu)_{\nu \in \mathbb{N}}$ and set

$$M := \sup_{\boldsymbol{x} \in U} |f(\boldsymbol{x})|.$$

Since f is Riemann integrable, it is bounded, and therefore $M < \infty$. Fix $\varepsilon > 0$.

Since U is Jordan measurable, its boundary is negligible. We can thus cover ∂U with finitely many open boxes such that their union Δ_ε has volume $< \frac{\varepsilon}{M}$. The set $S_\varepsilon := U \setminus \Delta_\varepsilon$. Observe that $S_\varepsilon \subset U$ is closed[8] and bounded so it is compact. The increasing collection of open sets $\boldsymbol{int}(K_\nu)$, $\nu \in \mathbb{N}$ is an open cover of the *compact* set S_ε and thus there exists $N = N(\varepsilon)$ such that $S_\varepsilon \subset K_\nu$ for all $\nu \geq N(\varepsilon)$. Note that this implies

$$U \setminus K_\nu \subset U \setminus S_\varepsilon \subset \Delta(\varepsilon)$$

so that

$$\operatorname{vol}_n(U \setminus K_\nu) \leq \operatorname{vol}_n(\Delta_\varepsilon) < \frac{\varepsilon}{M},\quad \forall \nu \geq N(\varepsilon).$$

For $\nu \geq N(\varepsilon)$ we have

$$\left| \int_U f(\boldsymbol{x})|d\boldsymbol{x}| - \int_{K_\nu} f(\boldsymbol{x})|d\boldsymbol{x}| \right| = \left| \int_{U \setminus K_\nu} f(\boldsymbol{x})|d\boldsymbol{x}| \right| \leq \int_{U \setminus K_\nu} |f(\boldsymbol{x})||d\boldsymbol{x}|$$
$$\leq \int_{U \setminus K_\nu} M|d\boldsymbol{x}| = M \operatorname{vol}_n(U \setminus K_\nu) < \varepsilon.$$

This proves (15.63). □

[8] Why?

15.4.2. *Absolutely integrable functions*

Let $U \subset \mathbb{R}^n$ be an open set. Recall that for any $A \subset \mathbb{R}^n$ we denoted by $\mathcal{J}_c(A)$ the collection of compact, Jordan measurable subsets of A.

Definition 15.62. A locally integrable function $f : U \to \mathbb{R}$ is called *absolutely integrable* if

$$\sup_{K \in \mathcal{J}_c(U)} \int_K |f(\boldsymbol{x})|\,|d\boldsymbol{x}| < \infty.$$

We will denote by $\mathcal{R}_a(U)$ the collection of absolutely integrable functions $f : U \to \mathbb{R}$. □

Proposition 15.63. *Let $f : U \to \mathbb{R}$ be a locally integrable function. Then the following statements are equivalent.*

(i) The function f is absolutely integrable.
(ii) For any Jordan measurable compact exhaustion $(K_\nu)_{\nu \in \mathbb{N}}$ of U the sequence

$$\int_{K_\nu} |f(\boldsymbol{x})|\,|d\boldsymbol{x}|$$

is bounded.
(iii) There exists a Jordan measurable compact exhaustion $(K_\nu)_{\nu \in \mathbb{N}}$ of U such that the sequence

$$\int_{K_\nu} |f(\boldsymbol{x})|\,|d\boldsymbol{x}|$$

is bounded.

Moreover, if any of the above conditions is satisfied, then

$$\lim_{\nu \to \infty} \int_{K_\nu} |f(\boldsymbol{x})|\,|d\boldsymbol{x}| = \sup_{K \in \mathcal{J}_c(U)} \int_K |f(\boldsymbol{x})|\,|d\boldsymbol{x}|.$$

Proof. Clearly (i) ⇒ (ii) ⇒ (iii). We only have to prove (iii) ⇒ (i). Suppose that $(K_\nu)_{\nu \in \mathbb{N}}$ is a Jordan measurable compact exhaustion of U such that the sequence

$$\int_{K_\nu} |f(\boldsymbol{x})|\,|d\boldsymbol{x}|$$

is bounded. We denote by L its supremum. We have to prove that,

$$\sup_{K \in \mathcal{J}_c(U)} \int_K |f(\boldsymbol{x})|\,|d\boldsymbol{x}| = L.$$

Since for every ν we have

$$\int_{K_\nu} |f(\boldsymbol{x})|\,|d\boldsymbol{x}| \leq \sup_{K \in \mathcal{J}_c(U)} \int_K |f(\boldsymbol{x})|\,|d\boldsymbol{x}|,$$

we deduce

$$L = \sup_\nu \int_{K_\nu} |f(\boldsymbol{x})|\,|d\boldsymbol{x}| \leq \sup_{K \in \mathcal{J}_c(U)} \int_K |f(\boldsymbol{x})|\,|d\boldsymbol{x}| =: L^*.$$

To prove that $L^* \leq L$ it suffices to show that, for any $K \in \mathcal{J}_c(U)$ we can find $\nu \in \mathbb{N}$ such that $K \subset \boldsymbol{int}(K_\nu)$. Indeed, if this were the case we would have

$$\int_K |f(\boldsymbol{x})|\,|d\boldsymbol{x}| \leq \int_{K_\nu} |f(\boldsymbol{x})|\,|d\boldsymbol{x}| \leq L.$$

To prove the claim note that the *increasing* collection of open sets $\boldsymbol{int}(K_\nu)$, $\nu \in \mathbb{N}$ is an open cover of U and thus also of the *compact* set K. Thus there exist natural numbers $\nu_1 < \cdots < \nu_\ell$ such that the finite collection $\boldsymbol{int}(K_{\nu_1}), \dots, \boldsymbol{int}(K_{\nu_\ell})$ covers K. Since

$$\boldsymbol{int}(K_{\nu_1}) \subset \cdots \subset \boldsymbol{int}(K_{\nu_\ell})$$

we deduce $K \subset \boldsymbol{int}(K_{\nu_\ell})$.

The last conclusion of the proposition follows by observing that the sequence

$$\int_{K_\nu} |f(\boldsymbol{x})|\,|d\boldsymbol{x}|, \;\; \nu \in \mathbb{N}$$

is nondecreasing, so

$$\lim_{\nu\to\infty} \int_{K_\nu} |f(\boldsymbol{x})|\,|d\boldsymbol{x}| = \sup_{\nu\in\mathbb{N}} \int_{K_\nu} |f(\boldsymbol{x})|\,|d\boldsymbol{x}| = L = L^*.$$

□

Definition 15.64. Suppose that $f : U \to [0,\infty)$ is a *nonnegative* locally integrable function. We set

$$\int_U^* |f(\boldsymbol{x})|\,|d\boldsymbol{x}| := \sup_{K\in\mathcal{J}_c(U)} \int_K |f(\boldsymbol{x})|\,|d\boldsymbol{x}|.$$ □

Remark 15.65. Note that if U is Jordan measurable and $f : U \to [0,\infty)$ is integrable, then it is absolutely integrable. Moreover, Propositions 15.61 and 15.63 show that

$$\int_U^* f(\boldsymbol{x})|d\boldsymbol{x}| = \int_U f(\boldsymbol{x})dx.$$ □

To proceed we need to introduce a useful trick. For any $x \in \mathbb{R}$ we set

$$x_+ := \max(x,0), \;\; x_- := \max(-x,0) = (-x)_+.$$

We will refer to $x_\pm$ as the *positive/negative part* of x. For example,

$$(2)_+ = 2, \;\; (2)_- = 0, \;\; (-3)_+ = 0, \;\; (-3)_- = 3.$$

Note that

$$x = x_+ - x_-, \;\; |x| = x_+ + x_-, \;\; \forall x \in \mathbb{R}.$$

The functions $x \mapsto x_\pm$ can be given the alternate descriptions

$$\boxed{x_+ = \frac{|x|+x}{2}, \;\; x_- = \frac{|x|-x}{2}, \;\; \forall x \in \mathbb{R}}.$$

This shows that the functions $x \mapsto x_\pm$ are Lipschitz since they are linear combinations of the Lipschitz functions $x \mapsto x$ and $x \mapsto |x|$.

Suppose now that U is an open set and $f : U \to \mathbb{R}$ is *absolutely integrable*. Theorem 15.12 implies that the functions $f_\pm : U \to \mathbb{R}$, $f_\pm(\boldsymbol{x}) = f(\boldsymbol{x})_\pm$, are locally integrable. From the equality

$$|f(\boldsymbol{x})| = f_+(\boldsymbol{x}) + f_-(\boldsymbol{x}), \ \ \forall \boldsymbol{x} \in U,$$

we deduce that the functions $f_\pm$ are also absolutely integrable. The equality $f = f_+ - f_-$ suggests the following concept.

Definition 15.66. Suppose that the function $f : U \to \mathbb{R}$ is absolutely integrable. We set

$$\int_U^{**} f(\boldsymbol{x})|d\boldsymbol{x}| := \int_U^{*} f_+(\boldsymbol{x})\,|d\boldsymbol{x}| - \int_U^{*} f_-(\boldsymbol{x})\,|d\boldsymbol{x}|.$$

We say that $\int_U^{**} f(\boldsymbol{x})|d\boldsymbol{x}|$ is the *improper integral* over U of the absolutely integrable function $f : U \to \mathbb{R}$. □

Remark 15.67. (a) If $f : U \to \mathbb{R}$ is absolutely integrable, then, for any Jordan measurable compact exhaustion $(K_\nu)_{\nu\in\mathbb{N}}$ of U we have

$$\begin{aligned}\int_U^{**} f(\boldsymbol{x})|d\boldsymbol{x}| &= \lim_{\nu\to\infty}\int_{K_\nu} f_+(\boldsymbol{x})\,|d\boldsymbol{x}| - \lim_{\nu\to\infty}\int_{K_\nu} f_-(\boldsymbol{x})\,|d\boldsymbol{x}| \\ &= \lim_{\nu\to\infty}\int_{K_\nu} f(\boldsymbol{x})|d\boldsymbol{x}|.\end{aligned} \tag{15.64}$$

(b) If $f : U \to [0,\infty)$ is absolutely integrable, then we deduce from the equality $f = f_+$ that

$$\int_U^{**} f(\boldsymbol{x})\,|d\boldsymbol{x}| := \int_U^{*} f_+(\boldsymbol{x})\,|d\boldsymbol{x}| = \int_U^{*} f(\boldsymbol{x})\,|d\boldsymbol{x}|.$$

(c) If U is Jordan measurable, and $f : U \to \mathbb{R}$ is Riemann integrable, then we deduce from Remark 15.65 that

$$\int_U^{**} f(\boldsymbol{x})\,|d\boldsymbol{x}| = \int_U f(\boldsymbol{x})\,|d\boldsymbol{x}|.$$ □

✍ *In view of the above remarks, and in order to control the proliferation of notations for closely related concepts, we will continue to use the notation $\int_U f(\boldsymbol{x})\,|d\boldsymbol{x}|$ when referring to the improper integral of f over U.*

Theorem 15.68 (Comparison Principle). *Suppose that $f, F : U \to \mathbb{R}$ are locally integrable functions such that*

$$|f(x)| \leq |F(\boldsymbol{x})|, \ \ \forall \boldsymbol{x} \in U.$$

(i) If F is absolutely integrable, then f is also absolutely integrable.

(ii) If f is not absolutely integrable, then neither is F.

Proof. Clearly (i) $\iff$ (ii) so it suffices to prove (i). We have

$$\int_K \big|\,f(\boldsymbol{x})\,\big|\,|d\boldsymbol{x}| \leq \int_K \big|\,F(\boldsymbol{x})\,\big|\,|d\boldsymbol{x}|, \ \ \forall K \in \mathfrak{J}_c(U)$$

so

$$\sup_{K\in\mathfrak{J}_c(U)}\int_K \big|\,f(\boldsymbol{x})\,\big|\,|d\boldsymbol{x}| \leq \sup_{K\in\mathfrak{J}_c(U)}\int_K \big|\,F(\boldsymbol{x})\,\big|\,|d\boldsymbol{x}| < \infty.$$

□

15.4.3. *Examples*

We want to discuss a few simple but important examples.

Example 15.69. Fix $n \in \mathbb{N}$. For $\alpha > 0$ define

$$p_\alpha : \mathbb{R}^n \setminus \{\mathbf{0}\} \to \mathbb{R}, \;\; p_\alpha(\boldsymbol{x}) = \|\boldsymbol{x}\|^{-\alpha}.$$

Suppose that U is the punctured unit ball in $\mathbb{R}^n$,

$$U = \big\{\, \boldsymbol{x} \in \mathbb{R}^n; \;\; 0 < \|\boldsymbol{x}\| < 1 \,\big\}.$$

The collection of annuli

$$K_\nu := \left\{ \boldsymbol{x} \in \mathbb{R}^n; \;\; \frac{1}{\nu} \leq \|\boldsymbol{x}\| \leq 1 - \frac{1}{\nu} \right\}, \;\; \nu \in \mathbb{N}$$

is a Jordan measurable compact exhaustion of U. Using (15.56) we deduce that

$$\int_{K_\nu} p_\alpha(\boldsymbol{x})|d\boldsymbol{x}| = n\boldsymbol{\omega}_n \int_{\frac{1}{\nu}}^{1-\frac{1}{\nu}} \rho^{n-1-\alpha} d\rho = n\boldsymbol{\omega}_n \times \begin{cases} \frac{1}{n-\alpha}\rho^{n-\alpha}\Big|_{1/\nu}^{1-1/\nu}, & \alpha \neq n, \\ \log \rho \Big|_{1/\nu}^{1-1/\nu}, & \alpha = n. \end{cases}$$

The last quantity has a finite limit as $\nu \to \infty$ if and only if $\alpha < n$, so the function $p_\alpha(\boldsymbol{x})$ is absolutely convergent on U if and only if $\alpha < n$.

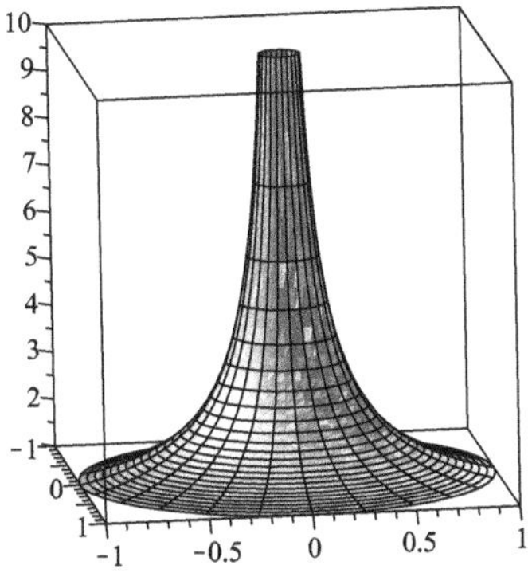

Fig. 15.14 *The graph of* $p_1(\boldsymbol{x}) = \|\boldsymbol{x}\|^{-1}$, $\boldsymbol{x} \in \mathbb{R}^2 \setminus \{\mathbf{0}\}$.

Fix $\beta > 0$. Consider now the complement in $\mathbb{R}^n$ of the closed unit ball,

$$V := \big\{\, \boldsymbol{x} \in \mathbb{R}^n; \;\; \|\boldsymbol{x}\| > 1 \,\big\}.$$

The collection of annuli

$$C_\nu := \left\{ \boldsymbol{x} \in \mathbb{R}^n;\ 1 + \frac{1}{\nu} \leq \|\boldsymbol{x}\| \leq \nu \right\},\ \ \nu \in \mathbb{N}$$

is a Jordan measurable compact exhaustion of V. We deduce as above that

$$\int_{C_\nu} p_\beta(\boldsymbol{x})|d\boldsymbol{x}| = n\boldsymbol{\omega}_n \times \begin{cases} \frac{1}{n-\beta}\rho^{n-\beta}\Big|_{1+1/\nu}^{\nu}, & \beta \neq n, \\ \\ \log\rho\Big|_{1+1/\nu}^{\nu}, & \beta = n. \end{cases}$$

The last quantity has a finite limit as $\nu \to \infty$ if and only if $\beta > n$, so the function $p_\beta(\boldsymbol{x})$ is absolutely convergent on V if and only if $\beta > n$. □

Example 15.70 (Gaussian integrals). Consider the function

$$f : \mathbb{R}^2 \to \mathbb{R},\ \ f(x,y) = e^{-x^2-y^2}.$$

It is locally integrable since it is continuous. To investigate if it is absolutely integrable consider the disks

$$D_\nu := \left\{ (x,y) \in \mathbb{R}^2;\ \sqrt{x^2+y^2} \leq \nu \right\},\ \ \nu \in \mathbb{N}.$$

Using polar coordinates $x = r\cos\theta$, $y = r\sin\theta$ we deduce

$$\begin{aligned}\int_{D_\nu} f(x,y)|dxdy| &= \int_0^{2\pi}\left(\int_0^\nu e^{-r^2} r dr\right) d\theta \\ &= 2\pi\int_0^\nu e^{-r^2} r dr = \pi\int_0^\nu e^{-r^2} d(r^2) \stackrel{u=r^2}{=} \pi\int_0^{\nu^2} e^{-u} du \\ &= \pi\left(1 - e^{-\nu^2}\right).\end{aligned}$$

Since $(D_\nu)_{\nu\in\mathbb{N}}$ is a Jordan measurable compact exhaustion of $\mathbb{R}^2$ we deduce that

$$\int_{\mathbb{R}^2} e^{-x^2-y^2} dxdy = \lim_{\nu\to\infty}\int_{D_\nu} e^{-x^2-y^2} dxdy = \pi.$$

On the other hand, if we set $S_\nu = [-\nu,\nu] \times [-\nu,\nu]$ we deduce

$$\begin{aligned}\pi &= \int_{\mathbb{R}^2} e^{-x^2-y^2}|dxdy| = \lim_{\nu\to\infty}\int_{S_\nu} e^{-x^2-y^2}|dxdy| \\ &= \lim_{\nu\to\infty}\int_{-\nu}^{\nu}\left(\int_{-\nu}^{\nu} e^{-x^2}e^{-y^2} dx\right) dy = \lim_{\nu\to\infty}\left(\int_{-\nu}^{\nu} e^{-x^2} dx\right)\left(\int_{\nu}^{\nu} e^{-y^2} dy\right) \\ &= \lim_{\nu\to\infty}\left(\int_{-\nu}^{\nu} e^{-x^2} dx\right)^2 = \left(\int_{\mathbb{R}} e^{-x^2} dx\right)^2.\end{aligned}$$

We have thus obtained the following famous result

$$\boxed{\int_{\mathbb{R}} e^{-x^2} dx = \sqrt{\pi}}. \tag{15.65}$$

□

Example 15.71 (The volume of the unit n-dimensional ball). We want to have another look at $\boldsymbol{\omega}_n$, the volume of the unit ball in $\mathbb{R}^n$. We will obtain a new description using an elegant trick of H. Weyl that is based on computing the integral

$$\boldsymbol{I}_n := \int_{\mathbb{R}^n} e^{-\|\boldsymbol{x}\|^2}\,|d\boldsymbol{x}|$$

in two different ways. Note first that

$$\boldsymbol{I}_n = \int_{\mathbb{R}^n} e^{-x_1^2-\cdots-x_n^2}|dx_1\cdots dx_n| = \int_{\mathbb{R}^n} e^{-x_1^2}\cdots e^{-x_n^2}|dx_1\cdots dx_n|$$

(use Fubini)

$$= \Big(\int_{\mathbb{R}} e^{-x_1^2}dx_1\Big)\cdots\Big(\int_{\mathbb{R}} e^{-x_n^2}dx_n\Big) = \Big(\int_{\mathbb{R}} e^{-x^2}dx\Big)^n \overset{(15.65)}{=} \pi^{\frac{n}{2}}.$$

On the other hand, the function $e^{-\|\boldsymbol{x}\|^2}$ is radially symmetric and we deduce from (15.56) that

$$\boldsymbol{I}_n = n\boldsymbol{\omega}_n \int_0^\infty e^{-\rho^2}\rho^{n-1}d\rho \overset{\rho=\sqrt{t}}{=} \frac{n\boldsymbol{\omega}_n}{2}e^{-t}t^{\frac{n}{2}-1}dt.$$

At this point we want to recall the definition of the Gamma function (9.70)

$$\Gamma(x) := \int_0^\infty e^{-t}t^{x-1}dt,\ \ x>0.$$

Thus

$$\pi^{\frac{n}{2}} = \boldsymbol{I}_n = \frac{n\boldsymbol{\omega}_n}{2}\Gamma\Big(\frac{n}{2}\Big).$$

Using the identity (9.72)

$$\Gamma(x+1) = x\Gamma(x),\ \ \forall x>0,$$

we deduce

$$\frac{n}{2}\Gamma\big(\frac{n}{2}\big) = \Gamma\big(\frac{n}{2}+1\big),$$

$$\boldsymbol{\omega}_n = \frac{\pi^{\frac{n}{2}}}{\Gamma\big(\frac{n}{2}+1\big)}. \tag{15.66}$$

To see how this relates to (15.54) we use again the identity (9.72) and we deduce

$$\Gamma(m) = m!\ \ \forall m\in\mathbb{N}$$

$$\Gamma\Big(\frac{2n+1}{2}\Big) = \frac{2n-1}{2}\Gamma\Big(\frac{2n-1}{2}\Big) = \cdots = \frac{(2n-1)!!}{2^{n-1}}\Gamma(1/2).$$

On the other hand

$$\Gamma(1/2) = \int_0^\infty e^{-t}t^{-1/2}dt \overset{t=x^2}{=} 2\int_0^\infty e^{-x^2}dx \overset{(15.65)}{=} \sqrt{\pi}.$$

□

Example 15.72 (Euler's Beta function). For $x, y > 0$ we set

$$\boxed{B(x,y) := \int_0^1 t^{x-1}(1-t)^{y-1}dt}. \tag{15.67}$$

This integral is convergent since $x-1, y-1 > -1$. The resulting function

$$(0,\infty)\times(0,\infty) \ni (x,y) \mapsto B(x,y) \in (0,\infty)$$

is known as *Euler's Beta function.*

If in the integral (15.67) we make the change in variables

$$u = \frac{t}{1-t},$$

then we observe that $u = 0$ when $t = 0$ and $u \to \infty$ as $t \nearrow 1$. Moreover, we have

$$(1-t)u = t \Rightarrow u = t(1+u) \Rightarrow t = \frac{u}{1+u} = 1 - \frac{1}{1+u}$$

$$\Rightarrow 1-t = \frac{1}{1+u}, \quad dt = \frac{1}{(1+u)^2}du,$$

$$t^{x-1}(1-t)^{y-1}dt = \left(\frac{u}{1+u}\right)^{x-1}\left(\frac{1}{1+u}\right)^{y-1}\frac{1}{(1+u)^2}du = \frac{u^{x-1}}{(1+u)^{x+y}}du,$$

so that

$$\boxed{B(x,y) = \int_0^\infty \frac{u^{x-1}}{(1+u)^{x+y}}du, \ \ \forall x,y > 0.} \tag{15.68}$$

Fix $\lambda > 0$. In the definition of Gamma function,

$$\Gamma(x) = \int_0^\infty t^{x-1}e^{-t}dt,$$

we make the change of variables $t = \lambda s$ and we deduce

$$\Gamma(x) = \int_0^\infty \lambda^{x-1}s^{x-1}e^{-\lambda s}\lambda ds = \lambda^x \int_0^\infty s^{x-1}e^{-\lambda s}ds.$$

Hence

$$\frac{1}{\lambda^x} = \frac{1}{\Gamma(x)}\int_0^\infty s^{x-1}e^{-\lambda s}ds, \ \ \forall x,\lambda > 0. \tag{15.69}$$

Using (15.69) we deduce

$$\frac{1}{(1+u)^{x+y}} = \frac{1}{\Gamma(x+y)}\int_0^\infty s^{x+y-1}e^{-(1+u)s}ds.$$

Using this in (15.68) we deduce

$$\begin{aligned} B(x,y) &= \frac{1}{\Gamma(x+y)}\int_0^\infty u^{x-1}\left(\int_0^\infty s^{x+y-1}e^{-(1+u)s}ds\right)du \\ &= \frac{1}{\Gamma(x+y)}\underbrace{\int_0^\infty\left(\int_0^\infty u^{x-1}s^{x+y-1}e^{-(1+u)s}ds\right)du}_{=:I}. \end{aligned} \tag{15.70}$$

At this point we want to invoke Fubini's theorem which allows us to conclude that we can interchange the order of integration so that

$$
\begin{aligned}
\boldsymbol{I} :&= \int_0^\infty \left(\int_0^\infty u^{x-1} s^{x+y-1} e^{-(1+u)s} ds \right) du \\
&= \int_0^\infty \left(\int_0^\infty s^{x+y-1} u^{x-1} e^{-(1+u)s} du \right) ds \\
&= \int_0^\infty s^{x+y-1} \left(\int_0^\infty u^{x-1} e^{-(1+u)s} du \right) ds \\
&= \int_0^\infty s^{x+y-1} \left(\int_0^\infty u^{x-1} e^{-s} e^{-su} du \right) ds \\
&= \int_0^\infty e^{-s} s^{x+y-1} \underbrace{\left(\int_0^\infty u^{x-1} e^{-su} du \right)}_{\overset{(15.69)}{=} \frac{\Gamma(x)}{s^x}} ds \\
&= \int_0^\infty e^{-s} s^{x+y-1} \frac{\Gamma(x)}{s^x} ds = \Gamma(x) \int_0^\infty s^{y-1} e^{-s} ds = \Gamma(x)\Gamma(y).
\end{aligned}
$$

Thus

$$\boldsymbol{I} = \Gamma(x)\Gamma(y).$$

Using the last equality in (15.70) we conclude that

$$\boxed{B(x,y) = \frac{\Gamma(x)\Gamma(y)}{\Gamma(x+y)}, \quad \forall x, y > 0}. \tag{15.71}$$

For example, the above equality shows that

$$\int_0^1 x^5(1-x)^5 = \frac{5!5!}{11!} = \frac{5!}{11 \cdot 10 \cdot 9 \cdot 8 \cdot 6} = \frac{1}{11 \cdot 9 \cdot 4} = \frac{1}{396}.$$

Even in this very simple case, it takes a bit of work proving this using the obvious method. □

Remark 15.73. The Gamma and Beta functions are ubiquitous in mathematics. They belong to the category of so-called *special functions*. The facts we have presented barely scratch the surface of the beautiful theory of these functions. There are many sources from which to learn about the Gamma function but, as entry point in this theory, no source comes close to the little gem [1] by Emil Artin.[9] First published 1931 in German, it remains to the day an example of beautiful mathematical writing. □

[9] Emil Artin (1898–1962) was one of the most influential mathematicians of the 20th century. He was a Professor at the University of Hamburg Germany until 1937 when he was forced to emigrate to US due to the political tensions in Germany at that time. His first US position was at the University of Notre Dame where, coincidently, the present book was also born.

15.5. Exercises

Exercise 15.1. Prove Lemma 15.2.

Hint. The proof is not hard, but it takes some thinking to write down a short and precise exposition. Start with the case $n = 2$ and then argue by induction on n. □

Exercise 15.2. Consider the triangle

$$T := \{ (x, y) \in \mathbb{R}^2;\ x, y \geq 0,\ x + y \leq 1 \}$$

and the function

$$f : [0,1] \times [0,1] \to \mathbb{R},\ f(x,y) = \begin{cases} 1, & (x,y) \in T, \\ 0, & (x,y) \notin T. \end{cases}$$

For each $n \in \mathbb{N}$ denote by $\boldsymbol{P}_n$ the partition of $[0,1] \times [0,1]$,

$$\boldsymbol{P}_n = (\boldsymbol{U}_n^x, \boldsymbol{U}_n^y),$$

where $\boldsymbol{U}_n^x$ is the uniform partition of order n of the interval $[0,1]$ on the x-axis (see Example 9.2) and $\boldsymbol{U}_n^y$ is the uniform partition of order n of the interval $[0,1]$ on the y-axis.

Compute $\omega(f, \boldsymbol{P}_n)$ and then show that

$$\lim_{n\to\infty} \omega(f, \boldsymbol{P}_n) = 0.$$

Hint. To understand what is going on investigate first the partition $\boldsymbol{P}_4$ depicted in Figure 15.15. □

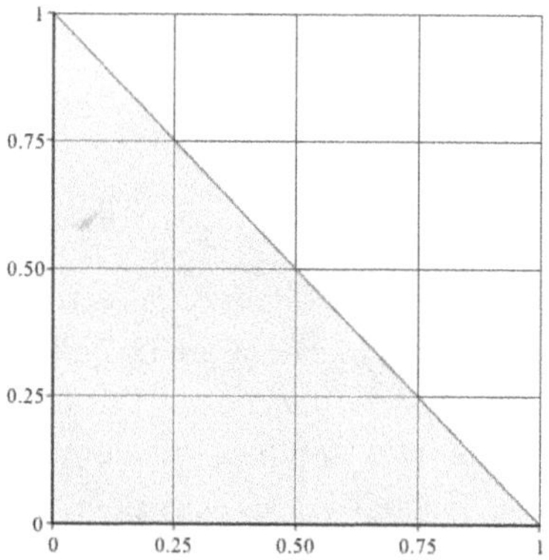

Fig. 15.15 *The partition $\boldsymbol{P}_4$ of the square $[0,1] \times [0,1]$. The triangle T is described by the shaded area.*

Exercise 15.3. Let $f, g : [0,1] \to \mathbb{R}$ be Riemann integrable functions. Define

$$s, p : [0,1] \times [0,1] \to \mathbb{R},\ s(x,y) = f(x) + g(y),\ p(x,y) = f(x)g(y),\ \forall x, y \in [0,1].$$

Prove that the functions s and p are also Riemann integrable. □

Exercise 15.4. Let $n \in \mathbb{N}$ and suppose that $B := [a_1, b_1] \times \cdots\cdots \times [a_n, b_n] \subset \mathbb{R}^n$ is a closed nondegenerate box. For any Riemann integrable function $f : B \to \mathbb{R}$ we define its mean on B to be the real number

$$\operatorname{Mean}(f) := \frac{1}{\operatorname{vol}_n(B)} \int_B f(\boldsymbol{x})|d\boldsymbol{x}|.$$

(i) Prove that for any Riemann integrable function $f : B \to \mathbb{R}$ we have

$$\inf_{\boldsymbol{x} \in B} f(\boldsymbol{x}) \leq \operatorname{Mean}(f) \leq \sup_{\boldsymbol{x} \in B} f(\boldsymbol{x}).$$

(ii) Prove that for any continuous function $f : B \to \mathbb{R}$ there exists $\bar{\boldsymbol{x}} \in B$ such that

$$f(\bar{\boldsymbol{x}}) = \operatorname{Mean}(f).$$

Hint. (ii) Use (i) and Corollary 12.49. □

Exercise 15.5. Denote by $\overline{C_r^n}$ the closed cube in $\mathbb{R}^n$ of radius r centered at $\mathbf{0}$,

$$\overline{C_r^n} = \{ \boldsymbol{x} \in \mathbb{R}^n;\ |x^i| \leq r,\ \forall i = 1, \dots, n \}.$$

Suppose that $f : \overline{C_1^n} \to \mathbb{R}$ is a continuous function. Prove that

$$\lim_{r \searrow 0} \frac{1}{\operatorname{vol}_n\left(\overline{C_r^n}\right)} \int_{\overline{C_r^n}} f(\boldsymbol{x})|d\boldsymbol{x}| = f(\mathbf{0}).$$

Hint. Use Exercise 15.4(ii). □

Exercise 15.6. Suppose that $B \subset \mathbb{R}^n$ is a nondegenerate closed box and $f, g : B \to \mathbb{R}$ are Riemann integrable functions. Fix $p, q \in (1, \infty)$ such that

$$\frac{1}{p} + \frac{1}{q} = 1.$$

(i) Prove that there exists $C > 0$ such that, for any partition $\boldsymbol{P}$ of B, we have

$$\omega(|f|^p, \boldsymbol{P}) < C\omega(f, \boldsymbol{P}),\ \ \omega(|g|^q, \boldsymbol{P}) < C\omega(g, \boldsymbol{P}).$$

(ii) Prove that there exists a sequence of partitions $(\mathbb{P}_\nu)_{\nu \in \mathbb{N}}$ of B such that

$$\lim_{\nu \to \infty} \omega(f, \mathbb{P}_\nu) = \lim_{\nu \to \infty} \omega(g, \mathbb{P}_\nu) = 0.$$

(iii) Prove that

$$\left| \int_B f(\boldsymbol{x})g(\boldsymbol{x})|d\boldsymbol{x}| \right| \leq \left(\int_B |f(\boldsymbol{x})|^p |d\boldsymbol{x}| \right)^{\frac{1}{p}} \left(\int_B |g(\boldsymbol{x})|^q |d\boldsymbol{x}| \right)^{\frac{1}{q}}. \tag{15.72}$$

Hint. (i) Have a look at the proof of (15.8). (iii) Use Proposition 15.9 coupled with (i) and (ii) to reduce (15.72) to (8.23). □

Exercise 15.7. Fix $n \in \mathbb{N}$

(i) Show that a subset $S \subset \mathbb{R}^n$ is negligible (see Definition 15.15) if and only if, for any $\varepsilon > 0$, there exists a sequence $(B_\nu)_{\nu \geq 1}$ of *closed* boxes such that

$$S \subset \bigcup_{\nu \geq 1} \boldsymbol{int}(B_\nu) \ \text{ and } \sum_{\nu \geq 1} \operatorname{vol}_n(B_\nu) < \varepsilon.$$

(ii) Show that if the *compact* subset $S \subset \mathbb{R}^n$ is negligible, then for any $\varepsilon > 0$ there exists *finitely many* closed boxes $B_1, \dots, B_N$ such that

$$S \subset \bigcup_{\nu=1}^{N} \boldsymbol{int}(B_\nu) \text{ and } \sum_{\nu=1}^{N} \operatorname{vol}_n(B_\nu) < \varepsilon.$$

Hint. (i) Observe that for any closed box B and any $\hbar > 0$ there exists a closed box B' such that $B \subset \boldsymbol{int}(B')$ and $\operatorname{vol}_n(B') - \operatorname{vol}_n(B) < \hbar$. □

Exercise 15.8. Prove that the following sets are negligible.

(i) A subset of a negligible subset.
(ii) The union of a sequence $(\mathcal{N}_\nu)_{\nu \geq 1}$ of negligible subsets of $\mathbb{R}^n$.
(iii) The coordinate hyperplane

$$H_t^i := \{ (x^1, x^2, \dots, x^n) \in \mathbb{R}^n \;\; x^i = t \} \subset \mathbb{R}^n,$$

where $i \in \{1, \dots, n\}$ and t is a fixed real number.

Hint. (ii) For $\nu = 1, 2, \dots$ cover $\mathcal{N}_\nu$ by a countable family of boxes $B_{\nu,1}, B_{\nu,2}, \dots$, such that

$$\sum_{k=1}^{\infty} \operatorname{vol}_n(B_{\nu,k}) < \frac{\varepsilon}{2^\nu}.$$

Then use the fact that the set $\mathbb{N} \times \mathbb{N}$ is countable; see Example 3.17. (iii) Use (ii). □

Exercise 15.9. Suppose that $f : [a, b] \to [0, \infty)$ and $g : [c, d] \to [0, \infty)$ are two nonnegative Riemann integrable functions. Define

$$h : [a, b] \times [c, d] \to \mathbb{R}, \;\; h(x, y) = f(x)g(y).$$

Prove that h is Riemann integrable and

$$\int_{[a,b]\times[c,d]} h(x, y)|dxdy| = \left(\int_a^b f(x)dx \right) \left(\int_c^d g(y)dy \right).$$

Hint. Choose a partition $\boldsymbol{P} = (\boldsymbol{P}_x, \boldsymbol{P}_y)$ of the box $[a, b] \times [c, d]$ and express the Darboux sum $\boldsymbol{S}_*(h, \boldsymbol{P})$ in terms of $\boldsymbol{S}_*(f, \boldsymbol{P}_x)$ and $\boldsymbol{S}_*(g, \boldsymbol{P}_y)$. Do the same for the upper Darboux sums. □

Exercise 15.10. Let $B = [0, 1] \times [1, 2] \subset \mathbb{R}^2$. Using Theorem 15.18 compute

$$\int_B \frac{1}{(x^1 + x^2)^2} |dx^1 dx^2|.$$ □

Exercise 15.11. Consider a nondegenerate box $B \subset \mathbb{R}^n$, a continuous function $f : B \to \mathbb{R}$ and a continuous convex function $\Phi : \mathbb{R} \to \mathbb{R}$. Prove that

$$\Phi\left(\frac{1}{\operatorname{vol}_n(B)} \int_B f(\boldsymbol{x}) \, |d\boldsymbol{x}| \right) \leq \frac{1}{\operatorname{vol}_n(B)} \int_B \Phi\big(f(\boldsymbol{x}) \big) |d\boldsymbol{x}|.$$

Hint. Use Riemann sums and Jensen's inequality (8.16). □

Exercise 15.12. (a) Let $a, b \in \mathbb{R}$, $a < b$. For any $\varepsilon > 0$ construct *explicitly* a continuous function $g_\varepsilon : [a, b] \to \mathbb{R}$ satisfying the following properties.

(i) $g_\varepsilon(a) = g_\varepsilon(b) = 0$.
(ii) $0 \leq g_\varepsilon(x) \leq 1, \forall x \in [a,b]$.
(iii)

$$\int_a^b dx - \varepsilon \leq \int_a^b g_\varepsilon(x)dx \leq \int_a^b dx.$$

(b) Let $n \in \mathbb{N}$ and suppose that $B = [a_1, b_1] \times \cdots \times [a_n, b_n] \subset \mathbb{R}^n$ is a closed nondegenerate box. Prove that, for any $\varepsilon > 0$ there exists a continuous function $h_\varepsilon : B \to \mathbb{R}$ satisfying the following conditions.

(i) $h_\varepsilon(\boldsymbol{x}) = 0$ for any point $\boldsymbol{x}$ the boundary of B, i.e., a point $\boldsymbol{x} = (x^1, \dots, x^n)$ such that $x^i = a_i$ or $x^i = b_i$ for some $i = 1, \dots, n$.
(ii) $0 \leq h_\varepsilon(\boldsymbol{x}) \leq 1, \forall \boldsymbol{x} \in B$.
(iii)

$$\int_B |d\boldsymbol{x}| - \varepsilon \leq \int_B h_\varepsilon(\boldsymbol{x})|d\boldsymbol{x}| \leq \int_B |d\boldsymbol{x}|.$$

Hint. (a) Think of a function whose graph looks like a trapezoid. (b) Seek h_ε of the form

$$h_\varepsilon(x^1, \dots, x^n) = g_\varepsilon^1(x^1) \cdots g_\varepsilon^n(x^n),$$

where $g_\varepsilon : [a_i, b_i] \to \mathbb{R}$ are chosen as in (a). Use Fubini to reach the desired conclusion. □

Exercise 15.13. Let $n \in \mathbb{N}$ and suppose that $B = [a_1, b_1] \times \cdots \times [a_n, b_n] \subset \mathbb{R}^n$ is a closed nondegenerate box. Suppose that $f : B \to [0, \infty)$ is a *nonnegative* Riemann integrable function. Prove that, for any $\varepsilon > 0$ there exists a continuous function $h_\varepsilon : B \to \mathbb{R}$ satisfying the following conditions.

(i) $h_\varepsilon(\boldsymbol{x}) = 0$ for any point $\boldsymbol{x}$ on the boundary of B, i.e., a point $\boldsymbol{x} = (x^1, \dots, x^n)$ such that $x^i = a_i$ or $x_i = b_i$ for some $i = 1, \dots, n$.
(ii) $0 \leq h_\varepsilon(\boldsymbol{x}) \leq f(\boldsymbol{x}), \forall \boldsymbol{x} \in B$.
(iii)

$$\int_B f(\boldsymbol{x})|d\boldsymbol{x}| - \varepsilon \leq \int_B h_\varepsilon(\boldsymbol{x})|d\boldsymbol{x}| \leq \int_B f(\boldsymbol{x})|d\boldsymbol{x}|.$$

Hint. Choose a partition $\boldsymbol{P}$ of B such that the mean oscillation $\omega(f, \boldsymbol{P})$ is very small. Next use Exercise 15.12(b) on each chamber of the partition $\boldsymbol{P}$. □

Exercise 15.14. Let $n, N \in \mathbb{N}$ and suppose that $S_1, \dots, S_N \subset \mathbb{R}^n$ are Jordan measurable sets. Prove that their union is also Jordan measurable and

$$\text{vol}_n\left(\bigcup_{k=1}^N S_k\right) \leq \sum_{k=1}^N \text{vol}_n(S_k).$$

Hint. Argue by induction using Proposition 15.30. □

Exercise 15.15. (a) Suppose that $K \subset \mathbb{R}^n$ is compact. Prove that K is negligible if and only if it is Jordan measurable and $\mathrm{vol}_n(K) = 0$.

(b) Suppose that $B \subset \mathbb{R}^n$ is a *closed* nondegenerate box. Prove that $\boldsymbol{int}(B)$ is Jordan measurable and

$$\mathrm{vol}_n\big(\,\boldsymbol{int}(B)\,\big) = \mathrm{vol}_n(B).$$

Hint. (a) If K is negligible use Corollary 15.29 to prove that it is Jordan measurable. To show that $\mathrm{vol}_n(K) = 0$ use Exercises 15.7 and 15.14. Conversely, suppose that K is Jordan measurable and $\mathrm{vol}_n(K) = 0$. Choose a box B containing K. Then $\int_B I_K(\boldsymbol{x})|d\boldsymbol{x}| = 0$. Thus for $\varepsilon > 0$ one can choose a partition $\boldsymbol{P}_\varepsilon$ of B such that $S^*(I_K, \boldsymbol{P}_\varepsilon) < \varepsilon$. Use this partition to show that you can cover K by finitely many boxes whose volumes add up to less than ε. (b) Invoke Proposition 15.21 and Lebesgue's Theorem to conclude that ∂B is negligible and then use (a). □

Exercise 15.16. Suppose that $f : [a,b] \times [a,b] \to \mathbb{R}$ is a continuous function. Prove that

$$\int_a^b dy \int_a^y f(x,y)\,dx = \int_a^b dx \int_x^b f(x,y)\,dy.$$

Hint. Compute the double integral $\int_C f(x,y)dxdy$ in two different ways for a suitable Jordan measurable set $C \subset [a,b] \times [a,b]$. □

Exercise 15.17. Denote by T the triangle in the plane determined by the lines

$$y = 2x, \quad y = \frac{x}{2}, \quad x + y = 6.$$

(i) Find the coordinates of the vertices of this triangle.
(ii) Draw a picture of this triangle.
(iii) Compute the integral

$$\int_T xy\,|dxdy|.$$

□

Exercise 15.18. Find the area of the region $R \subset \mathbb{R}^2$ defined as the intersection of the disks of radius 1 centered at the vertices of the square $[0,1] \times [0,1]$; see Figure 15.16.

Exercise 15.19. Consider the box $B = [a_1, b_1] \times [a_2, b_2] \subset \mathbb{R}^2$ and suppose that $f : B \to \mathbb{R}$ is continuous function. Define

$$F : B \to \mathbb{R}, \quad F(x,y) = \int_{[a_1,x]\times[a_2,y]} f(s,t)|dsdt|.$$

Show that the restriction of F to the interior of B is C^1 and then compute the partial derivatives $\frac{\partial F}{\partial x}, \frac{\partial F}{\partial y}$. □

Exercise 15.20. Suppose that $B \subset \mathbb{R}^n$ is a closed nondegenerate box and $f : B \to \mathbb{R}$ is a continuous function such that $f(\boldsymbol{x}) \geq 0$, $\forall \boldsymbol{x} \in B$. Prove that the following statements are equivalent.

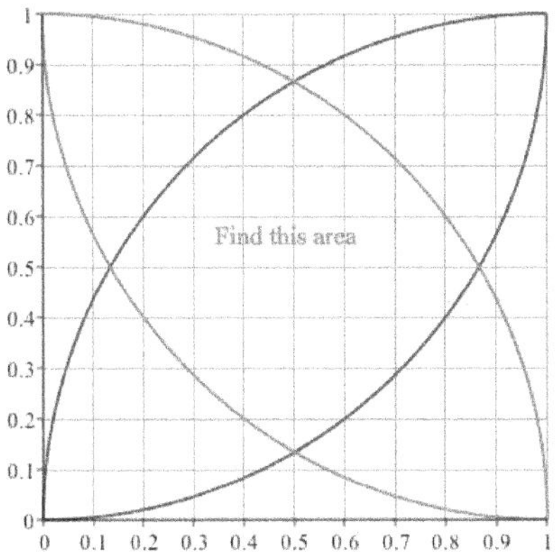

Fig. 15.16 *The overlap of 4 disks centered at the vertices of a square.*

(i) $f(\boldsymbol{x}) = 0, \forall \boldsymbol{x} \in B$.
(ii)

$$\int_B f(\boldsymbol{x})\,|d\boldsymbol{x}| = 0.$$

□

Exercise 15.21. Let $n \in \mathbb{N}$ and $r > 0$.

(i) Prove that the n-dimensional closed ball

$$\overline{B_r^n(\mathbf{0})} := \left\{\, \boldsymbol{x} \in \mathbb{R}^n;\ \ \|\boldsymbol{x}\| \leq r \,\right\}$$

is Jordan measurable.

(ii) Prove that the sphere

$$\Sigma_r(\mathbf{0}) := \left\{\, \boldsymbol{x} \in \mathbb{R}^n;\ \ \|\boldsymbol{x}\| = r \,\right\}$$

is Jordan measurable and $\operatorname{vol}_n(\Sigma_r(\mathbf{0})) = 0$.

(iii) Prove that the n-dimensional open ball

$$B_r^n(\mathbf{0}) := \left\{\, \boldsymbol{x} \in \mathbb{R}^n;\ \ \|\boldsymbol{x}\| < r \,\right\}$$

is Jordan measurable and

$$\operatorname{vol}_n\left(\, B_r^n(\mathbf{0}) \,\right) = \operatorname{vol}_n\left(\, \overline{B_r^n(\mathbf{0})} \,\right).$$

Hint. (i) Use induction on n and Proposition 15.40. (ii) Use Corollary 15.29. Conclude using Exercise 15.15 and Proposition 15.30(i). (iii) Follows from (i) and (ii). □

Exercise 15.22. For any $n \in \mathbb{N}$ and $r > 0$ denote by $\boldsymbol{\omega}_n(r)$ the n-dimensional volume of the n-dimensional ball $B_r^n(\mathbf{0}) \subset \mathbb{R}^n$. For simplicity, set $\boldsymbol{\omega}_n := \boldsymbol{\omega}_n(1)$, so that $\boldsymbol{\omega}_n$ is the volume of the unit ball in $\mathbb{R}^n$.

(i) Show that $\boldsymbol{\omega}_n(r) = \boldsymbol{\omega}_n r^n$.
(ii) Use Cavalieri's Principle to show that

$$\boldsymbol{\omega}_n := 2\boldsymbol{\omega}_{n-1} \int_0^1 \left(1 - x^2\right)^{\frac{n-1}{2}} dx.$$

(iii) Prove that $\boldsymbol{\omega}_n$ satisfies the equalities (15.54).

Hint. (i) Use the change-in-variables formula for the diffeomorphism $\Phi : \mathbb{R}^n \to \mathbb{R}^n$, $\Phi(\boldsymbol{x}) = r\boldsymbol{x}$ that sends $B_1(\mathbf{0})$ onto $B_r(\mathbf{0})$. (iii) Relate the integral in (ii) to the integrals (9.44) and (9.45). Then proceed by induction on n. □

Exercise 15.23. Prove that Lebesgue's Theorem 15.17 implies Corollary 15.38. □

Exercise 15.24. Fix a continuous function $f : \mathbb{R} \to \mathbb{R}$ and define recursively the sequence of functions $F_n : \mathbb{R} \to \mathbb{R}$, $n = 0, 1, 2 \ldots,$

$$F_n(x) = \frac{1}{n!} \int_0^x (x - y)^n f(y) dy.$$

(i) Show that, for all $n \in \mathbb{N}$, $F_n \in C^1(\mathbb{R})$, $F_n(0) = 0$ and $F_n'(x) = F_{n-1}(x)$, $\forall x \in \mathbb{R}$.
(ii) Show that if $(G_n)_{n \geq 1}$ is a sequence of C^1-functions on $\mathbb{R}$ such that

$$G_n(0) = 0, \ \ \forall n \geq 1, \ \ G_1'(x) = f(x),$$

$$G_{n+1}'(x) = G_n(x), \ \ \forall n \geq 1, \ \ x \in \mathbb{R},$$

then $G_n(x) = F_n(x)$, $\forall n \geq 1$, $x \in \mathbb{R}$.
(iii) Show

$$\int_0^x dx^1 \int_0^{x^1} dx^2 \cdots \int_0^{x^{n-1}} f(x^n) dx^n = \frac{1}{n!} \int_0^x (x - y)^n f(y) dy.$$

□

Exercise 15.25. Suppose that $K \subset \mathbb{R}^2$ is a compact Jordan measurable set that is symmetric with respect to the y-axis, i.e.,

$$(x, y) \in K \Longleftrightarrow (-x, y) \in K.$$

Let $f : \mathbb{R}^2 \to \mathbb{R}$ be a continuous function that is odd in the x-variable, i.e.,

$$f(-x, y) = -f(x, y), \ \ \forall (x, y) \in K.$$

(i) Prove that

$$\int_K f(x, y) |dxdy| = 0.$$

(ii) Let

$$D := \left\{ (x, y) \in \mathbb{R}^2; \ \ x^2 + y^2 \leq 1 \right\}.$$

Compute

$$\int_D x^{23} y^{24} |dxdy| \ \text{ and } \ \int_D (xy)^{24} |dxdy|.$$

Hint. (i) Use the change-in-variables formula. For (ii) you need to use the equalities (9.47) in Example 9.50. □

Exercise 15.26. Let $n \in \mathbb{N}$ and suppose that $K \subset \mathbb{R}^n$ is a compact, Jordan measurable set.

(i) Suppose that $L : \mathbb{R}^n \to \mathbb{R}^n$ is a linear isomorphism. Prove that

$$\mathrm{vol}_n\big(L(K)\big) = |\det L|\, \mathrm{vol}_n(K).$$

(ii) Suppose that $A : \mathbb{R}^n \to \mathbb{R}^n$ is a linear isometry, i.e., A is a linear operator and

$$\langle A\boldsymbol{x}, A\boldsymbol{y}\rangle = \langle \boldsymbol{x}, \boldsymbol{y}\rangle, \;\; \forall \boldsymbol{x}, \boldsymbol{y} \in \mathbb{R}^n.$$

Show that

$$\mathrm{vol}_n\big(A(K)\big) = \mathrm{vol}_n(K).$$

(iii) Let $\boldsymbol{v} \in \mathbb{R}^n$ and define $T_{\boldsymbol{v}} : \mathbb{R}^n \to \mathbb{R}^n$, $T_{\boldsymbol{v}}(\boldsymbol{x}) = \boldsymbol{v} + \boldsymbol{x}$. Show that

$$\mathrm{vol}_n\big(T_{\boldsymbol{v}}(K)\big) = \mathrm{vol}_n(K).$$

Hint. (ii) Use Exercise 11.25 and (i). □

Exercise 15.27. Let $n \in \mathbb{N}$ and suppose that $a_1, a_2, \dots, a_n > 0$. Consider the ellipsoid

$$\Sigma(a_1, \dots, a_n) := \Big\{ \boldsymbol{x} \in \mathbb{R}^n;\; \sum_{j=1}^n \frac{(x^j)^2}{a_j^2} \leq 1 \Big\}.$$

(i) Prove that

$$\mathrm{vol}_n\big(\Sigma(a_1, \dots, a_n)\big) = \boldsymbol{\omega}_n a_1 \cdots a_n,$$

where $\boldsymbol{\omega}_n$ is the volume of the unit ball in $\mathbb{R}^n$.

(ii) Show that

$$\int_{\Sigma(a_1,\dots,a_n)} x^1 x^2 \cdots x^n \, |dx^1 \cdots dx^n| = 0.$$

Hint. (i) Make the change in variables $x^i = a_i y^i$. (ii) Make the change in variables $x^1 = -y^1, x^2 = y^2, \dots, x^n = y^n$. □

Exercise 15.28. Let $m, n \in \mathbb{R}$ and suppose that $\rho : \mathbb{R}^m \times \mathbb{R}^n \to \mathbb{R}$ is a continuous function. We denote by $\boldsymbol{t} = (t^1, \dots, t^m)$ the Euclidean coordinates in $\mathbb{R}^m$ and by $\boldsymbol{x} = (x^1, \dots, x^n$ the Euclidean coordinates on $\mathbb{R}^n$. Fix a Riemann integrable function $f : \mathbb{R}^n \to \mathbb{R}$. (Note in particular that f must have compact support.) Define

$$\hat{f} : \mathbb{R}^m \to \mathbb{R}, \;\; \hat{f}(\boldsymbol{t}) = \int_{\mathbb{R}^n} \rho(\boldsymbol{t}, \boldsymbol{x}) f(\boldsymbol{x}) |d\boldsymbol{x}|.$$

(i) Show that $\hat{f}$ is continuous.

(ii) Suppose additionally that ρ is C^1. Show that $\hat{f}$ is also C^1 and

$$\frac{\partial \hat{f}}{\partial t^k}(\boldsymbol{t}) = \int_{\mathbb{R}^n} \frac{\partial \rho}{\partial t^k}(\boldsymbol{t}, \boldsymbol{x}) f(\boldsymbol{x}) |d\boldsymbol{x}|.$$

Hint. Fix (closed) box B such that $\operatorname{supp} f \subset B$. (i) Prove that if $\boldsymbol{t}_\nu \to \boldsymbol{t}$ as $\nu \to \infty$, then the functions $\boldsymbol{x} \mapsto \rho(\boldsymbol{t}_\nu, \boldsymbol{x})f(\boldsymbol{x})$ converge uniformly on B to the function $\boldsymbol{x} \mapsto \rho(\boldsymbol{t}, \boldsymbol{x})f(\boldsymbol{x})$. Conclude using Proposition 15.14. (ii) Use Lagrange's Mean Value Theorem 13.30 and the fact that the partial derivatives of ρ are bounded on the compact subsets of $\mathbb{R}^m \times \mathbb{R}^n$. □

Exercise 15.29. Suppose that $\rho : \mathbb{R}^n \to \mathbb{R}$ is a *nonnegative, compactly supported* continuous function such that

$$\int_{\mathbb{R}^n} \rho(\boldsymbol{x})\,|d\boldsymbol{x}| = 1. \tag{15.73}$$

Given a Riemann integrable function f and $\varepsilon > 0$ we define $f_\varepsilon : \mathbb{R}^n \to \mathbb{R}$,

$$f_\varepsilon(\boldsymbol{x}) = \varepsilon^{-n} \int_{\mathbb{R}^n} \rho\big(\varepsilon^{-1}(\boldsymbol{x} - \boldsymbol{y})\big) f(\boldsymbol{y})\,|d\boldsymbol{y}|.$$

(i) Prove that

$$f_\varepsilon(\boldsymbol{x}) = \varepsilon^{-n} \int_{\mathbb{R}^n} \rho\big(\varepsilon^{-1}\boldsymbol{z}\big) f(\boldsymbol{x} - \boldsymbol{z})\,|d\boldsymbol{z}| = \int_{\mathbb{R}^n} \rho(\boldsymbol{z}) f(\boldsymbol{x} - \varepsilon\boldsymbol{z})\,|d\boldsymbol{z}|.$$

(ii) Prove that the function f_ε is continuous and compactly supported.

(iii) Show that

$$\int_{\mathbb{R}^n} f_\varepsilon(\boldsymbol{x})\,|d\boldsymbol{x}| = \int_{\mathbb{R}^n} f(\boldsymbol{x})\,|d\boldsymbol{x}|.$$

(iv) Show that if, additionally, f is continuous, then

$$\lim_{\varepsilon \to 0} f_\varepsilon(\boldsymbol{x}) = f(\boldsymbol{x}), \quad \forall \boldsymbol{x} \in \mathbb{R}.$$

Hint. (i) Use the change in variables formula. (ii) Use Exercise 15.28(i). (iii) Use (i), (15.73) and Fubini. (iv) Use (15.73) and Proposition 15.14. □

Exercise 15.30. Show that the improper integral

$$\int_{0<x<y} (y^2 - x^2)e^{-y}\,dxdy$$

is absolutely convergent and then compute its value.

Hint. First draw the region $R := \{0 < x < y\}$. Note that in the region R the function $f(x, y) = (y^2 - x^2)e^{-y}$ is positive. Consider the compact exhaustion

$$K_\nu := \big\{\,(x, y) \in \mathbb{R}^2;\ 1/\nu \le x,\ y - x \ge 1/\nu,\ \ y \le \nu\,\big\},\ \ \nu \in \mathbb{N},$$

compute the integrals

$$I_\nu = \int_{K_\nu} (y^2 - x^2)e^{-y}|dxdy|,$$

and study the limit of I_ν as $\nu \to \infty$. □

Exercise 15.31. Fix $n \in \mathbb{N}$ and a continuous function $f : \mathbb{R} \to \mathbb{R}$. For $c > 0$ we denote by T_n^c the simplex

$$T_n^c := \big\{\,\boldsymbol{x} \in \mathbb{R}^n;\ x^1, \dots, x^n \ge 0,\ \ x^1 + \cdots + x^n \le c\,\big\}.$$

(i) Prove that

$$\operatorname{vol}_n(T_n^c) = \frac{c^n}{n!}.$$

(ii) Show that

$$\int_{T_n^c} f(x^1+\cdots+x^n)\,|d\boldsymbol{x}| = \frac{1}{(n-1)!}\int_0^c f(t)t^{n-1}dt.$$

(iii) Show that the integral

$$\int_{x^1,\dots,x^n\geq 0} \sin\big(\pi(x^1+\cdots+x^n)\big)\,|d\boldsymbol{x}|$$

is not absolutely convergent.

Hint. (i) Reduce to Example 15.45 via the change in variables $x^i = cy^i$, $i=1,\dots,n$. (ii) Make the change in variables $u^1=x^1,\dots,u^{n-1}=x^{n-1}, u^n = x^1+\cdots+x^n$ and then use Fubini coupled with (i). For (iii) use (ii). □

Exercise 15.32. Let $n\in\mathbb{N}$, $n>1$. Show that, for any $R>0$ the n-dimensional improper integral

$$\int_{0<\|\boldsymbol{x}\|\leq R} \ln\|\boldsymbol{x}\|\,|d\boldsymbol{x}|$$

is absolutely convergent and then compute its value. □

Exercise 15.33. Prove the equalities (15.51). □

Exercise 15.34. Prove that

$$\sum_{k=0}^n (-1)^k\frac{\binom{n}{k}}{n+k+1} = \frac{1}{(n+1)\binom{2n+1}{n}}.$$

Hint. Use (15.71). □

15.6. More challenging problems

Problem 15.1. (a) Consider a degree $(n-1)$ polynomial

$$P(x) = a_{n-1}x^{n-1}+a_{n-2}x^{n-2}+\cdots+a_1x+a_0,\quad a_{n-1}\neq 0.$$

Compute the determinant of the following matrix.

$$V = \begin{bmatrix} 1 & 1 & \cdots & 1\\ x_1 & x_2 & \cdots & x_n\\ \vdots & \vdots & \vdots & \vdots\\ x_1^{n-2} & x_2^{n-2} & \cdots & x_n^{n-2}\\ P(x_1) & P(x_2) & \cdots & P(x_n)\end{bmatrix}.$$

(b) Compute the determinants of the following $n\times n$ matrices

$$A = \begin{bmatrix} 1 & 1 & \cdots & 1\\ x_1 & x_2 & \cdots & x_n\\ \vdots & \vdots & \vdots & \vdots\\ x_1^{n-2} & x_2^{n-2} & \cdots & x_n^{n-2}\\ x_2x_3\cdots x_n & x_1x_3x_4\cdots x_n & \cdots & x_1x_2\cdots x_{n-1}\end{bmatrix},$$

and

$$B = \begin{bmatrix} 1 & 1 & \cdots & 1 \\ x_1 & x_2 & \cdots & x_n \\ \vdots & \vdots & \vdots & \vdots \\ x_1^{n-2} & x_2^{n-2} & \cdots & x_n^{n-2} \\ \left(\sum_{j\neq 1} x_j\right)^{n-1} & \left(\sum_{j\neq 2} x^j\right)^{n-1} & \cdots & \left(\sum_{j\neq n} x^j\right)^{n-1} \end{bmatrix}.$$ □

Problem 15.2. Suppose that $A = (a_{ij})_{1\leq i,j\leq n}$ is an $n \times n$ matrix with complex entries.
(a) Fix complex numbers $x_1, \dots, x_n$, $y_1, \dots, y_n$ and consider the $n \times n$ matrix B with entries

$$b_{ij} = x_i y_j a_{ij}.$$

Show that

$$\det B = (x_1 y_1 \cdots x_n y_n) \det A.$$

(b) Suppose that C is the $n \times n$ matrix with entries

$$c_{ij} = (-1)^{i+j} a_{ij}.$$

Show that $\det C = \det A$. □

Problem 15.3. (a) Suppose we are given three sequences of numbers $\underline{a} = (a_k)_{k\geq 1}$, $\underline{b} = (b_k)_{k\geq 1}$ and $\underline{c} = (c_k)_{k\geq 1}$. To these sequences we associate a sequence of tridiagonal matrices known as *Jacobi matrices*

$$J_n = \begin{bmatrix} a_1 & b_1 & 0 & 0 & \cdots & 0 & 0 \\ c_1 & a_2 & b_2 & 0 & \cdots & 0 & 0 \\ 0 & c_2 & a_3 & b_3 & \cdots & 0 & 0 \\ \vdots & \vdots & \vdots & \vdots & \vdots & \vdots & \vdots \\ 0 & 0 & 0 & 0 & \cdots & c_{n-1} & a_n \end{bmatrix}. \tag{J}$$

Show that

$$\det J_n = a_n \det J_{n-1} - b_{n-1} c_{n-1} \det J_{n-2}. \tag{15.74}$$

(b) Suppose that above we have

$$c_k = 1,\ \ b_k = 2,\ \ a_k = 3,\ \ \forall k \geq 1.$$

Compute J_1, J_2. Using (15.74) determine J_3, J_4, J_5, J_6, J_7. Can you detect a pattern? □

Problem 15.4. Suppose we are given a sequence of polynomials with complex coefficients $(P_n(x))_{n\geq 0}$, $\deg P_n = n$, for all $n \geq 0$,

$$P_n(x) = a_n x^n + \cdots,\ \ a_n \neq 0.$$

Denote by $\boldsymbol{V}_n$ the space of polynomials with complex coefficients and degree $\leq n$.
(a) Show that the collection $\{P_0(x), \dots, P_n(x)\}$ is a basis of $\boldsymbol{V}_n$.

(b) Show that for any $x_1, \ldots, x_n \in \mathbb{C}$ we have

$$\det \begin{bmatrix} P_0(x_1) & P_0(x_1) & \cdots & P_0(x_n) \\ P_1(x_1) & P_1(x_2) & \cdots & P_1(x_n) \\ \vdots & \vdots & \vdots & \vdots \\ P_{n-1}(x_1) & P_{n-1}(x_2) & \cdots & P_{n-1}(x_n) \end{bmatrix} = a_0 a_1 \cdots a_{n-1} \prod_{i<j} (x_j - x_i). \qquad \square$$

Problem 15.5. To any polynomial $P(x) = c_0 + c_1 x + \ldots + c_{n-1} x^{n-1}$ of degree $\leq n-1$ with complex coefficients we associate the $n \times n$ *circulant matrix*

$$C_P = \begin{bmatrix} c_0 & c_1 & c_2 & \cdots & c_{n-2} & c_{n-1} \\ c_{n-1} & c_0 & c_1 & \cdots & c_{n-3} & c_{n-2} \\ c_{n-2} & c_{n-1} & c_0 & \cdots & c_{n-4} & c_{n-3} \\ \cdots & \cdots & \cdots & \cdots & \cdots & \cdots \\ c_1 & c_2 & c_3 & \cdots & c_{n-1} & c_0 \end{bmatrix}.$$

Set

$$\rho := e^{\frac{2\pi \boldsymbol{i}}{n}}, \quad \boldsymbol{i} := \sqrt{-1},$$

so that $\rho^n = 1$. Consider the $n \times n$ Vandermonde matrix $V_\rho = V(1, \rho, \ldots, \rho^{n-1})$, where

$$V(x_1, \ldots, x_n) = \begin{bmatrix} 1 & 1 & \cdots & 1 \\ x_1 & x_2 & \cdots & x_n \\ \vdots & \vdots & \vdots & \vdots \\ x_1^{n-1} & x_2^{n-1} & \cdots & x_n^{n-1} \end{bmatrix}. \tag{15.75}$$

(a) Show that for any $j = 1, \ldots, n-1$ we have

$$1 + \rho^j + \rho^{2j} + \cdots + \rho^{(n-1)j} = 0.$$

(b) Show that

$$C_P \cdot V_\rho = V_\rho \cdot \operatorname{Diag}\big(P(1), P(\rho), \ldots, P(\rho^{n-1}) \big),$$

where $\operatorname{Diag}(a_1, \ldots, a_n)$ denotes the diagonal $n \times n$ matrix with diagonal entries $a_1, \ldots, a_n$.

(c) Show that

$$\det C_P = P(1) P(\rho) \cdots P(\rho^{n-1}). \qquad \square$$

(d)* Suppose that $P(x) = 1 + 2x + 3x^2 + 4x^3$ so that C_P is a 4×4 matrix with integer entries and thus $\det C_P$ is an integer. Find this *integer*. Can you generalize this computation?

Problem 15.6. Let $B_1(\mathbf{0})$ denote the unit ball in $\mathbb{R}^n$ centered at the origin. Compute

$$\int_{B_1(\mathbf{0})} (x^1 \cdots x^n)^2 |dx^1 \cdots dx^n|.$$

$\square$

Chapter 16

Integration over Submanifolds

We begin here the study of integration over "curved" regions of $\mathbb{R}^n$. This is a rather elaborate theory belonging properly to the area of mathematics called *differential geometry*. Time constraints prevent us from covering it in all its details and generality. Think of this as a first and low dimensional encounter with the subject.

16.1. Integration along curves

16.1.1. *Integration of functions along curves*

We start with a simpler situation. Fix $n \in \mathbb{N}$ and suppose that we are given a *convenient* C^1-curve $C \subset \mathbb{R}^n$, i.e., a curve (1-dimensional submanifold) that is the image of an *injective, immersion* $\boldsymbol{\alpha} : (a,b) \to \mathbb{R}^n$. Recall that $\boldsymbol{\alpha}$ is an immersion if $\boldsymbol{\alpha}'(t) \neq \mathbf{0}$, $\forall t \in (a,b)$. We think of $\boldsymbol{\alpha}(t)$ as describing the position at time t of a moving particle. The fact that $\boldsymbol{\alpha}$ is an immersion signifies that the particle does not double back. We will refer to a map $\boldsymbol{\alpha}$ with the above properties as a *parametrization* of the convenient curve.

During a tiny time interval $[t_0, t_0 + dt]$ the particle travels from $\boldsymbol{\alpha}(t_0)$ to $\boldsymbol{\alpha}(t_0 + dt)$. The arc of the curve C from $\boldsymbol{\alpha}(t_0)$ to $\boldsymbol{\alpha}(t_0 + dt)$ "does not bend too much" during the infinitesimal period of time dt so the distance $ds(t_0)$ covered by the particle along C can be approximated by the length of the line segment joining $\boldsymbol{\alpha}(t_0)$ to $\boldsymbol{\alpha}(t_0 + dt)$, i.e.,

$$ds(t_0) \approx \|\boldsymbol{\alpha}(t_0 + dt) - \boldsymbol{\alpha}(t_0)\| \approx \|\boldsymbol{\alpha}'(t_0)\|\,|dt|.$$

Thus, the length of C, viewed as the total distance covered by the particle ought to be

$$\operatorname{length}(C) \stackrel{?}{=} \int_a^b \|\boldsymbol{\alpha}'(t)\|\,|dt|.$$

Why the question mark? Intuition tells us that the length of a curve should be independent of the way a particle travels along it without backtracking. The right-hand side seems to depend on such travel as expressed by the parametrization $\boldsymbol{\alpha}$.

To deal with this issue we need to answer the following concrete question. Suppose that $\boldsymbol{\beta} : (c,d) \to \mathbb{R}^n$ is another parametrization of the curve C; see Figure 16.1. Can we conclude that

$$\int_a^b \|\boldsymbol{\alpha}'(t)\|\,|dt| = \int_c^d \|\boldsymbol{\beta}'(\tau)\|\,|d\tau|? \tag{16.1}$$

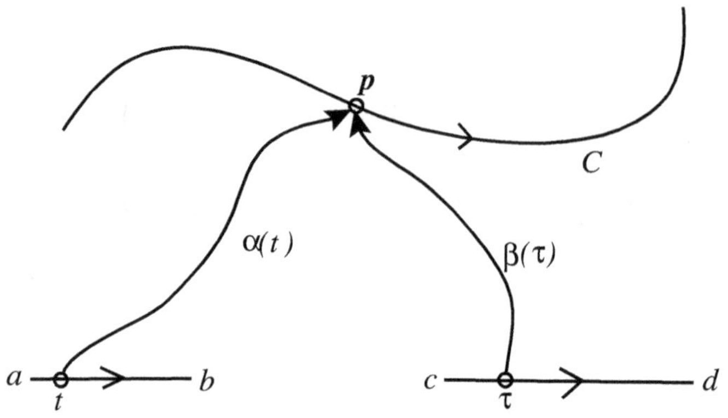

Fig. 16.1 *Different parametrizations of the same convenient curve C.*

A point $\boldsymbol{p} \in C$ corresponds uniquely via $\boldsymbol{\alpha}$ to a point $t \in (a, b)$. Via $\boldsymbol{\beta}$ the point $\boldsymbol{p}$ corresponds uniquely to a point $\tau \in (c, d)$. More precisely we have

$$\boldsymbol{\alpha}(t) = \boldsymbol{p} = \boldsymbol{\beta}(\tau).$$

The correspondence

$$t \mapsto \boldsymbol{p} \mapsto \tau = \tau(t)$$

produces a bijection $(a, b) \to (c, d)$ that can be described formally by the equality $\tau = \beta^{-1}\big(\alpha(t)\big)$. Proposition 14.34(B) implies that the function $(a, b) \ni t \mapsto \tau(t) \in (c, d)$ is C^1.

The change in variables formula for the Riemann integral implies

$$\int_c^d \|\boldsymbol{\beta}'(\tau)\| d\tau = \int_a^b \|\boldsymbol{\beta}'(\,\tau(t)\,)\| \cdot \left|\frac{d\tau}{dt}\right| dt.$$

Derivating with respect to t the equality $\boldsymbol{\beta}\big(\,\tau(t)\,\big) = \boldsymbol{\alpha}(t)$ we deduce

$$\boldsymbol{\beta}'(\,\tau(t)\,)\frac{d\tau}{dt} = \boldsymbol{\alpha}'(t). \tag{16.2}$$

Using the equalities

$$\boxed{ds(\tau) = \|\boldsymbol{\beta}'(\tau)\||d\tau|, \;\; ds(t) = \|\boldsymbol{\alpha}'(t)\||dt|, \;\; |d\tau| = \left|\frac{d\tau}{dt}\right| |dt|},$$

we deduce from (16.2) that $ds(\tau) = ds(t)$. For this reason we will rewrite the equality

$$\text{length}(C) = \int_a^b \|\boldsymbol{\alpha}'(t)\| \, |dt| = \int_c^d \|\boldsymbol{\beta}'(\tau)\| \, |d\tau|$$

in the simpler form

$$\boxed{\text{length}(C) = \int_C ds}.$$

More generally, the same argument as above shows that, if $f : C \to \mathbb{R}$ is a continuous function, then we have the equality

$$f\big(\boldsymbol{\alpha}(t)\big)\|\boldsymbol{\alpha}'(t)\|dt = f\big(\boldsymbol{\beta}(\tau)\big)\|\boldsymbol{\beta}'(\tau)\|d\tau$$

and we set

$$\boxed{\int_C f(\mathbf{p})ds := \int_a^b f\big(\boldsymbol{\alpha}(t)\big)\|\boldsymbol{\alpha}'(t)\|\,|dt| = \int_c^d f\big(\boldsymbol{\beta}(\tau)\big)\|\boldsymbol{\beta}'(\tau)\|\,|d\tau|}. \tag{16.3}$$

The integral in the left-hand side of (16.3) is called the *integral of f along the curve C*. This type of integral is traditionally known as *line integral of the first kind*.

☞ *The integrals in the right-hand side of (16.3) could be improper integrals and, as such, they may or may not be convergent. We say that the integral over the convenient curve C is well defined if the integrals in the right-hand side of (16.3) are convergent.*

Example 16.1. Consider the arc of helix H in $\mathbb{R}^3$ described by the parametrization (see Figure 16.2)

$$\boldsymbol{\alpha} : (0,1) \to \mathbb{R}^3, \quad \boldsymbol{\alpha}(t) = \big(\cos(4\pi t), \sin(4\pi t), 2t\big).$$

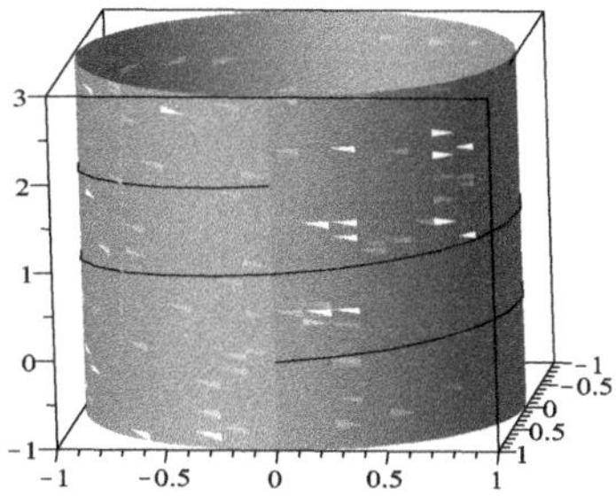

Fig. 16.2 *The helix described by the parametrization* $\big(\cos(4\pi t), \sin(4\pi t), 2t\big)$ *is winding up a cylinder of radius* 1.

It is not hard to see that $\boldsymbol{\alpha}$ is an injective immersion. Moreover

$$\dot{\boldsymbol{\alpha}}(t) = (-4\pi\sin(4\pi t), 4\pi\sin(4\pi t), 2),$$

$$\|\dot{\boldsymbol{\alpha}}(t)\| = \sqrt{(4\pi\sin 4\pi t)^2 + (4\pi\cos 4\pi t)^2 + 2^2} = \sqrt{16\pi^2+4} = 2\sqrt{4\pi^2+1}.$$

We deduce that

$$\text{length}(H) = \int_H ds = \int_0^1 \|\dot{\boldsymbol{\alpha}}(t)\|dt = 2\sqrt{4\pi^2+1}. \qquad \square$$

Example 16.2. Suppose that $f : [a,b] \to \mathbb{R}$ is a C^1 function. Its graph with endpoints removed is the set

$$\Gamma_f = \big\{ (x,y) \in \mathbb{R}^2;\ \ x \in (a,b),\ \ y = f(x) \big\}.$$

This is a curve that admits the parametrization

$$\boldsymbol{\alpha} : (a,b) \to \mathbb{R}^2,\ \ \boldsymbol{\alpha}(x) = \big(x,\ f(x)\big),\ \ x \in (a,b).$$

Then

$$\boldsymbol{\alpha}'(x) = \big(1, f'(x)\big),\ \ \|\boldsymbol{\alpha}'(x)\| = \sqrt{1+f'(x)^2}.$$

We deduce that the length of Γ_f is given by

$$\operatorname{length}(\Gamma_f) = \int_a^b \sqrt{1+f'(x)^2}dx.$$

This is in perfect agreement with our earlier definition (9.74). □

If the curve C is the union of pairwise disjoint convenient curves $C_1, \dots, C_\ell$, then, for any continuous function $f : C \to \mathbb{R}$ we set

$$\boxed{\int_C f(\boldsymbol{p})ds = \int_{\bigcup_{j=1}^{\ell} C_j} f(\boldsymbol{p})ds = \sum_{j=1}^{\ell}\int_{C_j} f(\boldsymbol{p})ds}.$$

Remark 16.3 (A few points do not matter). Suppose that $C \subset \mathbb{R}^n$ is a convenient curve and $\boldsymbol{p}_0$ is a point on C. If we remove the point $\boldsymbol{p}_0$ we obtain a new curve C' consisting of two arcs C_0, C_1 of the original curve; see Figure 16.3.

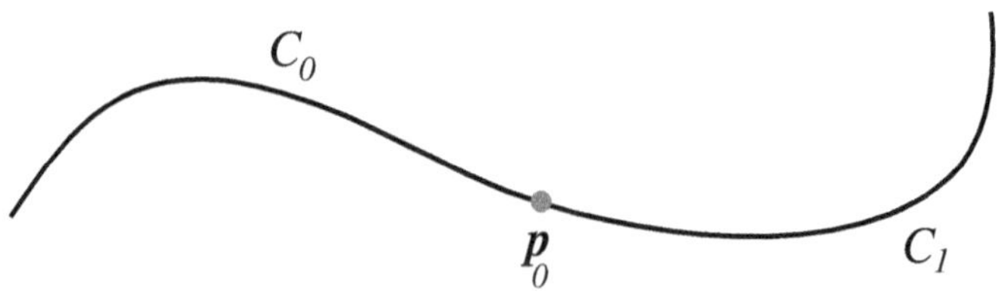

Fig. 16.3 *Cutting a convenient curve C in two parts by removing a point $\boldsymbol{p}_0$.*

We want to show that the removal of one point does not affect the computation of an integral, i.e., if $f : C \to \mathbb{R}$ is a continuous function, then

$$\int_C f(\boldsymbol{p})ds = \int_{C'} f(\boldsymbol{p})ds := \int_{C_0} f(\boldsymbol{p})ds + \int_{C_1} f(\boldsymbol{p})ds.$$

Indeed, fix a parametrization $\boldsymbol{\alpha} : (a,b) \to \mathbb{R}^n$ of C. Then, there exists $t_0 \in (a,b)$ such that $\boldsymbol{\alpha}(t_0) = \boldsymbol{p}_0$. Assume that C_0 is the curve swept by the moving point $\boldsymbol{\alpha}(t)$ as t runs from a

to t_0 and C_1 is swept when t runs from t_0 to b. Then

$$\begin{aligned}\int_C f(\boldsymbol{p})ds &= \int_a^b f\big(\boldsymbol{\alpha}(t)\big)\|\dot{\boldsymbol{\alpha}}(t)\|dt\\ &= \int_a^{t_0} f\big(\boldsymbol{\alpha}(t)\big)\|\dot{\boldsymbol{\alpha}}(t)\|dt + \int_{t_0}^b f\big(\boldsymbol{\alpha}(t)\big)\|\dot{\boldsymbol{\alpha}}(t)\|dt\\ &= \int_{C_0} f(\boldsymbol{p})ds + \int_{C_1} f(\boldsymbol{p})ds =: \int_{C'} f(\boldsymbol{p})ds.\end{aligned}$$

□

Definition 16.4. A curve C is called *quasi-convenient* if it admits a *convenient cut*, i.e., a finite subset $Z = \{\boldsymbol{p}_1, \dots, \boldsymbol{p}_\ell\} \subset C$ whose complement $C \setminus \{\boldsymbol{p}_1, \dots, \boldsymbol{p}_\ell\}$ is a union of finitely many pairwise disjoint convenient curves. □

Example 16.5. The unit circle in $\mathbb{R}^2$,

$$C := \{(x,y) \in \mathbb{R}^2;\ x^2 + y^2 = 1\}$$

is quasi-convenient. Indeed, the set consisting of the point $(1,0)$ is a convenient cut because its complement admits the parametrization

$$\boldsymbol{\alpha} : (0, 2\pi) \to \mathbb{R}^2,\ \ \boldsymbol{\alpha}(t) = \big(\cos t, \sin t\big).$$

□

Suppose now that C is a quasi-convenient curve and $f : C \to \mathbb{R}$ is a continuous function. The integral of f along C, denoted by

$$\int_C f(\boldsymbol{p})ds,$$

is defined as follows. Choose a convenient cut Z. Then the complement $C_Z = C \setminus Z$ is a union of finitely many pairwise disjoint convenient curves and then we set

$$\boxed{\int_C f(\boldsymbol{p})ds := \int_{C_Z} f(\boldsymbol{p})ds}.$$

To show that this definition is independent of the convenient cut Z, suppose that Z_0, Z_1 are two convenient cuts. Then $Z := Z_0 \cup Z_1$ is also a convenient cut and $Z_0, Z_1 \subset Z$. Note that C_Z is obtained from either C_{Z_0} or C_{Z_1} by removing a few points. From Remark 16.3 we deduce

$$\int_{C_{Z_0}} f(\boldsymbol{p})ds = \int_{C_Z} f(\boldsymbol{p})ds = \int_{C_{Z_1}} f(\boldsymbol{p})ds.$$

Example 16.6. Suppose that C is the unit circle in $\mathbb{R}^2$

$$C := \{(x,y) \in \mathbb{R}^2;\ x^2 + y^2 = 1\}.$$

Denote by Z the convenient cut $Z = \{(1,0)\}$. Then

$$\operatorname{length}(C) = \operatorname{length}(C_Z).$$

The complement C_Z admits the parametrization

$$\boldsymbol{\alpha} : (0, 2\pi) \to \mathbb{R}^2, \quad \boldsymbol{\alpha}(t) = \big(\cos t, \sin t \big).$$

We deduce

$$\dot{\boldsymbol{\alpha}}(t) = \big(-\sin t, \cos t \big),$$

$$\operatorname{length}(C_Z) = \int_0^{2\pi} \|\dot{\boldsymbol{\alpha}}(t)\| dt = \int_0^{2\pi} \sqrt{\sin^2 t + \cos^2 t} dt = 2\pi. \qquad \square$$

Definition 16.7 (Curves with boundary). Let $k, n \in \mathbb{N}$. A C^k*-curve with boundary* in $\mathbb{R}^n$ is a compact subset $C \subset \mathbb{R}^n$ such that, for any point $\boldsymbol{p}_0 \in C$, there exists an open neighborhood $\mathcal{U}$ of $\boldsymbol{p}_0$ in $\mathbb{R}^n$ and a C^k-diffeomorphism $\Psi : \mathcal{U} \to \mathbb{R}^n = \mathbb{R} \times \mathbb{R}^{n-1}$ with the following property: either the image of $\mathcal{U} \cap C$ is an interval (a, b) on the x^1-axis, $a < 0 < b$,

$$\Psi\big(\mathcal{U} \cap C\big) = (a, b) \times \mathbf{0}_{n-1} \in \mathbb{R} \times \mathbb{R}^{n-1}, \quad \Psi(\boldsymbol{p}_0) = (0, \mathbf{0}_{n-1}) \in \mathbb{R} \times \mathbb{R}^{n-1}, \tag{I}$$

or, the image of $\mathcal{U} \cap C$ is an interval $(a, 0]$ on the x^1-axis, $a < 0$,

$$\Psi\big(\mathcal{U} \cap C\big) = (a, 0] \times \mathbf{0}_{n-1} \in \mathbb{R} \times \mathbb{R}^{n-1}, \quad \Psi(\boldsymbol{p}_0) = (0, \mathbf{0}_{n-1}) \in \mathbb{R} \times \mathbb{R}^{n-1}. \tag{B}$$

The pair $(\mathcal{U}, \Psi)$ is called a (local) *straightening diffeomorphism* at $\boldsymbol{p}_0$, or s.d. for brevity.

If the alternative (B) occurs for some choice of straightening diffeomorphism, then we say that $\boldsymbol{p}_0 \in C$ is a *boundary point*. The *boundary* of a C^k-curve with boundary C, denoted by ∂C, is the (possibly empty) subset of C consisting of the boundary points. The curve C is called *closed* if $\partial C = \emptyset$.

A point $\boldsymbol{p}_0 \in C$ is called an *interior point* if it is not a boundary point, i.e., the alternative (I) holds for any choice of straightening diffeomorphism. The *interior* of C, denoted by C°, is the collection of interior points,

$$C^0 = C \setminus \partial C. \qquad \square$$

Example 16.8. (a) Suppose that $f : [a, b] \to \mathbb{R}$ is a C^1 function. Then its graph

$$\Gamma_f := \big\{ (x, y) \in \mathbb{R}^2; \;\; x \in [a, b], \;\; y = f(x) \big\}$$

is a curve with boundary. Its boundary consists of the endpoints of the graph

$$\partial\Gamma_f = \big\{ (a, f(a)), \; (b, f(b)) \big\}.$$

(b) More generally, suppose that

$$\boldsymbol{\alpha} : [a, b] \to \mathbb{R}^n, \quad \boldsymbol{\alpha}(t) = \begin{bmatrix} \alpha^1(t) \\ \alpha^2(t) \\ \vdots \\ \alpha^n(t) \end{bmatrix}$$

is an *injective* map such that each of the components α^i is a C^1-function and, $\forall t \in [a, b]$

$$\dot{\boldsymbol{\alpha}}(t) \neq \mathbf{0}.$$

Then the image of $\boldsymbol{\alpha}$ is a curve C with boundary consisting of two points

$$\partial C = \big\{ \boldsymbol{\alpha}(a), \; \boldsymbol{\alpha}(b) \big\}.$$

The proof of this fact is a variation of the proof of Proposition 14.34 and we will skip it. A curve with boundary obtained in this fashion is called *convenient*. A map $\boldsymbol{\alpha}$ as above is called a *parametrization* of the convenient curve (with boundary).

(c) The unit circle in $\mathbb{R}^2$ is a closed curve. $\square$

We have the following result whose proof we omit.

Theorem 16.9 (Classification of curves with boundary). *(a) Any curve with boundary $C \subset \mathbb{R}^n$ is the union of finitely many pairwise disjoint path connected curves with boundary called the* connected components *of* C

(b) If $C \subset \mathbb{R}^n$ is a path connected *C^k-curve with boundary, then it is either a convenient curve with boundary if $\partial C \neq \emptyset$, or, if C is closed, there exist $T > 0$ and a C^k-immersion $\boldsymbol{\alpha} : \mathbb{R} \to \mathbb{R}^n$ with the following properties.*

- $\boldsymbol{\alpha}(\mathbb{R}) = C$.
- $\boldsymbol{\alpha}(t) = \boldsymbol{\alpha}(t+T)$, $\forall t \in \mathbb{R}$.
- *The restriction to $[0,T)$ is injective.*

*A map with the above properties is called a T-*periodic parametrization *of the closed connected curve C.* □

Example 16.10. The closed curve winding around the torus in Figure 16.4 admits the 2π-periodic parametrization

$$\boldsymbol{\alpha} : \mathbb{R} \to \mathbb{R}^3, \;\; \boldsymbol{\alpha}(t) = \big(\,(3+\cos(3t)\,)\cos(2t), (3+\cos(3t)\,)\sin(2t), \sin(3t)\,\big). \quad \square$$

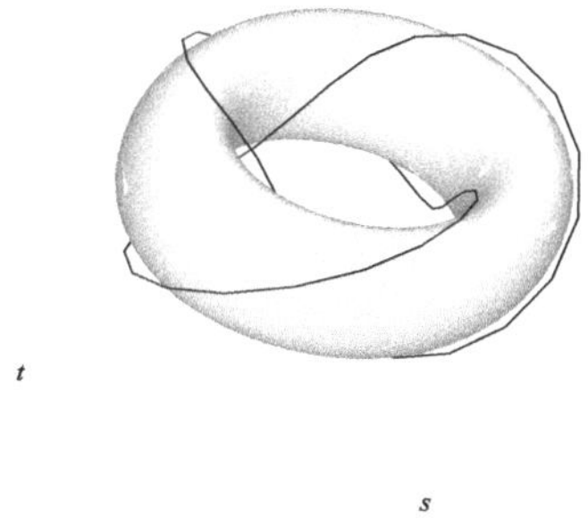

Fig. 16.4 *A torus knot.*

Remark 16.11. Suppose that $C \subset \mathbb{R}^n$ is a closed C^1-curve and $\boldsymbol{\alpha} : \mathbb{R} \to \mathbb{R}^n$ is a T-periodic parametrization of C. Set $\boldsymbol{p}_0 := \boldsymbol{\alpha}(0)$. Then $C' := C \setminus \{\boldsymbol{p}_0\}$. Then C' is a convenient curve and the restriction of $\boldsymbol{\alpha}$ to the open interval $(0,T)$ is a parametrization of C'. □

From the above classification theorem we deduce that if C is a curve with boundary, then its interior C° is a quasi-convenient curve. If $f : C \to \mathbb{R}$ is a continuous function, then we define

$$\boxed{\int_C f(\boldsymbol{p})ds := \int_{C^\circ} f(\boldsymbol{p})ds}.$$

Remark 16.12. We can think of a connected curve with boundary in $\mathbb{R}^n$ as a "bent wire". A function $f : C \to (0, \infty)$ can be thought of as a linear density: the quantity $f(\boldsymbol{p})ds$ would be the mass of an infinitesimal arc C of length ds starting at $\boldsymbol{p}$. The integral

$$\int_C f(\boldsymbol{p})ds$$

would then represent the mass of that "bent wire". □

Example 16.13. Consider the curve $C \subset \mathbb{R}^3$ obtained by intersecting the unit sphere $\{x^2 + y^2 + z^2 = 1\}$ with the cone spanned by the rays at the origin that make an angle of $\frac{\pi}{4}$ with the positive z-semiaxis: see Figure 16.5. We want to compute the integral

$$I = \int_C xyds. \tag{16.4}$$

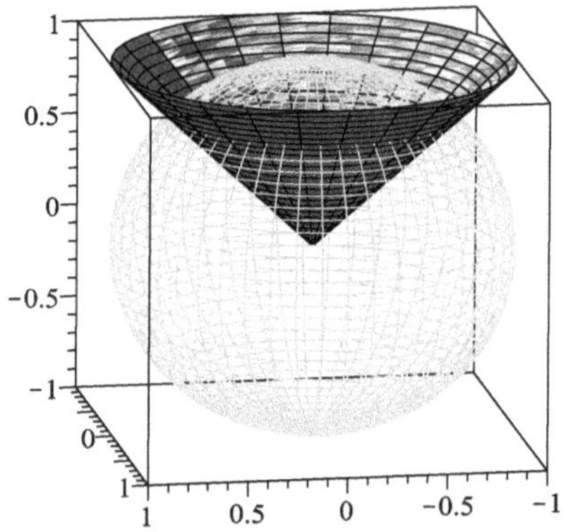

Fig. 16.5 *A cone intersecting a sphere.*

The resulting curve is a circle. To find a periodic parametrization for this circle we use spherical coordinates, (ρ, θ, φ),

$$\begin{gathered} x = \rho \sin\varphi \cos\theta, \;\; y = \rho \sin\varphi \sin\theta, \;\; z = \rho\cos\varphi, \\ \rho > 0, \;\; \theta \in [0, 2\pi], \;\; \varphi \in [0, \pi]. \end{gathered} \tag{16.5}$$

In these coordinates the unit sphere is described by the equation $\rho = 1$ and the cone is described by the equation $\varphi = \frac{\pi}{4}$. Taking into account that

$$\cos\frac{\pi}{4} = \sin\frac{\pi}{4} = \frac{\sqrt{2}}{2},$$

we deduce from (16.5) that a parametrization for C is described by

$$x = \frac{\sqrt{2}}{2}\sin\theta, \quad y = \frac{\sqrt{2}}{2}\cos\theta, \quad z = \frac{\sqrt{2}}{2}, \quad \theta \in [0, 2\pi]. \tag{16.6}$$

The arclength element ds on C is then

$$ds = \sqrt{x'(\theta)^2 + y'(\theta)^2 + z'(\theta)^2}\, d\theta = \frac{\sqrt{2}}{2}\, d\theta.$$

To compute the integral (16.4) we use the parametrization (16.6) and we deduce

$$\int_C xyds = \int_0^{2\pi} \left(\frac{\sqrt{2}}{2}\right)^3 \cos\theta \sin\theta d\theta = \left(\frac{\sqrt{2}}{2}\right)^3 \int_0^{2\pi} \frac{1}{2}\sin 2\theta\, d\theta = 0.$$

□

The next result shows that integral along a curve with boundary has several features in common with the Riemann integrals.

Proposition 16.14. *Suppose that $\Gamma \subset \mathbb{R}^n$ is a curve with boundary. (We recall that, by our definition, the curves with boundary are* compact.*) We denote by $C^0(\Gamma)$ the vector space of continuous functions $\Gamma \to \mathbb{R}$. Then the first kind integral along C defines a* linear map

$$C^0(\Gamma) \ni f \mapsto \int_\Gamma fds \in \mathbb{R}$$

satisfying the monotonicity property:

$$\int_\Gamma fds \leq \int_\Gamma gds$$

if $f(\mathbf{p}) \leq g(\mathbf{p})$, $\forall \mathbf{p} \in \Gamma$. □

We omit the simple proof of the above result.

16.1.2. *Integration of differential 1-forms over paths*

Let $n \in \mathbb{N}$ and suppose that $U \subset \mathbb{R}^n$ is an open set. A *differential form of degree* 1 or a *differential* 1*-form* on U is an expression ω of the form

$$\omega = \omega_1 dx^1 + \cdots + \omega_n dx^n,$$

where $\omega_1, \dots, \omega_n : U \to \mathbb{R}$ are continuous functions. The precise meaning of a 1-form is a bit more complicated to explain at this point, but for the goals we have in mind, it is irrelevant. We will denote by $\Omega^1(U)$ the collection of 1-forms on U.

Example 16.15. (a) If $f \in C^1(U)$, then its total differential as described in (13.21) is a 1-form

$$df = \partial_{x^1} f dx^1 + \cdots + \partial_{x^n} f dx^n.$$

A 1-form of this type is called *exact*.

(b) Suppose $\boldsymbol{F} : U \to \mathbb{R}^n$ is a continuous vector field on U,

$$\boldsymbol{F}(\boldsymbol{p}) = \begin{bmatrix} F^1(\boldsymbol{p}) \\ F^2(\boldsymbol{p}) \\ \vdots \\ F^n(\boldsymbol{p}) \end{bmatrix},$$

then the *infinitesimal work* is the 1-form

$$W_{\boldsymbol{F}} := F^1 dx^1 + \cdots + F^n dx^n.$$

Traditionally, in classical mechanics the infinitesimal work is denoted by $\boldsymbol{F} \cdot d\boldsymbol{p}$ or $\langle \boldsymbol{F}, d\boldsymbol{p} \rangle$, where "$\cdot$" is short-hand for inner product and $d\boldsymbol{p}$ denotes the "infinitesimal displacement"

$$d\boldsymbol{p} = \begin{bmatrix} dx^1 \\ dx^2 \\ \vdots \\ dx^n \end{bmatrix}.$$

Note that if $f \in C^1(U)$, then $df = W_{\nabla f}$.

(c) The *angular form* on $\mathbb{R}^2 \setminus \{\boldsymbol{0}\}$ is the infinitesimal work associated to the angular vector field (see Figure 16.6)

$$\Theta : \mathbb{R}^2 \setminus \{0\} \to \mathbb{R}^2, \;\; \Theta(x, y) := \begin{bmatrix} -\frac{y}{x^2+y^2} \\ \\ \frac{x}{x^2+y^2} \end{bmatrix}.$$

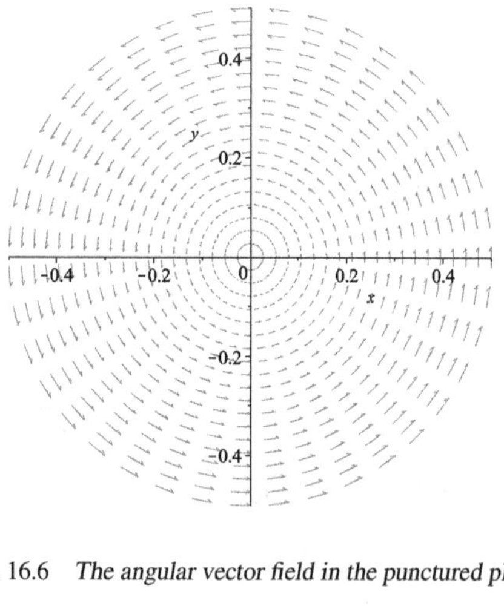

Fig. 16.6 *The angular vector field in the punctured plane.*

More explicitly

$$W_\Theta = -\frac{y}{x^2 + y^2} dx + \frac{x}{x^2 + y^2} dy. \qquad \square$$

Definition 16.16. Let $n \in \mathbb{N}$ and suppose that $U \subset \mathbb{R}^n$ is an open set. For any differential 1-form

$$\omega = \omega_1 dx^1 + \cdots + \omega_n dx^n \in \Omega^1(U),$$

and any C^1-path

$$\gamma : [a,b] \to U,\ \ \gamma(t) = \big(\gamma^1(t), \dots, \gamma^n(t)\big),\ \ \forall a \leq t \leq b,$$

we define the *integral of* ω *along the path* γ to be the real number

$$\int_\gamma \omega := \int_a^b \Big(\omega_1\big(\gamma(t)\big)\dot{\gamma}^1(t) + \cdots + \omega_n\big(\gamma(t)\big)\dot{\gamma}^n(t)\Big)dt.$$

The integral $\int_\gamma \omega$ is traditionally known as the *line integral of the second kind.* □

When ω is the infinitesimal work of a vector field $\boldsymbol{F}$, $\omega = W_{\boldsymbol{F}}$, then following the physicists' tradition, one uses the notation

$$\int_\gamma \boldsymbol{F} \cdot d\boldsymbol{p} := \int_\gamma W_{\boldsymbol{F}}.$$

Example 16.17. Fix a natural number N, a real number $R > 0$ and consider the path

$$\gamma_N : [0, 2\pi N] \to \mathbb{R}^2 \setminus \{\boldsymbol{0}\},\ \ \gamma_N(t) = \big(x(t), y(t)\big) := \big(R\cos t, R\sin t\big).$$

Let $W_\Theta \in \Omega^1(\mathbb{R}^2 \setminus \{0\})$ be the angular form defined in Example 16.15(c), i.e.,

$$W_\Theta = \frac{-y}{x^2+y^2}dx + \frac{x}{x^2+y^2}dy.$$

Then

$$\int_{\gamma_N} W_\Theta = \int_{\gamma_N} \left(\frac{-y}{x^2+y^2}\,dx + \frac{x}{x^2+y^2}\,dy\right)$$

$(dx = \dot{x}dt,\ dy = \dot{y}dt)$

$$= \int_0^{2\pi N} \frac{-y}{x^2+y^2}\,\dot{x}dt + \int_0^{2\pi N} \frac{x}{x^2+y^2}\,\dot{y}dt.$$

Observing that along γ_N we have $x^2 + y^2 = R^2$, $\dot{x} = -R\sin t$, $\dot{y} = R\cos t$, we deduce

$$\int_{\gamma_N} W_\Theta = \int_0^{2\pi N} \left(\frac{-R\sin t}{R^2}(-R\sin t) + \frac{R\cos t}{R^2}R\cos t\right)dt$$

$$= \int_0^{2\pi N} \big(\sin^2 t + \cos^2 t\big)dt = 2\pi N.$$

□

Theorem 16.18 (1-dimensional Stokes' formula). *Let* $n \in \mathbb{N}$ *and suppose that* $\mathcal{O} \subset \mathbb{R}^n$ *is an open subset. Then, for any* $f \in C^1(\mathcal{O})$*, and any* C^1*-path* $\gamma : [a,b] \to \mathcal{O}$ *we have*

$$\boxed{\int_\gamma df = f\big(\gamma(b)\big) - f\big(\gamma(a)\big)}. \tag{16.7}$$

Proof. Denote by $(x^1(t), \dots, x^n(t))$ the coordinates of the point $\gamma(t)$. Then

$$\int_\gamma df = \int_\gamma \left(\partial_{x^1} f dx^1 + \cdots + \partial_{x^n} f dx^n \right)$$

$$= \int_a^b \left(\partial_{x^1} f(x^1(t), \dots, x^n(t))\, \dot{x}^1 + \cdots + \partial_{x^n} f(x^1(t), \dots, x^n(t))\, \dot{x}^n \right) dt$$

(use the chain rule)

$$= \int_a^b \frac{d}{dt} f(\gamma(t))\, dt = f(\gamma(b)) - f(\gamma(a)).$$

□

Remark 16.19. (i) Although very simple, the above result has very important consequences. First, let us observe that df is the infinitesimal work of the vector field ∇f

$$df = W_{\nabla f}.$$

In classical mechanics the function $U = -f$ is often called the *potential* of the gradient vector field $\boldsymbol{F} = \nabla f$.

Think of $\gamma(t)$ as describing the motion of a particle interacting with the force field ∇f. For example ∇f can be the gravitational field and the particle is a "heavy" particle, i.e., a particle with positive mass. The integral

$$\int_\gamma df = \int_\gamma W_{\nabla f}$$

is interpreted in classical mechanics as the total energy required to generate the travel of the particle described by the path γ. Stokes' formula (16.7) shows that, when the force field is a gradient vector field, then this total energy depends only on the endpoints of the travel and not on what happened in between. In particular, if the path γ is closed, $\gamma(b) = \gamma(a)$, so the particle ends at the same point where it started, this total energy is trivial!

(ii) Let $U \subset \mathbb{R}^n$ be an open set. As we have mentioned earlier a 1-form $\omega \in \Omega^1(U)$ is *exact* if there exists $f \in C^1(U)$ such that $\omega = df$. A function f such that $df = \omega$ is called an antiderivative of ω.

If

$$\omega = \sum_{i=1}^n \omega_d x^i,$$

then ω is exact if there exists $f \in C^1(U)$ such that

$$\omega_i = \frac{\partial f}{\partial x^i}, \quad \forall i = 1, \dots, n.$$

Since $\partial_{x^i} \partial_{x^j} f = \partial_{x^j} \partial_{x^i} f$, $\forall i, j$, we deduce that, if ω is exact then

$$\boxed{\frac{\partial \omega_i}{\partial x^j} = \frac{\partial \omega_j}{\partial x^i}, \quad \forall i, j.} \tag{16.8}$$

A 1-form satisfying (16.8) is called *closed*. In other words,

$$\omega \text{ exact} \Rightarrow \omega \text{ closed}.$$

A famous result, known by the name of *Poincaré Lemma* [18, Thm. 4.11], states that the converse is true if U is *convex*,

$$U \text{ convex}, \omega \text{ closed} \Rightarrow \omega \text{ exact}.$$

The result is not true without the convexity assumption. Consider for example the 1-form

$$W_\Theta = \underbrace{\frac{-y}{x^2+y^2}}\, dx + \underbrace{\frac{x}{x^2+y^2}}\, dy \in \Omega^1\big(\mathbb{R}^2 \setminus \{0\}\big).$$

In this case $U = \mathbb{R}^2 \setminus \{0\}$ is not convex and it is the largest open set where W_Θ is well defined: the denominators in the definition of W_Θ become 0 exactly at the origin, the missing point.

As we will see soon in Example 16.35 we have $P'_y = Q'_y$, so the form W_Θ is closed. On the other hand, it is not exact because, as shown in Example 16.17, its integral over the counterclockwise oriented unit circle centered at the origin is 2π. The differential form W_Θ senses the hole in its domain!

This curious phenomenon is the beginning of a rather deep story called *cohomology*. □

Definition 16.20. Let $n \in \mathbb{N}$. A *piecewise C^1-path* in $\mathbb{R}^n$ is a continuous map $\gamma : [a, b] \to \mathbb{R}^n$ such that, there exists a partition

$$a = t_0 < t_1 < \cdots < t_\ell = b$$

of $[a, b]$ with the property that, for any $i = 1, \dots, \ell$, the restriction of γ to the subinterval $[t_{i-1}, t_i]$ is C^1. A partition of $[a, b]$ with the above properties is said to be *adapted* to γ. □

Example 16.21. Consider the continuous path $\gamma : [0, 4] \to \mathbb{R}^2$ defined by

$$\gamma(t) = \begin{cases} \big(t, 0\big), & t \in [0, 1], \\ \big(1, t-1\big), & t \in (1, 2], \\ \big(3-t, 1\big), & t \in (2, 3], \\ \big(0, 4-t\big), & t \in (3, 4]. \end{cases}$$

The image of this path is the boundary of the unit square $[0, 1] \times [0, 1] \subset \mathbb{R}^2$; see Figure 16.7. □

One can integrate differential forms over piecewise C^1-paths. Suppose that $U \subset \mathbb{R}^n$ is an open set,

$$\omega = \omega_1 dx^1 + \cdots + \omega_n dx^n \in \Omega^1(U)$$

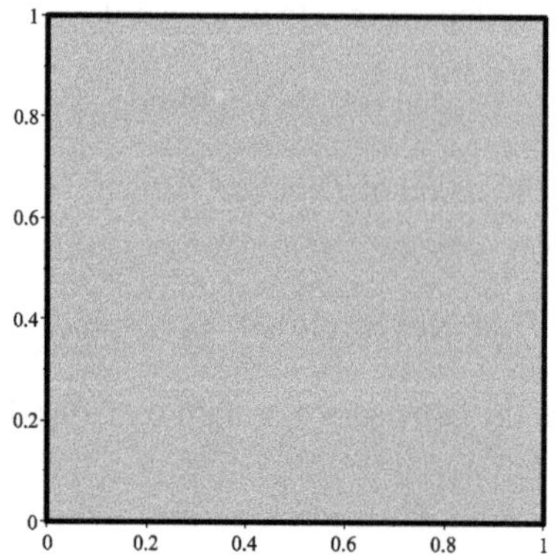

Fig. 16.7 *The unit square* $[0,1]^2$ *and its boundary.*

and $\gamma : [a,b] \to U$ is a C^1-path. If $a = t_0 < t_1 < \cdots < t_\ell = b$ is *any* partition of $[a,b]$ adapted to γ then we set

$$\boxed{\int_\gamma \omega = \sum_{i=1}^{\ell} \int_{t_{i-1}}^{t_i} \Big(\omega_1\big(\gamma(t)\big)\dot{\gamma}^1 + \cdots + \omega_n\big(\gamma(t)\big)\dot{\gamma}^n \Big) dt}.$$

One can show that the right-hand side is independent of the choice of the partition adapted to γ.

Example 16.22. Let γ denote the piecewise path described in Example 16.21. Suppose that

$$\omega = -ydx + xdy.$$

If we denote by $(x(t), y(t))$ the moving point $\gamma(t)$, then we deduce

$$\int_\gamma (-ydx + xdy) = \int_0^1 \big(-y\dot{x} + x\dot{y} \big) dt + \int_1^2 \big(-y\dot{x} + x\dot{y} \big) dt$$
$$+ \int_2^3 \big(-y\dot{x} + x\dot{y} \big) dt + \int_3^4 \big(-y\dot{x} + x\dot{y} \big) dt.$$

Observing that on the intervals $[0,1]$ and $[2,3]$ we have $\dot{y} = 0$ while on the others we have $\dot{x} = 0$, we deduce

$$\int_\gamma (-ydx + xdy) = -\underbrace{\int_0^1 y\dot{x}dt}_{y=0,} + \underbrace{\int_1^2 x\dot{y}dt}_{x=1,\,\dot{y}=1} - \underbrace{\int_2^3 y\dot{x}dt}_{y=1,\,\dot{x}=-1} + \underbrace{\int_3^4 x\dot{y}dt}_{x=0}$$
$$= \int_1^2 dt + \int_2^3 dt = 2.$$

□

16.1.3. *Integration of 1-forms over oriented curves*

There is a simple way of producing a path given a connected curve, namely to assign an orientation to that curve. Loosely speaking, an orientation describes a direction of motion along the curve without specifying the speed of that motion; see Figure 16.8.

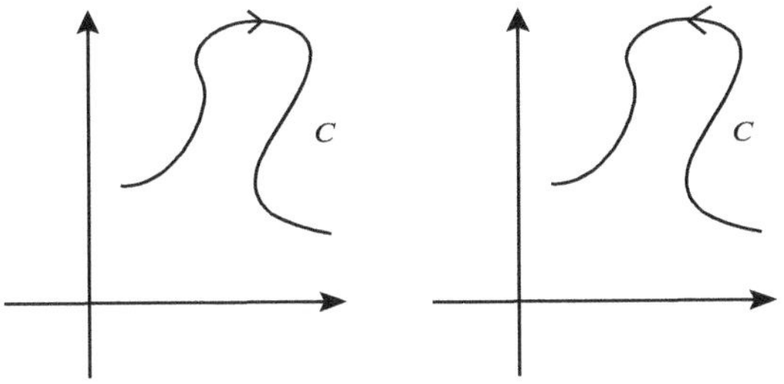

Fig. 16.8 *Two orientations on the same planar curve C.*

The direction of motion at a point on the curve C would be given by a unit vector tangent to the curve at that point. At each point $\boldsymbol{p} \in C$ there are exactly *two* unit vectors tangent to C and an orientation would correspond to choosing one such vector at each point, and the choices would vary continuously from one point to another. Here is a precise definition.

Definition 16.23. Let $n \in \mathbb{N}$.

(i) An *orientation* on a C^1-curve $C \subset \mathbb{R}^n$ is a continuous map $\boldsymbol{T} : C \to \mathbb{R}^n$ such that, $\forall \boldsymbol{p} \in C$, the vector $\boldsymbol{T}(\boldsymbol{p})$ has length 1 and it is tangent to C at $\boldsymbol{p} \in C$.

(ii) An *oriented curve* is a pair $(C, \boldsymbol{T})$, where C is a curve and $\boldsymbol{T}$ is an orientation on C.

(iii) An *orientation* on a (compact) C^1-curve with boundary C is an orientation on its interior C°.

(iv) Suppose that C is a convenient curve and $\boldsymbol{T}$ is an orientation of C. A parametrization $\gamma : (a, b) \to \mathbb{R}^n$ is said to be *compatible with the orientation* if, $\forall t \in (a, b)$, the tangent vectors

$$\dot{\gamma}(t), \ \ \boldsymbol{T}\big(\gamma(t)\big) \in T_{\gamma(t)}C$$

point in the same direction, i.e., $\big\langle \dot{\gamma}(t), \boldsymbol{T}\big(\gamma(t)\big) \big\rangle > 0$.

□

Let us observe that if $C \subset \mathbb{R}^n$ is a convenient curve, then any parametrization $\gamma(a, b) \to \mathbb{R}^n$ of C defines an orientation on $\boldsymbol{T} : C \to \mathbb{R}^n$ on C according to the rule

$$\boldsymbol{T}\big(\gamma(t)\big) = \frac{1}{\|\dot{\gamma}(t)\|}\dot{\gamma}(t). \ \ \forall t \in (a, b).$$

This is called the *orientation induced by the parametrization.* Clearly the parametrization is compatible with the orientation it induces since

$$\big\langle \dot{\gamma}(t), \boldsymbol{T}\big(\gamma(t)\big) \big\rangle = \|\dot{\gamma}(t)\| > 0.$$

It turns out that any orientation on a convenient curve is induced by a parametrization.

Lemma 16.24. *Suppose that C is a convenient curve in $\mathbb{R}^n$. For any parametrization $\gamma : (a,b) \to \mathbb{R}^n$ of C we denote by γ_- the parametrization*

$$\gamma_- : (-b,-a) \to \mathbb{R}^n, \quad \gamma_-(t) := \gamma(-t).$$

If $\boldsymbol{T}$ is an orientation on C, then exactly one of the parametrizations γ and γ_- is compatible with the orientation.

Proof. Observe that for any $t \in (a,b)$, the *nonzero* vectors $\boldsymbol{T}\big(\gamma(t)\big)$ and $\dot{\gamma}(t)$ belong to the 1-dimensional space $T_{\gamma(t)}C$. They are therefore collinear and thus

$$\big\langle \boldsymbol{T}\big(\gamma(t)\big), \dot{\gamma}(t) \big\rangle \neq 0, \quad \forall t \in (a,b).$$

Since the function

$$(a,b) \ni t \mapsto \big\langle \boldsymbol{T}\big(\gamma(t)\big), \dot{\gamma}(t) \big\rangle \in \mathbb{R}$$

and is nowhere zero, we deduce that either

$$\big\langle \boldsymbol{T}\big(\gamma(t)\big), \dot{\gamma}(t) \big\rangle > 0, \quad \forall t \in (a,b),$$

or

$$\big\langle \boldsymbol{T}\big(\gamma(t)\big), \dot{\gamma}(t) \big\rangle < 0, \quad \forall t \in (a,b).$$

In the first case we deduce that γ is compatible with $\boldsymbol{T}$, while in the second case we deduce that γ_- is compatible with $\boldsymbol{T}$ □

Suppose now that $U \subset \mathbb{R}^n$ is an open set,

$$\omega = \omega_1 dx^1 + \cdots + \omega_n dx^n \in \Omega^1(U),$$

$C \subset \mathbb{R}^n$ is a convenient curve and $\boldsymbol{T}$ is an orientation on C.

To define the integral of ω on the oriented convenient curve $(C, \boldsymbol{T})$ we need to first choose a parametrization $\boldsymbol{\alpha} : (a,b) \to \mathbb{R}^n$ of C compatible with the orientation. We then set

$$\int_C \omega = \int_{(C,\boldsymbol{T})} \omega := \int_{\boldsymbol{\alpha}} \omega = \int_a^b \Big(\omega_1\big(\boldsymbol{\alpha}(t)\big)\dot{\alpha}^1(t) + \cdots + \omega_n\big(\boldsymbol{\alpha}(t)\big)\dot{\alpha}^n(t) \Big) dt.$$

For the above definition to be consistent, the right-hand side has to be independent of parametrization compatible with the given orientation.

Lemma 16.25. *The above definition is independent of the choice of the parametrization of C compatible with the orientation.*

Proof. Indeed, if $\boldsymbol{\beta} : (c,d) \to \mathbb{R}^n$ is another such parametrization, then, as argued in Subsection 16.1.1 (see Figure 16.1), there exists a continuous C^1-bijection $(a,b) \ni t \mapsto \tau(t) \in (c,d)$ such that

$$\boldsymbol{\beta}\big(\tau(t)\big) = \boldsymbol{\alpha}(t), \quad \frac{d\boldsymbol{\beta}}{d\tau}\big(\tau(t)\big)\frac{d\tau}{dt} = \dot{\boldsymbol{\alpha}}(t), \quad \forall t \in (a,b).$$

Since the tangent vectors

$$\frac{d\boldsymbol{\beta}}{d\tau}\big(\tau(t)\big), \quad \frac{d\boldsymbol{\alpha}}{dt}(t)$$

point in the same directions we deduce

$$\frac{d\tau}{dt} > 0, \ \ \forall t \in (a, b).$$

The 1-dimensional change-in-variables formula implies that

$$\int_c^d \omega_i\big(\,\beta(\tau)\,\big)\frac{d\beta^i}{d\tau}d\tau = \int_a^b \omega_i\big(\,\beta(\tau(t))\,\big)\frac{d\beta^i}{d\tau}\frac{d\tau}{dt}dt = \int_a^b \omega_i\big(\,\boldsymbol{\alpha}(t)\,\big)\frac{d\alpha^i}{dt}dt, \ \ \forall i = 1, \dots, n.$$

Hence

$$\int_{\boldsymbol{\alpha}} \omega = \sum_{i=1}^n \int_a^b \omega_i\big(\,\boldsymbol{\alpha}(t)\,\big)\frac{d\alpha^i}{dt}dt = \sum_{i=1}^n \int_c^d \omega_i\big(\,\beta(\tau)\,\big)\frac{d\beta^i}{d\tau}d\tau = \int_{\boldsymbol{\beta}} \omega.$$

□

Proposition 16.26. *Let $U \subset \mathbb{R}^n$ be an open set, and $(C, \boldsymbol{T})$ an oriented convenient curve inside U. Then, for any 1-form $\omega \in \Omega^1(U)$ we have*

$$\int_{(C,-\boldsymbol{T})} \omega = -\int_{(C,\boldsymbol{T})} \omega.$$

Proof. Let

$$\omega = \omega_1 dx^1 + \cdots + \omega_n dx^n.$$

Fix a parametrization

$$\gamma : (a, b) \to \mathbb{R}^n, \ \ \gamma(t) = \begin{bmatrix} x^1(t) \\ x^2(t) \\ \vdots \\ x^n(t) \end{bmatrix},$$

compatible with the orientation $\boldsymbol{T}$. Then $\gamma_- : (-b, -a) \to \mathbb{R}^n$ is a parametrization compatible with $-\boldsymbol{T}$. We have

$$\begin{aligned}
\int_{(C,-\boldsymbol{T})} \omega &= \int_{\gamma_-} \omega \\
&= \int_{-b}^{-a} \Big(-\omega_1\big(\gamma(-t)\big)\dot{x}^1(-t) - \cdots - \omega_n\big(\gamma(-t)\big)\dot{x}^n(-t) \Big)dt \\
&\overset{t=-\tau}{=} \int_b^a \Big(\omega_1\big(\gamma(\tau)\big)\dot{x}^1(\tau) + \cdots + \omega_n\big(\gamma(\tau)\big)\dot{x}^1(\tau) \Big)d\tau \\
&= -\int_a^b \Big(\omega_1\big(\gamma(\tau)\big)\dot{x}^1(\tau) + \cdots + \omega_n\big(\gamma(\tau)\big)\dot{x}^1(\tau) \Big)d\tau \\
&= -\int_\gamma \omega = -\int_{(C,\boldsymbol{T})} \omega.
\end{aligned}$$

□

If C is an oriented quasi-convenient curve, choose a convenient cut $\{\boldsymbol{p}_1, \dots, \boldsymbol{p}_\ell\}$ so that C becomes a disjoint union of convenient curves $C_1, \dots, C_N$ equipped with orientations. Then define

$$\boxed{\int_C \omega := \sum_{i=1}^N \int_{C_i} \omega}.$$

One can show that the right-hand side of the above equality is independent of the choice of the convenient cut.

Let us finally observe that the interior of an oriented (compact) curve with boundary C is oriented and quasi-convenient and we define

$$\boxed{\int_C \omega := \int_{C^\circ} \omega}.$$

Example 16.27. Suppose that C is the quarter of the unit circle contained in the first quadrant and equipped with the counter-clockwise orientation; see Figure 16.9.

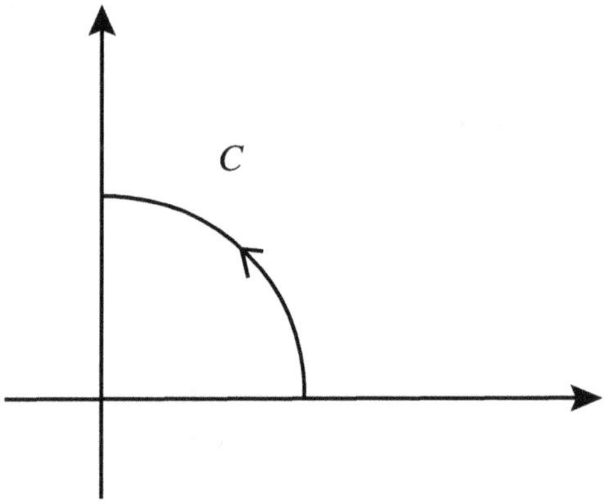

Fig. 16.9 *An arc of the unit circle equipped with the counterclockwise orientation.*

Its interior is a convenient curve and the map

$$(0, \pi/2) \ni t \mapsto (\cos t, \sin t) \in \mathbb{R}^2$$

is a parametrization compatible with the counterclockwise orientation. We have

$$\int_C W_\Theta = \int_C \left(\frac{-y}{x^2+y^2}dx + \frac{x}{x^2+y^2}dy \right)$$

(along C we have $dx = -\sin t dt$, $dy = \cos t dt$, $x^2 + y^2 = 1$)

$$= \int_0^{\frac{\pi}{2}} \Big((-\sin t)(-\sin t)dt + (\cos t)(\cos t)dt \Big) = \int_0^{\frac{\pi}{2}} dt = \frac{\pi}{2}. \qquad \square$$

Definition 16.28. A 1-*dimensional chain* in $\mathbb{R}^n$ is a collection

$$\mathscr{C} := \{ (C_1, m_1), \dots, (C_\nu, m_\nu) \}$$

where $C_1, \dots, C_n$ are compact, *oriented*, curves (with or without boundary) and $m_1, \dots, m_\nu$ are integers called the *local multiplicities* of the chain.

The integral of a 1-form ω over the above chain $\mathscr{C}$ is the real number

$$\int_{\mathscr{C}} \omega := \sum_{k=1}^{\nu} m_k \int_{C_k} \omega. \qquad \square$$

16.1.4. *The 2-dimensional Stokes' formula: a baby case*

The integrals of 1-forms over closed curves can often be computed as certain double integrals over appropriate regions. This is roughly speaking the content of Stokes' formula.

Definition 16.29. Let $k, n \in \mathbb{N}$. A *domain* of $\mathbb{R}^n$ is an open, path connected subset of $\mathbb{R}^n$. A domain $D \subset \mathbb{R}^n$ is called C^k if its boundary is an $(n-1)$-dimensional C^k-submanifold of $\mathbb{R}^n$. □

Example 16.30. In Figure 16.10 we have depicted two C^k-domains in $\mathbb{R}^2$. The domain D_1 is a closed disk and its boundary is a circle. The boundary of D_2 consists of three closed curves. □

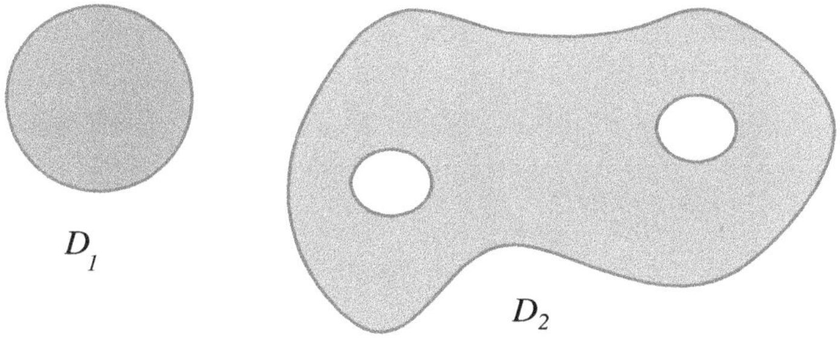

Fig. 16.10 *Two domains in* $\mathbb{R}^2$.

Suppose that $D \subset \mathbb{R}^2$ is a bounded C^k domain. As one can see from Figure 16.10, its boundary is a disjoint union of compact C^k curves in $\mathbb{R}^2$. Each of these components is equipped with a natural orientation called the *induced orientation*. It is determined by the right-hand rule; see Figure 16.11 and 16.12.

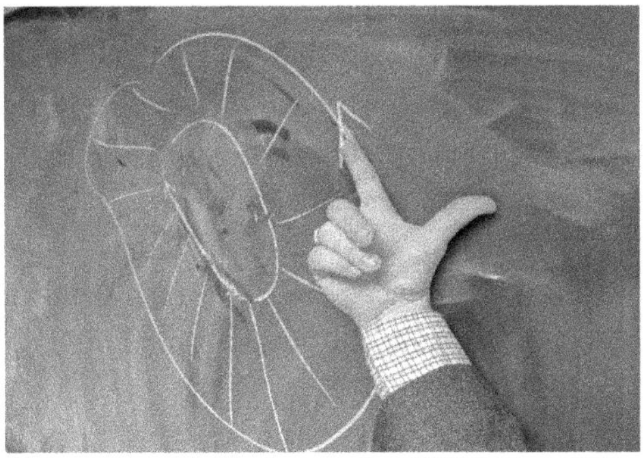

Fig. 16.11 *Right-hand rule.*

Fig. 16.12 *Right-hand rule.*

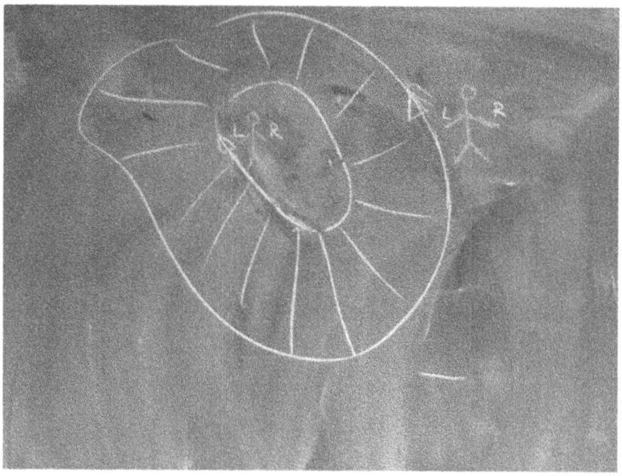

Fig. 16.13 *Walking around the boundary.*

Here is the explanation for the above figures: place your right hand on the boundary component, palm-up, so that the thumb points to the exterior of your domain. The index finger will then point in the direction given by the orientation of that boundary component. Another way of visualizing is as follows: if we walk along ∂D in the direction prescribed by the induced orientation, then the domain D will be to our left; see Figure 16.13.

Here is a more rigorous explanation. Given a point $\boldsymbol{p} \in \partial D$, there are exactly two unit vectors that are perpendicular to the tangent line $T_{\boldsymbol{p}}\partial D$. These vectors are called the *normal vectors* to ∂C at $\boldsymbol{p}$. Denote by $\boldsymbol{\nu}$ or $\boldsymbol{\nu}(\boldsymbol{p})$ the *outer normal vector at* $\mathbf{p} \in \partial D$: this vector is perpendicular to $T_{\boldsymbol{p}}\partial C$ and, when placed at $\boldsymbol{p}$, it points towards the exterior of D. The induced orientation at $\boldsymbol{p}$ is obtained from $\boldsymbol{\nu}(\boldsymbol{p})$ after a 90 degree counterclockwise rotation.

Thus, the boundary of a bounded C^1-domain $D \subset \mathbb{R}^2$ is a disjoint union of closed curves carrying orientations. It is therefore a 1-dimensional chain in the sense of Definition 16.28. Thus, we can integrate 1-forms on such boundaries.

Theorem 16.31 (Planar Stokes' theorem). *Let $U \subset \mathbb{R}^2$ be a bounded C^1-domain. Suppose that $\boldsymbol{F}$ is a C^1-vector field defined on an open set $\mathcal{O}$ that contains $\mathbf{cl}(U)$,*

$$\boldsymbol{F} : \mathcal{O} \to \mathbb{R}^2, \;\; \boldsymbol{F}(x,y) = \big(P(x,y), Q(x,y) \big).$$

Let $\boldsymbol{\nu} : \partial U \to \mathbb{R}^2$ be the outer normal vector field. We denoted by $\partial_+ U$ the boundary of U equipped with the induced orientation. Then the following hold.

$$\int_{\partial_+ U} (Pdx + Qdy) = \int_U \left(\frac{\partial Q}{\partial x} - \frac{\partial P}{\partial y} \right) |dxdy|, \tag{16.9a}$$

$$\int_{\partial U} \langle \boldsymbol{F}, \boldsymbol{\nu} \rangle ds = \int_U \left(\frac{\partial P}{\partial x} + \frac{\partial Q}{\partial y} \right) |dxdy|. \tag{16.9b}$$

□

Remark 16.32. It turns out that the equality (16.9b) follows rather easily from (16.9a). The hard part is proving (16.9a). □

Definition 16.33. Let $U \subset \mathbb{R}^n$ be an open set and

$$\boldsymbol{F} : U \to \mathbb{R}^n, \;\; \boldsymbol{F}(\boldsymbol{p}) = \begin{bmatrix} F^1(\boldsymbol{p}) \\ F^2(\boldsymbol{p}) \\ \vdots \\ F^n(\boldsymbol{p}) \end{bmatrix}$$

be a C^1 vector field. The *divergence* of $\boldsymbol{F}$ is the function

$$\operatorname{div} : U \to \mathbb{R}, \;\; \operatorname{div} \boldsymbol{F}(\boldsymbol{p}) = \frac{\partial F^1}{\partial x^1}(\boldsymbol{p}) + \cdots + \frac{\partial F^n}{\partial x^n}(\boldsymbol{p}).$$ □

Using the concept of divergence we can rephrase (16.9b) in the traditional form

$$\boxed{\int_{\partial D} \langle \boldsymbol{F}, \boldsymbol{\nu} \rangle ds = \int_D \operatorname{div} \boldsymbol{F} \, |dxdy|}. \tag{16.10}$$

The integral in the left-hand side is usually called the *the outer flux of $\boldsymbol{F}$ through ∂D*. The last equality is a special case of the *flux-divergence formula.*

Example 16.34. The *radial* vector field $\mathcal{R} : \mathbb{R}^2 \to \mathbb{R}^2$ is given by

$$\mathcal{R}(x,y) = \begin{bmatrix} x \\ y \end{bmatrix}.$$

Thus the vector field $\mathcal{R}$ associates to the point $\boldsymbol{p} \in (x,y)$, the point $\boldsymbol{p}$ itself but viewed as a vector in $\mathbb{R}^2$; see Figure 16.14. In other words, $\mathcal{R}$ is the identity map under another guise. Note that

$$\operatorname{div} \mathcal{R} = \partial_x x + \partial_y y = 2.$$

If D is a bounded domain with C^1-boundary, then the flux-divergence formula shows that the outer flux of $\mathcal{R}$ through the boundary ∂D is equal to twice the area of D. Thus we can compute the area of D by performing computations involving only boundary data and no interior data. □

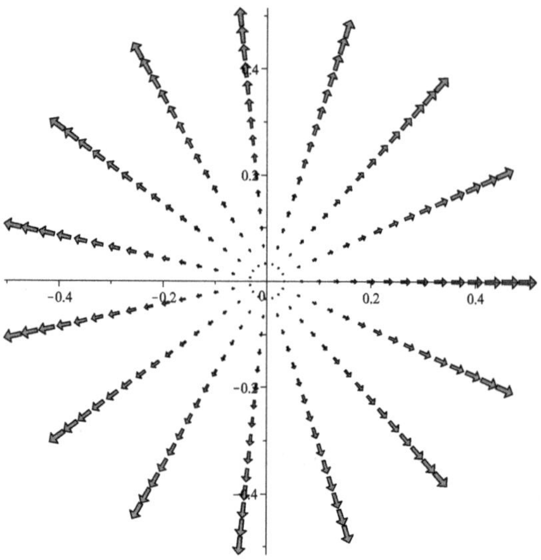

Fig. 16.14 *The radial vector field in the plane.*

Example 16.35. Suppose that $U \subset \mathbb{R}^2$ is a bounded C^1-domain whose boundary $C = \partial U$ is a closed connected curve. We assume that the origin $\mathbf{0}$ is not on the boundary of U. The induced orientation on ∂U is the counterclockwise orientation; see Figure 16.15. We denote by $\partial_+ U$ the boundary ∂U equipped with this orientation. We want to compute

$$\int_{\partial_+ U} W_\Theta = \int_{\partial_+ U} \Bigg(\underbrace{-\frac{y}{x^2+y^2}}_{P}\, dx + \underbrace{\frac{x}{x^2+y^2}}_{Q}\, dy \Bigg). \tag{16.11}$$

We distinguish two cases.

1. $\mathbf{0} \notin U$. To compute (16.11) we use (16.9a). To find $Q'_x - P'_y$ we set $r := \sqrt{x^2+y^2}$, so

$$P = -\frac{y}{r^2},\;\; Q = \frac{x}{r^2}.$$

We have

$$r'_x = \frac{x}{r},\;\; r'_y = \frac{y}{r},\;\; Q'_x = \frac{1}{r^2} - \frac{2x}{r^3}\cdot\frac{x}{r} = \frac{1}{r^2} - \frac{2x^2}{r^4},$$

$$P'_y = -\frac{1}{r^2} + \frac{2y}{r^3}\cdot\frac{y}{r} = -\frac{1}{r^2} + \frac{2y^2}{r^4}.$$

Thus

$$Q'_x - P'_y = \frac{2}{r^2} - \frac{2(x^2+y^2)}{r^4} = 0,\;\; \forall (x,y) \neq (0,0).$$

Using (16.9a) we deduce

$$\int_{\partial_+ U} W_\Theta = \int_U \big(Q'_x - P'_y \big)\, |dxdy| = 0.$$

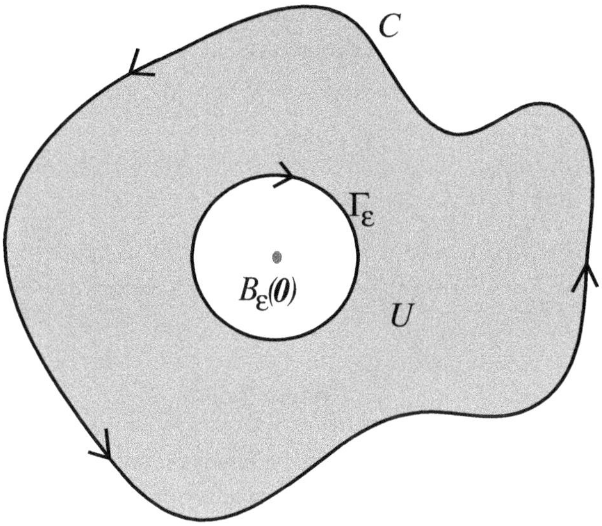

Fig. 16.15 *Integrating the angular form.*

2. $\mathbf{0} \in U$. The above approach does not work. Theorem 16.31 requires that the vector field Θ be defined on an open set containing U. This is not the case. The vector field Θ is not defined at the origin $\mathbf{0}$, and in fact the components P and Q do not have a limit as $(x, y) \to (0, 0)$ so they cannot be the restrictions of some continuous functions on U. Although this issue may seem trivial, it has enormous consequences.

To deal with this issue we will tread lightly around the singular point $\mathbf{0}$. Observe first that there exists $\varepsilon > 0$ such that the closed ball $\overline{B_\varepsilon(\mathbf{0})}$ is contained in U. Denote by U_ε the domain obtained by removing this ball from U,

$$U_\varepsilon := U \setminus \overline{B_\varepsilon(\mathbf{0})}.$$

The boundary of the domain U_ε has two components: the boundary ∂U and the boundary Γ_ε of $B_\varepsilon(\mathbf{0})$. Each of these components is equipped with an induced orientation: on ∂U this is the counterclockwise orientation, while on Γ_ε this is the clockwise orientation; see Figure 16.15. Observe that the clockwise orientation on Γ_ε is the opposite of the orientation induced as boundary of $B_\varepsilon(\mathbf{0})$. We denote by $\partial_+ B_\varepsilon(\mathbf{0})$ the curve Γ_ε equipped with the *counter*clockwise orientation. We have

$$\int_{\partial_+ U_\varepsilon} W_\theta = \int_{\partial_+ U} W_\Theta - \int_{\partial_+ B_\varepsilon(\mathbf{0})} W_\Theta.$$

Now observe that Θ is defined and C^1 on the open set $\mathbb{R}^2 \setminus \mathbf{0}$ that contains U_ε. We can now safely invoke Theorem 16.31 to conclude that

$$0 = \int_{U_\varepsilon} (Q'_x - P'_y) = \int_{\partial_+ U_\varepsilon} W_\Theta = \int_{\partial_+ U} W_\Theta - \int_{\partial_+ B_\varepsilon(\mathbf{0})} W_\Theta.$$

Hence

$$\int_{\partial_+ U} W_\Theta = \int_{\partial_+ B_\varepsilon(\mathbf{0})} W_\Theta.$$

To compute the right-hand side of the above equality observe that a parametrization of Γ_ε compatible with the counterclockwise orientation is

$$\gamma(t) = \big(\varepsilon\cos t, \varepsilon\sin t\big), \;\; t\in[0,2\pi].$$

Arguing exactly as in Example 16.27 we deduce

$$\int_{\partial_+ B_\varepsilon(\mathbf{0})} W_\Theta = \int_\gamma \left(\frac{-y}{x^2+y^2}dx + \frac{x}{x^2+y^2}dy\right)$$
$$= \int_0^{2\pi}\Big((-\sin t)(-\sin t)dt + (\cos t)(\cos t)dt\Big) = \int_0^{2\pi} dt = 2\pi.$$

This proves that

$$\int_{\partial_+ U} W_\Theta = 2\pi.$$

To put things in perspective let us mention that a famous result of Jordan[1] states that if C is a connected compact C^1-curve, then it is the boundary of a bounded C^1-domain U. We equip it with the orientation as boundary of U. If $\mathbf{0}\notin\partial U$, then $\int_C$ is well defined and the above computation shows

$$\frac{1}{2\pi}\int_{\partial_+ U} W_\Theta = I_U(\mathbf{0}) = \int_C W_\Theta = \begin{cases} 1, & \mathbf{0}\in U,\\ 0, & \mathbf{0}\notin U.\end{cases}$$

This "quantization" result has profound consequences in mathematics. □

The Planar Stokes' Theorem 16.31 extends to more general domains.

Definition 16.36. A domain $D\subset\mathbb{R}^2$ is called *piecewise* C^k if there exists a finite subset F of the boundary ∂D such that each component of $\partial D\setminus F$ is the interior of a C^k-curve with boundary. □

Example 16.37. Suppose that $\beta,\tau:[a,b]$ are C^1-functions such that

$$\beta(x) < \tau(x), \;\; \forall x\in(a,b).$$

Then the simple type domain

$$D(\beta,\tau) = \big\{\,(x,y)\in\mathbb{R}^2;\;\; x\in(a,b),\;\; \beta(x)<y<\tau(x)\,\big\} \tag{16.12}$$

is a piecewise C^1-domain. In particular an open rectangle $(a,b)\times(c,d)$ is a piecewise C^k domain. □

Suppose that $D\subset\mathbb{R}^2$ is a bounded piecewise C^1 domain. Remove a finite subset F of the boundary to obtain a union of convenient C^1-curves. In fact, each of these components is the interior of a (compact) curve with boundary. We denote by $\partial^* U$ the complement of this finite set. Using the right-hand rule we obtain orientations on each component of $\partial^* U$. We denote by $\partial^*_+ U$ the chain obtained this way, where the multiplicity of each oriented component of $\partial^* U$ is equal to 1. We have the following generalization of Theorem 16.31.

[1] I recommend T. Hales' very lively discussion of this result in [11].

Theorem 16.38 (Planar Stokes' theorem). *Let $U \subset \mathbb{R}^2$ be a bounded piecewise C^1-domain. Suppose that $\boldsymbol{F}$ is a C^1-vector field defined on an open set $\mathcal{O}$ that contains $\mathbf{cl}(U)$,*

$$\boldsymbol{F} : \mathcal{O} \to \mathbb{R}^2, \ \ \boldsymbol{F}(x, y) = \big(P(x, y), Q(x, y) \big).$$

Then

$$\int_{\partial_+^* U} (Pdx + Qdy) = \int_U \left(\frac{\partial Q}{dx} - \frac{\partial P}{dy} \right) |dxdy|. \tag{16.13}$$

□

16.2. Integration over surfaces

The various concepts of integrals over curves have higher dimensional counterparts, called integrals over submanifolds of a Euclidean space. In this section we will explain this concept only in the special case of 2-dimensional submanifolds. The proper presentation of the more general case requires a more complicated formalism that might bury the geometry of the construction. The restriction to the 2-dimensional situation will afford us more geometric transparency and a lighter algebraic burden. The extension to higher dimensions involves few new geometric ideas, but requires more algebraic travails.

The most basic example of integral over a surface is the concept of area. To define it we consider first the simplest of situations, when the surface in question is contained in a 2-dimensional vector subspace.

16.2.1. *The area of a parallelogram*

Suppose that we are given two linearly independent, i.e., non-collinear vectors

$$\boldsymbol{v}_1, \boldsymbol{v}_2 \in \mathbb{R}^2, \ \ \boldsymbol{v}_1 = \begin{bmatrix} v_1^1 \\ \\ v_1^2 \end{bmatrix}, \ \ \boldsymbol{v}_1 = \begin{bmatrix} v_2^1 \\ \\ v_2^2 \end{bmatrix}.$$

We denote by V the 2×2 matrix with columns $\boldsymbol{v}_1, \boldsymbol{v}_2$, i.e.,

$$V = \begin{bmatrix} v_1^1 & v_2^1 \\ \\ v_1^2 & v_2^2 \end{bmatrix}.$$

The parallelogram spanned by $\boldsymbol{v}_1, \boldsymbol{v}_2$ coincides with the parallelepiped spanned by these vectors defined in (15.34). More precisely, it is the set

$$P(\boldsymbol{v}_1, \boldsymbol{v}_2) = \big\{ x^1\boldsymbol{v}_1 + x^2\boldsymbol{v}_2; \ \ x^1, x^2 \in [0, 1] \big\} \subset \mathbb{R}^2. \tag{16.14}$$

According to (15.35), the area of this parallelogram is equal to the absolute value of the determinant of V,

$$\text{area}\big(P(\boldsymbol{v}_1, \boldsymbol{v}_2) \big) = \text{vol}_2 \big(P(\boldsymbol{v}_1, \boldsymbol{v}_2) \big) = \big| \det V \big|.$$

This equality has one "flaw": we need to know the coordinates of the vectors $\boldsymbol{v}_1, \boldsymbol{v}_2$. For the applications we have in mind we would like a formula that does involve this knowledge. To achieve this we consider the product of matrices

$$G = G(\boldsymbol{v}_1, \boldsymbol{v}_2) := V^T \cdot V = \begin{bmatrix} \langle \boldsymbol{v}_1, \boldsymbol{v}_1 \rangle & \langle \boldsymbol{v}_1, \boldsymbol{v}_2 \rangle \\ \langle \boldsymbol{v}_2, \boldsymbol{v}_1 \rangle & \langle \boldsymbol{v}_2, \boldsymbol{v}_2 \rangle \end{bmatrix}. \tag{16.15}$$

Note two things.

(i) The matrix $G(\boldsymbol{v}_1, \boldsymbol{v}_2)$ called the *Gramian* of $\boldsymbol{v}_1, \boldsymbol{v}_2$ is symmetric, and it is determined only by the scalar products $\langle \boldsymbol{v}_i, \boldsymbol{v}_j \rangle$.

(ii) We have

$$\det G = \det V^{\top} \det V = \big(\det V \big)^2.$$

We deduce the following very important formula

$$\operatorname{area}\big(P(\boldsymbol{v}_1, \boldsymbol{v}_2) \big) = \sqrt{\det G(\boldsymbol{v}_1, \boldsymbol{v}_2)}. \tag{16.16}$$

Now suppose that $n \in \mathbb{N}$, $n \geq 2$, and $\boldsymbol{u}_1, \boldsymbol{u}_2 \in \mathbb{R}^n$ are two linearly independent vectors. They define an $n \times 2$ matrix

$$U = [\boldsymbol{u}_1 \; \boldsymbol{u}_2]$$

whose columns are the vectors $\boldsymbol{u}_1, \boldsymbol{u}_2$. We define their Gramian $G(\boldsymbol{u}_1, \boldsymbol{u}_2)$ according to formula (16.15)

$$G(\boldsymbol{u}_1, \boldsymbol{u}_2) := U^{\top} U = \begin{bmatrix} \langle \boldsymbol{u}_1, \boldsymbol{u}_1 \rangle & \langle \boldsymbol{u}_1, \boldsymbol{u}_2 \rangle \\ \langle \boldsymbol{u}_2, \boldsymbol{u}_1 \rangle & \langle \boldsymbol{u}_2, \boldsymbol{u}_2 \rangle \end{bmatrix}.$$

The parallelogram spanned by $\boldsymbol{u}_1, \boldsymbol{u}_2$ is defined as in (16.14),

$$P(\boldsymbol{u}_1, \boldsymbol{u}_2) := \big\{ x^1 \boldsymbol{u}_1 + x^2 \boldsymbol{u}_2;\; x^1, x^2 \in [0,1] \big\} \subset \mathbb{R}^n.$$

We define the area of $P(\boldsymbol{u}_1, \boldsymbol{u}_2)$ by the formula

$$\boxed{\operatorname{area}\big(P(\boldsymbol{u}_1, \boldsymbol{u}_2) \big) := \sqrt{\det G(\boldsymbol{u}_1, \boldsymbol{u}_2)}}. \tag{16.17}$$

For example, if

$$\boldsymbol{u}_1 = (1,1,1) \in \mathbb{R}^3, \quad \boldsymbol{u}_2 = (1,0,-1) \in \mathbb{R}^3,$$

then $\langle \boldsymbol{u}_1, \boldsymbol{u}_1 \rangle = 3$, $\langle \boldsymbol{u}_1, \boldsymbol{u}_2 \rangle = 0$, $\langle \boldsymbol{u}_2, \boldsymbol{u}_2 \rangle = 2$,

$$G(\boldsymbol{u}_1, \boldsymbol{u}_2) = \begin{bmatrix} 3 & 0 \\ 0 & 2 \end{bmatrix}, \quad \det G(\boldsymbol{u}_1, \boldsymbol{u}_2) = 6, \quad \operatorname{area}\big(P(\boldsymbol{u}_1, \boldsymbol{u}_2) \big) = \sqrt{6}.$$

Remark 16.39. Let us point one other feature of (16.17) that adds extra plausibility to our definition by diktat of the area of a parallelogram in a higher dimensional space.

Recall (see Exercise 11.25) that an orthogonal operator $S : \mathbb{R}^n \to \mathbb{R}^n$ is a linear map S such that

$$\langle S\boldsymbol{u}, S\boldsymbol{v} \rangle = \langle \boldsymbol{u}, \boldsymbol{v} \rangle, \quad \forall \boldsymbol{u}, \boldsymbol{v} \in \mathbb{R}^n.$$

Orthogonal operators preserve distances between points. In particular, an orthogonal operator is bijective and it is natural to expect that it will map a parallelogram to another parallelogram with the same area. The area as defined in (16.14) displays this orthogonal invariance.

To see this, note that the definition of an orthogonal operator implies that, for any orthogonal operator S, we have

$$G(\boldsymbol{u}_1, \boldsymbol{u}_2) = G(S\boldsymbol{u}_1, S\boldsymbol{u}_2).$$

Note that the image via S of the parallelogram spanned by $\boldsymbol{u}_1, \boldsymbol{u}_2$ is the parallelogram spanned by $S\boldsymbol{u}_1, S\boldsymbol{u}_2$, i.e.,

$$S\big(P(\boldsymbol{u}_1, \boldsymbol{u}_2)\big) = P(S\boldsymbol{u}_1, S\boldsymbol{u}_2).$$

In particular, we deduce that

$$\operatorname{area}\big(SP(\boldsymbol{u}_1, \boldsymbol{u}_2)\big) = \operatorname{area}\big(P(S\boldsymbol{u}_1, S\boldsymbol{u}_2)\big) = \operatorname{area}\big(P(\boldsymbol{u}_1, \boldsymbol{u}_2)\big).$$

Let us also observe that if $\boldsymbol{u}_1, \boldsymbol{u}_2 \in \mathbb{R}^n$ are two linearly independent vectors then there exists an orthogonal operator $S : \mathbb{R}^n \to \mathbb{R}^n$ such that the vectors $\boldsymbol{v}_1 := S\boldsymbol{u}_1$ and $\boldsymbol{v}_2 := S\boldsymbol{u}_2$ belong to the subspace $\mathbb{R}^2 \times \mathbf{0} \subset \mathbb{R}^n$.[2] As we have explained at the beginning of this subsection the area of the parallelogram spanned by the vector $\boldsymbol{v}_1, \boldsymbol{v}_2 \subset \mathbb{R}^2$, defined in terms of Riemann integral, *must be given by* (16.17). Hence, due to the orthogonal invariance of the Gramian, the area of the parallelogram spanned by $\boldsymbol{u}_1, \boldsymbol{u}_2$ must also be given by (16.17). □

16.2.2. *Compact surfaces (with boundary)*

The concept of curve with boundary has a 2-dimensional counterpart.

Definition 16.40. Let $k, n \in \mathbb{N}$, $n \geq 2$. A *C^k-surface with boundary* in $\mathbb{R}^n$ is a *compact* subset $\Sigma \subset \mathbb{R}^n$ such that, for any point $\boldsymbol{p}_0 \in \Sigma$, there exists an open neighborhood $\mathcal{U}$ of $\boldsymbol{p}_0$ in $\mathbb{R}^n$ and a C^k-diffeomorphism $\Psi : \mathcal{U} \to \mathbb{R}^n$ such that $\overline{U} := \Psi(\mathcal{U} \cap \Sigma)$ is contained in the subspace $\mathbb{R}^2 \times \mathbf{0} \subset \mathbb{R}^n$, and it is either

(I) an open disk in $\mathbb{R}^2$ centered at $\Psi(\boldsymbol{p}_0)$ or
(B) the point $\Psi(\boldsymbol{p}_0)$ lies on the y-axis and $\overline{U}$ is the intersection of a disk as above with the half-plane

$$H_- := \big\{ (x, y) \in \mathbb{R}^2;\ \ x \leq 0 \big\}.$$

The pair $(\mathcal{U}, \Psi)$ is called a *straightening diffeomorphism* at $\boldsymbol{p}_0$. The pair $\big(\mathcal{U} \cap \Sigma, \Psi\big|_{\mathcal{U}\cap\Sigma}\big)$ is called a *local coordinate chart* of X at $\boldsymbol{p}_0$.

In the case (B), the point $\boldsymbol{p}_0 \in X$ is called a *boundary point* of X. Otherwise $\boldsymbol{p}_0$ is called an *interior* point.

The set of boundary points of X is called the *boundary* of X and it is denoted by ∂X. The set of interior points of X is called the *interior* of X and it is denoted by X°. The surface with boundary is called *closed* if its boundary is empty, $\partial X = \emptyset$. □

Remark 16.41. Before we present several examples of surfaces with boundary we want to mention without proofs a few technical facts.

- If $X \subset \mathbb{R}^n$ is a surface with boundary and $\boldsymbol{p}_0$ is a boundary point, then, *for any* straightening diffeomorphism $(\mathcal{U}, \Psi)$ at $\boldsymbol{p}_0$, the image $\Psi(U \cap X)$ is a half-disk B_r^-.
- The boundary of a surface with boundary is a *closed curve*. □

[2]This follows from a simple application of the Gram–Schmidt procedure.

Example 16.42 (Bounded C^k-domains in the plane). Suppose that $D \subset \mathbb{R}^2$ is a bounded C^k domain. For $n \geq 2$ we regard $\mathbb{R}^2$ as a subspace of $\mathbb{R}^n$. One can show that

- the closure $\boldsymbol{cl}(D)$ of a bounded C^k domain in $\mathbb{R}^2$ is a C^k-surface with boundary in $\mathbb{R}^n$;
- as a subset of $\mathbb{R}^2$, the closure $\boldsymbol{cl}(D)$ is Jordan measurable.

We will not present the proofs of the above claims. □

Example 16.43 (Graphs). Suppose that $D \subset \mathbb{R}^2$ is a bounded C^k domain and f is a C^k function defined on some open set U containing the closure of D, $f : U \to \mathbb{R}$. Then the graph of f over D

$$\Gamma_f(D) := \{ (x, y, z) \in \mathbb{R}^3; \ (x, y) \in D, \ z = f(x, y) \}$$

is a surface with boundary. Figure 16.16 depicts such a graph.

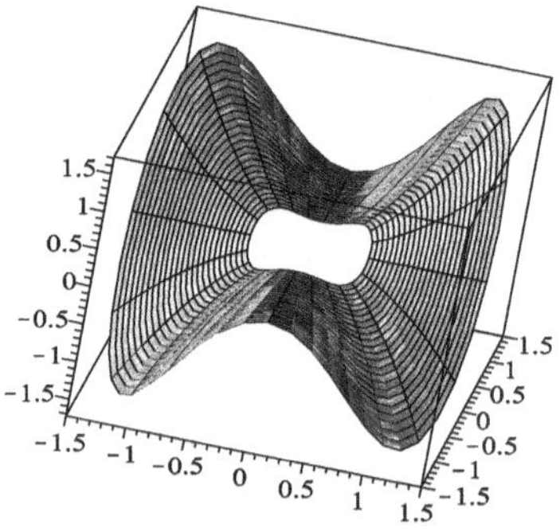

Fig. 16.16 *The graph of $f(x, y) = 2xy^2$ over the annular domain $0.5 \leq \sqrt{x^2 + y^2} \leq 1.5$ in $\mathbb{R}^2$.*

Example 16.44 (Surfaces of revolution). Given a C^1-function $f : [a, b] \to (0, \infty)$, then by rotating its graph about the x-axis we obtain a surface with boundary S_f whose boundary consists of two circles; see Figure 16.17. □

Example 16.45 (Cutting surfaces transversally by a hypersurface). Suppose that $S \subset \mathbb{R}^n$ is a closed C^k-surface and $f : \mathbb{R}^n \to \mathbb{R}$ is a C^k-function. Suppose that

$$\forall \boldsymbol{x} \in \mathbb{R}^n \ \ f(\boldsymbol{x}) = 0 \Rightarrow \nabla f(\boldsymbol{x}) \neq 0.$$

The zero set

$$Z = \{ \boldsymbol{x} \in \mathbb{R}^n; \ f(\boldsymbol{z}) = 0 \}$$

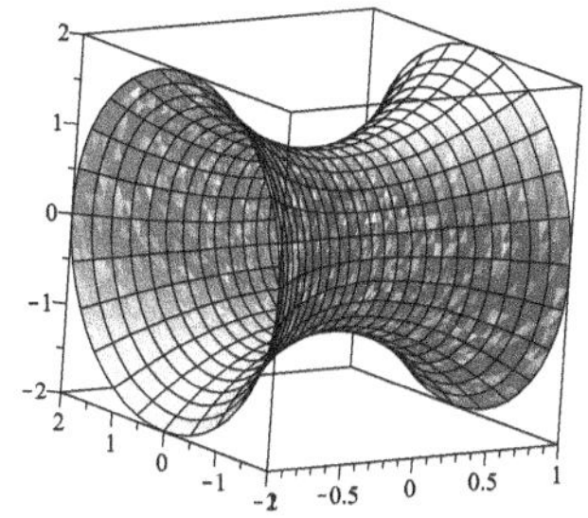

Fig. 16.17 *The surface of revolution generated by the graph of the function* $f : [-2, 1] \to (0, \infty)$, $f(x) = x^2 + 1$.

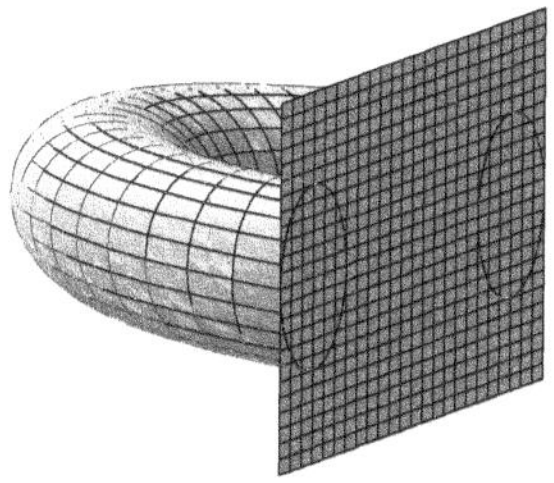

Fig. 16.18 *A half-torus in* $\mathbb{R}^3$.

is a hypersurface of $\mathbb{R}^n$. One expects that the intersection of the surface S with the hypersurface Z is a curve; think for example of a plane Z in $\mathbb{R}^3$ intersecting a surface S in $\mathbb{R}^3$.

This intuition is indeed true if this intersection is *transversal*. This means that, for any $\boldsymbol{p} \in S \cap Z$, the gradient vector $\nabla f(\boldsymbol{p})$ is *not perpendicular* to the tangent plane $T_{\boldsymbol{p}}S$. This claim follows from the implicit function theorem, but we will omit the details.

If this transversality condition is satisfied then

$$S_+ := \{\, \boldsymbol{p} \in S;\ \ f(\boldsymbol{p}) \geq 0 \,\}$$

is a surface with boundary $\partial S_+ = S \cap Z$. To get a better hold of this fact, think that S is the surface of a floating iceberg and $f(\boldsymbol{p})$ denotes the altitude of a point. Then, S_+ consists of the points on the iceberg with altitude ≥ 0, i.e., the points on the surface above the water level.

For example we cut the torus in Figure 14.6 with the vertical plane $y = 0$, we obtain the half-torus depicted in Figure 16.18.

Also, if we cut the sphere $S = \{x^2 + y^2 + z^2 = 1\}$ with the plane $z = \frac{1}{2}$, then the part above the level $z = \frac{1}{2}$ is the polar cap depicted in Figure 16.19. □

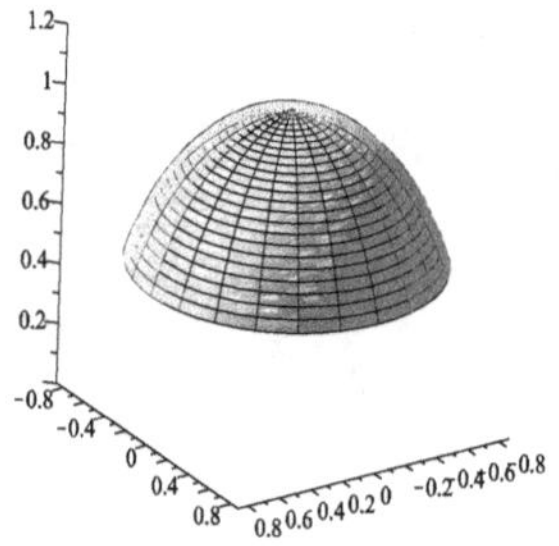

Fig. 16.19 *A half-sphere in* $\mathbb{R}^3$.

16.2.3. *Integrals over surfaces*

Let $n \in \mathbb{N}$, $n \geq 2$, and suppose that $\Sigma \subset \mathbb{R}^n$ is a C^1-surface. Fix a point $\boldsymbol{p}_0$ and a straightening diffeomorphism (U, Ψ) near $\boldsymbol{p}_0$; see Definition 14.31.

The restriction of Ψ to $U \cap \Sigma$ is a continuous bijection onto an open subset

$$V \subset \mathbb{R}^2 = \mathbb{R}^2 \times \mathbf{0} \subset \mathbb{R}^n.$$

Its inverse is given by a C^1-map $\Phi : V \to \mathbb{R}^n$ such that $\Phi(V) = U \cap \Sigma$. The map Φ is called the *local parametrization* of the surface Σ associated to the straightening diffeomorphism (U, Ψ).

The local parametrization Φ deforms the flat planar region V to a patch $\Phi(V)$ of the curved surface Σ; see Figure 16.20. The region V lies in the 2-dimensional vector space $\mathbb{R}^2$, and we denote by (s, t) the coordinates in this space. Moreover, it may help to think of the plane $\mathbb{R}^2$ as made of horizontal and vertical lines woven together. These lines form a grid dividing the plane into infinitesimally small rectangular patches of size $ds \times dt$; see Figure 16.21.

The parametrization Φ takes the rectangular (s, t)-grid to a curvilinear grid on the surface Σ; see Figure 16.20. For any point $\boldsymbol{p}$ in the patch $\Phi(V) \subset \Sigma$ there exists a *unique* point $(s, t) \in V$ such that $\boldsymbol{p} = \Phi(s, t)$. We will write this in a simplified form as

$$\boldsymbol{p} = \boldsymbol{p}(s, t).$$

If we keep t fixed and vary s we get a (red) horizontal line in the (s, t)-plane; see Figures 16.20 and 16.21. This (red) line is mapped by Φ to a (red) curve on Σ. Similarly, if we keep s fixed and vary t we get a vertical line in the (s, t)-plane mapped to a curve on Σ.

Now take a tiny rectangular patch of size $ds \times dt$ with lower left-hand corner situated at some point (s, t). The map Φ sends it to a tiny (infinitesimal) patch of the surface. Because this is so small we can assume it is almost flat and we can approximate it with the parallelogram spanned by the vectors (see Figure 16.20).

$$\boldsymbol{p}'_s ds = \Phi'_s ds \;\text{ and }\; \boldsymbol{p}'_t dt = \Phi'_t dt.$$

The area of this infinitesimal tangent parallelogram is usually referred to as the *area element*. It is denoted by $|dA|$ and we have

$$|dA| = \sqrt{\det G(\boldsymbol{p}'_s ds, \boldsymbol{p}'_t dt)} = \sqrt{\det G(\boldsymbol{p}'_s, \boldsymbol{p}'_t)}\; |dsdt|.$$

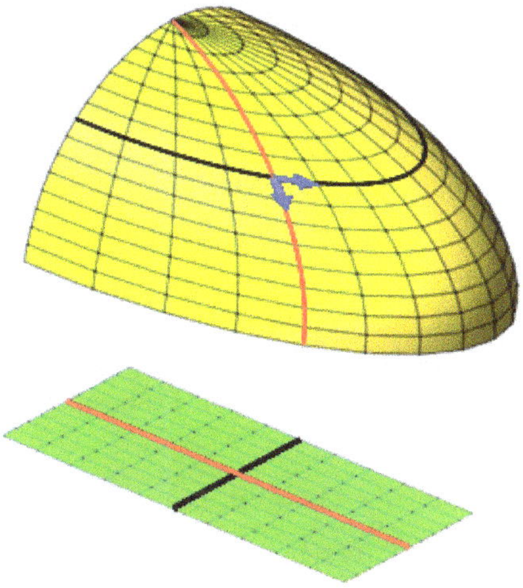

Fig. 16.20 *A local parametrization deforms a (flat) planar region to a patch of (curved) surface.*

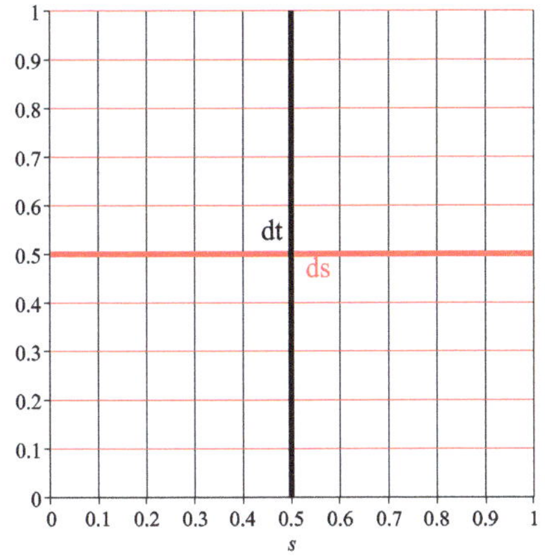

Fig. 16.21 *A planar infinitesimal grid.*

More intuitively, the parametrization Φ identifies a point $\boldsymbol{p} \in U \cap \Sigma$ with a point $(s,t) \in V \subset \mathbb{R}^2$. We write this $\boldsymbol{p} = \boldsymbol{p}(s,t)$ and we rewrite the above equality

$$|dA| = \sqrt{\det G(\boldsymbol{p}'_s, \boldsymbol{p}'_t)}\ |dsdt|.$$

The quantity $\sqrt{\det G(\boldsymbol{p}'_s, \boldsymbol{p}'_t)}\ |dsdt|$ describes the area element in the s,t coordinates. We provisorily define the area of $U \cap \Sigma$ to be

$$\boxed{\operatorname{area}(U \cap \Sigma) = \int_{U\cap\Sigma} |dA| := \int_V \sqrt{\det G(\boldsymbol{p}'_s, \boldsymbol{p}'_t)}\ |dsdt|}. \tag{16.18}$$

Suppose that $(U, \hat{\Psi})$ is another straightening diffeomorphism of S near $\boldsymbol{p}_0$. The restriction of $\hat{\Psi}$ to $U \cap \Sigma$ is a continuous bijection onto an open subset

$$\hat{V} \subset \mathbb{R}^2 = \mathbb{R}^2 \times \mathbf{0} \subset \mathbb{R}^n.$$

Its inverse is given by a C^1-map $\hat{\Phi} : \overline{V} \to \mathbb{R}^n$ such that $\hat{\Phi}(\overline{V}) = U \cap \Sigma$. We denote by (u, v) the Euclidean coordinates in the plane $\mathbb{R}^2$ that contains $\overline{V}$. A point $\boldsymbol{p} \in U \cap \Sigma$ can now be identified either with a point $(s, t) \in V$ or with a point $(u, v) \in \overline{V}$. We write this $\boldsymbol{p} = \boldsymbol{p}(s, t)$ and respectively $\boldsymbol{p} = \boldsymbol{p}(u, v)$.

We obtain in this fashion a map $V \mapsto \hat{V}$ that associates to the point $(s, t) \in V$ the unique point $(u, v) \in \hat{V}$ such that $\boldsymbol{p}(s, t) = \boldsymbol{p}(u, v)$. We will indicate this correspondence $(s, t) \mapsto (u, v)$ by writing $u = u(s, t)$ and $v = v(s, t)$. Proposition 14.34(B) implies that the correspondence is C^1. To recap

$$u = u(s, t), \quad v = v(s, t) \Longleftrightarrow \boldsymbol{p}(s, t) = \boldsymbol{p}(u, v).$$

We now have two possible definitions of area$(U \cap \Sigma)$

$$\operatorname{area}(U \cap \Sigma) = \int_V \sqrt{G(\boldsymbol{p}'_s, \boldsymbol{p}'_t)}\, |dsdt| \;\text{ or }\; \operatorname{area}(U \cap \Sigma) = \int_{\overline{V}} \sqrt{G(\boldsymbol{p}'_u, \boldsymbol{p}'_v)}\, |dudv|.$$

We want to show that they both yield the same result.

Lemma 16.46.

$$\int_V \sqrt{\det G(\boldsymbol{p}'_s, \boldsymbol{p}'_t)}\, |dsdt| = \int_{\overline{V}} \sqrt{\det G(\boldsymbol{p}'_u, \boldsymbol{p}'_v)}\, |dudv|.$$

Proof. We argue as in the proof of the equality (16.1). The key fact is the equality

$$\boldsymbol{p}(s, t) = \boldsymbol{p}(u, v).$$

Differentiating with respect to s, t we deduce

$$\boldsymbol{p}'_s = \boldsymbol{p}'_u u'_s + \boldsymbol{p}'_v v'_s, \quad \boldsymbol{p}'_t = \boldsymbol{p}'_u u'_t + \boldsymbol{p}'_v v'_t \tag{16.19}$$

Let us introduce the $n \times 2$ matrices

$$P_{s,t} := [\boldsymbol{p}'_s\ \boldsymbol{p}'_t], \quad P_{u,v} := [\boldsymbol{p}'_u\ \boldsymbol{p}'_v].$$

Thus the columns of $P_{s,t}$ consist of the vectors $\boldsymbol{p}'_s, \boldsymbol{p}'_t \in \mathbb{R}^n$, and the columns of $P_{u,v}$ consist of the vectors $\boldsymbol{p}'_u, \boldsymbol{p}'_v \in \mathbb{R}^n$. We can rewrite (16.19) as

$$P_{s,t} = P_{u,v} \cdot \underbrace{\begin{bmatrix} u'_s & u'_t \\ v'_s & v'_t \end{bmatrix}}_{J}.$$

We note that J is the Jacobian matrix of the transformation $(s, t) \mapsto (u, v)$

$$J = \frac{\partial(u, v)}{\partial(s, t)}.$$

Then

$$G(\boldsymbol{p}'_s, \boldsymbol{p}'_t) = P_{s,t}^T P_{s,t} = J^T P_{u,v}^T P_{u,v} J = J^T G(\boldsymbol{p}'_u, \boldsymbol{p}'_v) J,$$

so

$$\det G(\boldsymbol{p}'_s, \boldsymbol{p}'_t) = (\det J^T) \det G(\boldsymbol{p}'_u, \boldsymbol{p}'_v)(\det J) = \det G(\boldsymbol{p}'_u, \boldsymbol{p}'_v)(\det J)^2,$$

and thus

$$\sqrt{\det G(\boldsymbol{p}'_s, \boldsymbol{p}'_t)} = \sqrt{\det G(\boldsymbol{p}'_u, \boldsymbol{p}'_v)} \cdot |\det J|.$$

The change in variables formula then implies

$$\begin{aligned} \int_{\overline{V}} \sqrt{\det G(\boldsymbol{p}'_u, \boldsymbol{p}'_v)}\, |dudv| &= \int_V \sqrt{\det G(\boldsymbol{p}'_u, \boldsymbol{p}'_v)} \left| \det \frac{\partial(u, v)}{\partial(s, t)} \right| |dsdt| \\ &= \int_V \sqrt{\det G(\boldsymbol{p}'_u, \boldsymbol{p}'_v)} \cdot |\det J|\, |dsdt| = \int_V \sqrt{\det G(\boldsymbol{p}'_s, \boldsymbol{p}'_t)}\, |dsdt|. \end{aligned}$$

□

Definition 16.47 (Convenient surfaces with boundary). A *parametrized surface with boundary* is a triplet (Σ, D, Φ) with the following properties.

- $D \subset \mathbb{R}^2$ is a bounded domain in $\mathbb{R}^2$ with C^1-boundary;
- Φ is an injective immersion $\Phi : U \to \mathbb{R}^n$, where U is an open subset of $\mathbb{R}^2$ containing the closure $\boldsymbol{cl}(D)$ of D.
- $\Sigma = \Phi\big(\, \boldsymbol{cl}(D)\,\big)$.

The induced map $\Phi : \boldsymbol{cl}(D) \to \Sigma$ is called a *parametrization* of Σ. A surface with boundary is called *convenient* if it can be parametrized as above. □

For example, the surfaces depicted in Figures 16.16, 16.18, 16.19 and 16.17 are convenient.

Suppose now that $\Sigma \subset \mathbb{R}^n$ is a convenient surface with boundary with a parametrization $\Phi : \boldsymbol{cl}(D) \to \Sigma$. Here $D \subset \mathbb{R}^2$ is a bounded domain with C^1-boundary. If we denote by s, t the Euclidean coordinates in the plane $\mathbb{R}^2$ containing D, then we can describe the parametrization Φ as describing point $\boldsymbol{p}$ on Σ depending on the coordinates,

$$\boldsymbol{p} = \boldsymbol{p}(s,t).$$

If $f : \Sigma \to \mathbb{R}$ is a function on Σ, then we say that it is integrable over Σ if the function

$$D \ni (s,t) \mapsto f\big(\boldsymbol{p}(s,t)\big)\sqrt{\det G(\boldsymbol{p}'_s, \boldsymbol{p}'_t)}$$

is Riemann integrable. If this is the case, then we define *the integral of f over Σ* to be the number

$$\boxed{\int_{\Sigma} f(\boldsymbol{p})|dA(\boldsymbol{p})| := \int_D f\big(\boldsymbol{p}(s,t)\big)\sqrt{\det G(\boldsymbol{p}'_s, \boldsymbol{p}'_t)}\,|dsdt|}\,.$$

The left-hand side of the above equality makes no reference of any parametrization while the right-hand side is obviously described in terms of a concrete parametrization. We want to show that, if we change the parametrization, then the resulting right-hand side does not change its value, although it might look dramatically different.

If $\overline{\Phi} : \overline{D} \to \Sigma$ is another parametrization of Σ,

$$\overline{D} \in (u,v) \mapsto \boldsymbol{p}(u,v) = \overline{\Phi}(u,v),$$

then Lemma 16.46 shows that

$$\int_D f\big(\boldsymbol{p}(s,t)\big)\sqrt{\det G(\boldsymbol{p}'_s, \boldsymbol{p}'_t)}\,|dsdt| = \int_D f\big(\boldsymbol{p}(u,v)\big)\sqrt{\det G(\boldsymbol{p}'_u, \boldsymbol{p}'_v)}\,|dudv|.$$

Example 16.48 (Integration along graphs). Suppose that $D \subset \mathbb{R}^2$ is a bounded domain with C^1-boundary, and $h : \boldsymbol{cl}(D) \to \mathbb{R}$ is a C^1-function. Its graph

$$\Gamma_f := \big\{(x,y,z);\ \ (x,y) \in \boldsymbol{cl}(D),\ z = h(x,y)\,\big\}$$

is a convenient surface with boundary. The map

$$(x,y) \mapsto \boldsymbol{p}(x,y) := \big(\,x, y, h(x,y)\,\big) \in \mathbb{R}^3$$

is a parametrization of Γ_h. We have

$$\boldsymbol{p}'_x = \big(1, 0, h'_x(x, y)\big), \;\; \boldsymbol{p}'_y = \big(0, 1, h'_y(x, y)\big),$$

$$\langle \boldsymbol{p}'_x, \boldsymbol{p}'_x\rangle = 1 + \left|h'_x\right|^2, \;\; \langle \boldsymbol{p}'_y, \boldsymbol{p}'_y\rangle = 1 + \left|h'_y\right|^2, \;\; \langle \boldsymbol{p}'_x, \boldsymbol{p}'_y\rangle = h'_x h'_y,$$

$$\begin{aligned}\det G(\boldsymbol{p}'_x, \boldsymbol{p}'_y) &= \langle \boldsymbol{p}'_x, \boldsymbol{p}'_x\rangle \cdot \langle \boldsymbol{p}'_y, \boldsymbol{p}'_y\rangle - \langle \boldsymbol{p}'_x, \boldsymbol{p}'_y\rangle^2\\ &= \big(1 + \left|h'_x\right|^2\big)\big(1 + \left|h'_y\right|^2\big) - \big(h'_x h'_y\big)^2\\ &= 1 + \left|h'_x\right|^2 + \left|h'_y\right|^2 = 1 + \left\|\nabla h\right\|^2.\end{aligned}$$

Hence, the area element on the graph of h is

$$\boxed{|dA| = \sqrt{1 + \left\|\nabla h\right\|^2}\,|dxdy|}.$$

In particular, the area of Γ_h is

$$\operatorname{area}(\Gamma_h) = \int_D \sqrt{1 + \left\|\nabla h\right\|^2}\,|dxdy|. \qquad \square$$

Example 16.49 (Revolving graphs about the z-axis). Suppose that $0 < a < b$ and $f : [a, b] \to \mathbb{R}$ is a C^1-function. We denote by S_f the surface in $\mathbb{R}^3$ obtained by revolving the graph of f in the (x, z) plane,

$$\Gamma_f = \big\{(x, z);\;\; x \in [a, b],\;\; z = f(x)\,\big\}$$

about the z-axis. Denote by D the annulus in the (x, y)-plane described by the condition

$$a \leq r \leq b, \;\; r = \sqrt{x^2 + y^2}.$$

Then S_f is the graph of the function

$$\hat{f} : D \to \mathbb{R}, \;\; \hat{f}(x, y) = f(r) = f\big(\sqrt{x^2 + y^2}\,\big).$$

Note that

$$\hat{f}'_x = f'(r)\frac{x}{r}, \;\; \hat{f}'_y = f'(r)\frac{y}{r}, \;\; 1 + \|\nabla\hat{f}\|^2 = 1 + |f'(r)|^2.$$

Thus

$$\boxed{|dA| = \sqrt{1 + |f'(r)|^2}\,|dxdy|}.$$

$\square$

Suppose in general that Σ is a compact surface with, or without boundary, and

$$f : \Sigma \to \mathbb{R}$$

is a continuous function. We want to define the integral of f over Σ. We distinguish two cases.

Case 1. Let $\boldsymbol{p}_0 \in \Sigma$ and suppose that (U, Ψ) is a straightening diffeomorphism at $\boldsymbol{p}_0$ (see Definition 16.40). This induces a homeomorphism

$$\Psi : U \cap \Sigma \to D = \Psi(U \cap \Sigma) \subset \mathbb{R}^2,$$

where D is either an open disk or an open half-disk. We denote by (s,t) the Euclidean coordinates on $\mathbb{R}^2$. If

$$\boxed{\operatorname{supp} f \subset U}, \tag{16.20}$$

then we define

$$\int_\Sigma f\,|dA| = \int_D f\big(\boldsymbol{p}(s,t)\big)\sqrt{\det G(\boldsymbol{p}'_s,\boldsymbol{p}'_t)}\,|dsdt|.$$

Case 2. Let us explain how to deal with the general case, when the support of f is not necessarily contained in the domain of a straightening diffeomorphism as in (16.20).

Fix an *atlas* of Σ, i.e., a collection

$$\big\{\,(U_\alpha,\Psi_\alpha)\,\big\}_{\alpha\in\mathcal{A}}$$

where each (U_α,Ψ_α) is a straightening diffeomorphism of Σ at a point $\boldsymbol{p}_\alpha\in\Sigma$ and

$$\Sigma\subset\bigcup_{\alpha\in\mathcal{A}}U_\alpha.$$

Now choose a continuous partition of unity $\chi_1,\dots,\chi_N$ subordinated to the open cover $(U_\alpha)_{\alpha\in\mathcal{A}}$ of Σ. Thus, each χ_i is a compactly supported continuous function $\mathbb{R}^n\to\mathbb{R}$ and, for each $i=1,\dots,N$, there exists $\alpha(i)\in\mathcal{A}$ such that

$$\operatorname{supp}\chi_i\subset U_{\alpha(i)}. \tag{16.21}$$

Moreover

$$\chi_1(\boldsymbol{p})+\cdots+\chi_N(\boldsymbol{p})=1,\ \ \forall\boldsymbol{p}\in\Sigma.$$

In particular

$$f(\boldsymbol{p})=\chi_1(\boldsymbol{p})f(\boldsymbol{p})+\cdots+\chi_N(\boldsymbol{p})f(\boldsymbol{p}),\ \ \forall\boldsymbol{p}\in\Sigma.$$

Due to the inclusions (16.21), each of the functions $\chi_i f$ satisfies the support condition (16.20). We define

$$\boxed{\int_\Sigma f|dA| := \sum_{i=1}^N\int_\Sigma \chi_i f|dA|},$$

where each of the integrals in the right-hand side are defined as in **Case 1**.

Clearly, the definition used in Case 2 depends on several choices.

- A choice of atlas, i.e., a collection $\big\{\,(U_\alpha,\Psi_\alpha);\ \ \alpha\in\mathcal{A}\,\big\}$ of local charts covering Σ.
- A choice of a partition of unity subordinated to the open cover $(U_\alpha)_{\alpha\in\mathcal{A}}$ of Σ.

One can show that the end result is independent of these choices. We omit the proof of this fact. This type of integral over a surface is traditionally known as *surface integral of the first kind.*

The above definition of the integral of a continuous function over a compact surface is not very useful for concrete computations. In concrete situations surfaces are *quasi-parametrized.*

Definition 16.50. Suppose that $\Sigma\subset\mathbb{R}^n$ is a compact C^1-surface with or without boundary. A *quasi-parametrization* of Σ is an injective C^1-immersion $\Phi: D\to\mathbb{R}^n$, where $D\subset\mathbb{R}^2$ is a bounded domain, such that $\Phi(D)$ *almost covers* Σ, i.e., $\Phi(D)\subset\Sigma$ and $\Sigma\setminus\Phi(D)$ is a finite union of compact C^1-curves with or without boundary. □

The next result extends Theorem 15.50 to "curved" situations.

Theorem 16.51. *Let $n \in \mathbb{N}$ and suppose that $\Sigma \subset \mathbb{R}^n$ is a compact surface, with or without boundary. Suppose that*

$$\Phi : D \to \mathbb{R}^n, \;\; (s,t) \mapsto \boldsymbol{p}(s,t) := \Phi(s,t) \in \mathbb{R}^n$$

is quasi-parametrization of Σ. Then, for any continuous function $f : \Sigma \to \mathbb{R}$ we have

$$\int_\Sigma f(\boldsymbol{p})\,|dA(\boldsymbol{p})| = \int_D f\big(\boldsymbol{p}(s,t)\big)\sqrt{\det G(\boldsymbol{p}'_s, \boldsymbol{p}'_t)}\,|dsdt|. \qquad \square$$

Example 16.52. Consider the unit sphere

$$S := \big\{\,(x,y,z) \in \mathbb{R}^3;\;\; x^2+y^2+z^2 = 1\,\big\}.$$

In spherical coordinates (ρ, φ, θ) this sphere is described by the equation $\rho = 1$.

The spherical coordinates define an injective immersion

$$(0,\pi) \times (0, 2\pi) \ni (\varphi, \theta) \mapsto \boldsymbol{p}(\varphi,\theta) = (\sin\varphi\cos\theta, \sin\varphi\sin\theta, \cos\varphi)$$

that almost covers S: its image is the complement in the sphere of the "meridian" obtained by intersecting the sphere with the half-plane

$$y = 0, x \geq 0.$$

Thus this map almost covers S. Note that

$$\boldsymbol{p}'_\varphi = (\cos\varphi\cos\theta, \cos\varphi\sin\theta, -\sin\varphi), \;\; \boldsymbol{p}'_\theta = (-\sin\varphi\sin\theta, \sin\varphi\cos\theta, 0).$$

Then

$$\langle \boldsymbol{p}'_\varphi, \boldsymbol{p}'_\varphi\rangle = (\cos\varphi\cos\theta)^2 + (\cos\varphi\sin\theta)^2 + \sin^2\varphi = 1,$$

$$\langle \boldsymbol{p}'_\varphi, \boldsymbol{p}'_\theta\rangle = 0, \;\; \langle \boldsymbol{p}'_\theta, \boldsymbol{p}'_\theta\rangle = (\sin\varphi\sin\theta)^2 + (\sin\varphi\cos\theta)^2 = \sin^2\varphi,$$

so that

$$\sqrt{\det G(\boldsymbol{p}'_\varphi, \boldsymbol{p}'_\theta)} = \sin\varphi.$$

This shows that, *in spherical coordinates, the area element of the sphere is*

$$\boxed{|dA(\boldsymbol{p})| = \sin\varphi |d\varphi d\theta|}.$$

If $f : S \to \mathbb{R}$ is a continuous function, then

$$\int_S f(\boldsymbol{p})\,|dA(\boldsymbol{p})| = \int_{\substack{0\leq\theta\leq,2\pi,\\ 0\leq\varphi\leq\pi}} f(\varphi,\theta)\sin\varphi\,|d\varphi d\theta|.$$

In particular

$$\operatorname{area}(S) = \int_{\substack{0\leq\theta\leq 2\pi,\\ 0\leq\varphi\leq\pi}} \sin\varphi\,|d\varphi d\theta| = \underbrace{\left(\int_0^{2\pi} d\theta\right)}_{=2\pi}\underbrace{\left(\int_0^{\pi} \sin\varphi d\varphi\right)}_{=2} = 4\pi. \qquad \square$$

Example 16.53. Consider again the unit sphere

$$S := \{ (x,y,z) \in \mathbb{R}^3;\ \ x^2+y^2+z^2=1 \}.$$

Fix a real number $c \in (0,1)$ and consider the polar cap

$$S_c := \{ (x,y,z) \in S;\ \ z \geq c \}.$$

We want to compute the area of this surface with boundary. This time we will use cylindrical coordinates (r,θ,z). In cylindrical coordinates the sphere is described by the equation

$$r^2+z^2=1.$$

In the northern hemisphere $z>0$ we have

$$z=\sqrt{1-r^2}=\sqrt{1-x^2-y^2}.$$

Along the polar cap we have $z \geq c$ so

$$\sqrt{1-r^2} \geq c \Rightarrow 1-r^2 \geq c^2 \Rightarrow r^2 \leq 1-c^2 \Rightarrow s \leq \sqrt{1-c^2}.$$

Thus the polar cap admits the quasi-parametrization

$$\boldsymbol{p} = \begin{bmatrix} x \\ y \\ z \end{bmatrix} = \begin{bmatrix} r\cos\theta \\ r\sin\theta \\ \sqrt{1-r^2} \end{bmatrix},\ \ 0 \leq r \leq \sqrt{1-c^2},\ \ \theta \in [0,2\pi].$$

Then

$$\boldsymbol{p}'_r = \begin{bmatrix} \cos\theta \\ \sin\theta \\ -\frac{r}{\sqrt{1-r^2}} \end{bmatrix},\ \ \boldsymbol{p}'_\theta = \begin{bmatrix} -r\sin\theta \\ r\cos\theta \\ 0 \end{bmatrix},$$

$$\langle \boldsymbol{p}'_r, \boldsymbol{p}'_r \rangle = 1+\frac{r^2}{1-r^2} = \frac{1}{1-r^2},\ \ \langle \boldsymbol{p}'_\theta, \boldsymbol{p}'_\theta \rangle = r^2,\ \ \langle \boldsymbol{p}'_r, \boldsymbol{p}'_\theta \rangle = 0,$$

$$\det G(\boldsymbol{p}'_r, \boldsymbol{p}'_\theta \rangle = \langle \boldsymbol{p}'_r, \boldsymbol{p}'_r \rangle \cdot \langle \boldsymbol{p}'_\theta, \boldsymbol{p}'_\theta \rangle = \frac{r^2}{1-r^2}.$$

Then, in polar coordinates (r,θ) the area on the unit sphere is given by

$$|dA| = \frac{r}{\sqrt{1-r^2}}|drd\theta|,$$

and we deduce

$$\begin{aligned} \operatorname{area}(S_c) &= \int_{\substack{0 \leq r \leq \sqrt{1-c^2} \\ \theta \in [0,2\pi]}} \frac{r}{\sqrt{1-r^2}} drd\theta \\ &= 2\pi \int_0^{\sqrt{1-c^2}} \frac{r}{\sqrt{1-r^2}} dr = -2\pi \int_0^{\sqrt{1-c^2}} \frac{d(1-r^2)}{2\sqrt{1-r^2}} \\ &= -2\pi \left(\sqrt{1-r^2}\Big|_{r=0}^{r=\sqrt{1-c^2}} \right) = 2\pi(1-c). \end{aligned}$$

□

Example 16.54. Suppose that $g : [a,b] \to (0,\infty)$ is a C^1-function. We denote by $S_g \subset \mathbb{R}^3$ the surface obtained by rotating the graph of g about the x-axis. Using polar coordinates in the (y,z)-plane

$$y = r\cos\theta,\quad z = r\sin\theta$$

we can describe S_g via the equation $r = g(x)$. The injective immersion

$$(0,2\pi)\times(a,b) \ni (\theta,x) \mapsto \boldsymbol{p}(\theta,x) = \big(\, x, g(x)\cos\theta, g(x)\sin\theta \,\big) \in \mathbb{R}^3$$

almost covers S_g. We have

$$\boldsymbol{p}'_\theta = \big(\, 0, -g(x)\sin\theta, g(x)\cos\theta \,\big),\;\; \boldsymbol{p}'_x = \big(\, 1, g'(x)\cos\theta, g'(x)\sin\theta \,\big),$$

$$\langle \boldsymbol{p}'_\theta, \boldsymbol{p}'_\theta\rangle = g(x)^2,\;\; \langle \boldsymbol{p}'_\theta, \boldsymbol{p}'_x\rangle = 0,$$

$$\langle \boldsymbol{p}'_x, \boldsymbol{p}'_x\rangle = 1 + |g'(x)|^2,$$

$$\det G(\boldsymbol{p}'_\theta, \boldsymbol{p}'_x) = g(x)^2\big(\, 1 + |g'(x)|^2 \,\big),$$

so

$$|dA(\boldsymbol{p})| = g(x)\sqrt{1+|g'(x)|^2}\,|dx d\theta|.$$

In particular,

$$\begin{aligned}\operatorname{area}(S_g) &= \int_{\substack{0\le\theta\le 2\pi,\\ a\le x\le b}} g(x)\sqrt{1+|g'(x)|^2}\,|dx d\theta \\ &= \int_0^{2\pi} d\theta \int_a^b g(x)\sqrt{1+|g'(x)|^2} dx \\ &= 2\pi\int_a^b g(x)\sqrt{1+|g'(x)|^2} dx.\end{aligned}$$

This is in perfect agreement with the equality (9.77) obtained by alternate means.

For example, if $g(x) = R > 0$, $\forall x \in [a,b]$, then S_g is the cylinder

$$C_R := \big\{ (x,y,z) \in \mathbb{R}^3;\;\; y^2 + z^2 = R^2,\;\; x \in [a,b] \,\big\}.$$

Thus

$$\int_{C_R} f(x,y,z)|dA| = \int_{\substack{a\le x\le b\\ 0\le\theta\le 2\pi}} f\big(\, x, R\cos\theta, R\sin\theta \,\big)\, |dx d\theta|. \qquad \square$$

Example 16.55. Consider the hypersurface in $\mathbb{R}^3$ given by the equation

$$x^2 + y^2 = 2x.$$

Note that we can rewrite this as

$$x^2 - 2x + 1 + y^2 = 1 \Longleftrightarrow (x-1)^2 + y^2 = 1,$$

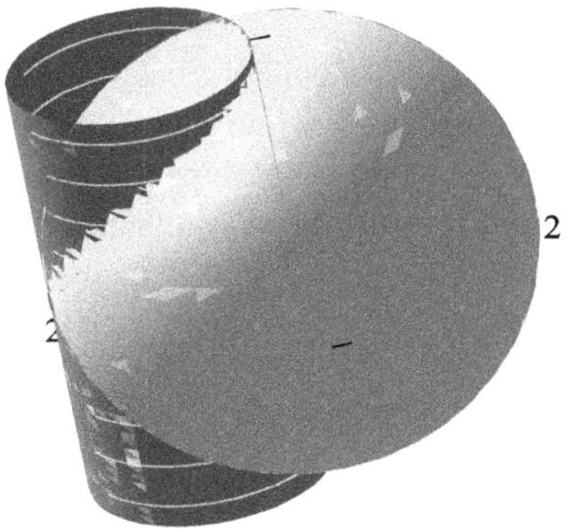

Fig. 16.22 *Intersecting a cylinder with a sphere.*

showing that this is a cylinder with radius 1 and vertical axis passing through the point $(1,0,0)$. We want to compute the area of the portion Σ of this cylinder that lies inside the sphere

$$x^2 + y^2 + z^2 = 4.$$

We observe first that Σ is symmetric with respect to the plane (x,y). We set

$$\Sigma_+ = \{ (x,y,z) \in \Sigma;\ \ z \geq 0 \}\ \ \Sigma_- = \{ (x,y,z) \in \Sigma;\ \ z \leq 0 \}.$$

Due to the above symmetry of Σ we have

$$\operatorname{area}(\Sigma_+) = \operatorname{area}(\Sigma_-) = \frac{1}{2}\operatorname{area}(\Sigma).$$

Using polar coordinates about $(1,0)$ we obtain the quasi-parametrization of Σ_+

$$x = 1 + \cos\theta,\ \ y = \sin\theta, z = z,$$

where

$$\theta \in (0, 2\pi),\ \ 0 < z < \sqrt{4 - x^2 - y^2} = \sqrt{4 - 2x} = \sqrt{2 - 2\cos\theta} = 2|\sin\theta|.$$

We have

$$\boldsymbol{p}'_\theta = (-\sin\theta, \cos\theta, 0),\ \ \boldsymbol{p}'_z = (0,0,1),$$

$$\det G(\boldsymbol{p}'_\theta, \boldsymbol{p}'_z) = 0,\ \ \operatorname{area}(\Sigma_+) = \int_0^{2\pi} |\sin\theta|\, d\theta = 4.$$

Hence $\operatorname{area}(\Sigma) = 8$. □

16.2.4. *Orientable surfaces in* $\mathbb{R}^3$

Suppose that $\Sigma \subset \mathbb{R}^3$ is a surface in $\mathbb{R}^3$. Roughly speaking a surface is orientable if it has two sides. For example, the xy-plane in $\mathbb{R}^3$ is orientable: it has one side facing the positive part of the z-axis, and one side facing the negative part of the z-axis. *Not all surfaces are two-sided.* The most famous and arguably the most important example is the *Möbius strip* (or band) depicted in Figure 16.23. It can be described by the parametrization

$$\begin{aligned} x &= \big(\, 3 + r\cos(t/2)\,\big)\cos(t)\,, \\ y &= \big(\, 3 + r\cos(t/2)\,\big)\sin(t)\,, \\ z &= r\sin(t/2)\,, \\ &-1 < r < 1,\;\; 0 \le t \le 2\pi. \end{aligned}$$

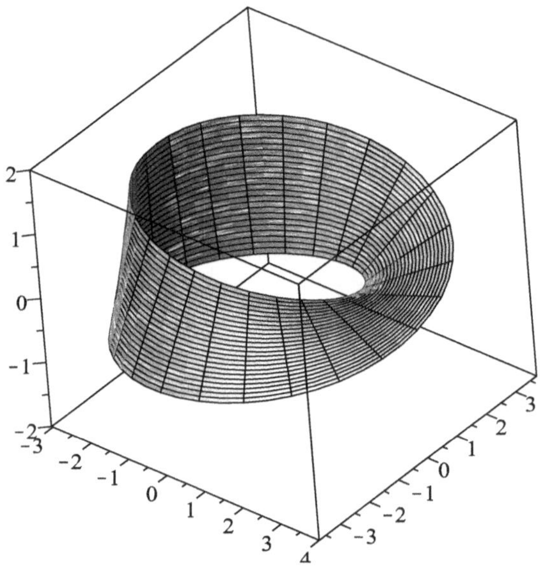

Fig. 16.23 *The Möbius strip (band) is the prototypical example of non-orientable surface.*

Definition 16.56. Let $\Sigma \subset \mathbb{R}^3$ be a surface in $\mathbb{R}^3$, with or without boundary. An *orientation on* Σ is a choice of a continuous, unit normal vector field along Σ, i.e., a continuous map $\boldsymbol{\nu} : \Sigma \to \mathbb{R}^3$ satisfying the following conditions:

(i) $\|\boldsymbol{\nu}(\boldsymbol{p})\| = 1, \forall \boldsymbol{p} \in \Sigma$.
(ii) $\boldsymbol{\nu}(\boldsymbol{p}) \perp T_{\boldsymbol{p}}\Sigma, \forall \boldsymbol{p} \in \Sigma^{\circ}$.[3]

The surface Σ is called *orientable* if it admits an orientation. An *oriented surface* is a pair $(\Sigma, \boldsymbol{\nu})$ consisting of a surface $\Sigma \subset \mathbb{R}^3$ and an orientation $\boldsymbol{\nu}$ on Σ. □

[3] We recall that Σ^0 denotes the interior of Σ.

Intuitively, the unit normal vector field defining an orientation points towards one side of the surface.

Example 16.57. Suppose that $f : \mathbb{R}^3 \to \mathbb{R}$ is a C^1-function. We denote by Z its zero set,

$$Z = \{ \boldsymbol{p} \in \mathbb{R}^3;\ f(\boldsymbol{p}) = 0 \}.$$

Suppose that

$$\nabla f(\boldsymbol{p}) \neq \boldsymbol{0},\ \ \forall \boldsymbol{p} \in Z.$$

The implicit function theorem shows that Z is a surface in $\mathbb{R}^3$. It is orientable because the vector field

$$\boldsymbol{\nu} : Z \to \mathbb{R}^3,\ \ \boldsymbol{\nu}(\boldsymbol{p}) = \frac{1}{\|\nabla f(\boldsymbol{p})\|} \nabla f(\boldsymbol{p})$$

is an orientation on Z. □

Example 16.58. Suppose that $U \subset \mathbb{R}^3$ is a bounded domain with C^1-boundary. Then its boundary is orientable. The *induced orientation* of the boundary is that defined by the unit normal vector field that points towards *the exterior* of U. We will use the notation $\partial_+ U$ when referring to the boundary of U equipped with this induced orientation. □

Important orientation convention. Suppose that $\Sigma \subset \mathbb{R}^3$ is a compact C^1-surface with nonempty boundary $\partial\Sigma$. Fix an orientation on Σ described by the normal unit vector field $\boldsymbol{\nu} : \Sigma \to \mathbb{R}^3$. Then we can equip the boundary with an orientation as follows: a person traveling on $\partial\Sigma$ according to this orientation while the toe-to-head direction is given by $\boldsymbol{\nu}$ will notice the surface Σ to her left-hand side; see Figure 16.24. This orientation of $\partial\Sigma$ is called the *orientation induced by the orientation of* Σ.

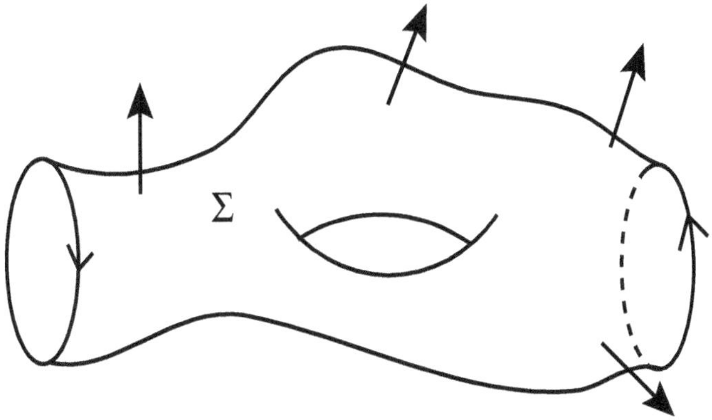

Fig. 16.24 *An orientation on a surface induces in a natural way an orientation on its boundary.*

Remark 16.59. Observe that if $D \subset \mathbb{R}^2$ is a bounded C^1-domain, then it is also a surface with boundary. The constant unit vector field $\boldsymbol{k}$ along D defines an orientation on D which in turn induces an orientation on the boundary ∂D. This orientation coincides with the induced orientation as described on page 593. □

16.2.5. *The flux of a vector field through an oriented surface in* $\mathbb{R}^3$

Definition 16.60 (Flux a vector field). Suppose that $\Sigma \subset \mathbb{R}^3$ is compact surface (with or without boundary) and $\boldsymbol{\nu}$ is an orientation on Σ. Suppose that $\boldsymbol{F} : \Sigma \to \mathbb{R}^3$ is a continuous vector field along Σ. The *flux* of $\boldsymbol{F}$ in the direction defined by the orientation is the scalar

$$\operatorname{Flux}(\boldsymbol{F}, \Sigma, \boldsymbol{\nu}) := \int_\Sigma \langle \boldsymbol{F}(\boldsymbol{p}), \boldsymbol{\nu}(\boldsymbol{p}) \rangle \, |dA(\boldsymbol{p})|. \qquad \square$$

Remark 16.61 (Computing the flux of a vector field). Suppose $\Sigma \subset \mathbb{R}^3$ is a compact surface (with or without boundary), $\boldsymbol{\nu}$ is an orientation on Σ and $\boldsymbol{F}$ is a continuous vector field along Σ,

$$\boldsymbol{F} : \Sigma \to \mathbb{R}^3,$$

$$\boldsymbol{F}(\boldsymbol{p}) = P(x,y,z)\boldsymbol{i} + Q(x,y,z)\boldsymbol{j} + R(x,y,z)\boldsymbol{k} = \begin{bmatrix} P(x,y,z) \\ Q(x,y,z) \\ R(x,y,z) \end{bmatrix}. \tag{16.22}$$

To compute the flux $\operatorname{Flux}(\boldsymbol{F}, \Sigma, \boldsymbol{\nu})$ one typically proceeds as follows.

Step 1. Parametrizing. Fix a quasi-parametrization of Σ, $\Phi : D \to \mathbb{R}^3$, where D is a bounded open domain of $\mathbb{R}^2$,

$$D \ni (s,t) \mapsto \Phi(s,t) = \boldsymbol{p}(s,t) = \begin{bmatrix} x(s,t) \\ y(s,t) \\ z(s,t) \end{bmatrix}. \tag{16.23}$$

Step 2. Understanding the orientation. The vectors $\boldsymbol{p}_s'$, $\boldsymbol{p}_t'$ are linearly independent and tangent to Σ. Thus $\boldsymbol{p}_s' \times \boldsymbol{p}_t'$ is perpendicular to the tangent space of Σ at $\boldsymbol{p}(s,t)$. In particular, exactly one of the vectors $\boldsymbol{p}_s' \times \boldsymbol{p}_t'$ or $\boldsymbol{p}_t' \times \boldsymbol{p}_s' = -\boldsymbol{p}_s' \times \boldsymbol{p}_t'$ points in the same direction as the normal $\boldsymbol{\nu}\big(\boldsymbol{p}(s,t)\big)$ defining the orientation. Find $\epsilon(s,t) \in \{\pm 1\}$ such that $\epsilon(s,t)(\boldsymbol{p}_s' \times \boldsymbol{p}_t')$ and $\boldsymbol{\nu}$ point in the same direction, i.e.,

$$\epsilon(s,t) = \begin{cases} 1, & \big\langle \boldsymbol{p}_s' \times \boldsymbol{p}_t', \boldsymbol{\nu} \big\rangle > 0, \\ -1, & \big\langle \boldsymbol{p}_s' \times \boldsymbol{p}_t', \boldsymbol{\nu} \big\rangle < 0. \end{cases}$$

Note that

$$\boxed{\epsilon(s,t) = -\epsilon(t,s)}.$$

Step 3. Integrating. We have

$$\boldsymbol{\nu} = \frac{\epsilon(s,t)}{\|\boldsymbol{p}_s' \times \boldsymbol{p}_t'\|} \boldsymbol{p}_s' \times \boldsymbol{p}_t'.$$

On the other hand (see Exercise 16.8)

$$|dA| = \|\boldsymbol{p}_s' \times \boldsymbol{p}_t'\| \, |dsdt|,$$

$$\boxed{\langle \boldsymbol{F}, \boldsymbol{\nu} \rangle \, |dA| = \epsilon(s,t) \langle \boldsymbol{F}, \boldsymbol{p}_s' \times \boldsymbol{p}_t' \rangle \, |dsdt|}.$$

Observe that

$$\langle \boldsymbol{F}, \boldsymbol{p}_s' \times \boldsymbol{p}_t' \rangle = \det \begin{bmatrix} P & x_s' & x_t' \\ Q & y_s' & y_t' \\ R & z_s' & z_t' \end{bmatrix}.$$

Then

$$\boxed{\operatorname{Flux}(\boldsymbol{F}, \Sigma, \boldsymbol{\nu}) = \int_D \epsilon(s,t) \det \begin{bmatrix} P & x_s' & x_t' \\ Q & y_s' & y_t' \\ R & z_s' & z_t' \end{bmatrix} |dsdt|} \tag{16.24}$$

or, equivalently,

$$\boxed{\operatorname{Flux}(\boldsymbol{F}, \Sigma, \boldsymbol{\nu}) = \int_D \epsilon(t,s) \det \begin{bmatrix} P & x_t' & x_s' \\ Q & y_t' & y_s' \\ R & z_t' & z_s' \end{bmatrix} |dsdt|}. \tag{16.25}$$

Expanding along the first column of the determinant in (16.24) we deduce

$$\det \begin{bmatrix} P & x_s' & x_t' \\ Q & y_s' & y_t' \\ R & z_s' & z_t' \end{bmatrix} = P \det \begin{bmatrix} y_s' & y_t' \\ z_s' & z_t' \end{bmatrix} + Q \det \begin{bmatrix} z_s' & z_t' \\ x_s' & x_t' \end{bmatrix} + R \det \begin{bmatrix} x_s' & x_t' \\ y_s' & y_t' \end{bmatrix}. \tag{16.26}$$

The above steps are best remembered using the language of *differential forms of degree* 2 or 2-*forms*.

To the vector field $\boldsymbol{F} = P\boldsymbol{i} + Q\boldsymbol{j} + R\boldsymbol{k}$ we associate the degree 2 differential form

$$\boxed{\Phi_{\boldsymbol{F}} := Pdy \wedge dz + Qdz \wedge dx + Rdx \wedge dy}. \tag{16.27}$$

Above, the symbol "$\wedge$" is called the *exterior product* or the *wedge*. Do not worry about its meaning yet. For now you only need to know that it differs from a usual product in that it is *anti-commutative*, i.e., *for any differential* 1-*forms* ω *and* η,

$$\boxed{\omega \wedge \eta = -\eta \wedge \omega, \quad \omega \wedge \omega = 0}.$$

The product $\omega \wedge \eta$ is a differential form of degree 2.

In (16.26) we think of x, y, z as functions depending on the variables s, t as in (16.23). Then dx, dy, dz are the (total) differentials of these functions as defined in (13.21). We have

$$\begin{aligned} dy \wedge dz &= (y_s' ds + y_t' dt) \wedge (z_s' ds + z_t' dt) \\ &= \underbrace{y_s' z_s' ds \wedge ds}_{=0} + y_s' z_t' \, ds \wedge dt + y_t' z_s' \underbrace{dt \wedge ds}_{=-ds \wedge dt} + \underbrace{y_t' z_t' dt \wedge dt}_{=0} \\ &= \big(y_s' z_t' - z_s' y_t' \big) ds \wedge dt \\ &= \det \begin{bmatrix} y_s' & y_t' \\ z_s' & z_t' \end{bmatrix} ds \wedge dt. \end{aligned}$$

We can write this as

$$\det\begin{bmatrix} y_s' & y_t' \\ z_s' & z_t' \end{bmatrix} = \frac{dy\wedge dz}{ds\wedge dt}. \tag{16.28}$$

Arguing in a similar fashion we deduce from (16.26)

$$\det\begin{bmatrix} P & x_s' & x_t' \\ Q & y_s' & y_t' \\ R & z_s' & z_t' \end{bmatrix} = P\frac{dy\wedge dz}{ds\wedge dt} + Q\frac{dz\wedge dx}{ds\wedge dt} + R\frac{dx\wedge dy}{ds\wedge dt},$$

so

$$Pdy\wedge dz + Qdz\wedge dz + Rdz\wedge dy = \det\begin{bmatrix} P & x_s' & x_t' \\ Q & y_s' & y_t' \\ R & z_s' & z_t' \end{bmatrix} ds\wedge dt.$$

Now observe that

$$\langle \boldsymbol{F}, \boldsymbol{\nu} \rangle\, |dA| = \epsilon(s,t)\langle \boldsymbol{F}, \boldsymbol{p}_s', \boldsymbol{p}_t' \rangle\, |dsdt| = \det\begin{bmatrix} P & x_s' & x_t' \\ Q & y_s' & y_t' \\ R & z_s' & z_t' \end{bmatrix} \epsilon(s,t)\,|dsdt|.$$

It is now time to give an idea of what $ds\wedge dt$ is. We "define"

$$\boxed{ds\wedge dt := \epsilon(s,t)\,|dsdt|}. \tag{16.29}$$

This is not an entirely satisfying definition because it is not clear what the nature of $|dsdt|$ is. Intuitively, it is the area of an "infinitesimal curvilinear parallelogram" on Σ, but the concept of "infinitesimal parallelogram" is a rather nebulous one. This will have to do for a while. Note that (16.28) implies

$$dy\wedge dz = \det\begin{bmatrix} y_s' & y_t' \\ z_s' & z_t' \end{bmatrix} ds\wedge dt = \epsilon(s,t)\det\begin{bmatrix} y_s' & y_t' \\ z_s' & z_t' \end{bmatrix} |dsdt|.$$

The issue of the nature of $\wedge$ aside, we deduce that

$$\langle \boldsymbol{F}, \boldsymbol{\nu} \rangle\, |dA| = Pdy\wedge dz + Qdz\wedge dx + Rdx\wedge dy.$$

For this reason we set

$$\boxed{\int_{\Sigma,\boldsymbol{\nu}} \Phi_{\boldsymbol{F}} := \operatorname{Flux}(\boldsymbol{F}, \Sigma, \boldsymbol{\nu})},$$

where we recall that

$$\Phi_{\boldsymbol{F}} = Pdy\wedge dz + Qdz\wedge dx + Rdx\wedge dy.$$

The integral

$$\int_{\Sigma,\boldsymbol{\nu}} Pdy\wedge dz + Qdz\wedge dx + Rdx\wedge dy$$

is traditionally called a *surface integral of the second kind.* It depends on a choice of orientation specified by the unit normal vector field $\boldsymbol{\nu}$. The differential form $\Phi_{\boldsymbol{F}}$ is sometimes referred to as the *infinitesimal flux of* $\boldsymbol{F}$. □

Example 16.62. Let us see how the above strategy works in a special case. Suppose that Σ is unit sphere

$$\Sigma = \{ (x,y,z) \in \mathbb{R}^3;\ x^2+y^2+z^2=1 \}.$$

We fix on Σ the orientation defined by the unit normal vector field pointing towards the exterior of the unit ball bounded by this sphere. Let $\boldsymbol{F}$ be the vector field $\boldsymbol{F} = \boldsymbol{i} + \boldsymbol{j}$.

The spherical coordinates provide a quasi-parametrization of this sphere

$$(\theta,\varphi) \mapsto \boldsymbol{p}(\theta,\varphi) = \begin{bmatrix} x = \sin\varphi\cos\theta \\ y = \sin\varphi\sin\theta \\ z = \cos\varphi, \end{bmatrix},\quad \theta \in (0,2\pi),\quad \varphi \in (0,\pi).$$

If we keep φ fixed and we let θ vary increasingly, the moving point $\theta \mapsto \boldsymbol{p}(\theta,\varphi)$ runs West-to-East along a parallel; see Figure 16.25. The vector $\boldsymbol{p}'_\theta$ is tangent to this parallel and points East. If we keep θ fixed and we let φ vary increasingly, the moving point $\theta \mapsto \boldsymbol{p}(\theta,\varphi)$ runs North-to-South along a meridian; see Figure 16.25. The vector $\boldsymbol{p}'_\varphi$ is tangent to this meridian and points South.

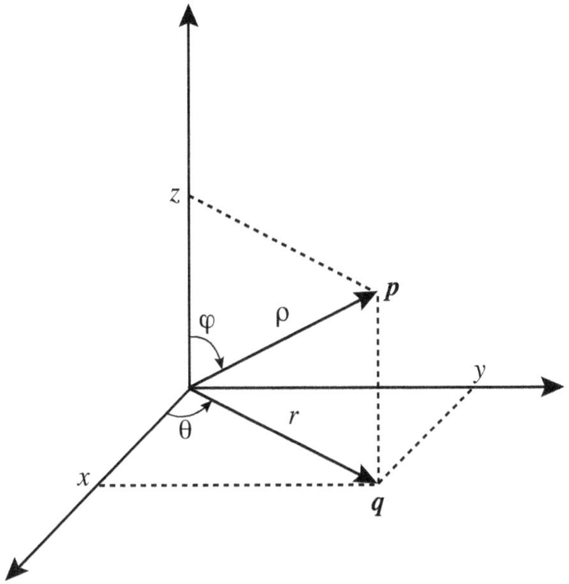

Fig. 16.25 *If we vary θ keeping φ fixed the point $\boldsymbol{p}$ runs along a parallel, while if we vary φ keeping theta fixed the point $\boldsymbol{p}$ runs along a meridian.*

The right-hand-rule for computing cross products (see page 354) shows that $\boldsymbol{p}'_\varphi \times \boldsymbol{p}'_\theta$ points towards the exterior of the sphere, i.e., in the same direction as the normal $\boldsymbol{\nu}$ defining the chosen orientation on the sphere. Thus, in this case

$$\epsilon(\varphi,\theta) = 1 = -\epsilon(\theta,\varphi).$$

We have

$$\langle \boldsymbol{F}, \boldsymbol{p}'_\varphi \times \boldsymbol{p}'_\theta \rangle = \det \begin{bmatrix} P & x'_\varphi & x'_\theta \\ Q & y'_\varphi & y'_\theta \\ R & z'_\varphi & z'_\theta \end{bmatrix}$$
$$= \det \begin{bmatrix} 1 & \cos\varphi\cos\theta & -\sin\varphi\sin\theta \\ 1 & \cos\varphi\sin\theta & \sin\varphi\cos\theta \\ 0 & -\sin\varphi & 0 \end{bmatrix}$$
$$= \det \begin{bmatrix} \cos\varphi\sin\theta & \sin\varphi\cos\theta \\ -\sin\varphi & 0 \end{bmatrix} - \det \begin{bmatrix} \cos\varphi\cos\theta & -\sin\varphi\sin\theta \\ -\sin\varphi & 0 \end{bmatrix}$$
$$= \sin^2\varphi\big(\sin\theta + \cos\theta \big),$$

and

$$\int_\Sigma \langle \boldsymbol{F}, \boldsymbol{\nu} \rangle \, |dA| = \int_{\substack{0\leq\theta\leq 2\pi,\\ 0\leq\varphi\leq\pi}} \sin^2\varphi\big(\sin\theta + \cos\theta \big)\, |d\varphi d\theta|$$
$$= \left(\int_0^\pi \sin\varphi \, d\varphi \right) \cdot \underbrace{\left(\int_0^{2\pi} (\sin\theta + \cos\theta)\, d\theta \right)}_{=0} = 0.$$

□

16.2.6. *Stokes' Formulæ*

To state these formulæ we need to introduce new concepts. Suppose that $U \subset \mathbb{R}^3$ is an open set and $\boldsymbol{F} : U \to \mathbb{R}^3$ is a C^1 vector field

$$\boldsymbol{F} = P\boldsymbol{i} + Q\boldsymbol{j} + R\boldsymbol{k}.$$

The *curl* of $\boldsymbol{F}$ is the continuous vector field $\operatorname{curl} F : U \to \mathbb{R}^3$

$$\boxed{\operatorname{curl} F := \big(\partial_y R - \partial_z Q\big)\boldsymbol{i} + \big(\partial_z P - \partial_x R\big)\boldsymbol{j} + \big(\partial_x Q - \partial_y P\big)\boldsymbol{k}}. \tag{16.30}$$

The right-hand side of the above equality looks intimidating, but there are cleverer ways of describing it.

Consider the formal[4] vector field

$$\nabla := \partial_x \boldsymbol{i} + \partial_y \boldsymbol{j} + \partial_z \boldsymbol{k}.$$

We then have

$$\boxed{\operatorname{curl} \boldsymbol{F} = \nabla \times \boldsymbol{F}}, \tag{16.31}$$

where "$\times$" denotes the cross product of two vectors in $\mathbb{R}^3$. Recall that it is uniquely determined by the anti-commutativity equalities

$$\boldsymbol{i} \times \boldsymbol{j} = -\boldsymbol{j} \times \boldsymbol{i} = \boldsymbol{k}, \;\; \boldsymbol{j} \times \boldsymbol{k} = -\boldsymbol{k} \times \boldsymbol{j} = \boldsymbol{i}, \;\; \boldsymbol{k} \times \boldsymbol{i} = -\boldsymbol{i} \times \boldsymbol{k} = \boldsymbol{j},$$

[4]In mathematics the term *formal* usually refers to objects that exist on paper and whose natures are not important for a particular argument. In this particular case ∇ can be given a precise rigorous meaning.

$$\boldsymbol{i} \times \boldsymbol{i} = \boldsymbol{j} \times \boldsymbol{j} = \boldsymbol{k} \times \boldsymbol{k} = \boldsymbol{0}.$$

Let us point out that if we use the classical interpretation of the inner product on $\mathbb{R}^3$ as a "dot product"

$$\boldsymbol{x} \cdot \boldsymbol{y} := \langle \boldsymbol{x}, \boldsymbol{y} \rangle, \;\; \boldsymbol{x}, \boldsymbol{y} \in \mathbb{R}^3,$$

then the divergence of $\boldsymbol{F}$ can be given by the more compact description

$$\boxed{\operatorname{div} \boldsymbol{F} = \nabla \cdot \boldsymbol{F} = \langle \nabla, \boldsymbol{F} \rangle}. \tag{16.32}$$

Theorem 16.63 (2D Stokes). *Suppose that $\Sigma \subset \mathbb{R}^3$ is a compact C^1-surface with boundary oriented by a unit normal vector field $\boldsymbol{\nu} : \Sigma \to \mathbb{R}^3$. Denote by $\partial_+\Sigma$ the boundary of Σ equipped with the orientation induced by the orientation of Σ as described at page 615. If*

$$\boldsymbol{F} := P\boldsymbol{i} + Q\boldsymbol{j} + R\boldsymbol{k}$$

is a C^1 vector field defined on an open set $\mathcal{O}$ containing Σ, then

$$\int_{\partial_+\Sigma} W_{\boldsymbol{F}} = \int_{\Sigma} (\operatorname{curl} \boldsymbol{F}) \cdot \boldsymbol{\nu} \, dA = \operatorname{Flux}(\nabla \times \boldsymbol{F}, \Sigma, \boldsymbol{\nu}) = \int_{\Sigma} \Phi_{\operatorname{curl} \boldsymbol{F}}. \tag{16.33}$$

□

The equality (16.33) is often referred to as *Green's formula*

Theorem 16.64 (3D Stokes). *Suppose that $U \subset \mathbb{R}^3$ is a bounded C^1-domain. Let $\boldsymbol{\nu} : \partial U \to \mathbb{R}^3$ denote the outer unit normal vector field along ∂U; see Example 16.58. If*

$$\boldsymbol{F} := P\boldsymbol{i} + Q\boldsymbol{j} + R\boldsymbol{k}$$

*is a C^1 vector field defined on an open set $\mathcal{O}$ containing **cl** U, then*

$$\int_{\partial_+ U} \Phi_{\boldsymbol{F}} = \operatorname{Flux}(\boldsymbol{F}, \partial U, \boldsymbol{\nu}) = \int_U \operatorname{div} \boldsymbol{F} \; |dxdydz|. \tag{16.34}$$

□

Often (16.34) is referred to as *divergence formula.*

Example 16.65. Suppose that $D \subset \mathbb{R}^3$ with a bounded C^1-domain and $\boldsymbol{R} = x\boldsymbol{i} + y\boldsymbol{j} + z\boldsymbol{k}$. Denote by $\boldsymbol{\nu}_{out}$ the outer unit normal vector field. Note that

$$\operatorname{div} \boldsymbol{R} = 3.$$

The divergence formula then implies

$$\operatorname{Flux}(\boldsymbol{R}, \partial D, \boldsymbol{\nu}_{out}) = \int_D 3 \, |dxdydz| = 3 \operatorname{vol}(D).$$ □

Example 16.66. Consider the vector field $\boldsymbol{V} : \mathbb{R}^3 \setminus \boldsymbol{0} \to \mathbb{R}^3$

$$\boldsymbol{V}(x,y,z) = \frac{x}{\rho^3}\boldsymbol{i} + \frac{y}{\rho^3}\boldsymbol{j} + \frac{z}{\rho^3}\boldsymbol{k},\;\; \rho = \sqrt{x^2+y^2+z^2}.$$

Let $B_r(0)$ the open ball of radius r centered at $0 \in \mathbb{R}^3$. Its boundary is the sphere $\Sigma_r(0)$ of radius r centered at 0. The outer unit vector field along $\partial B_r(0)$ is

$$\boldsymbol{\nu}_{out}(x,y,z) = \frac{x}{r}\boldsymbol{i} + \frac{y}{r}\boldsymbol{j} + \frac{z}{r}\boldsymbol{k}.$$

Thus, along $\partial B_r(0)$ we have $\rho = r$ and

$$\langle\, \boldsymbol{V}(x,y,z), \boldsymbol{\nu}_{out}(x,y,z)\,\rangle = \frac{x^2+y^2+z^2}{r^4} = \frac{r^2}{r^4} = \frac{1}{r^2}.$$

Thus

$$\operatorname{Flux}(\boldsymbol{V}, \partial B_r(0), \boldsymbol{\nu}_{out}\rangle = \int_{\Sigma_r} \frac{1}{r^2} dA = \frac{1}{2}\operatorname{area}(\Sigma_r) = 4\pi.$$

Let us compute the divergence of $\boldsymbol{V}$. From the equality $\rho^2 = x^2+y^2+z^2$ we deduce

$$\rho'_x = \frac{x}{\rho},\;\; \rho'_y = \frac{y}{\rho},\;\; \rho'_z = \frac{z}{\rho},$$

$$\partial_x\left(\frac{x}{\rho^3}\right) = \frac{1}{\rho^3} - 3\frac{x^2}{\rho^5},\;\; \partial_y\left(\frac{y}{\rho^3}\right) = \frac{1}{\rho^3} - 3\frac{y^2}{\rho^5},\;\; \partial_z\left(\frac{z}{\rho^3}\right) = \frac{1}{\rho^3} - 3\frac{z^2}{\rho^5}.$$

Thus

$$\operatorname{div}\boldsymbol{V} = \frac{3}{\rho^3} - 3\frac{x^2+y^2+z^2}{\rho^5} = 0$$

so that

$$\int_{B_r(0)} \operatorname{div}\boldsymbol{V}\, |dxdydz| = 0.$$

This seems to contradict the divergence theorem. The problem with this is that the vector field $\boldsymbol{V}$ has a singularity at the origin and it is not defined there. The divergence formula requires that the vector field be defined *everywhere* in the domain.

Suppose now that D is a bounded C^1 domain such that $0 \in D$. We want to compute $\operatorname{Flux}(\boldsymbol{V}, \partial D, \boldsymbol{\nu}_{out})$.

There exists $r_0 > 0$ such that $B_{r_0}(0) \subset D$. For any $0 < \varepsilon < r_0$ we set

$$D_\varepsilon := D \setminus \boldsymbol{cl}\,(\,B_\varepsilon(0)\,).$$

The vector field is well defined everywhere on D_ε and the divergence formula implies

$$\operatorname{Flux}(\boldsymbol{V}, \partial D_\varepsilon, \boldsymbol{\nu}_{out}) = \int_{D_\varepsilon} \operatorname{div}\boldsymbol{V}\, |dxdydz| = 0.$$

Now observe that

$$\partial D_\varepsilon = \partial D \cup \partial B_\varepsilon(0).$$

The normal vector field along $\partial B_\varepsilon(0)$ that points towards the exterior of D_ε is the normal vector field that points towards the *interior* of $B_\varepsilon(0)$. We denote it with $\boldsymbol{\nu}_{in}(B_\varepsilon)$. Thus

$$\begin{aligned} 0 &= \operatorname{Flux}(\boldsymbol{V}, \partial D_\varepsilon, \boldsymbol{\nu}_{out}) \\ &= \operatorname{Flux}(\boldsymbol{V}, \partial D, \boldsymbol{\nu}_{out}) + \operatorname{Flux}(\boldsymbol{V}, \partial B_\varepsilon, \boldsymbol{\nu}_{in}(B_\varepsilon)) \\ &= \operatorname{Flux}(\boldsymbol{V}, \partial D, \boldsymbol{\nu}_{out}) - \operatorname{Flux}(\boldsymbol{V}, \partial B_\varepsilon, \boldsymbol{\nu}_{out}(B_\varepsilon)) \\ &= \operatorname{Flux}(\boldsymbol{V}, \partial D, \boldsymbol{\nu}_{out}) - 4\pi, \end{aligned}$$

so that

$$\operatorname{Flux}(\boldsymbol{V}, \partial D, \boldsymbol{\nu}_{out}) = 4\pi. \qquad \square$$

Remark 16.67. Suppose that $D \subset \mathbb{R}^2$ is a bounded C^1-domain, and

$$\boldsymbol{F} : U \to \mathbb{R}^2, \quad \boldsymbol{F} = P(x,y)\boldsymbol{i} + Q(x,y)\boldsymbol{j}$$

is a C^1-vector field defined on an open set $U \subset \mathbb{R}^2$ that contains $\boldsymbol{cl}(D)$. Then D can be viewed as a surface with boundary in $\mathbb{R}^3$. If we write

$$\boldsymbol{F} = P(x,y)\boldsymbol{i} + Q(x,y)\boldsymbol{j} + 0\boldsymbol{k}$$

we see that we can view $\boldsymbol{F}$ as a 3-dimensional C^1-vector field defined on the open set $\mathcal{O} = U \times \mathbb{R} \subset \mathbb{R}^3$. The constant vector field $\boldsymbol{k}$ defines an orientation on D, and the induced orientation on ∂D defined by this unit normal vector field coincides with the orientation of ∂D as boundary of a planar domain. Note that

$$\operatorname{curl} \boldsymbol{F} = \big(Q'_x - P'_y\big)\boldsymbol{k},$$

so we deduce from (16.33)

$$\int_{\partial_+ D} \boldsymbol{F} \cdot \boldsymbol{p} = \operatorname{Flux}(\operatorname{curl} \boldsymbol{F}, D, \boldsymbol{k}) = \int_D \big(Q'_x - P'_y\big)\, |dxdy|.$$

This shows that (16.9a) is a special case of (16.33). $\square$

16.3. Differential forms and their calculus

16.3.1. *Differential forms on Euclidean spaces*

So far we have encountered differential forms of degree 1

$$Pdx + Qdy + Rdz,$$

and differential forms of degree 2,

$$Pdy \wedge dz + Qdz \wedge dx + Rdx \wedge dy,$$

where we recall that the operation "$\wedge$" satisfies the unusual anti-commutativity conditions

$$dx \wedge dy = -dy \wedge dx, \;\; dy \wedge dz = -dz \wedge dy, \;\; dz \wedge dx = -dx \wedge dz,$$

$$dx \wedge dx = dy \wedge dy = dz \wedge dz = 0.$$

A *differential form of degree* 3 or *3-form* on an open set $\mathcal{O} \in \mathbb{R}^3$ is an expression of the form

$$\rho \, dx \wedge dy \wedge dz,$$

where $\rho : \mathcal{O} \to \mathbb{R}$ is a continuous function. Again, we avoid explaining the meaning of the quantity $dx \wedge dy \wedge dz$, but we want to point out a few oddities.

$$\begin{aligned} dx \wedge dy \wedge dz &= dx \wedge \big(dy \wedge dz \big) = dx \wedge \big(-dz \wedge dy \big) \\ &= -\big(dx \wedge dz \big) \wedge dy = \big(dz \wedge dx \big) \wedge dy = dz \wedge dx \wedge dy. \end{aligned}$$

This is a bit surprising! The anti-commutative operation "$\wedge$" becomes commutative when 2-forms are involved,

$$dx \wedge dy \wedge dz = dz \wedge dx \wedge dy.$$

We still have not explained the meaning of the quantities $dx \wedge dy$, $dx \wedge dy \wedge dz$, etc., and we will not do so for a while.

Definition 16.68. Given an open set $\mathcal{O} \subset \mathbb{R}^n$ and $k = 1, \dots, n$ we denote by $\Omega^k(\mathcal{O})$ the space of *differential forms of degree* k on $\mathcal{O}$, i.e., expressions of the form

$$\omega = \sum_{1 \le i_1 < \cdots < i_k \le n} \omega_{i_1, \dots, i_k} dx^{i_1} \wedge \cdots \wedge dx^{i_k},$$

where the coefficients $\omega_{i_1, \dots, i_k}$ are continuous functions on $\mathcal{O}$. We set

$$\Omega^0(\mathcal{O}) := C^0(\mathcal{O}).$$

A differential form is called C^m if its coefficients are C^m functions. We denote by $\Omega^k(\mathcal{O})_{C^m}$ the space of C^m-forms of degree k. □

To simplify the exposition we introduce the following conventions.

- We set $[m] := \{1, \dots, m\}$.
- We denote by $[n]^{[m]}$ the set of maps $[m] \to [n]$, $k \mapsto i_k$. We will refer to such maps as multi-indices. We describe such a map I using the notation $I = (i_1, \dots, i_m)$. We denote by $\{I\}$ the range of I, $\{I\} := \{i_1, \dots, i_m\}$.
- We denote by $\operatorname{Inj}(m, n)$ the set of *injections* $[m] \to [n]$.
- We denote by $\mathfrak{S}_n$ the group of permutations of n objects, $\mathfrak{S}_n = \operatorname{Inj}(n, n)$.
- We denote by $\operatorname{Inj}^+(m, n)$ the subset of increasing multi-indices $I : [m] \to [n]$.
- If $\sigma \in \mathfrak{S}_m$ and $I = (i_1, \dots, i_m) \in \operatorname{Inj}(m, n)$ we set

$$I_\sigma = (i_{\sigma(1)}, \dots, i_{\sigma(m)}) \in \operatorname{Inj}(m, n).$$

- For $I \in \operatorname{Inj}(m, n)$ we set

$$d\boldsymbol{x}^{\wedge I} := dx^{i_1} \wedge \cdots \wedge dx^{i_m}. \tag{16.35}$$

  The terms $d\boldsymbol{x}^{\wedge I}$, $I \in \operatorname{Inj}(m, n)$ are called *exterior monomials*.

Thus, any $\omega \in \Omega^k(\mathcal{O})$ has the form

$$\omega = \sum_{I \in \text{Inj}^+(k,n)} \omega_I d\boldsymbol{x}^{\wedge I}, \;\; \omega_I \in C^0(\mathcal{O}). \tag{16.36}$$

The space $\Omega^k(\mathcal{O})$ is a vector space. The addition is defined by

$$\left(\sum_I \alpha_I d\boldsymbol{x}^{\wedge I}\right) + \left(\sum_I \beta_I d\boldsymbol{x}^{\wedge I}\right) = \sum_I (\alpha_I + \beta_I) d\boldsymbol{x}^{\wedge I},$$

and the scalar multiplication is defined in a similar fashion.

The elementary monomials dx^i satisfy the anti-commutativity rules

$$dx^i \wedge dx^j = \begin{cases} -dx^j \wedge dx^i, & i \neq j, \\ 0, & i = j. \end{cases} \tag{16.37}$$

This implies that if $I \in \text{Inj}(m,n)$ and $\sigma \in \mathfrak{S}_m$, then

$$d\boldsymbol{x}^{\wedge I_\sigma} = \epsilon(\sigma) d\boldsymbol{x}^{\wedge I}, \tag{16.38}$$

where $\epsilon(\sigma) \in \{-1, 1\}$ is the *signature* or *parity* of the permutation σ,

$$\epsilon(\sigma) = \prod_{1 \leq i < j \leq m} \frac{\sigma(j) - \sigma(i)}{j - i}.$$

We can define $d\boldsymbol{x}^{\wedge I}$ in the obvious way for any $I \in [n]^{[m]}$ with the understanding that $d\boldsymbol{x}^{\wedge I} = 0$ if I is not an injection. For example, if $I = (2,3,2)$, then

$$d\boldsymbol{x}^{\wedge I} = dx^2 \wedge dx^3 \wedge dx^2 = -dx^2 \wedge dx^2 \wedge dx^3 = 0.$$

For $I \in [n]^{[k]}$ and $J \in [n]^{[\ell]}$ we define

$$I * J := (i_1, \dots, i_k, j_1, \dots, j_\ell) \in [n]^{[k+\ell]}.$$

Then

$$d\boldsymbol{x}^{\wedge I} \wedge d\boldsymbol{x}^{\wedge J} = d\boldsymbol{x}^{\wedge I * J}.$$

This allows us to define a product

$$\wedge : \Omega^k(\mathcal{O}) \times \Omega^\ell(\mathcal{O}) \to \Omega^{k+\ell}(\mathcal{O}), \;\; k + \ell \leq n,$$

$$\left(\sum_{I \in \text{Inj}^+(k,n)} \alpha_I d\boldsymbol{x}^{\wedge I}\right) \wedge \left(\sum_{J \in \text{Inj}^+(\ell,n)} \beta_J d\boldsymbol{x}^{\wedge J}\right)$$

$$:= \sum_{\substack{I \in \text{Inj}^+(k,n), \\ J \in \text{Inj}^+(\ell,n)}} \alpha_I \beta_J d\boldsymbol{x}^{\wedge I} \wedge d\boldsymbol{x}^{\wedge J} = \sum_{\substack{I \in \text{Inj}^+(k,n), \\ J \in \text{Inj}^+(\ell,n)}} \alpha_I \beta_J d\boldsymbol{x}^{\wedge I * J}.$$

As we know, the 1-forms can be integrated over *oriented* curves, and the 2-forms can be integrated over *oriented* surfaces. A 3-form $\rho dx \wedge dy \wedge dz \in \Omega(\mathcal{O})$ can also be integrated and we set

$$\int_{\mathcal{O}} \rho \, dx \wedge dy \wedge dz := \int_{\mathcal{O}} \rho \, |dxdydz|,$$

whenever the integral on the right-hand side is absolutely convergent.

Let us point out a curious but important fact: since $dx \wedge dy \wedge dz = -dx \wedge dz \wedge dy$, we have

$$\int_{\mathcal{O}} \rho \, dx \wedge dz \wedge dy = -\int_{\mathcal{O}} \rho \, dx \wedge dy \wedge dz.$$

There is also a notion of derivative of a differential form called *exterior derivative*.

Definition 16.69 (Exterior derivative). Let $\mathcal{O} \subset \mathbb{R}^n$ be an open subset. The *exterior derivative* is the linear operator

$$d : \Omega^k(\mathcal{O})_{C^1} \to \Omega^{k+1}(\mathcal{O}), \;\; k = 0, 1, \dots, n-1,$$

defined as follows.

- If $k = 0$ so that $\Omega^0(\mathcal{O})_{C^1} = C^1(\mathcal{O})$, then

$$df = \sum_{i=1}^n \frac{\partial f}{\partial x^i}\, dx^i \in \Omega^1(\mathcal{O}), \;\; \forall f \in C^1(\mathcal{O}).$$

- If $0 < k < n$, then for any $\alpha \in \Omega^k(\mathcal{O})_{C^1}$ we set

$$d\alpha := d\left(\sum_{I \in \mathrm{Inj}^+(k,n)} \alpha_I d\boldsymbol{x}^{\wedge I} \right) = \sum_{I \in \mathrm{Inj}^+(k,n)} d\alpha_I \wedge d\boldsymbol{x}^{\wedge I},$$

where $d\alpha^I \in \Omega^1(\mathcal{O})$ is the (total) differential of the C^1-function α^I.
- If $k = n$, then for any $\alpha \in \Omega^k(\mathcal{O})_{C^1}$ we set $d\alpha = 0$.

□

Example 16.70. Let $\mathcal{O} \subset \mathbb{R}^3$ be an open set. The exterior derivative of a 0-form on $\mathcal{O}$ is a 1-form, the exterior derivative of a 1-form is a 2-form, the exterior derivative of a 2-form is a 3-form, and the exterior derivative of a 3-form on $\mathcal{O}$ is defined to be zero. Here is how one computes these exterior derivatives.

The differential of a C^1 form of degree zero, i.e., a C^1-function $f : \mathcal{O} \to \mathbb{R}$ is its total differential

$$df = f'_x dx + f'_y dy + f'_z dz = W_{\nabla f}.$$

If ω is a C^1 differential form of degree 1 on $\mathcal{O}$,

$$\omega = Pdx + Qdy + Rdz = W_{\boldsymbol{F}}, \;\; \boldsymbol{F} = P\boldsymbol{i} + Q\boldsymbol{j} + R\boldsymbol{k},$$

then

$$\begin{aligned} d\omega = dW_{\boldsymbol{F}} &= dP \wedge dx + dQ \wedge dy + dR \wedge dz \\ &= (P'_x dz + P'_y dy + P'_z dz) \wedge dx + (Q'_x dx + Q'_y dy + Q'_z dz) \wedge dy \\ &\quad + (R'_x dx + R'_y dy + R'_z dz) \wedge dz \\ &= P'_y\, dy \wedge dx + P'_z\, dz \wedge dx + Q'_x\, dx \wedge dy \\ &\quad + Q'_z\, dz \wedge dy + R'_x\, dx \wedge dz + R'_y\, dy \wedge dz \\ &= \left(R'_y - Q'_z \right) dy \wedge dz + \left(P'_z - R'_z \right) dz \wedge dx + \left(Q'_x - P'_y \right) dx \wedge dy \end{aligned}$$

(use (16.27) and (16.30))

$$= \Phi_{\mathrm{curl}\, \boldsymbol{F}}.$$

Thus

$$\boxed{dW_{\boldsymbol{F}} = \Phi_{\mathrm{curl}\, \boldsymbol{F}}}. \tag{16.39}$$

Suppose finally that η is a C^1 form of degree 2 on $\mathcal{O}$

$$\eta = \Phi_{\boldsymbol{F}} = P dy \wedge dz + Q\, dz \wedge dx + R\, dx \wedge dy.$$

Then

$$\begin{aligned} d\eta &= dP \wedge dy \wedge dz + dQ \wedge dz \wedge dx + dR \wedge dx \wedge dy \\ &= \big(P'_x dx + P'_y dy + P'_z dz \big)\, dy \wedge dz \\ &\quad + \big(Q'_x dx + Q'_y dy + Q'_z dz \big) dz \wedge dz \\ &\quad + \big(R'_x dx + R'_y dy + R'_z dz \big)\, dx \wedge dy \\ &= P'_x\, dx \wedge dy \wedge dz + Q'_y\, dy \wedge dz \wedge dx + R'_z\, dz \wedge dx \wedge dy \\ &= \big(P'_x + Q'_y + R'_z \big)\, dx \wedge dy \wedge dz = \big(\operatorname{div} \boldsymbol{F} \big)\, dx \wedge dy \wedge dz. \end{aligned}$$

Thus

$$\boxed{d\Phi_{\boldsymbol{F}} = \big(\operatorname{div} \boldsymbol{F} \big)\, dx \wedge dy \wedge dz}. \tag{16.40}$$

□

In view of the computations in Example 16.70 we can give the following equivalent reformulations of Theorem 16.63 and Theorem 16.64.

Theorem 16.71. *Suppose that $\boldsymbol{F}$ is a C^1-vector field defined on the open set $\mathcal{O} \subset \mathbb{R}^3$.*

(i) If $\Sigma \subset \mathcal{O}$ is a compact, oriented *surface with boundary, then*

$$\int_{\partial_+ \Sigma} W_{\boldsymbol{F}} = \int_{\Sigma} dW_{\boldsymbol{F}}.$$

(ii) If $U \subset \mathbb{R}^3$ is a bounded C^1 domain such that $\mathbf{cl}\, U \subset \mathcal{O}$, then

$$\int_{\partial_+ U} \Phi_{\boldsymbol{F}} = \int_{U} d\Phi_{\boldsymbol{F}}.$$

□

The above result is not a low dimensional accident. In the remainder of this section we will describe how it generalizes to higher dimensions. The key to this process is a more subtle operation on differential forms.

Suppose that $\mathcal{O}_0 \subset \mathbb{R}^{n_0}$ and $\mathcal{O}_1 \subset \mathbb{R}^{n_1}$ are open sets. We denote by $\boldsymbol{x} = (x^i)$ the Cartesian coordinates in $\mathbb{R}^{n_0}$ and by $\boldsymbol{y} = (y^j)$ the Cartesian coordinates in $\mathbb{R}^{n_1}$. Let

$$\Phi : \mathcal{O}_0 \to \mathcal{O}_1$$

be a C^1-map described in the above coordinates by the functions

$$y^j = \Phi^j\big(x^1, \dots, x^{n_0}\big), \;\; j = 1, \dots, n_1.$$

For each $k \leq \min(n_0, n_1)$, the *pullback* via Φ of a k-form on $\mathcal{O}_1$ (to a k-form on $\mathcal{O}_0$) is the linear operator

$$\Phi^* : \Omega^k(\mathcal{O}_1) \to \Omega^k(\mathcal{O}_0),$$

$$\Phi^*\eta = \Phi^*\left(\sum_{I\in \mathrm{Inj}^+(k,n_1)} \eta_I d\boldsymbol{y}^{\wedge I}\right) \tag{16.41}$$

$$:= \sum_{1\leq i_1<\cdots<i_k\leq n_1} \eta_{i_1,\dots,i_k}\big(\Phi(\boldsymbol{x})\big) d\Phi^{i_1}(\boldsymbol{x})\wedge\cdots\wedge d\Phi^{i_k}(\boldsymbol{x}),\ \ \forall \eta\in\Omega^k(\mathcal{O}_1), \tag{16.42}$$

where $d\Phi^i$ is the (total) differential of Φ^i viewed as a 1-form,

$$d\Phi^i = \sum_{j=1}^{n_0} \frac{\partial \Phi^i}{\partial x^j} dx^j.$$

Example 16.72. (a) Consider the map

$$\Phi: \mathbb{R}^2 \to \mathbb{R}^2,\ \ (r,\theta)\mapsto (x,y) = (r\cos\theta, r\sin\theta).$$

Then

$$\begin{aligned}\Phi^*(dx\wedge dy) &= d(r\cos\theta)\wedge d(r\sin\theta)\\ &= (\cos\theta dr - r\sin\theta d\theta)\wedge(\sin\theta dr + r\cos\theta d\theta)\\ &= r\cos^2\theta dr\wedge d\theta - r\sin^2\theta\underbrace{d\theta\wedge dr}_{=-dr\wedge d\theta} = (r\cos^2\theta + r\sin^2\theta)dr\wedge d\theta\\ &= rdr\wedge d\theta = \det J_\Phi\, dr\wedge d\theta.\end{aligned}$$

More generally, given open sets $U,V\subset\mathbb{R}^n$ and $\Phi: U\to V$ a C^1-map, we have

$$\Phi^*\big(dv^1\wedge\cdots\wedge dv^n\big) = (\det J_\Phi)du^1\wedge\ldots\wedge du^n, \tag{16.43}$$

where (v^i) are the Cartesian coordinates on V and (u^j) are the Cartesian coordinates on U.

(b) Consider the map

$$\Phi: (0,\infty)\times\mathbb{R}\to\mathbb{R}^2\setminus\{\boldsymbol{0}\},\ \ (r,\theta)\mapsto(x,y) = (r\cos\theta, r\sin\theta).$$

Let

$$\omega = -\frac{y}{x^2+y^2}dx + \frac{x}{x^2+y^2}dy.$$

Then

$$\Phi^*\omega = \frac{-r\sin\theta\, d(r\cos\theta) + r\cos\theta\, d(r\sin\theta)}{r^2} = d\theta.$$

□

Example 16.73. Suppose that $U\subset\mathbb{R}^n$ is an open set that nontrivially intersects the subspace

$$\mathbb{R}^m\times\boldsymbol{0} = \big\{\,(u^1,\dots,u^n)\in\mathbb{R}^n:\ \ u^i = 0,\ \ \forall i>m\,\big\}.$$

We set $\overline{U} := U\cap\mathbb{R}^m\times\boldsymbol{0}$ and we denote by i the inclusion map

$$\overline{U}\ni\bar{\boldsymbol{u}}\mapsto(\bar{\boldsymbol{u}},\boldsymbol{0})\in U.$$

For any $k \leq m$ and any $\alpha \in \Omega^k(U)$ we set

$$\alpha\big|_{\overline{U}} := i^*\alpha. \tag{16.44}$$

For example, if $k = m$ and

$$\alpha = \sum_{I \in \mathrm{Inj}^+(m,n)} \alpha_I(u) d\boldsymbol{u}^{\wedge I},$$

then

$$\alpha\big|_{\overline{U}} = \alpha_{1,2,\dots,m}(\overline{\boldsymbol{u}}, \boldsymbol{0}) du^1 \wedge \cdots \wedge u^m \in \Omega^1(\overline{U}).$$ □

Proposition 16.74. *Suppose that $U_i \subset \mathbb{R}^{n_i}$, $i = 0, 1$ are open sets and $\Phi : U_0 \to U_1$ is a C^1-map. Then the following hold.*

(i) For any $0 \leq k, \ell \leq n_1$, $\alpha \in \Omega^k(U_1)$, $\beta \in \Omega^\ell(U_1)$ we have the product formula

$$d(\alpha \wedge \beta) = (d\alpha) \wedge \beta + (-1)^k \alpha \wedge (d\beta). \tag{16.45}$$

(ii) For any $0 \leq k, \ell \leq n_1$, $\alpha \in \Omega^k(U_1)$, $\beta \in \Omega^\ell(U_1)$ we have

$$\Phi^*(\alpha \wedge \beta) = \Phi^*\alpha \wedge \Phi^*\beta. \tag{16.46}$$

(iii) For any $k = 0, 1, \dots, n-1$ and any $\alpha \in \Omega^k(U_1)_{C^1}$ we have

$$d(\Phi^*\alpha) = \Phi^*(d\alpha). \tag{16.47}$$

(iv) For any $k = 0, 1, \dots, n-1$ and any $\alpha \in \Omega^k(U_1)_{C^2}$ we have

$$d(d\alpha) = 0. \tag{16.48}$$

□

Remark 16.75. Let us comment a bit about the proof. First, observe that it suffices to prove the results in the special case when α, β are monomials

$$\alpha = f d\boldsymbol{x}^{\wedge I}, \quad \beta = g d\boldsymbol{x}^{\wedge J}.$$

In this case (16.45) follows immediately from the definition of exterior derivative. The equality (16.46) follows from the definition of pullback after an elementary but rather strenuous computation.

As for (16.47), the crucial case is $k = 0$ so α is C^1-function. Fortunately this was discussed in great detail in Remark 13.26. The general case follows by combining the case $k = 0$ with (16.46).

Finally, let us remark that, using (16.45) we can reduce (16.48) to the case $k = 0$. In this case the identity (16.48) is equivalent with the known fact

$$\frac{\partial^2 f}{\partial x^i \partial x^j} = \frac{\partial^2 f}{\partial x^j \partial x^i}, \quad \forall u \in C^2(U), \;\; 1 \leq i, j \leq n.$$ □

The top degree forms on $\mathbb{R}^n$ can be integrated. Let $U \subset \mathbb{R}^n$ be an open set. Denote by $\Omega^k_{\text{cpt}}(U)$ the space of degree k-forms on U with compact support, i.e., forms η such that there exists a compact set $K \subset U$ such that all the coefficients η_I are zero outside K.

Observe first that a degree n form ω defined on open set U in $\mathbb{R}^n$ has the form

$$\omega = \rho_\omega dV_n := \rho_\omega du^1 \wedge \cdots \wedge du^n$$

where ρ_ω is a continuous function on U called the *density* of ω, and $u^1, \dots, u^n$ are the canonical Cartesian coordinates on $\mathbb{R}^n$. The top degree form dV_n is called the *canonical volume form* on $\mathbb{R}^n$.

Example 16.76. Suppose that U, V are open subset of $\mathbb{R}^n$ and $\Phi : U \to V$ is a C^1-map. Then the density of the top degree form $\omega = \Phi^* dV_n \in \Omega^n(U)$ is

$$\rho_\omega(\boldsymbol{u}) = \det J_\Phi(\boldsymbol{u}). \qquad \square$$

Definition 16.77. Let $U \subset \mathbb{R}^n$ be an open set.

- An *orientation* on U is a choice of a nowhere vanishing form η of maximum degree n.
- Two orientations defined by the top degree forms η_0 or η_1 are to be considered *equivalent* if there exists a positive continuous function $\rho : U \to (0, \infty)$ such that $\eta_1 = \rho\eta_0$.

$\square$

Remark 16.78. Observe that if U is an open subset in $\mathbb{R}^n$, then any nowhere vanishing degree n form ω can be described explicitly as a product

$$\omega = \rho_\omega dV_n,$$

where density ρ_ω is a nowhere vanishing continuous function on U. We have a well defined continuous function[5]

$$\epsilon = \epsilon_\omega : U \to \{-1, 1\}, \ \ \epsilon_\omega(\boldsymbol{u}) = \operatorname{sign} \rho_\omega(\boldsymbol{u}) = \frac{\rho_\omega(\boldsymbol{u})}{|\rho_\omega(\boldsymbol{u})|}.$$

The orientation defined by ϵ is therefore equivalent with the orientation defined by $\epsilon_\omega dV_n$.

For this reason we will identify the set of possible nonequivalent orientations on U with the set $\mathbb{O}(U)$ of continuous functions $\epsilon : U \to \{-1, 1\}$. The orientation defined by ϵ is, by definition, the orientation defined by the nowhere vanishing top degree form $\epsilon(\boldsymbol{u}) dV_n$.

If U is a *path connected* open subset of $\mathbb{R}^n$, then there are precisely two nonequivalent orientations, one defined by the form

$$dV_n := dx^1 \wedge \cdots \wedge dx^n,$$

called the *canonical orientation*, and one defined by $-dV_n$.

If U has k connected components, then there are 2^k nonequivalent choices of orientation, each determined by a continuous function

$$\epsilon : U \to \{-1, 1\}. \qquad \square$$

[5] Such a function is constant on the connected components of U.

Definition 16.79. An *oriented* open subset of $\mathbb{R}^n$ is a pair (U, ϵ), where U is an open set and $\epsilon \in \mathbb{O}(U)$ is an orientation on U. □

Any oriented open set (U, ϵ) in $\mathbb{R}^n$ defines a linear map

$$\int_{U,\epsilon} : \Omega^n_{\mathrm{cpt}}(U) \to \mathbb{R},$$

$$\int_{U,\epsilon} \omega = \int_{U,\epsilon} \rho_\omega du^1 \wedge \cdots \wedge du^n := \int_U \epsilon(\boldsymbol{u}) \rho_\omega(\boldsymbol{u}) \, |du^1 \cdots du^n|. \tag{16.49}$$

When ϵ is identically equal to 1 we omit it from the notion.

Let us point out a simple but confusing fact. For any $\sigma \in \mathfrak{S}_n$ we have

$$\epsilon(\sigma) \rho_\omega du^{\sigma(1)} \wedge \cdots \wedge du^{\sigma(n)} = \omega,$$

$$\int_U \rho_\omega du^{\sigma(1)} \wedge \cdots \wedge du^{\sigma(n)} = \epsilon(\sigma) \int_U \rho_\omega du^1 \wedge \cdots \wedge du^n.$$

Because of this it is important to keep in mind the following naive but important advice.

☛ ***When integrating differential forms, the order in which we write the coordinates is crucial!***

Definition 16.80. Let (U_i, ϵ_i), $i = 0, 1$ be oriented open sets in $\mathbb{R}^n$, $i = 0, 1$ and $\Phi : U_0 \to U_1$ a C^1-diffeomorphism. We say that Φ is *orientation preserving* if

$$\epsilon_0(\boldsymbol{u}) \epsilon_1\big(\Phi(\boldsymbol{u})\big) \det J_\Phi(\boldsymbol{u}) > 0, \ \ \forall \boldsymbol{u} \in U_0. \qquad □$$

Proposition 16.81. *Suppose that (U_i, ϵ_i), $i = 0, 1$ are oriented open sets in $\mathbb{R}^n$, $i = 0, 1$ and $\Phi : U_0 \to U_1$ an orientation preserving diffeomorphism such that $U_1 = \Phi(U_0)$, then for any $\eta \in \Omega^n_{\mathrm{cpt}}(U_1)$ we have*

$$\int_{U_1, \epsilon_1} \eta = \int_{U_0, \epsilon_0} \Phi^* \eta. \tag{16.50}$$

□

Statement (i) follows from the chain rule, while statement (ii) is essentially the change in variables formula for multiple integrals.

We close this subsection with a technical result which will play a key role in extending to higher dimensions the concept of integration of a differential form.

Proposition 16.82. *Suppose that $V_i \subset \mathbb{R}^n$, $i = 0, 1$, are open sets such that*

$$\overline{V}_i := V_i \cap \mathbb{R}^m \times \mathbf{0} \neq \emptyset, \ \ i = 0, 1.$$

Let $\Phi : V_0 \to \mathbb{R}^n$ be a C^1-diffeomorphism such that (see Figure 16.26)

$$\Phi(V_0) = V_1, \ \ \Phi(\overline{V}_0) = \overline{V}_1.$$

Then the following hold.

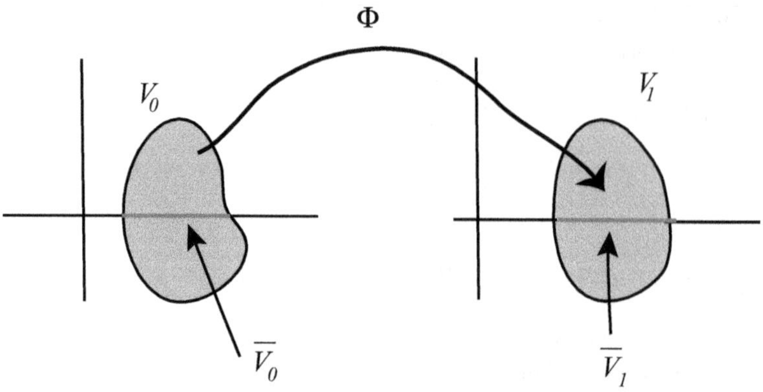

Fig. 16.26 *A transition map.*

(i) The induced map $\overline{\Phi} : \overline{V}_0 \to \overline{V}_1 \subset \mathbb{R}^m$ is a C^1-diffeomorphism.
(ii) If $\omega_1 \in \Omega^m(V_1)$ and $\omega_0 := \Phi^*\omega_1$, then (see (16.44))

$$\omega_0\big|_{\overline{V}_0} = \overline{\Phi}^*\omega_1\big|_{\overline{V}_1}.$$

Proof. Denote by $\boldsymbol{y} = (y^1, \dots, y^n)$ the Cartesian coordinates on V_1 and by $\boldsymbol{x} = (x^1, \dots, x^n)$ the Cartesian coordinates on V_0. We set

$$\overline{\boldsymbol{x}} := (x^1, \dots, x^m), \ \ \overline{\boldsymbol{y}} = (y^1, \dots, y^m),$$

$$\boldsymbol{x}_\perp := (x^{m+1}, \dots, x^n), \ \ \boldsymbol{y}_\perp = (y^{m+1}, \dots, y^n).$$

For simplicity we set

$$\overline{\omega}_i := \omega_i\big|_{\overline{V}_i}, \ \ i = 0, 1.$$

The diffeomorphism Φ is described by a collection of functions

$$y^i = y^i(\boldsymbol{x}), \ \ i = 1, \dots, n.$$

Since $\Phi(\overline{V}_0) = \overline{V}_1$, we deduce

$$y^j(\overline{\boldsymbol{x}}, \boldsymbol{0}) = 0, \ \ \forall \overline{\boldsymbol{x}} \in \overline{V}_0, \ \ j > m.$$

We write this

$$\frac{\partial \boldsymbol{y}_\perp}{\partial \overline{\boldsymbol{x}}}(\overline{\boldsymbol{x}}, \boldsymbol{0}) = 0. \tag{16.51}$$

(i) The map $\overline{\Phi}$ is described by the functions

$$y^j = y^j(\overline{\boldsymbol{x}}, \boldsymbol{0}), \ \ j = 1, \dots, k.$$

We write this succinctly

$$\overline{\boldsymbol{y}} = \overline{\boldsymbol{y}}(\overline{\boldsymbol{x}}, \boldsymbol{0}).$$

The map $\overline{\Phi} : \overline{V}_0 \to \overline{V}_1$ is a homeomorphism since $\Phi : V_0 \to V_1$ is such. We have to show that if $(\overline{\boldsymbol{x}}, \boldsymbol{0}) \in \overline{V}_0$, then

$$\det J_{\overline{\Phi}}(\overline{\boldsymbol{x}}, \boldsymbol{0}) \neq 0.$$

The Jacobian $J_{\overline{\Phi}}(\overline{\boldsymbol{x}}, \boldsymbol{0})$ is given by the $m \times m$ matrix

$$\frac{\partial \overline{\boldsymbol{y}}}{\partial \overline{\boldsymbol{x}}}(\overline{\boldsymbol{x}}, \boldsymbol{0}).$$

We know that $\det J_\Phi(\overline{\boldsymbol{x}},0) \neq 0$ since Φ is a diffeomorphism. Now observe that $J_\Phi(\overline{\boldsymbol{x}},0)$ has the block decomposition

$$J_\Phi(\overline{\boldsymbol{x}},0) = \begin{bmatrix} \partial\overline{\boldsymbol{y}}/\partial\overline{\boldsymbol{x}} & \partial\boldsymbol{y}_\perp/\partial\overline{\boldsymbol{x}} \\ \partial\boldsymbol{y}_\perp/\partial\overline{\boldsymbol{x}} & \partial\boldsymbol{y}_\perp/\partial\boldsymbol{x}_\perp \end{bmatrix}_{(\overline{\boldsymbol{x}},\mathbf{0})} \stackrel{(16.51)}{=} \begin{bmatrix} \partial\overline{\boldsymbol{y}}/\partial\overline{\boldsymbol{x}} & 0 \\ \partial\boldsymbol{y}_\perp/\partial\overline{\boldsymbol{x}} & \partial\boldsymbol{y}_\perp/\partial\boldsymbol{x}_\perp \end{bmatrix}_{(\overline{\boldsymbol{x}},\mathbf{0})}.$$

Hence

$$0 \neq \det J_\Phi(\overline{\boldsymbol{x}},0) = \det\frac{\partial\overline{\boldsymbol{y}}}{\partial\overline{\boldsymbol{x}}}\cdot\det\frac{\partial\boldsymbol{y}_\perp}{\partial\boldsymbol{x}_\perp} \Rightarrow \det J_{\overline{\Phi}}(\overline{\boldsymbol{x}},\mathbf{0}) = \det\frac{\partial\overline{\boldsymbol{y}}}{\partial\overline{\boldsymbol{x}}} \neq 0.$$

(ii) Let

$$\omega_1 = \sum_{I\in\mathrm{Inj}^+(m,n)} \omega_I(\boldsymbol{y})d\boldsymbol{y}^{\wedge I}.$$

Then

$$\omega_0 = \sum_{I\in\mathrm{Inj}^+(m,n)} \omega_I\big(\boldsymbol{y}(\boldsymbol{x})\big)dy^{i_1}(\boldsymbol{x})\wedge\cdots\wedge dy^{i_m}(\boldsymbol{x}),$$

$$\overline{\omega}_1 = \omega_{1,\dots,m}(\overline{\boldsymbol{y}},\mathbf{0})dy^1\wedge\cdots\wedge dy^m,$$

$$\overline{\omega}_0 = \sum_{I\in\mathrm{Inj}^+(m,n)} \omega_I\big(\boldsymbol{y}(\overline{\boldsymbol{x}},\mathbf{0})\big)dy^{i_1}(\overline{\boldsymbol{x}},\mathbf{0})\wedge\cdots\wedge dy^{i_m}(\overline{\boldsymbol{x}},\mathbf{0}).$$

From (16.51) we deduce that for $j > m$ we have

$$dy^j(\overline{\boldsymbol{x}},\mathbf{0}) = \sum_{i=1}^m \frac{\partial y^j}{\partial x^i}(\overline{\boldsymbol{x}},\mathbf{0})dx^i = 0.$$

Hence

$$\overline{\omega}_0 = \omega_{1,2,\dots,m}\big(\boldsymbol{y}(\overline{\boldsymbol{x}},\mathbf{0})\big)dy^1(\overline{\boldsymbol{x}},\mathbf{0})\wedge\cdots\wedge dy^m(\overline{\boldsymbol{x}},\mathbf{0}) = \overline{\Phi}^*\overline{\omega}_1.$$

□

16.3.2. *Orientable submanifolds*

Suppose that X is an m-dimensional C^1-submanifold of $\mathbb{R}^n$, $0 < m < n$. Every point $\boldsymbol{p}\in X$ admits (at least) a *straightening diffeomorphism* $(\mathcal{U},\Psi)$. For brevity we will use the acronym *s.d.* when referring to straightening diffeomorphisms. We recall what this entails (see Definition 14.31).

- $\mathcal{U}$ is an open neighborhood of $\boldsymbol{p}\in\mathbb{R}^n$.
- $\Psi:\mathcal{U}\to\mathbb{R}^n$ is a C^1-diffeomorphism with image $U := \Psi(\mathcal{U})$.
- $\Psi(\mathcal{U}\cap X) = \overline{U} := U\cap\mathbb{R}^m\times\mathbf{0}$.
- We denote by $\overline{\Phi}$ the induced map $\Psi^{-1}:\overline{U}\to\mathcal{U}$.

An *orientation* for the s.d. $(\mathcal{U},\Psi)$ is a choice of orientation $\epsilon\in\mathbb{O}(\overline{U})$. An *oriented* s.d. is a triplet $(\mathcal{U},\Psi,\epsilon)$, where $(\mathcal{U},\Psi)$ is an s.d. and ϵ is an orientation of that s.d.

If $(\mathcal{U}_0,\Psi_0)$ and $(\mathcal{U}_1,\Psi_1)$ are two straightening diffeomorphism near $\boldsymbol{p}\in X$, we set $\mathcal{V} := \mathcal{U}_0\cap\mathcal{U}_1$, and we get open sets (see Figure 16.27)

$$U_i = \Psi_i(\mathcal{U}_i)\subset\mathbb{R}^n,\ \ V_i = \Psi_i(\mathcal{V})\subset U_i,\ \ i=0,1,$$

$$\overline{U}_i := U_i\cap\mathbb{R}^m\times\mathbf{0}\subset\mathbb{R}^m,\ \ \overline{V}_i = V_i\cap\mathbb{R}^m\times\mathbf{0}\subset\overline{U}_i,\ \ i=0,1,$$

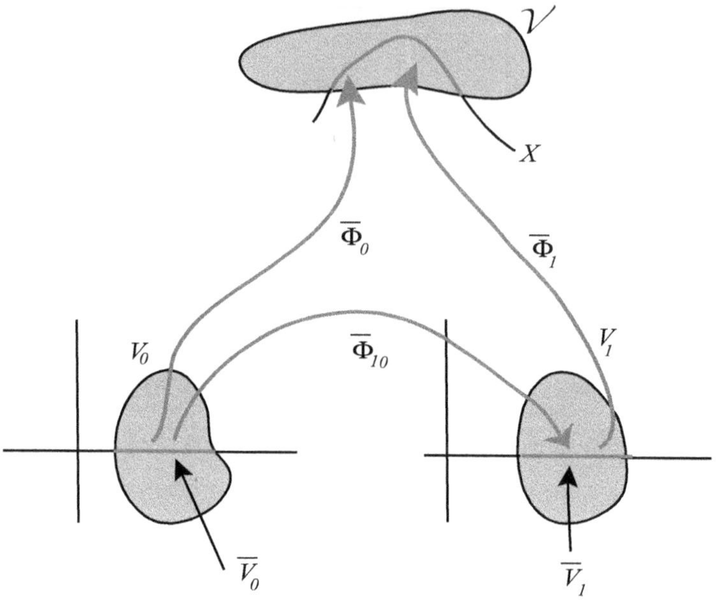

Fig. 16.27 *The transition map determined by two overlapping straightening diffeomorhisms.*

and C^1-maps

$$\overline{\Phi}_i : \overline{V}_i \to \mathcal{V}.$$

The composition

$$\Phi_{10} : \Psi_1 \circ \Psi_0^{-1} : V_0 \to V_1$$

is a diffeomorphism. It induces a homeomorphism

$$\overline{\Phi}_{10} : \overline{V}_0 \to \overline{V}_1.$$

Note that

$$\overline{\Phi}_{10} = \Psi_1 \circ \overline{\Phi}_0.$$

According to Proposition 16.82(i) the induced map $\overline{\Phi}_{10}$ is a diffeomorphism with image $\overline{V}_1$. We will refer to $\overline{\Phi}_{10}$ as the *transition diffeomorphism* associated to the pair of overlapping s.d.-s $(\mathcal{U}_i, \Psi_i)$, $i = 0, 1$.

Definition 16.83. Let $X \subset \mathbb{R}^n$ be an m-dimensional C^1-submanifold, $1 \leq m \leq n$.

(i) An *atlas* of X is a collection of s.d.-s $\{ (\mathcal{U}_i, \Psi_i) \}_{i \in I}$ such that the collection $(\mathcal{U}_i)_{i \in I}$ is an open cover of X. For $i, j \in I$ we set $\mathcal{U}_{ij} := \mathcal{U}_i \cap \mathcal{U}_j$.

(ii) An *orientation* of an atlas $\{ (\mathcal{U}_i, \Psi_i) \}_{i \in I}$ of X is a choice of orientations $\epsilon_i \in \mathbb{O}(\overline{U}_i)$, $\overline{U}_i = \Psi_i(\mathcal{U}_i \cap X) \subset \mathbb{R}^m$, $i \in I$. The orientation is called *coherent* if, for any $i, j \in I$ such that $\mathcal{U}_{ij} \neq \emptyset$, the associated transition map

$$\overline{\Phi}_{ji} : (\overline{V}_i, \epsilon_i) \to (\overline{V}_j, \epsilon_j)$$

is orientation preserving. Above $\overline{V}_i = \Psi_i(\mathcal{U}_{ij})$, $\overline{V}_j = \Psi_j(\mathcal{U}_{ij})$.

(iii) The submanifold X is called *orientable* if it admits a *coherently* oriented atlas.
(iv) An *orientation* on X is a choice of a coherently oriented atlas.
(v) Two orientations on X given by the coherently oriented atlases

$$\mathcal{A} := \left\{ (\mathcal{U}_i, \Psi_i, \epsilon_i) \right\}_{i \in I}, \quad \mathcal{B} := \left\{ (\mathcal{V}_j, \Psi_j, \epsilon_j) \right\}_{j \in J}$$

are to be considered equivalent if their union is also a *coherently* oriented atlas.

□

Example 16.84. Suppose that X is described by $n - m$ equations

$$X := \left\{ \boldsymbol{x} \in \mathbb{R}^n : \ F^1(\boldsymbol{x}) = \cdots = F^{n-m}(\boldsymbol{x}) = 0 \right\}, \ F^1, \dots, F^{n-m} \in C^1(\mathbb{R}^n)$$

such that, for any $\boldsymbol{x} \in X$ the vectors $\nabla F^1(\boldsymbol{x}), \dots, \nabla F^{n-m}(\boldsymbol{x})$ are linearly independent.

Suppose that $\mathcal{A} := (\mathcal{U}_i, \Psi_i)$ is an atlas for X. Set as usual

$$\overline{U}_i := \Psi_i(\mathcal{U}_i \cap X) \subset \mathbb{R}^m, \ \ \overline{\Phi}_i := \Psi_i^{-1}\big|_{\overline{U}_i}.$$

For simplicity we set $\boldsymbol{x}(\boldsymbol{\nu}) := \overline{\Phi}_i(\boldsymbol{u})$. Denote by $u^1, \dots, u^m$ the canonical Cartesian coordinates on $\overline{U}_i$ and we set

$$T_k(\boldsymbol{u}) = \frac{\partial \overline{\Phi}_i}{\partial u^i}(\boldsymbol{u}), \ \ k = 1, \dots, m.$$

The collection $\{T_1(\boldsymbol{u}), \dots, T_m(\boldsymbol{u})\}$ is a basis of the tangent pace $T_{\boldsymbol{x}(\boldsymbol{u})}X$. The vectors $\nabla F^j(\boldsymbol{x}(\boldsymbol{u}))$ are perpendicular to this space. It follows that the $n \times n$ matrix $B_i(\boldsymbol{u})$ with columns

$$\nabla F^1(\boldsymbol{x}(\boldsymbol{u})), \dots, \nabla F^{n-m}(\boldsymbol{x}(\boldsymbol{u})), T_1(\boldsymbol{u}), \dots, T_m(\boldsymbol{u})$$

is nonsingular. We obtain an orientation ϵ_i on $\overline{U}_i$ given by

$$\epsilon_i(\boldsymbol{u}) = \operatorname{sign} \det B_i(\boldsymbol{u}).$$

One can show that the collection $\left\{ (\mathcal{U}_i, \Psi_i, \epsilon_i) \right\}$ is a coherently oriented atlas and thus defines an orientation on X.

The natural proof of this fact is based on a bit more differential geometry than I can safely assume you, the reader, may know at this point in time. There exist proofs of this claim that use essentially only linear algebra, but the geometric meaning will be lost in the heap of computations. For this reason I have decided not to include a proof of this claim. Instead, I encourage you to supply a proof in the special case $m = 2$, $n = 3$ and compare this with the arguments in Remark 16.61. □

16.3.3. *Integration along oriented submanifolds*

Suppose that $X \subset \mathbb{R}^n$ is an orientable m-dimensional C^1-submanifold. We denote by $\Omega^m_{\mathrm{cpt}}(X)$ the subspace of $\Omega^m_{\mathrm{cpt}}(\mathbb{R}^n)$ consisting of compactly supported degree m forms ω such that $X \cap \operatorname{supp} \omega$ is a compact subset of X. We want to associate to any orientation $\boldsymbol{or}_X$ on X an integration map

$$\int_{X, \boldsymbol{or}_X} : \Omega^m_{\mathrm{cpt}}(X) \to \mathbb{R}.$$

We will build this integral in stages.

Fix an orientation $\vec{\epsilon}$ on X defined by a coherently oriented atlas

$$\mathcal{A} = \{ (\mathcal{U}_i, \Psi_i, \epsilon_i) \}_{i \in I}.$$

We denote by $\Omega^m_{\mathrm{cpt}}(X, \mathcal{A})$ the subspace of $\Omega^m_{\mathrm{cpt}}(\mathbb{R}^n)$ consisting of compactly supported degree m forms ω such that

- The support of ω is contained in the union
$$\mathcal{U}_{\mathcal{A}} := \bigcup_{i \in I} \mathcal{U}_i.$$
- $X \cap \operatorname{supp} \omega$ is a compact subset of X.

We will define a canonical linear map

$$\int_X = \int_{X,\mathcal{A}} : \Omega^m_{\mathrm{cpt}}(X, \mathcal{A}) \to \mathbb{R}, \;\; \Omega^m_{\mathrm{cpt}}(X, \mathcal{A}) \ni \omega \mapsto \int_{X,\mathcal{A}} \omega.$$

We achieve this in several steps.

Step 1. *The form* ω *has small support, i.e.,* $\exists i \in I$ *such that* $\operatorname{supp} \omega \cap X \subset \mathcal{U}_i \cap X$.

Consider the map

$$\overline{U}_i \ni \boldsymbol{u} \mapsto \boldsymbol{x}(\boldsymbol{u}) = \overline{\Phi}_i(\boldsymbol{u}) \in \mathcal{U}.$$

Set

$$\bar{\omega}_i := \overline{\Phi}_i^* \omega \in \Omega^m(\overline{U}_i).$$

In this case we provisorily define

$$\int_{X,\mathcal{A}} \omega := \int_{\overline{U}_{i,\epsilon_i}} \bar{\omega}_i = \int_{\overline{U}_{i,\epsilon_i}} \overline{\Phi}_i^* \omega, \tag{16.52}$$

where $\int_{U,\epsilon}$ is defined in (16.49).

The above definition is provisory because it depends on the choice of a chart $(\mathcal{U}_i, \Psi_i, \epsilon_i)$ such that

$$\operatorname{supp} \omega \cap X \subset X \cap \mathcal{U}_i.$$

What happens if, for another j, we have

$$\operatorname{supp} \omega \cap X \subset \mathcal{U}_j \cap X?$$

In this case we could use the chart $(\mathcal{U}_j, \Psi_j, \epsilon_j)$ to define $\int_{X,\mathcal{A}}$ and we cannot dismiss off hand that we might get a different answer than in (16.52). Let us show that this is not the case.

Suppose that we also have $\operatorname{supp} \omega \cap X \subset \mathcal{U}_j \cap X$ for a different $j \in I$. Then, we could propose a new definition of $\int_X \omega$

$$\int_{X,\mathcal{A}} \omega := \int_{\overline{U}_{j,\epsilon_j}} \bar{\omega}_j.$$

Set

$$\overline{V}_i := \Psi_i(\mathcal{U}_i \cap \mathcal{U}_j \cap X),\ \ \overline{V}_j := \Psi_j(\mathcal{U}_i \cap \mathcal{U}_j \cap X).$$

Thus

$$\operatorname{supp} \bar{\omega}_i \subset \overline{V}_i,\ \ \operatorname{supp} \bar{\omega}_j \subset \overline{V}_j.$$

From Proposition 16.82 we deduce that

$$\bar{\omega}_i = \overline{\Phi}_{ji}^* \bar{\omega}_j.$$

Since

$$\overline{\Phi}_{ji} : (\overline{V}_i, \epsilon_i) \to (\overline{V}_j, \epsilon_j)$$

is orientation preserving we have

$$\int_{\overline{U}_{i,\epsilon_i}} \bar{\omega}_i = \int_{\overline{V}_{i,\epsilon_i}} \bar{\omega}_i \stackrel{(16.50)}{=} \int_{\overline{V}_{j,\epsilon_j}} \bar{\omega}_j = \int_{\overline{U}_{j,\epsilon_j}} \bar{\omega}_j.$$

Set $X_i := X \cap \mathcal{U}_i$. We have thus defined an integration map

$$\int_{X_i} : \Omega^m_{\mathrm{cpt}}(X_i) \to \mathbb{R}.$$

This map is linear and it is independent of any other choice of local coordinates we could choose on X_i.

Step 2. *Extension to* $\Omega^m_{\mathrm{cpt}}(X, \mathcal{A})$. Let $\omega \in \Omega^m_{\mathrm{cpt}}(X, \mathcal{A})$. Choose a continuous partition of unity on $\operatorname{supp} \omega$

$$\psi_1, \dots, \psi_k : \mathbb{R}^n \to \mathbb{R}$$

subordinated to the open cover $(\mathcal{U}_i)_{i \in I}$. For each $a = 1, \dots, k$ choose $i(a) \in I$ such that $\operatorname{supp} \psi_a \subset \mathcal{U}_{i(a)}$. We have

$$\omega = \sum_{a=1}^{l} \psi_a \omega.$$

Note that $\omega_a = \psi_a \omega \in \Omega^m_{\mathrm{cpt}}(X_{i(a)})$. We define

$$\int_{X,\mathcal{A}} \omega = \sum_a \int_{X_{i(a)}} \psi_a \omega.$$

A priori, this definition depends on the choice of the partition of unity. Let us show that this is not the case.

Choose another partition of unity on $\operatorname{supp} \omega$, $\phi_1, \dots, \phi_\ell$, subordinated to the cover $(\mathcal{U}_i)_{i \in I}$. For each $b = 1, \dots, \ell$ choose $j(b) \in I$ such that $\operatorname{supp} \phi_b \subset \mathcal{U}_{j(b)}$. Note that

$$\omega_a = \sum_b \underbrace{\phi_b \omega_a}_{\omega_{ab}}.$$

Since $\operatorname{supp} \omega_{ab} \subset \mathcal{U}_{i(a)} \cap \mathcal{U}_{j(b)}$ we have

$$\int_{X_{i(a)}} \omega_a = \sum_b \int_{X_{i(a)}} \omega_{ab} = \sum_b \int_{X_{j(b)}} \omega_{ab},$$

so

$$\sum_a \int_{X_{i(a)}} \psi_a \omega = \sum_a \int_{X_{i(a)}} \omega_a = \sum_b \int_{X_{j(b)}} \underbrace{\sum_a \omega_{ab}}_{=\phi_b \omega} = \sum_b \int_{X_{j(b)}} \phi_b \omega.$$

This proves that the definition of $\int_{X,\mathcal{A}}$ is independent of the choices of partitions of unity.

Step 3. *Different atlases yield identical integrals.*

The argument in the previous step shows that if $\mathcal{A}$ and $\mathcal{B}$ are two coherently oriented atlases such that $\mathcal{A} \subset \mathcal{B}$, then

$$\Omega^m_{\text{cpt}}(X, \mathcal{A}) \subset \Omega^m_{\text{cpt}}(X, \mathcal{B}),$$

and, for any $\omega \in \Omega^m_{\text{cpt}}(X, \mathcal{A})$, we have

$$\int_{X,\mathcal{A}} \omega = \int_{X,\mathcal{B}} \omega.$$

In particular, this shows that if the coherently oriented atlases $\mathcal{A}$ and $\mathcal{B}$ define equivalent orientation, then for any $\omega \in \Omega^m_{\text{cpt}}(X, \mathcal{A}) \cap \Omega^m_{\text{cpt}}(X, \mathcal{B})$ we have

$$\int_{X,\mathcal{A}} \omega = \int_{X,\mathcal{A}\cup\mathcal{B}} \omega = \int_{X,\mathcal{B}} \omega.$$

We have thus constructed an integration map defined over

$$\bigcup_{\mathcal{A}} \Omega^n_{\text{cpt}}(X, \mathcal{A}) \subset \Omega^n_{\text{cpt}}(X),$$

where the above union is over all the coherently oriented atlases of X defining a given orientation.

Step 4. *Extend the integral to* $\Omega^n_{\text{cpt}}(X)$.

Let us remark that the proof in **Step 3** shows that if $\omega_1, \omega_2 \in \Omega_{\text{cpt}}(X, \mathcal{A})$ coincide in an open neighborhood of X in $\mathbb{R}^n$, then

$$\int_{X,\mathcal{A}} \omega_1 = \int_{X,\mathcal{A}} \omega_2.$$

Using partitions of unity one can show (but we will skip the details) that for any $\omega \in \Omega^m_{\text{cpt}}(X)$ and for any coherently oriented atlas $\mathcal{A}$ defining an orientation $\boldsymbol{or}_X$ on X, there exists a form $\tilde{\omega} \in \Omega^m_{\text{cpt}}(X, \mathcal{A})$ such that $\omega = \tilde{\omega}$ in an open neighborhood of X. We then set

$$\int_{X,\boldsymbol{or}_X} \omega := \int_{X,\mathcal{A}} \tilde{\omega}.$$

Clearly the right-hand side does not depend on any particular choices of $\tilde{\omega}$.

16.3.4. *The general Stokes' formula*

We first need to introduce the concept of manifolds with boundary. This is a simple generalization of the concept of surface with boundary introduced in Definition 16.40. We will skip many technical details.

Definition 16.85. Let $k, m, n \in \mathbb{N}$, $n \geq m \geq 1$. An m-dimensional C^k-*submanifold with boundary* in $\mathbb{R}^n$ is a *compact* subset $X \subset \mathbb{R}^n$ such that, for any point $\boldsymbol{p}_0$, there exists an open neighborhood $\mathfrak{U}$ of $\boldsymbol{p}_0$ in $\mathbb{R}^n$ and a C^k-diffeomorphism $\Psi : \mathfrak{U} \to \mathbb{R}^n$ such that the image $\overline{U} = \Psi(\mathfrak{U} \cap X)$ is contained in the subspace $\mathbb{R}^m \times \mathbf{0} \subset \mathbb{R}^n$ and it is either

(I) an open ball in $\mathbb{R}^m$ centered at $\Psi(\boldsymbol{p}_0)$ or
(B) the point $\Psi(\boldsymbol{p}_0)$ lies in plane $\{x^1 = 0\} \subset \mathbb{R}^m$ and $\overline{U}$ is the intersection of an open ball $B_r(\boldsymbol{p}_0)$ with the half-plane

$$\boldsymbol{H}^m_- := \{ (x^1, x^2, \dots, x^m) \in \mathbb{R}^2;\ x^1 \leq 0 \}.$$

The pair $(\mathfrak{U}, \Psi)$ is called a *straightening diffeomorphism* (abbreviated s.d.) at $\boldsymbol{p}_0$. The pair $\left(\mathfrak{U} \cap X, \Psi|_{\mathfrak{U} \cap X} \right)$ is called a *local coordinate chart* of X at $\boldsymbol{p}_0$.

The situations (I) and (B) are mutually exclusive. In the case (B), the point $\boldsymbol{p}_0 \in X$ is called a *boundary point* of X. In the case (I) the point $\boldsymbol{p}_0$ is called an *interior* point.

The set of boundary points of X is called the *boundary* of X and it is denoted by ∂X. The set of interior points of X is called the *interior* of X and it is denoted by X°. The submanifold with boundary is called *closed* if its boundary is empty, $\partial X = \emptyset$. □

An atlas of a manifold with boundary X is a collection of s.d.-s $\{ (\mathfrak{U}_i, \Psi_i) \}_{i \in I}$ such that

$$X \subset \bigcup_{i \in I} \mathfrak{U}_i.$$

An *orientation* of an s.d. $(\mathfrak{U}, \Psi)$ is an orientation on the interior of $\Psi(X \cap \mathfrak{U})$. The transition maps are defined in a similar fashion which leads as in the boundary-less case to the concept of orientation of a manifold with boundary. Equivalently, an orientation on a manifold with boundary is equivalent to a choice of orientation on its interior.

There is a new phenomenon. Namely, an orientation on X induces in a natural fashion an orientation on its boundary ∂X.

Suppose that $\mathcal{A} := \{ (\mathfrak{U}_i, \Psi_i, \epsilon_i) \}_{i \in I}$ is a coherently oriented atlas of X. The s.d.-s $(\mathfrak{U}_i, \Psi_i)$ are of two types.

- *interior*, i.e., $\mathfrak{U} \cap \partial X = \emptyset$ and
- *boundary*, i.e., $\mathfrak{U} \cap \partial X \neq \emptyset$.

Consider the subcollection $\mathcal{A}_\partial := \{ (\mathfrak{U}_a, \Psi_a) \}_{a \in A \subset I}$ consisting of all the boundary type s.d.-s in $\mathcal{A}$. The collection $\mathcal{A}_\partial$ is also an atlas for the manifold ∂X. For any $a \in A$ we set

$$\overline{V}_a := \Psi_a(\mathfrak{U}_A \cap \partial X) \subset \partial \boldsymbol{H}^m_-.$$

Note that for $a \in A$ we have

$$\overline{U}_a := \Psi_a(\mathfrak{U}_a \cap X) = B_{r(a)}(\boldsymbol{p}_a) \cap \boldsymbol{H}_-^m, \;\; \boldsymbol{p}_a \in \partial \boldsymbol{H}_-^m.$$

Denote by $u^1, \dots, u^m$ the Cartesian coordinates on the space $\mathbb{R}^m$ where the half-space $\boldsymbol{H}_-^m$ lives. An orientation ϵ_a on the interior $\boldsymbol{int}\overline{U}_a$ is given by the top degree form

$$\omega_a := \epsilon_a du^1 \wedge du^2 \wedge \cdots \wedge du^m.$$

The boundary $\partial \boldsymbol{H}_-^m$ is the subspace $\mathbb{R}^{m-1}$ with Cartesian coordinates $u^2, \dots, u^m$. The induced orientation on $\overline{V}_a$, denoted by $\partial\epsilon_a$, is described by the top degree form

$$\omega_a^\partial := \epsilon_a du^2 \wedge \cdots \wedge du^m.$$

There is a simple mnemonic device to help you remember this construction. It is called the *outer conormal first* convention. Let us explain.

Note that traveling in $\boldsymbol{H}_-^m$ in the direction of increasing u^1 one eventually exits $\boldsymbol{H}_-^m$. Equivalently, observe that along $\partial \boldsymbol{H}_-^m$ the vector field $\boldsymbol{e}_1 = (1, 0, \dots, 0) \in \mathbb{R}^m$ is an outer pointing normal vector field. Note that

$$\omega_a = du^1 \wedge \omega_a^\partial,$$

or,

$$\text{orientation interior} = \boxed{\text{outer conormal}} \wedge \text{orientation boundary},$$

whence the terminology *outer-conormal-first* convention.

One can show that if

$$\mathcal{A} := \big\{ (\mathfrak{U}_i, \Psi_i, \epsilon_i) \big\}_{i \in I}$$

is a coherently oriented atlas of the manifold with boundary X, then the collection

$$\mathcal{A}_\partial = \big\{ (\mathfrak{U}_a, \Psi_a, \partial\epsilon_a) \big\}_{a \in A}, \;\; A = \big\{ a \in I;\; U_a \cap \partial X \neq \emptyset \big\},$$

is a coherently oriented atlas of ∂X. While we will not present all the tedious details, we want to explain the simple fact behind this. Its proof is left to you as an exercise.

Lemma 16.86. *Let $(\overline{U}_i, \epsilon_i)$, $i = 0, 1$ be two oriented open subsets of $\mathbb{R}^m$ such that $\overline{U}_i \cap \partial \boldsymbol{H}_-^m \neq \emptyset$ for all $i = 0, 1$. If $\Phi : (\overline{U}_0, \epsilon_0) \to (\overline{U}_1, \epsilon_1)$ is an orientation preserving diffeomorphism such that*

$$\Phi\big(\overline{U}_0 \cap \partial \boldsymbol{H}_-^m\big) \subset \partial \boldsymbol{H}_-^m,$$

then the induced map

$$\Phi_\partial : (\overline{U}_0 \cap \partial \boldsymbol{H}_-^m, \partial\epsilon_0) \to (\overline{U}_1 \cap \partial \boldsymbol{H}_-^m, \partial\epsilon_1)$$

is also orientation preserving. □

Just like in the case of submanifolds without boundary, an orientation ϵ on an m-dimensional manifold with boundary defines an integration map

$$\int_{X,\epsilon} : \Omega^m(\mathbb{R}^n) \to \mathbb{R}.$$

The submanifolds with boundary are, by definition, compact so we no longer need to work with compactly supported forms.

The construction follows the same four steps as in the bondary-less case so we can safely omit the details.

At the same time, we have another oriented submanifold $(\partial X, \partial\epsilon)$. In particular, we also have an integration map

$$\int_{\partial X,\partial\epsilon} : \Omega^{m-1}(\partial X) \to \mathbb{R}.$$

Theorem 16.87 (General Stokes' formula). *Suppose that (X,ϵ) is an m-dimensional oriented C^1-submanifold with boundary of $\mathbb{R}^n$. Then for any $\omega \in \Omega^{m-1}(\mathbb{R}^n)_{C^1}$ we have*

$$\int_{\partial X,\partial\epsilon} \omega = \int_{X,\epsilon} d\omega, \tag{16.53}$$

where $d\omega \in \Omega^m(\mathbb{R}^n)$ is the exterior derivative of ω.

Proof. Fix a coherently oriented atlas

$$\mathcal{A} := \big\{\, (\mathcal{U}_i, \Psi_i, \epsilon_i) \,\big\}_{i\in I}$$

that defines the orientation ϵ. Set

$$\mathcal{U}_{\mathcal{A}} := \bigcup_{i\in I} U_i.$$

Fix a compact set K such that[6]

$$X \subset \boldsymbol{int}\, K,\;\; K \subset \mathcal{U}_{\mathcal{A}}. \tag{16.54}$$

Using the results in Exercise 13.14 we can find a C^1-partition of unity along K and subordinated to the open cover $(U_i)_{i\in I}$. Recall that this is a *finite* collection $(\chi_s)_{s\in S}$ of compactly supported C^1-functions $\chi_s : \mathbb{R}^n \to \mathbb{R}$ satisfying the following properties.

- For all $s \in S$ there exists $i = i(s) \in I$ such that supp $\chi_s \subset \mathcal{U}_{i(s)}$.
- $$\sum_{s\in S} \chi_s(\boldsymbol{x}) = 1, \;\; \forall \boldsymbol{x} \in K.$$

Let $\omega \in \Omega^{m-1}(\mathbb{R}^n)_{C^1}$. For $s \in S$ set

$$\eta_s := \chi_s\omega \in \Omega^{m-1}_{\text{cpt}}(\mathbb{R}^n),$$

[6] Can you see why a compact set K satisfying (16.54) exists?

and define

$$\eta := \sum_s \eta_s.$$

Note that on $\boldsymbol{int}\, K \supset X$ we have

$$\omega = \eta, \ \ d\omega = d\eta = \sum_s d\eta_s,$$

so it suffices to prove (16.53) for η. On the other hand,

$$\int_{\partial X, \partial\epsilon} \eta = \sum_s \int_{\partial X, \partial\epsilon} \eta_s,$$

$$\int_{X,\epsilon} d\eta = \sum_s \int_{X,\epsilon} d\eta_s,$$

so it suffices to prove (16.53) for each of the individual η_s. Thus we have to prove that (16.53) holds for forms $\eta \in \Omega^{m-1}_{\text{cpt}}(\mathbb{R}^n)_{C^1}$ satisfying the additional property that there exists an oriented s.d. $(\mathcal{U}, \Psi, \epsilon)$ such that $\operatorname{supp} \eta \subset \mathcal{U}$. We distinguish two cases.

Interior case, i.e., $\mathcal{U} \cap \partial X = \emptyset$. In this case η is identically zero in a neighborhood of ∂X, so

$$\int_{\partial X, \partial\epsilon} \eta = 0.$$

We have to prove that

$$\int_{X,\epsilon} d\eta = 0.$$

We set $\overline{U} = \Psi(\mathcal{U} \cap X)$ so that $\overline{U}$ is an open subset of $\mathbb{R}^m$. Denote by Φ the inverse of Ψ and by $\overline{\Phi}$ the restriction of Φ to $\overline{U}$. Then, according to (16.52), we have

$$\int_{X,\epsilon} d\eta = \int_{\overline{U},\epsilon} \overline{\Phi}^* d\eta.$$

We set

$$\bar{\eta} := \overline{\Phi}^* \eta.$$

From (16.47) we deduce that

$$\overline{\Phi}^* d\eta = \overline{\Phi}^* \eta d\bar{\eta}.$$

Thus we have to show that

$$\int_{\overline{U},\epsilon} d\bar{\eta} = 0$$

for any $\bar{\eta} \in \Omega^{m-1}_{\text{cpt}}(\overline{U})_{C^1}$.

Let $\bar{\eta}$ be such a degree $(m-1)$ form. Fix a positive number R such that

$$\overline{U} \subset C^m_R := [-R, R]^m \subset \mathbb{R}^m.$$

We have

$$\bar{\eta} = \eta_1 du^2 \wedge du^3 \wedge \cdots \wedge du^m + \eta_2 du^1 \wedge du^3 \wedge \cdots \wedge du^m + \cdots + \eta_m du^1 \wedge du^2 \wedge \cdots \wedge du^{m-1}.$$

This can be written in a more compact form as

$$\bar{\eta} = \sum_{k=1}^{m} \eta_k du^1 \wedge \cdots \wedge \widehat{du^k} \wedge \cdots \wedge du^m, \tag{16.55}$$

where a hat $\widehat{\ }$ indicates a missing entry. We deduce

$$d\bar{\eta} = \left(\frac{\partial \eta_1}{\partial u^1} - \frac{\partial \eta_2}{\partial u^2} + \cdots + (-1)^{m-1}\frac{\partial \eta_m}{\partial u^m}\right) du^1 \wedge du^2 \wedge \cdots \wedge du^m \tag{16.56}$$

$$= \left(\sum_{k=1}^{m}(-1)^{k-1}\frac{\partial \eta_k}{\partial u^k}\right) du^1 \wedge du^2 \wedge \cdots \wedge du^m. \tag{16.57}$$

Then

$$\int_{\overline{U},\epsilon} d\bar{\eta} = \sum_{k=1}^{m}(-1)^{k-1}\epsilon \int_{\overline{U}} \frac{\partial \eta_k}{\partial u^k}\, |du^1 \cdots du^m|.$$

We will prove that

$$\int_{\overline{U}} \frac{\partial \eta_k}{\partial u^k}\, |du^1 \cdots du^m| = 0, \quad \forall k = 1, \ldots, m.$$

For simplicity we discuss only the case $k = 1$. The other cases are completely similar. We have

$$\int_{\overline{U}} \frac{\partial \eta_1}{\partial u^1}\, |du^1 \cdots du^m| = \int_{C_R^m} \frac{\partial \eta_1}{\partial u^1}\, |du^1 \cdots du^m|$$

(use Fubini)

$$= \int_{C_R^{m-1}} \left(\int_{-R}^{R} \frac{\partial \eta_1}{\partial x^1}\, |dx^1|\right) |du^2 \cdots du^m|$$

$$= \int_{C_R^{m-1}} \Big(\underbrace{\eta_1(R, u^2, \ldots, u^m)}_{=0} - \underbrace{\eta_1(-R, u^2, \ldots, u^m)}_{=0} \Big) = 0.$$

Boundary case, i.e., $\mathcal{U} \cap \partial X = \emptyset$. In this case $\overline{U} := \Psi(\mathcal{U} \cap X)$ is a half-ball (see Figure 16.28)

$$\overline{U} = \boldsymbol{H}_-^m \cap B_r(\boldsymbol{p}_0), \quad \boldsymbol{p}_0 \in \partial \boldsymbol{H}_-^m.$$

We can find $L > 0$ such that $\overline{U}$ is contained in the closed box

$$B^- = [-L, L]^m \cap \boldsymbol{H}_-^m \subset \mathbb{R}^m.$$

We define

$$B^0 := [-L, L]^m \cap \partial \boldsymbol{H}_-^m = \{0\} \times [-L, L]^{m-1}.$$

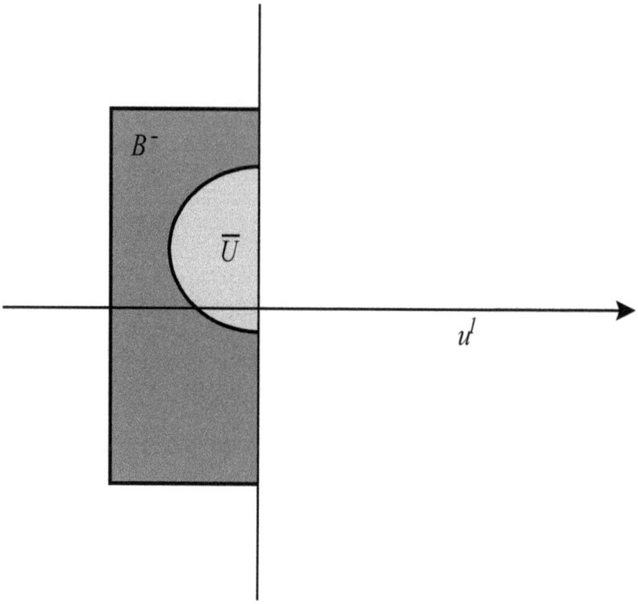

Fig. 16.28 *Integrating over a half-ball.*

Arguing as in the previous case we deduce that

$$\int_{X,\epsilon} d\eta = \int_{\overline{U},\epsilon} d\bar{\eta} = \int_{B^-,\epsilon} d\bar{\eta}, \quad \int_{\partial X,\partial\epsilon} \eta = \int_{B^0,\partial\epsilon} \bar{\eta}.$$

If

$$\bar{\eta} = \sum_{k=1}^{m} \eta_k du^1 \wedge \cdots \wedge \widehat{du^k} \wedge \cdots \wedge du^m,$$

then

$$\int_{B^-,\epsilon} d\bar{\eta} = \sum_{k=1}^{m} (-1)^{k-1}\epsilon \int_{B^-} \frac{\partial \eta_k}{\partial u^k}\, |du^1 \cdots du^m|,$$

and

$$\int_{B^0,\partial\epsilon} = \epsilon \int_{B^0} \eta_1(0,u^2,\dots,u^m)\, |du^2 \cdots du^m|.$$

We will show that

$$\int_{B^-} \frac{\partial \eta_1}{\partial u^1}\, |du^1 \cdots du^m| = \int_{B^0} \eta_1(0,u^2,\dots,u^m)\, |du^2 \cdots du^m|, \tag{16.58a}$$

$$\int_{B^-} \frac{\partial \eta_k}{\partial u^k}\, |du^1 \cdots du^m| = 0, \quad \forall k = 2,\dots,m. \tag{16.58b}$$

To prove (16.58a) we use Fubini's theorem and we deduce

$$\int_{B^-} \frac{\partial \eta_1}{\partial u^1}\, |du^1 \cdots du^m| = \int_{|u^k|\leq L,\ 2\leq k\leq m} \left(\int_{-L}^{0} \frac{\partial \eta_1}{\partial u^1} du^1 \right) |du^2 \cdots du^m|$$

(use the Fundamental Theorem of Calculus)

$$= \int_{|u^k|\leq L,\ 2\leq k\leq m} \left(\eta_1(0,u^2,\dots,u^m) - \underbrace{\eta_1(-L,u^2,\dots,u^m)}_{=0} \right) |du^2\cdots du^m|$$

$$= \int_{B^0} \eta_1(0,u^2,\dots,u^m)\,|du^2\cdots du^m|.$$

The equality (16.58b) also follows from Fubini's theorem. We prove only the case $k=m$ which involves simpler notations.

$$\int_{B^-} \frac{\partial \eta_m}{\partial u^m}\,|du^1\cdots du^m|$$

$$= \int_{[-L,0]\times[-L,L]^{m-2}} |du^1du^2\cdots du^{m-1}| \left(\int_{-L}^{L} \frac{\partial \eta_m}{\partial u^m}(u^1,\dots,u^{m-1},u^m)\,du^m \right)$$

(use the Fundamental Theorem of Calculus)

$$= \int_{[-L,0]\times[-L,L]\times\cdots\times[-L,L]} |du^1du^2\cdots du^{m-1}| \underbrace{\left(\eta_m(u^1,\dots,u^{m-1},u^m)\Big|_{u^m=-L}^{u^m=L} \right)}_{=0} = 0.$$

□

16.3.5. *What are these differential forms anyway*

In lieu of epilogue to this book, I will try to crack open the door to another world to which the considerations in this last section properly belong.

We have developed a theory of integration of objects whose nature was left nebulous. What are these differential forms?

Suppose that U is an open subset of $\mathbb{R}^n$, $n\geq 2$. The equality (13.20) of Example 13.18 explained that the terms dx^i should be viewed as *linear forms*. A linear form, as you know, is a "beast" that, when fed a vector, it spits out a number. If you feed a vector $\boldsymbol{v}$ to the "beast" $dx^i=\boldsymbol{e}^i$, it will return the number v^i, the i-th coordinate of the vector $\boldsymbol{v}$.

The exterior monomials

$$dx^i\wedge dx^j = \boldsymbol{e}^i\wedge\boldsymbol{e}^j,\, dx^i\wedge dx^j\wedge dx^k = \boldsymbol{e}^i\wedge\boldsymbol{e}^j\wedge\boldsymbol{e}^k,$$

etc., are more sophisticated "beasts": they are multilinear maps with certain additional properties. To explain their nature we consider a slightly more complicated situation.

Let $m\in\mathbb{N}$ and consider m linear functionals

$$\alpha^1,\dots,\alpha^m \to \mathbb{R}.$$

Their exterior product is m-linear map

$$\alpha^1\wedge\cdots\wedge\alpha^m : \underbrace{\mathbb{R}^n\times\cdots\times\mathbb{R}^n}_{m} \to \mathbb{R},$$

$$\alpha^1\wedge\cdots\wedge\alpha^m(\boldsymbol{v}_1,\dots,\boldsymbol{v}_m):=\det\begin{bmatrix}\alpha^1(\boldsymbol{v}_1) & \alpha^1(\boldsymbol{v}_2) & \cdots & \alpha^1(\boldsymbol{v}_m)\\ \alpha^2(\boldsymbol{v}_1) & \alpha^2(\boldsymbol{v}_2) & \cdots & \alpha^2(\boldsymbol{v}_m)\\ \vdots & \vdots & \ddots & \vdots\\ \alpha^m(\boldsymbol{v}_1) & \alpha^m(\boldsymbol{v}_2) & \cdots & \alpha^m(\boldsymbol{v}_m)\end{bmatrix}. \tag{16.59}$$

The exterior monomial $\alpha^1\wedge\cdots\wedge\alpha^m$ is thus a "beast" that, when fed m vectors, it outputs a number.

Note that the m-linear form $\alpha^1\wedge\cdots\wedge\alpha^m$ satisfies the skew-symmetry conditions

$$\alpha^{\sigma(1)}\wedge\cdots\wedge\alpha^{\sigma(m)}=\epsilon(\sigma)\alpha^1\wedge\cdots\wedge\alpha^m,$$

$$\alpha^1\wedge\cdots\wedge\alpha^m(\boldsymbol{v}_{\sigma(1)},\dots,\boldsymbol{v}_{\sigma(m)})=\epsilon(\sigma)\alpha^1\wedge\cdots\wedge\alpha^m(\boldsymbol{v}_1,\dots,\boldsymbol{v}_m),$$

for any permutation $\sigma\in\mathfrak{S}_m$. Let us observe that if $m>n$, then any collection of m vectors $\boldsymbol{v}_1,\dots,\boldsymbol{v}_m\in\mathbb{R}^n$ is linearly dependent, and we deduce from (16.59) that

$$\alpha^1\wedge\cdots\wedge\alpha^m=0$$

for any linear forms $\alpha_1,\dots,\alpha_m:\mathbb{R}^n\to\mathbb{R}$.

When $m=n$ and $\alpha^i=\boldsymbol{e}^i$, then

$$\boldsymbol{e}^1\wedge\cdots\wedge\boldsymbol{e}^n(\boldsymbol{v}_1,\dots,\boldsymbol{v}_n)=\det\big[\boldsymbol{v}_j^i\big]_{1\le i,j\le n}\overset{(15.35)}{=}\pm\operatorname{vol}\big(\boldsymbol{P}(\boldsymbol{v}_1,\dots,\boldsymbol{v}_n)\big),$$

where we recall that $\boldsymbol{P}(\boldsymbol{v}_1,\dots,\boldsymbol{v}_n)$ denotes the parallelepiped spanned by $\boldsymbol{v}_1,\dots,\boldsymbol{v}_n$.

More generally, if $m<n$, then for any vectors $\boldsymbol{v}_1,\dots,\boldsymbol{v}_m$, the number

$$\boldsymbol{e}^1\wedge\cdots\wedge\boldsymbol{e}^m(\boldsymbol{v}^1,\dots,\boldsymbol{v}_m)$$

is equal, up to a sign, with the volume of the parallelepiped spanned by the orthogonal projections of the vectors $\boldsymbol{v}_1,\dots,\boldsymbol{v}_m$ onto the m-dimensional subspace of $\mathbb{R}^n$ spanned by $\boldsymbol{e}_1,\dots,\boldsymbol{e}_m$.

In general, and exterior form of degree m is an m-linear map

$$\omega:\underbrace{\mathbb{R}^n\times\cdots\times\mathbb{R}^n}_{m}\to\mathbb{R},$$

satisfying the skew-symmetry condition

$$\omega(\boldsymbol{v}_{\sigma(1)},\dots,\boldsymbol{v}_{\sigma(m)})=\epsilon(\sigma)\omega(\boldsymbol{v}_1,\dots,\boldsymbol{v}_m),\ \ \forall\boldsymbol{v}_1,\dots,\boldsymbol{v}_m\in\mathbb{R}^n,\ \ \sigma\in\mathfrak{S}_m. \tag{16.60}$$

We denote by $\Lambda^m\mathbb{R}^n$ the vector space of skew-symmetric, m-linear forms on $\mathbb{R}^n$. Such a form can be thought of as "gauging m-dimensional" parallelepipeds. This gauging is of a special kind: its output depends on the order in which we "feed" the vectors spanning the parallelepiped according to (16.60).

Take for the example the case $m=2$. Think of a parallelogram as having two faces: a white face and a black face. When a 2-form gauges a white-face-up parallelogram, it outputs a number, but when it gauges the same parallelogram but with the black face up, it outputs the opposite number.

We can use the canonical basis $e_1, \dots, e_n$ to express an m-form ω as a linear combination (compare with (16.36))

$$\omega = \sum_{I \in \mathrm{Inj}^+(m,n)} \omega_I e^{\wedge I},$$

where, for any $I = (i_1, \dots, i_m) \in \mathrm{Inj}^+(m,n)$, we have

$$e^{\wedge I} = e^{i_1} \wedge \cdots \wedge e^{i_m}, \quad \omega_I := \omega(e_{i_1}, \dots, e_{i_m}) \in \mathbb{R}.$$

In other words the collection

$$\left(e^I \right)_{I \in \mathrm{Inj}^+(m,n)}$$

is a basis of $\Lambda^m \mathbb{R}^n$. In particular

$$\dim \Lambda^m \mathbb{R}^m = \binom{n}{m}.$$

A differential form of degree m on an open set $U \subset \mathbb{R}^m$ is a continuous assignment of an m- form $\omega_{\boldsymbol{x}}$ to each point $\boldsymbol{x} \in U$. More precisely this means that

$$\omega_{\boldsymbol{x}} = \sum_{I \in \mathrm{Inj}^+(m,n)} \omega_I(\boldsymbol{x}) e^{\wedge I},$$

where $\omega_I(\boldsymbol{x})$ depends continuously on $\boldsymbol{x}$. More compactly, $\Omega^1(U)$ is the space of continuous maps

$$U \to \Lambda^m \mathbb{R}^n = \mathbb{R}^{\binom{n}{m}}.$$

Intuitively we can think that we have a continuous family of "gauges" $\omega_{\boldsymbol{x}}$, where $\omega_{\boldsymbol{x}}$ is to be used to gauge parallelepipeds originating at $\boldsymbol{x}$.

In the case $m = 2$ such a differential form gauges parallelograms. In particular, given a surface $S \subset \mathbb{R}^n$, such a form gauges infinitesimal parallelograms on S, i.e., parallelograms spanned by a pair of vectors tangent to the same point $\boldsymbol{x} \in S$. We use the form $\omega_{\boldsymbol{x}}$ to gauge such a parallelogram. An orientation on S is essentially a rule we use to determine the order in which we feed infinitesimal parallelograms to the differential form because we know that the output is order sensitive.

The above intuitive interpretation of differential forms gives a pretty accurate idea on the nature of differential forms. Unfortunately, it is essentially useless if we want to perform meaningful mathematical computations with them.

At this point a deeper look at the concept of differential form is needed and this requires substantial algebraic and analytic considerations. Fortunately you now have all knowledge you need to digest the classic little book [18] of M. Spivak on this topic. Although it is more than half a century old as I write these lines, it remains very actual and a gem of mathematical writing. However, do not let the tiny size of [18] fool you: it has a high density of subtle ideas per square inch of page.

You ought to be very familiar with the first half of [18]. The second half, on integration of differential forms and various Stokes' formulæ, is a rather steep, but very rewarding intellectual climb.

16.4. Exercises

Exercise 16.1. Denote by C the line segment in the plane $\mathbb{R}^2$ that connects the points $\boldsymbol{p}_0 := (3,0)$ and $\boldsymbol{p}_1 := (0,4)$.

(i) Compute the integrals

$$\int_C ds,\ \int_C f(\boldsymbol{p})ds,\ f(x,y) = x^2 + y^2.$$

(ii) Compute the integral of the angular form

$$W_\Theta = \frac{-y}{x^2+y^2}dx + \frac{x}{x^2+y^2}dy$$

along the segment C equipped with the orientation corresponding to the travel from $\boldsymbol{p}_1$ to $\boldsymbol{p}_0$.

□

Exercise 16.2. Denote by S the open square $(-1,1)\times(-1,1)$ in $\mathbb{R}^2$. Suppose that $P, Q : S \to \mathbb{R}$ are continuous functions. We set

$$\omega = Pdx + Qdy.$$

Prove that the following statements are equivalent.

(i) There exists $f \in C^1(S)$ such that $df = \omega$, i.e.,

$$P = \frac{\partial f}{\partial x},\ Q = \frac{\partial f}{\partial y}.$$

(ii) For any piecewise C^1 path $\gamma : [a,b] \to S$ such that $\gamma(a) = \gamma(b)$ we have

$$\int_\gamma \omega = 0.$$

(iii) For any piecewise C^1 paths $\gamma_i : [a_i, b_i] \to S$, $i = 1, 2$, such that

$$\gamma_1(a_1) = \gamma_2(a_2),\ \gamma_1(b_1) = \gamma_2(b_2)$$

we have

$$\int_{\gamma_1} \omega = \int_{\gamma_2} \omega.$$

Hint. (iii) ⇒ (i) Define

$$f(x,y) = \int_0^x P(s,0)ds + \int_0^y Q(x,t)dt$$

and use (iii) to prove that

$$f(x+h,y) = f(x,y) + \int_x^{x+h} P(s,y)ds,\ f(x,y+k) = f(x,y) + \int_y^{y+k} Q(x,t)dt,$$

and $\partial_x f = P$, $\partial_y f = Q$.

□

Exercise 16.3. Let $U \subset \mathbb{R}^n$ be an open set and suppose that $f, g : U \to \mathbb{R}$ are C^2-functions. Prove that

$$\Delta g = \operatorname{div} \nabla g, \;\; \operatorname{div}(f \nabla g) = \langle \nabla f, \nabla g \rangle + f \Delta g,$$

where Δg is the Laplacian of g,

$$\Delta g := \sum_{k=1}^{n} \partial^2_{x^k} g.$$

□

Exercise 16.4. Let $D \subset \mathbb{R}^2$ be a bounded domain with C^1-boundary, U an open set containing $\boldsymbol{cl}\, D$ and $f, g : U \to \mathbb{R}$ are C^2 function. Denote by $\boldsymbol{\nu}$ the outer normal vector field along ∂D. Prove that

$$\int_D f \Delta g \, |dxdy| = \int_{\partial D} f(\boldsymbol{p}) \frac{\partial g}{\partial \boldsymbol{\nu}}(\boldsymbol{p}) |ds| - \int_D \langle \nabla f, \nabla g \rangle \, |dxdy|,$$

$$\int_{\partial D} \frac{\partial g}{\partial \boldsymbol{\nu}} \, |ds| = \int_D \Delta g \, |dxdy|,$$

$$\int_D f \Delta g |dxdy| = \int_{\partial D} \left(f(\boldsymbol{p}) \frac{\partial g}{\partial \boldsymbol{\nu}}(\boldsymbol{p}) - g(\boldsymbol{p}) \frac{\partial f}{\partial \boldsymbol{\nu}}(\boldsymbol{p}) \right) |ds| + \int_D g \Delta f \, |dxdy|,$$

where we recall that, for $\boldsymbol{p} \in \partial D$, we have

$$\frac{\partial g}{\partial \boldsymbol{\nu}}(\boldsymbol{p}) := \langle \nabla g(\boldsymbol{p}), \boldsymbol{\nu}(\boldsymbol{p}) \rangle.$$

Hint. Use Exercise 16.3 and the flux-divergence formula (16.10). □

Exercise 16.5. Consider the function $K : \mathbb{R}^2 \to \mathbb{R}$,

$$K(x, y) = \begin{cases} \frac{1}{2\pi} \ln r, & (x, y) \neq (0, 0), \\ 0, & (x, y) = (0, 0), \;\; r = \sqrt{x^2 + y^2}. \end{cases}$$

Suppose that $f : \mathbb{R}^2 \to \mathbb{R}$ is a C^2-function with *compact support.*

(i) Show that $\Delta K = 0$ on $\mathbb{R}^2 \setminus \{\boldsymbol{0}\}$.

(ii) Show that the integral

$$\int_{\mathbb{R}^2} K \Delta f \, |dxdy|$$

is absolutely integrable.

(iii) Show that

$$\int_{\mathbb{R}^2} K \Delta f \, |dxdy| = -f(0).$$

Hint. (ii) Have a look back at Exercise 15.32. (iii) Fix $R > 0$ sufficiently large so that the support of f is contained in the disk $D_R = \{ r < R \}$. For $\varepsilon > 0$ small we consider the disk $D_\varepsilon = \{ r < \varepsilon \}$ and the annulus $A_{\varepsilon, R} := \{ \varepsilon < r < R \}$. Use Exercise 16.4 to show that

$$\int_{A_{\varepsilon,R}} K \Delta f \, |dxdy| = - \int_{\partial D_\varepsilon} \left(f \frac{\partial K}{\partial \boldsymbol{\nu}} - K \frac{\partial f}{\partial \boldsymbol{\nu}} \right) |ds|,$$

and then prove that

$$\lim_{\varepsilon \searrow 0} \int_{\partial D_\varepsilon} K \frac{\partial f}{\partial \boldsymbol{\nu}} \, ds = 0, \;\; \lim_{\varepsilon \searrow 0} \int_{\partial D_\varepsilon} f \frac{\partial K}{\partial \boldsymbol{\nu}} \, |ds| = f(0).$$

□

Exercise 16.6. Suppose that $\beta, \tau : [a, b]$ are C^1-functions such that

$$\beta(x) < \tau(x), \ \ \forall x \in (a, b).$$

Consider the simple type domain

$$D(\beta, \tau) = \{ (x, y) \in \mathbb{R}^2; \ x \in (a, b), \ \beta(x) < y < \tau(x) \}. \tag{16.61}$$

Prove Stokes' formula (16.13) when the piecewise C^1 domain is $U = D(\beta, \tau)$.

Hint. To compute $\int_D P'_y |dxdy|$ use Fubini's Theorem 15.41. To compute $\int_D Q'_x |dxdy|$ use the change of variables

$$x = u, \ \ y = \beta(u) + (\tau(u) - \beta(u))v, \ \ u \in [a, b], \ \ v \in [0, 1],$$

and the chain rule

$$\frac{\partial}{\partial x} = \frac{\partial u}{\partial x}\frac{\partial}{\partial u} + \frac{\partial v}{\partial x}\frac{\partial}{\partial v} = \partial_u - \frac{\beta'(u) + (\tau'(u) - \beta'(u))v}{\tau(u) - \beta(u)}\partial_v.$$

(You have to justify the second equality above.) We set

$$f(u, v) := Q\big(x(u, v), y(u, v) \big).$$

Show that

$$\int_D \frac{\partial Q}{\partial x} |dxdy| = \int_{\substack{a \leq u \leq b \\ 0 \leq v \leq 1}} \underbrace{\left(\partial_u f - \frac{\beta'(u) + (\tau'(u) - \beta'(u))v}{\tau(u) - \beta(u)} \partial_v f \right) \big(\tau(u) - \beta(u) \big)}_{=:g(u,v)} |dudv|.$$

On the other hand, if we write Qdy in (u, v) coordinates we get

$$Qdy = f(u, v)y'_u du + f(u, v)y'_v dv = \underbrace{f(u, v)\big(\beta'(u) + v(\tau'(u) - \beta'(u))\big)}_{=:A(u,v)} du + \underbrace{f(u, v)\tau(u)}_{=:B(u,v)} dv.$$

Show that

$$g(u, v) = \frac{\partial B}{\partial u} - \frac{\partial A}{\partial v}$$

and then compute

$$\int_{\substack{a \leq u \leq b \\ 0 \leq v \leq 1}} \left(\frac{\partial B}{\partial u} - \frac{\partial A}{\partial v} \right) |dudv|$$

using Fubini. □

Exercise 16.7. Suppose that $D_1, D_2 \subset \mathbb{R}^2$ are two bounded piecewise C^1 domains that intersect only along portions of their boundaries and $\partial D_1 \cap \partial D_2$ is a piecewise C^1 connected curve. Set $D := D_1 \cup D_2$.

(i) Show that D is also piecewise C^1.

(ii) Assume that $\mathcal{O} \subset \mathbb{R}^2$ is an open set containing the closure of D and $\boldsymbol{F} : \mathcal{O} \to \mathbb{R}^2$ is a continuous vector field, $F(x, y) = \big(P(x, y), Q(x, y) \big)$. Show that

$$\int_{\partial_+^* D} Pdx + Qdy = \int_{\partial_+^* D_1} Pdx + Qdy + \int_{\partial_+^* D_2} Pdx + Qdy.$$

(iii) Conclude that if Stokes' formula (16.13) holds for D_1 and D_2 then it also holds for $D = D_1 \cup D_2$.

□

Exercise 16.8. Suppose that $\boldsymbol{u}, \boldsymbol{v} \in \mathbb{R}^3$ are two linearly independent vectors. Prove that

$$\|\boldsymbol{u} \times \boldsymbol{v}\| = \operatorname{area}\big(P(\boldsymbol{u}, \boldsymbol{v})\big),$$

where the cross product "$\times$" is defined by (11.22) and $\operatorname{area}\big(P(\boldsymbol{u}, \boldsymbol{v})\big)$ is defined by (16.17). □

Exercise 16.9. Fix $a, b, r, R \in \mathbb{R}$, $a, R > r > 0$. Consider the map $\Phi : \mathbb{R}^2 \setminus \{\mathbf{0}\} \to \mathbb{R}^3$

$$\Phi(x, y) = \begin{bmatrix} \frac{R}{r}x \\ \\ \frac{R}{r}y \\ \\ ar + b \end{bmatrix}, \quad \text{where } r = r(x, y) = \sqrt{x^2 + y^2}.$$

Denote by D the annulus

$$D = \big\{ (x, y) \in \mathbb{R}^2;\ 1 < \sqrt{x^2 + y^2} < 2 \big\}.$$

(i) Show that Φ satisfies all the conditions (i)–(iii) in Proposition 14.34.
(ii) Show that the image S of Φ is a cylinder of radius R with the z-axis as symmetry axis.
(iii) Describe the area element on S in terms of the coordinates x, y induced by Φ.
(iv) Set $\Sigma := \Phi\big(\, \boldsymbol{cl}(D)\big)$. Show that Σ is a convenient surface with boundary and Φ defines a parametrization of Σ.
(v) Denote by f the restriction to Σ of the function $f(x, y, z) = z$. Compute

$$\int_\Sigma f(\boldsymbol{p})\, |dA(\boldsymbol{p})|.$$

□

Exercise 16.10. Suppose that $S \subset \mathbb{R}^n$ is a *compact* surface, with or without boundary.

(i) Show $\operatorname{area}(S) < \infty$.
(ii) If $f : S \to \mathbb{R}$ is continuous and $f(\boldsymbol{p}) \geq 0$, $\forall \boldsymbol{p} \in S$, then

$$\int_S f(\boldsymbol{p})\, |dA(\boldsymbol{p})| \geq 0.$$

(iii) Prove that if $L : \mathbb{R}^n \to \mathbb{R}^n$ is an orthogonal linear operator, i.e., $L^\top L = \mathbb{1}_n$, then

$$\operatorname{area}\big(L(S)\big) = \operatorname{area}(S).$$

(iv) If $f, g : S \to \mathbb{R}$ are continuous and $f(\boldsymbol{p}) \geq g(\boldsymbol{p})$, $\forall \boldsymbol{p} \in S$, then

$$\int_S f(\boldsymbol{p})\, |dA(\boldsymbol{p})| \geq \int_S g(\boldsymbol{p})\, |dA(\boldsymbol{p})|.$$

(v) If $f : S \to \mathbb{R}$ is continuous, $C > 0$ and $|f(\boldsymbol{p})| \leq C$, $\forall \boldsymbol{p} \in S$, then

$$\left| \int_S f(\boldsymbol{p})\, |dA(\boldsymbol{p})| \right| \leq C \operatorname{area}(S).$$

□

Exercise 16.11. For each $r > 0$ we denote by S_r the sphere of radius r in $\mathbb{R}^3$ centered at the origin, i.e.,

$$S_r := \{ (x, y, z) \in \mathbb{R}^3;\ x^2 + y^2 + z^2 = r^2 \}.$$

(i) Show that $\text{area}(S_r) = 4\pi r^2$.

(ii) Prove that if $f : \mathbb{R}^3 \to \mathbb{R}$ is a continuous function, then

$$\lim_{r\to 0} \frac{1}{4\pi r^2} \int_{S_r} f(\boldsymbol{p})\, |dA(\boldsymbol{p})| = f(0).$$

□

Exercise 16.12. Let S denote the unit sphere in $\mathbb{R}^3$

$$S = \{ (x, y, z) \in \mathbb{R}^3;\ x^2 + y^2 + z^2 = 1 \}.$$

Denote by $\boldsymbol{N}$ the North Pole, i.e., the point on S with coordinates $(0, 0, 1)$. The *stereographic projection* is the map

$$\boldsymbol{F} : S \setminus \{\boldsymbol{N}\} \to \mathbb{R}^2 \times \boldsymbol{0} = \{(x, y, z) \in \mathbb{R}^3 :\ \ z = 0\},$$

$$\boldsymbol{F}(\boldsymbol{p}) = \text{intersection of the line } \boldsymbol{Np} \text{ with the plane } \mathbb{R}^2 \times \boldsymbol{0}.$$

(i) For $\boldsymbol{p} \in S \setminus \{\boldsymbol{N}\}$ compute the coordinates (u, v) of $\boldsymbol{F}(\boldsymbol{p})$ in terms of the coordinates (x, y, z) of $\boldsymbol{p}$ and conversely, compute the coordinates (x, y, z) of $\boldsymbol{p}$ in terms of the coordinates (u, v) of $\boldsymbol{F}(\boldsymbol{p})$.

(ii) Prove that $\boldsymbol{F}$ is a homeomorphism and the inverse map $\Phi = \boldsymbol{F}^{-1} : \mathbb{R}^2 \to S \setminus \{\boldsymbol{N}\}$ is an immersion so Φ is a parametrization of $S \setminus \{\boldsymbol{N}\}$.

(iii) Describe the area element dA on $S \setminus \{\boldsymbol{N}\}$ in terms of the coordinates (u, v) defined by Φ.

□

Exercise 16.13. Suppose that $\mathcal{O} \subset \mathbb{R}^3$ is an open set, $f : \mathcal{O} \to \mathbb{R}$ is a C^2-function and $\boldsymbol{F} : \mathcal{O} \to \mathbb{R}^3$ is a C^2-vector field. Compute

$$\text{curl}\,(\nabla f),\ \ \text{div}\,(\nabla f),$$

$$\text{curl}\,(\text{curl}\,\boldsymbol{F}),\ \ \text{div}\,(\text{curl}\,\boldsymbol{F}),\ \ \nabla(\text{div}\,\boldsymbol{F}).$$ □

Exercise 16.14. Let $D \subset \mathbb{R}^3$ be a bounded domain with C^1-boundary, U an open set containing $\boldsymbol{cl}\, D$ and $f, g : U \to \mathbb{R}$ are C^2 function. Denote by $\boldsymbol{\nu}$ the outer normal vector field along ∂D. Prove that

$$\int_D f\Delta g\, |dxdydz| = \int_{\partial D} f(\boldsymbol{p}) \frac{\partial g}{\partial \boldsymbol{\nu}}(\boldsymbol{p}) |dA| - \int_D \langle \nabla f, \nabla g \rangle\, |dxdydz|,$$

$$\int_{\partial D} \frac{\partial g}{\partial \boldsymbol{\nu}}\, |dA| = \int_D \Delta g\, |dxdydz|,$$

$$\int_D f\Delta g|dxdydz| = \int_{\partial D}\left(f(\boldsymbol{p})\frac{\partial g}{\partial \boldsymbol{\nu}}(\boldsymbol{p}) - g(\boldsymbol{p})\frac{\partial f}{\partial \boldsymbol{\nu}}(\boldsymbol{p})\right)|dA| + \int_D g\Delta f\,|dxdydz|,$$

where we recall that, for $\boldsymbol{p} \in \partial D$, we have

$$\frac{\partial g}{\partial \boldsymbol{\nu}}(\boldsymbol{p}) := \langle \nabla g(\boldsymbol{p}), \boldsymbol{\nu}(\boldsymbol{p})\rangle.$$

Hint. Use Exercise 16.3 and the flux-divergence formula (16.34). □

Exercise 16.15. Consider the function $K : \mathbb{R}^3 \to \mathbb{R}$,

$$K(x,y,z) = \begin{cases} \frac{1}{4\pi\rho}, & (x,y,z) \neq (0,0,0),\\ 0, & (x,y,z) = (0,0,0), \end{cases} \quad \rho = \sqrt{x^2+y^2+z^2}.$$

Suppose that $f : \mathbb{R}^3 \to \mathbb{R}$ is a C^2-function with *compact support.*

(i) Show that $\Delta K = 0$ on $\mathbb{R}^3 \setminus \{\mathbf{0}\}$.
(ii) Show that the integral

$$\int_{\mathbb{R}^3} K\Delta f\,|dxdydz|$$

is absolutely integrable.
(iii) Show that

$$\int_{\mathbb{R}^3} K\Delta f\,|dxdydz| = -f(0).$$

Hint. (ii) Have a look back at Example 15.69. (iii) Fix $R > 0$ sufficiently large so that the support of f is contained in the ball $B_R = \{\rho < R\}$. For $\varepsilon > 0$ small we consider the ball $B_\varepsilon = \{\rho < \varepsilon\}$ and the annulus $A_{\varepsilon,R} := \{\varepsilon < \rho < R\}$. Use Exercise 16.14 to show that

$$\int_{A_{\varepsilon,R}} K\Delta f\,|dxdydz| = -\int_{\partial B_\varepsilon}\left(f\frac{\partial K}{\partial \boldsymbol{\nu}} - K\frac{\partial f}{\partial \boldsymbol{\nu}}\right)|dA|,$$

and then prove that

$$\lim_{\varepsilon\searrow 0}\int_{\partial B_\varepsilon} K\frac{\partial f}{\partial \boldsymbol{\nu}}\,|dA| = 0,\quad \lim_{\varepsilon\searrow 0}\int_{\partial B_\varepsilon} f\frac{\partial K}{\partial \boldsymbol{\nu}}\,|dA| = f(0).$$

□

Exercise 16.16. Suppose that $f : \mathbb{R}^n \to \mathbb{R}$ is a C^1-function with compact support. Denote by H^- the half-space

$$H^- := \left\{ (x^1,\dots,x^n) \in \mathbb{R}^n;\ x^1 \leq 0 \right\}.$$

Prove that

$$\int_{H^-} \frac{\partial f}{\partial x^1}(\boldsymbol{x})|dx^1\cdots dx^n| = \int_{\mathbb{R}^{n-1}} f(0,x^2,\dots x^n)\,|dx^2\cdots dx^n|,$$

and

$$\int_{H^-} \frac{\partial f}{\partial x^k}(\boldsymbol{x})|dx^1\cdots dx^n| = 0,\ \ \forall k \geq 2.$$

Hint. Use Fubini. □

Exercise 16.17. Let $m, n \in \mathbb{N}$, $m \leq n$. Consider

$$\omega_1, \dots, \omega_m \in \Omega^1(\mathbb{R}^n),$$

$$\omega_i = \sum_{j=1}^{n} \omega_{ij} dx^j, \;\; i = 1, \dots .$$

Prove that

$$\omega_1 \wedge \cdots \wedge \omega_m = \sum_{J \in \mathrm{Inj}^+(m,n)} \det \big(\omega_{ij_k}\big)_{1 \leq j,k \leq m} dx^{\wedge J}.$$

In particular, if $m = n$

$$\omega_1 \wedge \cdots \wedge \omega_n := \Big(\det \big(\omega_{ij}\big)_{1 \leq i,j \leq n} \Big)\, dx^1 \wedge \cdots \wedge dx^n.$$

□

Exercise 16.18. Let $m, n \in \mathbb{N}$, $m \leq n$. Consider

$$\omega_1, \dots, \omega_m \in \Omega^1(\mathbb{R}^n),$$

$$\omega_i = \sum_{j=1}^{n} \omega_{ij} dx^j, \;\; i = 1, \dots, n.$$

Prove that

$$\omega_1 \wedge \cdots \wedge \omega_m = \sum_{J \in \mathrm{Inj}^+(m,n)} \det \big(\omega_{ij_k}\big)_{1 \leq j,k \leq m} dx^{\wedge J}.$$

In particular, if $m = n$

$$\omega_1 \wedge \cdots \wedge \omega_n = \Big(\det \big(\omega_{ij}\big)_{1 \leq i,j \leq n} \Big)\, dx^1 \wedge \cdots \wedge dx^n.$$

□

Exercise 16.19. Prove (16.43). □

Exercise 16.20. Prove Proposition 16.81.

Hint. For part (i) consider first the case when α is a monomial $\alpha = \alpha_I dv^{\wedge I}$, $\alpha_I \in C^1(V)$, where $I \in \mathrm{Inj}^+(k,n)$. Start with the case $I = (1, 2, \dots, k)$. □

Exercise 16.21. Prove Lemma 16.86. □

16.5. More challenging problems

Problem 16.1. (a) Suppose that $f : [a, b] \to \mathbb{R}$ is a C^1-function. Prove that the length of its graph is not smaller than that of the length of the line segment that connects the endpoints of the graph.

(b) Suppose that $C \subset \mathbb{R}^n$ is a compact, connected C^1-curve with *nonempty* boundary. Prove that its length is not smaller than that of the line segment that connects its endpoints.

□

Problem 16.2. Let $n \in \mathbb{N}$, $n \geq 2$. Suppose that $S \subset \mathbb{R}^n$ is a closed C^1-surface and $f : \mathbb{R}^n \to \mathbb{R}$ is a C^1-function satisfying the following *transversality condition*: if $\boldsymbol{p} \in S$ and $f(\boldsymbol{p}) = 0$, then $\nabla f(\boldsymbol{p}) \not\perp T_{\boldsymbol{p}} S$. Prove that the set

$$\big\{ \boldsymbol{p} \in S;\;\; f(\boldsymbol{p}) = 0 \big\}$$

is a closed C^1-curve.

Hint. Use the implicit function theorem. □

Bibliography

[1] E. Artin: *The Gamma Function*, Holt, Rinehart & Winston, 1964.

[2] W. A. Coppel: *Number Theory. An Introduction to Mathematics*, 2nd Edition, Universitext, Springer Verlag, 2009.

[3] H. Davenport: *The Higher Arithmetic*, 7th Edition, Cambridge University Press, 1999.

[4] B. Demidovich: *Problems in Mathematical Analysis*, Mir Publishers, Moscow, 1989.

[5] J. J. Duistermaat, J. A. Kolk: *Mutidimensional Real Analysis I. Differentiation*, Cambridge University Press, 2004.

[6] J. J. Duistermaat, J. A. Kolk: *Mutidimensional Real Analysis II. Integration*, Cambridge University Press, 2004.

[7] P. Duren: *Invitation to Classical Analysis*, Pure and Applied Undergraduate Texts, Amer. Math. Soc., 2012.

[8] W. Feller: *An Introduction to Probability Theory and Its Applications*, Vol. 1, 3rd Edition, John Wiley & Sons, 1968.

[9] A. Friedman: *Advanced Calculus*, Dover, 2007.

[10] B. R. Gelbaum, J. M. H. Olmstead: *Counterexamples in Analysis*, Dover, 2003.

[11] T. Hales: *The Jordan curve theorem, formally and informally*, Amer. Math. Monthly, **114**(2007), 882–894.

[12] G. H. Hardy: *Weierstrass's non-differentiable function*, Trans. A.M.S., **17**(1916), 301–325.

[13] G. H. Hardy, J. E. Littlewood, G. Polya: *Inequalities*, 2nd Edition, Cambridge University Press, 1952.

[14] D. Hilbert, P. Bernays: *Foundations of Mathematics I*, C. P. Wirth, J. Siekmann, M. Gabbay, Editors, College Publications, 2011.

[15] K. Hrbacek, T. Jech: *Introduction to Set Theory*, 3rd Edition, Marcel Dekker, 1999.

[16] S. Lang: *Undergraduate Analysis*, 2nd Edition, Springer Verlag, 1997.

[17] J. H. Silverman: *A Friendly Introduction to Number Theory*, Prentice Hall, 1997.

[18] M. Spivak: *Calculus on Manifolds*, Perseus Books, Cambridge MA, 1965.

[19] J. M. Steele: *The Cauchy–Schwarz Master Class*, Cambridge University Press, 2004.

[20] A. Tarski: *Introduction to Logic and to the Methodology of Deductive Sciences*, Dover Publications, 1996.

[21] S. Treil: *Linear Algebra Done Wrong*, `https://www.math.brown.edu/~treil/papers/LADW/LADW.html`

[22] E. T. Whittaker, G. N. Watson: *A Course of Modern Analysis*, 4th Edition, Cambridge University Press, 1927.

[23] V. A. Zorich: *Mathematical Analysis I*, Springer Verlag, 2004.
[24] V. A. Zorich: *Mathematical Analysis II*, Springer Verlag, 2004.

Index

∇f, 425
$(2k)!!$, 269
$(2k-1)!!$, 269
$A \setminus B$, 8
$A \times B$, 9
BW property, *see* Bolzano–Weierstrass property
$B_r^n(\boldsymbol{p})$, 356
$B_r(\boldsymbol{p})$, 356
$C(X)$, 385
$C^0(I)$, 160
$C^1(U)$, 422
C°, 580
$C^\infty(I)$, 160
$C^\infty(U)$, 437
$C^k(U)$, 437
$C^n(I)$, 160
$C_{\text{cpt}}(\mathbb{R}^n)$, 398, 517
$C_r(\boldsymbol{p})$, 357
$D(K, \beta, \tau)$, 523
$G(\boldsymbol{v}_1, \boldsymbol{v}_2)$, 600
$H_{\boldsymbol{p},\boldsymbol{N}}$, 352
I_S, 516
$I_k(\boldsymbol{P})$, 241
$J_{\boldsymbol{F}}(\boldsymbol{x}_0)$, 414
$P(\boldsymbol{v}_1, \ldots, \boldsymbol{v}_n)$, 531
$R \bullet C$, 341
$T_{\boldsymbol{x}_0} X$, 481
$V^\perp$, 354, 484
$W_{\boldsymbol{F}}$, 584
$X \sim Y$, 36
$\#X$, 36
$\Delta_k(\boldsymbol{P})$, 241
$\text{Hom}(\mathbb{R}^n, \mathbb{R}^m)$, 340
$\Leftrightarrow$, 2
$\text{Mat}_n(\mathbb{R})$, 341
$\text{Mat}_{m \times n}(\mathbb{R})$, 341
$\text{Mean}(f)$, 263
$\Omega^1(U)$, 583
$\Omega^k(\mathcal{O})$, 624
$\text{Proj}_C\, \boldsymbol{x}_0$, 405
$\boldsymbol{R}(f)$, 11
$\Rightarrow$, 2
$\Sigma_r(\boldsymbol{p})$, 404
$\|A\|_{HS}$, 384
$\|\boldsymbol{P}\|$, 241
$\|\boldsymbol{x}\|$, 348
$\|\boldsymbol{x}\|_\infty$, 357
arccos, 145
arcsin, 145
arctan, 145
$\arg z$, 313
$\mathbb{C}$, 311
$\mathbb{I}_n$, 36
$\mathbb{N}$, 33
$\mathbb{N}_0$, 34
$\mathbb{Q}$, 44
$\mathbb{R}$, 17
$\mathbb{S}^1$, 479
$\mathbb{S}^2$, 480
$\mathbb{Z}$, 42
$\boldsymbol{e}^k$, 336
$\binom{n}{k}$, 39
$\boldsymbol{\omega}_n$, 543
$\boldsymbol{pq}$, 335
$\boldsymbol{F}'_{x^i}$, 416
$\boldsymbol{H}(f, \boldsymbol{x}_0)$, 451
$\boldsymbol{P}' \succ \boldsymbol{P}$, 247
$\boldsymbol{P} \vee \boldsymbol{P}'$, 248
$\boldsymbol{S}(f, \boldsymbol{P}, \underline{\xi})$, 242

$\boldsymbol{S}^*(f, \boldsymbol{P})$, 501
$\boldsymbol{S}_*(f), \boldsymbol{S}^*(f)$, 249
$\boldsymbol{S}_*(f, \boldsymbol{P})$, 501
$\boldsymbol{x} \perp \boldsymbol{y}$, 350
$\boldsymbol{x}^\downarrow$, 351
$\boldsymbol{\xi}^\dagger$, 351
$\cap$, 8
$\boldsymbol{cl}(X)$, 397
cos, 122
cosh, 182, 236
cot, 123
$\cup$, 8
curl F, 620
diam(S), 395
dist(z_1, z_2), 317
div $\boldsymbol{F}$, 595
$\mathscr{C}(\boldsymbol{P})$, 500
$\mathcal{J}(A)$, 551
$\mathcal{J}_c(A)$, 551, 554
$\mathcal{N}_a$, 101
$\mathcal{R}[0, 1]$, 367
$\mathcal{R}[a, b]$, 243
$\mathcal{R}_n(S)$, 521
$\mathcal{S}(\boldsymbol{P})$, 242
$\mathcal{SN}_a$, 101
$\ell_{\boldsymbol{p},\boldsymbol{v}}$, 332
$\emptyset$, 8
$\epsilon(\sigma)$, 625
$\exists$, 5
$\forall$, 5
$\frac{\partial \boldsymbol{F}}{\partial \boldsymbol{x}}$, 414
$\frac{\partial \boldsymbol{F}}{\partial x^i}$, 416
$\frac{\partial \boldsymbol{F}}{\partial x^j}$, 416
$\frac{\partial \boldsymbol{F}}{\partial \boldsymbol{v}}$, 416
$\frac{d^n f}{dx^n}$, 159
$\boldsymbol{i}$, 311
$\mathbf{Im}\, z$, 311
$\in$, 7
inf, 26
$\int_B f(\boldsymbol{x})|d\boldsymbol{x}|$, 503
$\int_B f(\boldsymbol{x}) dx^1 \cdots dx^n$, 503
$\int_C f(\mathbf{p}) ds$, 576, 577
$\int_\Sigma f(\mathbf{p})|dA(\mathbf{p})|$, 607
$\int_a^b f(x)dx$, 243
$\boldsymbol{int}(X)$, 397
ker A, 346
$\wedge$, 2
$\lfloor x \rfloor$, 42
$\lim_{n\to\infty} x_n$, 60
lim inf, 95
lim sup, 95
$\longleftrightarrow$, 4
$\neg$, 2
$\vee$, 2
$\ni$, 7
$\notin$, 7
$\omega(f, \boldsymbol{P})$, 245
$\mathbb{1}_X$, 11
osc, 145, 395
$\overline{B_r(\mathbf{p})}$, 359
$\overrightarrow{\boldsymbol{pq}}$, 334
∂X, 397
$\partial_{\boldsymbol{v}} \boldsymbol{F}$, 416
$\partial_{x^i} \boldsymbol{F}$, 416
$\partial_{x^j} \boldsymbol{F}$, 416
lg, 109
ln, 109
log, 109
$\log_a$, 109
$\mathbf{Re}\, z$, 311
sin, 122
sinh, 182, 236
$\sqrt[n]{a}$, 48
$\subset$, 8
$\subsetneq$, 8
sup, 26
supp(f), 398
tan, 123
$|X|$, 36
$|x|$, 22
$|z|$, 312
$\bar{z}$, 312
$\{f \leq r\}$, 407
a^x, 106
$a^{\frac{1}{n}}$, 48
dV_n, 630
$d\boldsymbol{F}(\boldsymbol{x}_0)$, 413
e, 70
$f'(x_0)$, 156
$f : X \to Y$, 10
$f^{(n)}$, 159
f^{-1}, 13
i_A, 12
$m|n$, 43
$n!$, 39
$n!!$, 269
$x > y$, 21
$x \geq y$, 21
x^{-1}, 20

$x_{\pm}$, 555

$\measuredangle(\boldsymbol{x}, \boldsymbol{y})$, 349
$\overline{C_r(\boldsymbol{p})}$, 359

Abel's trick, 98, 131
absolute value, 22
AM-GM inequality, 212
angular form, 584
antiderivative, 221
area, 295, 296
area element, 604
arithmetic progression, 57
  initial term, 57
  ratio of, 57
atlas, 634
  coherent orientation, 634
  coherently oriented, 635
  orientation, 634
automorphism, 370
axes, 28

Banach space, 367
Bernstein polynomial, 189
Beta function, 560
binary operation, 18
binomial
  coefficient, 39
  formula, 40, 82
Bolzano–Weierstrass property, 387
bounded
  map, 395
  sequence, 58
  set, 25
box
  closed, 387
  nondegenerate, 499
  open, 387
  volume, 499

canonical basis, 331
Cartesian
  coordinates, 29
  plane, 28, 120
    axes, 28
Cauchy product, 98
Cauchy–Schwarz inequality, 214
center of mass, 201
chain, 592
  local multiplicities, 592
chain rule, *see* rule
chord, 202
closed
  ball, 359
  cube, 359
  in X, 366
  set, 358, 367
closed curve, 580
closure, *see* set
cluster point, 101, 365
compact exhaustion, 552
  Jordan measurable, 552
compact set, 393
compactness, 393
complex number, 311
  argument, 313
  conjugate, 312
  norm, 312
  trigonometric representation, 314
complex plane, 313
connected component, 581
conservation law, 436
convenient cut, 579
convergence, 59, 367
  pointwise, 136, 385
  uniform, 136, 385
convex
  function, 202
  set, 335
coordinate axes, 334
coordinate neighborhood, 473
coordinates
  Cartesian, 330
  cylindrical, 480, 536
  polar, 479, 480, 532
  spherical, 480, 537
    n-dimensional, 540
critical point, 175, 452
cross product, 352
curl, 620
curve, 473, 575
  convenient, 575
    parametrization, 575
  length of a, 575
  orientation on a, 589
  oriented, 589
  quasi-convenient, 579
  with boundary, 580
    convenient, 580

Darboux
  integrable, 249
  integral
    lower, 249, 503
    upper, 249, 503
  sum
    lower, 246, 501
    upper, 246, 501
decreasing
  function, 108
dense subset, 366
derivative, *see* function
  along a path, 431
derivative along a vector, 416
diameter, 395
diffeomorphism, 456
  elementary, 548
  orientation preserving, 631
difference quotients, 158
differential, 158, 413
  Fréchet, 413
differential equation, 186, 231
  linear, 231
differential form, 583, 624
  degree 1, 583
  degree 2, 617
  degree 3, 624
  exact, 583, 586
  integral along a path, 585
differential from
  exterior derivative, 625
direction, 426
distance
  Euclidean, 355
divergence, 595
domain, 593
  C^k, 593
  piecewise C^k, 598
dual, 351

Einstein's convention, 330, 337
elementary diffeomorphism, 548
Euclidean
  space, 329
Euler
  Beta function, 560
  formula, 326
  Gamma function, 292
  identity, 432
  number, 70, 71
exact form, 586
exterior
  derivative, 625
  monomial, 624
extremum
  local, 173

facet, 499
factorial, 39
Fermat Principle, 452
Fermat's Principle, 174
flow line, 434
flux, 595, 616
  infinitesimal, 618
formula, 1
  change in variables, 530
  change of variables, 273
  Euler, 326
  flux-divergence, 595
  Green, 621
  Moivre's, 314
  Newton's binomial, 40, 161, 183, 315
  Pascal, 183
  Pascal's, 40
  Stirling, 277
  Stokes', 593
  Wallis, 270, 280
function, 10
  C^n, 160
  n-times differentiable, 159
  absolutely integrable, 554
  average value of a, 263
  bijective, 11, 108
  codomain, 11
  concave, 203
  continuous, 133
  continuous at a point, 133
  convex, 202, 491
  critical point of a, 175
  Darboux integrable, 249
  decreasing, 108
  derivative of a, 156
  differentiable, 156
  domain of, 11
  even, 123, 178, 302
  fiber of, 11
  graph of, 11, 29, 217, 220, 378
  homogeneous, 406
  hyperbolic, 236
  image, 11

implicit, 469
increasing, 108
indefinite integral of a, 221
indicator, 516
injective, 11
integrable, 503, 516
inverse of, 13
linearizable, 155
Lipschitz, 128, 135, 186
locally integrable, 551
mean of a, 263
monotone, 108
nondecreasing, 107, 108
nonincreasing, 108
odd, 123, 302
one-to-one, 11
onto, 11
oscillation of a, 145
periodic, 123
piecewise C^1, 294
piecewise constant, 260
range, 11
Riemann integrable, 242, 503, 516
smooth, 160, 437
stationary point of a, 175
strictly monotone, 108, 143
support of, 398
surjective, 11
trigonometric, 122
uniformly continuous, 146, 396

Gamma function, 292
Gauss bell, 235
geometric progression, 58
initial term, 58
ratio, 58
gradient, 425
Gramian, 600
graph, 11, 29, 109, 122, 217, 220, 378

Hölder's inequality, 213
Heine–Borel property, 390
weak, 390
Hermite polynomial, 235
Hessian, 451
homeomorphic sets, 397
homeomorphism, 397
hyperbolic
cosine, 182
functions, 182
sine, 182
hyperplane, 338
normal vector, 352
hypersurface, 496

iff, 3
imaginary part, 311
immersion, 474
increasing
function, 108
inequality
AM-GM, 212
Bernoulli, 39
Cauchy–Schwarz, 214
Hölder's, 213
Jensen's, 211
Minkowski, 215, 367
triangle, 355
Young's, 179, 234
infimum, 26
infinitesimal work, 584
inner product
canonical, 347
integer, 42
integer part, 42
integrable function, 503
integral
improper, 556
along a curve, 577
along a path, 585
improper, 282
absolutely convergent, 290
convergent, 282
indefinite, 221
iterated, 511
repeated, 511
Riemann, 243
integral curve, 434
integration
by parts, 223
by substitution, 226
interior, 397
interval, 22
closed, 22
open, 22

Jacobi matrix, 572
Jacobian
matrix, 414
Jordan measurable, 518

Kronecker symbol, 331, 350

L'Hôpital's rule, *see* rule
Lagrange remainder, *see* Taylor approximation
Landau notation, 126
Laplacian, 463
  polar coordinates, 463
Legendre polynomial, 184, 307
Leibniz rule, *see* rule
length, 292, 575
limit, 379
limit point, 74, 95
line, 332
  parametric equations of a, 334
  segment, 335
line direction vector of, 332
line integral
  first kind, 577
  second kind, 585
linear
  form, 336
    basic, 336
  functional, 336
  map, 340
  operator, 340
    ker, 346, 483
    kernel of, 346
linear approximation, 155, 414
linearization, 155, 414
Lipschitz
  constant, 381
  function, 128, 135, 186
  map, 381
local
  extremum, 173, 452
    strict, 173
  maximum, 173, 452
    strict, 173
  minimum, 173, 451
    strict, 173
local chart, 473
local coordinate chart, 601, 639
logarithm, 109
  natural, 109
lower bound, 25

Möbius strip, 614
Maclaurin
  polynomial, 191
map
  continuous, 380
  differentiable, 412
mapping, *see* function
marginal, 511, 514
matrix, 341
  associated to linear operator, 342
  column, 341
  diagonal, 345
  invertible, 370
  multiplication, 344
  nilpotent, 370
  orthogonal, 372
  product, 344
  row, 341
  square, 341
  symmetric, 345
    indefinite, 453
    negative definite, 453
    positive definite, 453
  trace, 371
  transpose of a, 372
maximum
  global, 140
  local, 173
  strict local, 173
mean oscillation, 246, 501
minimum
  global, 140
  local, 173
  strict local, 173
Minkowski's inequality, 215
multi-index, 441
  size, 441

natural basis, 331
negligible, 510
neighborhood, 60, 101
  deleted, 102
  open, 356
  symmetric, 101
Newton's method, 208
norm, 216, 366
  Euclidean, 348
  Frobenius, 384
  Hilbert–Schmidt, 384
  sup-, 357, 367
normed space
  complete, 367
normed space, 366
number

Euler, 70, 81
natural, 33

open
ball, 356, 367
cube, 357
disk, 318
in X, 366
set, 318, 356, 367
open cover, 390
open neighborhood, 356
operator
linear, 340
orthogonal, 372, 600
orientable surface, 614
orientation, 614, 630, 639
induced, 593, 615
oriented open set, 631
oriented surface, 614
orthogonal operator, 372, 600
oscillation, 395

pairing, 341
parallelepiped, 531
parametrization, 334, 474, 478
local, 473, 604
partial derivatives, 416
partial sum, 76, 320
partition, 241, 500
interval of a, 241
mesh size, 241, 500
node of a, 241
order of a, 241
refinement of a, 247
sample, 242
uniform, 242, 562
partition of unity, 399, 547
continuous, 399
subordinated to, 399
path
continuous, 380
differentiable, 431
piecewise C^1, 587
path connected, 386
permutation
signature of a, 625
Poincaré Lemma, 587
potential, 435
power series, 87, 323
coefficients, 323
domain of convergence, 87
radius of convergence, 88, 96, 324
predicate, 1
preimage, 11
prime integral, 436
primitive, 221
principle
Archimedes', 41
Cavalieri, 520, 527
inclusion-exclusion, 519
induction, 34
squeezing, 61, 382
well ordering, 36
product rule, *see* rule
pullback, 627

quantifier, 5
existential, 5
universal, 5
quasi-parametrization, 609
quotient rule, *see* rule

radially symmetric, 545
radius of convergence, 88, 96, 324
ratio test, 84
rational
function, 229
number, 44
real
line, 27
real number, 17
negative, 21
nonnegative, 21
positive, 20
real part, 311
Riemann
integrable, 242
integral, 243, 503
sum, 242, 504
zeta function, 80
right-hand rule, 354
roots of unity, 316
rule
chain, 167, 427
inverse function, 171
L'Hôpital's, 197
Leibniz, 164
product, 164
quotient, 164

s.d., 633
  orientation of a, 633
  oriented, 633
s.t., 5
segment, 335
sequence, 57
  bounded, 58, 318, 361, 389
  Cauchy, 74, 363, 367
  convergent, 59, 318, 360
  decreasing, 58
  divergent, 59
  Fibonacci, 58
  fundamental, 74, 363
  increasing, 58
  limit point of, 74
  monotone, 58
  nondecreasing, 58
  nonincreasing, 58
series, 76, 374
  absolutely convergent, 83, 322
  Cauchy product, 98
  conditionally convergent, 86
  convergent, 76, 320
  geometric, 76
  ratio test, 84
  sum of the, 76, 320
set
  boundary of a, 397
  bounded, 25, 388
  bounded above, 25
  bounded below, 25
  closed, 318
  closure of a, 397
  countable, 37
  finite, 36
    cardinality, 36
  inductive, 33
  interior of a, 397
  open, 318
simple type, 295, 523, 526
simplex, 527
space
  Euclidean, 329
sphere
  Euclidean, 404
stationary point, 175
stereorgraphic projection, 652
Stokes' formula
  1-dimensional, 585
straightening chart, 473
straightening diffeomorphism, 473
sublevel set, 407
submanifold, 473
  boundary of a, 639
  boundary point, 639
  closed, 639
  explicit description, 476
  implicit description, 477
  interior of a, 639
  local chart, 473
  local parametrization, 473
  orientable, 635
  orientation of a, 635
  parametric description, 474
  parametrization, 474
  with boundary, 639
submersion, 477, 478
subsequence, 58
subset, 8
  proper, 8
subspace
  affine, 339
  coordinate, 467
  linear, 339
  vector, 339
support, 398
supremum, 26
surface, 473, 480
  boundary of a, 601
  boundary point, 601
  closed, 601
  convenient, 607
    parametrization, 607
  interior of a, 601
  local parametrization, 604
  orientable, 614
  orientation of a, 614
  oriented, 614
  with boundary, 601
    parametrized, 607
surface integral
  first kind, 609
  second kind, 618

tangent
  space, 481
  vector, 481
tangent line, 158
tautology, 4, 6
Taylor

approximation, 193
integral remainder, 270
Lagrange remainder, 194
remainder, 193
polynomial, 191
series, 191
Taylor approximation, 449
test
ratio, 84
Weierstrass, 84
theorem
absolute convergence, 83
Banach's fixed point, 374
Bolzano–Weierstrass, 73, 388, 389
Cauchy, 75, 83, 284
Cauchy's finite increment, 181
chain rule, 167, 427
classification of curves with boundary, 581
comparison principle, 80
continuity of uniform limits, 136
D'Alembert test, 84
Darboux, 181
Fermat's Principle, 174
Fubini, 511
fundamental theorem of calculus, 265, 266
fundamental theorem of arithmetic, 43
Heine–Borel, 390
implicit function, 465, 468
integral mean value, 263
intermediate value, 394
intermediate value theorem, 140
inverse function, 457
Jensen's inequality, 211, 564
Lagrange, 175, 432
Lagrange multipliers, 488
mean value, 175, 432
nested intervals, 72
planar Stokes', 594, 599
Pythagoras, 350
Pythagoras', 293
Ratio Test, 84, 87
Riemann–Darboux, 249, 504
Rolle, 175
Stokes, 621
Taylor approximation, 193
Weierstrass, 69, 139, 395, 396
Weierstrass M-test, 84
Well Ordering Principle, 36
total differential, 423, 617
trace, 371
transformation, *see* function
transition diffeomorphism, 634
transversal intersection, 603
transversality, 478
trigonometric
circle, 121
function, 122

uniform convergence, *see* convergence
upper bound, 25

vector
length, 348
vector field, 434
vector space, 330
dual, 336
normed, 366
vectors, 329
addition of, 330
angle, 349
Cartesian coordinates, 330
collinear, 332, 368
orthogonal, 350
velocity, 431
vertex, 499
volume, 499

Wronskian, 186

www.ingramcontent.com/pod-product-compliance
Lightning Source LLC
Chambersburg PA
CBHW081210220326
41598CB00037B/6735